THE AGRICULTURAL SKY

A Concept to Revolutionize Farming

THE AGRICULTURAL SKY

A Concept to Revolutionize Farming

K. R. Krishna, PhD

AAP | APPLE ACADEMIC PRESS

First edition published 2023

Apple Academic Press Inc.
1265 Goldenrod Circle, NE,
Palm Bay, FL 32905 USA

760 Laurentian Drive, Unit 19,
Burlington, ON L7N 0A4, CANADA

CRC Press
6000 Broken Sound Parkway NW,
Suite 300, Boca Raton, FL 33487-2742 USA

4 Park Square, Milton Park,
Abingdon, Oxon, OX14 4RN UK

Library and Archives Canada Cataloguing in Publication

Title: The agricultural sky : a concept to revolutionize farming / K.R. Krishna, PhD
Names: Krishna, K. R. (Kowligi R.), author.
Description: First edition. | Includes bibliographical references and index.
Identifiers: Canadiana (print) 20220490074 | Canadiana (ebook) 20220490112 | ISBN 9781774911648 (hardcover) | ISBN 9781774911655 (softcover) | ISBN 9781003328247 (ebook)
Subjects: LCSH: Plant-atmosphere relationships. | LCSH: Field crops. | LCSH: Agriculture.
Classification: LCC QK754.4 .K75 2023 | DDC 630.2/515—dc23

Library of Congress Cataloging-in-Publication Data

CIP data on file with US Library of Congress

ISBN: 978-1-77491-164-8 (hbk)
ISBN: 978-1-77491-165-5 (pbk)
ISBN: 978-1-00332-824-7 (ebk)

About the Author

K. R. Krishna, PhD, is an agricultural scientist. He was formerly a Visiting Professor and Research Scholar at the Soil and Water Science Department at the University of Florida, Gainesville, USA. He is retired from the International Crops Research Institute for the Semi-Arid Tropics (ICRISAT), India. Dr. Krishna has authored several books on international agriculture, encompassing topics such as agroecosystems, field crops, soil fertility and crop management, precision farming, and soil microbiology. His more recent books deal with agricultural robotics, drones, and satellite guidance to improve soil fertility and crop productivity. He is a member of the International Society of Precision Farming, American Society of Agronomy, Soil Science Society of America, Ecological Society of America, Indian Society of Agronomy, and Soil Science Society of India.

OTHER BOOKS BY THE AUTHOR

Aerial Robotics in Agriculture: Parafoils, Blimps, Aerostats, and Kites

Unmanned Aerial Vehicles in Crop Production: A Compendium

Agricultural Drones: A Peaceful Pursuit

Push Button Agriculture: Robotics, Drones, and Satellite-Guided Soil and Crop Management

Agricultural Prairies: Natural Resources and Crop Productivity

Agroecosystems: Soils, Crops, Nutrient Dynamics, and Productivity

Precision Farming: Soil Fertility and Productivity Aspects

Maize Agroecosystem: Nutrient Dynamics and Productivity

Contents

The theme of the book:

We have mended soils, genetically modified crops, reorganized water resource, then, why not modify "Agricultural sky" to improve crop productivity.

By Kowligi Krishna

Abbreviations

AI	artificial intelligence
AWT	airborne wind turbine
BAT	buoyant airborne turbines
BC	black carbon
BNF	biological nitrogen fixation
BrC	brown carbon
CA	conservation agricultural
CABI	Centre for Agriculture and Biosciences
CO	carbon monoxide
CO_2	carbon dioxide
CSP	concentrated solar power
CWSI	crop's water stress index
DAPI	4,6-diamidino2-phenylindole
DCD	dicyandiamide
DLIS	Desert Locust Information System
DSM	digital surface models
EC	electrical conductivity
FAO	Food and Agriculture Organization
FHB	Fusarium head blight
GCS	ground control station
GHG	greenhouse gas
GWP	global warming potential
HM	heavy metals
ICBN	International Code of Botanical Nomenclature
IPEV	insect pollination economic value
IPM	integrated pest management
ISS	International Space Station
LAI	leaf area index
LDD	long distance dispersal
LTA	lighter-than-air
MHM	millet head miner
MSRI	Modified System of Rice Intensification
NASA	National Aeronautics and Space Agency
NMHC	non-methane hydrocarbons

NMVOC	non-methane volatile organic compounds
NO	nitric oxide
NOx	nitrogen oxides
NOx	oxides of nitrogen
NO_2	nitrogen dioxide
PAR	photosynthetically active radiation
PGPR	plant growth promoting bacteria
PM	particulate matter
PV	photovoltaic
RH	relative humidity
SDS	sand and dust storms
SLCP	short-lived climate pollutants
SOA	secondary organic aerosol
SO_2	sulphur dioxide
SRI	system of rice intensification
SRM	satellite-based rice monitoring
TPM	total particulate matter
UAVs	unmanned aerial vehicles
VOC	volatile organic compounds
VTOL	vertical take-off and lift
WRB	World Reference Base

Acknowledgments

This book has been prepared with generous gift of copies of research papers, chapters, and reports from several scientists and farm experts. They have provided key research papers, photographs, and other published material. A few of them have offered permission to utilize the images developed at their institutions. A few CEOs, marketing officers, and industrial professionals have provided important photographs. Following is a list of CEOs, industry specialists, research scientists, and their organizations to whom I wish to express my appreciation.

Mrs. Aditi Munshi, Amplus Solar Energy Solutions Ltd, Bengaluru 560001, Karnataka, India

Dr. Clyde Beaver, Creative Service Manager, International Maize and Wheat Centre, El Baton, Mexico

Food and Agricultural Organization of the United Nations, Vielle de Terme, Caracalla, Rome, Italy

Dr. Hans-Peter Thamm, Geo Technic, Neustr. 40, 53545 LinzAm Rhein, Germany

Dr. Janet Allen and James Irons, National Aeronautics and Space Agency, Washington D.C. USA

Dr. Kianian Shahryar, Cereal Disease Laboratory, United States Department of Agriculture, Minnesota, USA

Mrs. Lincy Prasanna, Vinvelli Unmanned Systems Inc. Cedar Rapids, Iowa, USA

Mrs. Sandra Allsopp, Allsopp Helikites Ltd. Fordingbridge, Hampshire, United Kingdom

Mr. Sharath Kowligi, Mrs. Roopashree, B. N. and Family, Eindhoven, Nord Brabandt, Netherlands,

United States Department of Agriculture (USDA), Beltsville, USA

United State Geological Survey (USGS), Washington, D.C., USA

I wish to thank Dr. (Mrs.) Uma Krishna, Mr. Sharath Kowligi, and Mrs. Roopashree, B. N. I wish to offer my best wishes to Ms Tara Kowligi.

Preface

Agricultural sky has molded several agrarian regions since the earliest efforts by humans (farmers) to cultivate crops in the backyards of their dwellings, small fields, or in large expanses. The blue sky above crops that is seemingly clear, filled with tranquil or sometimes clouds, is really a repository of large number of gases, mineral or organic particulate matter, dust, mist, turbulent wind, innumerable species of microorganisms, tiny biotic flora/fauna, seeds, insects, etc. At any moment, agrarian sky supports complex interactions of biotic and abiotic aspects with perhaps immediate and/or delayed influence on crops sown on the ground. *Perhaps, we should have understood farm skies much better than it has been done so far.*

Agricultural sky is perhaps an important natural entity that has initiated, controlled, and led the evolution of agriculture for past 12 millennia. Most agroecosystems, (i.e., large cropping expanses) traced in different continents are the result of massive influence imparted by the natural and man-made factors that operate in the agricultural sky. Agricultural evolution *per se* must have been intricately connected with factors that operate above the crop's canopy. We can trace innumerable treatises, research reports, and papers that deal with soils, soil fertility, soil moisture, soil-borne pathogens, and soil-dwelling insect pests. Soil factors that affect crops are too many. Soil characteristics could affect crops and their genetic selection programs, the agronomic procedures adopted, and final harvest size. Soil factors may exert influence either individually or by acting in a composite manner. Perhaps, similarly, we need to accumulate detailed and accurate knowledge about agrarian sky.

In the present context, the "Agricultural Sky" is a dominant natural entity that has influenced, interacted with, and guided the evolution of crops, farming practices, and cropping systems. For example, season, diurnal pattern, precipitation, ambient temperature, wind, photosynthetic radiation, aerial microbes, insects, birds may have decided which crop and its genotype would be preferred. Similarly, in the same location, biotic factors in the sky are important. An agricultural sky that spreads diseases or insect pests determines that only crops tolerant to disease/insect pressure can evolve into larger expanses. Susceptible crops get progressively rejected due to factors operating from the agricultural sky.

Agricultural sky is the primary abode and medium for transit and spread for a diverse set of microbes, including microbial pathogens and beneficial species such as N-fixers, N-transformers, P solubilizers, mycorrhizal propagules, etc. Phyllosphere microbes depend almost entirely on the aerospace above crops for aerial distribution. On several occasions in history, the agricultural sky has mediated pandemics/epidemics and devastated food crops in the vast expanses. The spread of fungal or bacterial spores/propagules or insects or birds has been mediated via low- or high-speed wind or turbulence generated in the sky. How accurately can we judge or forecast or take remedial measures for such a situation? Perhaps we are still not sure of a good answer. Agricultural sky harbors several types of insects. Dreaded pests like locusts, foliage-destroying cut worms, army worms, pod/seed eaters are also aided and abetted by the agricultural sky and its climatic conditions. Birds of diverse genera/species are easily traceable in any of the agroecosystems. Here, again, aerospace above crops and its operating factors must have affected the selection of crops cultivars and their productivity. Modifying the whole region of sky above each farm to become hostile to pests/diseases is not a possibility, still.

Agricultural sky is, in reality, a composite of several factors. Each factor has influenced the soils, crops, and farm procedures to different extents. For example, if agricultural sky is prone to offer high-intensity precipitation, then contouring, mulching, and erosion control methods are apt. Also, planting crops needing wetland condition or relatively more water has been preferred. Precipitation that emanates from sky determines farm procedures. Drylands suffer soil moisture scarcity. Droughts bestowed via sky affects crop productivity. Agricultural sky also influences nutrient shifts through dust storms, fogs, storms, inundation, etc. Overall agricultural sky potentially dictates farming, in one way or other. *The question here is whether we have recognized this glaring fact with appropriate seriousness.*

Often, among several aspects of "agricultural sky" we have confined our investigations to climatic and few other factors. When we study entire "agroecosystems," aspects related to agricultural sky may also need exclusive attention. Not just weather but all others too. For example, crop residue burning directly spoils the atmosphere. It induces deterioration of sky. The emissions, particulate matter, and pollution need perhaps immediate and a steady application of remedies. We have known this fact for several decades. *Farmers have no business to spoil the sky that belongs to every one of us and all of biotic flora and fauna too!* Agrarian sky over the entire landmass needs greater attention. This is to avoid emissions and spoilage via pollution. To do this, we need a good understanding of the nutrient dynamics in the soil the inside of crop tissue, and interactions of emissions in the atmosphere. *Agronomic practices detrimental to upkeep of agrarian sky need to be carefully reduced.* Farm belts of each nation may have to take responsibility to an appropriate extent to reduce the damage to agrarian sky. Emissions of $NH4^+$, NO_2, N_2O, CH_4, CO_2 into the atmosphere (agrarian sky) requires careful monitoring and management.

We may, in the future, use agrarian sky to supply a large share of nutrient needs of crops via a foliar mode. This aspect needs detailed attention by crop scientists and environmentalists. In the future, foliage may be responsible for nutrient absorption by crops, to a greater extent. This is in addition to gaseous exchange, nutrient accumulation, and biomass production (C-fixation), which are the usual functions of foliage.

This treatise is titled *The Agricultural Sky: A Concept to Revolutionize Farming*. It aims to explain a concept termed "Agricultural Sky." Adoption of such a concept could revolutionize agronomic procedures, leading to higher yields. Agricultural sky is considered an important aspect of a farm, just like soils, water, and crop species. The most important portion in a cropping expanse could, after all, be the sky. This is because it is vital to a crop's growth and sustenance in an ecosystem. Also, agricultural sky has immense influence on a crop's survival, growth, and productivity. We may not have realized the entire gamut of effects mediated by the agricultural sky. This book suggests and tries to convince that "agricultural sky" along with its natural and man-made factors is a concept of great utility to farm world in the future years. Agricultural sky needs to be emphasized to a greater extent, during farming. Appropriately, greater investment of capital, research time, and perseverance by farmers is the need of the hour. After all, we have not utilized "agricultural sky" in the best possible way and to its full extent. *There is a gap between what we have done so far with agricultural sky and what is possible.* Simple steps such as integration of various aspects of aerospace are useful.

The Earth, its features on the surface, terrain, natural vegetation, and crops have all affected the sky. Agricultural sky has been dynamic with regard to several traits. *Perhaps, the agricultural sky noticed by the early farmer several millennia ago was different from what we know of it at present. We could make a good study of the evolutionary changes of agrarian sky, as impacted by farming.* For example, during recent decades we know that farming has emitted a relatively larger quantity of gaseous compounds and deteriorated the agrarian sky. We should accrue knowledge about evolutionary trends of agricultural sky. We should also be ready with appropriate forecasts regarding the nature of agricultural sky. Physical intrusions into agrarian sky by farmers have been really feeble in the past. However, with, advent of foliar sprays, aerial vehicles such as small drones, sprayer drones, helikites, aerostats, and blimps, we will modify the agricultural skyline. We may notice tethered aerostats in large numbers. They may be seen projecting from each farm. How does this affect agrarian procedures and productivity? This aspect needs attention. Aerial application of irrigation via sprayer drones may completely replace surface, furrow, or even drip irrigation. So, agricultural sky may face a certain evolutionary change.

Agricultural sky supports both abiotic and biotic factors. Agricultural sky is a dynamic region with regard to abiotic atmospheric constituents. It includes even the aerial vehicles such as drones, parafoils, aerostats, etc. Biotic aspects include microbes, insects, birds, etc. In fact, the skyline above crops is said to change drastically globally, when farmers tend to use robotic aerial vehicles (e.g., drones, sprayer drones, parafoils, aerostats, helikites). However, legislation about agricultural sky or the region above farms/fields is unclear. Farmers are not sold the sky (i.e., aerospace or atmosphere) when they purchase land to cultivate. Agrarian sky is not to be utilized for anything other than for foliar sprays of nutrients, fungicides, herbicides, and pesticides or drones to image the crops. We are not sure if upkeep and regulation of agricultural sky above each small or large farm is an individual farmer's responsibility. This is because a farm legally does not include the rights to manipulate agrarian sky above it. Legally, we have no knowledge about to what height the agrarian sky extends above a farmer's field, to what height farmers can reach into the atmosphere and manipulate natural conditions to their advantage. Agricultural sky is still not classified based on a set of characters like we do with soils, crops, microbes, etc. We need to prepare a strong basis for classification of agrarian sky and inform farmers to react accordingly regarding crop selection and their production procedures. In simple yet glaring terms, agrarian sky above Sahel is entirely different from that encountered in the Great Plains of North America or Northern European Temperate plains or tropical crop production zones in the Indo-Gangetic Plains. Agrarian sky differs even between fields located at short distance apart. We can classify agrarian sky as sickly if propagules of pathogens are encountered in higher density above a farm or as "deleterious" if cold fronts or tornadoes are rampant. Perhaps, we can rate the agrarian sky on a scale from 1 to 10 based on its influence on productivity of food grains/fruits. *The commercial value of agrarian land should then depend to a certain extent on the characteristics of agrarian sky immediately above it.*

Agricultural sky is a dynamic entity regarding its constituents and influence on global farming enterprises, natural flora, and even fauna. Agricultural sky is a boon to farmers when we think of benefits that a farmer reaps. Agricultural sky is a detrimental aspect of farming when we consider the devastation that it causes to farms, crops, domestic animals, natural vegetation, and other inputs.

Agricultural sky has not been studied in great detail. Still, most farmers do have a semblance of understanding about what rain, thunderstorms, or deposits and erosion can do to their farms. Farm procedures aimed at modifying the agricultural sky itself or to avoid its detrimental effects need to be highlighted. We should also enhance our ability to modify procedures so that we may accentuate the beneficial effects of agricultural sky. We should derive the best possible results out of agricultural sky in the future. For unknown reasons, we have neglected this portion of an agroecosystem. Perhaps we assume that agricultural sky is taken care when we deal with an entire field and its problems. It is definitely not so in practical situations. Agricultural sky has its specific share of good and bad effects on global farming. So, it needs special attention like we do with soils or crops.

Agricultural sky is a natural gift like soil or water. Let us strive to get the best out of it. At the same time, exceedingly high levels of precautions are needed to retain its pristine qualities. Agricultural sky is an important region offering innumerable advantages to human civilization and food-generating tendencies of farmers residing in different agrarian regions. To a farmer, soil, water, crops, and sky too should be on the same level of importance. Agricultural sky cannot be just meteorology or other aspects seen in isolation. An integrated approach is needed. The sum total effect of sky should be made positive during farming.

Phyllosphere is the region surrounding the leaves of any crop. This small area with 0.2 mm thickness on leaf's surface is important. Overall, there is good reason to understand the phylloplane/phyllosphere better than we do at present. Phyllosphere is in direct contact with the atmosphere immediately above the crop's canopy. It is no doubt a portion of agrarian sky. *They say, phyllosphere is the largest area in an agroecosystem.* Larger than the land surface itself. Phyllosphere is perhaps the most crucial region that affects crop productivity in the future, whenever aerial sprays of moisture, nutrients, and herbicide become more common than now. Phyllosphere is a key region that regulates moisture, nutrients, microbes, and biotic factors in an agroecosystem. Foliar sprays of nutrients are becoming popular. Therefore, in the future, foliage, leaf area index, and phyllosphere interactions will regulate crop productivity. It is a fact that crop production using foliar supply of nutrients, water, and microbial sprays depend immensely on phyllosphere. In any region with natural vegetation and/or crops, phyllosphere is the region through which all aerial nutrients, moisture, systemic pesticides, etc. have to pass.

Plant geneticists may have to focus more on the leaf, its surface characteristics, and cellular physiology. We may have to obtain cultivars that thrive more on aerial/foliar sprays of water/nutrients. Leaf surface features such as cuticles, palisade layers, stomata, lenticels, and hydathodes are important. Farmers may prefer to channel nutrients via foliage mode more than they do so, now using soil/roots. Therefore, crops with a foliage suitable for aerial sprays will become preferred in the future.

Agricultural sky also dictates the existence, proliferation, and severity of diseases/pests surrounding the cropping zones. It decides to a certain extent the types of pests that crops encounter. A crop's resistance to pests/diseases is influenced to different extents by the agrarian sky. If we consider the evolution of genetic resistance in cultivated crops, perhaps agricultural sky had its share of influence on it. The disease/pest escape phenomenon is immensely controlled by the weather patterns, seasons, and cropping systems adopted. Farmers have to just read the season/weather to know the pathogen/insect pests and to modify the planting dates and cropping sequences.

Agricultural sky literally dictates the agronomic procedures adopted on soil. Such an influence begins with the selection of land, tillage, interculture, etc. Fertilizer inputs, pruning, training, girdling, spraying hormones, pesticides, fungicides, etc. are all dependent on the type of sky that farmers encounter above their fields. The crop harvest dates are perhaps stringently controlled by the sky and its manifestation when the crops are mature. A day's delay or a day too early will show its impact on productivity. Agricultural sky is indeed important to farmers, much more than several other aspects of the farm world although we have not made it to look so. We mended soils, water, crops more than we have done with agricultural sky. Why was the sky and its factors neglected?

Now, let us consider this treatise, in particular. There are seven chapters. The first chapter offers an elaborate introduction, full of definitions and explanation about the concept and related terms. Agricultural sky and its composition particularly various gaseous and particulate matter, nutrient dynamics including greenhouse gas emissions have been described in detail in Chapter 2. Chapter 3 includes elaborate discussions about biotic factors like aerial microbes, plant pathogens in the aerospace, insects, pests, locusts, fungal pandemics, aerial transmission, and disease severity. Chapter 4 deals with latest methods of crop production that adopt unmanned aerial vehicles (i.e., robots, drones, autonomous centre-pivot sprayers, etc.). Chapter 5 makes it clear that the aerospace above crops is useful in producing energy via wind and solar radiation. Chapter 6 describes an aerial approach to produce crops. It is called "aeroponics" and at present suits best to grow vegetables. Chapter 7 is a relatively elaborate exposition of various features relevant to crop production in the major food-generating systems of the world. A few examples dealt with are North American Great Plains, Pampas of Argentina, Sahelian production zones of West Africa, Indo-Gangetic Plains, etc.

This book on agricultural sky and its various aspects is useful to students and professors in universities as well as to researchers in industries dealing with aerial aspects of farming. Libraries meant for general public should possess a copy. Several universities may eventually develop departments of research/teaching on this concept. Faculties treating agrarian sky separately may evolve. Just like soil or water science or crop science, it is time that we monitored and studied all aspects of "Agricultural Sky" exclusively.

September 2021
Krishna Kowligi
Bangalore, India

CHAPTER 1

The Agricultural Sky: An Introduction

ABSTRACT

The theme of this treatise is to describe the importance of "Agricultural Sky." "Agricultural Sky" is a recent concept. It pertains to the aerospace above crops. The intention of this chapter is to first introduce, define, explain, and depict salient features of the sky in general. Hitherto, the word "Agricultural sky" has not been defined satisfactorily. Perhaps it is not used frequently by farm experts. Hence, first, definitions for the key words namely, "Agriculture," "Sky," and "Agricultural Sky" have been offered. "Agrosphere" that is an ecological concept has also been included. Initially, explanations about earth's atmospheric layers such as troposphere, stratosphere, mesosphere, ionosphere, and even space above cropping expanses have been provided. The "agricultural sky" is literally the area above the ground and/or crop's canopy till border layers of ionosphere/space. Agricultural sky includes both natural and man-made factors. Such factors could be characterized further into biotic and abiotic in nature. These components of agrarian sky may affect agricultural farming individually or in innumerable combinations. The spread and intensity of their influence on crops in space and time needs attention. The altitude above crop's canopy and volume of the agrarian sky also need consideration. The agrarian sky actually imparts diverse effects on crops, vegetation, and other natural resources.

There is vast literature explaining the nature of sky. A gist of it relevant to global farming has been enlisted. Its various layers such as troposphere, stratosphere, mesosphere, and ionosphere have been defined. Agricultural sky extends from the ground surface and/or the crop canopy till the fringes of ionosphere and parts of space immediately above the agroecosystem. The altitude of agricultural sky, like sky in general, may extend from ground

The Agricultural Sky: A Concept to Revolutionize Farming. K. R. Krishna, PhD (Author)

surface to 11 km of immediate atmosphere (troposphere) till 41 km, that is, upper stratosphere and reaching heights up to 71 km from the cropped fields. This zone within atmosphere, which is termed "agricultural sky," has played a crucial role, dictating the evolution of crops and cropping systems, etc.

Agricultural sky is not just about ambient weather that crops have to negotiate. Agricultural sky encompasses a few climatic indices. They are abiotic, natural and of immense value to crop production on the earth's surface. The climatic indices such as photosynthetic radiation, relative humidity, temperature, diurnal length, air and gaseous matter, precipitation, storms etc. are of concern to farming enterprises. Photosynthetic radiation influences carbon fixation trends. Light *per se* affects chlorophyll pigmentation, several enzymes, morphogenesis, and biomass formation. Relative humidity (RH) is indicative of presence of water vapor in the agrarian sky. The RH dictates the type of natural vegetation and crop species supported by the region. For example, a predominantly lowland, flooded rice needs consistently high RH in the atmosphere. A dry Sahelian aerospace allows only a hardy drought-tolerant crop like millets. The RH influences several aspects of crops like foliage, gaseous exchange, duration, maturity, and quality of grains. The ambient temperature has marked effects on crops. The selection of crop species, season, sowing time, and cropping systems depends immensely on temperature bestowed by the agrarian sky on the crop's canopy. Extremely high or low temperature in the sky means we have to select appropriately tolerant cultivars and adopt suitable agronomic measures.

Pollutants of diverse origin and potentially toxic nature are reaching conspicuous levels in the agricultural sky. Therefore, agricultural sky can literally be a prominent detriment if it harbors pollutants at above the threshold levels. Farming procedures such as crop burning, application of pesticides at high concentration and at short intervals, gaseous emissions, and ozone can severely damage crop quality and productivity. Pollutants in the sky can literally wipe out crops under certain circumstances. Smog, fog, acid fumes, and particulate carbonaceous matter that throng the agrarian sky affect photosynthetic efficiency.

We may note that greenhouse gas emissions and related global warming are of great concern to experts dealing with global agriculture. It is said that 23% of GHG get emitted from cropping expanses in different continents. This is primarily due to excessive application of fertilizer-based nutrients like N and S. Carbon emissions are severe in many locations due to excessive ploughing and oxidative conditions. These agronomic procedures spoil the quality of the agrarian sky.

Wind is created due to differential pressure above crop's canopy. Wind could be a detriment causing soil erosion, dust bowls, storms, tornadoes, heat waves. Wind may induce movement of gaseous and particulate matter over large distances. Intercontinental drifts of dust, microbes, and disease propagules are common in certain agrarian regions. Wind that mediates drift of pollen and induces pollination is a useful aspect.

Clouds are major sources of precipitation that occurs as rainfall. Precipitation may also happen in the form of first, snow, fog, or mist. Clouds and other forms of precipitation may affect several physiological process relevant foliage and grain yield formation. We may realize that individual and interactive effects of "agricultural sky" have been conspicuous since ages. Such abiotic factors have indeed guided evolution of agrarian belts. The genetic selection of crops might have depended, to a great extent, on such aerial factors.

Several definitions and explanations about the sky in general and agricultural sky specifically have been mentioned. A concept proposed some years ago is called "agrosphere." "Agrosphere" is an ecological concept meant to enhance our ability to understand the agricultural biomes. Agricultural sky could be considered a part of this ecological entity on earth. Aspects such as topography, soil, water crops, cropping systems, and importantly the aerospace agricultural sky have been revisited. The agrarian sky extends into vast space above the global agricultural zones. It differs markedly and sometimes feebly based on geographic location and agricultural farming trends. We need to understand this aspect clearly and utilize shrewdly in the future.

Like agricultural soils, water, and crops, agricultural sky too is an important factor during generation of food grains, fiber, wood, and other necessities derived from vegetation. Perhaps we ought to realize this fact a trifle better than at present.

Agricultural sky and its various natural components have been discussed briefly. They include topics such as the climatic indices namely photosynthetic radiation, relative humidity, temperature, etc. The air in the atmosphere, gaseous compounds, and greenhouse gas emissions have been stressed as an important factor that influences crop production. Wind as an atmospheric phenomenon has been discussed in detail along with other aspects such as clouds, precipitation (i.e. rainfall, frost snow, fog, smog, rainstorms), heat waves, dust storms, droughts are part of agricultural sky. Clearly, agrarian sky must have played havoc with crops and farm enterprises since several millennia. Pollution of agrarian sky with particulate matter, ozone, gaseous compounds etc are equally important detriments that influence evolution, sustenance, and productivity of crops.

Heat waves are atmospheric phenomena that are harsh on crops, fauna, and humans. Heat waves may occur in conjunction with droughts and dust storms. Such combinations destroy standing crops. They reduce crop productivity. In some instances, crops get wiped out entirely. Heat waves occur regularly in certain agrarian regions like Sahel in West Africa (e.g., Harmattan), Haboob in Sudan, etc. They are no doubt detrimental to natural vegetation and crops. Even heat-tolerant and hardy millets may suffer when the agrarian sky is uncongenial. Crop genetic stocks with the ability to tolerate heat wave effects are needed. Farm procedures and cultivars that escape heat wave need priority.

Major biotic components of the agrarian sky discussed in this chapter and elsewhere in others in this treatise are the micro-organisms, insects, and aves. These biotic aspects could cause detriments as pests. At times, they are devastating when they occur as epidemics. Microbes and insect pests could spread all over the globe as pandemics. There are several beneficial aspects attributable to them such as biological N fixing microbes, honeybees, predator insects, birds of prey acting as biological control agents, pollinator birds, etc. Dispersal of microbes, mainly fungal pathogens, is worth noting. Fungal spores are disseminated via dust storms and high-altitude wind streams (jets) (e.g. rust fungi). There is vast information on wind and its role in trans-Atlantic transport of dust, sand particles, microbes especially disease propagules (e.g. *Puccinia graminis tritici*). There are several fungal, bacterial, viral diseases that are dispersed via wind. Both short- and long-distance dispersal of spores are observed in major agrarian belts.

We have bestowed relatively lower levels of inquisitiveness, intellect, and funds to understand the diseases mediated via agricultural sky. Of course, a couple of exceptions exist. Agricultural sky and its biotic aspects have not been pooled, integrated, and analyzed prior to prescribing appropriate responses. We tend to answer disease/insect pest-related aspects individually. Agricultural sky is perhaps the most important aspect when we think of airborne diseases, insects, and even birds that are pests.

There are also several benefits from biotic segment of agrarian sky. Pollination that literally determines the crop productivity via its influence on seed set is manifested via agrarian sky. Honey production by bees is also an aeolian activity. Aves and their influence on faming need greater attention and better management. Several examples of aves and their role in biological control, pollination, and as pests are described.

We have bestowed great interest and improved our ability to study soils, water resources, crops, and genetic stocks. However, our effort to study agrarian sky in detail and utilize it is relatively feeble. It has almost confined to knowing the seasons, weather, and a few factors like precipitation pattern. So, effort has been made to describe various factors in the agrarian sky that have immediate impacts on crops. Several of the agrarian sky-related aspects impose detrimental effects. A few others offer beneficial effects all through the season.

Man-made factors operative in the agricultural sky are the foliar sprays of irrigation water, nutrients, pesticides, and herbicides. During past several millennia, we have not utilized the agricultural sky, that is, aerospace above crops to any worthwhile extent. Particularly to surveillance, obtain arial imagery, analyze and treat the crops, accordingly. Unmanned aerial vehicles (UAVs) are recent invasions into agrarian sky. They provide excellent vantage points to collect digital data of crops. They could be of immense utility to farmers who opt to harness agrarian sky better. Such versatile aerial vehicles transit above crops swiftly and provide useful digital information regarding growth, biomass accumulation, chlorophyll content, N status, disease, pest attack, and so on. In fact, drones and tethered helikites are expected to revolutionize the way we accrue data about cropped fields and respond during farming.

Satellites are useful in deciphering spread of natural vegetation, agricultural crops, soils, water bodies, and so on. Their role in monitoring vast agrarian biomes and forewarning any impending disasters has been mentioned. Finally, this chapter provides a unique comparative discussion dealing with the agricultural soils, water, crops, and sky. It tries to coax us to study the agricultural sky in greater detail, obtain data accurately, and act swiftly. After all, agricultural sky is beneficial to crops—we may realize this fact.

Overall, this chapter tries to pool vast information about agricultural sky. It provides introductory information. Also, the following six chapters offer detailed discussions on agrarian sky and its influence on global cropping belts.

1.1 INTRODUCTION

1.1.1 DEFINITION AND EXPLANATIONS FOR AGRICULTURE

Agriculture involves the resources and services of different ecospheres such as the lithosphere, hydrosphere, atmosphere, and biosphere. The agricultural enterprise anywhere on earth is dependent on natural resources, man-made resources, and human activity relevant immediately to food crop production, rearing domestic animals and generation of other products such as firewood and fiber. During past several millennia man has bestowed great interest and inquisitiveness to study the various factors that determine the productivity of agricultural farms. Crop production experts have made enormous strides in understanding the importance of soil, water, air, and inputs such as nutrients, plant protection chemicals, and herbicides. They have also made detailed studies on the effect of seasons, weather patterns, and ambient atmosphere in general, on the crop production trends. *At this juncture, we may note that, the sky per se above the soil surface and crop's canopy in any given agrarian region influences the crop enormously.* Roughly, the region above the agricultural field soils and crop's canopy could be called the "Agricultural Sky." "Agricultural sky" is a highly heterogenous zone in terms of its contents and many manifestations. The physico-chemical traits of agricultural sky vary in space, time, and intensity. The atmosphere and layers above the soil surface/crop's canopy such as troposphere and stratosphere offer several different inputs to crops. A few examples are precipitation, particulate matter, nutrients, radiation, gases such as N_2, CO_2, O_2, and SO_2. The "agricultural sky" also supports large posse of biotic factors such as the micro-organisms, insects, birds, and flying mammals (e.g. bats, vampires) of diverse abilities, microbes with the ability for nitrogen fixation, plant disease causation, and putrefaction etc.). The biotic factors affect soils and crops either directly or indirectly. Actually, agricultural sky is in constant interaction with ground (soil) factors in an agrarian belt. The atmosphere influences crop production immensely through the seasonal variations in weather patterns. The photosynthetic radiation needed for carbon fixation transits through the agricultural sky and is made available to vegetation on earth. Precipitation is an exceedingly important atmospheric factor that affects agrarian regions and their productivity. Generally, the phrase "agricultural sky" has meant the blue sky and atmospheric components. However, in practical agricultural situations, the "agricultural sky" is a region with many more variable factors, both abiotic and biotic. Agricultural sky has immense influence on crop's productivity.

Agricultural sky means the transparent clear region or sometimes slightly hazy one or that with moisture laden clouds is deceptive in appearance. In reality, it has a marked influence on the large expanse of crops. We may have to understand the influence of this "agricultural sky" and its array of components in greater detail and accurately. We may then modify this important region to our advantage. We have to note that, "agricultural sky" has not been treated comprehensively as a concept that includes several factors affecting vegetation and crops on the ground. *The interaction among various factors that operate in the "agricultural sky" is important. Yet, it has received only meagre attention. At the same time, as a caution, our inquisitiveness and efforts to utilize agricultural sky in a better way should not spoil this serene looking or at times turbulent part of the earth.* During recent years, there have been concerns expressed vehemently about the gaseous emissions, excessive particulate matter, haze, smog, and climate change that affect the sky. In the near future, we may be flying agricultural drones, parafoils, blimps, and aerostats in a greater intensity into the agricultural sky. Yet, they may not clog the agricultural skyline. Also, they will not distract other agricultural activities that we have conducted through the sky since many years.

Now, let us consider the meaning of the key terms used in the title of the chapter. Of course, depending on the present context , that is, The "Agricultural Sky." Agriculture is a worldwide phenomenon invented, devised, and mastered, approximately for 12000 years, through human ingenuity. It deals with production of food crops including vegetables, fruit and fiber, forest plantations, fuel wood, pastures, and livestock. There are several definitions for agriculture. Overall, most dictionaries state that agriculture is a process that offers food grains.

Agriculture involves components such as crops, soil, water, and atmosphere. Soils utilized for crop production are termed generally as "agricultural soils." Such agricultural soils are studied in great detail for traits such as texture, structure, nutrient contents, and their availability to crop roots, soil's moisture holding characteristics, soil-borne plant pathogens, insect pests, nematodes, etc. Soils are amended accordingly to derive better crop yield. Crops used for human consumption and to support industrial productivity are grouped as "Agricultural crops." Most of these were domesticated since Neolithic age. They are produced and consumed by the populace. Agricultural Crops are those species grown mostly for human consumption in large expanses. There are food crops, fruit-bearing orchards, vegetables, fiber crops, wood plantations, medicinal species, etc. Agricultural crops constitute only a portion of the large diversity of plant species available on earth. Water resources utilized to irrigate crops are augmented through a variety of means such as precipitation, rivers, lakes, ponds, irrigation projects, and canals (GAO, 2019). The portion of water resource utilized for crop production is termed "Agricultural Water" resources. We may note here that all types of water resources are not utilized for agricultural crop production. There are also exclusive "Agricultural inputs." For example, fertilizers are specialized items derived from natural resources through mining, for example, phosphatic rocks. Such inputs are manufactured through industrial procedures into products like nitrogen fertilizers, super phosphate, pesticides, bactericides, fungicides, herbicides, etc.

We have studied several aspects of agriculture that has direct or indirect bearing on the atmosphere and the sky above the cropping expanses. Primarily, aspects such as ambient atmosphere, its weather patterns, precipitation, relative humidity, temperature, wind, and photosynthetic radiation for the entire cropping season have been understood in detail. The influence of factors such as the greenhouse gas emissions, atmospheric deposits, plant pathogen density in the air above crops, insects, and their pattern of devastation too have been studied. The sky above crops has been utilized efficiently to sprinkle irrigation water, spray fertilizer-based nutrients, pesticides, herbicides, and several of the amendments. Recently, we have made great strides in utilizing the agricultural sky to conduct spectral analysis of crops. We are flying drone aircrafts with multi-spectral cameras, periodically, to assess the crops. Such aerial spectral analysis helps in application of agricultural inputs, accurately. Matching the crop's needs in time and space is important. These are essential aspects to an agricultural enterprise. Hence, a concept that encompasses several aspects of agricultural sky is needed. The concept "Agricultural sky" thus involves innumerable factors that operate in the region immediately above the soil surface and/or crop canopy. Some of these factors interact within the "agricultural sky."

Let us consider various definitions and explanations offered for the word "Agriculture" (see Table 1.1). Primarily, it relates to human activity relevant to food crop generation that was invented some 12,000 years ago, in the Fertile Crescent region. Historically, mending field soil using implements that were still rudimentary in design, sowing land races and domesticated species of crops, and collecting dehiscent and/or non-dehiscent seeds were the hallmark of activities included under the word "agriculture." Harris and Fuller (2014) have offered an extended definition and overview of the word "agriculture" itself and the way we have adopted it to connote generation of food crops and other products, through the ages. According to them, "Agriculture" is the most comprehensive word used to denote the many ways in which crop plants and domestic animals sustain the global human population, by providing food and other products. Let us consider the etymology of the English word "agriculture." It is derived from the Latin word "Ager" that means field and "colo" means to cultivate. The combined word that is "agricultura" in Latin means, to till the field with an aim to produce crops. However, the word "agriculture" has evolved to mean a variety of activities that generate a range of products in open fields and farms. It describes activities related to adaptation of crops and their genotypes including crop husbandry procedures, rearing domesticated farm animals like cattle, sheep, poultry etc. Harris and Fuller (2014) further stated that agriculture currently is an all-encompassing word. It includes subactivities such as horticulture, forest plantation, mixed cropping that include food crops, pastures, and cattle rearing etc. They summarize that, through the ages the word "agriculture" has meant a broad spectrum of activities because of the multi-disciplinary nature of the profession (see Helbaum, 2010). Yet, overall, it has meant cropping the landscapes involved in food generation.

TABLE 1.1 A Few Definitions of the Word "Agriculture."

Following are the definitions offered by different dictionaries that are in use commonly. They are:
Agriculture is farming and the methods that are used to raise and look after crops and animals (Collins Dictionary, 2019).
Agriculture is the practise or work of farming (Cambridge Dictionary, 2019)
The Merriam-Webster Dictionary states that, agriculture is the science, art, or practice of cultivating the soil, producing crops, and raising livestock and in varying degrees, the preparation and marketing of the resulting products (Merriam-Webster Dictionary, 2019)
Agriculture is the art and science of cultivating the soil, growing crops and raising livestock (National Geographic, 2019)
The science or practice of farming, including growing crops and raising animals for the production of food, fiber, fuel, and other products (NAL, 2019).
Oxford English Dictionary (2020) defines agriculture very broadly as "The science and art of cultivating the soil including tillage and other crop husbandry procedures and rearing livestock.
Yet another definition states that, agriculture is the science and art of cultivating plants and livestock. The major agricultural products can be grouped into foods, fibers, fuels and raw materials (such as rubber). Food classes include cereals (grains), legumes, vegetables, fruits, oils, meat, milk, fungi (e.g. mushroom) and eggs. Over one-third of the world's workers are employed in agriculture. It is second only to the service sector. However, in the developed nations, the number of agricultural workers has decreased significantly over the centuries (Wikipedia, 2019a)

There are indeed several variants of agricultural practices adopted by farmers. They are named accordingly by highlighting the salient feature (s). A few examples are as follows.

Mixed farming is common in many agrarian regions. Mixed farming that includes production of crop, fodder, livestock, and any other essential product is among the most conspicuous variants of agriculture (Krishna 2003, 2008, 2016). Agropastoral pursuits have also been a common variant until the sedentary agriculture took shape in larger areas (Harris and Fuller, 2014). Here, the word agriculture encompasses pastoralism that is adopted to feed the cattle or other domestic animals surrounding the villages. Nomadic or migratory practices of food crop generation have also been adopted. Such practices too are classified as agriculture.

"Regenerative Agriculture" describes farming and grazing practices that, among other benefits, reverse climate change. This set of practice helps in rebuilding soil organic matter and restoring degraded soil biodiversity. It leads to both carbon drawdown and improvement of the water cycle (Institute for Sustainable Development, 2019). "Regenerative Agriculture" is a holistic land management practice. It aims at efficient photosynthesis in plants to complete the carbon cycle. Also, it helps to build soil fertility and health. Regenerative agriculture supposedly improves soil organic matter content. It aims to improve soil biota and its diversity. Most importantly, regenerative agricultural practices aim at reducing the ill effects of climate change on crop productivity.

Precision Agriculture is a relatively recent concept. It is propounded to overcome the variation in crop productivity caused by various lacunae like soil nutrient or moisture availability. It also considers the variations in the impact of detrimental factors such as disease, insect pest, or drought (Khosla, 2010; Krishna, 2013, Stafford, 2000, 2005; Zhang, 2015, Zhang and Pierce, 2013). The procedures involved stipulate supplying inputs such as fertilized-based nutrients, water, herbicides, pesticides, and fungicides in quantities, to match the variation in space and time. Digital maps depicting variations regarding each factor (e.g. soil-N fertility, soil moisture, weed) that affects the crop are essential. Presently, aerial robotics, that is, agricultural drones are being touted and even utilized to a certain extent. Drones are used to derive such digital data and maps depicting variations in soil fertility factors or disease/pest infestation. Such digital data are then utilized by the farm vehicles fitted with variable-rate dispensers (nozzles) (Krishna, 2013, 2016, 2018; SenseFly, 2015, 2016; Precision Hawk, 2014).

Subsistence farming involves production of the crops and/or livestock. The entire produce raised are used to support the farmer and the farmer's family. It leaves no surplus for sale or trade. We should note that subsistence farmers were common in the preindustrial agriculture, throughout the world

(Editors, 2020). Subsistence Agriculture involves cultivating soils using inherent fertility of soil. The extraneous supply of nutrient, irrigation, and plant protection chemicals is meagre. In such farms, total productivity of food grains, fruits, or other products is relatively low. Often, farmers thriving on subsistence farming are also constrained economically. They lack in ability for high capital investment. The role of soil, its inherent organic matter, mineral nutrients, moisture, and crop genotypes are crucial. The soil nutrients have to be utilized efficiently and recycled, to attain yield stability. Since plant protection chemicals are used sparingly, if at all, the crop genotypes selected have to manifest high degree of genetic resistance, to pests/diseases. Subsistence farming is common in the semiarid and arid tracts. Also, in drylands where inputs are feeble or not at all envisaged by farmers.

The term "subsistence agriculture" is used synonymously with such concepts as traditional, small scale, peasant, low income, resource poor, low-input, or low technology farming (Heidhaues and Bruntrup, 2003).

Reports suggest that subsistence agriculture (or family farming) is generally confined to areas with low level of economic development. The subsistence agriculture often involves low-external input level and low productivity (per land and/or per labor).

Intensive Agriculture aims at utilizing agricultural land efficiently. It aims to harvest appreciably high levels of produce such as grains, fruits, vegetables, fiber, and wood. The farmers supply high levels of fertilizer-based nutrients, irrigation, and plant protection chemicals. Proportionately, farmers reap higher harvests. Whatever the type of farming is adopted, we should recognize that agricultural sky bestows several advantages and plays a vital role in the development of crops in the fields and food grain generation. So, we should understand the importance of agricultural sky a bit better than what is known at present.

There are many other types of agricultural enterprises. Many of them are specialized in terms of crops, genotypes, inputs, and agronomic procedures adopted. For example, arboriculture involves cultivation of trees. Wetland agriculture involves crop production practices suited to raise crops in flooded condition (e.g. rice, sugarcane). Wetland agriculture involves frequent irrigation, flooding, and stagnating water in the crop fields. Irrigated crop production is common across different agrarian regions. This type of agriculture involves supply of water to crops through furrows or sprinklers. This is in addition to that received via precipitation. Agriculture that depends predominantly or almost exclusively on precipitation received naturally is called “Rainfed Agriculture.” Rainfed agriculture may result in lower grain production due to erratic precipitation patterns. Intermittent droughts are common during rainfed agriculture. Droughts often result in reduction of biomass and grain formation. Silviculture involves production of forest trees that offer a range of products. Viticulture involves production of grape vines that offer grapes for the distilleries and human consumption etc. Citriculture involves production of citrus trees.

We should note that, in all the above cases the ingredients required for raising crops are the soil (lithosphere), water (hydrosphere), crops (biosphere), and atmosphere with carbon-dioxide for photosynthesis and oxygen for respiration (Agricultural Sky). The agricultural sky is an important aspect of farming although not recognized appropriately well, so far. It supplies various requirements to the crop. The agricultural sky exhibits extra-ordinary diversity in terms of precipitation, gaseous components, emissions, particulate matter, biotic components, etc. No doubt, "agricultural sky" is a highly heterogenous entity. Agricultural sky above each of the agrarian regions may differ leading to differential crop productivity. It is blatantly clear that agricultural sky above the subsistence farming regions of Sahel or the rice-producing tracts of South Asia or the Wheat/soybean cultivation regions of Cerrados/Pampas or Forest plantations in temperate North America and Europe are not at all similar. Particularly, if we analyzed them in greater detail. Let us quote an example. The meagre nutrients received as sand/dust particles plus the morning dew and scanty precipitation received in Sahel are important. Also, such inputs received from sky on wet land rice regions may go undetected or deemed negligible in terms of its effects on crop sustenance. It means that impact of agricultural sky differs based on the prevailing agro-environment.

There is a concept touted that is titled as "vertical farming" or "sky farming." It involves production of vegetable crops, small annual flower species, and few others on the terraces of a high-rise building right in the urban zones. Often, they are denoted as sky farming enterprises. This form of agricultural production described has no bearing on the central theme of this volume. The word "sky," here, only suggests that crop production is conducted in locations perched in the high-rise buildings or on terrace gardens etc. Regarding factors concerned with crop production under the "sky farming" on high rise urban locations, we may note that they are similar to farming under controlled conditions, using green houses on terraces or open space (Chamberlain, 2007). A few advantages suggested about such "sky farming" in high rise apartments relate to reduction of energy loss from buildings, efficient recycling of nutrient inputs, possible reduction of emissions from soil that otherwise occurs in open field cropping systems. A single high-rise tower or sky farm may be equivalent to several hectares of land occurring in open expanses. At this juncture, we may realize that "sky farming" mentioned here is costly. It involves high capital investment for supplying inputs and recycling them. It needs sophisticated gadgets and trained technicians. There are control systems needed to circulate water and nutrients in carefully measured quantities, into closed enclosures, if it is a greenhouse enterprise.

Also, we should note that there are companies with names that include "sky" as suffix. They have large holding of farmland. There are also brand names of agro-industries that manufacture tractors, seed drill, fertilizer inoculators etc. They use the word "sky." Yet, they have no bearing on the central theme of the book. These products are not in any way directly or indirectly suggestive of the happening in the sky above the cropland. Particularly, in terms of atmosphere, weather patterns, aerial inputs, physico-chemical reactions, emissions, photosynthetic carbon fixation or growth, and yield formation (see Barnett, 2015; Sky Drill Inc, 2019).

Whatever the advantages and disadvantages quoted in the literature, we may note here that the theme of this book titled, "Agricultural sky" has no bearing on the topic stated in the above paragraph. The theme of this book relates to several different phenomena and factors that operate in the "Agricultural sky" , that is, the sky (troposphere) or atmospheric space above soil surface and crop's canopy. It also involves a variety of ways by which we can mend the "agricultural sky" or utilize it efficiently, to produce crops in the vast expanses termed usually as agroecosystems.

1.1.1.1 WHAT IS AGROSPHERE

There is a term "Agrosphere" coined to refer exclusively to zones utilized for agricultural crop production on earth (Krishna, 2003). In terms of geography, agrosphere is an exclusive area on earth that supports agricultural practices, with a purpose of generating food, fodder, fuel, and fiber. No doubt, generation of crops is conducted within a specified zone carved out of the lithosphere. Currently, about a third of earth's land surface is being utilized for farming. Therefore, agrosphere is that much expansive. It extends to 13,500 km^2 (13.5 billion ha) on earth (Wood et al., 2000; FAO, 1998, 2019). The sky above the cropping expanses within the agrosphere is the center piece of the discussions, in this volume. There is still 2.7 billion ha of land that could be potentially utilized for crop production (FAO, 2019). Clearly, agrosphere could be extended proportionately into more area on land surface. In that case, the agricultural sky and its influence on global crop production gets proportionately extended into newer areas. We should also get conversant about the influence of agroecosystems, crops, and cropping intensity on agricultural sky.

We may note that the above definitions for the word "Agriculture" "consider predominantly the components (crops, soil, water, atmosphere, etc.), procedures (seeding, interculture, fertilizer application, pesticides spray, etc.), and products (food grains, fiber wood, etc.) concerned. However, we may also define and try to understand term "agriculture" considering the geographic features that it involves, also the agroecosystems (cropping expanses) and their peculiarities. Agrosphere is a term that has been used to depict the areas on earth that support agricultural activity. It is an exclusive area. Physically, it is supposed to extend into 30% of the land surface of the world (Krishna, 2003). Agrosphere depends on the natural resources and services of other ecospheres such as the lithosphere, atmosphere, hydrosphere, and biosphere (Krishna, 2003; Wood et al., 2000). The major ingredients that lead to the formation of a highly productive agrosphere are the soil, water, crops, their genotypes plus human activity.

The demarcation of boundary of agrosphere as suggested by Krishna (2003) includes the space from the tip of the root system in the underground soil, its spread, then from soil surface till the tip of shoot system or canopy and atmosphere. Now, considering the context of this book—the agricultural sky is actually that portion of agrosphere that extends from soil surface till ionosphere. It includes all the agricultural cropping belts. The various portions of atmosphere such as the water vapour, gases, particulate matter, volatile organic substances, plus physico-chemical features such as the temperature, moisture, relative humidity, precipitation, and other weather parameters are all included in the agricultural sky. So, in this volume, we are basically concerned with various manifestations of agricultural sky portion of agrosphere. It deals exclusively with influence of factors that operate in the agricultural sky and affect crop production. Aspects related to agricultural sky (above the agrosphere) and their influence on vegetation, productivity, and nutrient cycling need greater attention.

The peculiarities of agricultural sky as touted this book are not yet fully understood. The reason is that there is no previous literature specifically pointing to the details of "agricultural sky" as a concept. Agricultural sky has not been characterized and classified using different criteria so far.

Agricultural sky, no doubt, is part of the agrosphere as defined by Krishna (2003). Further, Krishna (2003) prefers to demarcate the agrosphere into agroecosystems. Typically, these are vast cropping expanses (e.g. Great Plains of North America cropping belts; Pampas cropping zones; Cerrados Maize/Soybean cropping zones, Indo-Gangetic plains with wheat/rice agro-ecosystem etc). Such a treatment may help us in understanding the impact of factors operating through the agricultural sky, on a relatively uniform crop belt. However, we may realize that global agricultural zones also include complex and multiple cropping sequences and intercrops. Overall, the above proposition to study influence of agricultural sky on agroecosystems, at large, seems a good idea. It may help us to identify certain problems common to subregions of an agroecosystem. After all, crops may have to negotiate factors operative in the agricultural sky, throughout the season (Krishna, 2003, 2005, 2008, 2016, 2018).

1.1.2 DEFINITIONS FOR THE "SKY"

Now, let us consider the other key word in the title of this chapter—the "Sky." Sky by definition is a portion above earth's surface. It includes atmosphere, troposphere, stratosphere, and ionosphere leading to space as the next layer (see Table 1.2; Plate 1.1). Sky is definitely not a uniform entity. It is highly heterogenic spatially and temporally. It varies enormously based on location. Physico-chemical composition and biological activities also vary. Here, a new term "Agricultural Sky" is introduced to suggest that sky above cropping expanses, pastures, or dairy cattle zones is different from the general term "sky." The agricultural sky exhibits certain distinct aspects. Agricultural sky extends into wherever cropping has been practised. The agricultural sky has not been studied in great detail. It has not been modified, to any great extent (exception is cloud seeding, sprinklers, and sprayers) and utilized advantageously, by agricultural experts, so far.

Let us consider the etymology and usage of the word "sky" within the context of agriculture and related activities. As a caution, there are several private companies that generate agricultural produce in open farmland. There are also enterprises that generate horticultural fruits and ornamental commodities, such as passion fruits, strawberries, and flowers on high-rise buildings. They use the word "sky" perhaps to denote that crop production occurs in high rise space of urban locations. They are located high above the land surface. However, it has no relevance to the theme of this book.

TABLE 1.2 Definitions of "Sky" in General.

Following are the definitions for the word "Sky':
Sky is the area above the earth in which clouds, the sun, and the stars can be seen.
(Cambridge Dictionary, 2019)
Sky is the upper atmosphere or the clear expanse that constitutes an apparent great vault or arch over the earth (Merriam-Webster Dictionary, 2019)
The sky (or celestial dome) is everything that lies above the surface of the Earth, including the atmosphere and outer-space (Wikipedia, 2019b)
The atmosphere above a given point, especially as visible from the ground during the day (Definitions and Translations, 2019)
The space above the Earth where clouds, sun and stars appear (Longman's Dictionary, 2019)

1.1.3 EARTH'S ATMOSPHERIC LAYERS

Let us consider the details about the layers of atmosphere (i.e. the gaseous envelope) or Sky above the earth's surface, including agrarian regions. Troposphere is that part of the atmosphere that is closest to earth's surface. It extends from earth's surface to a distance of 8–14 km high (NASA, 2019a; Zell, 2017). The troposphere is heated by solar radiation and conduction. The tropopause is the boundary layer between the troposphere and stratosphere. The thickness of troposphere depends upon a number of atmospheric variables specific to that altitude. The troposphere ranges from a thickness of approximately 5.5 miles (9 km) in the polar regions, to a thickness of approximately 10 miles (16 km) in equatorial regions. Troposphere is usually the densest of the layers of sky (also agriculture sky) above the soil surface/crop's canopy (Plate 1.1). Manifestation of weather processes and parameters are intense within layer. Such weather parameters are easily discernible using the specialized weather gadgets. Air pressure drops and temperatures get colder as we climb higher in the troposphere. Further, we may note that tropospheric constituents and processes are in close proximity to earth's surface. Therefore, they interact closely with surface vegetation and crops.

Troposphere harbors large regions of natural vegetation. Also, agrarian regions are filled with diverse crops species, a large number of domesticated animals, and human beings. Troposphere also holds clouds. Almost 99% of the rain-bearing clouds are traced in troposphere. Such clouds are important for recycling water from sky to earth in the form of precipitation. The aspects of weather that operate within the troposphere are indeed crucial to optimal growth and productivity of agricultural crops, also, sustenance of life on earth in general.

Stratosphere is the immediate next layer of atmosphere above the troposphere (Plate 1.1). It begins with the end of troposphere. It extends till 50 km distance high from earth's surface (Plate 1.1). We find the occurrence of "ozone layer" in the stratosphere. The ozone layer absorbs and scatters the ultra-violet rays received. The ozone molecules in the stratosphere absorb ultra-violet radiation and convert it to heat. When present at the lowest layer in the atmosphere , that is, troposphere (8–13 km) distance, the ozone layer could be a challenge, to optimal crop growth and food grain generation (Emberson et al., 2018a, 2018b; EPA, 2019). The stratosphere actually gets warmer with altitude from earth's surface. The lower stratosphere is said to be less turbulent and hence passenger airplanes prefer to reach this height of around 30,000–33,000 ft. from sea level.

Mesosphere is the region above the stratosphere. It extends from 50 km up to 85 km high from earth' surface. Meteors that enter the earth's atmosphere burn away in this zone. Mesosphere is a cooler region with low air pressure. Air pressure is not congenial to breathe and reduces further as we transit to higher altitude within this layer. Temperature drops to –90°C (–130°F or 280°K). These temperatures recorded are among the lowest, for different layers of atmosphere above the sea level.

Thermosphere extends from 65 km to 600 km high from earth's surface. It begins just above the Mesosphere (Plate 1.1). Thermosphere has very thin air. It is more like a space with vacuum. High-energy X-rays and ultraviolet rays are absorbed in this layer. This phenomenon increases temperature enormously. In the upper thermosphere, temperature ranges from about 500°C (932°F) to 2000°C (3,632°F) or higher. The aurora, the Northern Lights, and Southern Lights occur in the thermosphere. Most of the satellites meant for military and civilian uses such as agriculture, transport, and GPS guidance occur in this layer of the earth's sky.

Ionosphere is a layer that extends from 48 km to 965 km high from the earth's surface. This sphere overlaps with mesosphere and thermosphere. Actually, the ionosphere is not a definitive layer unlike others mentioned above. It is a series of regions in the layers of mesosphere, and thermosphere. These regions harbor electrons from the parent atoms and molecules that are actually knocked off by the high energy radiation from the sun. The ionosphere is a layer that is abundant in electrons and ionized atoms (Zell, 2017). The richness of electrically charged ions gives it the name "ionosphere." The outer fringes of ionosphere are close to space. Ionosphere is a region that grows and shrinks depending on solar conditions. It can be divided into subregions: D, E, and F. Such a division is based on wavelength of solar radiation that gets absorbed. The ionosphere is a critical link in the chain of Sun-Earth interactions. This region is what makes radio communications possible (Horneck et al., 2010; Zell, 2017).

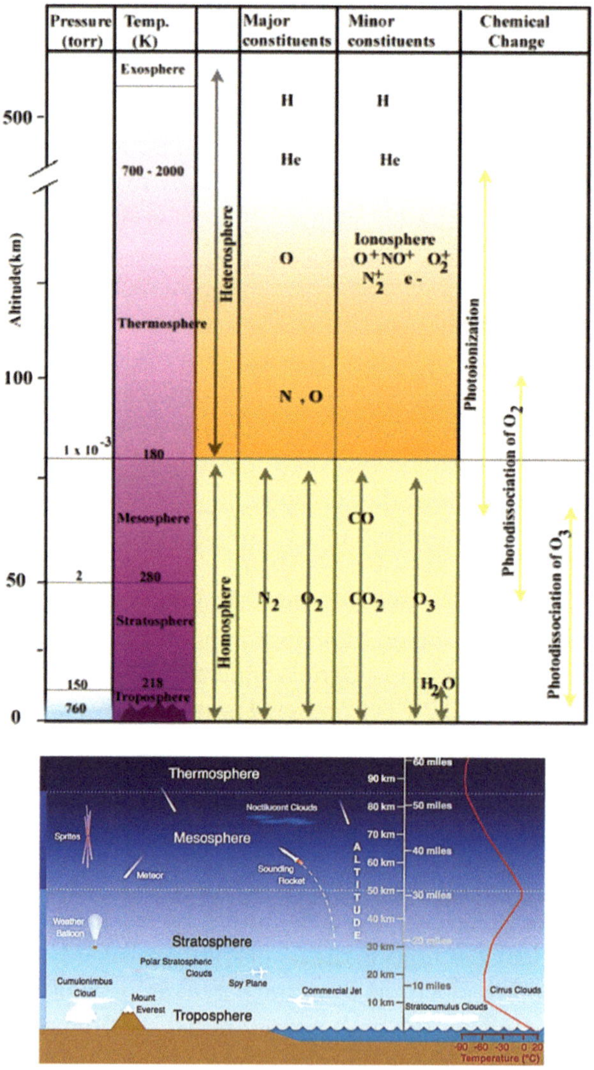

PLATE 1.1 The Sky and its regions.
Source: NASA/Goddard Space Centre, Washington D.C. USA; US Environment Protection Agency, Washington D.C.; see EPA, 2019; Zell, H. 2017

Exosphere is the upper limit of our atmosphere. It extends from the top of the thermosphere up to 100,000 km high from the earth's surface. The exosphere is supposedly the outermost layer of earth's gaseous envelope—atmosphere. However, many consider that earth's atmosphere ends with the Mesosphere layer. The air is extremely thin in the exosphere. It is almost absent giving a vacuum condition. Physically, the exosphere extends to 100,000 km to 120,000 km from sea level. We have no accurate idea about how these farther layers of gaseous envelope affect the agricultural weather or crops.

The earth's atmosphere also acts as traps for electrons (i.e. negatively charged particles) and protons (i.e. positively charged particles). They occur abundantly at 3000 to 16000 km altitude above the earth's surface. The zone is called "Van Allen Radiation Belt." The region is also called "Magnetosphere."

Now, where is the edge of the earth's atmosphere? A point where atmosphere ends and space begins. Generally, such a delineation is known to occur at a point 100 km from earth's surface. It is known as "Karman Line." About 99.7% of the earth's atmosphere is said to lie beneath this "Karman line." However, some reports from NASA's observatories state that a cloud of hydrogen atoms called the "geocorona" may actually occur at 391,000 km away from earth's surface (Buis, 2019). This is much beyond the moon's orbit. We have no idea whether such a phenomenon has any bearing on other atmospheric layers or earth's surface features.

At this juncture, we may note that the explanations provided for different layers of the sky, in general, also apply if one wants to define and describe the "Agricultural Sky." The "agricultural sky" refers to the atmospheric layers above the surface of earth, specifically, in the region where agricultural crop production is in vogue.

Sky, in general, has been identified over each location as a specific zone with definite traits. Sky above a region has often been identified and named, based on the earth's geological features, or other manifestations. For example, sky above high oceans and seas is termed "Marine sky" with its specific traits. Sky above coastal belts is termed the "Coastal Sky." Sky above urban locations and city with skyscrapers is termed "Urban Sky," for example, a highly polluted urban sky is a common phrase during recent years. Sky above deserts and arid regions is termed the "Desert sky." Sky above wet tropical regions such as in Amazonia or Congo region is termed the "Tropical sky." Sky above temperate regions that support vegetation and large-scale farming is called "Temperate sky." Sky above arctic locations with predominantly snow laden regions is termed "Arctic sky." We can classify further and call the skies as forest sky, agricultural sky, those above natural prairies and cereal stretches as "Prairie sky" etc. Similarly, we should be able to study in detail the sky/atmosphere, above the major food generating regions of the world, such as the Great Plains of North America. We may call it Great Plains Sky. The sky above Cerrados shrub vegetation and farm belt could be termed "Cerrados Sky; that above Pampas farm belts as "Pampas Sky"; that above European Plains as European Agrarian sky etc (see chapter 7 for greater details). In the future, we should be able to characterize the types of skies (atmosphere) and relate them to farming. So far, we have not identified and described the general characteristics of agricultural sky, in any detail. We should be able to characterize the agricultural sky. Perhaps even quantify its relevance to farm productivity. Farms should be monitored for status of agricultural sky immediately above the crop fields. We have to perhaps study agricultural sky in the same way as we did with agricultural soils. *No doubt, agricultural sky is as important as soils on which crops are generated. We should also note that, sky (in general) that occurs anywhere in nonagricultural belt may also influence farm activities. The drifting wind, precipitation, and several atmospheric factors do shift from one location in the sky to another, as time lapses.*

1.2 THE "AGRICULTURAL SKY" —A BRIEF EXPLANATION OF NATURAL COMPONENTS

So far, the word "agricultural sky" has not been defined satisfactorily, although the agricultural sky and its manifestations have been identified by the farmers. Since the invention of "agriculture," in other words seeding and cultivation of crops some 10000–12000 years ago, farmers have reaped the benefits of the "Agricultural Sky." The agricultural sky has bestowed a variety of services and advantages to the farming community. Agricultural sky along with its components has played an immensely important role in the farming belts of the world, since past several millennia, particularly, in the choice of crops, their genotypes, evolution of cropping systems, sequences, rotations etc. It has influenced the agricultural terrain, its topography, the soil, and its suitability for cropping. Agronomic procedures adopted by farmers too have depended on the agricultural sky, to a great extent. Plus, it has affected the productivity of soils in various agroecosystems. Yet, generally, farmers have considered agricultural sky or the region above their crop's canopy as a natural resource without paying great attention to it. First, farmers try to match their cropping based on the seasons created predominantly by the factors operating in the agricultural sky. Farmers receive services such as photosynthetic radiation, precipitation, the nutrient laden dust, the microbial components, beneficial insects and pests, aves, etc. In other words, agricultural sky has played a predominant role in providing several resources needed for farming. It has helped in soil conditioning. Agricultural sky offers nitrogen to crops via the biological N fixing organisms that are resident in soil/air. Of course, it has also induced soil deterioration whenever wind speed is beyond threshold or precipitation is excessive. Severe soil erosion has been the result. Overall, there is a need for a concerted effort to understand the agricultural sky in greater detail, then characterize it, classify and demarcate as accurately as possible.

Here, we may note that, definitions offered for the word "sky," in general, do not suffice. The dictionary meanings mostly stress on the physical demarcations of atmosphere and celestial manifestations of the sky, in general. Physically, "agricultural sky" too could be defined as the seemingly clear region beginning just above the soil surface (earth), then including the crop's canopy and the atmosphere above earth's surface and extending till space. The atmosphere includes the extended region called troposphere that begins immediately above the earth's surface plus crop's canopy if it is a cropped area. It is followed by a layer termed "stratosphere," then followed by one known as "mesosphere." There is a mesosphere pause above which is the thermosphere and ionosphere in sequence. Several physico-chemical constituents such as the gaseous matter (e.g. water vapor, N_2, NO_2, N_2O, NH_3, CO_2, CH_4, SO_2, VOCs, and traces of several other gases); particulate matter emanating from dust storms, industries, agricultural operations, and other events on earth may all be traced in the agricultural sky. Agricultural sky also harbors the gases and particulate matter that are derived through cropping operations, such as spraying pesticides, fungicides, herbicides, foliar fertilizer formulation (e.g. NH_4, KNO_3, Urea, $ZnCl_2$, B as Borax etc.), and crop residue burning. The crops also experience the consequences of several physico-chemical reactions between the constituents within the agricultural sky. Often, precipitation may carry the constituents of the agricultural sky on to crop's canopy and earth's surface (soil). Agricultural sky is also the source of atmospheric-N. Atmospheric-N is fixed by certain soil microbes. The process improves soil-N fertility. Agricultural sky also harbors and mediates dissemination of disease propagules and insect pests.

Agricultural sky often induces changes in the agricultural terrain. Here, we include several abiotic and biotic factors that operate within agricultural sky. Also, how they affect the land/soil formation processes, expression of soil fertility, biodiversity of a vegetation, cropping systems, and their productivity. It also includes man-made factors such as foliar sprays, pesticide, and fungicide sprays.

Within the present context, that is, "agricultural sky," we may note that various biotic factors are important. For example, birds, insects, and microbes that throng the agricultural sky could be beneficial, detrimental, or sometimes stay as commensals and perish. Agricultural sky supports wind velocities that could be utilized to generate electricity and support irrigation of crops through lift irrigation. Solar radiation received through the atmosphere could be utilized to generate electricity. Such electricity supports several different agricultural operations/activities (Plate 1.2). There are also clear reports and efforts to generate crops without the soil. It is termed the "aeroponics." It involves the creation of aerosol with all essential nutrients needed for crop production. Such exclusively aerosol-supported crop production is still confined to controlled glasshouse conditions, not practiced in the open in expanses of any size.

PLATE 1.2 Solar panels help to generate electricity by receiving radiation from the agrarian sky.
Source: Amplus Solar Energy Solutions Ltd, Bangalore, Karnataka, India
Note: Both solar cell panels and crops could be located in the same field and utilized simultaneously.

The details in the above paragraphs make it clear that the definition for agricultural sky includes several other abiotic and biotic phenomena that occur in the atmosphere. Agricultural sky is not the same as sky that we refer to in general discussions. The sky generally envelopes all of the earth's surface (globe). It occurs above oceans, mountains, arid deserts, tropical forests, agrarian regions, barren high-altitude snow mountains (e.g. Alps, Himalayas), and ice-capped polar regions, etc. On the contrary, "Agricultural sky" that we consider here just confines to region above the crop fields and large expanses of monotonous agrarian regions. In nature, the agricultural sky may mix with other portions of the sky based on turbulence in the sky, wind currents, and various atmospheric phenomena such as precipitation.

The agricultural sky is composed of several components that may operate individually and affect the crop production procedures and trends. Several of the components may also exhibit physico-chemical interaction within the atmosphere. Then, crops on earth's surface may actually experience their interactive effects. Broadly, natural components that make up the agricultural sky could be easily classified into abiotic and biotic factors. Abiotic factors include the photosynthetic radiation, clouds, precipitation, relative humidity, ambient temperature, wind, storms, dust bowls, gaseous emissions, particulate matter, acid rains, pollutants etc. The biotic factors include the microbial load of atmosphere, insect pests, and aves. The man-made factors that are related to the agricultural sky also affect the crop production trends. Man-made factors include aerial irrigation that creates mists, pesticide/fungicide sprays, industrial emissions, crop residue burning, wind power generation using low altitude turbines and/or high-altitude aerostat-supported turbines, smoke screens to raise ambient temperature under the tree canopy within plantations during winter etc (Koppmann et al., 2005).

First, let us consider the abiotic factors that may vary enormously in terms of their influence on agrarian regions and crops, in particular.

1.2.1 THE CLIMATIC INDICES

Thus far, we have collected enormous data about the atmospheric parameters that have direct relevance to crop production. Among the abiotic factors studied we have understood salient aspects about the climatic indices such as photosynthetic radiation, temperature, precipitation including storms, frost, snow, hail, gaseous emissions, particulate matter, acid rain, and pollution, if any. Each climatic factor has its own effect individually or in interaction with others operating in the atmosphere. Agricultural sky mediates a large share of the climatic effects that agrarian regions experience.

Historical climate data that is focused on agriculture has been studied, to assess climate-related risk. It indicates acceleration in warming trends. They say, extreme heat waves have occurred since 1980 over the major growing regions of wheat, maize, soybean, and rice. Warming trends are often accompanied by drought and vapor pressure deficit. They have deleterious impacts on crop yields (Lobell et al., 2011a; 2011b, 2005). In certain geographical regions, the climate-related effects on crops have been severe. Detailed analysis of data on climate in West African Sahelian indicates that climate change effects such as temperature increase and deviation in precipitation pattern can lead to perceptible loss of grain yield loss of major crops, such as millet, sorghum, cowpea. The extent of biomass and grain loss due to climate change effects is uncertain. Yet, estimates reveal that mean yield of cereal could decrease by 8% in the entire African continent. In Sahel, in particular, it could be 11% reduction by 2050. Hence, the planners need to invest and aim at procedures that fully consider the possible reduction in yield, caused due to enhanced population of humans and farm animals (Shepard, 2019; Sultan et al., 2019).

The recent increase in the mean temperature in the Sahel has been attributed to human activities. The decade 2000–2009 is approximately 1°C warmer in West Africa. Frequency of hot days is much higher. They say, annual rainfall in West Africa during the 20 century, which is characterized by the succession of wet periods with droughts does not seem to be driven entirely by human activities. They found a significant increase in the intensity of rainy events (Sultan et al., 2019).

In the immediately following sections, it is intended to discuss the climatic components/indices as related to agrarian regions, with emphasis on the agricultural sky.

1.2.1.1 PHOTOSYNTHETIC RADIATION

Basically, a set of congenial climatic factors operating in the agricultural sky is a necessity for full expression of crop's genetic potential. We may note that climatic factors that naturally operate in the sky above the crop's canopy affect them both favorably and detrimentally. The intensity and duration of impact of climate-related factors should be optimum. Climatic factors or indices could be detrimental to crop production, if they reach beyond the extremes. At the same time, we should adopt crop production tactics that maximize the advantages from climatic factors. There are a few climatic factors such as light, temperature, humidity, precipitation, and wind that affect growth and productivity of the crops.

Light is a climatic factor that affects crops. Photosynthetic radiation is essential for the formation and functioning of leaf chlorophyll. Chlorophyll is related to photosynthesis, that is, carbon fixation and biomass formation by the crop. Subsequently, several different types of organic compounds such as enzymes, vitamins, hormones, and phenolics are synthesized, utilizing the basic organic compounds (Bareja, 2011; Devlin, 1975).

The light quality (wavelength measured as nm on electromagnetic spectrum), intensity (degree of brightness (i.e. lux), and diurnal pattern, that is, day length for which the crops are exposed to sun's radiation affects the chlorophyll and C-fixation. Light also affects plant processes such as phototropism, photomorphogenesis, translocation of biochemical compounds, and mineral absorption. Leaf etiolation due to lack of light radiation or nutritional disorders too affect reception of photosynthetic radiation. Lack of chlorophyll reduces photosynthetic processes and efficiency of carbon fixation.

Light interception by natural vegetation, agricultural crops, and pastures could be reduced through several means. Natural factors that operate in the sky and reduce the impingement of light radiation on ground crops are clouds, precipitation, haze, dust, and smog. If the vegetation has upper and lower story plant species such as in thick forests and commercial plantations, then shadows caused by the main stand reduce interception of light. There are crop's canopy-related factors that reduce photosynthetic light interception. They are reduced foliage formation due to insufficient input, loss of foliage due to drought, inundation, diseases, pests, dust cover over canopy/leaves, physiological senescence of leaves etc. Intercropping practices too affect light interception. Biotic factors such as disease that creates pustules or variegation in color or reduces chlorophyll formation reduce photosynthetic light interception. Similarly, insect pests that reduce foliage (e.g. cut worms, leaf rollers, leaf miners) could reduce reception of photosynthetic light radiation from the sky. Any reduction in leaf area, defoliation, or damage to leaf tissue can result in reduced photosynthetic efficiency (C-fixation) (Abellanosa and Pava, 1987; Devlin, 1975; Poincelot, 1980; Tiaz et al., 2014)

Photosynthetic light impinges on crop's canopy via the agricultural sky. Light is absorbed by the natural vegetation and crops in the open expanses. Artificial lighting or diffused sunlight through the greenhouse is another possibility, if the crop is cultivated in entirely or partially controlled green houses. There is no doubt that the photosynthetic efficiency of crops varies for a variety of reasons. There is a genetic basis for differences of photosynthetic efficiency. In most previous reports, the photosynthetic efficiency of major cereals, for example rice has been studied under conditions when the impinging radiation is optimum. However, in a practical field situation, light impinging on crops fluctuates based on, say, cloudiness, canopy structure, leaf formation trend (LAI) etc. (FFAR, 2020). Further, investigations have revealed that in rice, the leaf and canopy formation are important. The extent of diffused light trapped by lower leaves in the canopy may affect photosynthetic light interception and CO_2 fixation. We ought to realize that crop's canopy structure and its under storey is important in terms of CO_2 fixation, that is, biomass formation.

1.2.1.2 RELATIVE HUMIDITY

Relative humidity (RH) is indicative of the water vapor that the air (agricultural sky) holds, at a given temperature. The RH is the ratio of actual water vapor content to the saturated water vapoulr content, at a given temperature. RH is expressed as percent of the maximum water vapor that air or soil can hold, at a given temperature. Mean maximum RH is usually encountered in the morning sky and mean minimum RH in the early afternoon. RH is low during afternoon when ambient temperature is higher. Mean RH is generally high above the crops cultivated in equatorial regions/tropics. RH decreases as we traverse toward polar regions. Warm air holds more water vapor compared to cool air. RH ranges from 0.01, say in the frigid polar region to 5% in humid tropics. Relative humidity (RH) in the ambient atmosphere immediately above the crop's canopy is no doubt important. RH in the agricultural sky affects several aspects of crops. Primarily, it affects the functioning of stomata that has direct bearing on moisture and gas exchange, and photosynthesis. RH affects cell turgidity, cell and leaf growth, also stomatal functioning. Therefore, it has immense influence on CO_2 fixation and transpiration. RH may also affect pollination, pollen viability, and seed set. RH affects bacterial, fungal, and insect activity in the crop's canopy and surroundings. Hence, RH has immediate influence on disease/pest attack on the crop.

Relative humidity in the ambient atmosphere does affect the type of crop that can flourish in a given region. Most crops thrive well at a relative humidity of 40–60% in the ambient atmosphere during the crop season.

Clearly, cultivation of wetland crops such as rice or sugarcane is possible only in tropics with high humidity (RH 60–80%). Similarly, vegetable crops such as cucumber too need high relative humidity during growth and formation of fruits. However, high RH of 90% reduces quality of fruits. In addition, fungal diseases may set in on crops. In the arid regions, where RH during crop season is low, crops such as small millets (*Setaria italica*), sorghum, pearl millet could thrive better. No doubt, RH as a climatic factor must have influenced the selection of crops and cropping systems. It may also act on crops in conjunction with other factors such as diurnal temperature fluctuations, photosynthetic radiation. We need greater details and clearer examples about RH as a major factor of agricultural sky that influences the crops.

Let us consider a few examples. The wheat belt in Northeast China is an important grain-generating region. It encounters a series of weather-related maladies. Therefore, wheat genotypes sown should be highly versatile and adapted to several climate change factors. The wheat cultivars grown were insensitive to climatic factors like precipitation, wind speed, and evapotranspiration. Temperature, solar radiation, and relative humidity of air above canopy were major meteorological factors affecting wheat yield. Such factors affected the wheat crop in the eastern and western Inner Mongolia. The increased air relative humidity would make the western spring wheat yield increase and the eastern spring wheat yield decrease.

Chepfer et al. (2019) point out that natural vegetation and crops are perpetually exposed to diurnal fluctuations in clouds, their intensity, and relative humidity. Effect of diurnal cycle of solar radiation on the climate system is well defined. But diurnal evolutions of water vapor and clouds induced by the solar radiation are not yet deciphered in detail, across the tropics. We need to study the cloud patterns and relative humidity profiles of the agricultural sky above, perhaps, each agrarian location/field. The relative humidity profiles may change swiftly based on vertical distribution of clouds and their thickness/thinness. As a consequence, in the tropics, where clouds are frequent, the relative humidity as well as solar radiation that impinges on the crops vary with time, within a day. These manifestations in the agricultural sky do have a certain clear impact on crop growth, seed set, grain filling, grain maturity, and even grain quality. To study the diurnal variations in relative humidity, researchers have utilized the Megha-Tropiques satellite. It obtains data from a low orbit, mostly, about the cloud profiles, solar radiation profiles, and relative humidity profiles. It is important to take note of profiles of relative humidity in the agricultural sky above the cropping belts. Then, understand its influence on the crop productivity.

Relative humidity is an important factor operating in the agricultural sky above crops such as rice. The RH and its interaction with canopy/leaf temperature seem to influence the grain productivity. The RH has impact on flag leaf and panicle initiation and its fertility. Higher RH can affect seed set detrimentally (Yan et al., 2010). They say, decreasing RH under high temperature conditions is useful to attain better seed set. High temperature along with more than optimum RH leads to droopy panicles. Such panicles are also prone to fungal disease attack. Adjusting planting dates and selecting genotypes that mature at apt dates is required. In nature, we often encounter environmental factors operating in clusters that is, a few at a time. So, their interactive effects could be more important. We may guess that, through the ages, RH, ambient temperature, canopy temperature, and panicle physiology might have interacted consistently, to arrive at the rice genotypes that we now cultivate. Surely, RH as a natural factor related to agricultural sky needs emphasis. The influence of divergent botanical species and crops in monocrops on the relative humidity due to transpiration may also need some attention. A large patch of crop too has its influence on the climate-related parameters above the canopy (agricultural sky).

Relative humidity has direct influence on pathogen/insect pests. Sorghum is an important cereal crop in the central and southern Great Plains of USA. High humidity during seed set and maturation sometimes induces ergot, smut, and other fungal diseases on sorghum panicles (Stack, 2000). High RH is said to be congenial for production and dispersal of secondary conidia of these fungi that affect the panicle.

Relative humidity also affects the insect vectors of viral diseases. Reports suggest that increased temperature and low relative humidity reduce the survival of vector and the pathogen (Wosula et al., 2015).

High temperature can affect the morphogenetic expression and physiological functions that lead to normal seed set and maturation. High temperatures are usually detrimental to grain formation by hybrid rice (Yan et al., 2017).

Relative humidity levels between 55% and 90% (1.0 kPa to 0.2 kPa vapor pressure deficit) have no effect on growth and productivity of horticultural crops (Grange and Hand, 2015). Relative humidity values less than it create water deficit and crops could get stunted. Therefore, yield realized could be lower. Relative humidity higher than the above-mentioned optimum range leads to fungal infections. It could encourage physiological disorders and microbial infections. Clearly, RH of the agricultural sky has its impact on crops.

A few reports suggest that a climatic factor such as relative humidity affects both the host and pathogen, simultaneously, perhaps to different extents. For example, relative humidity is said to influence, both host plant and fungal pathogen, in case of soybean-Phytophthora interaction. The expression of resistance genes and physiological manifestation of soybean at 63% relative humidity helped pathogen, to infect and establish. In the host, transcript level of the gene that imparts resistance to Phytophthora was minimum, at 63% RH (Xu et al., 2016).

1.2.1.3 TEMPERATURE

Temperature is an important climatic index. It affects several aspects of natural vegetation and crops.

Ambient temperature partly decides the kind of vegetation and crop species that can flourish in a geographic region. Ambient temperature changes with altitude and topography. Ambient temperature in interaction with altitude affects the type of vegetation and crop species cultivated. If we begin from the soil surface at sea level, the atmospheric temperature that hovers at an average of 90°F (32.2°C), usually, supports tropical/temperate trees, crops, and natural vegetation. The temperature drops by 3°F for every 1000 ft. altitude as we traverse to stratosphere. At 4000 ft. altitude above the sea level, we encounter a slight drop in ambient temperature to 78°F (25.5°C). At this altitude we may trace deciduous trees. Crops acclimatized to higher altitudes thrive well. For example, certain cereal genotypes accustomed to mountainous terrain etc. At 8000 ft. above sea level, the terrain usually supports forest, but agricultural crops get sparse. The ambient temperature at this altitude is about 66°F (18.9°C). At 12000 ft. above sea level, the temperature encountered is 54°F (12.2°C). The vegetation changes to Tundra type cold-tolerant shrub or grassy vegetation. No doubt, the temperature drops as we reach higher altitude. At 16000 ft. altitude, temperature on ground surface hovers at 42°F and at 20,000 ft altitude, it may reach 30°F (−1.1°C). At 32 000 ft. altitude, ambient temperature above the crop (i.e. in the agricultural sky) is usually 18°F (−7.7°C). The naturally cool temperatures that prevail at high altitudes hardly have any impact on crops that flourish on the low plain region. If it is a mountainous region, then such low temperature (18–12°F or −7.7 to −11.1°C) allows only snow-capped terrain and no crops are possible.

Overall, in the troposphere, the ambient temperature drops by 6.5°C per each km altitude, up to 10 km. In this region, ambient temperature has direct impact on the type of vegetation that flourishes its growth and productivity. In the Stratosphere, temperature hovers at –45 to –50°C. In the Mesosphere, that is between 30 to 50 km above the ground surface, the temperature rises by 0.5°C per km height reaching a warm 25°C at 50 km altitude.

Globally, agricultural crops thrive from 50° N to 50°S latitude. They experience different temperature ranges (Krishna, 2003; Wood et al., 2000). Crops flourish and offer optimum harvest in areas with congenial temperature such as tropics, subtropics, or temperature zones. They also thrive but with low productivity in areas with temperature ranges not so congenial throughout the year/season. Extreme temperature detrimental to crop growth may occur in earth's polar region or deserts where both terrain and sky are harsh for crop growth.

In general, plants may survive from 0 to 50°C in the outfield. Extreme temperatures within this range are often tolerated only for a certain short span of period, in a crop season. However, each crop species has an optimum temperature, beyond which its growth and grain yield formation get affected detrimentally. Ambient temperature affects a series of physiological manifestations of crops. They are, braking of seed dormancy and induction of germination, protein synthesis, photosynthesis, respiration, transpiration, translocation of photosynthates and water etc. Warmer temperature may stimulate higher rate of photosynthesis, enzyme activity, and growth. They say, for 10°C increase, the rate of enzyme activity in the plant tissue doubles. However, at extremely higher temperature, the growth retards. Low ambient and soil temperatures reduce the rate of several physiological processes. Low temperatures induce hibernation and dormancy. No doubt the temperature is an important climatic index.

Higher temperature affects crops in the field in different ways. It reduces photosynthetic efficiency, induces leaf senescence, decreases pollen production and its viability, reduces grain number and weight. Crops may complete cycle in a shorter duration impairing proper expression of physiological stages and restricting the period for grain maturity (Challinor et al., 2005; Deryng et al., 2014; Prasad et al., 2006; Ugarte et al., 2007).

Higher temperature for longer duration (heat waves) may affect crop yield. Crop's productivity may increase due to warmer conditions if it is a cold region. Warming avoids sever cold stress. While in warm tropics higher temperature beyond 2°C to 3°C may affect photosynthesis and carbon fixation. Most congenial temperature for carbon fixation is around 20°C–25°C. So, if heat waves enhance average temperature beyond threshold of 30°C–32°C, then photosynthetic efficiency decreases. They say, heat waves and global warming affect Northern hemisphere to a greater extent because this portion of earth supports enormously large agrarian stretches of diverse crops. There are also forecasts that continuously warmer temperature or heat waves may extend natural vegetation and crops into higher latitudes (Bralower and Bice, 2019).

Further, Lobell et al. (2011a; 2011b) state that warming trends have been experienced by almost all major crop species. It has been felt in all agrarian regions. Growing season temperature has increased by 30% in some regions. Clearly, global agricultural crop production is vulnerable to in-season deviations of temperature from optimum. They say a 1°C rise in temperature due to climate change processes may affect crop yield both ways. It could induce biomass formation in locations at higher latitudes, but in tropics and lower latitudes it could reduce biomass formation.

Siebert and Ewert (2014) have stated that severity of extreme temperature may get accentuated in future years. Such extreme temperatures actually interact with several different climate change factors. It results in complex patterns of effects on crops. The positive effects of enhanced CO_2 are clear, but simultaneously higher temperature decreases crop growth and yield. Modeling the effects of heat waves has its advantages, particularly, while prescribing remedial measures (Deryng et al., 2014; Korzekwa, 2018; Rotter et al., 2011).

Ambient temperature is a factor intrinsic to the present concept, that is, "Agricultural Sky." Ambient temperature affects a series of aspects of any agrarian region. Particularly, it affects soil and various physico-chemical and biological activities. Regarding crops in an agroecosystem, it influences first the seed germination and seedling establishment. Growth stages such as foliage formation, boot leaf/panicle initiation, and grain formation are stringently influenced by the seasonal fluctuations of temperature. The choice of cropping system and sowing time is highly influenced by the ambient temperature.

It is said that climate change does not always affect agrarian regions and field crops detrimentally. For example, temperature increase has often shown a stimulation in crop growth and productivity (Lobell et al., 2005). In Mexico, between 1980 and 2001, such alterations in temperature have increased wheat yield by 25%. In case of maize, crop productivity has shown an increase in grain yield, due to increase in ambient temperature. However, after a certain threshold temperature, further increase in ambient temperature reduces crop yield (Zhu and Troy, 2018). The threshold temperature for soybean is 30°C and for maize it is 29°C. In India, wheat exposed to temperature beyond threshold has reduced grain yield by 6% for every degree increase beyond 34°C. The influence of interactive effects of temperature and precipitation on crop is complicated.

Zhao et al. (2017) have tried to remind us that crops such as wheat, rice, maize, and millets provide us with 66% of calorie requirements, on a daily basis. It is therefore too important for us to know a great deal about influence of temperature on these very crops, first. Based on a detailed global survey of productivity trends they have concluded that temperature above normal is detrimental to crops. For example, in case of wheat, each degree C above normal depreciates grain yield, by 6.0% of the normal. About 3.2% decrease in rice grain yield, 7.4% decrease of maize grains, and 3.1% of soybean grains is to be expected, for every °C rise above optimum temperature. The above depreciations of grain yield occur despite selecting best genotypes and agronomic procedures. Yet, such depreciation in grain yield due to higher temperature depends on several characteristics of the geographic location (Zhao et al., 2017).

1.2.2 AIR, ITS GASEOUS COMPOSITION, AND GREENHOUSE GAS EMISSIONS

Agrarian regions are bestowed with a mixture of gases that is collectively called "air." The air above agrarian terrains is composed of 78% nitrogen gas, 21% oxygen, 1% argon, and 0.036% carbon dioxide. Other gases such as carbon monoxide, methane, sulfur di-oxide, chloroflourocarbons, and propane also occur in traces. In addition, agrarian regions show up dust particles, carbon soot, asbestos, lead, ozone, and a few other particulate matters. Fluctuations in the gaseous mixture have its impact on crop growth and productivity. During recent years, enhanced concentrations of CO_2, CO, NO_2 have been perceived in agrarian regions. Locations closer to industries often show higher levels of carbon particles and SO_2 concentration. They affect crop's physiological manifestation adversely, if found beyond threshold (Grewer et al., 2016). Air above the coastal cropping zones may get influenced by the evaporation of sea water and breeze.

One of the basic facts to be recognized is that between 1880 and 2018 the earth's temperature increased by 0.85°C, on an average. Greenhouse gas emissions and resulting alterations in the contents of each gas in the atmosphere have been attributed as the cause, for increased temperature. Among gaseous emissions from farming zones, methane (CH_4) has 25 times the warming potential of CO_2 and N_2O has 298 times more warming potential. At this juncture, we may note that "Paris Agreement" intends to fix the increase in temperature to < 2°C. It asks us to reduce GHG emissions by 20% by 2020 and by 30% by 2030. *Here, we should note that agricultural emissions account for 23% of global GHG emissions.*

Forecasts suggest that enhanced population levels by the year 2050 may mean intensive cultivation of crops. It definitely involves higher fertilizer inputs. Regarding fertilizer-N, improved management is necessary. A combination of organic-N and legume cultivars will help to increase nitrogen efficiency. It may reduce nitrous oxide emissions. Low N-emission fertilizers need to be utilized. Use of nitrification inhibitors should be made mandatory. In the future, emissions from dairy, particularly CH_4 with greater potential of increasing atmospheric temperature, may be generated, at 30–40% levels more than at present (Lanigan–Teagasc, 2017; USEPA, 2006).

Lanigan–Teagasc (2017) state that we have to stabilize GHG emission rates, particularly, CH_4, CO_2, and NO_2. Further, we have to adopt methods that enhance carbon sequestration in the soil, instead of releasing it to atmosphere. Also, we have to reduce burning of fossil fuel and crop residue wherever feasible. Livestock farming is a source of greenhouse gas (GHG) emissions. As stated above, it emits one of the most potent greenhouse gases—methane (CH_4) (Food and Agricultural Organization of the United Nations, 2006; USEPA, 2006). However, through manure anaerobic digestion technology, a portion of these emissions can be mitigated (Pronto and Gooch, 2009).

The bottom-line suggestion for reducing emissions from agricultural zones is to produce crops with least possible input levels. This will ensure that emissions of GHG such as CO_2, CH_4, N_2O that contribute to climate change will automatically reduce. Stabilizing CH_4 emissions from dairy industry is essential. Lanigan–Teagasc (2017) suggest that an integrated approach involving farmers, industry experts, and environmental specialists is essential to mitigate GHG emissions (Goodland and Anhang, 2009; International Centre for Tropical Agriculture, 2019).

Taub (2010) points out that atmospheric concentrations of carbon dioxide have been steadily rising, from approximately 315 ppm (parts per million) in 1959 to a current atmospheric average of approximately 385 ppm. Current forecasts indicate that it might rise to as much as 500–1000 ppm by the year 2100 (IPCC, 2007). So, high concentrations of atmospheric CO_2 predicted for the year 2100 will have major implications for crop's physiology. Enhanced CO_2 in the atmosphere induces higher rates of photosynthesis.

For example, at CO_2 concentration ranging from 475 to 600 ppm most plant species show higher rates of photosynthesis. It increases by an average of 40% over normal (Ainsworth and Rogers, 2017). The CO_2 concentration in the ambient atmosphere also regulates the stomatal activity. No doubt, photosynthesis and stomatal behavior are central to plant growth, biomass, and yield formation. It seems biomass increased on average by 17% for the aboveground, and more than 30% for the belowground portions of plants (Ainsworth and Long, 2005; de Graaff and Van Gronigen, 2006). This increased growth is also reflected in the harvestable yield of crops. Reports indicate increases in grain yield of 12–14% under elevated CO_2 (Ainsworth, 2008; Long and Ainsworth, 2006). As a consequence, leaf carbohydrates per unit leaf area increase, up to 20–30% of original levels. Also, nitrogen and mineral concentration in the leaf tissue decreases due to dilution (Ainsworth and Long, 2005; Taub and Wang, 2008). The effects of elevated CO_2 vary based on crop species. Crops that utilize the C_4 photosynthetic pathway respond relatively feebly to elevated CO_2 than do other types of plants that adopt C_3 mechanisms. Rising CO_2 is therefore likely to have complex effects on the growth and composition of natural plant communities.

1.2.2.1 GREENHOUSE GAS EMISSIONS (GHG)

Reports by the Environmental Protection Agency of USA suggest that, overall, several economic sectors emanate GHG. Agriculture is among major GHG emitters. Electricity generation leads at 33.5% of total GHG emission. Transportation contributes 27.7%, Industries emanate 18.6%, and Residential users cause 5.0% GHG emission. Fossil fuel use leads to 7.2% GHG emissions, while other sources account for 4% GHG emission. Agriculture emanates 8.1% GHG. Again, out of this 8.1% GHG emitted, there are several sectors within agriculture that contribute to the emissions. Agricultural soils are the major emitters of CO_2 and other GHGs. Agricultural soils account for 61% of total agriculture-related GHG emission. Enteric fermentation (dairy) results in 18% GHG, manure management releases 9%. Fossil fuel burning accounts for 7% GHG emissions and other sources emit 4% of GHG (Faulkner and Easton, 2015; Grewer et al., 2016; Herrero et al., 2009, 2011; U.S. Environmental Protection Agency, 2011, 2014). It is clear that soil management procedures need to be stringently regulated to curtail GHG emissions from agrarian regions. Reports by FAO of the United Nations suggest that animal feed production, enteric fermentation, and manure management are major sources of GHG from the livestock sector (Food and Agricultural Organization, 2007). Ammonia volatilization from livestock production systems is also responsible for indirect emissions of N_2O. The pastures and cattle need to be maintained efficiently to manage GHG at lowest levels. The GHG emissions from livestock sector are estimated at 17.5% of the total anthropogenic emissions of GHG (Eyshi et al., 2014; IPCC, 2007, 2014). There are opportunities for adoption of mitigation strategies in order to minimize such emissions.

Atmosphere, which surrounds the earth, is an important region that supports life and its activities. The atmosphere is composed of several gases. The atmosphere is a life-giving blanket of air that surrounds our Earth (NASA, 2016). These gases protect natural vegetation, forest stands, crops, animals, and humans from the Sun's intense ultraviolet radiation. This is a crucial aspect since it allows life to flourish on earth. Many of the recent reports highlight that greenhouse gases such as carbon dioxide, ozone, and methane are increasing from year to year. These gases trap infrared radiation (heat) emitted from Earth's surface and atmosphere, causing the atmosphere to warm. On the other hand, clouds and aerosol that are composed of minute solid particles such as dust, smoke, and pollutants actually reflect the sun's radiative energy. So, it leads to cooling. The balance between the incoming and reflected solar radiation and emitted infra-red energy, helps in regulating the earth's climate and sustaining life (NASA, 2016).

Considering the present context, we can say that greenhouse gas (GHG) emission is a natural phenomenon that regulates the temperature of earth's surface and atmosphere. The sun's heat that hits the earth is reflected to a certain extent. This heat is trapped by GHG such as water vapor and CO_2. In the absence of GHG, earth's temperature could be much colder. Agricultural activity is among major GHG generators. Agricultural crop production methods lead to emission of GHG such as CO_2, CO, CH_4, N_2O, NO_2, VOC. The different GHGs have different potencies in the atmosphere. The global warming potential of GHGs differs enormously. It is expressed as a "carbon dioxide equivalent or CO_2e." As stated earlier, for methane (CH_4) it is 21 and for nitrous oxide (NO_2) it is 320. It means CH_4 is 21 times more potent in causing global warming and NO_2 is 320 times more potent in causing global warming compared with CO_2. So, such GHGs entrap higher quantities of heat compared to CO_2. It is said that levels of these GHGs are increasing at a faster pace. It is higher than at any time previously during past centuries. Hence, earth's surface temperatures are increasing at several locations. Such changes could result in higher evapo-transpiration from agrarian regions. It affects precipitation pattern, induces flash floods, and storms above earth's surface.

Nitrous oxide emissions contribute to the greenhouse effect. Further, such NO_2 emissions also induce depletion of stratospheric ozone. We may note that 90% of the atmospheric N_2O is generated, during the microbial transformation of nitrate (NO_3) and ammonia (NH_4^+), in soils and water (FAO, 2015).

Ammonia is absorbed through the leaves via stomata. This physiological mechanism enhances the N uptake by crops supplied with fertilizer-N using foliar sprays. The N uptake efficiency of foliar fertilizer-N is generally higher than that applied to soil. Soil injection of granules/powder results in series of transformations due to soil chemical and microbial processes. There is also fixation of N into organic fraction of soil that reduces fertilizer-N efficiency. However, consequences of foliar uptake and processing of an alkaline gas on terrestrial vegetation or crops need to be understood, in greater detail.

Ammonia is applied in a liquid form dissolved in water or as aerosol. Such a method is utilized by many farms in North America and Europe, to supply N to soil and crops. Ammonia is a highly reactive gas, and it is alkaline in reaction. Ammonia is generated due to natural processes as well as anthropogenic reasons. Agricultural crop production that utilizes excessive dosages of N to achieve higher yield goals is among the major reasons for ammonia liberated into sky. Ammonia is released into air due to breakdown and volatilization of urea (fertilizer N) applied to fields.

The agricultural sector in the African continent is among the major GHG emitters. The GHG emission is increasing rapidly as time lapses. It is attributed to high inputs that are channeled into farmland. They say, Eastern African farmland emits the highest amount of GHG among African nations (Tongwane and Moeletsi, 2018). The crop residue burning is a major cause of such emissions (see Plate 1.4).

Atmospheric ammonia has impacts on both local and international (transboundary) scales. In the atmosphere, ammonia reacts with acid pollutants such as the products of SO_2 and NO_2 emissions, to produce fine ammonium (NH_4^+) containing aerosol. While the lifetime of NH_3 is relatively short (<10 to 100 km), NH_4^+ may be transferred much longer distances (100 to >1000 km).

Ammonia (NH_4^+) is absorbed by the leaves via stomata. Ammonia volatilizes rapidly. So, it is emitted into the atmosphere when the surface concentration exceeds that of the surrounding air. Losses of NH_3 by volatilization of fertilizer-N formulation range from negligible amounts to >50% of the applied fertilizer-N. The extent of N loss due to volatilization depends on chemical fertilizer/manure type. For a chemical fertilizer such as urea, volatilization rates are higher than ammonium nitrate. Method of application (e.g. foliar spray or injection or surface application) and environmental conditions also affect emission rates (Freney and Denmead, 1992; Freney et al., 1991; Peoples et al., 1997; Peoples and Herridge, 1990). Solubility and dissolution processes primarily drive the magnitude of NH_3 emissions. It is higher in warm drying conditions and smaller in cool wet conditions.

1.2.3 WIND AND RELATED PHENOMENON IN THE AGRICULTURAL SKY

Wind above crop fields is caused due to differential pressure in the atmosphere. The wind could be due to movement of large mass, or it could be small stream or jet flow. Wind that flows close to ground cools, contracts, and then the pressure rises. When it warms, the air mass expands, and the pressure reduces. On a large scale, winds and differential pressures in the atmosphere cause seasonal wind patterns. Such patterns affect weather over large agrarian belts (e.g. monsoon). Wind is useful to agriculturists in different ways. For example, it hastens pollination of flowers in fields. Wind pollinated crop species depend on it. Wind causes dispersal of propagules of disease-causing microbes (Isard and Russo, 2011). Wind could be utilized at low (e.g. windmills) and high altitude over crops, to generate power (tethered aerostats with turbines) (Boyer, 2012; Cherubini et al., 2015; Glass, 2018; Krishna, 2020a, 2020b; MIT, 2013; Stanford Report, 2009; Wikipedia, 2020a). Wind also affects dispersal of weed seeds, insects, microbes etc.

Wind shows horizontal movement based on atmospheric pressure differences. It transits from areas with high-pressure to low-pressure zones. Further, variations in ambient temperature and solar radiation too cause wind. As a consequence, air moves rapidly from one location to another, in the atmosphere. With regard to effect of wind on crops, we may note that rapid movement of wind (breeze) increases evapotranspiration. This leads to high requirement of water by the crops (Singh, 2017). High-speed wind (or dust storms) also disperses nutrients held on dust and soil/sand particles (Sterk et al., 1996).

Tornadoes are a weather-related phenomenon caused by ambient temperature, atmospheric pressure, and wind. The frequency, number, intensity, and destructive power of tornadoes may vary. There are also certain areas to which tornadoes are endemic. In North America, for example, over 1200 tornadoes occur annually. Most tornadoes in the United States occur east of the Rocky Mountains. The Great Plains, the Midwest, the Mississippi Valley, and the southern United States are all areas that are vulnerable to tornadoes (also called "Tornado Alley'). They occur seasonally. For example, in the Southern and Central Great Plains of North America (Kansas, Oklahoma, Texas), Central Florida, and Great Lakes (Michigan, Minnesota, Wisconsin) region, tornadoes occur periodically. It causes devastation to standing cereal crops, timber trees (Coblentz, 2019), farms, and urban properties etc (Farmlife, 2019).

Agricultural sky perceives tornado whenever moist air lashes. Thunderstorms capable of creating tornadoes usually occur in the afternoon, that is, when atmospheric temperature is highest (i.e. 16.00–19.00 hr. of a day). Tornadoes peak immediately after spring and as summer begins. It coincides with March to June end (Tornado season) in Southern Great Plains. However, we should note that tornadoes occur during any period of the year. For example, Florida's violent tornadoes occur in the winter. It is attributed to cold air that lashes the region.

Tornadoes are usually identified and classified based on the extent of damage. Then, the speed of the wind is gauged. For example, a low-intensity low damage causing tornado is classified as "EF0." The other extreme in the "Fujita scale" of destructive power of a tornado is classified as "EF5." Its wind speed is usually above 200 mph or 320 kmph. Tornadoes grouped under EF4 and EF5 using Fujita scale are termed "Violent Tornadoes." These tornadoes shift large quantities of surface soil and other organic matter from one location and spread it to longer distances. Tornadoes are forecasted using data received from weather satellites and Doppler Radars (Kingsfield and DeBeers, 2017; Wikipedia, 2019c). Tornadoes could naturally disperse seeds of weeds, crops, microbes, insects, etc.

In West Africa, wind and dust storms are observed frequently. Heat transfer from soil surface to atmosphere also generates turbulence. In Sahelian region (Senegal, Mali, Burkina Faso, Niger, and Chad) we can observe the "conical tornadoe." Locally, they are called "dust devils." Such dusty tornadoes in Sahel could be small. They are usually couple of meters in width and height. Sometimes, tornadoes may reach 1000 m in height. They distribute soil, dust particles, biotic fraction such as microbes, insects etc. from one location to another in the sandy region. Such tornadoes appear for short duration and disappear. Often, they get generated in the afternoons when soil/atmosphere gets heated and then cools suddenly later. They are given different names in different languages (UNCCD, 2001).

Tornadoes affect both the soil surface and crops. The soil particles, dust, and even larger debris are disturbed and lashed into atmosphere. The soil and dust matter are transported to large distances of many kilometers and deposited. Therefore, a tornado mixes the air, soil particles, dust, microbes, insects etc. from one location with others. Localized tornadoes may hurl sand/dust particle only for shorter distances. However, dust storms that transit horizontally may distribute sand/dust particles along with nutrients and micro-organisms resident on them, to long distances. It could cover continental distances before settling at a place.

Airborne wind power generation is a recent technology. It is yet to make a mark with the general population. Wind power is sufficiently a large source of energy (Bull and Phillips, 2018; Cassimally, 2012; Inman 2012). Wind power generation at low and high altitudes is affected by the velocity. (Glass, 2018). We should take note that agricultural sky is not always a region that is abode to intense precipitation, storms of high strength and destructive power, or tornadoes. Agricultural sky is not just the region that harbors disease-causing organisms (e.g. Rust fungi, bacterial pathogens, viruses) and disperses them, or the one that spreads aerial insect pests (e.g. Locusts, flies, bugs, aphids, etc.) and detrimental bird pests.

The agricultural sky also supports uniform breeze of wind above the cropping belts. The low altitude winds get interrupted by ground features (e.g. trees, wind brakes, undulated terrain, and buildings). Therefore, wind could be of low velocity. At high altitude (200–2000 m.a.s.l.) winds are uniform, less turbulent, and with least interruption. They could be suited best to generate electricity in the sky, above the farms. Almost each large farm can own a tethered aerostat with turbine and generate the electric power! They say, usually, a single turbine placed aloft by an aerostat suffices to supply electricity for 7 to 8 farms on the ground (Glass, 2018; Krishna, 2020a, 2020b).

The wind above the agricultural expanses and other locations such as coastal region could be utilized efficiently by generating wind power (electricity). Windmills of variety of sizes, capacities, and efficiencies are available. Windmill could be adopted for lift irrigation. Wind power generation above the agrarian zones could be classified at least into two types. A. low-altitude wind power generation usually generates electricity of lower order. It is not very consistent. Interruptions due to variations in the wind velocity are common. High altitude wind power generation utilizes the stronger and uniform wind currents that blow at altitudes above 200–2000 ft. from the ground or crop's canopy level. Recently, aerostats have been effectively employed to generate electricity from high altitude. They have constructed helium-filled aerostats with light weight turbines. Such turbines are floated at high altitudes. High-altitude wind power is generally more consistent since the wind currents are uniform. Winds rarely get interrupted at high altitudes (Glass, 2018; Krishna, 2020a, 2020b).

Wind-aided pollen transfer is an important phenomenon related to agricultural sky. Insufficient production and liberation of pollen and accompanied by low winds may affect normal pollen dispersal. It leads to reduced seed set and yield.

1.2.4 CLOUDS ABOVE AGRICULTURAL FIELDS

The "agricultural sky" could be clear or cloudy depending on the geographic region and season of the year. Clouds are most conspicuous aspects of an agricultural sky. Clouds, their timely formation, and causation of precipitation are perhaps the crucial factors. Clouds influence crop production immensely in any given agrarian region. Clouds bring precipitation that is crucial for crop growth. On a yearly basis, clouds determine the extent of total rain that impinges the agrarian regions. The agrarian sky supports formation of several types of clouds. Each has its special effects on the atmosphere and vegetation on the ground.

Crops absorb maximum photosynthetic radiation in the range of 400–700 nm. Clouds have a direct role in the process of radiation capture by crops and its use efficiency (Murchie and Reynolds, 2018). Fluctuating cloud covers often affect the crop productivity of large expanses. In nature, crops have to adapt themselves rapidly to changing intensities and durations of photosynthetic radiation. Light from Sun, as stated earlier, is affected by cloud cover, smog, haze, precipitation events, drizzle for long durations, shades etc. To maintain optimum rates of photosynthesis, it seems crops utilize a protein known as "thioredoxin." This protein helps plants to adapt to the variations of light intensities (Kromdijk et al., 2017; Smith, 2019).

Cloud cover obstructs and reduces solar radiation from reaching the crop's canopy. Hence, it affects photosynthetic efficiency of a crop. During a season, if cloud cover occurs for a long duration, then, carbon fixation, biomass, and grain yield formation may all get affected, negatively. However, we may note that cloud cover that occurs transitorily for short duration over the crops has certain beneficial effects. Crops that loose water due to excessive transpiration may derive benefits, due to cloud cover. A good cloud cover reduces loss of water from the canopy. For example, cloud cover that occurred for 10% of the time during day helped forest species by reducing loss of moisture (Young and Smith, 1983).

Actually, atmospheric phenomena above the crop's canopy lead to formation of clouds. Such clouds have a definite role. They help in the modulation of solar radiation and loss of infrared radiation. Such clouds are also the locus for the formation of precipitation. Precipitation events and pattern of seasonal rains are dependent on cloud formation process and their transit. According to observations by National Aeronautics and Space Agency (USA), for the past few decades, clouds have a few crucial functions in the atmosphere. They are clouds that cool the earth's surface by reflecting incoming sunlight. Clouds warm earth's surface by absorbing heat emitted from earth's surface. Clouds could either cool or warm the surface of cropping zones (Tselioudis, 2019).

Satellite imagery using high-resolution sensors could be obscured by cloudiness. Mere hazy images with low resolution are possible under cloudy conditions. Such hazy images do not provide necessary digital information about the crops. Lack of clear images of agricultural terrain is listed among the disadvantages of an extended cloud cover (Witchcraft et al., 2015a, 2015b). The cloud cover interferes with regular optical observation of global agricultural regions. The problem due to cloud cover is greater if the revisit-time of satellites is longer. In fact, reports suggest that, obtaining cloud-free images using LANDSAT or MODIS satellite with longer revisit period is really difficult.

In tropical and semiarid regions there is evidence that, at regional scales, a moist, cooler, vegetated surface increases cloudiness. It induces rainfall in even larger areas based on wind velocity (DeAngelis et al., 2010; Harding and Snyder, 2012, 2014; Puma and Cook, 2010; Spracklen et al., 2012; Spracklen and Garcia-Carreras, 2015). Cloud cover is significantly correlated with inter-annual variations in vegetation. Reports about cloud feedback suggest that, in arid lands, it is the opposite of what has been previously observed in tropical and semiarid regions. Clearly, different sets of processes might be involved in cloud formation and consequent rainfall in arid regions. Such cloudiness, diffuse radiation, and reduction of ambient and soil temperature may benefit vegetative growth in arid lands (Carreras, 2016).

Rice is cultivated under luxuriant sunlight available in tropical sky. Yet, the extent of cloudiness and the number of cloudy days can affect the crop's productivity. The photosynthetically active radiation (PAR) intercepted depends on season, planting time, plant density, and genotype. Cloudy weather and extended rainy period affect crop growth. Cloudiness at critical stages often resulted in reduced grain yield. The quality of grains may also get reduced. Chaffy grains may increase if it is cloudy during seed-fill. Low light stress has rigorously constrained the rice production in various rice-growing regions, especially in Southeast Asia (Gautam et al., 2019). Selecting suitable planting dates to avoid clouds in the agricultural sky during critical stages of the crop is no doubt essential.

Clouds in the agricultural sky are no doubt a positive factor when we count on the precipitation. They bring about increased grain yield. However, clouds for extended periods do create shaded conditions resulting in reduced interception of sunlight. Like other cereal crops, maize has critical stages when photosynthetic radiation should be available. Maize needs sunlight during silking and grain fill stages. Indeed, some reports during 2019 showed that paucity of sunlight due to cloudiness during August month did affect maize grain productivity. Studies on relationships between shade created artificially or via clouds have shown that productivity reduces, particularly, if cloudy days are more during silking (Nelson, 2015; Yunshan et al., 2019). Shading during reproductive stages can cause a 30% reduction in grain yield (Schmidt and Colville, 1967; Yunshan et al., 2019).

Let us consider an example. Here, the importance of sunlight and role of clouds on tree crops such as apple orchards have been focussed. Basically, interception of solar energy is crucial to the development and fruiting pattern of the apple orchards. Light interception may depend on geographic region, season, cloudiness, planting pattern and density of orchard, leaf area indices of cultivars, pruning methods adopted etc. Sunlight bestowed via "agricultural sky" has to be intercepted with greater efficiency. We have to note the number of sunny versus cloudy days. Yield potential of orchards depends on sunlight harnessed by the trees. The relationship between sunlight intercepted and apple productivity is a thoroughly studied aspect. Such data has been accrued for several cultivars of apples planted in different geographic locations (Lakso and Robinson, 2015). Clouds that can reduce interception of sunlight are important factors affecting the productivity of apple orchards. Clouds may reduce light interception during stages when it is most crucial to apple tree. Generally, in apple trees, carbohydrates are synthesized during seasons when sunlight is luxuriant.

First, clouds reduce photosynthetically active radiation from reaching the crop's canopy. Clouds and cloudiness have a tendency to enhance relative humidity in the atmosphere. Such levels of RH are known to encourage fungal growth. Fungal growth and its multiplication may get enhanced leading to diseases on crops.

As stated earlier, clouds and cloudiness affect the accuracy of satellite imagery obtained about agricultural zones. There are several studies involving remote sensing of Brazilian agricultural regions. They depict the difficulties and inaccuracies that cloud cover over the farm belts cause (Asner, 2001; Eberhardt et al., 2016; Witchcraft et al., 2014, 2015a, 2015b; Wilson et al., 2014). In some cases, correction factors could be used, to arrive at clearer maps of agrarian regions. Therefore, clouds in the agricultural sky that bring in precipitation may be detrimental if aerial imagery of high accuracy is required. Hence, during past few years, agricultural experts are consistently recommending use of unmanned aerial vehicles (UAV). Such small drone aircrafts or tethered aerostats are adopted, to obtain sharp and highly accurate maps of each farm, the terrain, crops, their boundaries, and growth status (see Krishna, 2016, 2018, 2020a, 2020b).

1.2.5 PRECIPITATION

1.2.5.1 RAINFALL

Precipitation of water occurs in different forms. Rainfall is the conversion of water vapor held in clouds to droplets that fall on the ground surface and vegetation. It is the most common form of precipitation. Precipitation of water may also occur as ice, frost, snowflakes, or hailstorms depending on geographic location. The seasonality, diurnal temperature changes, and clouds induce such precipitations. The interaction between the precipitation and temperature is of utmost importance, to agriculturists. It almost decides the type of vegetation, pattern of biomass formation, seasonal changes in vegetation, diversity of vegetation, crop species cultivated, and crop productivity. Lack of precipitation that induces drought can be severe, if temperature too is unfavorable. *No doubt, the agricultural sky and its manifestations literally hold sway over agrarian regions,* particularly, in terms of cropping patterns, crop species, and productivity.

Lobell et al. (2011a, 2011b) point out that there is considerable literature and data about the seasonal changes in climatic parameters such as precipitation. Historical trends of precipitation for each of the major food-generating agroecosystems could be easily consulted. Especially, if one wants to draw useful inferences. For example, we have daily precipitation and temperature data in great detail from 1950 till 2018. Our inferences about the impact of changes in agricultural sky and its influence on cropping systems could be based on such data pool.

Next, growing season total rainfall is an independent variable. It is a proxy for the potential water supply for crop growth. Lobell et al. (2011a, 2011b) state that regional trends for precipitation pattern could be discerned. But on a global scale there was no clear trend attributable to climate change effects, during the past 6 decades. A few other reports state that precipitation and temperature trends have decreased European wheat yield by 2.5% but have slightly increased maize yields (Lobell and Burke, 2010; Moore and Lobell, 2015). Generally, climate trends have caused a decline of 3.8% and 5.5% in global maize and wheat yields, from 1980 to 2008 (Lobell et al., 2011a, 2011b). But, yields have declined with increasing night-time temperatures (Lobell et al., 2008).

Let us consider an example from wheat farming regions of Central and Southern Great Plains of North America. The agrarian region here is deemed as dryland with annual precipitation lower than that required for cropping systems adopted. A wheat/legume cropping system is receiving about 400–450 mm of rainfall from the agricultural sky. It is clearly deficient. The evapotranspiration in this cropping belt is about 1460 to 1975 mm. Hence, Peterson and Westfall (2005) recommend adoption of as many methods and improvements that maximize precipitation use efficiency. Precipitation as a phenomenon is regulated through agricultural sky. This in turn stipulates the crops, cropping systems, and yield levels in an agroecosystem. Therefore, we need to procure data, fit accurate models, and adopt agronomic procedures that enhance precipitation use efficiency. For example, No-till systems, or wheat-sunflower cropping systems etc. can be adopted (Patrignani et al., 2018). Total precipitation and its pattern clearly decide the wheat genotype that dominates the agrarian landscape. Cultivars tolerant to low precipitation levels get preferred. Also, we must realize that precipitation pattern as a factor must have guided the genetic diversity of natural vegetation and wheat-fallow cropping system, by acting as an important selection pressure. It means agricultural sky acts as an important "selection pressure" of crops. Low rainfall received through the sky selects only crops/natural vegetation tolerant to scanty precipitation. Further, a good example is the selection of drought-tolerant crops that suit dryland areas. Here, the agricultural sky stipulates that only drought-tolerant crop species/genotypes will be suited. We must realize that agricultural sky, particularly, has immense influence on genetic diversity of crops and natural vegetation that flourishes in a given region.

1.2.5.2 FROST IN AGRARIAN REGIONS

Frost, its formation and distribution are phenomena that occur in the agricultural sky. Frost may manifest in the sky and affect crops cultivated in certain locations within a continent. In the temperate regions, crops are prone to suffer damage due to frost (i.e., frost injury). Frosts can have severe effect on seed germination. Frost also affects seedlings. It retards leaf formation, growth, and maturity of crops. Frosts affect the crops to different degrees based on location, topography, and general vegetation pattern of the field. For example, in California, citrus orchards located in the eastern flank of San Joaquin valley are least affected by frost damage. The wind direction and velocity are important aspects to consider (FAO, 2003). Frost damage also depends on the extent to which a particular crop or its organs are susceptible. The stage of crop at which frost deposition/damage occurs is also important. We can identify at least two types of frosts that afflict the crops. They are advection frost and radiation frost.

Frost damage to crops results not only from cold temperature but mainly from extracellular (i.e. not inside the cells) ice formation inside plant tissue, which draws water out, dehydrates the cells, and causes injury to the cells. Following mild cold periods, plants tend to harden against freeze injury, and they lose the hardening after a warm spell. A combination of these and other factors determine the temperature at which ice forms inside the plant tissue and when damage occurs. The amount of frost injury increases as the temperature falls and the temperature corresponding to a specific level of damage is called a "critical temperature" or "critical damage temperature", and it is given the symbol "Tc" (De Mello-Areu, 2018).

Both cloud cover and fog increase downward long-wave radiation and lessen net radiation losses. The occurrence of a "radiation frost" is rare during cloudy or foggy conditions. It is because the net radiation losses are reduced. On rare occasions, sub-zero temperatures occur during cloudy conditions associated with an advection frost. However, radiation frosts are considerably more common than "advection frosts" in certain areas. According to FAO (2003), there are several precautionary aspects while developing a farm. Also, while producing food crops or trees in an area prone to frost damage. They are appropriate selection of site considering the topography and other related aspects; managing cold air drainage; selection of cold tolerant crop species for production; planting canopy trees; following correct fertilizer schedules; adopting proper agronomic procedures; applying plastic mulches; heaters, helicopters, sprinklers, proper pruning; applying chemicals to delay bloom if it coincides with frost appearance; growing cover crops; applying truck covers if it is a tree crop; etc. Above all, farmers should be consulting the weather forecasts rather regularly and adopt procedures, accordingly. Again, here, the agricultural sky stipulates use of cold/frost tolerant crop species. Genetic selection of cereals such as wheat and barley are being conducted in many agricultural stations located in temperate climate. This is to match the crop with the temperate sky.

1.2.5.3 SNOW AFFECTS AGRARIAN REGION

Snow is frozen precipitation composed of white or translucent ice particles in complex hexagonal patterns. Mist is a phenomenon caused by small droplets of water suspended in air. Physically, it is an example of a dispersion. It is most commonly seen where warm, moist air meets sudden cooling, such as in exhaled air in the winter, or when throwing water onto the hot stove of a sauna. It can be created artificially with aerosol canisters if the humidity and temperature conditions are right. It can also occur as part of natural weather when humid air cools rapidly. For example, when the air comes into contact with surfaces that are much cooler than the air.

Dew is the moisture deposited in the form of water droplets on the surface of natural vegetation, crops and other objects located near the ground level. It forms when nocturnal (night-time) terrestrial temperature falls below normal. Mist and dew play a vital role in drylands where rainfall is meagre. The little moisture that settles on canopy/leaves assures that plants survive with small amounts of moisture. Such moisture from dew is absorbed via stomata and hydathodes on the leaves. Often, in semiarid and arid cropping zones, dew provides lifesaving moisture to crops.

1.2.5.4 FOG AND SMOG IN THE AGRICULTURAL SKY

Fogs are fairly common occurrence in agricultural zones. Fog is made of water droplets. These droplets disperse the sunlight. As a result, it reduces visibility on the ground. Concurrently, fog reduces the net photosynthetic radiation that impinges the crop's canopy. It can affect the quality (wave-length), intensity, and duration of light interception by crops. Fog layers form when moist air is cooled to its dew point, that is, saturation point. We can easily identify a few different types of fogs. All of them are not experienced by an agrarian region. Their formation depends on the geographic location and its weather patterns, seasons, etc.

Radiation fog seen above the crop's canopy is caused when surface heat is radiated back into sky, during the night times. As the earth's surface cools, the air reaches high humidity which eventually transforms into fog (water droplets).

Advection fog is formed when warm moist air moves over a cold surface. It results in condensation of moisture into droplets and hence fog is created. Such fogs are common above sea surface or in the coastal agricultural regions.

Unslope fog is experienced in the mountainous or hilly regions. It is caused when winds push the moist air on the sloppy terrain. The moist air condenses to form the fog.

Ice fogs are traceable in temperate agrarian regions. They are formed when the temperature of air that blows over the crop fields is below freezing point.

Freezing fog is formed when the air that lashes above the pastures and crop's canopy has "super cooled" water droplets. The ground surface and crop's canopy could be covered with ice crystals (Kim, 2018)

All of the above types of fogs are detrimental to optimum physiological activity of the crops. Fogs first can reduce photosynthetic light reaching the crop's foliage. Therefore, if fogs occur during crop season it reduces net carbon fixation and biomass accumulation rates. In addition, fogs that bring in cold weather and moisture droplets/ice affect the optimum physiological activity of crops. Crop species and their genotypes selected for the season have to be "cold tolerant." At the minimum, crop genotypes should survive the entire duration of cold season. Overall, fogs can reduce both optimum light interception (i.e. photosynthetic rates) and metabolic processes. It depends on the type of fog that hits above the agrarian region.

Smog is a mixture of smoke and fog. The smoke often includes carbonaceous particles plus pollutants.

Smogs occur above the cities and industrial zones that experience high ambient temperatures. Similarly, large- scale crop-residue burning within an agrarian region can cause high temperatures. Resultant wind currents induce smog formation. For example, smog above Chinese cities that are blown into the adjoining richly cultivated agrarian regions may reduce crop yield. The smog reduces photosynthetic light interception. Smog also reduces biomass/ grain formation due to pollution. It is common to see malformed leaves and canopy if the pollution effects due to smog are severe. Smog detection and analysis of gaseous and particulate fraction is important. Forecasts using data from aerial observation, for example, using drones and weather balloons are useful (Krishna, 2018).

There are a few terms related to smog and fog. Haze, for example, is an atmospheric phenomenon closely related to smog. It occurs when dust, smoke, and dry particulate contaminants mix in the atmosphere. It can affect light interception by crops. Haze is again an atmospheric phenomenon that results due to excessive sulfur di-oxide or other similar gases in the air. This is a situation common to agrarian regions close to volcanic eruptions or industrial emissions.

1.2.6 RAINSTORMS, HEAT WAVES, DUST STORMS, AND DROUGHTS

Let us define and explain a few terms relevant to storms. First, storm means it is a violent disturbance in atmosphere concerning the wind, cloud, precipitation, and sometimes even dust. Storms mostly have to do with the weather conditions with an accentuation of strong winds. There are indeed several types of storms identified by common man and expert weatherman. They are hailstorm, ice storm, rainstorm, snowstorm, thunderstorm, windstorm, cyclone, hurricane, and tornado. We should note that all of these storms, any kind, any intensity, or any duration, are basically a phenomenon related to and occurring in the sky. Many of the above kinds of storms get manifested in the "Agricultural sky." It impacts crops, from seeding stage to maturity/ harvest. They can be severe on crops. Soil erosion and loss of fertility is a clear possibility if the storms are severe. A hailstorm is accompanied by hailstones that is, ice pellets larger than 5 mm. Ice storm is almost similar to hails. But the precipitation converts to ice as it lands on ground surface. The rain freezes to collect as ice. In agrarian regions prone to cold fronts, snow, or ice sheets, the food crops have to possess genes that bestow a certain degree of cold tolerance. For example, wheat cultivars grown in Northern Great Plains in Canada, Dakotas, or those cultivated in Scotland, Ukraine, or Greater Mongolia or southern Argentina need to be cold tolerant.

A rainstorm is characterized by large quantities of precipitation. It is sometimes denoted as heavy rains. A snowstorm is generally windy with ice crystals falling incessantly on the ground. A blizzard that affects standing crops is actually a violent snowstorm with wind speed reaching above 56 kmph. In the temperate regions standing crops with seedlings and even tree plantations do suffer from snow damage, sometimes rather severely.

There are windstorms that occur in agrarian regions. These windstorms may occur at harvest period. Then, they cause damage to panicles/grains. A fruit plantation, say, with mangoes, apples, or even grapes may suffer loss of harvest.

A cyclone is a violent storm with rotating winds. The cyclones rotate clockwise in the Southern Hemisphere and anticlockwise in the Northern Hemisphere. A hurricane is a severe form of cyclone. A tornado is a cyclonic storm with highly concentrated winds and destructive power. Cyclone of larger dimensions may affect the entire agrarian region and bring about destruction to crops both via wind and floods caused by the rainfall. Overall, storms do occur in the sky immediately above the crops and they could be detrimental.

1.2.6.1 HEAT WAVES IN AGRICULTURAL SKY AND ITS CONSEQUENCES ON CROP PRODUCTIVITY

Heat waves are natural phenomena entirely related to ambient atmosphere, that is, "agricultural sky," if we considered the agrarian regions, exclusively. According to Intergovernmental Panel on Climate Change (2007), agricultural terrain in general is supposed to experience an increase in temperature by 2–4°C, in the 21st century. Such enhanced ambient temperature may cause severe detrimental effects on crop production zones in tropics and arid regions that are entirely rainfed (Jha et al., 2014). Heat waves affect almost every stage of the crop from seed germination till grain maturity (Jha et al., 2014). Heat waves generated in a location may even hit farms situated in the adjacent vicinity.

Let us define heat wave. A "heat wave" is described as existence of maximum temperatures above a certain threshold for a specified period of time. Such thermal extremes were rare in occurrence in the past. But they are now becoming more common, year-by-year. Heat waves are grouped into categories as defined below: "Moderate Heat Wave" means five consecutive days with "Daily Maximum Temperature" $\geq$ 35°C and < 40°C. "Mild Heat Wave" means five consecutive days with "Daily Maximum Temperature" $\geq$ 30°C and < 35°C. Without exception, heat waves affect the crops, their growth, and productivity (Rasul et al., 2019)

Now, how severe is the effect of heat stress and periodic heat waves on establishment of crops in fields. Also, how severe is the effect of heat wave on productivity of different agricultural regions. Forecasts suggest that increased ambient temperatures in North American continent could reduce soybean yield by 16%, by the year 2100 (Kucharik and Serbian, 2018). In case of barley, which is a temperate cereal crop, it is said the grain yield depreciated by 8 million tonnes, during 1976–2006. In terms of exchequer, it is equivalent to loss of US$ 1.0 billion (Lobell and Field, 2017). Let us consider an example related to maize. The grain yield records of over 20,000 maize trials spreading into all regions of the world indicated that, it decreased by 1- 1.7 t/ha exclusively due to heat stress. It was deduced that increase of each degree of ambient temperature above 30°C resulted in grain yield loss. Again, in Europe, crops grown in temperate conditions suffered due to heat stress. If heat stress occurred during growing season, they say, increase of each 1°C in ambient temperature resulted in perceptible decrease in grain yield of cereals, such as wheat, barley, and maize (Lobell et al, 2011a; Olesen et al., 2011; Stott et al., 2004).

No doubt, heat wave is a phenomenon related to agricultural sky. It is a natural phenomenon. It is getting more frequent due to climate change. First, we have to practice procedures that reduce climate change. While producing crops we may select suitable agronomic procedures. We should avoid seasons that expose crops to high temperatures. Crops with heat wave escape mechanism should be preferred. Further, Jha et al. (2014) state that there is knowledge accrued regarding inherent (genetic) ability of crops for heat tolerance. So, we should select genotypes that possess genes for heat tolerance. Planting crop species and genotypes that tolerate heat waves/ stress may be a good idea in subsistence farming zones exposed to semiarid or arid weather conditions.

Simulation studies at Cornell University indicate that it is heat stress caused during a drought that causes greater depression in crop growth and yield, than dearth of soil moisture, although, together, dearth for moisture and heat wave can be highly detrimental (Garris, 2019). Further, they have used 3–4 scenarios of climate change and tried to predict the consequences on grain yield of major cereals such as the wheat, maize, rice. They have predicted that climate change related to heat stress may result in 8% to 19% grain yield decrease by 2050. Under most severe heat waves/drought, the grain yield of cereals could reduce by 20–50% compared to present levels by 2100. Only hardy cereal species such as sorghum and pearl millet may tide over the heat stress a bit better than others like rice (Garris, 2019; Ortiz-Obea, 2019). Clearly, heat units received from the agricultural sky, radiant heat, and reflected heat from soil surface all affect the crop's performance. We may have to monitor climate change effects more stringently in areas prone to heat stress, particularly, in semiarid and arid agrarian regions.

Regarding detrimental effects, it is said heat waves often reduce soil moisture leading to drought/dust bowl conditions. It induces loss of crops, domestic animals, and even farmers. Often such heat waves that occur for over a month may induce farmer migrations and abandoning of fields. There are a few examples to note. In Russia, heat waves during 2010 induced loss of crops, cattle, and exchequer about US$ 15 billion. In 1995, in the Great Lakes region of USA, warmer conditions beyond threshold affected farmers and crops severely. Wheat grain yield reduced by 25–36% due to heat waves. Heat waves actually caused elevation of temperatures beyond 3–4°C from the normal. Fruit orchards too reported reductions in yield (Bralower and Bice, 2019).

A few reports on influence of heat waves and drought suggest that such maladies have been affecting various agrarian regions, periodically. They have caused about 10% annual yield decrease due to heat waves. It is said that nations with well-developed agriculture may get affected more due to high temperature, heat waves, and periodic drought. Agrarian regions in America, Europe, and Australasia suffered a 19.9% decrease in cereal yield, during past few decades. While in Africa the grain yield dropped by 9% due to heat waves/drought (Korzekwa, 2018; Leff et al., 2004; McSweeney, 2018). According to Challinor et al. (2005), Lesk (2016) and Ramankutty (2002), monocropping zones may suffer greater yield loss compared to a mixed-cropping enterprise in response to heat waves. Further, they say, heat waves and droughts have affected grain yield by 6–7% during 1967 to 1984. However, it was much higher at 13.7% from 1985 to 2014. Clearly, heat waves/droughts have been severe. They caused greater grain yield loss in recent years. These heat waves are natural phenomena related to agricultural sky. No doubt we have to bestow greater research attention to the happenings in the agricultural sky.

Based on data from over 30 years since the 1970s, researchers at Joint Research Centre of European Commission simulated the effect of extremes of heat, water, and resultant drought conditions on wheat grain yield. It has been concluded that occurrence of heat waves can be detrimental. In fact, yearly wheat yield variations in the European region could be explained, using heat waves related data. About 40% of yield variability could be attributed to extremes of heat and water availability (JRC, 2019).

Simulations using the commonly adopted crop modeling programs such as EPIC or others suggested that, in the European plains, France in particular, weather-related grain yield depreciation could become perceptible in the future (Van der Velde, 2012). Wheat and maize have responded slightly differently to extreme heat waves in the European plains. Since the frequency of heat waves has been forecasted to increase, we should expect grain yield depreciation of both wheat and maize. In fact, in certain years such as in 2003 and 2007, grain yield of cereals reported were far too less compared to simulations, based on data from 2 decades.

Experts dealing with climate change effects on crop production in the African continent have stated that frequency of heat waves and drought could become greater than ever before. What we know as unusual hot weather could be a routine event in about 20 years later. Also, heat waves may occur in any season of the year. They have analyzed weather data for the stretch between 1979 and 2015 and have concluded that hot spell every year and for longer duration is to be expected. Also, a rise of 2°C in ambient temperature will be perceived by crops all over the African continent (Institute of Physics, 2016). No doubt, agricultural sky above the African continent will affect crop production more conspicuously, than perceived currently. Heat waves in conjunction with other factors of sky may have to be tackled carefully. An integrated approach is required. If not, we have to face definite decrease in crop productivity. Severity of heat waves plus frequency of occurrence of such events will decide the extent of crop loss and yield depreciation in the agrarian zones of Africa.

In addition to crops in the field, heat waves do affect soil, its water, air, and microbes in the ecosystem.

Researchers at Wheat Research Centre in Victoria, Australia have applied latest observational techniques involving spectral analysis and computer modeling. They aimed to assess the ability of wheat genotypes to withstand heat waves. Heatwaves do affect wheat productivity in Australia. Heat waves affect wheat severely if it occurs at critical stages. Temperatures above normal for more than five days before "boot leaf stage" reduced the number of wheat grains on a plant. It also affected seed set and maturation. Crop modeling could be used to assess crop's response and design remedial measures appropriately.

Mehrabi and Ramankutty (2017) have opined that since heat waves and droughts are getting ever more frequent and affecting the crop productivity, we need to understand such disasters on a suitable scale of intensity. There could be heat waves causing greater depreciation than known earlier. We have to quantify the heat waves. That is characteristics versus the crop loss recorded at a location. They further state that our knowledge about crop loss due to heat waves/droughts could be focussed on the major food crops. For example, loss in grain productivity of crops such as wheat, maize, rice, oilseed crops, and pulses in relation to heat wave events needs to be known. They have reported that for the 6 major grain crops tested in 131 different countries, grain yield depreciation due to heatwave/drought differed. Cereal grain loss due to heat wave ranged from 4.1% (in Angola) to 5.7% (in Botswana) compared to normal years. It resulted in 4.4% grain yield decrease in USA and 5.2% in Australia. We should also consider the loss in agricultural exchequer suffered by different countries due to heat waves. Reports suggest that in 2017, it was about 28 billion US$ in India, 38 billion US$ in Russia, 116 billion US$ in USA etc. Overall, understanding the cause of grain yield loss and quantifying seems important. It helps us to adopt remedial measures quickly and appropriately if we have some authentic forecasts. Overall, heat wave is a phenomenon mediated via agricultural sky. It affects crop production.

1.2.6.2 DROUGHTS AND THEIR IMPACT ON CROP PRODUCTIVITY

Drought is a complex phenomenon that affects various agroecosystems in different ways, year after year. Drought is partly an agricultural sky-related phenomenon that affects crops immensely in certain regions of the world. Several factors related to agricultural sky operate for variable period and intensities. Droughts vary enormously with regard to the extent of crop loss caused by them. Top five nations in the world whose crop belts suffer perceptible loss of food grains due to droughts are China, United States, France, South Korea, and India (Kim et al., 2019). The decrease in food grain productivity in these countries ranges from 10% to 48% of normal, for major crops such as maize, rice, wheat, and soybean.

According to Kim et al. (2019), droughts represent extreme climate. Droughts are detrimental to crops and their sustenance in the outfield. Drought reduces crop production and food security. Droughts and their influence on agrarian regions have been studied since several decades. We have bestowed lot of research time and funds to know the implications of mild/severe droughts on various crops. However, Kim et al. (2019) state that, globally, extent of its spread and loss to crop productivity due to droughts is poorly understood. They state that, during 1983 till 2010, globally, about 454 million ha of cropped area suffered drought of different intensities and for different lengths of time. The cumulative grain yield loss amounted to 166 billion US$ per year. Each drought event, it seems, reduced exchequer by 0.8%, during past 2 decades. Grain yield depreciation occurred on crops such as wheat, maize, rice, soybean, and millets (Kim et al., 2019; Worland, 2016). A few other studies deal with analysis of previous reports on droughts and heat waves and their impact on crops. They indicate that during past 50 years, drought-related maladies have reduced grain productivity of major cereals by 10%, each year, rather routinely (Puiu, 2016).

1.2.6.3 DUST STORMS IN AGRARIAN REGIONS

Dust storms and Sandstorms are phenomena generated by an interaction of soil surface and high-speed winds that blow over on agrarian regions. They occur conspicuously during certain seasons of the year. Dust storms are also called sandstorms, particularly, when they occur on sandy plains regions. Dust storms are defined as events in which visibility gets reduced to 1 km or less, as a result of blowing dust (Sterk et al., 1996; Wikipedia, 2020b).

Dust/sandstorms are meteorological phenomena more frequently observed in the drylands or sandy farmlands, for example, in Sahelian West Africa, West Asia, Drylands of Gobi region of Mongolia, in Western Australia, and Northwest India (see Kumar et al., 2014; Ulbrich, 2012). Many of the above regions experience typical monsoon dust storms that distribute large quantities of nutrient-laden sand/dust, microbes, and other minute biotic forms. Dust storms lift and carry dry dirt and dust material from the plains. They may carry dry dusty soil, if they occur in agrarian regions with soft soil surface (e.g. Central Plains of North America, say Oklahoma, Kansas). They are termed "Dust Bowls." Storms are called sandstorms when they lift sand particles with mineral nutrients attached to them. Desert winds aerosolize billions of tons of dust (Kellog and Griffin, 2006). It has been estimated that the annual quantity of desert dust that makes regional or global airborne migrations is 0.5–5.0 billion tons each year (Perkins, 2001). In addition to their detrimental effects on large expanses of crops, dust storms can be harmful to animals and humans. Dust storms are one of the major contributors of allergic reactions, breathing problems, and lung disease (Kohfeld and Tegen, 2007). Atmospheric dust can have substantial regional impacts on agriculture and human health. Dust storms affect agricultural crop production, by causing erosion of the soil surface. Dust can also affect the quality of the air that we breathe. Minute particles of soil dust with diameters smaller than 10 μm in the atmosphere have been shown to be damaging to human respiratory health.

Let us consider an example from Australia. Recent dust storms were created due to strong winds, low pressure trough and months of dry warm weather. The dry weather had resulted in parched soil. So, the wind easily carried surface soil particles. Clay particles of < 4 μm, silt particles of 4 to 62.5 μm, and sand particles of 62.5 μ to 2 mm were lifted into atmosphere. It created a dust storm of significant size. First, such dust storms reduce the topsoil fertility. It results in soil erosion that needs capital to correct the situation. Such dust storms, it seems are common in the Australian continent, particularly, during spring and summer. The blowing sandstorms can uproot seedlings and affect crop stand (De Deckker, 2019; EOS 2015; Julie, 2020; Middleton, 2018).

The term sandstorm is used most often in the context of desert storms, especially in the Sahara and Sahel regions or places where sand is more prevalent than silt. Such sandstorms are common in sandy terrain (e.g. West Africa). Worldwide, areas in North and West Africa, particularly, sub-Sahara, Central, and Southern regions of North American Great Plains, Arabian Peninsula, Western Australia, Gobi region of North China, and Mongolia are prone to crop loss due to such storms. Often, they occur in junction with prolonged droughts and dry conditions. They say, poor management of drylands and lack of cover crops or other suitable conservation procedures during "fallow" period causes loss of topsoil and sand.

Recent reports suggest that dust storms are the major source of nanoparticles in the sky. High-speed winds that lash the agrarian regions generate mineral dust and pollutants. It is said that, about 50% of aerosols in the troposphere could carry mineral dusts originating from major deserts of the earth. For example, mineral dust (nanoparticles) emanating from Gobi Desert of Mangolia can effectively get transported across large distances. They can be deleterious to atmosphere in the region. The nanoparticles carried by dust storms could also get concentrated at certain geographic locations. Again, they can be immensely detrimental to crops and other biota (Manna and Bandyopadhyay, 2019).

"Harmattan" in sub-Saharan West Africa, "Haboob" in Sudan in North Africa, are examples of heat waves accompanied with windstorms and hanging sand haze (Plate 1.3). The cropping pattern adopted usually avoids the heat waves and sand/dust storms, by leaving the fields fallow. However, in case seeds are sown, their emergence and survival of seedlings are affected by high temperatures of soil/sand crystals. Sand particles get hot. They often reach 45°C surrounding the seedlings. It leads to scorching and loss of leaf tissue of emerging saplings. Such storms often carry large quantities of sand particles from one location to other. The surface soil fertility in one location where the sand is picked gets reduced. While, at location of deposition of sand dust, the fertility gets enhanced (Romheld et al., 2000; Sterk et al., 1996; Stahr and Herrmann, 1996). Again, heavy precipitations, storms, snowfall, dust storms, and heat waves are all phenomena that engulf agrarian regions. They may force farmers to avoid the seasons in which they are accentuated. If not, a standing crop may suffer damage. We need to consider the agricultural sky above such crop belts. Then, classify the sky above the crop, accordingly. Dust storms of different kinds should be considered while characterizing and classifying the agricultural sky. It helps researchers and farmers alike while conducting crop production trials. In simplest terms, agricultural sky above pearl millet fields in Sahel is prone to Harmattan or Haboob (AgWeb, 2020; Weather Online, 2020). They need to be thoroughly characterized and classified. Further, we should note that agricultural sky may differ at short distances. So, demarcations need to be clearly noted with GPS coordinates on maps.

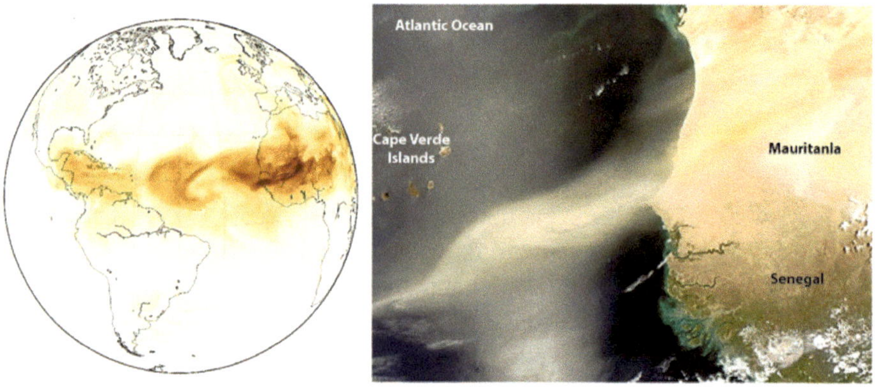

PLATE 1.3 The Sahelian/Saharan dust plumes.

Note: Storms such as Harmattan or Haboob from West/North Africa blow into South and Central American regions. Plumes could be large reaching a size of 200 km in the Western Sahara and Mauritania. The above images were obtained on April 30, 2003, as per NASA, USA. Such dust periodically carries large mass of mineral rich sand particles, organic particles, and most importantly lots of biotic species, mainly, microbes (fungal propagules, bacteria, virus, insect and their eggs, etc.). Such sandstorms are among most important agricultural sky-related phenomena. They deplete/add to nutrients in West African Sahelian farming belts. Sandstorms along with heat waves can be highly detrimental to establishment of seedling of major cereals such as pearl millet, sorghum, or even a legume such as cowpea. Wind-mediated soil erosion is an important phenomenon related to agricultural sky above the Sahelian cropping zones.

Source: National Aeronautics and Space Agency (NASA), Washington, D.C. USA; https://eoimages.gsfc.nasa.gov/images/imagerecords/92000/92358/dust_geo_2018179_lrg.png;

Weather Online- Espana https://www.woespana.es/reports/wind/Haboob.htm; also see NASA's Earth Observatory-https://earthobservatory.nasa.gov/images/83260/cape-verde-under-dust;

1.2.7 OZONE AND SMOG IN THE ATMOSPHERE ABOVE CROPS

Ozone is not directly emitted to the atmosphere from human activities. Instead, it is formed within the atmosphere through a series of complex chemical reactions. The chemical processes involve volatile organic compounds (VOCs) and nitrogen oxides (NOx), such as nitric oxide (NO) and nitrogen dioxide (NO_2) (NASA, 2019a). Clearly, it is a problem created in the agricultural sky (Andreae et al., 2001; Andreae and Merlet, 2001; Cao et al., 2008; Koppmann et al., 2005; Yang et al., 2008). Ozone and peroxyacetyl nitrate travel close to the ground as a result of smog. They have damaging effects on plants, resulting in the leaf discoloration and yield reduction. It reduces the amount of photosynthesis possible by the plants. As stated earlier, basically, a smog is a combination of smoke and fog. The initial reaction of photochemical smog is the photolysis of nitrogen dioxide (NO_2) into nitrogen (NO) monoxide and ozone (O_3). In addition, unburnt particulate hydrocarbons liberated during crop residue burning also react with atomic oxygen. It results in a variety of aldehydes and ketones. Such aldehydes and ketone along with ozone and nitrogen dioxide react to generate series of derivatives known as Peroxyacyl nitrates and Peroxybenzyl nitrate. These compounds are scorching. They exhibit "tear gas" like properties in the presence of free radicals. Such chemical processes cause a yellowish-brown colored smog. Such smogs are often termed the "Photochemical smog."

Let us consider an example from the Canadian plains where crops such as wheat, canola, and soybean flourish. Reports indicate that respiration and photosynthesis in the leaves was restricted. Stomatal functions were affected. So, normal gas exchange process got altered. Crop growth too got retarded, due to smog. Smog that occurs for extended periods during the growth phase can be deleterious to crops. The maturity of grains gets delayed. Often, the grain yield was lowered due to smog (Ainsworth, 2017; Avnery 2011a, 2011b; Whetter, 2018). Ozone in the atmosphere is monitored continuously by the aerospace agencies (NASA, 2019b). This is because ozone can reduce the crop growth and yield formation severely. Such effects due to ozone may get accentuate, if unattended.

Ozone and aerosols with particulate organic pollutants contaminate the agricultural sky if wildfires occur for longer durations. Such pollutants are detrimental to normal growth pattern of natural vegetation as well as farmers' crops. Productivity of crops decreases significantly due to pollutants liberated by wildfires. Interestingly, Yue and Nadine (2018) stated that wildfires do not just affect the crops in surrounding vicinity. The smoke and pollutants created by wildfires are known to affect crop productivity even 100s of km away from the "hot spot" that experienced the wildfire. It has been found that forest that stands several km away from the wildfire-affected area too got affected by ozone and pollutants. Further, it has been pointed out that such wildfire-related pollution of atmosphere and resultant damage to normal ecological processes and biomass productivity may affect other regions too.

Let us consider an example from a food grain-producing region in South Asia. Reports from Indus region state that cropping zones have got exposed to smog and fog mix, since a decade. The smog contained ozone, peroxyacetyl nitrate, and hydrogen sulfide cocktails. It affected the natural vegetation, wheat fields, vegetables, and pastures severely. Bhatti (2017) and Mati (2019) state that smog that pollutes the agricultural sky is detrimental to a greater extent on certain crops, such as sweet corn, sunflower, tomatoes, potatoes, beans, gourds, turnip, and grape vines grown in the Indus region. Smog has actually reduced photosynthetic efficiency of the crops, resulting in lower yield.

The serene, congenial, and pollutant-free agricultural sky that supports crop production luxuriantly could be spoilt. It could deteriorate either transitorily or even for long durations due to smog. Such a smog filled with black smoke and carbonaceous matter plus other toxic gases, usually, results after an extended period of forest fire. Such forest fires were observed in California, USA, or Southeast Australia, recently. A nuclear catastrophe too can cause extended burning of forests, crops, and other vegetation. Crutzen et al. (1984) studied the results of nuclear catastrophe. They forecast that forest tree and bush fires will last for months. Fires liberate smog filled with pollutants, CO_2, CO, NO, SO_2 at high rates. The hanging smoke in the post nuclear war period reduces photosynthetic light interception by trees and field crops for extended periods, say a few months. The above effects are in addition to radiation-related pollution of agricultural sky.

Let us consider an example. The agrarian regions of India generate large amounts of crop residue. The crop residue gets burnt in the open field, periodically. The process leads to release of greenhouse gases and particulate matter. Jain et al. (2014) reported that, in 2009, total amount of residue generated annually was 620 Mt. Out of which ~15.9% residue was burnt *in situ* on farm. Cereal crop residue accounted for 58%, fibers 28%, and sugarcane stubbles 17%. Among cereals, rice straw contributed 40% of the total residue burnt followed by wheat straw (22%) and sugarcane trash (20%). Burning of crop residues emitted 8.57 Mt of CO, 141.15 Mt of CO_2, 0.037 Mt of SO_x, 0.23 Mt of NO_x, 0.12 Mt of NH_3, 1.46 Mt non-methane Volatile Organic Carbon, 0.65 Mt of non-methane hydrocarbons emissions, and 1.21 Mt of particulate matter, for the year 2008–09. Crop residue burning does affect atmospheric quality because of emissions (Badarinath et al., 2006; Bhuvaneshwari et al., 2019; Mittal et al., 2009; Sahai et al., 2011; Plate 1.4).

Smog filled with dust particles could have certain beneficial effects on soils/crops. Smoke created by the burning crop residue could affect the atmospheric composition. The smog that carries ash and organic dusts that settle on cropped fields, adds to soil nutrients. Usually, the ash particles contain nutrients such as Mg, Ca, S etc. So, a smoke screen in the sky could be useful too. It may protect crops from harmful radiation (UV rays). It can also protect the crop by thwarting a temperature rise around the crop's canopy (Kohls, 2019).

PLATE 1.4 Stubble burning immediately after harvest is a common practice in cereal fields in many parts of the world.

Source: Dr Clyde Beaver, Creative Services Manager, International Centre for Maize and What, Mexico, CYMMIT Photo archives.

Note: Crop residue/stubble burning is a routine agronomic procedure practiced in the Indo-Gangetic plains where wheat/rice are the major field crops. Stubble burning adds to mineral content of the soil and reduces expenditure on removal of crop residues from the fields. Crop residue burning becomes a large-scale atmospheric phenomenon immediately after a crop is harvested. It causes environmental problem. Crop residue burning adds to particulate matter, organic compounds, sulfur etc. in the atmosphere.

1.2.8 POLLUTION IN THE SKY ABOVE AGRARIAN REGIONS

Here, we start with the assumption that clear sky above the crops and the entire agrarian region may actually be a repository of innumerable types of pollutants. Each of these pollutants may occur at different stages of crop period, at different intensities, and for different lengths of period. Agricultural crops can be injured when exposed to high concentrations of various pollutants in the air. Injury ranges from visible discolorations on the foliage to reduced growth and yield, to premature death of the plant. The severity of the injury depends not only on the particular pollutant plus other environmental factors. Other factors include the length of exposure to the pollutant, the plant species, and its stage of development. Environmental factors conducive to a build-up of the pollutant could aggravate detrimental effects (Ministry of Environment, 2003).

We can identify deleterious effects of air pollution by looking at the symptoms on the crop's canopy that each of the pollution factors causes. The injury to foliage, discoloration of chlorophyll pigment, appearance of lesions on leaf tissue, and dead tissue are common symptoms. Plants in open field are often stunted. Plus, pollutant-affected plants do not keep pace with healthy ones, regarding growth stages and maturity.

Ozone is among the important air pollutants that affects crops. The deleterious effect of ozone was discovered in 1944 in the crops grown around Los Angles in California, in USA. At present, ozone pollution effects have been noticed in several agrarian regions of the world. Symptoms of ozone toxicity are first seen on the upper surface of leaves. The leaves show discoloration, flecking, bronzing, and bleaching. Yield reduction too occurs.

Sulfur pollution in the air occurs mainly due to coal burning industries. Sulfur di-oxide emissions can also occur due to burning of petroleum and other fuels. Sulfur dioxide (SO_2) contamination in air results from industrial activities such as copper smelting. The SO_2 formed readily converts to other compounds such as sulfuric acid, sulfates, and sulfur trioxide (Poonia, 2012). Sulfur di-oxide is easily recognized by its pungent "egg" smell. For example, industries involved in production of fertilizers, paper products, sulfuric acid, food preservatives may realize SO_2. In nature, it gets released into the atmosphere through volcanic eruptions (Merck, 2020; Poonia, 2012). Sulfur enters leaf tissue via stomata. So, concentration higher than the threshold can be toxic to crops. Lesions on leaves and necrotic patches are common, if SO_2 pollution is severe in the air above crops. Susceptibility to sulfur is dependent on factors such as crop species, its stage, period for which the crop gets exposed to sulfur pollution etc.

Fluoride pollution occurs due to combustion of coal in the industries that produce, ceramics, glass, fertilizers, steel, aluminum, and bricks. Fluoride emissions are also observed above industries that produce phosphate fertilizers, hydrofluoric acids, and other chemicals that require use of fluorides (Al Attar et al., 2012; Dartan et al., 2017; Ministry of Environment, 2003). Major effects of fluoride pollution in the air are retardation of leaf growth and elongation, appearance of grayish patches and loss of chlorophyll, appearance of reddish-brown lesions, etc. Fluoride emissions may reach soil and crops when precipitation occurs. Fluoride-contaminated water may percolate to lower horizons of soil in agricultural fields (Bhat et al., 2015; Kassir et al., 2012).

We encounter particulate pollution in the air. Such particles could be organic or inorganic in chemical composition. Particulate matter generated during crop residue burning is an important pollutant. Such particulate pollutants when they mix with other atmospheric constituents and water vapor can generate smog. Particulate matter that occurs as haze for longer durations above the crop's canopy can reduce photosynthetic carbon fixation, by crops. Crop's gross biomass productivity lessens due to the hanging particulate matter in the atmosphere (Andreae and Merlet, 2001; Andreae et al., 2001)

Farm operations such as application of nitrogenous fertilizers, organic manure, and burning crop residues *in situ* are routine. It seems these are among the major causes of gaseous and particulate pollution of the atmosphere above the crop's canopy (Krajick and Lee 2016). In certain geographical regions, emissions from agricultural procedures and those from industries interact. Gaseous ammonia applied to fields may often combine with other industrial emissions and solid particulate pollution of the atmosphere. This results in the formation of certain "Aerosols." The particulate pollution levels are expected to increase as we opt for higher amounts of fertilizer-N inputs into the fields. World Meteorological Organisation's standards allow only 10 μg particulate matter m^3 in the atmosphere.

Indonesia, for example, adopts a mixed farming trend that has farming patches interspersed with forests. This is in addition to larger thick forest plantation regions. Forest burning does occur intermittently due to various causes. Such fires emanate pollutants, carbonaceous pellets, CO_2, CO, and SO_2 (Davies and Unam, 2019; Rossi et al., 2016). There is no doubt that a polluted sky above the agricultural zone or forest does affect photosynthetic rate and respiration. Plantations, fruit crops, wetland, and arable crop species such as rice, legumes, and vegetables too suffer, due to smog that carries pollutants.

1.2.8.1 ACID RAINS ARE FORMED IN THE SKY ABOVE AGRARIAN REGIONS

Acid rain is a phenomenon generated in the sky. It is precipitation that is highly acidic in reaction. It is caused due to elevated levels of hydrogen ions (pH). In the agrarian regions, acid rain is created whenever gaseous and particulate emissions are caused because of crop residue burning. Such emissions contain sulfur di-oxide (SO_2) and nitrous oxide (NO_2). These gases mix with water vapor in the atmosphere to produce acids. Acid rain refers to acidic precipitation in the form of rain (water drops), snow, mist, frost, sleet, and hail. Acid rain has been reported to occur in the agrarian regions of all continents. Forests may receive acidic precipitation and suffer detrimental effects due to it. Burning fossil fuel, forest fires, fertilizer and coal industries, thermal plants, and homes are some examples of sulfur di-oxide generators. The gaseous SO_2 oxidizes to SO_3 and forms sulfurous and sulfuric acid in combination with water.

Acid rains are also termed wet aerosols or depositions. In such acidic aerosols, water drops are mixed with nitrates, sulfates, and organic acids. Rain drops usually possess a pH of 5.4 to 5.7, that is, in normal situations. In an acidic rain the pH drops. Acid rain is one of the problems while producing crops and forest plantations. It is known to cause "brown clouds" that affect photosynthetic efficiency of crops (Molina and Molina, 2002).

Natural vegetation, crops, and soils are the major receptors of acid rain. Reports suggest that productivity of agrarian regions is affected detrimentally, due to acid rains. The crops species, of course, differ in their ability to tolerate acidic precipitation that falls on canopy (leaves) and soil surface. Cereal crops tolerate acidic deposits on their canopy better than dicotyledons (Lal, 2016; Yang, 1989). Acid rain penetrates leaf tissue via cuticle. It diffuses into soil profile via the pores and affects soil pH and composition of soil solute. Acid rain affects soil micro-organisms and their activity levels. Acid, no doubt, causes abnormalities in the plant tissue. It affects normal physiological functions. Acid rain retards crop growth, delays boot leaf formation and grain maturity. During crop season, acidic precipitation may mix with clouds and acidify them further. Usually, increased emissions of SO_2, N_2O, NO_2 cause such acid rains. Compounds with sulfate and nitrate moiety are largely responsible for acid rains in the agrarian regions. In addition to gaseous emissions, particulate matter, both organic and inorganic that emanate due to crop residue burning may cause acidic rains. Particulate matter too may mix with clouds and form precipitations that are highly acidic, say of pH 2.5 to 4.5 (Jacobson et al., 1988).

Smog induces formation of acid rain. Such an acid rain has severe detrimental effects on humans, animals, dairy cattle, and most importantly crops grown in large expanses. Crops are prone to retarded growth, wilting, loss of photosynthetic ability, and reduced biomass/grain harvest. Smog can cause plants to lose 10 to 40 percent growth. In addition, according to University of California researchers, smog weakens the plants. Therefore, it makes them more susceptible to diseases and pests that can cause further damage. It is said that in California alone, agricultural farmers lose 2 to 6 billion US dollars a year. It is attributed to reduced productivity caused by smog and acid rains.

1.2.9 CLIMATE CHANGE: A MAJOR AGRICULTURAL SKY-RELATED MANIFESTATION

Agricultural crop production is immensely dependent on the expression of climatic indices, either individually or in conjunction. The net influence of the several of the climatic indices is felt by the crops. The agricultural sky actually allows for constant interaction of climatic indices. The climate change that results, makes farmers adopt matching changes in crop production procedures. During a season, crops have to adapt to the climate change. Therefore, farmers have to select crop species and genotypes that suit the altered weather patterns. They say, agricultural producers have an excellent track-record of making matching alterations in crop production tactics, particularly, in response to what happens to agricultural sky (Faulkner and Easton, 2015). Yet, it is said crops are vulnerable to constantly changing climatic patterns. During recent years, crop productivity in many of the major food grain production zones has suffered. Weather-related disasters such as drought, dust storms, floods, cold fronts, and disease have reduced crop yield. During 2012, in USA, it seems about 15.7 billion US$ was lost in agricultural exchequer. It could be attributed to climate change effects. Droughts that are mild or harsh have affected agricultural productivity. This has occurred despite farmers exhibiting greater flexibility in the selection of crop species and cropping systems. Overall, it is clear that agricultural sky is a dominant aspect of any agrarian region. The natural climatic factors related to it have direct effects on crops. Such inputs could be positive or deleterious to crops (Perarnaud et al., 2005).

There is actually a spike in the GHG emission during recent years. It is caused indirectly by the enhanced global population. To meet increased demand for food grain, agricultural inputs have been increased. Natural resources have experienced a depletion. Incessant release of GHG has led to uneven warming of atmosphere above land mass. The climate change effects, such as dust storms and heat waves, have impacted agrarian regions of North America, West Asia, and Europe. The productivity of crops such as maize, wheat, and legumes has shown decline in lower latitudes. In higher latitudes, it has affected food grain generation by the crops favorably (Hirji, 2019a, 2019b: UNCCD, 2001; Walthall et al., 2012). Diversifying crops has been suggested as a method to overcome the drastic effects of GHG. In countries such as India and China, diversifying crops may reduce the impact of GHG, perceptibly (IANS, 2019).

In the tropical Africa, climate change is marked with increased frequency of storms, drought, and flooding. Climate change induces altered hydrological cycles and precipitation. Simulations suggest that climate change effects may differ based on the crop species and location within the tropics. It affects the yield levels of presently grown major food crops such as yam, maize, sorghum, cassava, cowpea, and other vegetables (Adewuyi et al., 2014, 2015; Chikezie et al., 2015; Food and Agricultural Organization, 2007; Jones and Thronton, 2003). Overall, crop diversification is a possible remedy to overcome ill effects of GHG emissions, into agricultural sky.

Fossil fuel burning is a major source of GHG. The greenhouse gases get released into atmosphere. Such GHG may accumulate in agricultural sky. We may note that agricultural farming includes repeated ploughing of land. It induces soil oxidation and microbial respiration. Application of nitrogenous fertilizers and dairy cattle rearing induces GHG emission. During 2007 to 2018, agricultural farming accounted for 23% of global GHG emissions attributable to human activity (Hirji, 2019 a, 2019b; Wood et al., 2000).

Climate change is a manifestation related to entire agroecosystem. It includes the soils, crops, and ambient atmosphere. The agricultural sky plays a vital role in causing climate change. Climate change effects on crops could be severe. It could lead to small or massive loss of biomass and grain productivity (Chloupek et al., 2004; Cuculeanu et al., 1999; Perarnaud et al., 2005). Farmers may often incur higher costs, to overcome the anomalies caused by climate change. In fact, climate change is among top rated risks that farmers in different agrarian regions have to negotiate. It is still not easy to forecast and develop methods, to overcome climate change effects. In some countries, such as Australia, the agricultural agencies are alert to climate change. They try to identify the various deleterious effects that climate change may inflict on to the crops. They identify specific risks and ask the farmers, to take preventive measures, to ensure grain productivity at optimum levels (CSIRO, 2020).

In a farm, response of crops to seasonal variations is important. No doubt, diversification of crops and cropping systems can overcome the climate change effects to a certain extent. Whole farm planning to compensate for the losses incurred due to climate change is a good idea. Material inputs and agronomic procedures could be modified, to suit the climate change effects. The climate change is predominantly an agricultural sky-related manifestation. We ought to know the extent and duration of its impact on crops. The type of effect such as loss of foliage, lack of flowering, or poor seed set etc. that occur due to water stress should be understood, in greater detail. It then helps us to design appropriate remedial measures. In general, aspects of climate that operate through the agricultural sky could manifest only for short duration. Sometimes, climate change may extend for lengthier period, say, for years. For example, a change in the monsoon pattern may affect crops for a year. Agricultural crops may show up the effects of monsoon on large expanses of thousands of kms. Global warming may actually affect the crops and entire agrarian region of thousands of km^2, for several decades. A cold front during the winter affects for 100s of km, for a day or a week. For example, cold front in Florida citrus belt is felt for 50 to 100 km each year during winter. A severe storm affects crops within about 10–15 km^2 area, for a few hours. A cloud mass may affect the crop canopy for hours and for a distance of 1–2 km. A dust storm affects the conduct of agronomic procedures. It causes losses of topsoil. It may affect area of 10–100 km^2, but for a small period (duration) of time. It could be 1 hr or even less. Overall, the climate change effects differ in their impact in terms of intensity, period, and extent of devastation to farmland. So, we need to read the agricultural sky accurately and be ready with remedies (CSIRO, 2020; UNCCD, 2001). There are currently innumerable reports about the climate change effects on agricultural crop production. Such reports also alert us about the climate change effects on frequency of "dust bowl like" conditions, droughts, and floods in the major crop production zones. Let us consider an example. They say, if unchecked, GHG emissions will cause a clear 1.5°C to 2°C increase in ambient temperature. In particular, the temperature over land, that is, farming belts is increasing rapidly and at twice the speed of global average. As a consequence, it could induce dust storms and droughts more frequently (Hirji, 2019a).

Higher temperatures have several direct impacts on crop production. They are a) Higher temperature will lead to enhanced evapotranspiration. It dries up soils more rapidly, ultimately, raising the humidity of the atmosphere. It decreases crop's water uptake. It leads to variable soil moisture and relative humidity. b) Temperatures higher than threshold will reduce organic carbon levels in the soil via oxidation. Higher soil temperature also induces soil microbial activity and their respiration leading to higher evolution of CO_2. Such a situation can further reduce soil moisture levels and subsequently affect crop productivity. c) Higher temperature may affect germination, d) Higher temperature reduces frost risk in temperate regions. Warmer winters in many regions could allow earlier planting. However, warmer temperature could also induce a range of various agricultural pests and diseases.

We should note that increased atmospheric CO_2 levels have the potential to increase crop productivity.

Ambient temperature increases as GHG emissions (e.g. CO_2) get pronounced. The resulting warmer temperatures could induce many crops to grow more rapidly. However, it could also consequently reduce yields of some crops. Crops tend to grow faster in warmer conditions. Therefore, it reduces length of vegetative periods as well as grain maturity period. Crops may complete life cycle even before optimum nutrient, water, and CO_2 has been imbibed. Actually, crop duration gets reduced if ambient weather is warmer.

Crop plants use photosynthetic radiation to synthesize carbohydrate from CO_2. No doubt, greater CO_2 concentrations can result in greater carbohydrate production. A small increase in ambient temperature coupled with increasing CO_2 could benefit certain crops. Of course, water and nutrients should not be limiting. Overall, biomass and grain productivity of agrarian regions does depend on climate change effects. Such consequences could be negative or positive depending on geographic location.

Let us consider a few facts about organic farming and its relation to climate change. They say, crop burning causes the greenhouse gas emissions (CO_2, CH_4, N_2O). They have direct impact on warming and climate change. Fossil fuel burning is an important source of carbon emissions worldwide. As stated earlier, agricultural farming is responsible for 20–23% of global greenhouse gas emissions (FAO, 2002). Organic agriculture enables ecosystems to adjust better to the effects of climate change. It also offers a possibility to reduce the emissions of agricultural GHG. Reports suggest that CO_2 emissions per hectare of organic agriculture systems are 48 to 66 percent lower than in conventional systems. In organic agriculture, almost 70 percent of the CO_2 emissions were due to fuel consumption and the production of machinery. Moreover, in conventional systems 75 percent of the CO_2 emissions are ascribed to N-fertilizers, feedstuff, and fuels (FAO, 2002). During past decades, soil carbon levels have generally decreased under agricultural land use. Agronomic strategies that improve recycling of organic matter, internal nutrient cycles are useful. Low- or no-tillage practices may stabilize organic matter content.

Nitrogen fertilizer inputs into phosphorus-limited ecosystems often generate N_2O (and NO) fluxes. Such N fluxes are 10 to 100 times greater than that, from P-non-limiting crop fields. Excessive fertilizer-N is the main cause of accumulation of N in soil. Accumulated-N becomes vulnerable and gets emitted into agricultural sky as N_2O or NO_2 or NH_4. There are several reasons for lowered levels of N_2O emission, during organic farming. A few of them are a) lowered levels of fertilizer-N supplied to fields under organic farming; b) optimal C: N rations in organic manure applied; and c) use of cover crops and organic matter recycling procedures.

Agriculture accounts for major share of methane emissions at 66% of total man-made source. Methane emissions occur during rice production. Methane is also produced when crop residue is burnt. Ruminant and cattle farming emit more of CH_4. About two-thirds of the total man-made CH_4 arises mainly from paddy fields, burning of biomass and ruminants. However, we need accurate data on the effects of organic agriculture on methane emissions.

As stated earlier, North American Great Plains is among the major food grain-generating systems. This region has experienced climate change effects. Climate change effects on the major crops of the US Great Plains have been studied in detail, by several researchers. Reports suggest that temperature changes during the years 1968 till 2018 have been congenial to maize, but detrimental to sorghum and soybean. Observed precipitation trends have been beneficial to all three crops namely maize, sorghum, and soybean (Kukal and Irmak, 2018). Supplemental irrigation has generally mitigated the ill effects of climate change on the crops. Obviously, factors operating in the agricultural sky above the vast plains have induced alterations in crop growth and productivity, to different extents. Each of the factors had differing effects on crops. Perhaps, in due course, computer programs that assess the impact of each climatic factor and arrive at suggestions to farmers will become common, so that farmers can select the best crop and genotype, and also best agronomic procedures.

Let us consider an example that depicts combined effect of temperature, relative humidity, and solar radiation on wheat productivity. The climate change effect noticed from 1981 to 2014 was conspicuous. The reduction of spring wheat yield in response to the climate warming from 1981 to

2014 was significant. It reduced to an average of 3564 kg·ha^{-1}. The regional differences in yields were significant. The maximum potential yield of spring wheat was traced in the western region. The air temperature and soil surface temperature were the climatic factors that affected the key phenophases of spring wheat in Inner Mongolia (Zhao et al., 2017). Clearly, here, our preferences for wheat genotypes get influenced by their ability to negotiate climatic factors and changes that occur, in parameters such as temperature, and relative humidity and even others.

Agricultural sky is a good repository of pathogens affecting crops. The airborne fungi, bacteria, and viruses may all be influenced to different levels. Agricultural sky also affects their dispersal, infectivity, and extent of damage, that is, in terms of crop's growth and grain productivity. Climate change has its impact on several aspects of these very pathogens that inflict damage to crops (Helfer, 2014). No doubt, climate change affects both the host and the fungal pathogens. Each of the factors may have different impacts on fungi. Further, Helfer (2014) stated that graminaceous rust fungi were affected detrimentally by an increase in ambient temperature and CO_2. A few studies on major crops of the world such as wheat, maize, rice, soybean, legumes, and oilseeds indicate that climate change could accentuate the damage caused by viral pathogens. For example, Trabicki et al. (2015, 2017) and Serfling et al. (2016) have reported that viral pathogens of wheat such as Wheat Streak Mosaic virus, Barley Yellow Dwarf Virus, and a few others that cause severe loss of yield could gain in severity, due to climate change. Plants/crops exposed to higher CO_2 and temperature in the ambient atmosphere, generally, showed higher viral titer (e.g. Barley Yellow Dwarf Virus). The virus disease symptoms were pronounced in crops experiencing higher CO_2 in the air. Climate change effects mediated by higher CO_2, temperature, and relative humidity do enhance Brown rust incidence. It is not just the virus caused diseases of wheat that get accentuated due to climate change. The pathogen profiles in the atmosphere may vary enormously from the normal (IPCC, 2014). The patterns of fungal diseases get affected by climate change (IPCC, 2014; Barford, 2013; Bregaglio et al., 2013; Chen et al., 2011; Fisher et al., 2012; Helfer, 2014). For example, Junk et al. (2016) reported that simulations indicate that more common fungal diseases such as Brown rust disease caused by *Puccinia graminis tritici,* Wheat leaf rust *caused by P. triticiana,* Stripe rust *caused by P. striformis* also become prominent. It ultimately reduces wheat grain yield. Major agricultural sky-related factors such as relative humidity, precipitation, and air temperature have a great influence on the spread of fungal spores in the air and disease causation. Obviously, we can use occurrence of fungal spores in the sky and climate change-related parameters to classify the agricultural sky. Such a classification may help us to assess the feasibility of wheat cultivation. The classification of agricultural sky based on such biotic factors is rear or not done at all. This is despite the fact that agricultural sky must have stipulated and influenced the selection of wheat genotypes and evolution of cropping systems. In addition to vast stretches of cereals, RH in the sky above pastures and turf grasses too has similar effects. For example, there are fungal diseases such as Grey Leaf spot that affects rye grass. The causal agent *Magnothorpe oryzae* is sensitive to RH in the ambient atmosphere. At RH 88% the fungus infects the host grass. At RH 96% conidiation occurs. Higher RH induces formation of hyphae. Therefore, fungal biomass gets enhanced. Clearly, pasture quality is affected by the RH and its impact on fungal diseases (Li et al., 2014). Greater details about agricultural sky-mediated transmission of dreaded rust fungi on wheat or other cereals is available in Chapters 4 and 7.

1.3 BIOTIC FACTORS IN THE AGRICULTURAL SKY

Agricultural sky is the abode for biotic factors such as micro-organisms (i.e., bacteria, fungi, viruses, archaea, insects, and aves). Each of these aerial biotic aspects has its impact on crop productivity. An overview is provided in the following paragraphs. However, detailed discussions are available in Chapter 3.

1.3.1 MICROBES IN AGRICULTURAL SKY

The agricultural sky harbors variable population densities of comensalistic microbial flora. It harbors mainly the bacteria, their resting stages such as spores, fungal hyphae, spores, and other propagules that float just like dust particles, actinomycete species, and their propagules, archae, virus particles, etc. The microbial diversity encountered in the agricultural sky is indeed enormous. The species diversity may often depend on the type of agrarian region, crops that dominate the agroecosystem and agronomic procedures adopted by farmers. Closeness of the specific farm location to other features, such as water bodies, industries, mines too affects microbial diversity and population. Geographic location and weather-related parameters such as wind, temperature, RH, precipitation may all affect the microbial population.

Aeromicrobiology is the study of microbes which are suspended in the air. These microbes are referred to as bio-aerosols (Al-Dagal and Fung, 1990; Brandl et al., 2008; Pepper and Gerba, 2015). Microbial population in the atmosphere is relatively less, if we compared it with soil or oceans at different depths (Amato, 2012). Generally, bacterial population in atmosphere may range from 10^{4} to 10^{6} m^{-3}; fungal propagules range from 10^{3} to 10^{4} m^{-3}, viruses range from 10^{6} to 10^{7} m^{-3}, and pollen particles may range from 10–10^{4} m^{-3}. Aerobiology researchers often try to assess ratio of bacterial to fungal propagules or bacterial to viral particles etc. It helps them to trace the fluctuations in the dominant fraction of microbial load in the air above crops.

Generally, the bioaerosol particles encountered in the agricultural sky include pollen, bacterial, fungal spores/hyphal fragments, viruses, archaea, yeasts, lichens, small seeds, etc. (Desperés et al., 2012; Frohlich-Novoisky et al., 2012; Nunez et al., 2016). Further, it has been suggested that fluctuations in population and diversity of bioaerosols are affected by land-use pattern in the agrarian regions. Temporal variability of bioaerosol composition (microbial species) could also be attributed to climate factors.

Atmosphere, including troposphere immediately above the crop's canopy, then, in stratosphere up to above 21–42 km height above the crops do harbor microbes (Smith et al., 2011; Wade, 2013). As stated above, the microbial diversity above the cropland perhaps depends on too many factors related to atmosphere, earth's surface features, soil types, water bodies, crops etc. The causes for fluctuation of microbial population too were not known. Perhaps seasonal variation could occur and crop species that flourish may affect microbes in the air immediately above their canopy. The distribution of microbes vertically in the regions above crops too varies immensely (Griffin 2004; Griffin et al., 2017; Smith et al., 2018). Smith et al. (2018) reported that microbial signatures (i.e. bioaerosols) could be detected up to 38 km height from surface of agrarian regions. Airborne microorganisms in the upper troposphere and lower stratosphere remain elusive, due to a lack of reliable sample collection systems (Horneck et al., 2010; Smith et al., 2018). There is a report that stratosphere zone above a semiarid region of South India clearly showed that microbes reside at heights between 20 and 41 km above cropped land. The microbial diversity in the stratosphere was detected using 16s RNA sequences. The three new bacterial species detected were *Bacillus hoylei, B. isronensis*, and *B. aryabhata* (ISRO, 2009). The mechanism of survival was not studied. Reports by NASA (2019c) and Khodadad et al. (2016) suggest that bacterial species (e.g. *B. pumilus*) transported from earth's surface and released into stratosphere do not survive for periods longer than 6–8 hr. Smith et al. (2011) too have stated that, under simulated stratosphere conditions *B. subtilis* did not survive for more than 6 hr. This is because of harsh conditions and radiations. However, it was reported that some of these stratosphere microbes are tolerant to UV radiation (ISRO, 2009). Biochemical transformation brought about by the stratosphere microbes was not known. There are reports that earth's surface, including the vast agrarian regions, are incessantly impinged with microbial species namely, bacteria, fungal propagules, and really large number of viral particles (Wehner, 2018). Microbes could be freely transmitted between various layers of atmosphere and earth's surface. Such transmission is dependent on the physical conditions, including harsh radiation. We ought to study the role of atmospheric microbes and their ability to inoculate the earth's surface periodically. They could be replenishing the earth's surface, particularly, soils, water bodies, and vegetation with microbes. The *vice versa*, meaning that soil microbes could be kicked up along with dust into higher reaches of atmosphere. Microbiology of the space and spaceship environments has also been studied and reviewed (Horneck et al., 2010). However, considering the context and theme of this volume, it is out of purview. The intention here is to highlight the role of microbes (e.g. dissemination of biological nitrogen fixers, fungal pathogens, viral particles), their movement in the agricultural sky, mainly, in troposphere region and its influence on cropping belts.

Microbes float naturally in the atmosphere due to light weight and buoyancy. They transit in the atmosphere just like dust particles. They may land on soil surface at any place after a certain distance and may get blown again into "agricultural sky." Long distance travel, even across continents, is also possible (Acosta-Martinez et al., 2015; Bowers et al., 2009, 2013; Prospero et al., 2005). Schmale and Ross (2015) have opined that microbial highways do occur in the atmosphere. It definitely includes agricultural sky. Several aerosol specialists have mentioned that we need methods that accurately gauge and enumerate different micro-organisms, in the atmosphere (Smith and Griffin, 2013). Winds and dust storms aid long distance dispersal of micro-organisms. Acosta-Martinez et al. (2015) report that each year several terra tons of soil particles and dust from agricultural zones in the Sahel and sandy Sahara travels to South and North American agrarian regions (see Acosta-Martinez et al., 2015; Smith et al., 2012, 2013; Plate 1.3). Trans-pacific dispersal of microbes has been reported by Smith et al. (2013). Such intercontinental dispersion of microbes should be monitored periodically. Asian dust/microbial movement into North American pacific region is known to enrich the soil, in that region. Obviously, large population of different microbes, including commensals, pathogens, free living and symbiotic N-fixing microbes, NO_3 transforming microbes, and mycorrhizal fungal spores may reach the soil, in the distant agrarian region. Such a phenomenon of dust storm-aided movement of microbes could be occurring in other parts of earth. For example, in deserts and surrounding locations of Gobi, in Manchuria, Thar in Western India, Atacama in South America, Kalahari in Southern Africa, and desert in Western Australia etc. These regions might be aiding microbial transit in large scale into farming belts via desert sand particles. Microbial movement (floating) in agricultural sky need not always be related to loose-textured soil, dust, or sandy deserts. Thickly cropped or forested tropical zones too may release lots of microbial species into atmosphere, whenever wind blows. Moderately loamy soils are also prone to release dust/microbes into air. Several of the microbes that transit withstanding harsh atmospheric conditions, it seems, are also resistant to UV radiation effects. Incidentally, microbial population in the atmosphere (sky) may have important role to play in the physico-chemical transformations of nutrient-enriched organic debris. They may take part in the nutrient cycling processes in the agricultural sky. These aspects, particularly, nitrogen fixation induced by bioaerosols could be beneficial to farmers who receive the aerial dust position. Quantification of N added through bioaerosols into a farm soil is worthwhile.

Microbial population in the agrarian sky in turn could be potentially deleterious to crops, if the density of virulent forms of plant pathogens is higher. At this juncture, we should realize that agricultural sky plays a vital role in the dispersal of harmful microbes that potentially cause diseases in the crops. Let us consider as examples the dispersal of plant pathogenic fungi through the agricultural sky. Several different bacterial, fungal, and viral diseases of crops are disseminated aerially. Wind currents above crop fields or agrarian regions are known to aid transmission of disease propagules. This can lead to epidemics of crop disease in agrarian regions. No doubt, agricultural sky has a major role in dispersal and dissemination of dreadful rust fungi. They affect cereal crops worldwide. Understanding dispersal of major plant pathogenic fungi across fields, regions, and even continents through the atmosphere (agricultural sky) has been a major pre-occupation of plant scientists, particularly, since at least 5–6 decades. We have accrued vast knowledge about how and to what extent atmospheric transmission plays its vital role, in disseminating pathogenic fungi (Brown and Hovmoller, 2002; Pepper and Gerba, 2015). Yet, we are left with vast gaps in our knowledge about rust fungi that cause devastations on major cereals, such as wheat, maize, and millets. For example, Isard and Russo (2011) have provided details about the various routes that rust fungi attacking wheat adopt, to reach different agrarian zones of the world. We encounter cereal rusts in good intensity. They are being mainly dispersed through atmospheric wind/ dust storms. At present, there are wheat rust disease monitoring networks that monitor the crop and sometimes even the "agricultural sky" above the cropped fields. They try to detect propagules of rust fungi. It is worthwhile to alert wheat farmers about the dissemination of rust fungi. We can forecast the extent of devastations the "agricultural sky" may mediate if prophylaxis is not adopted. Satellite imagery and drones that offer verifiable images of rust attack could be used effectively in gathering data about crop disease, at an early stage. Spectral signatures of crop disease obtained from drones could be of immense value (See Krishna, 2018, 2020 a; Huang et al., 2008, Khot et al., 2014; Garcia-Ruiz et al., 2016).

Let us consider another example of plant pathogenic fungus that affects wheat crop. Fungal spores spread because of their ability to transit rapidly. In Australia, wheat crop is affected by powdery mildew. The agrarian sky is humid and warm. Plus, it has winds to disperse the fungal spores from infected crops. This leads to devastation. The fungus has to be sprayed with plant protection chemicals. Sprays should be done to stubbles/stem before they expose the fluffy powdery mildew infection (spores) to wind. The secondary infection that is predominantly mediated via a windy farm sky causes greater damage (Thomas and Jayasena, 2018).

Fungal spores do migrate to long distances from one agroecosystem to other. The migratory fungi may reach new field with crops located, at a short or even long distance. Reports suggest that timing of spore release, weather conditions, wind speed, and its direction are important (Oneto et al., 2020). In fact, such migration of spores of plant pathogenic fungi in the air is the cause for rapid spread of dreaded crop diseases. They may cause total failures of crops. A few examples of airborne plant pathogenic fungi that get dispersed via agricultural sky are Rust fungi (*Puccinia spp*); *Mycosphaerella spp* (Leaf blotch of wheat and Bananas) (Burt, 2002; Meredith et al., 1973; Zhang et al., 2005); *Gibrella zeae* (rots in cereals) (Paulitz, 1996); *Venturia inequalis* (Apple Scab) (Aylor and Sutton 1992); *Helminthosporium maydis, Leptosprhaerria maculans, Alternaria spp* etc. Obviously, fungal life cycle events, crop stages, and most importantly, the weather/wind conditions play vital role in the spread of crop disease (Oneto et al., 2020; Pepper and Gerba, 2015). Further, Oneto et al. (2020) have provided a list of plant pathogenic fungi along with their spore release pattern and longevity in the atmosphere during dispersal. Such information is useful while conducting prophylaxis, to avoid crop damage. In fact, we should be able to demarcate and classify the agricultural sky above the set of fields or small farming zone or even agrarian belts into those prone to disease based on the fungal propagules, for example, as "disease prone agricultural sky" or "healthy agricultural sky." The sky could be graded based on intensity of fungal spores recorded.

1.3.2 INSECTS AND AGRICULTURAL SKY

Insects, in general, are well distributed in the troposphere. Troposphere is the lowest and closest portion of atmosphere to earth's vegetation. Insects are distributed at different altitudes above the earth's surface up to 15,000 ft from the ground surface. Some of the early attempts to understand the distribution of eggs, nymphs, larvae, and adult insects in the atmosphere at different altitudes were made in USA in the 1930s. Glick (1939) states that insects were actually detected, using traps placed on mountains, light houses, balloons, higher floors of buildings, tall trees, etc. Later came the use of airplanes fitted with insect traps. One of the detailed reports by USDA, Washington D.C. USA points out that much of the insect activity is localized between 0 and 200 ft above crop's canopy. They traced innumerable genera of insects. Many of them were pests on crops. Insect distribution was less dense between 200 and 5000 ft. above ground surface. However, beyond 5000 ft. and up to 15,000 ft altitude, insect population was feeble. Some of the genera traced at different altitudes in the agricultural sky over farming zones in southern United States of America and Mexico were Coleoptera, Diptera, Ephemeraptera, Heteroptera, Homoptera, Lepidoptera, Neuroptera, Mecoptera, Siphonoptera, and Thysanopera. Reports about dispersal of specific insect orders of different continents are also available (Guppy, 1925). Regarding distribution, Glick (1939) states that insects are well distributed in the atmosphere during May–June month of the year in Northern hemisphere. The population is really feeble in December–January months. Ambient temperature is the most important factor that affects insect population at different altitudes in the agricultural sky. Most congenial temperatures were at 23.8°C to 27.6°C. Other atmospheric factors such as vapor pressure, barometric pressure, wind speed and direction, convection, and turbulence all had their share of effect on insect distribution in the air above crops (Guppy, 1925; Parman, 1920). Most importantly, day and night too had its effect on insect movement and distribution in the sky.

The above studies quoted are among the earliest ones and almost century old. At present, particularly, with the advent of drones, we may trace the insect population as many times by flying them at various altitudes and collecting the samples from the traps. There are also airplanes that have the ability to collect samples from up to 30–35000 ft altitude. At such an altitude the prevailing temperature is most uncongenial for any insects to survive. Space shuttles, satellites too can be of use to trace insects at different altitudes in the agricultural sky.

According to the Royal Society of Entomology of Great Britain, there are over one million species of insects that have been described and classified. A relatively small number of them are pests on cultivated crops (Royal Society of Entomology of Great Britain, 2020). Further, within the context of this book, we are concerned with only those insect species that utilize space above the crop's canopy, that is, the agricultural sky. Agricultural sky supports innumerable species of insects and their eggs and even early stages. A sizeable number of them are regular pests that spend a portion of life cycle in the atmosphere. Aerial stages could be devastating on crops. These agricultural pests are transmitted both through wind (eggs, nymphs etc.) and through their flight across the vast stretches of crop fields. The agricultural sky is no doubt a repository large number of pests that attack crops. Many of them appear seasonally as the precipitation, natural vegetation, and crops appear in the agrarian regions. Agricultural scientists are engrossed with several pests that adopt to conditions in the agricultural sky. They have tried to understand how insects perpetuate, disseminate, and cause devastations on crops and their productivity.

There are indeed millions of publications dealing with major insect pests that occur on crops. The Centre for Agriculture and Biosciences (CABI) lists at least top 20 aerial insects that are of great concern to global agriculture (CABI, 2019). They are *Helicoverpa armigera* that affects cotton (cotton boll worm), cereals such as maize, sorghum, wheat, legumes like pigeon pea, chickpea and vegetables like beans, tomatoes, brinjal etc (see Bailey, 2019; CABI, 2019; CESAR, 2019; Queensland Department of Agriculture and Fisheries, 2019; UCIPM, 2020). The insect pest *Helicopverpa* has been reported in over 100 nations, each year during the past decade. Next, *Bemisia tebaci* also called tobacco white fly is widespread on the field crops (e.g. tobacco, cotton, cassava) of tropics. It has been reported in over 56 nations. It is devastating on tobacco and other Solanaceous species. It is also a major vector for plant viruses. So, as the white fly moves across the sky it also transmits viruses. A few other lepidopterous insect pests of greater consequences to crops are: *Spodaptera litura, S. frugiera,* and *S. exigua.* The army worms are widespread in all the continents. *Plutella xylostella,* the diamond back moth that attack vegetables, particularly, Brassicas are severe in Asia, Europe, and Oceania. The spotted mites *Tetranychus sp, Aphis gossipy, Nilaparvata lugens* Brown leaf hopper are together top pests that farmers encounter through the agricultural sky. There are thrips of several genera that occupy the canopy of crops and affect yield. In addition, the list includes top pests of tree crops. It includes several fruit fly species, Psillids, Jassids, mealy bugs, leaf miners, and plant hoppers. They cause damages on fruit plantations. There might be many soil insects such as root worms, etc. that are equally devastating. But they are not dealt with here considering the context of the book. Insect pests relevant to agricultural sky and mainly canopy are only discussed here. Reports suggest that annually we spray pesticides worth 15 billion US$ to regulate these aerial pests. In some cases, the loss, if unattended, could reach 300 billion (e.g. cereals, legumes, oilseeds, gourds, cucurbits etc.). Clearly, agricultural pests that throng the sky and spread at rapid rate are a major agricultural sky-related phenomenon (CABI, 2019).

Pearl millet is a staple cereal food to humans and domestic animals of the Sahelian West Africa. Pearl millet grown in the subsistence farming zones in the Sahel is affected severely by the head miner (*Heliocheilus albipunctella*). It is a major constraint to millet production in the Sahelian zone. The larvae feed on the panicle, leading to 40–85% yield losses. It seems use of biological control *Habrobracon* (=Bracon) *hebetor* is effective against the millet head miner. The release of the parasitoid wasp has led to over 70% parasitism of the millet head miner larvae (ICRISAT, 2008). This is a good example of an aerial phenomenon that controls the pest or keeps it at threshold.

In certain parts of world, locusts are major biotic phenomenon in the agrarian sky. Locusts (short-horned grasshoppers) have devastating effect on general vegetation and field crops (Lecoq, 2005). Their movement in the sky could be monitored, using satellite imagery. Unmanned Aerial Vehicles, specialized farm scouts, and their networks and tribal farmers in the area are employed. This is to get advance information about locusts. At present, locusts are frequently reported in Northwest India, East and North Africa, and West Africa. The agricultural sky that offers easy transit for large clouds of locusts need to be made uncongenial and suppressive.

Locusts or short-horned grasshoppers are among the most dreaded insects. They possess potential to devastate agrarian regions. Locusts are actually "short horned grasshoppers" belonging to the order Orthoptera and family Acridoidea. Locusts are worldwide in distribution. Locusts move through the sky in large numbers. Most often, locusts move along the agricultural sky in discrete bands known as "swarm." They alight to cause devastation of vegetation/crops in vast stretches in perhaps shortest period. Locust swarms or clouds of grasshoppers contain hundreds of millions of individuals (Arthur, 2008; Simon, 2020; Song et al., 2018). Locusts are gregarious and their swarms travel large distances. To quote an example, population dynamics studies conducted by CIRAD researchers indicate that West African Sahel is actually invaded by locusts via Chad. Such locusts usually arrive in Chad from West Asia, North Africa (Sudan), and East Africa (Kenya, Eritrea) (CIRAD, 2020).

Reports suggest that there were at least 8 different Desert locust plagues. It was caused by several different grasshopper species (e.g. *Schistocerca gregaria, Anacridium sp.*) that occurred in Africa (Egypt, Sudan, South Sudan, Ethiopia, Libya, Northern Chad, Tunisia, Algeria) and Asia during the previous century. Entire Sahelian belt in West Africa is vulnerable to locust attack that destroys millets and legume crops. The recent locust attack that devastated North Africa and West and South Asia damaged crops, mainly cereals/millets in about 29 km^2 area (not ha). Desert locusts bring about severe damage to cereal/legume intercrops in West Africa. Mauritania is a Sahelian nation that already suffers from low grain productivity. Reports state that 36–43% grain loss for a stretch of 5 years until 2004 could be attributed to desert locusts (Hebei et al., 2004; Simon, 2020.). Such desert locusts are an aerial phenomenon that usually appears immediately after the onset of rainy season (June/July) and lasts till crops senescence. The swarms forage through the entire stretch of crop production zones. Locusts do interact with livestock grazing zones and can potentially devastate pastures. They can devour pasture grasses in a short while. Locusts and their foraging behavior are dependent on native vegetation plus the weather pattern. For example, a rainfall event that stimulates greenery in Sahel and North African cropping zones recently induced high locust incidence (Chauffin, 2020). Such surge in aerial phenomenon of locusts has been often reported from Somalia, Eritrea, Sudan, and even Kenya. Further, we may note that locusts are known to be carried by winds though long distances. They cross Atlantic and reach Caribbean and South American farming regions. Sahelian soil/dust laced with grasshopper eggs may easily float across Atlantic Ocean (Ritchie and Pedlegey, 1989; Rosenburg and Burt, 1999).

Certain tree grasshopper species (e.g. *Calliptamus italicus*-Italian Stout bodied tree locust; *Anacridium melanhordon*-Sahelian tree Locust) are well spread out into entire Sahelian crop production zones. They are primarily tree leaf eaters. These locust species do not compete strongly for pasture grasses (Elamin et al., 2008; LeGall et al., 2019; Sergeev and Van'Kova, 2008). However, tree productivity gets affected. For example, in Sudan, tree locusts could reduce gum Arabic crop yield from 273 kg ha to 93 kg ha through mild attack on trees (Elamin et al., 2008).

Lecoq (2005) states that crop loss due to insect pests is not a new phenomenon in subsistent and poorer regions of African continent. For example, in addition to the usual insect infestation, the desert locust in arid regions of Africa has been occurring regularly, since the historical times. During recent past, desert locusts have devastated African cropping periodically since the 1960s (FAO, 1968, 1972; Steedman, 1990). Prevention using advanced information about locusts has been useful. However, desert locust persists sometimes peaking again as severe pests. In the most recent past, desert locust attack that occurred between 2003 and 2005 has been the most devastating aerial phenomenon. Monitoring the agricultural sky is therefore mandatory in areas prone to such devastating pests (Lecoq, 2005).

There are specialized agencies such as Desert Locust Information System (DLIS). They disseminate accurate information about locust attacks. Usually, satellite imagery is utilized to track the locust clouds. Then, forewarn the farmers and farm agencies about the impending damage that lurks through the farm sky. Currently, there is a trend to utilize agricultural drone to obtain aerial imagery of locusts. Radar sightings are also utilized to inform the farm population about locusts, their size, and possible extent of damage by the grasshoppers (Atherton, 2017; CIRAD, 2020; Krishna 2016, 2018; Le Gall et al., 2019; Yue et al., 2012).

The high intensity of individual grasshoppers that make up the locust swarms in the agricultural sky is not easy to control. Particularly, using the traditional pesticide application methods, logistics, and the sheer large quantity of pesticide sprays required make it difficult to adopt. Biological control of short-horned grasshoppers has also been attempted (Carruthers and Onsager, 1993; Chapman and Joerne, 1990; Goettel and Johnson, 1997; Krishna, 2018; Lomer et al., 2001). Such a measure depends on the virulence of the biological control agent, usually a pathogenic microbe or a predator insect that kills the grasshopper. Recently, drone air crafts filled with chemical pesticides or biological control formulations have been suggested, for use against locusts (Atherton, 2017; Joerne and Gaines, 1990; Krishna, 2018; Kreutz, 2017).

A report from the Desert Locust Information System (2019), a center for locust monitoring, states that recently mobile/satellite-based applications to map the movement of locusts are made available. For example, "*eLocust3*" is an application that tracks the daily movement of locusts in Sudan, Chad, etc. They say, at least a few million grasshoppers move 150 km distance daily. They devour 2 g of green leaves per insect. If left unattended, they say, locust swarms may spread into 20% cropping zones on earth. Satellite imagery has been utilized consistently for the past few years, to monitor locust attacks and movement in Kenya, Ethiopia, Sudan, Libya, and Chad.

Now, let us consider a beneficial activity of insects that throng the agricultural sky, that is, insect-mediated pollination. Pollination is a phenomenon entirely accomplished in the agricultural sky. Pollination is the transfer of pollen between plants (canopies). It induces fertilization and sexual reproduction. The two types of pollination known are abiotic and biotic. Abiotic pollination takes place without the involvement of living organisms, for example, where pollen is transported by wind. Biotic pollination is the result of the movement of pollen aided by living organisms (e.g. bees, insects, birds). The biotic pollination is the dominant form accounting for 90% of pollen transfer mechanism in the agrarian fields (Abramovitz, 1998).

Insect pollinators influence the stability, diversity, and function of natural and agricultural plant communities. There are indeed several reports suggesting that pollinator population and activity is essential for optimum seed set. In nature, insect pollinators may directly influence crop's reproductive function and provide ecosystem services, by influencing the quality and quantity of crop yield. They say, globally, about 35% of all crop species cultivated depend on insect-mediated pollination to a certain extent. About 70% of fruit and vegetable crop species depend on insect for pollination (Ricketts, 2008; Saunders, 2017). Reports by European Institute of Sustainable Agriculture state that insect species other than honeybee too need greater attention. Methods adopted in farm should induce their population and activity. For example, there are about 240,000 species of dipteran insects that are capable of pollination in the cropped fields, mainly flies, midges, etc. (Kunast et al., 2019). Rader et al. (2016) state that well-managed bee species and their population plus the wild bees do aid crop pollination. They are among most efficient pollinators. However, there are several non-bee pollinators such as flies, beetles, butterflies, wasps, ants, birds, and bats. To compensate for the lower pollination efficiency, we can adopt non-bee insect species for pollination. Non-bee species, it seems, may visit the flowers often, say 25% more than the bees (Rader et al., 2013, 2016).

Pollen collection, its carriage along the fields, and release in a nearby crop field are perhaps best examples of beneficial role that insects play in an agricultural sky. In many fields, bee population is a key indicator of effective pollination of crops. The pollinator function is carried out by a large number of different bee species along with other pollinator insects. Bees do aid pollination of crops that are otherwise, in general course, wind-pollinated species (Rader et al., 2016; Saunders, 2017; Wheelock, 2014). Garibaldi (2013) state that, irrespective of bees offering eco-systematic functions, there are crops that are naturally wind pollinated. Here, wind has its share of influence on seed formation and productivity. Saunders (2017) has provided a long list of crop species and natural vegetation, along with several bee, moth, wasps, and Scirphid fly species that pollinate wind-pollinated plants.

Plantations too depend immensely on insect pollinators. Coffee plantations are traced often along with native tree species. Trees provide the shade needed for the coffee bushes. It is said that native bee species are known to support geneflow by aiding pollination. Often, long distance pollination within the plantation ecosystem is achieved via native bee species. They travel over 1800 m from the place where they pick pollen (Jha and Dick, 2010).

Let us consider a case study about insect pollinators, which includes mostly honeybees. Reports from farm research agencies in United Kingdom state that pollination through bees and other insects is important for optimum productivity of crops such as brassicas, apples, strawberries, onions, carrots beans, and other vegetable species (Potts et al., 2014). Clearly, we have to manage the agricultural sky above the crops with high levels of pollinators. High levels of insect population help in stabilizing grain yield. It supposedly reduces variability within the cropped field. There are tree species such as apples whose production depends immensely (85%) on optimum pollination levels prior to fruit set. Most Brassicas too depend on pollinators to achieve good harvest. Overall, about 37% of crops in UK are dependent on bees/insect pollinators. Agricultural entomologists have identified over 1500 insect species that can help in pollination of crops. The species diversity of aerial pollinators of course depends on location, season, natural vegetation, and crops that flourish in a given area. Flower-rich field margins could be helpful in enhancing bee/insect pollinator population in the agricultural sky. Therefore, we have to reduce cutting of natural vegetation and hedges with flowers.

Insect pollinators are under threat from a variety of human pressures, particularly, habitat loss through land use intensification. There is an urgent need for research and management programmes that enhance pollinator populations (Bartemeous, 2014; Ricketts et al., 2008). As such avoiding use of harmful pesticides and the deterrents during flowering period and until seed set is prescribed. This is to improve bee-aided pollination of crop flowers (European Food Safety Agency, 2013).

Honeybees are among most important pollinator insect species in any agrarian region. Although there are other insects such as butterflies, moths, scirphid flies, wasps, and bugs that accomplish a similar task (Torchio, 1987). Some agricultural landscapes provide suitable habitat for pollinators. However, there are many agricultural practices that negatively impact native pollinator communities. Agricultural landscapes can also have high densities of wild and exotic honeybees (*Apis mellifera scutellata*) (Kennedy, 2013), for example, *Xylocopa spp*, *Scaptogtrigona spp*, and *Trigano spp*. A report by the European Institute for Sustainable Agriculture states that among the 20,000 or more species of honeybees that throng the agrarian sky over the European plains, *Apis mellifera* the common European Honeybees is most prolific pollinator. Its population directly affects the pollination efficiency and agricultural crop productivity. In fact, about 70% of the crops grown in the European plains depend, to a get extent, on *Apis mellifera* and other bee species. During recent years, there has been a declining trend in honeybee population and activity in many locations. This has been attributed to overuse of pesticides, honeybee parasites getting accentuated and negligence of farmers who do not encourage bee population in the fields (Biesmeijer and Roberts, 2006; European Food Safety Authority, 2013). The European farm agency insists that honeybee protection is essential. After all, European food grain generation that is valued at 153 billion US$ each year is highly dependent on pollinators, such as *Apis mellifera* and other insects (Abramowitz, 1998; Kunast et al., 2019). Overall, it has been pointed out that factors that induce bee population and pollination activity have to be understood in detail and accentuated. Agrarian regions world over should adopt procedures that induce pollinator activity. The agrarian sky anywhere has to be most congenial for pollinator activity. This will ensure higher grain harvests (Kunast et al., 2019).

1.3.3 AVES IN THE AGRICULTURAL SKY

Foremost, the agricultural sky is abode to aves of great diversity. Their population is again variable depending on several factors. Aves operate from the sky with great impunity and versatility. They do affect a series of aspects related to crops. Across the agrarian regions of the world, birds of diverse species act as pests on crops. They attack crops in fields, at different stages of the growth beginning from seeds that are sown in the field. Bird pests are no doubt an aerial agricultural phenomenon of detrimental consequence, if they are not warded off or eradicated. Cereal grain crops at panicle stage are among most susceptible to bird damage. Let us consider an example. In the Indo-Gangetic plains, wheat-rice agroecosystem has stabilized through the past five decades. It is forecasted to last for several more. The wheat crop is supposedly an excellent foraging zone of several hundred species of birds. These birds feed on grains. Many of the bird species are migratory too. They cause severe damage to crop at the ripe panicle maturity stage, bringing down the productivity. At the same time, there are several bird species noted on the wheat field that are insectivorous. They help in biological control of insect pests. Relative abundance of bird pest and those beneficial need to be ascertained, in each region (Borad and Parasharya, 2018). We can trace examples of bird pest as well as those species that reduce insect attack, in almost every major agrarian region of different continents. Bird pests are an example of aerial phenomenon which needs greater attention. Agricultural sky has a major role in it.

Sunflower is a crop prone to severe bird damage, particularly, at the grains set stage of the head. Bird damage to crop actually begins right at 4 leaf-stage. Pigeons that attack sunflower can be devastating, leading to almost total loss of grains. To control this agricultural sky-related phenomenon, farmers are adopting an aerial method. It involves bird scaring using helikites (Allsopp Helikites, 2019). Such helikites are lofted into sky above the crops. They possess recorded sounds of birds of prey that ward off the pests (see Allsopp et al., 2013; Krishna, 2020a, 2020b). Several millet species such as pearl millet, proso millet, foxtail millet, finger millet are prone to bird pest damage. Incidentally, a few small millet species, particularly, certain genotypes are suited as bird feed (Patel, 2016). Hence, no doubt, bird species relish feeding on millet fields. It can be disastrous if no precautions are taken to scare them off. Fruit damage to plantations can get severe. This is again an aerial phenomenon of great consequence to farmers' exchequer. Birds usually reduce the value of the fruits or totally damage them. Overall, we may note that crop/plantations may differ in their susceptibility to bird species. Bird species that infest crops may differ. Yet, a certain damage could be avoided, if prophylactic bird scaring is practiced (see Krishna, 2020). Helikites with recorded sound of raptors are known to reduce pigeons, particularly, pigeon species that feed on sunflower grains and pulse seeds (Allsopp Helikites, 2019). These helikites are also part of the agricultural sky. We could use them effectively to reduce grain loss. There are companies that produce helikites specifically meant for scaring bird pests.

Several bird species conduct an aerial life for most part of the day. They are attracted by ripening grain crops. At harvest they flock to drying and threshing floors. There are also birds that develop a liking for grain storage zones. Some have developed a close association with the more permanent sources of cereals and cereal products. They have become a nuisance in food stores and the vicinity of warehouses (Caliboso, 2013).

Birds that throng the agricultural sky above crop belts are highly beneficial as biological control agents of rodents and insect pests. Wet land crops such as rice are known to benefit from egrets, geese, waterfowls, and several other species that consume insects in the wetlands. Raptors, falcons, barn owls, small songbirds, and several other species have been found useful in controlling rodents and insects, in orchards/cereal farms of Europe. It seems just keeping a few nest boxes and allowing birds to colonize around apple orchards, reduces insect pest that attacked apple orchards, in Netherlands. Blue bird species kept insect populations at low level in the vineyards of California. Birds of prey are indeed effective in keeping the pest on threshold. Plus, they offer the advantage of not employing environmentally harmful chemicals (McGashen, 2018).

Seed dispersal by different species of birds that localize above crop fields is a useful phenomenon. The seed dispersal occurs through the agricultural sky. Birds usually disperse the seeds of a broad range of plant/crop species through bodies, beaks, and claws. Also, while feeding on fruits/grains and through droppings, the distance to which the seeds of crop/plant species get dispersed depends on a few factors. Reports suggest that it could range from 20 m to several kms depending on bird species and habitat (Levey et al., 2005). Basically, it involves landscapes filled with natural vegetation and crops. The diversity of vegetation and crop species cultivated in a habitat is important. It is of course related to the bird's habitat within which it transits. It is an important factor that decides the distance to which seeds could be dispersed. Therefore, habitat disruption, fragmentation due to logging, or removal of natural vegetation or loss of cropped zones affects the bird species diversity and effective dispersal of seeds (Levey et al., 2005). There are indeed several species of birds and bats that mediate aerial dispersal of seeds of forest, fruit plantations, and field crops (Levey, et al., 2005; Murray, 1988; Whittaker and Jones, 1994). Bats may disperse 5–8 seeds m^{-2} during night. Birds could spread out 8–12 seeds m^{-2} day^{-1} depending on source, for example, fruits, vegetables, field grain crops (Murray, 1988). Bird- mediated seed dispersal, it seems, could play a vital role in rebuilding and stabilizing natural vegetation surrounding the cropped field (Whittaker and Jones, 1994). Long-distance dispersal of vegetable and grain crop seeds by birds is also a possibility. There are reports that in Spain, long-distance migratory birds that picked seeds from natural vegetation and cropped field could have dispersed them to areas in North Africa. Birds might have played a role in the introduction of new crop species through their long-distance migration, year after year. Migratory birds do disperse seeds across continents (Viana et al., 2016).

There are also birds that thrive as commensals in the agroecosystems. They may not be detrimental to crops. They may offer no advantages either by reducing insect pest. Yet, they are part of ecosystem functions in a farm. The agrarian region offers them a niche to survive and perpetuate. Guanos is a fertilizer derived from bird's defecations. There are two types of Guanos. Fresh guanos applied to crops is rich in nitrogen, phosphorus, and calcium. The drier version is rich in phosphorus and calcium. The avian population actually cycles the nutrients efficiently in the agroecosystem. Some examples of brand names of guano fertilizers are "Peru-guano," "Guano Boom' etc (Schnug et al., 2019).

Overall, there are several useful roles for birds that flourish in the agro-ecosystems. We could accentuate those species. Perhaps, bird damage to crops could be reduced by warding off the detrimental species, using helikite scarers (see Plate 1.5). It is wise not to use pesticide or baits, to destroy the species. It is also clear that we have to characterize "agricultural sky" based on avian population, its diversity, and positive or negative effects the bird populations have on the crop productivity. We could use the avian-related traits while classifying the agricultural sky. Agricultural sky could be classified as "rich in pollinators," "rich in insect eaters," high in bird pest," etc. If one encounters sky classified as "high in pest birds" then we try to deploy helikite bird scarers etc. If one wants to plant sunflower, then he selects areas with sky classified as rich in pollinators, during different periods of the year. A sky (region) classified as rich in bee/avian pollinators is good for growing oilseed Brassicas (Rape, Mustard) etc. A region with its sky rich in pollinators is also the one best suited to be used in plant breeding programs. Plant breeders need bird's activity, particularly, if the crop species is cross-pollinated. Crop breeding is dependent on minimal levels of population of insect/bird pollinators.

1.4 MAN-MADE FACTORS IN THE AGRICULTURAL SKY

1.4.1 *FOLIAR FERTILIZATION*

Soil plays a mute witness to the above ground portions of crops and their ability to absorb the nutrients, hormones, pesticides, or fungicides applied, using aerial sprayers. Of course, soil absorbs several of these aerial inputs. These inputs undergo physico-chemical and microbial transformations in the soil phase. These aspects are not within the purview of this book. *Foliar fertilization is a procedure conducted entirely in aerial space above the crop's canopy*. The nutrients are channeled to crops via liquid formulations. The stomata and hydathodes on the leaves absorb the nutrients and translocate it to different parts of plant. Several different mineral nutrients such as N in the form of urea, $NH4^+$, KNO_3, $NHNO_3$, P as HPO_4, K as KNO_3 or KSO_4 and a few micronutrients such as B. Zn, Fe, Mn are channeled through the foliage (i.e. agricultural sky). At present, we are not sure how many different essential nutrients can be absorbed by crop species through leaves (canopy), for what duration, and to what extent, if applied as aerial sprays. It is worth experimenting and confirming prior to adopting foliar fertilizers in commercial farms. Are there crops that absorb all the essential nutrients and entire water requirements via their canopy (i.e. leaves). In that case, root systems will be an organ meant only for anchorage.

There are a few problems to be evaluated and solved prior to adopting foliar methods of supplying nutrients to crops. A few of them are as follows: What are the elements available in atmosphere? How many are useful to crops? How many are not absorbed by crops despite being supplied through agricultural sky? What is the mechanism for expelling or rejecting a nutrient element at the leaf surface? How much of a nutrient can be channeled using aerial sprays on the canopy?

Rhizosphere is a region surrounding the roots. It extends from root's surface (rhizoplane) to 0.2 mm thickness. At this stage, we may note that like "rhizosphere" surrounding roots in soil, there is also phyllosphere and phylloplane micro-regions, in the aerial canopy of crops/trees. Phylloplane and phyllosphere are important regions through which every ml of water and nutrients/pesticides/fungicides have to pass, before entering the plants' vascular system, if foliar irrigation and/or nutrition is practised. Actually, phylloplane and phyllosphere in a crop's canopy are concepts/regions comparable to "rhizoplane and rhizosphere" of underground roots system of a cropping belt. Indeed, canopy, its size, structure, leaf area, and stomatal distribution are among the important aspects, when we conduct aerial spraying of liquid fertilizer formulation or plant protection chemicals.

1.4.2 AERIAL SPRAYS OF PLANT PROTECTION CHEMICALS

Aerial sprays of plant protection chemicals like pesticides, herbicides, growth hormones, and even biological control agents are common, in many agrarian belts. Such an agronomic procedure does affect the agrarian sky just above the crop canopy, particularly, above the field that was sprayed. Indeed, plant protection chemicals while they avoid pest attack, the toxic chemicals could be detrimental to microbes, nontarget insects, and birds. Plant protection with long residual effect (half-life) needs careful handling. Aerial sprays of herbicides could affect non target natural vegetation. Aerial sprays of growth hormones have to be accurately done. If not, nontarget inter crops and adjacent fields with a different crop species could get affected. Such aerial sprays may drift to neighbor's fields or even to longer distances therefore spoiling the atmosphere. Plant protection chemicals sprayed could settle down on soil surface and percolate. Aerial sprays of plant protection chemicals have been generally done using sprayers operated by farm labor scouts. Farm vehicles fitted with sprayers are common. During recent years, low flying multi-copters and helicopters have been found to be too efficient in spraying pesticides, herbicides, hormones etc. Detailed discussions on aerial sprays have been made available in Chapter 4.

1.4.3 USAGE OF UNMANNED AERIAL VEHICLES IN THE AGRICULTURAL SKY ABOVE FARM ENTERPRISES

At the outset, we may note that agricultural drones offer us aerial imagery of crops from vantage locations in the sky. It was never possible for farmers since several millennia. In fact, analysis of crops using aerial imagery is itself, historically, a recent effort by farm researchers. *Agricultural drones are the most recent man-made factors to enter the agrarian regions, more specifically the "Agricultural Sky."* They have the potential to revolutionize the way we conduct several of the agronomic procedures (Krishna, 2018). Recent reports suggest that aerial drones have indeed found a niche above the crop's canopy. They fly briskly in the agricultural sky drawing most valuable spectral data. Unmanned Aerial Vehicle (UAV), particularly, multi-copters (e.g. MJ-AGRAS-1; DJI, 2016, 2019) and helicopters (e.g. Yamaha's RMAX, RMAX, 2015; HSE's Hercules, Sanders, 2017), have been adopted to spray plant protection chemicals and foliar nutrients. Agricultural drones primarily keep a vigil over the crop's progress in growth, and disease/insect/

weed infestation, if any, in cropped fields. Crop genotypes are evaluated using spectral data (e.g. NDVI, Leaf Area, Chlorophyll content). Farm surveillance through aerial photography is an added advantage derived from UAVs (FAO of the United Nations, 2017; Krishna, 2018; 2020a; Palilo 2014; Tattaris and Reynolds, 2015). Farm surveillance using UAVs in agricultural sky is supposedly efficient in terms of costs. We may note here that originally the UAVs were common in the general civilian sky above urban locations. Drones were used often for surveillance of traffic, public events etc. Otherwise, they were most common in the military air space of feuding nations. However, Krishna (2018; 2020a) clearly points out that there is a general trend among farming community worldwide, to infuse their "agricultural sky" with versatile UAVs (see Krishna, 2018, 2020a, 2020b). Not just the small aerial robots, that is, drone aircrafts, farmers are also evaluating other types of UAVs such as parafoils (Pudelko et al., 2012; Tetracam Inc, 2018; Thamm, 2011; Thamm et al., 2013;), microlights (see Krishna 2020b; Northwing, 2018), tethered blimps (CIMMYT, 2012; Haire, 2019), aerostats, helikites (Allsopp Helikites Ltd, 2019), and kites (Aber and Aber, 2016; Aber et al., 2009). As stated earlier, many of these aerial vehicles were first part of the civilian and military sky, but they are now being modified or tailored specifically to suit the agricultural purposes. Worldwide there are over 1200 UAV-producing companies that generate several models suitable for use above farms (Krishna, 2020a; Wikipedia, 2017). They will aid efficient use of agricultural sky.

Fixed-winged drones (e.g. SenseFly's *eBee*, Precision Hawk's Lancaster, Trimble's UX5) are popular aerial agricultural vehicles. They are used mainly to obtain aerial imagery (visual range) and multispectral data of agricultural terrain, soil types, surface features, crops, irrigation channels, etc. The spectral data obtained about healthy crops (NDVI, GNDVI, leaf chlorophyll, crop's water stress index, i.e. CWSI) is useful while deciding the sequence, timing, and intensity of agronomic procedures. Aerial sprays or soil injection of fertilizers and irrigation procedures are also often dependent on the digital data. Such digital data is obtained by fixed-winged drone, at a rapid pace. In the air, a pre-programmed fixed winged drone collects data of about 50–100 ac farm in an hour. The spectral data from fixed-winged drones are also useful while mapping the pest/disease affected patches. Such digital data could be utilized directly on the sprayers, during precision farming operations (Palilo, 2014). Fixed-winged drones are relatively easy to operate and affordable. They are expected to throng the agricultural sky in big number, in due course (Krishna, 2020a).

As stated above, agricultural drones are man-made aerial phenomena. They are being utilized to obtain spectral data about healthy crops, along with spots that are affected by different pests, if any. For example, damage to soybean crop by aphids could be identified and mapped, using spectral imagery (Marston et al., 2019). The spectral data from drones is usually referred to a data bank and compared to identify the pest species. The same data can then be utilized in guiding the variable-rate applicators of pesticides, using ground vehicles or drones fitted with aerial spray bar (Atherton, 2017; Filho et al., 2019; HSE, 2015, 2017; Krishna, 2018, 2020a, 2020b; United Soybean Board, 2013; Yue et al., 2012; RMAX, 2015). By using aerial robots (drones) we are actually pitting a man-made factor (UAV) to conquer a natural aerial insect pest attack. Agricultural sky will be utilized efficiently, if we adopt aerial robotics to a greater extent than at present. Agricultural drones reduce human labor involved in scouting large areas of cropped fields. They offer highly accurate spectral data with GPS tags. Agricultural drones are economically efficient. Aerial observations are made at rapid pace covering large area of crop fields. Therefore, drones reduce human drudgery in the field.

The agricultural sky is now being invaded by yet another type of drones called the "autonomous helicopters" and "autonomous multi-rotor copter" drones. Several models of copter UAVs are produced in North America, Europe, China, Japan, and other Fareast nations. Many of these drone models have already made a mark in the agrarian sky. They are most frequently traced in the sky above rice fields. In addition to obtaining visual images, videos, and digital data, copter drones are being employed in plant protection operations. For example, helicopters such as RMAX are being adopted regularly to apply plant protection chemicals from the sky. They are efficient both in terms of time required to conduct an aerial task and costs (see Khot et al., 2014; Krishna, 2018, 2020a). The copter drones, in particular, offer detailed close-up maps of agrarian terrain with GPS tags. Copter drones too provide explicit digital data for use on farm vehicles. In addition, these copters are adopted during foliar inputs of fertilizers. Applying herbicides, fungicides, and pesticides is easier, if we adopt copter drones (Atherton, 2017; Huang et al., 2008; Yue et al., 2012). Farmers try to utilize the "agricultural sky" to accomplish varied tasks (Huang et al., 2008; Khot et al., 2014). The fixed-winged and copter drones operate in the sky from close locations above the crop's canopy, say, a few feet to 300 ft above in the sky. At this juncture, we may take note that airplane campaigns are less efficient. Satellites used to obtain aerial imagery are less efficient in terms revisit time. Resolution of images is insufficient. Problems related to hazy pictures, due to clouds or haze in the sky, too exist (see Krishna, 2018).

There are several other types of aerial vehicles such as the parafoils, parachutes, blimps, tethered blimps, tethered aerostats, and kites. They could be operated in the agricultural sky (see Krishna, 2020b). Each type of aerial vehicle has its advantages and lacunae. Farmers need to select the best ones that suit their purpose and affordability. Autonomous parafoils are being evaluated stringently for use over crop fields, to collect aerial imagery and digital data (Thamm, 2011; Thamm et al., 2013; Pudelko et al., 2012). Blimps have been adopted to obtain aerial imagery. The tethered versions placed above the crops are known to provide farmers and research staff at experimental stations with continuous data about crops. They are useful in monitoring crops all through the season. Blimps could be utilized to transport agricultural goods, aerially. Helikites (Plate 1.5) and kites have been sought by farmers to ward off bird pests, to keep a vigilance over the farms and to obtain aerial imagery (Allsopp Helikites Ltd., 2019).

Overall, the agriculture sky is to experience innumerable incursions by the aerial robots, in near future. Like it or not, they are efficient. So, they may take over major portion of scouting activity in farms all over the world. Plus, they are destined to conduct a series of farm operations utilizing the agricultural sky.

PLATE 1.5 A Helikite with multi-spectral sensor for aerial imagery.

Source: Mrs Sandra Allsopp, Fordingbridge, United Kingdom.

Note: Often helikites are fitted with recordings of predatory birds. Thus, pest species of birds are scared and warded off from the cropped fields. Such scaring is done during seed-set and maturation or during seeding/germination. The same helikites could be utilized for aerial imagery. Note the sensor hanging from the helium balloon.

1.4.4 USAGE OF AGRICULTURAL SATELLITES

Satellites have been in use above the agricultural landscapes since past five decades, beginning with perhaps LANDSAT system launched by United States of America (NASA/USGS, 2020). Landsat satellite has been utilized effectively to monitor occurrence of thunderstorms and tornado, and also to generate images of clouds and precipitation patterns (Bentley et al., 2002; NASA/USGS, 2020). Satellites localize themselves in the thermosphere layer of the agriculture sky. They have offered details about earth's terrain, its salient features, topography of agricultural zones, soil types that are utilized for crop production, soil moisture distribution maps for the top 10 inches, cropping systems, and their margins. Crop species and even specific genotypes have been mapped in certain cases. Seasonal variation in natural vegetation and demarcation of genetic diversity has been a major contribution of the satellite technology. At present, there are innumerable satellites that provide aerial imagery of agricultural expanses. Satellites offer details about global weather patterns, for use by farming community (e.g., SPOT, LANDSAT, NASA's SMAP). Sharper images of agricultural terrain even individual farms are possible. It is dependent on the resolution of the imagery offered by the specific satellites. Agricultural sky with its satellites has been most efficiently used in mapping entire farms and farming belts. Such aerial maps are accurate since they provide GPS coordinates. Further, groups of satellites (GPS, GLONASS) have immensely helped the farm vehicles, during the conduct of agronomic procedures. Precision faming is a recent technique, which involves liberal use of GPS methods. Global Positioning System is adopted to mark the "management blocks," also, to conduct almost every other farm operation, such as land tillage, seeding, inter-culture, weeding, fertilizer application using variable-rate dispensers, harvesting adopting GPS-guided combines etc. (Khosla, 2010; Krishna, 2013, 2016; Zhang 2015). Satellites stationed in the sky (i.e. thermosphere layer of atmosphere) are almost inseparable from farming methods, in the future (e.g. LANDSAT 8, NASA's Soil Moisture Active Passive satellite). Agricultural satellites stationed away in the thermosphere offer only relatively low-resolution images. They may not be good enough, if small details of farms are required. Resolution of imagery ranges from few meters (e.g. 2.5 m to 5 m in case of SPOT satellite of France) to one km, based on the satellite. Satellites are best adopted when monitoring and mapping large agroecosystems and cropped zones. They are still not feasible to be adopted if clouds or haze interfere with imagery. In case we need aerial imagery and digital details repeatedly, then, the re-visit time of satellites is often uncongenial to farm community. Satellites need high capital investment during launch and use by farmers. In comparison, agricultural drones fly at low altitude above the crops" canopy. They offer excellent higher resolution images, if farmers need close-up details of crops (see Krishna, 2016, 2018). Overall, we may realize that satellites are part of agricultural sky. They are of great utility to global agriculture. A few examples of satellites utilized by agricultural agencies of different countries are as follows. Landsat 8, originally developed by National Aeronautics and Space Agency, USA in the 1970s. IKONOS satellite series maintained by a private satellite agency called DigitalGlobe Inc. situated in California. SPOT 7 maintained by System pour l'Observation de la Terre, CNES, France. IRS Indian Remote Sensing Satellite series. NigeriaSAT-2 NigeriaSAT-X and series, etc.

Drought, floods, dust storms, water resources, disease/pest attacks in large areas e.g. locusts in North Africa could be monitored using Landsat 8. They could be mapped using mobile applications (Desert Locust Information System, 2019), Satellites and internet connectivity are essential for robotic farm vehicles, for example, fertilizer inoculators, combine harvesters and grain transport truck employed in the large farms. Believe it or not, in the future, agricultural robotics may after all immensely depend on satellite guidance (See Krishna, 2016, 2018, 2020a). The information from data banks that are needed by the ground and aerial robotics is channeled, using the satellites placed in the sky. No doubt, in the future, "agriculture sky" will be utilized more efficiently to accomplish a wide variety of farm operations. This holds true while operating both semi-autonomous and entirely robotic vehicles (e.g. Kinze Autonomous Grain corn harvesting systems) on the ground and in guiding the aerial robots (e.g. drones, blimps, aerostats etc.) (see Krishna, 2016). *In the future, aerial robots in the agricultural sky may become indispensable.*

1.5 AGRICULTURAL SOILS, WATER, CROPS, AND SKY: A COMPARISON

First, we have studied a great deal about crops, soil, water, and even sky regarding several aspects relevant to agriculture. In case of soil, we have made great strides in understanding the geological processes that induce soil formation, its minerology, various natural traits, physico-chemical and biological phenomenon, soil loss and deterioration processes etc (Wilding et al., 1983a, 1983b). These aspects pertain to soils in general. Further, agricultural soils that specifically suit and are utilized for crop production are clearly defined, characterized, classified, demarcated (mapped), and studied (Beyer, 2003; Brady, 2000; ISSS-WRB 1998 etc). In case of agricultural soils, aspects such as seed germination, seedling establishment, soil fertility (nutrients), moisture content, aeration, and soil health are considered important. Most of the agricultural soils are clearly identified and classified. Similarly, if we consider plants, then we have plant species classified as agricultural crops. There are species that are part of a natural vegetation in an agroecosystem. Agricultural crops species provide food grains, vegetables, oil seeds, fibers and other products.

Agricultural water and its importance to the sustenance of natural vegetation and agrarian cropping expanse are well understood. Irrigated crops account for the 39% of cropped land. The phenomenon of precipitation (rainfall) has been studied in great detail. Voluminous data and reports are available about the rainfall trends in the previous years and decades. The major sources of agricultural water are the precipitation, surface water such as rivers, lakes, ponds and dams. The groundwater that is drawn using lift irrigation is another important source of agricultural water. The rainfed agriculture is supported by the natural precipitation caused by clouds in the sky. Globally, it accounts for 78% of cropped land. Rainfed agriculture is the mainstay for all of the major food grain generating agrarian expenses. Agricultural sky and water are perhaps inseparable factors when it comes to food grain generating regions (CAB International, 2009).

Agricultural crops are again a very well investigated, studied and publicised aspects. There are over 7000 agricultural crops known to us. Yet only a small fraction of those species is preferred by humans for regular cultivation. Agricultural crops have been studied in great detail about their botanical features. They are actually classified based on botanical characters. Of course, there are several ways of classification of agricultural crops. The classification of agricultural crops utilizes several characteristics such as preferences to seasons, climatic requirements, stature of the crops, water needs, the features of crop and produce that they generate, economic aspects etc. At present, major cereal crops such as wheat, rice and maize plus a few millets offer the grains. These cereal species, supply carbohydrates to human population and fodder to farm animals. A few legumes such as soybean, cowpea, chickpea, lentils, beans etc supply the proteins needed for human population and farm animals.

The sky, in general, has been studied in great detail about the atmosphere and its various phenomena, such as weather, precipitation patterns, gaseous emissions etc. However, so far, we have not compiled scientific basis to identify the "agricultural sky." We should know the minimum traits needed for it to support farming. Demarcations between types of "agricultural sky" too is absent. We have to first fix the criterion on which the agricultural sky could be identified, marked, classified, studied and explained. This aspect is difficult without proper basis. There seems to be certain advantage in knowing the "agricultural sky." The region above each farm or larger expanse or agroecosystems is the "agricultural sky." The physico-chemical nature of sky above each cropping belt needs to be characterised. Similarly, we should be able to offer clear data about the biological components of agricultural sky. For example, microbial flora, pathogens of crop plants sown in the area, insects that are harmful (pests) or beneficial (bees), birds that are useful in reducing the harmful insect population or those that help in pollination etc. Even the usage aerial vehicles such as drones, blimps, aerostats, could be recorded. The usage of sprinkler for water supply is an important aspect of agricultural sky. This procedure helps in channelling the water both through soil and via leaf surface. Similarly, aerial sprayers utilized to disperse pesticides, herbicides, fungicides are to be included, as part of phenomenon related to agricultural sky.

1.5.1 AGRICULTURAL SOILS

Agricultural land on earth extends into 13,500 km^2 (i.e. 13.5 billion ha). It is composed of several different types of soils. Soil is a natural entity that provides anchorage, mineral nutrients, organic fraction and water to agricultural crops and natural vegetation. These soils differ for several characteristics that have immediate relevance to crop production. It comprises of solids that could be organic or mineral in chemical nature, liquids and gases. Soil is characterized by distinguishable horizons and layers. The horizons are formed due to natural soil formation processes, weathering processes, and growth of natural vegetation (Soil Survey Staff, 1995). Soil is also defined using physico-chemical characteristics such as texture, structure color, clay content, organic matter, pH, redox potential, nutrient availability etc (Jenny, 1980; see Wilding, 1983a, 1983b). Soils are classified based on parent material from which they are derived and physico-chemical characteristics.

Some of the earliest efforts to classify soils were made in Russia by V. V. Dokuchaev in the late 19th century (see Dokuchaev, 1951; Lebedeva et al., 2008; Prosolov 1916, 1936). There are at least three major classification systems used to characterize and identify the global soils (See Beyer, 2003; also See Brady, 2000). They are, World Reference Base (WRB) (Bridges et al., 1998; ISSS-WRB 1998; Stolbovoi, 2020); USDA Soil taxonomy (Soil Survey Staff, 1998) and FAO Soil resource classification system (FAO, 1997). Regarding agricultural soils, WRB system emphasises on soil chemical constituents, particularly the organic fraction. The USDA Soil Taxonomy, in general, emphasises on pedogenic features, morphology, physical and chemical characteristics. No doubt, one of the important aims of soil characterization and taxonomy is to help farmers, in understanding and judging them. Particularly, to understand how useful the field soils are for agricultural crop production. It helps farmers in selecting apt crops species for each soil type. Not all soil types classified using above taxonomic systems are apt for crop production. A few of the soil types are more congenial for raising crops. Let us try to list the soil types encountered in the major food grain generating regions of the world. They are: US corn belt thrives on soil types classified as Ustolls, Udolls and Kastanozems. The major crops of Cerrados such as maize, wheat, soybean and others are grown on Alisols, Oxisols, Faerralsols, and Luvisols. Crops such as wheat, sorghum, soybean that dominate the Pampas are grown on Mollisols, Ultisolls, Entisols, and Aridisols. European plains support a large cropping stretch. Crop species such as wheat, maize, barley, lentils, brassicas are predominant. They thrive on highly fertile Chernozems that are rich in organic fraction. Several other soil orders are also encountered in the European plains. No doubt, we need to understand the soil type and its influence on crops (Food and Agricultural Organization of the United Nations, 1997).

We should also keep in mind, the importance of agricultural sky and its factors that have a say in crop production. After all, the agricultural sky above the chernozem belt in France may differ immensely from that encountered above the similar chernozem found in Netherlands, Germany, Poland, and Ukraine. The GHG emissions, dust, industrial pollution, precipitation patterns, dust storms may be different while the soil type encountered is the same chernozem. We have to attempt to characterize and classify the agricultural sky too. Perhaps, mention it along with soil classes and subclasses. The main food grain crops of West African Savannas like cereals (pearl millet, sorghum), legumes (cowpea, groundnut) and other crops thrive on Sandy Oxisols, Ultisols, Alfisols, Paleudalfs, Eutroprepts, Tropequents

(Buerkert et al., 2000; Charreau, 1974; Ducreax, 1984; Sene and Badiane, 2001; Yost et al., 2002). The agricultural sky above the Sahelian region impacts soil surface and its moisture status, immensely. In West Asia, crops such as wheat, chickpea and lentils thrive on Aridisols, Inceptisols, Mollisols and Vertisols. Major crops such as wheat, rice, cajanus, cowpea, and sugarcane are cultivated on Inceptisol in the Indo-Gangetic plains. But, in South India, crops such as rice, sorghum, maize, cajanus, cowpea, groundnut, and sunflower thrive on Alfisols, Vertisols, and Coastal sandy soil.

The agricultural sky above each soil type too may have to be emphasised in the future. We ought to realize that soils and atmosphere interact and influence the features relevant to crop production. The agricultural sky directly impacts the surface soils through radiation, precipitation event, wind, and soil erosion. Biotic components such as microbes, insects, and aves from the sky too affect the soil conditions, to a certain extent. The *vice versa,* that is, effects of soil on agricultural sky and its quality with reference to farming too have to be considered. For example, soils applied with excessive fertilizers may emanate GHG to a greater extent. A sandy or dusty soil may affect the sky by inducing dust bowl conditions. Perhaps, while we describe soil types and their utility for farming, we should also make a list of few features of "agricultural sky," particularly, one immediately above the terrain/soil. For example, Alfisols in dry land regions with agricultural sky that offers scarce precipitation, scorching heat, and sunlight and dusty microbial pathogen filled atmosphere could be detrimental to crops. On the other hand, Alfisols with agricultural sky that offers optimum precipitation pattern, and low in insect pests, and plant disease propagules is congenial to produce crops. Clearly, just understanding and classifying agricultural soil does not suffice. It is advisable to analyze the agricultural sky periodically, above a particular field along with soil characteristics. This could be done prior to selecting crop species and planting. In practical agriculture, commercial value of fertile may get affected by factors operative in the agricultural sky. However, agricultural sky is not given importance while judging the fields.

1.5.2 AGRICULTURAL WATER

Global water cycle that describes the movement of water into below ground, surface, and atmosphere portions of earth has its impact on agricultural water resources. This includes the precipitation caused via clouds in the agricultural sky (GAO 2019; NASA, 2020). Water occurs in different forms such as gaseous (vapor, steam), liquid (water), and solid (frost, snow, ice etc). Precipitation from clouds is a key source of rainfed agriculture. Portions of surface water too are utilized for crop production. Precipitation formed via cloud involves bulk movement of water from sky to earth's surface. Large quantities of water transit in the agricultural sky as clouds. It helps to feed large agroecosystems with requisite water. It replenishes several different sources of water such as surface lakes, ponds, and rivers and even groundwater.

Water is among key factors that decides the global crop production trends. Agricultural water is that portion of water available on earth which is specifically useable for crop production. Not all sources/types of water available on earth are congenial for use in agricultural expanses (Centre for Disease Control and Prevention USA, 2020, United States Geological Survey, 2004). Agricultural water is generally defined as that water which is adopted to grow field crops and fresh vegetable in the fields, fruit crops in plantations and to maintain livestock.

The share of fresh water available on earth for agriculture and human needs is low. Large amounts of fresh water are actually stored out of reach in the glaciers and ice caps. The freshwater lakes, ponds rivers, and dams constitute only 0.3% of total water on earth. A sizeable share of ground water in aquifers is not accessible. No doubt, rainfall received from the sky is a precious source of fresh water for all farmers. Further, we may note that the agricultural sector uses >80% of the developed freshwater supply of the world (Nair et al., 2014; Seckler et al., 2003; Shiklomanov 1998). As stated earlier, water that moves above crops as clouds is important.

Global agriculture is highly dependent on water resource. Water may get channeled through the agricultural sky as precipitation. Otherwise, it could be made available from surface and ground water sources. Reports by many United Nations Agencies state that, in recent years, agricultural crop production trends have consistently faced severe constraints. Such constraints are predominantly related to water resources. Atmospheric sources such as precipitation have gone erratic in many locations. Ground and surface water have become progressively scanty. Extreme weather patterns and El Nino conditions too have affected farming belts by causing storms, floods, and soil erosion. In the present context, we are concerned with managing water resources derived from atmosphere as precipitation. It is important to note that globally most crops are grown as rainfed crops. Therefore, managing precipitation to derive greater water use efficiency should receive high priority (Irrigation Association, 2019; OECD, 2019). There are suggestions that we can attempt to solve such problems by concentrating on hotspots.

Geographic locations that face water scarcity due to scanty precipitation and uncongenial pattern could be highlighted. We should also note that intensive withdrawal of ground water, contamination due to excessive usage of fertilizer and pesticides are processes caused by crop production.

Let us consider a few salient facts about the water received as rainfall from clouds, its usage in agriculture, and other aspects as reported by FAO (AQUASTAT, 2014). Globally, about 69% of water received from precipitation is withdrawn for agricultural purposes. Among continents, Asia utilizes 84% of fresh water received for agriculture, Africa utilizes 82%, Americas utilize 52%, and Europe utilizes 23% for crop production. The clouds supply about 814 mm rainfall (water) per year on land (110000 km^2 area). The precipitation is not evenly distributed across different geographic regions. Precipitation is garnered predominantly by the forests/natural vegetation. About 56% of rainfall is evapo-transpired back to atmosphere by forest/natural vegetation. About 5% water is utilized by rainfed agriculture and it is transpired. The rest 39% water received as rainfall is used as freshwater resources by humans for irrigating crops and other purposes (AQUASTAT, 2014). There are indeed large areas (stretches) of agricultural crop production that receive insufficient rainfall from clouds. Hence, they need extra water through irrigation. Farmers do utilize water from different sources to improve crop yield. They are fresh water sources like precipitation, surface water, renewable ground water, nonrenewable ground water. A few nonconventional water sources used in farming are desalinated water, treated wastewater from urban regions, and agricultural drainage water that is reused (AQUASTAT, 2014).

Rainfed agriculture is the most prevalent form of crop production tactics in the world. Rainfed agriculture where precipitation is the mainstay is predominantly practiced in developing countries. Rainfed farming is practiced in 95% of Sub-Saharan Africa, 90% of South America, 75% of North African drylands, 65% of East and South Asia. Agricultural sky that mediates precipitation events deserves greater attention. Farmers with poor resources often rely on rainfall. Rainfall is itself a phenomenon influenced by several factors that operate in the agricultural sky. Rainfed agriculture often suffers from scanty or deficient water received by the crops. Often the rainfall distribution too is un-congenial. This fact underlines the importance of agricultural sky and its role in channeling water to global agriculture via precipitation. Incidentally, there are hot spots identified by UN agencies that suffer from water scarcity. They mostly coincide with those that record extremely low cereal crop productivity. For example, dry savannas (e.g. Pampas of Argentina), steppes (Southern Russian Steppes), Sahelian zones (West Africa), and semiarid cropping belts (India) that experience low rainfall offer commensurately low grain yields. No doubt, we should bestow greater interest in understanding the "agricultural sky" above such savannas, prairies, steppes, semi-arid regions. This is to maximize water use efficiency and crop productivity. Perhaps, agricultural sky above drier and rainfed zones such as the Sahelian belt needs greater attention, than bestowed till date.

The primary source of water for agricultural production for most of the world is rainfall. Three main characteristics of rainfall are its amount, frequency, and intensity. Their values vary from place to place, day to day, month to month, and also year to year. We are prone to consider "effective rainfall" as a useful measure (FAO, 2012). Agriculturists consider as effective that portion of the total rainfall which directly satisfies crops' water needs. It includes surface run-off which can be used for crop production on their farms, by being pumped from ponds or wells. In dryland farms, when the land is left fallow, effective rainfall is that which can be conserved for the following crop. An individual farmer considers that effective rainfall is that quantity that is useful in raising crops planted on his soil, under his management. Water that moves out of the field by run-off or by deep percolation beyond the root zone of his crop is ineffective (FAO, 2012).

Most reports state that agriculture, meaning the crop production procedures withdraw major share of the water received as precipitation (rainfall, frost, snow) through the sky and that drawn out as ground water. About 65% of the total water withdrawals is accounted for by agricultural irrigation (Irrigation Association, 2019). Production of electricity using thermal/hydro stations too consumes large portions of fresh water. Agricultural water is predominantly utilized to irrigate crops, either using sprinklers or via surface methods. It is also used supply fertilizers in dissolved state, and to apply plant protection chemicals. Sometimes, aerosol and sprays are used to cool the crop canopy if the crop is sensitive to ambient heat. Water available for agricultural crop production is actually derived from several types of sources/storage areas. For example, precipitation from agricultural sky through clouds (rainfall), melting snow, rivers, streams, ponds, lakes, wells, ground water extracted from borewells, dams, and channels etc. In the present context, we confine our attention to agricultural water and only the portion of it that is derived from the sky, that is, precipitation classified as rainfall, frost, snow, mist dew etc. All of them are related to agricultural sky some way. Water applied using the aerial methods such as sprinklers, center pivot systems is also included for discussions here.

In USA, about 40% (137,000 million gallons per day) of freshwater resources is withdrawn for irrigating agricultural crops and rest 60% for other purposes. About 31% of surface water is withdrawn for agricultural crop production and other purposes account for 69%. Regarding ground water, about 68% of the available fresh water is used for irrigating crops and 32% is utilized for other purposes. Even in developed nations such as USA, out of 315 million ac of cropped land only 19.5% area is irrigated. Rest 80% is not irrigated, This agrarian zone is dependent on natural precipitation (Irrigation Association, 2019). At this juncture, we may note that major share of food grain generating areas in North America and even elsewhere thrive on precipitation via agricultural sky. They are termed the rainfed agriculture.

Farmers adopt either rainfed or irrigated crop production tactics. Agricultural sky is the main source of precipitation (rainfall) that forms the basis of rainfed agriculture. The timing, intensity, and area covered by such precipitation events differs based on geographic factors. About 78% of crop production occurs under rainfed conditions. About 20–23% crop land is irrigated. Irrigation is defined as controlled use of water resources to match the crop's need in time and space. Irrigation actually comprises water applied during growing season, as well as prior to it during field preparation (GAO, 2019; Malaker et al., 2019).

The quality of water utilized for crop production is important. There are possibilities that such quality stipulations are overlooked during rampant adoption irrigation. Historically, water salinization has been the major factor that depreciated quality of water for crop production (Malaker et al., 2019). At present, irrigation water is polluted with trace elements, pharmaceuticals, salts, steroids, plasticizers etc. Ground water sources are contaminated with trace elements. But the precipitation received through the sky is often contaminated with particulate matter, sulfur compounds, acidic chemicals, pesticides, and greenhouse gases etc.

Frequently used methods of application of agricultural water to crops are as follows: Surface irrigation involves distribution of water across land by gravity. Mechanical pumps are not utilized. Localized irrigation is distribution of water under low pressure using pipes. Water is applied to each plant. Drip irrigation involves application of water in drops to points nearest to roots. Such methods reduce loss of water via evaporation and runoff. Sprinkler irrigation is an aerial technique. It is a method that is mediated using agricultural sky, that is, aerospace above the crop's canopy. Here, water is distributed over-head using high-pressure sprinkler or guns from a central location in the field or using moving sprinklers (e.g. center pivot systems).

Water is distributed by a system of sprinklers that move on wheeled towers in a circular pattern. This system is common in the plains region of North America and Europe. There are also lateral move sprinklers. Water is distributed across land by raising the water table, through a system of pumping stations, canals, gates, and ditches. This type of irrigation is most effective in areas with high water tables.

In the present context, it is pertinent to emphasize on climate change effects on agricultural water. Global agriculture depends immensely on water cycle and its influence on immediate availability of water resources for crop production procedures (GAO, 2019). Climate change impacts on agriculture, it seems, are primarily felt due to disruption of global water cycle at various points. One other opinion expressed is that a major portion of climate change effects on agriculture *per se* is mediated via water cycle-related events. Global warming may cause glaciers to shrink. The precipitation patterns could change severely causing floods or droughts of different intensities. Fecht (2019) states that climate change actually affects water cycle at too many points. Alterations in air temperature and circulation can affect where exactly rainfall occurs.

Precipitation pattern that gets altered is among important points to consider. In the agrarian regions it could lead to loss of crop productivity. Therefore, while deciding on cropping systems, crop species, and water resources we have to invariably think of climate change effects on agricultural water. In fact, a few agricultural specialists have opined that, regarding water for crop production, farmers have to consider the climate change effects rather than standard weather parameters. There is literally nothing stationary about agricultural climate. Forecasts on precipitation pattern are not narrow in range because of climate change effects.

Rojas et al. (2019) have pointed out that climate change affects one of the major agricultural water resources, that is, precipitation. It also influences the precipitation pattern and its distribution across various cropping belts. Let us consider the major crop production zones that include maize, wheat, rice, and soybean cultivation areas. These crops may encounter precipitation changes from 1–13% decrease or 3–31% increase depending on the year. Forecasts suggest that these changes in precipitation will influence the grain harvests from these crops as early as 2040. Harvest changes may get conspicuous by 2071 (Challinor et al., 2014; Rojas et al., 2019). The suggestions are to breed crops accordingly and be ready to match the climate change effects and overcome it. Further, adopt agronomic procedures that allow better use of precipitation (Challinor et al., 2016).

While calculating water requirements for a field/cropping system, the equation should include a "constant value" accounting for climate change effects. Such climate change is attributable to greenhouse gas emissions, excessive use of fertilizers, irrigation, and frequent changes in precipitation pattern in a geographic location.

Climate change is expected to affect water availability in areas that are already at risk of shortages. Management of agricultural water is therefore important in rainfed drylands. Iglesias and Garrote (2015) state that, in Europe and even in other regions, procurement and management of agricultural water is becoming complex. Appearance of risks related to efficient use of agricultural water has become more frequent (Iglesias et al., 1996).

Droughts could become intense and continue for longer period due to altered precipitation patterns. This is attributed to higher average daily temperature caused by GHG emissions (IPCC, 2014). Researchers at Stanford University have noticed that droughts coinciding with higher temperature have altered precipitation pattern. It has also affected crop production procedures preferred by farmers (Grace Communication Foundation, 2020).

Climate change also induces frequent precipitation events of higher intensity. It leads to flooding and deluge in the farmer's fields. It causes loss of soil structure. It also causes massive damage to standing crops in the field. Several species of crops are vulnerable to stagnating conditions. Soil erosion is another malady related to high-intensity precipitation induced by climate change. Melting glaciers could affect surface water and precipitation pattern.

By 2040, rainfall on wheat, soybean, rice, and maize will have changed, even if "Paris Agreement" emission targets are met. Projections show parts of Europe, Africa, the Americas, and Australia will be drier, while the tropics and north will be wetter (IITA, 2019; Rojas et al., 2019). Researchers at IITA (International Institute for Tropical Agriculture, Nigeria) have pointed out that, despite efforts to reduce GHG, precipitation changes that occur during the next few years may affect production of major crops such as wheat, maize, rice, and soybean. Climate change could influence agricultural water resources to different extents, depending on geographic location and related factors (Xing-Guo et al., 2017). Incidentally, precipitation is entirely a phenomenon related to agricultural sky, that is, atmosphere above the crops' canopy, mainly troposphere. Drier conditions in wheat-producing regions, for example, may reduce precipitation by 27% in Australia. Percentage reduction of precipitation predicted for other nations is Algeria 90%, Morocco 90%, Mexico 74%, Spain 66%, Chile 40%, Italy 30%, Egypt 15% (IITA, 2019; Rojas et al., 2019).

The quality of irrigation water utilized during crop production is important. Generally, it is dependent on both natural and human-related factors. Such factors influence surface or ground water sources. They affect weathering of bedrock minerals, deposition of dust due to wind, leaching of organic and mineral matter into streams, ponds, and other water bodies etc. Precipitation water received as rainfall that collects as surface source too could be contaminated, by the same processes. Precipitation is prone to contamination by the atmospheric dust, particulate matter, gasses that could be miscible with precipitation water, toxic fumes, emissions such as SO_2 and NH_3. Volatile Organic Compounds (VOC) may also contaminate rainfall. They say, in many major agroecosystems, declining quality of irrigation water is an important constraint. It is potentially a factor that can get accentuated in due course. Atmospheric contamination is increasing each year (UNDESA, 2020; UNEP 2008a).

Agrarian regions are vulnerable to spoilage due to eutrophication. Eutrophication is caused by excessive leaching of soil nutrients into irrigation channels. Nutrients applied as fertilizer and industrial effluents also mix with irrigation channels. Atmospheric inputs such as dust, any particulate matter caused by crop residue burning, fossil fuels, and bush/forest fires also contaminate water resources. Agricultural sky could be a source of contamination if remedial measures are not adopted. Lessening of GHG, industrial fumes/effluents, and fuel burning is required. Precipitation events may not bring in heavily pollution affected water. However, later, as surface water it might get polluted with a variety of contaminant, nutrients, toxicants etc. There are several processes that divert the applied fertilizer-based nutrients into irrigation water. For example, farmers may apply higher dosages of fertilizer-N. However, excess N may creep into streams, channels and percolate into soil profile. Irrigation water applied to crops should possess <5 mg N per liter of water (UNDESA, 2020; UNEP 2008a, 2008b).

A few of the pollutants are introduced into atmosphere even as precipitation occurs. Several of them get included during surface runoff and storage. Ground water gets contaminated during percolation of water through the contaminated soil profile. According to Mateo-Sagasta et al. (2017) water pollutants in agricultural production systems are as follows:

a. Nutrients: Primarily, nitrogen and phosphorus present in chemical and organic fertilizers as well as animal excreta. They are normally found in water as nitrate, ammonia, or phosphate
b. Pesticides: Herbicides, insecticides, fungicides, and bactericides, including organophosphates, carbamates, pyrethroids, organochlorine

pesticides, and others (pesticides, such as DDT, are banned in most countries but are still being used illegally and persistently)

c. Salts: Ions of sodium, chloride, potassium, magnesium, sulphate, calcium, and bicarbonate. They are measured in water, either directly as total dissolved solids or indirectly as electric conductivity
d. Sediment: Measured in water as total suspended solids.
e. Organic matter: Chemical or biochemical oxygen demanding substances (e.g. organic materials such as plant matter and livestock excreta).
f. Pathogens: Bacterial indicators: e.g., Escherichia coli, total coliforms, faecal coliforms, and enterococci
g. Metals: Selenium, lead, copper, mercury, arsenic, and manganese. Geologic weathering and release of minerals into streams, channels, and irrigation water is an important aspect.
h. Emerging pollutants: Drug residues, hormones, and feed additives.

1.5.3 AGRICULTURAL CROPS

The natural vegetation around the globe exhibits extra-ordinarily wide morphological, physiological, and genetic diversity. Plant species traced in natural vegetation exhibit genetic diversity that allows them to grow, mature, and perpetuate. Plant species also show great versatility to adapt to terrain, its topography, soils, water, and atmospheric conditions. Plant species are no doubt versatile. Therefore, they thrive in different agro-environmental zones. Like natural vegetation, agricultural crops *per se* are equally versatile in flourishing in different types of agrarian environments. Several crop species show excellent adaptation to environment. Such crops grow and bear mature fruits/grains even in harsher climate (see FAO/IIASA, 1999; Krishna, 2003, 2008; Leff et al., 2004).

In this section on agricultural crops, let us also consider a few examples of agricultural crops that need soil, water, and agricultural sky to different extents, also those that are adapted to entirely hydroponic or aeroponic conditions. There are crops that form the fruits, tubers, rhizomes, or other modified organs under the soil. Several species form yield entirely under soil. A few crops grow entirely or float above the water bodies (submerged crops). Crops could be grown without soil but as hydroponic cultures. Aeroponic cultures thrive on aerosol. Aerosol provides the necessary nutrients, oxygen tension for roots to grow and crops to mature.

As stated earlier, agricultural crops are no doubt versatile. Yet, there are limitations imposed by nature. There are no species in natural vegetation zones (say, forests, shrubs, pastures, mixed vegetations) that grow entirely in subterranean conditions. It seems futile to think of plants with leaves hidden from sun and inside the soil. The purpose of converting CO_2 using photosynthetic radiation will not be served if leaves are formed/held away from sun. Crops are grown in soils with different texture, structure, fertility, and moisture-holding capacities. Arable crops are those that grow in soils with moderate texture, moisture, and aeration. Wetland crops need stagnating water. There are several crops that grow in submerged, micro-aerophilic conditions. For example, Wetland rice (*Oryza sativa*), Water Cress (*Nasturtium microphyllum, N, officinale;* Brassicaceae), water spinach (*Ipomea aquatica*), Water Caltrop (*Trapa natans*), Water Pepper (*Persicaria hydropiper)* etc. There are crop species adapted to arid sandy soils. Although not belonging to the group of edible crops, there are large number of plant species that are entirely immersed in water, during their entire life cycle. They grow under sea water or even in freshwater lakes.

However, there are indeed innumerable food crops that are adapted to form under-ground organs such as rhizomes, corms (Amorphophallus, Colacasia), bulbs (e.g. tulips, daffodils, lilies, onion, garlic), stolons (*Trifolium subterranium*) etc. Underground organs serve in storing food and in perpetuation/rejuvenation upon appearance of congenial conditions (e.g. turmeric, turnip, ginger, onion, garlic, amorphophallus, sub-terranean clover etc.) Globally, we grow and consume several crop species that form underground carbohydrate storage modification. Yet relatively very few stem and root tuber crops are cultivated and consumed, in comparison to cereal and pulse grain crops (see FAOSTAT 1999, 2000; Kipple and Ornelius; 2000; Krishna, 2003; QA International Collectif, 2007; Wood et al., 2000),

We can trace plants with underground stems. There are at least four types of underground stems. Rhizomes are modified underground stems. Usually, rhizomes have nodes and scaly leaves (e.g. ginger, turmeric). There are also stem tubers such as potatoes that are again subterranean in formation. Underground stems are easily identified and differentiated from underground roots (e.g. radish, carrot) due to presence of scaly leaves. The rhizomes, and stem tubers are actually nutrient (carbohydrate and protein) storage organs. They help the plant in overwintering and rejuvenating upon appearance of congenial conditions. There are several edible crops that are cultivated for their rhizomes and tuber. Bulbs are flattened, disc-shaped underground stems. Onions and garlic are examples of bulbs. Corms are portions of the plant's main stem that are found underground. When the above-ground stem dies at the end of the growing season, the corm often remains active so that a new above-ground stem can emerge in the spring. Examples of corms include Amorphophallus and Colocasia. There are crops that form underground pods (e.g. *Arachis hypogea).*

There are plants that grow and complete their life cycle under water. They produce flowers and seeds and perpetuate under water. The influence of sky on them and vice versa seems feeble, if any (Schloesser, 2010; Benson, 2016). There are several domesticated crops that are grown submerged in water. They utilize relatively excessive amounts of water and need stagnating conditions. The foliage, flowers, and fruits/grains are borne above the surface of water, into the sky. They may grow in flooded, well puddled fields, but their dependence on the various factors related to the agricultural sky above the crop's canopy is similar to any other species. A small alteration in cloudiness, precipitation pattern, temperature or diurnal length, pathogenic microbes or insect pests in the air can still immensely affect growth and panicle initiation and grain formation. Among the crops, rice species are most widely spread in the tropical agrarian regions of Asia and Africa. There two species of rice, *Oryza sativa* and *O. glaberrima* (West African variety). We may note here that over 1.5 billion people thrive on rice grown mostly as submerged crop. There are rice varieties that prefer growing as floating rice (e.g. Boro rice.). They are cultivated in the delta region of South Asia.

Wild aquatic rice species are traced in certain regions. For example, there are three aquatic submerged wild rice species that are native to North America. They are Northern Wild Rice (*Zizania palustris*). It is traced as submerged aquatic crop in the Great lake's region and in marshes of Boreal Forest of Alberta, Saskatchewan, and Manitoba. *Zizania aquatica* is found in the Atlantic and Gulf coasts. *Zizania. texana* is a wild submerge rice found in small areas of San Marcos River in Central Texas (Natural Resource Conservation Service-Plant Data Base, 2020; Wikipedia 2020z). *Zizania latifolia* (Manchuria rice) is a wild submerged rice. It is perennial and native to Manchuria and mainland China (PFAF Organization, 2020).

In addition, there are several other submerged crop species that serve as vegetable or offer edible rhizomes etc. Let us consider a few examples. Water Cress (*Nasturtium microphyllum, N, officinale;* Brassicaceae) is among the oldest leafy vegetable utilized by humans. It is native to Central Asia and Europe. It is semi-aquatic and well suited to be cultivated in a hydroponic system (Morgan, 2020). Leaves, sprouts, and stems are edible. Water Pepper (*Persicaria hydropiper*, Family Polygonaceae) is grown in locations with severe dampness and shallow water. It is grown in the temperate region mainly for their pungent flavor. In the Fareast (e.g. Japan) it is a vegetable grown in submerged and swampy region.

Water Spinach (*Ipomea aquatica*) is adopted to swamps, water channels, and damp soils. It is grown as edible vegetable under submerged water conditions in Southeast and East Asia. Its stem grows rapidly to 2–3 m length and stays floating above water. It is used by Malaysian and Chinese in cuisines.

Wasabi (*Wasabia japonica*; Family Brassicaceae) is found in the riverbeds in southeast Asia. It is a leafy vegetable with pungent flavor.

Chinese Water Chestnut (*Eleocharis dulcis*; Family Cyperaceae) is native to southern China and Philippines. It is grown in flooded conditions in paddy fields. It offers edible corms.

Water Caltrop (*Trapa natans*; Family Trapaceae). is a floating aquatic species. It grows in deep slow-moving water bodies. It is native to warm temperate regions of Africa and Eurasia. Chinese farmers have grown water Caltrop for over 3000 years. At present, it is not a frequently grown submerged crop.

Lotus (*Nelumbo nucifera*; Family Nelumbonaceae) is an aquatic perennial plant with edible parts. Flowers, seeds, rhizomes, and petals are consumed (Wikipedia, 2020c).

There are several crop species that have been successfully grown entirely as an aeroponic culture. In nature, there are orchids that thrive in atmosphere utilizing relative humidity and moisture in the atmosphere. They garner nutrients from atmosphere and grow. Aeroponic culture of leafy vegetables, fruits, and flowers is possible. In European nations, glasshouses with facility for hydroponic and aeroponic cultivation of vegetables are in vogue, in large numbers.

Let us consider a few highly relevant facts about the crops that have been domesticated and utilized by humans. *Agricultural sky, like soils, water, and other abiotic as well as biotic factors might have shaped this evolution with different levels of impact.* We realize that human preferences and perseverance have immensely helped in domestication and accentuation of different crop species and their genotypes. Here, we have to highlight the role "agricultural sky" and its components in the evolution of crops. Agricultural expanses and crop species that fill the vast agroecosystems might have been feebly or stringently affected by the "agricultural sky." Yet, we have not identified nor recorded such specific influence of the agricultural sky on crop utilization. Simplest examples that we can think of are sky above the crops in tropics has induced vast patches of rice, sorghum, pearl millet and other small millets, several tropical legumes (cowpea, pigeon pea), tropical fruit crops such as mangoes, plantains, pomegranate, and melons. Temperate sky has induced cold-tolerant wheat, barley, brassicas, lentils etc. Several factors must have operated along with those related to agricultural sky, while the crops were selected, domesticated, refined, and utilized by humans.

It seems, currently, we have identified, domesticated, and used over 7000 edible plant species that are available to be exploited, although about 10 to 50,000 plant species are edible. However, we have been regularly cultivating only few species to serve our nutritional requirements. Regarding trees, although there are 2632 species that could be exploited, again, we grow only few tree species in plantations (Kindt et al., 1997). The human preferences have further narrowed the crop species cultivated to mere 30 species (Wood et al., 2000). At present, these 30 species of food crops serve over 9 billion human beings. Clearly, in addition to human preferences, it is the natural resource such as the soil, water (irrigation), and factors in the sky that decide the crops that get cultivated. Atmospheric factors such as season, precipitation pattern, temperature, gaseous composition, and biotic factors such as disease-causing microbes, insect pests, and birds affect the crops. Our preferences for food crops have been influenced by such features of "agricultural sky." We have to characterize the agriculture sky above cropping zones in greater detail. Then, find out the specific preference of the crop species. Also, we have yet to find out how crop species that flourish affect the microclimate under the canopy, ambient atmospheric composition, weather parameters, and biotic factors. Overall, we should note that agricultural crops could be grouped based on the atmospheric factors. We should note that, through the ages, during their evolution and domestication, crop species cultivated have been influenced immensely, by the agricultural sky and its factors. We may have to identify specific features, if any, of atmosphere (agricultural sky) that induced the evolution of the crop, its domestication, and intense usage by humans. Wet land rice is a good example of a crop that evolved in regions with high water availability (precipitation or riverine) and is now grown as submerged paddy in southeast Asia and elsewhere. We may be able to pinpoint several crops whose evolution was predominantly influenced and guided by the factors operating in the agricultural sky. On the other extreme, we have innumerable germplasm lines/cultivars of hardy cereals such as pearl millet, sorghum, setaria, panicum, that thrive under harsh conditions of agricultural sky.

There are indeed several criteria used to classify agricultural crops. We can begin with their botanical features. According to International Code of Botanical Nomenclature (ICBN), plants are classified into taxa, orders etc. based on the following criteria (botanical features). We can classify them using criteria such as morphology, growth habit, maturity period, mode of reproduction, edible part etc. Crops can also be classified based on their ecological predisposition such as wetland crops, dryland crops, perennial grasses, creepers, epiphytotic twiners (e.g. pepper, beetle), understorey crops (e.g. coffee, tea, citrus in South India). We can classify crops based on the regions, preferred climatic conditions, and food product. Globally, major crop classifications are defined by the Food and Agriculture Organization (FAO) of the United Nations. Whatever is the crop species and its classification, basically, it does depend immensely on the agricultural sky, in addition to soil, water, and timely agronomic procedures. We should be able to characterize the agricultural sky using as many traits as possible. Most importantly, we should be able to arrive at well matched "crop versus agricultural sky" combinations. We need detailed knowledge about agricultural sky above each field/farm, in a given season and then match the crop species. Perhaps several of the prophylactic measures could be adopted using information relevant to agricultural sky. Satellite imagery-based classification of natural vegetation and crops is also in vogue. The reflectance patterns and NDVI data collected periodically are utilized to identify and classify cropping systems (Barret et al., 2000; Krishna, 2016).

The classification of plant species based on the botanical features and growth habit is among the earliest efforts. It aimed to group the plant species found on earth. A few other botanists employed the nativity or the region of maximum genetic diversity to group the plants (CIAT, 2020; Filatneko et al., 2020; Hawkes, 2020; Kipple and Ornelius, 2000; Pearson, 1992). Domesticated crops were all easily identified based on the origin (or nativity), in other words where they were domesticated and also location where they are relatively plentiful. Classification of edible crops is essential. We may utilize several criteria and group them. Such classifications usually aimed at helping a definite purpose. For example, we can select crops based on agrarian region and its resources (e.g. wetland, dry arable land, sandy semi-arid land etc); based on food preferences (e.g. cereals like wheat, rice, maize millets, cowpea, lentils, peas etc.); based on commercial value (cash crops) (e.g. sugarcane, cotton etc). During a recent census conducted on agricultural crops by the Food and Agricultural Organization of the United Nations Rome, Italy) they have suggested "Indicative Crop Classification" (FAO, 2010; United Nations Statistical Division, 2015). These edible species could be annuals, biennials, and permanently perennial crops. The indicative crop classification has the following major groupings:

Cereals (e.g. Wheat, Rice, Maize, Sorghum, Oats, Barley, Millets)

Fruits and Nuts (e.g. Tropical and Subtropical Fruits, Citrus Fruits, Grapes, Berries, Pome and Stone fruits, Nuts)

Oilseeds (e.g. Brassicas, Groundnuts etc.)

Root and Tuber crops with high starch and inulin (Sweet Potato, Potato, Cassava)

Beverage and Spice crops (Tea, Coffee, Cardamom, Cloves, Pepper etc.)

Legume crops (Soybean, Chickpea, Lentil, Cowpea, Pigeon pea, Beans, Field beans)

Sugar crops (e.g. Sugarcane, Beet, Dates)

Other crops (e.g. Grasses and Fodder, Fiber crops, Medicinal and Aromatic species, Rubber, Flower crops, Tobacco etc.).

The above classification is more relevant to farmers and commercial operators dealing with agricultural commodities. There are also attempts to classify/group the crops based on the soil fertility, also based on weather patterns encountered in the agricultural sky. However, we may note that agricultural sky itself has not yet been classified properly. We have to standardize the traits on which to base an authentic identification and classification of agricultural sky. In fact, names that depict characteristics of the agricultural sky too have to be decided. Then, based on authentic names and their primary characteristics of agricultural sky, we have to classify the crops. Ultimately, we have to be ready with a list of crop species that suit a particular type of agricultural sky. This is just like what we have done with soils and the crops that suit a particular soil type and its fertility level. Overall, we should learn to identify the major role that agricultural sky and its parameters play in allowing agricultural/commercial crops, to thrive. It is essential to match the crops with agricultural sky. *After all, agricultural sky is among the most important factors that has shaped agrarian evolution during past 1–12 millennia.* During past millennia, agricultural sky with all its factors might have induced genetic evolution of crops. In fact, the dominant crop species might have been selected by the factors operative in the sky, as much as those in soil or more recently the human preferences. In other words, genetic diversity of natural vegetation and crops grown might have been actually decided to a good extent by the factors that operate in the agricultural sky.

1.5.4 AGRICULTURAL SKY: A CONCEPT TO PONDER, EVALUATE, AND WORK OUT

Here, we intend to clearly distinguish between the "sky" in general and "agricultural sky." "Agricultural sky" is the region specifically above the agrarian region. In other words, "Sky," in general, engulfs earth's surface area *in toto*. It includes atmosphere, that is, all its layers such as troposphere, stratosphere, mesosphere, and ionosphere plus a small region of space. But, "Agricultural sky" is the exclusive area above agricultural terrain across different continents. "Agricultural sky" spreads only to the extent crops, plantations and forest stand occur on the earth. Since agricultural crop production zones extend only to one-third of the land mass, proportionately, agricultural sky, its area too extends, just above it. If agricultural area shrinks or expands in certain geographic regions, then agriculture sky follows suit in terms of area. Agricultural sky too fluctuates accordingly. Agricultural sky and its factors have immense impact on crops. Agricultural sky is the main theme of this book. The proposed concept could be adopted by farm experts and farmers all over the world. *It has not been done so far.*

Foremost, "agricultural sky" as a concept could be of great utility to agricultural experts, farmers, natural resources experts, ecologists, meteorologists, climate change enthusiasts, policy makers, etc. Agricultural sky could lead us to more accurate knowledge about agrarian regions. It might improve the efficiency of various inputs applied and agronomic procedures adopted to raise crops. We have to characterize the agricultural sky, classify and demarcate the zones appropriately. This situation is unlike agricultural soils, crops or water about which we have accrued detailed knowledge, during past century. Regarding, sky in general, there is enormous literature about atmosphere and its various manifestations. Our generalized view about "sky" does not suffice. It may not offer great advantages to farmers. We need to characterize "agricultural sky" demarcate and then utilize just the same way we did for soils, water and crops. Indeed, we are at rudimentary stages of using a concept such as "agricultural sky," if not, none so far. In future, we may also need to compare and contrast our knowledge about agricultural sky with that of soils, water, crops, in terms of their important effects on various aspects of farm production.

By concept "Agricultural sky" as stated earlier, includes the region just above the soil/crop's canopy or region above a farming belt, or that above vast patches of crops or a large agro-ecosystem itself. Let us consider the situation in reverse. The vice versa could be equally true with soils/crops affecting the ambient agricultural sky. How does a cropped field or cropping system influence the sky above it? Reports state that, at present, about 38% of Earth's terrain is already engulfed with agricultural crops and pastures (Foley et al., 2005). Any expansion or intensification affects the ecological aspects of the terrain (Eitleberg et al., 2015). Not just terrain, the agrarian procedures adopted and the standing crops *per se* may have mild to severe effects on the agricultural sky. Effect of crops and their diversity on the climatic change factors could be important. Influence of crops on agricultural sky also needs to be understood in greater detail. Particularly, its effects on biotic factors that operate in the agricultural sky. The man-made alterations to the sky may also depend to a great extent on cropping patterns adopted in a region. Therefore, we may have to characterize and classify the agricultural sky after studying crops and agronomic procedures adopted in a location. The classification of agricultural sky, based on several different factors that are operative in the terrain and in the sky, needs consideration. For example, agricultural sky above a vast cereal expanse needs attention.

Here, in this section on "agricultural sky," let us try to include the biotic fractions such as microbes, insects and aves along with the other usual parameters of the atmosphere. In the recent times, we have also introduced the Unmanned Aerial Vehicles (UAVs), i.e. small aircrafts, blimps, aerostats, helikites etc. They need special consideration under the section on agricultural sky. They have the potential to revolutionize agricultural production procedures and reduce drudgery in farms (See Krishna 2016, 2018, 2020a, 2020b).

Why have we not characterized agricultural sky in great detail? We have no good reason put forth for this question, so far. In fact, the agricultural sky above a cropped field of a single hectare may differ from the next one. The characteristics of agricultural sky may fluctuate based on season, geographic location, soil type, crop species, its stage and husbandry procedures adopted. No doubt, agricultural sky above large cropping belts such as Great Plains, European steppes, Cerrados, Pampas, Sahel, Indo-Gangetic plains are definitely distinct in their characteristics. Yet we have not made great strides in understanding its relevance. We have not utilized the peculiarities of agricultural sky to our advantage while producing the specific crops in these agro-ecosystems.

The overall agroclimatic condition of an agrarian region is important. The agricultural sky has its share of influence on the climate and productivity of agrarian regions. Farmers often try to select adapted crops, their species and genotypes. Agronomic procedures such as tillage its intensity and other intercultural operations are often highly dependent on the agroclimatic manifestations. Further, farm inputs such as fertilizers, plant protection chemicals are decided based on the yield potential of the crop/s in a given weather condition above the field (Eitzinger et al., 2007). Several types of climatic risks do emanate from the sky. For example, storms, high rainfall, dust storms, snow, frost or smog, pathogens, insect pest, lack of pollinators, adoption of UAVs, satellite data etc. may all affect farmers ability to manage high productivity of crops. Farmers have to carefully manage several factors including those related to agricultural sky prior to fixing, cropping systems and yield goals. Food grain production strategies for short and long term too depend on factors related to agricultural sky.

Farmers in each region may utilize the agricultural sky and its components to different extents, depending on the importance they have bestowed on aerial factors. For example, farmers in predominantly irrigated region may think of vagaries in precipitation pattern a trifle less important. While a farmer in rainfed semi-arid region considers precipitation quantity, its intensity and pattern most important. A tropical farmer may not worry about photosynthetic radiation. Whereas, in cloudy temperate climate, photosynthetic radiation and ambient temperature may reign supreme in terms of importance to crop productivity. Basically, a list of characteristics of the agricultural sky above a farm belt seems mandatory. It provides useful information for farm experts/farmers, to develop proper planting and crop husbandry schedules. They can also adopt plant protection procedures and time their harvest accurately.

The UAV (Unmanned Aerial Vehicle) is an aerial technology that has been adopted by farmers, during recent years. The agricultural drones that throng the sky offers specific advantages to farming experts and farmers. The UAV technology is capable of aerial imagery, collection of digital data about terrain and crops. UAVs also possess and ability to apply inputs from sky, using precision technology. There are farmers and aerial technology experts who believe that UAVs can bring about a perceptible transformation to the way we conduct agronomic procedures. Ultimately it may enhance economic efficiency and productivity of farming (Krishna, 2016, 2018, 2020a; Yanmar, 2020). These improvements are actually touted to transform agricultural crop production by managing farms from the sky. This aspect forms only a small part of the concept "Agricultural sky" being explained in this book. In future, while characterizing agricultural sky we may have to consider usage of agricultural drones for aerial imagery, aerial spraying of fertilizer formulations, pesticides, fungicides, and herbicides as criteria.

As state earlier, agricultural soil, water, and crops have been classified using a definite set of traits. Such classification has been useful to agricultural experts, farm technicians, and farmers all over the world. The same is not true about the concept proposed here, that is, "Agricultural Sky." Often agricultural sky or generally sky above agrarian regions has meant atmosphere only, mainly meteorological aspects, such as seasonal diurnal patterns, photosynthetic radiation levels, weather, precipitation trends wind, storms, pollution climate change etc. However, agricultural sky is not just atmospheric aspects. It includes several aspects related to biotic factors such as microbes, insects, birds, man-made factors such as sprinkler sprays, of farm chemical layers, UAVs, satellites usage etc.

We can characterize the agricultural sky above each small, cropped field, a larger farm or a vast area or cropping belt. The extent of details, particularly semi-quantitative or quantitative data collected for the said region of sky/atmosphere, is important. Our accuracy in depicting and distinguishing the agricultural sky and its type depends on the accuracy of the data. Agricultural sky could be characterized and classified using the data about following traits: they are:

Atmospheric traits: Ambient temperature, diurnal period, photosynthetic radiation, clouds, wind, precipitation pattern, relative humidity, CO_2 concentration, N_2O, NO_2, SO_2 and CO concentration and VOCs in the atmosphere, particulate matter, greenhouse gas emissions, atmospheric pollution; smog, fog, haze, storms, dust storms, heat waves, drought (scanty precipitation), nutrient dispersal and transport through high-speed wind, dust storms, and rainfall. There are indeed innumerable standardized methods, and instrumentation available to assess the atmospheric properties of the agricultural sky. Incidentally, agricultural meteorology is a well-developed faculty that helps us to classify the sky above crops.

Biological Traits: We have not yet assessed agricultural sky in terms of its biological prowess, to any great extent. The biotic features could be beneficial or detrimental to crop land below it. Actually, monitoring biotic components of the agricultural sky, at least those relevant immediately to crops, is a good idea. It forewarns of impending dangers from large-scale disease or insect spread, if any etc. Methods to assess spread of beneficial microbes through the sky need attention (e.g. Biological Nitrogen Fixation, Arbuscular mycorrhizae, Phyllosphere microbes etc.).

Microbes in the air above crops, microbial load, and diversity could be used to classify agricultural sky. A large number of crop diseases are caused by the propagules of microbes that float and settle down on crops to infect and cause devastations to cropland. Microbial plant pathogens and their propagule intensity/diversity in the air above crops (e.g. dissemination of rust fungi, mildews, late and early blights, leaf mosaic and mottle viruses) is a good criterion, to judge the health of the sky. Propagules of symbiotic N-fixing bacteria and mycorrhizas are easily traceable in the agricultural sky. This trait too could be used to classify agricultural sky. Agricultural sky disseminates diverse types of symbiotic microbes through wind.

Beneficial insects such as bees and other pollinators, insect predators offering biological control of other insect pest species (e.g. *Trichogramma sp, Bracon spp*, mealy bugs etc.) throng the agricultural sky above crops, at various stages of growth in different seasons. There are innumerable insect pests that are detrimental to crops. Insect pests that are localized in the sky above crops cause a wide range of deleterious effects. Sometimes, they are devastating, if prophylaxis is absent (e.g. locusts, army worms, leaf miners, leaf eaters, stem borers, grain eaters, vectors of viruses).

Aves are beneficial as biological control agents or pollinators. Aves act as crop destroyers at grain formation and ripening stage, that is, panicle/seed eaters. Birds are source of fertilizers (guanos). Birds act as dispersal agents of seeds from natural vegetation, crops, and weeds; dispersal of disease propagules affecting crops, farm animals and farmers; human pathogens in the air above farms.

There are now several man-made traits that function in the agricultural sky. They need due consideration by farmers prior to planning crop production. A few conspicuous processes that utilize agricultural sky are sprinkler irrigation, foliar fertilizer application, and plant protection sprays. When water and chemicals are channeled through the agricultural sky, we may note that leaves, phyllosphere, and stomata/hydathodes are key aspects. We should realize that every small amount of formulation supplied via canopy/leaves has to pass the phyllosphere region. Canopy, foliage, and phyllosphere all need detailed consideration when we try to take advantage of agricultural sky, to supply inputs. In the future, when farmers tend to supply more of nutrients and spray plant protection chemicals, using aerial methods, the foliage and phyllosphere may hold the key, regarding their effectivity. Phyllosphere in the aerial portion of crops is as important as rhizosphere in the soil. Every molecule of water, nutrients, pesticides, fungicides, hormones, or any other aerial input has to pass through the phyllosphere/phylloplane.

Unmanned Aerial Vehicle Usage: Agricultural drones that fly periodically in the sky above the experimental stations conduct vigilance, record spectral data about crops, and obtain real-time imagery, instantaneously. Small UAVs are utilized to study the influence of inputs and agronomic procedures on crops. We may note that farmers/researchers may also keep a blimp floating or place a tethered aerostat above the agricultural experimental station throughout the season. This is to obtain crucial data about farm and crops meant for the evaluation. In such a case, the agricultural sky will be in use for the entire season. In an agroecosystem, if tethered blimps and aerostats are floated for entire season or year-round, then the agricultural skyline itself may get modified. This seems a clear possibility because of low cost of tethered aerostats and helikites if we consider their performance per unit investment and/or unit time above the experimental farms.

Airplane campaigns are in vogue to assess agricultural terrain, natural resources, cropping expanses, cropping systems, to spray water, foliar nutrients, pesticides, and fungicides. Airplanes are utilized for aerial photography. However, during recent years, agricultural drones that are efficient, versatile and could be utilized repeatedly have replaced airplane campaigns. Airplanes need extra skills to manoeuver at low altitude above the crop. Yet, we may note that airplane campaigns have provided some excellent data about agricultural resources, since several decades.

Satellite usage to assess the agricultural terrain, soil types, water resources, natural vegetation, crops is ever increasing. Satellites have been utilized to monitor crop growth status (NDVI), moisture status, drought/flood situations, disease/pest-related disasters etc. satellite-guided farming is in vogue in many regions. The farm vehicles are controlled using the GPS coordinates and guided using radio/internet signals (See Krishna 2016, 2018, 2020a, 2020b). Satellites that monitor soil moisture in surface layers of farm belts are used by developed nations such as USA. Of course, satellites that monitor storms, rainfall patterns, floods, etc. are in use since several years. To a great extent agricultural sky has been put to efficient use if we consider satellites. The problems attached with such sue are the revisit times, low resolution, and cost of developing and stationing satellites in the space. Satellites serve exceptionally large number of farmers situated in far off areas.

Agricultural sky could become a major source of wind power in agrarian regions. Power generation through low- and high-altitude wind turbines seems important. High-altitude wind turbines placed on aerostats that float above the farms in the agricultural sky need greater attention. They could be a sought-after technology, in the future. In this case, agricultural sky serves farmer's needs for electricity, excellently. They say a set of few farms, say 7–8, could be supplied with electricity from turbines placed on aerostats at 200 m–2000 m above cropped fields. Sometimes, such aerostats have multiple functions in the agricultural sky. They are used in obtaining aerial imagery of crops. They can offer visual and digital data to precision vehicles operated in the farm Simultaneously, tethered aerostat can conduct general aerial surveillance, bird scaring etc. These are in addition to harnessing wind power thorough turbines (Glass, 2018; Krishna, 2020). Blimps filled with helium could be of great utility when lofted into the agricultural sky. They could be used for aerial imagery, surveillance, and even transport of agricultural goods. Kites are among the earliest of man-made aerial vehicles used to obtain imagery of agricultural terrain, and to scare birds from fields with ripe crops.

Agricultural crop production in the controlled glass house conditions has gained popularity in developed and developing countries. Such facilities also possess an atmosphere preserved within the glasshouse structure. We generally monitor the environment above glasshouse crops rather carefully. Usually, it is much easier to sample, measure, and compile data from greenhouses. We can assess the atmospheric parameters within the confined regions of a glasshouse. The atmospheric parameters may be altered based on the preferences of crops and yield goals set by the farmer. Overall, there are innumerable man-made methods by which we can utilize the agricultural sky, to our own great advantage. We have to consider each one on merit of efficiency and ease of operation, prior to adopting them.

Overall, "Agricultural Sky" is the concept touted here. We have to watch this concept's acceptance in practical agricultural. We have to assess the net advantages accrued by farmers. The benefits reaped may vary depending on the geographic location. Monitoring the events in the agricultural sky, particularly, those related to weather and its possible impact on crops sown in the fields, is important. It allows farmers a longer reaction time. Farmers may modify the agronomic procedure, wherever necessary. Modifying biotic fractions, particularly, microbial pathogens in the agrarian sky may hold the key to avoiding epidemics/pandemics on major cereals such as wheat, maize, or rice. These aspects are similar to that noticed with agricultural soils, water, or crops. We may be able to control and modify several parameters of agricultural sky to our requirements, but not all. Research toward total control of agricultural sky is still a far-fetched idea. Yet, it is worth pursuing. *The take home message here is that agricultural sky is important perpetually. We need to preserve its characteristics at congenial levels for better crop productivity.*

KEYWORDS

- **agricultural sky**
- **aves**
- **biotic factors**
- **climate indices**
- **dust storms**
- **insects**
- **microbes**
- **precipitation**
- **unmanned aerial vehicles**

REFERENCES

Abellanosa, A. L.; Pava, W. M. *An Introduction to Crop Science*; CMU Museum: Manila, Philippines, 1987, pp 245.

Aber, J. S.; Aber, S. W.; Buster, L.; Jensen, W. E.; Sleezer, R. L. Challenge of Infra-Red Kite Aerial Photography: A Digital Update. *Trans. Kans. Acad. Sci.* **2009,** *112*, 31–39.

Aber, S.; Aber, S. W. Kite Aerial Photography Equipment [Online] 2016, pp 1–15. http://www.geospectra.net/kite/equip/equip.htm/ (accessed Aug 27, 2019).

Abramovitz, J. N. *Putting a Value on Nature's 'Free' Services*; World Watch, 1998; pp 10–19.

Acosta-Martinez, V.; Van Pelt, S.; Moore-Kucera, J.; Baddock, M. C.; Zobeck, T. M. Microbiology of Wind-Eroded Sediments: Current Knowledge and Future Research Directions. *Aeolian Res.* **2015,** *18*, 99–113.

Adewuyi, S. A.; Okojie, L. O.; Folorunso, B.; Bada, B. S. Effect of Climate Change on Yam and Cassava Production in Oyo State, Nigeria: A Co-Integration Model Approach. *Int. J. Agric. Econ. Rural Dev.* **2014,** *6*, 53–59.

Adewuyi, S.; Fologrunso, B.; Okojie, L. O.; Akerele, S. Effect of Climate Change on Food Crops and Vulnerability Assessment in Oyo State. *J. Econ. Int. Finance* **2015,** *7*, 16–24.

–AgWeb, AgDay National Forecast: Weather. *Farm Journal* [Online] 2020, pp 1–3. https://www.agweb.com/weather/ (accessed Aug 15, 2021).

Ainsworth, E. A. Rice Production in a Changing Climate: A Meta-Analysis of Responses to Elevated Carbon Dioxide and Elevated Ozone Concentration. *Glob. Change Biol.* **2008,** *14*, 1642–1650.

Ainsworth, E. A. Understanding and Improving Global Crop Response to Ozone Pollution. *Plant J.* **2017,** *90*, 886–897.

Ainsworth, E. A.; Long, S. P. What have We Learned from 15 Years of Free-Air CO_2 Enrichment (FACE)? A Meta-Analytic Review of the Responses of Photosynthesis, Canopy Properties and Plant Production to Rising CO_2. *New Phytol.* **2005,** *165*, 351–372.

Ainsworth, E. A.; Rogers, A. The Response of Photosynthesis and Stomatal Conductance to Rising (CO_2): Mechanisms and Environmental Interactions. *Plant Cell Environ.* **2017,** *30*, 258–270.

Al Attar, L.; Al-Qudat, M.; Shamami, K.; Ghany, B. A.; Kanakri, S. 2012. Case Study: Heavy Metals and Fluoride Contents in the Materials of Syrian Phosphate Industry and in the Vicinity of Phosphogypsum Piles. *Environ. Technol.* **2017,** *33*, 143–152.

Al-Dagal, M.; Fung, D. Y. Aeromicrobiology—A Review. *Critic. Rev. Food Sci. Nutr.* **1990,** *29* (5), 333–340.

Allsopp Helikites Ltd. Aerial Photography. [Online] 2019, pp 1–7, Allsopp Helikite Ltd, Hampshire, England, United Kingdom. http://www.allsopphelikites.com/index.php?mod=page&id_pag=33 (accessed July 19, 2021).

Allsopp, S.; Reynaud, L.; Mohorcic, M. Integrated Project: ABSOLUTE-Aerial Base Stations with Opportunistic Links for Unexpected and Temporary Events [Online] 2013, pp 21–25. https://cordis.europa.eu/docs/projects/cnect/2/318632/080/deliverables/001-fp7ict20118318632absoluted23v10isa.pdf (accessed Mar 10, 2019).

Amato, P. Clouds Provide Atmospheric Oases For Microbes. Microbe Magazine: N. Page. American Society for Microbiology [Online] 2012. http://www.microbemagazine.org/index.php/02–2012-home/4547-clouds-provide-atmospheric-oases-for-microbes (accessed Aug 12, 2021).

Andreae, M. O.; Merlet, P. Emission of Trace Gases and Aerosols from Biomass Burning. *Glob. Biogeochem. Cycles* **2001,** *15*, 955–966.

Andreae, M. O.; Artaxo, P.; Fischer, H.; Freitas, S. R.; Grégoire, J. M.; Hansel, A.; Hoor, P.; Kormann, R.; Krejci, R.; Lange, L.; Lelieveld, J.; Lindinger, W.; Longo, K.; Peters, W.; De Reus, M.; Scheeren, B.; Silva Dias, M. A. F.; Strom, J.; Van Velthoven, P. F. J.; Williams, J. Transport of Biomass Burning Smoke to the Upper Troposphere by Deep Convection in the Equatorial Region. *Geophys. Res. Lett.* **2001,** *28*, 951–954.

AQUASTAT, Water from the Clouds. Food and Agricultural Organization of the United Nations, Rome, Italy [Online] 2014, pp 1–9. http://www.fao.org/nr/aquastat (accessed Feb 10, 2021).

Arthur, S. Grasshoppers and Locusts as Agricultural Pests. In *Encyclopaedia of Entomology*; Springer: Dordrecht, 2020; pp 1002–1004.

Asner, G. P. Cloud Cover in Landsat Observations of the Brazilian Amazon. *Int. J. Remot Sens.* **2001,** *22*, 3855–3862.

Atherton, K. Drone Sprays Pesticides around Crops [Online] 2017. http://www.popsci.com/agri-drone-is-precision-pesticide-machine (accessed Nov 19, 2017).

Avnery, S. Global Crop Yield Reductions due to Surface Ozone Exposure: 1. Year 2000 Crop Production Losses and Economic Damage. *Atmos. Environ.* **2011a,** *45* (13), 2284–2296.

Avnery, S. Global Crop Yield Reductions due to Surface Ozone Exposure: 2. Year 2030 Potential Crop Production Losses and Economic Damage under Two Scenarios of O_3 Pollution. *Atmos. Environ.* **2011b,** *45*, 2297–2309.

Aylor, D. E.; Sutton, T. B. Release of *Venturia Inaequalis* Ascospores during Unsteady Rain: Relationship to Spore Transport and Deposition. *Phytopathology* **1992,** *82*, 532–540.

Badarinath, K. V. S.; Kiran Chand, T. R.; Krishna Prasad, V. Agriculture Crop Residue Burning in: The Indo-Gangetic Plains: A Study Using IRS-P6 Awifs Satellite Data. *Curr. Sci.* **2006,** *91*, 1085–1089.

Bailey, P. T. *Pests of Field Crops and Pastures*; Department of Agriculture and Fisheries, University Of Queensland, Queensland: Australia, 2019; pp 1–4.

Bareja, B. G. Climatic Factors Can Promote or Inhibit Crop Growth, Development and Productivity [Online] 2011. https://www.cropsreview.com/climatic-factors.html/1=24 (accessed Dec 18, 2020).

Barford, E. Crop Pests Advancing with Global Warming. *Nature* [Online] 2013. https://www.nature.com/articles/nature.2013.13644; doi:10.1038/nature.2013.13644/ (accessed Aug 19, 2021).

Barnett, M. Blue Skies Leads Ghana's Agricultural Come Back [Online] 2015, p. 1. https://www.blueskies.com/blue-skies-leads-ghanas-agricultural-come-back/ (accessed Sept 10, 2021).

Barret, R.; Crowther, P.; Laurence, R.; Lincoln, R. Agricultural Crop Identification Using SPOT and LANDSAT Imagery in Tasmania. *Int. Arch. Photogramm. Remote Sens.* **2000,** *33*, 133–139.

Bartemeous, I. Contribution of Insect Pollinators to Crop Yield and Quality Varies with Agricultural Intensification. *PeerJ* **2014,** *2*, e328.

Benson, E. Bees of the Sea: Tiny Crustaceans Pollinate Under Water Plants. *Nat. Commun.* [Online] 2016, pp 1–3, https://www.newscientist.com/article/2114930-bees-of-the-sea-tiny-crustaceans-pollinate-underwater-plants/ (accessed Aug 2, 2021).

Bentley, M. L.; Mote, T. L.; Thabepanya, P. Using Landsat to Identify Thunderstorm Damage in Agricultural Regions. *Bull. Am. Meteorol. Soc.* **2002,** *23*, 363–374.

Beyer, L. Soil Geography and Sustainability of Cultivation. In *Soil Fertility and Crop Production*; Krishna, E. D. K. R., Ed., Science Publishers Inc.: New Hampshire, USA, 2003; pp 33–63.

Bhat, N.; Jain, S.; Asawa, K.; Tak, M.; Shinde, K.; Singh, A.; Gandhi, N.; Gupta, V. V. Assessment of Fluoride Concentration of Soil and Vegetable in the Vicinity of Zinc Smelter, Debari, Udaipur, Rajasthan, India. *J. Clin. Diagnos. Res.* **2015,** *9*, 63–66.

Bhatti, I. Smog Will Add To Farmer's Woes [Online] 2017, pp 1–4. The Nation, Rawalpindi, Pakistan. https://nation.com.pk/09-nov-2017/smog-will-add-to-farmers-woes/ (accessed Aug 22, 2021).

Bhuvaneshwari, S.; Hettiarachchi, H.; Meegoda, J. N. Crop Residue Burning in India: Policy Challenges and Potential Solutions. *Int. J. Environ. Res. Public Health* **2019,** *16*, 832–837.

Biesmeijer, J. C.; Roberts, S. E. A. Parallel Declines in Pollinators and Insect-Pollinated Plants in Britain and the Netherlands. *Science* **2006,** *313*, 351–354.

Borad, C. K.; Parasharya, B. M. Community Structure of Birds in Fields of Central Gujarat. *J. Entomol. Zool. Stud.* **2018,** *7*, 19–24.

Bowers, R. M.; Clements, N.; Emerson, J. B.; Wiedinmyer, C.; Hannigan, M. P.; Fierer, N. Seasonal Variability in Bacterial and Fungal Diversity of the Near-Surface Atmosphere. *Environ. Sci. Technol.* **2013,** *47*, 12097–12106.

Bowers, R. M.; Lauber, C. L.; Wiedinmyer, C.; Hamady, M.; Hallar, A. G.; Fall, R.; et al., Characterization of Airborne Microbial Communities at a High-Elevation Site and their Potential to act as Atmospheric Ice Nuclei. *Appl. Environ. Microbiol.* **2009,** *75*, 5121–5130.

Boyer, M. Altaeroes Energies Floating Wind Turbines Tap into Strong High-Altitude Winds [Online] 2012, pp 1–7. http://in habitat.com/altaeros-energies-floating-wind-turbines-tap-into-stron-high-altitude-winds/ (accessed Aug 12, 2021).

Brady, N. C. *Nature and Properties of Soils*; Prentice Hall: India, New Delhi, 2000; pp 565.

Bralower, T.; Bice, D. Temperature and Waves. *Internal Report* [Online], 2019, pp 1–73. College of Earth and Mineral Sciences, Pennsylvania State University: Pennsylvania, USA. https://www.e-education.psu.edu/earth103/node/914/ (accessed Aug 5, 2021).

Brandl, H.; Daniken, A.; Hit, C.; Krebs, W. Short-Term Dynamic Patterns of Bioaerosol Generation and Displacement in an Indoor Environment. *Aerobiologia* **2008,** *24*, 203–209.

Bregaglio, S.; Donatelli, M.; Confalonieri, R. Fungal Infections of Rice, Wheat, and Grape in Ein 2030–2050. *Agron. Sustain. Dev.* **2013,** *33*, 767–776.

Bridges, E. M.; Batjes, N. H.; Nachtergaele, F. O. *World Reference Base for Soils Resources-Atlas*; Acco Leuven, Ammersorft, Flemish Brabant: Belgium, 1998; pp 1–79.

Brown, J. K.; Hovmoller, M. S. Aerial Dispersal of Pathogens on the Global and Continental Scales and its Impact on Plant Disease. *Science* **2002,** *297*, 537–541.

Buerkert, A.; Bagayoko, M.; Bationo, A.; Romheld, V. Soil Fertility and Crop Production in Semi-Arid and Sub-Humid West Africa. In *Farming in West Africa: Issues, Potentials and Perspectives*; Graef, F., Lawrence, P., Von Oppen, M., Eds., Verlag, Stuttgart: Germany, 2000; pp 27–57.

Buis, A. Edge of the Outer Space. In *Earth's Atmosphere: A Multi-Layered Cake*[Online] 2019, pp 1–8. National Aeronautics and Space Agency: Global Climate Change Section: Washington DC. USA. pp https://climate.nasa.gov/news/2919/earths-atmosphere-a-multi-layered-cake/ (accessed Aug 17, 2021).

Bull, L.; Phillips, N. Towards the Design of Aerostat Wind Turbine Arrays Through AI [Online] 2018, pp 1–3. Department of Computer Science & Creative Technologies, University of the West of England: Bristol, United Kingdom. https://arxiv.org/ftp/arxiv/papers/1811/1811.05290.pdf/ (accessed Aug 30, 2021).

Burt, P. J. A. *"Airborne Dispersal of Mycosphaerella fijiensis" in Mycosphaerella Leaf Spot Diseases of Bananas: Present Status and Outlook*, Proceedings of 2nd International Workshop on Mycosphaerella Leaf Spot Diseases Held in San José, Costa Rica; Jacome, L, Eds.; The International Network for the Improvement of Banana and Plantain: Montpellier, France, 2002; pp 111–122.

CAB International. Rainfed Agriculture [Online] 2009. International Water Management Institute: Srilanka. http://www.iwmi.cgiar.org/publications/cabi_publications/ca_cabi_series/rainfed_agriculture/protected/rainfed_agriculture_unlocking_the_potential.pdf/ (accessed Aug 8, 2021).

CABI. Plant Health: The State of Research: Global Threat to Plants [Online] 2019. https://stateoftheworldsplants.org/2017/report/sotwp_2017_10_plant_health_state_of_research.pdf/ (accessed Aug 27, 2021).

Caliboso, F. M. Birds as Pests of Grain Storage. Food and Agricultural Organization of the United Nations [Online] 2013, pp 1–4. Rome, Italy. http://www.fao.org/3/x5048e/x5048e0g.htm/ (accessed Aug 10, 2021).

Cambridge Dictionary. Sky [Online] 2019. https://dictionary.cambridge.org/dictionary/english/sky (accessed July 4, 2021).

Cao, G. L.; Zhang, X. Y.; Wang, Y. Q.; Zheng, F. C. Estimation of Emissions from Field Burning of Crop Straw in China. *Chinese Sci. Bull.* **2008,** *53*, 784–790.

Carreras, L. G. Observations of Increased Cloud Cover Over an Irrigated Agriculture in an Arid Environment [Online], 2016. https://journals.ametsoc.org/doi/full/10.1175/jhm-d-16–0208.1/https://doi.org/10.1175/jhm-d-16–0208.1 (accessed Aug 12, 2021).

Carruthers, R. I.; Onsager, J. A. Perspective on the use of Exotic Natural Enemies for Biological Control of Pest Grasshoppers (Orthoptera: Acrididae). *Environ. Entomol.* **1993,** *22*, 885–905.

Cassimally, K. A. Forget Aeolians, We Need Airborne Wind Farms to Harness Maximum Wind Energy [Online] 2012, pp 1–4. https://www.nature.com/scitable/blog/labcoat-life/forget_aeolians_we_need_airborne (accessed July 15, 2021).

Centre For Disease Control and Prevention, USA, Agricultural Water [Online] 2020, pp 1–3. https://www.cdc.gov/healthywater/other/agricultural/index.html/ (accessed Aug 4, 2021).

CESAR. Russian Wheat Aphid [Online], 2019, pp 1–5. Department of Agriculture and Fisheries, University Of Queensland, Queensland, Australia. http://cesaraustralia.com/sustainable-agriculture/pestnotes/insect/russian-wheat-aphid/ (accessed Aug 28, 2021).

Challinor, A. J.; Koehler, A. K.; Ramirez-Villegas, J.; Whitfield, S.; Das, B. A Meta-Analysis of Crop Yield Under Climate Change and Adaptation. *Nat. Climate Change* **2014,** *4*, 287–291.

Challinor, A. J.; Koehler, A. K.; Ramirez-Villegas, J.; Whitfield, S.; Das, B. Current Warming will Reduce Yields Unless Maize Breeding and Seed Systems Adapt Immediately. *Nat. Clim. Chang.* **2016,** *6*, 954–958.

Challinor, A. J.; Wheeler, T. R.; Crauford, P. Q.; Slingo, J. M. Simulation of the Impact of High Temperature in Wheat, Barley, Triticale Field Crops Research **2005,** *100*, 240–248.

Chamberlain, L. Sky Farming-Turning Skyscrapers into Crop Farms. *New York Times* [Online] 2007, pp 1–18. New York, USA. http://nymag.com/news/features/30020/ (accessed Dec 25, 2019).

Chapman, R. F.; Joerne, A. Biology of Grasshoppers [Online] 1990, p. 576. Wiley: New York, USA. https://www.wiley.com/en-us/biology+of+grasshoppers-p-9780471609018/ (accessed Jun 24, 2021).

Charreau, C. Soils of Tropical Dry and Wet Climatic Areas of West Africa and their Use and Management. A Series of Lectures. In *Agronomy Mimeograph 74–127*; Department of Agronomy, Cornell University: Cornell, USA, 1974; pp 1–63.

Chauffin, M. Unprecedented Locust Swarms Devastating Several Countries in Africa Fueled by Multiple Weather Factors [Online] 2020, pp 1–6. Accuweather. https://www.accuweather.com/en/weather-news/unprecedented-locust-swarms-devastating-several-countries-in-africa-fueled-by-multiple-weather-factors/666385/ (accessed Jun 24, 2021).

Chen, I.-C.; Hill, J. K.; Ohlemüller, R.; Roy, D. B, Thomas, C. D. Rapid Range Shifts of Species Associated with High Levels of Climate Warming. *Science* **2011,** *333*, 1024–1026.

Chepfer, H.; Brogniez, H.; Noel, V. Diurnal Variations of Cloud and Relative Humidity Profiles Across the Tropics. *Nat. Res. Sci. Rep.* **2019,** *9*, 16045.

Cherubini, A.; Papini, A.; Vertechy, R.; Fontana, M. Airborne Wind Energy Systems: A Review of the Technologies. *Renew. Sust. Energ. Rev.* **2015,** *51*, 1461–1476.

Chikezie, C.; Ibekwe, U. C.; Ohajianya, D. O.; Orebiyi, J. S.; Ehirim, N. C.; Henri-Ukoha, A.; Nwaiwu, I. U. O.; Ajah, E. A.; Essien, U. A.; Anthony, G.; Oshaji, I. O. Effect of Climate Change on Food Crop Production in Southeast Nigeria: A Co-Integration Model Approach. *Int. J. Weather Clim. Conserv. Res.* **2015,** *2*, 47–56.

Chloupek, O.; Hrstkova, P.; Schweigert, P. Yield and Its Stability, Crop Diversity, Adaptability and Response to Climate Change, Weather and Fertilization Over 75 Years in the Czech Republic in Some European Nations. *Field Crops Res.* **2004,** *85*, 167–190.

CIAT. Where Our Food Crops Come From. An Interactive Map. Centro Internacional Agricultura Tropicale [Online] 2020, pp 1–5. Cali. Columbia. https://ciat.cgiar.org/project/where-our-food-crops-come-from/ (accessed Aug 28, 2020).
CIMMYT. Obregon Blimp Airborne and Eyeing Plots [Online] 2012, pp 1–4. http://www.cimmyt.org/obregon-blimp-airborne-and-eyeing-plots/ (accessed Aug 24, 2021).
CIRAD. West Africa is Threatened by a Desert Locust Invasion. CIRAD-Agricultural Research and Development: Paris, France [Online] 2020. https://www.cirad.fr/en/news/all-news-items/press-releases/2020/estimate-risks-swarmrefs-chad/1–3 (accessed Apr 25, 2020).
Coblentz, B. Tornadoes Destruction Brings Agricultural Loss. *MSU Communications* [Online] 2019, pp 1–5. https://www.farmprogress.com/management/tornadoes-destruction-brings-agricultural-loss/ (accessed Jul 10, 2021).
Collins Dictionary. Agriculture [Online] 2019, pp 1–2. https://www.collinsdictionary.com/dictionary/english/agriculture (accessed Jul 20, 2021).
Crutzen, P. J.; Galbaly, I. E.; Brühl, C. Atmospheric Effects From Post-Nuclear Fires. *Clim. Change* **1984,** *6*, 323–364.
CSIRO Weather and Climate Risk in Agriculture [Online] 2020, pp 1–23. Council of Scientific and Industrial Research Organization: Canberra, Australia. http://www.bom.gov.au/watl/about-weather-and-climate/risk/ (accessed Jan 14, 2020).
Cuculeanu, V.; Marica, A.; Simota, A. Climate Change Impact on Agricultural Crops and Adaptation Options in Romania. *Clim. Res.* **1999,** *12*, 153–160.
Dartan, G.; Taspinar, F.; Toroz, I. Analysis of Fluoride Pollution from Fertilizer Industry and Phosphor-Gypsum Piles in Agricultural Area. *J. Indus. Pollut. Control* **2017,** *33*, 662–668.
Davies, S. J.; Unam, L. Smoke-Haze from the 1997 Indonesian Forest Fires: Effects on Pollution Levels, Local Climate, Atmospheric CO_2 Concentrations, and Tree Photosynthesis [Online] 2019, pp 1–13. https://www.sciencedirect.com/science/article/abs/pii/s0378112799000602 (accessed Sept. 3, 2021).
De Deckker, P. An Evaluation of Australia as a Major Source of Dust. *Earth Sci. Rev.* **2019,** *194*, 536–567.
De Graaff, M. A.; Van Groenigen, K. J. Interactions between Plant Growth and Soil Nutrient Cycling Under Elevated CO_2: A Meta-Analysis. *Glob. Change Biol.* **2006,** *12*, 2077–2091.
De Mello-Areu, J. P. Protecting Fruit Production Damage from Frost Damage. European Commission, EIP AGRI Focus Group [Online] 2018. pp 1–20. https://ec.europa.eu/eip/agriculture/sites/default/files/fg30_frost_damage_starting_paper_2018_en.pdf (accessed Jun 24, 2021).
Deangelis, A.; Dominguez, F.; Fan, Y.; Robock, A.; Kustu, M. D.; Robinson, D. Evidence of Enhanced Precipitation due to Irrigation Over the Great Plains of the United States. *J. Geophys. Res.* **2010,** *115*, D15115–15121.
Definitions and Translations. Definitions For Sky [Online] 2019. https://www.definitions.net/definition/sky (accessed Aug 4, 2021).
Deryng, D.; Conway, D.; Ramankutty, N.; Price, J.; Warren, R. Global Crop Yield Response to Extreme Heat Stress Under Multiple Climate Change Futures. *Environ. Res. Lett.* **2014,** *9*, 034011.
Desert Locust Information System. Elocust3: Solving Age Old Problem with New Technology [Online] 2019, pp 1–3. FAO of the United Nations: Rome, Italy pp . http://www.fao.org/fao-stories/article/en/c/1244192/ (accessed Jul 6, 2021).

Desperés, V. R.; Huffman, J. A.; Burrows, S. M.; Hoose, C.; Safatov, A. S.; Buryak, G.; Fröhlich-Nowoisky, J.; Elbert, W.; Andreae, M. O.; Pöschl, U.; Jaenicke, R. Primary Biological Aerosol Particles in the Atmosphere: A Review. *Tellus. Ser. Biochem. Phys. Meteorol.* **2012,** 64.

Devlin, R. M. *Plant Physiology Van Nostrand Publishers*; New York, USA **1975,** pp 600.

DJI. Above The World. *UAS Magazine*, 2016, pp 238–239.

DJI. DJI AGRAS M-1 [Online] 2019, pp 1–8. DJI Company Ltd.: Shenzhen, China. https://dji.com/mg-1/info (accessed Jul 29, 2021).

Dokuchaev, V. V. Selected Works, Vols. I and II, Moscow-Leningrad. In *Description of Russian Sols*; 1951, p. 419.

Ducreax, A. 1984 Characterization Mecanique Des Dols Sableaux Et Sablo-Agrileux Zone Et Techniques De l'Africe De l'Oeust. Academic De Montpellier, Universite Des Sciences Et Techniques Du Landuedoc. Montpellier, Franc, (English: Ducreax, A. 1984 Mechanical Characterization Of Sandy Udols and Sandy-Argileux Zone and Techniques Of The Oeust Africe. Academic Of Montpellier, University Of Science and Technology Of The Landuedoc. Montpellier, France), pp 1–38.

Eberhardt, I.D.R.; Schultz, B.; Rizzi, R.; Sanchez, I.D.A.; Foormaggio, A. R.; Atzberger, C.; Mello, M.R.; Immitzer, M.; Trabequinni, K.; Foschiera, W.; Louis, A.J.; B. 2016 Cloud Cover Assessments For Operational Crop Monitoring Systems In Tropical Area. Remote Sensing 8: 219 DOI: 10.3390/Rs8030219 (July 20th, 2021).

Editors, 2020 Subsistence Farming. Encyclopaedia Britannica. Https://Www.Britannica.Com/Topic/Subsistence-Farming/pp. 1–2 (February 29th, 2020).

Eitleberg, D. A.; Van Vliet, J.; Verburg, P. H. 2015 A Review Of Global Potentially Available Crop Land Estimates and their Consequences For Model-Based Assessments. *Glob. Change Biol.* 21:1236–1248. Https://Doi.Org/10.1111/Gcb.12733 (July 25, 2021).

Eitzinger, J.; Utset, A.; Trnka, M.; Zalud, Z.; Nikolaev, M.; Uskov, I. Weather and Climate and Optimization of Farm Technologies at Different Input Levels. In *Managing Weather and Climate Risks in Agriculture*; Sivakumar, M. V. K., Motha, R. P., Eds.; Springer: Berlin, Heidelberg, 2007; pp. 141–170.

Elamin, H. M. A.; Roth, M.; Taha, M. E. In *The Consequences of Defoliation of Gum Arabic Tree (Acacia Senegal) by Sahelian Tree Locust (Anacridium Melanorhodon) for the Gum Producers in North Kordofan State, Sudan*, Conference on International Research on Food Security, Natural Resource Management and Rural Development, University of Hohenheim: Stuttgart, Germany, 2008; pp 7–9.

Emberson, L. D.; Pleijal, H.; Ainsworth, E. A.; Van Den Berg, M.; Ren, W.; Osborne, S.; Mills, G. Pandey, D.; Dentener, F.; Buker, P.; Ewert, T., Koeble, R.; Van Dingron, R. Ozone Effects on Crops and Considerations in Crop Models. *Eur. J. Agron.* **2018b,** *100*, 19–34.

Emberson, L.; Pandey, D.; Mortenson, S. Why Ozone Levels Pose a Challenge to Food Security. Down To Earth.Org [Online] 2018a, pp 1–14 https://www.downtoearth.org.in/news/agriculture/why-ozone-levels-pose-a-challenge-to-food-security-60170 https://www.downtoearth.org.in/news/agriculture/why-ozone-levels-pose-a-challenge-to-food-security-60170 (accessed Jul 16, 2021).

EOS. *Major Dust Storm Sweeps Across Australia. Earth Observatory Service* [Online] 2015, pp 1–2. National Aeronautics and Space Agency: Washington D.C. USA https://earthobservatory.nasa.gov/images/146146/major-dust-storm-sweeps-across-australia/ (accessed Jul 15, 2021).

EPA. *Basic Ozone Layer Science*[Online] 2019, pp 1–8. United States Environment Protection Agency: Washington, D. C. https://www.epa.gov/ozone-layer-protection/basic-ozone-layer-science accessed Dec 16, 2019).

European Food Safety Authority (EFSA). Pesticides and Bees: EFSA Finalises New Guidance. Guidance on the Risk Assessment of Plant Protection Products on Bees (*Apis Mellifera, Bombus Spp.;* Solitary Bees) [Online] 2013. https://www.efsa.europa.eu/en/efsajournal/pub/3295/ (accessed Jul 23, 2021).

Eyshi, R. E.; Gaiser, T.; Siebert, S.; Ewert, F. Adaptation of Crop Production to Climate Change by Crop Substitution. *Mitig. Adapt. Strateg. Glob. Change* **2015,** *20*, 1155–1174.

FAO of the United Nations. 2017 Drones, Data, Food Security, UAVs Offer New Perspectives on Agriculture [Online] 2017, pp 1–4. Rome, Italy. http://www.fao.org/news/podcast/drones/en/ (accessed Jan 1, 2017).

FAO, *Desert Locust Project*; Final Report. Report No FAO/SF:34/DLC. Food and Agriculture Organization of the United Nations: Rome. 1968, p 86.

FAO, *Projet Relatif au Criquet Pèlerin*; Rapport Complémentaire (Juillet 1966-Décembre 1970). Report No FAO/SF:34/DLC. Food and Agriculture Organization of the United Nations: Rome, 1972.

FAO. The Status of World's Soil Resources. Chapter 6 Global Soil Status, Processes and Trends [Online] 1997, pp 1–71. http://www.fao.org/3/bc595e/bc595e.pdf/ (accessed July 20, 2021).

FAO. *The State of World's Plant Genetic Resources for Food and Agriculture*; Food and Agricultural Organization of the United Nations: Rome, Italy, 1998; pp 1–328.

FAO. Organic Agriculture and Climate Change [Online] 2002, pp 1–11. Food and Agricultural Organization of the United Nations: Rome Italy, Repository article http://www.fao.org/3/y4137e/y4137e02b.htm/ (accessed July 28, 2019).

FAO. Passive Protection Methods for Frost Control in Farms [Online] 2003, pp 1–30. Food and Agricultural Organization of the United Nations: Rome, Italy. http://www.fao.org/3/y7223e/y7223e0c.htm/ (accessed July 24, 2019).

FAO. Indicative Crop Classification [Online] 2010, pp 1–4. World Program for the Census of Agricultural Crops-2010. http://www.stats.gov.cn/english/icas/papers/p020071115312159845171.pdf/ (accessed July 28, 2021).

FAO. Irrigation: Introduction [Online] 2012, pp 1–14. Food and Agricultural Organization of the United Nations: Rome, Italy, Document Repository. http://www.fao.org/3/x5560e/x5560e02.htm/ (accessed July 23, 2021).

FAO. Organic Agriculture and Climate Change [Online] 2015, pp 1–11. Food and Agricultural Organization of the United Nations: Rome Italy. http://www.fao.org/3/y4137e/y4137e02b.htm/ (accessed Jul 23, 2021).

FAO. Crop Production and Natural Resource [Online] 2019, pp 1–23. Food and Agricultural Organization of the United Nations: Rome, Italy. http://www.fao.org/3/y4252e/y4252e06.htm (August 5, 2021).

FAO/IIASA. Global Agroecological Zoning. FAO Land and Water Digital Media [Online] 1999, Series No 11 pp 1–8. Food and Agricultural Organization of the United Nations: Rome Italy. http://www.fao.org/nr/gaez/en/ (accessed June 24, 2021).

FAOSTAT. Statistical Data Base [Online] 1999. Food and Agricultural Organization of the United Nations: Rome, Italy. http://apps.fao.org/ (accessed June 24, 2021).

FAOSTAT. Statistical Data Base [Online] 2000. Food and Agricultural Organization of the United Nations: Rome, Italy. http://apps.fao.org (accessed April 20, 2021).

Farmlife, Tornadoes Cause Damage in South. *Progressive Farmer* [Online] 2019, pp 1–3. https://www.dtnpf.com/agriculture/web/ag/news/world-policy/article/2019/03/10/tornadoes-cause-damage-south/ (accessed July 10, 2021).

Faulkner, J. W.; Easton, Z. W. *Agricultural Adaptation to Climate Changes: Improving Resilience in Row Crops*; University of Vermont Centre for Sustainable Agriculture-Extension Service: Burlington, Vermont USA, 2015; pp 1–8.

Fecht, S. How Climate Change Impacts Our Water. *State of the Planet* [Online] 2019, pp 1–12. Earth Institute/Columbia University, USA. https://blogs.ei.columbia.edu/2019/09/23/climate-change-impacts-water/ (accessed April 8, 2020).

FFAR. Photosynthesis Varies Greatly Across Cultivars-Natural Diversity Could Yields. Realizing Increased Photosynthetic Efficiencies. RIPE. Foundation for Food and Agricultural Research Communications [Online] 2020. http://foundationfar.org (accessed July 10, 2020).

Filatneko, A. A.; Diederischsen, A.; Hammer, K. Vavilov's Theories of Crop Domestication in the Ancient Mediterranean Area [Online] 2020, pp 1–23. https://www.bioversityinternational.org/fileadmin/bioversity/publications/web_version/47/ch06.htm#bm2/ (accessed July 7, 2021).

Filho, F. H.; Haldens, W. B.; Kong, Z.; De Lange, E. S. Drones: Innovative Technology for Precision Pest Management. *J. Econ. Entomol.* **2019,** *20,* 1–25.

Fisher, M. C.; Henk, D. A.; Briggs, C. J.; Brownstein, J. S.; Madoff, L. C.; Mccraw, S. L.; Gurr, S. J. Emerging Fungal Threats to Animal, Plant and Ecosystem Health. *Nature* **2012,** *484*, 186–194.

Foley, J. A.; Defries, R.; Asner, G. P.; Barford, C.; Bonan, G.; Carpenter, S.; Chapin, F. S.; Coe, M. T.; Daily, G. C.; Gibbs, H. K.; Helkowski, J. H.; Holloway, T.; Howard, E.; Kucharik, C. J.; Monfreda, C.; Patz, J.; Prentice, I. C.; Ramankutty, N.; Snyder, P. K. Global Consequences of Land Use. *Science* **2005,** *309* (5734), 570–574.

Food and Agricultural Organization of the United Nations. *Soil Map of the World;* Technical Paper no 20 ISRIC Wageningen, Netherlands, Revised Print. 1997, pp 140. http://www.fao.org/soils-portal/soil-survey/soil-maps-and-databases/faounesco-soil-map-of-the-world/en// (accessed May 8, 2021).

Food and Agricultural Organization of the United Nations. Livestock's Long Shadow: Environmental Issues and Options [Online] 2006, pp 1–8. Rome, Italy.http://www.fao.org/3/a0701e/a0701e00.htm/(accessed July 25, 2021).

Food and Agricultural Organization. Adaptation to Climate Change in Agriculture, Forestry and Fisheries: Perspective, Framework and Priorities [Online] 2007, pp 1–32. http://www.fao.org/nr/climpag/pub/adaptation_to_climate_change_2007.pdf (accessed July 21, 2021).

Freney, J. R.; Denmead, O. T. Factors Controlling Ammonia and Nitrous Oxide Emissions from Flooded Rice Fields. *Ecol. Bull.* **1992,** *42*, 188–194.

Freney, J. R.; Denmead, O. T.; Saffina, P. J.; Wood, A. W.; Chapman, L.; Hurney, A. P. In *Ammonia Loss from Sugarcane Fields as Affected by Fertilizer Placement, Irrigation and Canopy Development*, Proceedings of the Australian Society of Sugarcane Technologists. 1991, pp 38–43.

Fröhlich-Novoisky, J.; Burrows, S. M.; Xie, Z.; Engling, G.; Solomon, A. A.; Fraser, M. N.; Mayol-Bracero, O. L.; Artaxo, S.; Begerow, D.; Conrad, R.; Andreae, M. O.; Després, V. R.; Pöschl, U. Biogeography in the Air: Fungal Diversity Over Land and Oceans. *Biogeosciences* **2012,** *9*, 1125–1136.

GAO. *Irrigated Agriculture: Technologies, Practices and Implications for Water Scarcity*; United States Government Accountability Office: Washington D.C. USA, 2019; pp 1–179.

Garcia-Ruiz, F.; Sankaran, S.; Maja, J. M.; Lee, W. S.; Rasmussen, J.; Ehsani, R. Comparison of Two Aerial Imaging Platforms for Identification of Huanglongbing Infected Citrus Trees. *Comp. Electron. Agric.* **2016,** *91*, 106–115.

Garibaldi, L. A. Wild Pollinators Enhance Fruit Set of Crops Regardless of Honey-Bee Abundance. *Science* **2013,** *339* (6127), 1608–1611.

Garris, A. Heat, Not Drought, will Drive Lower Crop Yields, Researchers Say [Online] 2019, pp. 1–4. Cornell University, Environmental Science Department, Cornell University: New York, USA. https://phys.org/news/2019–06-drought-crop-yields.html/ (accessed August 4, 2021).

Gautam, P.; Lal, B.; Nayak, A. K.; Raja, R.; Panda, B. B.; Tripathi, R.; Shahid, M.; Kumar, U.; Swain, C. K. Inter-Relationship between Intercepted Radiation and Rice Yield Influence by Transplanting Time, Method and Variety. *Int. J. Biometeorol.* **2019,** *63*, 337–349.

Glass, B. Soaring 'Super Towers' Aim to Bring Mobile Broadband to Rural Areas. (by Steadler, T. 2018) [Online] 2018, pp 1–7. ITU News. https://news.itu.int/soaring-supertowers-aim-to-bring-mobile-broadband-to-rural-areas/ (accessed February 15, 2019).

Glick, P. A. The Distribution of Insects, Spiders and Mites in the Air. Bureau of Entomology and Quarantine, United States of America, Washington D.C. Technical Bulletin No, 1939, 673, pp 1–153.

Goettel, M. S.; Johnson, D. L. Microbial Control of Grasshoppers and Locusts. *Mem. Ent. Soc. Can.* **1997,** *171*, 400–432.

Goodland, R.; Anhang, J. *Livestock and Climate Change. What if the Key Actors in Climate Change Were Pigs, Chickens and Cows*. World Watch Institute: Washington D.C. USA, 2009, vol. 1, pp 11–19. https://www.semanticscholar.org/paper/livestock-and-climate-change%3a-what-if-the-key-in-goodland-anhang/6704c7a0777c82357704d82b9ae8007c1197cb07/ (accessed May 11, 2021).

Grace Communication Foundation. The Impact of Climate Change on Water Resources. Climate and Environment Footprints [Online] 2020, pp 1–3. https://www.watercalculator.org/footprint/climate-change-water-resources/ (accessed July 24, 2021).

Grange, R. I.; Hand, D. W. A Review of the Effects of Atmospheric Humidity on the Growth of Horticultural Crops. *J. Hortic. Sci.* **2015,** *62*, 125–134.

Grewer, U.; Bockel, L.; Galford, G.; Gurwick, N.; Nash, J.; Pirolli, G.; Wollenberg, E. A Methodology for Greenhouse Gas Emission and Carbon Sequestration Assessments in Agriculture: Supplemental Materials for Info Note Series Analysing Low Emissions Agricultural Practices in USAID Development Projects, CGIAR Research Program on Climate Change, Agriculture and Food Security (CCAFS). Published by The International Centre For Tropical Agriculture (CIAT) and the Food and Agriculture Organization of the United Nations (FAO). CCAFS Working Paper No. 187, 2016; pp 1–30.

Griffin, D. W. Terrestrial Microorganisms at an Altitude of 20,000 m in Earth's Atmosphere. *Aerobiologia* **2004,** *20*, 135–140.

Griffin, D.; Gonzalez-Martin, C.; Hoose, C.; Smith, D. Global Scale Atmospheric Dispersion of Microorganisms. In *Microbiology of Aerosols*; Delort, A. M.; Amato, P. John Wiley and Sons, Inc.: Hoboken, NJ, 2017; pp 155–194.

Guppy, H. B. Dispersal of Butterflies and Other Insects. *Nature (London)* **1925,** *116*, 543.

Haire, B. Blimp Helps Fine-Tune Irrigation [Online] 2019, pp 1–6. University of Georgia Extension Service: Georgia, USA. https://georgiaforages.caes.uga.edu/news/story/2045/infrared-cotton.html (accessed July 10, 2021).

Harding, K. J.; Snyder, P. K. Modeling the Atmosphere Response to Irrigation in the Great Plains Part 1. General Impacts on Precipitation and Energy Budget. *J. Hydrometeorol.* **2012,** *13*, 1667–1686.

Harding, K. J.; Snyder, P. K. Examining Future Changes in the Character of Central US Warm-Season Precipitation Using Dynamical Downscaling. *J. Geophys. Res. Atmos.* **2014,** *119*, 131116–13136.

Harris, Fuller, D. X. Agriculture: Definition and Overview, 2014, pp 1–17. doi:10.1007/978–1-4419–0465–2_64/ (accessed June 24, 2021).

Hawkes, J. E. Back to Vavilov: Why Wee Plants Domesticated in Some Areas and Not in Others [Online] 2020. https://www.bioversityinternational.org/fileadmin/bioversity/publications/web_version/47/ch06.htm/1–18 (accessed May 7, 2020).

Hebei, S.; Beninati, N. Scaglia, J. F.; Egal, F.; Diuf, M. Special Repot: FAO/WFP Crop and Food Supply Assessment Mission to Mauritania with Special Focus on Losses due to Desert Locus [Online] 2004, pp 1–3. http://www.fao.org/qiews (accessed April 2, 2020).

Heidhaues, I.; Bruntrup, M. Subsistence Agriculture in Central and Eastern Europe: How to Break the Vicious Cycle [Online] 2003, pp 1–27. Institute of Agriculture Development in Central and Eastern Europe. http://www.fao.org/family-farming/detail/en/c/335462/ (accessed December 5, 2019).

Helbaum, N. The Bases for a Classification of World Agriculture. *Profess. Geographer* **2010,** *9*, 2–17.

Helfer, S. Rust Fungi and Global Change. *New Phytol.* **2014,** *201*, 770–780.

Herrero, M.; Thornton, P. C.; Garber, P.; Reid, R. S. Livestock Livelihoods and the Environmental Understandings: The Trade-Offs. *Curr. Opin. Environ. Sustain.* **2009,** *1*, 11–120.

Herrero, M.; Villinga, T. V.; Gerber, J.; Tara, G. Livestock and Greenhouse Gas Emissions: The Importance of Getting the Numbers Right. *Anim. Feed Sci. Technol.* **2011,** *166*, 771–782.

Hirji, Z. These Scientists are Radically Changing how they Live to Cope with Climate Change. Buzzfeed News [Online] 2019b, pp 1–3. https://www.buzzfeednews.com/article/zahrahirji/scientists-climate-change-action (accessed April 23, 2019).

Hirji, Z.; Climate Change is Triggering More Heat Waves, Droughts and Crop Yield Declines, Study Says [Online] 2019a, pp. 1–7. https://www.buzzfeednews.com/article/zahrahirji/climate-change-land-report-ipcc/ (accessed January 16, 2019).

Horneck, G.; Klaus, D. M.; Mancinelli, R. L. Space Microbiology. *Microbiol. Mol. Biolo. Rev.* **2010,** *74*, 121–156.

HSE. Intelligent Imaging Systems [Online] 2015, pp 1–4. http://www.uavcropdustersrayers.com/agriculture_delta_fw-70__fixed_wing_uav.htm (accessed May 9, 2015).

HSE. *HSE Agricultural and Conservation UA*; Homeland Surveillance and Electronics LLC: Illinois, USA, 2017; pp. 1–4.

Huang, W.; Lin, L.; Haung, M.; Wang, J.; Wan, H. Monitoring Wheat Yellow Rust with Dynamic Hyperspectral Data [Online] 2008, pp 1–12. http://www.researchgate.net/c/nveacx/javascript/lib/pdfjs/web/viewer. html?file=http (accessed September 29, 2015).

IANS. Diversifying Crops will Mitigate Climate Impact in India: Study [Online] 2019, pp 1–12. https://www.business-standard.com/article/news-ians/diversifying-crops-will-mitigate-climate-impact-in-india-study-119061800248_1.html (accessed January 16, 2020).

ICRISAT. Biological Control of Millet Head Miner (*Heliocheilus Albipunctella*) [Online] 2008, pp. 1–2. International Crops Research Centre Sahelian Centre: Niamey, Niger. http://www.icrisat.org/proven-technologies/ (accessed May 3, 2008).

Iglesias, A.; Garrote, L. Adaptation Strategies for Agricultural Water Management under Climate Change in Europe. *Agric. Water Manag.* **2015,** *155*, 113–134.

Iglesias, A.; Erda, L.; Rosenzweig, C. Climate Change in Asia: A Review of the Vulnerability and Adaptation of Crop Production. *Water Air Soil Pollut.* **1996,** *92*, 13–27.

IITA. Rainfall Changes for Key Crops Predicted Even with Reduced Greenhouse Gas Emissions, 2019.

Inman, M. Energy High in the Sky: Expert Perspectives on Airborne Wind Energy Systems [Online] 2012, pp. 1–15. Near Zero. http://nearzero.org/ (accessed February 15, 2019).

Institute for Sustainable Development. Regenerative Agriculture Initiatives [Online] 2019, pp 1–5. California State University: Chico, California, USA. https://thecarbonunderground.org/wp-content/uploads/2017/02/regen-ag-definition-7.27.17-1.pdf (accessed December 4, 2019).

Institute of Physics. Frequency of Extreme Heat Waves on the Increase in the African Continent: Could Occur Annually by 2040 [Online] 2016, pp 1–5. https://phys.org/news/2016–05-frequency-extreme-africa-annually.html/ (March 10, 2020).

Intergovernmental Panel on Climate Change (IPCC). The Physical Science Basis. Contribution of Working Group I to the Fourth Assessment Report of the Intergovernmental Panel on Climate Change. Solomon, S., Qin, D., Manning, M., Chen, Z., Marquis, M., Averyt, K. B., Tignor, M., Miller, H. L., Eds.; https://www.ipcc.ch/report/ar4/wg1/. p 996 (accessed August 1, 2021).

International Centre for Tropical Agriculture. Rainfall Changes for Key Crops Predicted Even with Reduced Greenhouse Gas Emissions [Online] 2019, pp 1–3. Sciencedaily. www.sciencedaily.com/releases/2019/03/190311173147.htm (accessed July 21, 2021).

International Panel On Climate Change. Climate Change 2014: Impacts, Adaptation, and Vulnerability Vol. 2 [Online] 2014. IPCC: Stanford, California. http://ipcc-wg2.gov/ar5/ (accessed December 25, 2019).

IPCC. Climate Change: The Physical Science Basis. *Contribution of Working Group I to the Fourth Assessment Report of the Intergovernmental Panel on Climate Change* [Online] 2007, pp 1–112. Cambridge, UK: Cambridge University Press, Cambridge, United Kingdom, https://archive.ipcc.ch/pdf/assessment-report/ar4/syr/ar4_syr_full_report.pdf/.

IPCC. Summary for Policymakers. In: Climatechange-2014: Impacts, Adaptation, and Vulnerability. Part A: Global and Sectoral Aspects. In *Contribution of Working Group II to the Fifth Assessment Report of the Intergovernmental Panel on Climate Change*; Cambridge: United Kingdom, and New York, NY, USA, Cambridge, UK, 2014, pp 1–32.

Irrigation Association. Principles of Efficient Agricultural Irrigation [Online] 2019, pp 1–9. Irrigation Association: Fairfax, Virginia, USA. http://www.irrigation.org/ (accessed July 20, 2020).

Isard, S. A.; Russo, J. M. In *Global Movements of Rust Fungi in the Atmosphere*, American Phytopathological Society: Field Crops Symposium, San Antonio, Texas, United States Of America, 2011 ; p 32. http://www.plantmanagementnetwork.org/proceedings/fcrs/2011/presentations/isard.pdf/(accessed July 21, 2021).

ISRO. New Microbes Discovered in Earth's Stratosphere [Online] 2009, pp 1–3. Indian Space Research Organization: Ahmedabad, India. https://www.sciencedaily.com/releases/2009/03/090318094642.htm/ (accessed Jully 12, 2021).

ISSS-WRB. *World Reference base for Soil Resources*, World Soil Resources Reports. Food and Agricultural Organization of the United Nations: ISRIC, and ISSS, Rome. p 88. http://www.fao.org/3/i3794en/i3794en.pdf/ (accessed July 12, 2021).

Jacobson, J. S.; Heller, L. I.; Yamada, K. E.; Osmeloski, J. F.; Bethard, T. Foliar Injury and Growth Response of Red Spruce to Sulphate and Nitrate Acidic Mist. *Can. J. For. Res.* **1988,** *20*, 58–65.

Jain, N.; Bhatia, A.; Pathak, H. Emission of Air Pollutants from Crop Residue Burning in India. *Aerosol Air Qual. Res.* **2014,** *14*, 422–435.

Jenny, H. *Factors of Soil Formation Ecological Studies*; Springer Verlag: New York, 1980; p. 377.

Jha, S.; Dick, C. W. Native Bees Mediate Long-Distance Pollination in Coffee Landscape. *Proc. Natl Acad. Sci. U S A.* **2010,** *107*, 13760–13764.

Jha, U. C.; Bohra, A.; Singh, N. P. Heat Stress in Crop Plants: Its Nature, Impact and Integrated Breeding Strategies to Improve Heat Tolerance. *Plant Breed.* **2014,** *133*, 679–701.

Joerne, A.; and Gaines, S. B. Population Dynamics and Regulation in Grasshoppers. In *Biology Of Grasshoppers*; Chapman, R. F., Joerne, A., Eds.; Wiley: New York, USA, **1990,** pp 415–460.

Jones, P. G.; Thornton, P. K. The Potential Impacts of Climate Change in Maize Production in Africa and Latin America in 2055. *Glob. Environ. Change* **2003,** *13*, 51–59.

JRC. Extreme Weather and Climate Change Account for 40% of Global Wheat Production Variability [Online] 2019, pp 1–4. Joint Research Centre, European Commission:Brussels. https://www.sciencedaily.com/releases/2017/07/170704094104.htm/ (accessed July 2, 2021).

Julie, C. Massive Dust Storms Across Australia Extends almost the Total Width of the Continent [Online] 2020. https://watchers.news/2020/01/11/massive-dust-storm-across-australia-extends-almost-the-total-width-of-the-continent/ pp. 1–5 (accessed July 25, 2021).

Junk, J.; Kauadio, L.; Delfosse, P.; El-Jarroudi, M. Effect of Regional Climate Change on Brown Rust Disease in Winter Wheat. *Clim. Change* **2016,** *135*, 439–451.

Kassir, L. N.; Lartiges, B.; Ouaini, N. Effect of Fertilizer Industry Emissions on Local Soil Contamination: A Case Study of a Phosphate Plant on the East Mediterranean Coast. *Environ. Technol.* **2012,** *33*, 873–885.

Kellog, C. A.; Griffin, E. W. Aerobiology and the Global Transport of Desert Dust. *Trends in Ecology and Evolution* [Online] 2006, pp 1–5. http://doi.org/10.1016/j.tree.2006.07.004/ (accessed July 19, 2021).

Kennedy, C. M. A Global Quantitative Synthesis of Local and Landscape Effects on Wild Bee Pollinators in Agroecosystems. *Ecol. Lett.* **2013,** *16*, 584–599.

Khodadad, C. L.; Wang, G. M.; Leando, J. M..; Thakar, P. J.; Lane, M. A.; Jhon, A. C.; Smith, A. J. Stratosphere Conditions Inactivate Bacterial Endospores. *Astrobiology* **2016,** *17*, 337–359.

Khosla, R. In *Precision Agriculture: Challenges and Opportunities in a Flat World*, 19th World Congress of Soil Science, Soil Solutions for a Changing World; Brisbane, 2010, pp 1–6.

Khot, L.; Sankaran, S.; Cummings, T.; Johnson, D.; Carter, A.; Serra, S.; Musacchi, S. In *Unmanned Aerial Systems Applications in Washington State Agriculture*, Proceedings of 12th International Conference on Precision Agriculture, Sacramento, California, USA, 2014; p 129.

Kim, J. 2018 The Difference Between Smog and Fog [Online] pp 1–8. https://sciencing.com/difference-between-smog-fog-8093163.html/ (accessed January 12, 2020).

Kim, W.; Iizumi, T.; Nishimori, M. Global Pattern of Crop Production Losses Associated with Drought from 1963 to 2009. *J. Appl. Meteorol. Climatol.* **2019,** *58*, 1233–1244.

Kindt, R.; Mausya, S.; Kimotho, J.; Waruhice, A. Tree Seed Suppliers-N Sources of Seeds and Microsymbionts [Online] 1997, pp 1–17. International Centre for Agroforestry Nairobi, Kenya. https://agris.fao.org/agris-search/search.do?recordid=qi9700056 (accessed Juky 28, 2021).

Kingsfield, D. M.; Debeers, K. M. Landsat Identification of Tornado Damage by Land Cover and Evaluation of Damage Recovery in Forests. *J. Appl. Meteorol. Climatol.* **2017,** *56*, 965–987.

Kipple, K. W.; Orneleas, K. C. *The Cambridge World History of Food*; Cambridge University Press Cambridge: United Kingdom, 2000. p 2153.

Kohfeld, K. E.; Tegen, I. The Atmosphere. Record of Mineral Aerosol and their Roles in the Earth System. *Treatise Geochem.* **2007,** *4*, 1–26.

Kohls, J. How does Wildfire Ash and Smoke Impact Crops [Online] 2019, pp 1–8. https://dutchopeners.com/how-does-wildfire-ash-and-smoke-impact-crops/ (accessed July 10, 2021).

Koppmann, R.; Czapiewski, K. V.; Reid, J. S. A Review of Biomass Burning Emissions Part I: Gaseous Emissions of Carbon Monoxide, Methane, Volatile Organic Compounds, and Nitrogen Containing Compounds. *Atmos. Chem. Phys. Discuss.* **2005,** *5*, 10455–10516.

Korzekwa, K. Whether Wheat Weathers Heat Waves [Online] 2018, pp 1–3. American Society of Agronomy. https://www.agronomy.org/science-news/whether-wheat-weathers-heat-waves/ (accessed July 23, 2021).

Krajick, K.; Lee, K. A Major Source of Air Pollution: Farms [Online] 2016, pp. 1–12. The Earth Institute, Columbia University, USA. https://www.earth.columbia.edu/articles/view/3281/ (accessed July 14, 2021).

Kreutz, L. Drones Helping to Fight Mites in Strawberry and Almond Fields [Online] 2017, pp 1–4. https://mae.ucdavis.edu/news/drones-helping-fight-mites-strawberry-almond-fields (accessed July 7, 2021).

Krishna, K. R. *Agrosphere: Nutrient Dynamics Crop Productivity and Ecology*; Science Publishers: Enfield, New Hampshire, USA, 2003; pp 1–27.

Krishna, K. R. *Peanut Agroecosystem. Nutrient Dynamics and Productivity*; Alpha Science Publishers: Oxford, United Kingdom, 2008; p 297.

Krishna, K. R. *Precision Farming: Soil Fertility and Productivity Aspects*; Apple Academic Press Inc.: Waretown, New Jersey, USA, 2013; p 178.

Krishna, K. R. *Push Button Agriculture: Robotics, Drones and Satellite-Guided Soil and Crop Management*. Apple Academic Press Inc.: Waretown, New Jersey, USA, 2016; p 400.

Krishna, K. R. *Agricultural Drones: A Peaceful Pursuit*; Apple Academic Press Inc.: Waretown, New Jersey, USA, 2018; p 394.

Krishna, K. R. *Unmanned Aerial Vehicle Systems in Crop Production: A Compendium*; Apple Academic Press Inc.: Palm Bay, Florida, USA, 2020a; pp 1–688.

Krishna, K. R. *Aerial Robots in Agriculture: Parafoils, Blimps, Aerostats, and Kites*; Apple Academic Press Inc.: Palm Bay, Florida, USA, 2020b; p 398.

Kromdijk, J.; Glowacka, K.; Long, S. Realizing Increased Photosynthetic Efficiency: for Sustainable. Increases in crop yield [Online] 2017, p 14. The Science Breaker https://ripe.illinois.edu/news/cloudy-days-cost-yield-scientists-hacked-photosynthesis/ (accessedJanuary 20, 2020).

Kreutz, L. Drones Helping to Fight Mites in Strawberry and Almond Fields [Online] 2017, pp 1–4. https://mae.ucdavis.edu/news/drones-helping-fight-mites-strawberry-almond-fields (accessed July 7, 2021).

Kucharik, C. J.; S. P. Serbian. Impacts of Recent Climate Change on Wisconsin Corn and Soybean Yield Trends. *Environ. Res. Lett.* **2018,** *3*, 034–053.

Kukal, M. S.; Irmak, S. Climate-Driven Crop Yield and Yield Variability Impacts on the US Great Plains Agricultural Production. *Sci. Rep.* **2018,** *8*, 3450 (accessed August 3, 2021).

Kumar, R.; Barth, M. C.; Pfister, G. G.; Naje, M.; Brassier, G. P. 2014 WRF-Chem Simulation of a Typical Pre-Monsoon Dust Storm in Northern India. Influence on Aerosol Optical Properties and Radiation Budget. *Atmos. Chem. Phys.* **2014,** *14*, 2431–2446.

Kunast, C.; Riffel, M.; De Graeff, R.; Whitmore, S. *Pollination and Agriculture: Agricultural Productivity and Pollinator Protection*; European Institute for Sustainable Agriculture: Brussels, Belgium, 2019, pp 1–55.

Lakso, A. N.; Robinson, T. I. Sunlight, Yield and Productivity of Apple. *N. Y. Fruit Q.* **2015,** *22*, 1–5.

Lal, N. Effects of Acid Rain on Plant Growth and Development. *e-J. Sci. Technol.* **2016,** 1–23.

Lanigan–Teagasc, G. Reducing Greenhouse Gas Emissions from Agriculture. *Soils, Environment and Land-Use Programme* [Online] 2017, pp 1–5, Johnstown Castle: Wexford, Great Britain. https://www.teagasc.ie/publications/2017/reducing-greenhouse-gas-emissions-from-agriculture.php/ (accessed July10th, 2021).

Le Gall, M.; Overson, R.; Cease, A. A Global Review on Locusts (Orthoptera: Acridoidea) and their Interactions with Livestock Grazing Practices. *Front. Ecol. Evol.* **2019,** *263*, 1–12.

Lebedeva, I. I.; Tonkonogov, L. I.; Gerasimonova, M. I. A New Classification of Soils of Russia. *Eurasian Soil Sci.* 2008, *41*, 91–99.

Lecoq, M. Desert Locust Management: From Ecology to Anthropology. *J. Orthoptera Res.* **2005,** *14*, 179–186.

Leff, B.; Raman Kutty, N.; Foley, S. Geographic Distribution of Major Crops Across the World. *Glob. Biogeochem. Cycles* **2004,** *18*, 1–27.

Lesk, C. Influence of Extreme Weather Disasters on Global Crop Production. *Nature* **2016,** *529*, 84–87.

Levey, B. J.; Bolker, B. M.; Tawksberg, J. J.; Sargent, S.; Haddad, N. M. Effects of Landscapes Corridors on Seed Dispersal by Birds. *Science* **2005,** *309*, 146–148.

Li, V.; Uddin, W.; Kanishiki, J. E. Effect of Relative Humidity on Infection, Colonization and Conidiation of *Magnathorpe Oryzae* on Perennial Rye Grass. *Plant Pathol.* **2014,** *63*, 590–597.

Lobell, D. B.; Burke, M. B. On the use of Statistical Models to Predict Crop Yield Responses to Climate Change. *Agric. Forest Meteorol.* **2010,** *150* (11), 1443–1452.

Lobell, D. B.; Field, C. B. Global Scale Climate-Crop Yield Relationships and the Impacts of Recent Warming. *Environ. Res. Lett.* **2017,** *2*, 014–022.

Lobell, D. B.; Burke, M. B.; Tebaldi, C.; Mastrandrea, M. D.; Falcon, W. P.; Naylor, R. L. Prioritizing Climate Change Adaptation Needs for Food Security in 2030. *Science* **2008,** *319* (5863), 607–610.

Lobell, D. B.; Banziger, M.; Magorokosho, C.; Vivek, B. Nonlinear Heat Effects On African Maize As Evidenced by Historical Yield Trials. *Nat. Clim. Change*. **2011a,** *1*, 42–45.

Lobell, D. B.; Ortiz Monasterial, J. I.; Asner, G. P.; Matson, P. A.; Naylor, R. L.; Falcon, W. P. Analysis of Wheat Yield and Climatic Trends in Mexico. *Field Crops Res.* **2005,** *94* (2), 250–256..

Lobell, D. B.; Schwenker, W.; Costa-Roberts, J. 2011b. Climate Trends and Global Crop Production Since. *Science* 1980, *333* (6042), 616–620.

Lomer, C. J.; Bateman, R. P.; Johnson, D. L.; Langewald, J.; Thomas, M. B. Biological Control of Locusts and Grasshoppers. *Annu. Rev. Entomol.* **2001,** *46*, 667–702.

Long, S. P.; Ainsworth, E. A. Food for Thought: Lower-Than-Expected Crop Yield Stimulation with Rising CO_2 Concentrations. *Science* **2006,** *312*, 1918–1921.

Longman's Dictionary. Sky [Online] 2019. https://www.ldoceonline.com/dictionary/sky (accessed December 4, 2019).

Malaker, A.; Snow, D. I.; Ray, C. Irrigation Water Quality: A Contemporary Perspective. *Water* **2019,** 1482–1501.

Manna, I.; Bandyopadhyay, M. Physicochemical Perturbation of Plants on Exposure to Metal Oxide Nanoparticle. *Nanomater. Plants Algae Microorgan.* **2019,** 323–373 .

Marston, Z. P. D.; Cira, T. M.; Hodgson, E. W.; Knight, J. F.; Macrae, I. V.; Koch, R. L. Detection of Stress Induced by Soybean Paid (Hemiptera: Aphididae) Using Spectral Imagery from Unmanned Aerial Vehicles. *J. Econ. Entomol.* **2019,** *113*, 779–786.

Mateo-Sagasta, S.; Zadeh, S. M.; Turral, H. *Water Pollution in Agriculture: A Global Review*; Food and Agricultural Organization of the United Nations: Rome, Italy and International Water Management Institute: Colombo, S, pp 23–34.

Mati, P. K. Smog can Cause Lower Yields of Vegetables, Fruits and Field Crops Say Agricultural Experts. DND Dispatch [Online] 2019, pp. 1–5, Lahore, Pakistan. https://dnd.com.pk/smog-can-cause-lower-yields-of-vegetables-fruits-crops-agri-experts/135178/ (accessed July 22, 2021).

Mcgashen, A. Protecting Crops with Predators Instead of Poisons [Online] 2018, pp 1–3. https://www.ehn.org/back-to-basics-tackling-farm-pests-with-predator-birds-2546940909.html/ (accessed July 21, 2020).

Mcsweeney, R. Food and Farming: Droughts and Heat Waves Cause 10% Drop in Annual Crop Production [Online] 2018, pp 1–8. https://www.carbonbrief.org/droughts-and-heatwaves-cause-10-drop-in-annual-crop-harvests/ (accessed July 5, 2021).

Mehrabi, Z.; Ramankutty, N. The Cost of Heat Waves and Droughts for Global Crop Production [Online] 2017, pp 1–8. https://www.biorxiv.org/content/10.1101/188151v1 (accessed July 21, 2021).

Merck. Air Application Monitoring-Petrochemical Industries: Sulphur Dioxide [Online] 2020, pp 1–2. Merch Kya, Darmstadt, Germany. https://www.sigmaaldrich.com/analytical-chromatography/air-monitoring/applications/petrochemical/sulfur-dioxide.html/ (accessed July 1, 2021).

Meredith, D. S.; Lawrence, J. S.; Firman, I. D. Ascospore Release and Dispersal in Black Leaf Streak Disease of Bananas. *Trans. Br. Mycol. Soc.* **1973,** *60*, 547–554.

Merriam-Webster Dictionary. Sky [Online] 2019. https://www.merriam-webster.com/dictionary/sky (accessed July 7, 2021).

Middleton, N. Sand and Dust Storms: Underrated Natural Hazards. *Disasters* **2018,** *43*, 390–409.

Ministry of Environment. Effects of Air Pollution on Agricultural Crops [Online] 2003, pp 1–7. http://www.omafra.gov.on.ca/english/crops/facts/01–015.htm/canada. (accessed July 23, 2021).

MIT. Two Alums Dream of Tethered Aerostat Wind Farms. *Slice of MIT* [Online] 2013, pp 1–4. Massachusetts Institute of Technology, USA. https://alum.mit.edu/slice/two-alums-dream-tethered-aerostat-wind-farms/ (accessed July 13, 2021).

Mittal, S. K.; Bushel, K.; Singh, N.; Agarwal, R.; Awasthi, A.; Gupta, P. K. Ambient Air Quality during Wheat and Rice Crop Stubble Burning Episodes in Patiala. *Atmosp. Environ.* **2009,** *43*, 238–244.

Molina, L.T.; Molina, M. *Air Quality in the Mexico Mega City*‘ Kluwer Academic Publishers, 2002; p 379.

Moore, F. C.; Lobell, D. B. The Finger Print of Climate Trends on European Crop Yields. *Proc. Natl Acad. Sci.* **2015,** *112* (9), 2670–2675.

Morgan, L. How to Grow Aquatic Crops Hydroponically [Online] 2020, pp 1–8. https://www.maximumyield.com/aquatic-crops/2/1321/ (accessed July 1, 2021).

Murchie, E.; Reynolds, M. *Crop Radiation Capture and use Efficiency*. 2018; pp 1–12.

Murray, K. G. Avian Seed Dispersal. *Ecol. Monogr.* **1988,** *58*. 271–298.

Nair, S.; Johnson, J.; Wang, C. Efficiency of Irrigation Water Use: A Review from the Perspectives of Multi-Disciplines. *Agronomy J.* **2014,** *105*, 351–363.

NAL. Agriculture [Online] 2019. National Agricultural Library Glossary of Terms. https://agclass.nal.usda.gov/ (accessed July 22, 2021).

NASA. AIR: Our Atmosphere [Online] 2016, pp 1–3. National Aeronautics and Space Agency: Washington, D.C., USA. http:/www.eos.nasa.gov (accessed July 12, 2020).

NASA. Earth's Atmosphere: A Multi-Layered Cake [Online] 2019a, pp 1–8. National Aeronautics and Space Agency: Global Climate Change Section Washington DC. USA. https://climate.nasa.gov/news/2919/earths-atmosphere-a-multi-layered-cake/ (accessed July 17, 2021).

NASA. NASA Food Security Program—2019 [Online] 2019b, pp 1–3. http://www.science.gsfc.nasa.gov/610/applied-sciences/food.html/ (accessed July 3, 2019).

NASA. In Exposing Microorganisms in Stratosphere, *Astrobiology Science Conference -2015 In Wikipedia*, 2019c, pp 1–4. https://en.wikipedia.org/wiki/exposing_microorganisms_in_the_stratosphere/ (accessed April 14, 2020).

NASA. Global Precipitation Measurement [Online] 2020, pp 1–2. NAA: USA. https://gpm.nasa.gov/science/global-water-cycle/ (accessed April 6, 2020).

NASA/USGS. LANDSAT Missions [Online] 2020. https://www.usgs.gov/land-resources/nli/landsat/1–3 (accessed May 7, 2020).

National Geographic. Agriculture [Online] 2019. https://www.nationalgeographic.org/encyclopedia/agriculture/

Natural Resource Conservation Service-Plant Data Base. *Zizania Texana*-Texas Wild Rice [Online] 2020, pp 1–3. United States Department of Agriculture: Beltsville, Maryland, USA. https://plants.usda.gov/core/profile?symbol=zite/ (accessed May 13, 2020).

Nelson, B. *Will Cloudy Weather Damage your Corn*; Purdue University News: Lafayette, Indiana, USA, 2015; pp 1–4.

Northwing. Northwing ATF Soaring Trike [Online] 2018, p 15. http://northwing.com/atf-trike.aspx (accessed June 7, 2018).

Nunez, A.; Amo De Paz, G.; Rastrojo, A.; Garcia, A. M.; Alcami, A.; Gutierrez-Bustillo, G.; Moreno, D. A. Monitoring of Airborne Biological Particles in Outdoor Atmosphere. 1. Importance, Variability and Ratios. *Int. Microbiol.* **2016,** *19*, 1–13.

OECD. Water and Agriculture Managing Water Sustainably is Key to the Future Food and Agriculture Organization for Economic Development and Cooperation [Online] 2019, pp 1–14. Paris, France. https://www.oecd.org/agriculture/topics/water-and-agriculture/ (April 9, 2020).

Olesen, J. E.; Trinka, M.; Kersabaum, K. C.; Shjely, A. O.; Seguin, A. B.; Peltonen-Sainio, P. Rossi, F.; Kozyra, J.; Micale, F. Impacts and Adaptation of European Crop Production Systems in Climate Change [Online] 2011.

Oneto, D. L.; Golan, J.; Mazino, A.; Pringle, A.; Seminara, A. Timing of Fungal Spore Release during Atmospheric Transport. *Proc. Natl Acad. Sci. U. S. A.* **2020,** *117*, 5134–5143.

Ortiz-Obea, A. Unpacking Climatic Drivers of US Agricultural Yields. *Environ. Res. Lett.* **2019,** *14* (6), 064003.

Oxford English Dictionary. Agriculture [Online] 2020, p 1. https://www.oxfordlearnersdictionaries.com/definition/english/agriculture/ (accessed July 8, 2021).

Palilo, A. Monitoring and Management of Maize Rust (*Puccinia Sorghi*) by a Drone Prototype in Southern Highlands, Tanzania [Online] 2014, pp 1–15. Sokoine University of Agriculture: Morogoro, Tanzania. http://www.academia.edu/8063999/monitoring_and_management of maize-rust disease (accessed July 18, 2021).

Parman, D. C. Observation of the Effect of Storm Phenomenon on Insect Activity. *J. Econ. Entomol.* **1920,** *13*, 339–343.

Patel, K. B. Studies on Control of Bird Pests that Damage Millet Crops in Semiarid Zone of North Gujarat, India. *J. Environ. Res. Develop.* **2016,** *11*, 132–150.

Patrignani, A.; Godsey, C. B.; Ocshner, T. E. No-Till Diversified Cropping System for Efficient Allocation of Precipitation in Southern Great Plains. *Agrosyst. Geosci. Environ.* **2018,** *2*, 1–8.

Paulitz, T. C. Diurnal Release of Ascospores by *Gibberella Zeae* in Inoculated Wheat Plots. *Plant Disease* **1996,** *80*, 674–678.

Pearson, C. J. 1992 Field Crop Ecosystem. Elsevier Science Publications Inc. Netherlands pp 576.

Peoples, M. B.; Herridge, D. F. Nitrogen Fixation by Legumes in Tropical and Subtropical Agriculture. *Adv. Agron.* **1990,** *44*, 15–52.

Peoples, M. B.; Turner, G. L.; Shah, Z.; Shah, S.; Aslam, M.; Ali, A.; Maskey, Afandi, F.; Shwenke, G. D.; Horridge, D. F. In *Evaluation of the N^{15} Abundance Technique for Measuring N2 Fixation in Plots and Farmer's Fields, Extending Nitrogen Fixation Research to Farmer's Fields. Proceedings of International Workshop on Managing Legume Nitrogen Fixation in Cropping Systems of Asia*; Rupela, O. P.; Johansen, C.; Horridge, Eds.; International Crops Research Institute for the Semi-Aid Tropics: Patancheru, Andhra Pradesh, India, 1997; pp 52–75.

Pepper, I. L.; Gerba, C. P. Aeromicrobiology. In *Environmental Microbiology*, 3rd ed.; Elsevier Inc.: New York, USA, 2015; pp 89–110.

Perarnaud, V.; Seguin, B.; Malezieaux, E.; Deque, M.; Loustau, D. Agrometeorological Research and Applications Needed to Prepare Agriculture and Forestry to 21st Century Climate Change. *Climate Change* **2005,** *70*, 319–340.

Perkins. Dust Storms [Online] 2001. https://www.sciencedirect.com/topics/agricultural-and-biological-sciences/dust-storms. pp 1–8 (accessed July 20th, 2021).

Peterson, G. A.; Westfall, D. G. Managing Precipitation use in Sustainable Dryland Agroecosystem [Online] 2005, pp 1–24. https://doi.org/10.1111/j.17348.2004.tb00326.x/ (accessed August 31, 2021).

PFAF Organization. *Zizania Latifolia.* Plants for a Future Organization [Online] 2020. https://plants.usda.gov/core/profile?symbol=zite/1–3 (accessed July 13, 2021).

Poincelot, R. P. *Horticulture: Principles and Practical Applications*; Prentice Hall Inc.: Wisconsin, USA, 1980; p 642.

Poonia, S. Sulphur Dioxide Pollution: A Review on Trends of Emission and Impact on Organisms and their Environment. *Int. J. Eng. Manag. Res.* **2012,** *2*, 57–60.

Potts, S.; Garret, M.; Senapathi, D.; Breeze, T. Insect Pollinator Initiative. Living with Environmental Change [Online] 2014, pp 1–7. http://www.lwec.org.uk (accessed April 20, 2021).

Prasad, P. V. V.; Boote, K. J.; Allen, L. H. Adverse High Temperature Effects on Pollen Viability, Seed Set, Seed Yield and Harvest Index of Rain Sorghum (*Sorghum Bicolor*) are Severe at Elevated Carbon Dioxide due to Higher Tissue Temperature. *Agric. Forest Meteorol.* **2006,** *139*, 237–251.

Prasolov, L. I. *About the Classification and Nomenclature of Soils*; Transactions of Pochvovedenic Institute, Academy of Nauk SSSR: Russia, vol. 1: 1936; pp. 101–111.

Precision Hawk LLC, 2014 Lancaster Platforms In Agriculture. Http://Www.Precisionhawk.Com/Index.Htm #Industries pp 1–4 (August 5, 2021).

Pronto, T.; Gooch, C. A. Greenhouse Gas Emission Reduction due to Anaerobic Digestion of Dairy Manure. *Am. Soc. Agric. Biol. Eng.* Paper No: 09–7015, 2009; pp. 1–19.

Prosolov, L. I. Chernozems of Azov Steppes. *Pochhvovedenic* **1916,** *1*, 23–46.

Prospero, J. M.; Blades, E.; Mathison, G.; Naidu, R. Interhemispheric Transport of Viable Fungi and Bacteria from Africa to the Caribbean with Soil Dust. *Aerobiologia* **2005,** *21*, 1–19.

Pudelko, R.; Stuczynski, T.; Borzecka-Walker The Suitability of an Unmanned Arial Vehicle (UAV) for the Evaluation of Experimental Fields and Crops. *Zemdirbyste Agriculture* **2012,** *99*, 431–436.

Puiu, T. In the Last 50 Years Droughts and Heat Waves Destroyed 1/10 of Crops [Online] 2016, pp. 1–3. https://www.zmescience.com/ecology/world-problems/droughts-heat-wave-crop-production-0532543/ (accessed July 14, 2021).

Puma, M. J.; Cook, B. I. Effects of Irrigation on Global Climate the 20 Century. *J. Geophys. Res.* **2010,** *115*, D161120.

QA International Collectif. The Visual Guide to Understanding Plants & the Vegetable Kingdom—Plants and the Vegetable Kingdom. In *Québec Amerique* 2007; 35 37.

Queensland Department of Agriculture and Fisheries. Insect Pests of Sunflower [Online] 2019, pp 1–4. University of Queensland: Queensland, Australia. https://www.daf.qld.gov.au/business-priorities/agriculture/plants/crops-pastures/broadacre-field-crops/insect-pest-management-specific-crops/insect-pest-management-sunflowers/ (accessed July 28, 2021).

Rader, R.; Bartemeous, I.; Garibaldi, L. A. Non-Bee Insects are Important Contributors to Global Crop Pollination. *Proc. Natl Acad. Sci. U. S. A.* **2016,** *113*, 146–151.

Rader, R.; Edwards, W.; Westcott, D. A.; Cunningham, S. A.; Howlett, B. G. Diurnal Effectiveness of Pollination by Bees and Flies in Agricultural *Brassica Rapa*: implications for Ecosystem Resilience. *Basic Appl. Ecol.* **2013,** *14* (1), 20–27.

Ramankutty, N.; Norman, J.; Foley, L. A.; Mcsweeney, L. The Global Distribution of Cultivable Land: Current Patterns and Sensitivity to Possible Climate Change. *Glob. Ecol. Biogeogr.* **2002,** *11* (5), 377–392.

Rasul, G.; Chaudhry, Q. Z.; Mahmood, A.; Hyder, K. W. Effect of Temperature Rise on Crop Growth and Productivity. *Pak. J. Meteorol.* **2019,** *8*, 53–62.

Ricketts, T. H. Landscape Effects on Crop Pollination Services: Are there General Patterns? *Ecol. Lett.* **2008,** *11*, 499–515.

Ritchie, J. M.; Pedlegey, D. E. 1989 Desert Locust Cross The Atlantic. Antenna 13: 10–13.

RMAX. RMAX Specifications [Online] 2015, pp 1–8. Yamaha Motor Company: Japan https://www.yamaha-motor.com.au/products/sky/aerial-systems/rmax (September 8th, 2021).

Rojas, M.; Lambert, F.; Ramirez-Villegarsd, J.; Challinor, A. J. Emergence of Robust Precipitation Changes across Crop Production Areas in the 21st Century. *Proc. Natl Acad. Sci. U.S.A.* **2019,** *116*, 6673–6678.

Romheld, D.; Marschner, H.; Becker, K.; Schecht, E.; Hulsebusch, C.; Buerkert, A. Effects of Site Factors on the Efficiency of Soil Amendments and on Nutrient Fluxes in Subsistence-Oriented Cropping of the West African Sahel [Online] 2000, pp 1–24. http://www.troz.uni-hohenheim.de/research/sfb308/ebcont/bcheck (accessed May 10, 2020).

Rosenburg, J.; Burt, P. J. A. Windborne Displacement of Desert Locust. Windborne Displacement of Desert Locust from Africa to the Caribbean and South America. *Aerobiologia* **1999,** *15*, 166–175.

Rossi, S.; Tubiello, F. N.; Prospero, P.; Salvatore, M.; Jacobs, H.; Biancalani, R.; House, J. I.; Boschetti, L. FAOSTAT Estimates of Greenhouse Gas Emissions from Biomass and Peat Fires. *Clim. Change* **2016,** *135*, 699–711.

Rotter, R. P.; Carter, T. R, Olesen, J. E.; Porter, J. R. Crop-Climate Models Need an Overhaul. *Nat. Clim. Change* **2011,** *1*, 175–177.

Royal Society of Entomology of Great Britain. Classification of Insects [Online] 2020, pp 1–4. https://www.royensoc.co.uk/insect-classification/ (accessed July 27, 2021).

Sahai, S.; Sharma, C.; Singh, S. K.; Gupta, P. K. Assessment of Trace Gases, Carbon and Nitrogen Emissions from Field Burning of Agricultural Residues in India. *Nutr. Cycl. Agroecosystems* **2011,** *89*, 143–157.

Sanders, T. HSE Hercules 30 [Online] 2017. HSE Inc.: Illinois, USA. https://hse-uav.com (accessed July 20, 2017).

Saunders, M. E. Insect Pollinators Collect Pollen from Wind Pollinated Plants: Implications for Pollination Ecology and Sustainable Agriculture. *Insect Conserv. Divers.* **2017,** *11*, 13–31.

Schloesser, D. W. A Field Guide to Valuable Underwater Aquatic Plants of the Great Lakes [Online] 2010. United States Geological Survey. https://pubs.er.usgs.gov/publication/93728/ (accessed July 25, 2021).

Schmale, D. G. III.; Ross, S. D. Highways in the Sky: Scales of Atmospheric Transport of Plant Pathogens. *Annu. Rev. Phytopathol.* **2015,** *53*, 591–611.

Schmidt, W. H.; Colville, W. L. Yield and Yield Components of *Zea mays* L. as Influenced by Artificially Induced Shade. *Crop Sci.* **1967,** *7*, 137–140.

Schnug, E.; Jacobs, F.; Stevens, K. Guano: The White Gold of Seabirds. In *Seabirds*; Intech open, 2019; pp 1–8.

Seckler, D.; Molden, D.; Sakthivadivel, R. The Concept of Efficiency in Water-Resources Management and Policy. In *Water Productivity in Agriculture: Limits and Opportunities for Improvement*; Kijney, J. W.; Baker, R., Eds.; CABI: Wallingford, UK, 2003; pp 37–51.

Sene, M.; Badiane, A. *Phosphogypsum Efficiency to Correct Soil P Deficiency and/or Soil Acidity*. Institute Senegalaise De Researches Agricoles. Ministere De L'Agriculture, Republic Du Senegal Report; 2001, pp 17–30.

Sensefly Inc. Ebee Sensefly: The Professional Mapping Drone [Online] 2016, pp 1–5. Sensefly Inc. A Parrot Company: Chesseaux-Lousanne: Switzerland. https://www.sensefly.com/drone/ebee.html (accessed July 29, 2021).

Sensefly. Drones for Agriculture. Sensefly—A Parrot Inc [Online] 2015, pp 1–5. Chessaeux-Lousanne: Switzerland. http://www.sensfly.com/applications/agricuture.html (accessed May 23, 2015).

Serfling, A.; Kopahnke, D.; Habekuss, A.; Novakazi, F.; Ordon, J. *Wheat Diseases: An Overview*; Burleigh Dodds Science Publishing, 2016; pp 1–32.

Sergeev, M. G.; and Van'kova, I. A. Dynamics of the Italian Locust *Calliptamus Italicus* L. Population in the Southeast of the West Siberian Plain. *Contemp. Prob. Ecol.* **2008,** *1*, 204–209.

Shepard, D. Global Warming: Severe Consequences For Africa [Online] 2019, pp 1–9. Africa Renewal. https://www.un.org/africarenewal/magazine/december-2018-march-2019/global-warming-severe-consequences-africa(accessed July 27, 2021).

Shiklomanov, I. *World Water Resources: A New Appraisal and Assessment for the 21st Century*[Online] 1998; pp 1–37. UNESCO, Paris. Https://Snia.Mop.Gob.Cl/Sad/PHI710.Pdf.

Shroeder, J. F. Hazards and Disasters in Afghanistan. In *Natural Resources In Afghanistan: Geographic and Geologic Perspective*; Elsevier: New York, USA, 2014; pp 1–235.

Siebert, S.; Ewert, F. Future Crop Production Threatened by Extreme Heat. *Environ. Res. Lett.* **2014,** *9*, 04100.

Simon, M. The Terrifying Science Behind Locus Plagues of Africa. *WIRED Science* [Online] 2020, pp 1–4. https://www.wired.com/story/the-terrifying-science-behind-the-locust-plagues-of-africa/ (accessed July25, 2021).

Singh, A. How does Weather and Climate Affect the Agriculture Productivity in India [Online] 2017, pp 1–5. https://www.myfarminfo.com/blog/how-does-weather-and-climate-affect-the-agricultural-productivity-in-india/ (accessed July 24, 2021).

Sky Drill Inc. Sky Agriculture-Seed Drills [Online] 2019, pp 1–3. OPICO Limited Lincolnshire: United Kingdom. https://www.sky-agriculture.co.uk/ (accessed July 20th, 2021).

Smith, D. J.; Griffin, D. W. Inadequate Methods and Questionable Conclusions in Atmospheric Life Study. *Proc. Natl Acad. Sci. U. S. A.* **2013,** *110*, E2084..

Smith, D. J.; Griffin, D. W.; Mcpeers, R. B.; Ward, P. B.; Schuerger, A. J. Microbial Survival in the Stratosphere and its Implications for Global Dispersal. *Aerobiologia* **2011,** *27*, 319–332.

Smith, D. J.; Jaffe, D. A.; Birmele, M. N.; Griffin, D. W.; Schuerger, A. C.; Hee, J.; et al., Free Tropospheric Transport of Microorganisms from Asia to North America. *Microb. Ecol.* **2012,** *64*, 973–985.

Smith, D. J.; Ravichander, J. M.; Jain, S.; Griffin, D. W.; Yu, H.; Tan, Q.; Thissen, J.; Lusby, T.; Nicoll, P.; Shedler, S.; Partinez, P.; Osorio, A.; Lechniak, J.; Choi, S.; Sabino, K.; Iverson, K.; Chan, L. Jiang, C.; Mcgrath, J. Airborne Bacteria in Earth's Lower Stratosphere Resemble Taxa Detected in the Troposphere: Results from a New NASA Aircraft Bioaerosol Collector (ABC). *Front. Microbiol.* **2018,** 1–21.

Smith, D. J.; Timonen, H. H.; Jaffe, D. A. Griffin, D. W.; Bimele, M. V.; Pery, P. J.; Ward, P. D.; Roberts, M. E. Intercontinental Dispersal of Bacteria and Archaea by Transpacific Winds. *Appl. Environ. Microbiol.* **2013,** *79*, 1134–1139.

Smith, J. How Plants Adjust Photosynthesis in Response to Fluctuating Light Intensities [Online] 2019, pp 1–6. Ludwig Maximilian University of Munich: Munich, Germany. https://phys.org/news/2016–12-adjust-photosynthesis-response-fluctuating-intensities.html (accessed July 20, 2021).

Soil Survey Staff. *Soil Survey Laboratory Information Manual*, Soil Survey Investigations Report 45 USDA-NRCD United States Department of Agriculture-Natural Resources Conservation Department Lincoln, Nebraska, USA, 1995; p 305.

Soil Survey Staff. *Keys to Soil Taxonomy*, 8th ed.; USDA-NRCS Beltsville: Maryland, USA, 1998; p 326.

Song, H.; Mariño-Pérez, R.; Woller, D. A.; Cigliano, M. M. Evolution, Diversification, and Biogeography of Grasshoppers (Orthoptera: Acrididae). *Insect Syst. Diversity* **2018,** *2*, 1–25..

Spracklen, D. V.; Garcia-Carreras, L. The Impact of Amazonian Deforestation on Amazon Basin Rainfall. *Geophys. Newslett.* **2015,** *42*, 9546–9552.

Spracklen, D. V.; Taylor, C. M.; Arnold, S. R. Observations of Increased Tropical Rainfall Preceded by Air Passage Over Forests. *Nature* **2012,** *489*, 282–285.

Stack, J. *Sorghum Ergot in Northern Great Plains*; University of Nebraska Cooperative Extension EC 00–1879-S 2000; pp 1–7.

Stafford, J. V. *Implementing Precision Agriculture in the 21st Century*; Wageningen Academic Publishers: Wageningen, The Netherlands, 2000; pp 733–737.

Stafford, J. V. *Precision Agriculture 05. Precision Agriculture*; Wageningen Academic Publishers: Wageningen, Netherlands, 2005; pp. 105.

Stahr, K.; Herrmann, L. Origin, Deposition and Composition of Dust and Consequences for Soil and Site Properties with Special Reference to the Semi-Arid Regions of West Africa. In *Wind Erosion in Niger: Implications and Control Measures in a Millet-Based Farming System*; Buerkert, B.; Allison, B. E.; Von Oppen, M., Eds.; Kluwer Academic Publishers: Dordrecht. The Netherlands, 1996; pp 56–65.

Stanford Report. High Altitude Winds: The Greatest Source of Concentrated Energy on Earth [Online] 2009, pp 1–4. http://news.stanford.edu/news/2009/june24/high-altitude-winds-062309.html (accessed July 12, 2021).

Steedman, A. *Locust Handbook*; Natural Resources Institute: Chatham, United Kingdom, 1990; p. 88.

Sterk, G.; Herman, L.; Bationo, A. Windblown Nutrient Transport and Soil Productivity Changes in Southwest Niger. *Land Degrad. Dev.* **1996,** *7*, 325–335.

Stolbovoi, V. Description of Russian Soils [Online] 2020, pp 1–5. National Institute for Snow and Ice: Moscow/Leningad. https://nsidc.org/data/docs/fgdc/ggd601_russia_soil_maps/russian_soil_desc.html/ (accessed Aug 6, 2021).

Stott, P. A.; Stone, D. A.; Allen, M. R. Human Contribution to the European Heat Wave of 2003. *Nature* **2004,** *432*, 610–614.

Sultan, B.; Defrance, D. Izumi, T. Evidence of Crop Production Losses in West Africa due to Historical Global Warming in two crop models. *Sci. Rep.* **2019,** *9* (1), 12834.

Tattaris, M.; Reynolds, M. In *Applications of an Aerial Remote Sensing Platform*, Proceedings of the International Trigo (Wheat) Yield Workshop. Reynolds, M.; Mollero, G.; Mollins, J.; Braun, H., Eds.; International Maize and Wheat Centre, CIMMYT: Mexico, 2015, pp 1–5.

Taub, D. R. Effects of Rising Atmospheric Concentrations of Carbon Dioxide on Plants. *Nat. Educ. Knowledge* **2010,** *3* (10), 21.

Taub, D. R.; Wang, X. Z. Why are Nitrogen Concentrations in Plant Tissues Lower Under Elevated CO_2? A Critical Examination of the Hypotheses. *J. Integrat. Plant Biol.* **2008,** *50*, 1365–1374.

Tetracam Inc. Hawkeye: Autonomous Aerial Imaging [Online] 2018, pp 1–8. http://www.tetracam.com/producthawkeyewindow2.htm (accessed Jul 21, 2021).

Thamm, H. P. SUSI 62 A Robust and Safe Parachute UAV with Long Flight Time and Good Payload International Archives of the Photogrammetry. *Remote Sens. Spat. Inform. Sci.* **2011,** *38*, 19–24.

Thamm, H. P.; Menz, G.; Becker, M.; Kuria, D. N.; Misana, S.; Kohn, D. The use of UAS for Assessing Agricultural Systems in AN Wetland in Tanzania, in the Wet Season, for Sustainable Agriculture and Providing Ground Truth For Terra Sar X Data. ISPRS International Archives of the Photogrammetry, Remote Sensing and Spatial Information Sciences [Online] 2013. XL-1/W2: 10.5194/Isprsarchives-XL-1-W2–401–2013 pp 401–406 (accessed Aug 7, 2021).

Thomas, G.; Jayasena, K. Managing Powdery Mildew in Wheat in Western Australia [Online] 2018, pp 1–7. Department of Primary Industries: Agriculture and Food. Western Australia. https://www.agric.wa.gov.au/spring/managing-powdery-mildew-wheat/ (accessed Jul 18, 2021).

Tiaz, L.; Ziegler, E.; Moller, I. M.; Murphy, A. *Plant Physiology and Development.* Sinauer Associates: Sunderland, Massachusetts, USA, 2014; p. 778.

Tongwane, M. I.; Moeletsi, M. E. A Review of Greenhouse Gas Emissions in the Agricultural Sector of Africa. *Agric. Syst.* **2018,** *166*, 124–134.

Torchio, P. In *Use of Non-Honeybee Species as Pollinators of Crops*, Proceedings of the Entomological Society of Gutario, Vol. 118, No. S, 1987, pp 111–124.

Trabicki, P.; Nancarrow, N.; Bosque-Pérez, N. A.; Rodoni, B.; Aftab, M.; Freeman, A.; Yen, A.; Fitzgerald, G. J. Virus Incidence in Wheat Increases Under Elevated CO_2: A 4-Year Study of Yellow Dwarf Viruses from A Free Air Carbon Dioxide Facility. *Virus Res.* **2017,** *241*, 137–144..

Trabicki, P.; Nancarrow, V.; Cole, E.; Bosque-Perez, N. A.; Constable, F.; Freeman, A. J.; Rodini, A.; Yen, A. L.; Luck, J. E.; Fitzegerald, G. J. Virus Disease in Wheat Predicted to Increase with Climate Change. *Glob. Change Biol.* **2015,** *21*, 3511–3519.

Tselioudis, G. *International Satellite Cloud Climatology Project*; National Aeronautics and Space Agency: Washington D.C. USA, 2019; pp 1–23.

U.S. Environmental Protection Agency. Sources of Greenhouse Gas Emissions [Online] 2011, pp 1–11. U.S. EPA Climate Change Web Page. www.epa.gov/climatechange/ghgemissions/sources.html (accessed Jul 12, 2021).

U.S. Environmental Protection Agency. Overview of Greenhouse Gases. U.S. EPA Web Page [Online] 2014, pp 1–3. www.epa.gov/climatechange/ghgemissions/gases.html (accessed Jul 10, 2021).

UCIPM. Integrated Pest Management [Online] 2020, pp 1–4. Agricultural and Natural Resources College, Davis, University of California, USA. http://ipm.ucanr.edu/pmg/selectnewpest.small-grains.html/ (accessed Jul 28, 2021).

Ugarte, C.; Calderinin, D. F.; Slafer, G. A. Grain Weight and Grain Number Responsiveness to Pre-Anthesis Temperature and Day Length. *Agric. Forest Meteorol.* **2007,** *152*, 44–57.

Ulbrich, U. Climate of the Mediterranean: Dust Storms. In *Climate of the Mediterranean Region*; Lionello, P., Eds.; Elsevier, 2012; pp 301–346.

UNCCD. Global Alarm: Dust and Sandstorm from the World's Dryland. United Nations Convention to Combat Desertification [Online] 2001 p. 343, United Nations Buildings: Bangkok, Thailand. https://www.droughtmanagement.info/literature/unccd_global_alarm_dust_and_sandstorms_2002.pdf (accessed Jul 23, 2021).

UNDESA. International Decade for Action 'Water For Life' 2005–2015 United Nations Department for Economic and Social Affairs [Online] 2020, pp 1–8. https://www.un.org/waterforlifedecade/quality.shtml/ (accessed Jul 4, 2021).

UNEP. Water Program: Vital Water Graphics [Online] 2008a, p. 1. United Nations Environment Program https://www.un.org/waterforlifedecade/quality.shtml/ (accessed Jul 7, 2021).

UNEP. *Water Quality for Ecosystem and Human Health,* 2nd ed.; United Nations Educational, Social and Cultural Organization: Paris, 2008b; pp 1–2.

United Nations Statistical Division. Central Product Classification. Indicative Crop Classification for the Agricultural Census (ICC) Version 1.1 [Online] 2015, p. 239. https://unstats.un.org/unsd/classifications/family/detail/1002/ (accessed July 28, 2021).

United Soybean Board. Farming's Newest Precision Agricultural Tool Takes Data to a New Horizon [Online] 2013. http://www.united soybean.org/article/new-precision-agriculture-could-revolutionize-farming (accessed Jul 26, 2021).

United States Geological Survey. Estimated Use of Water in the US in 2000 [Online] 2004. http://pubs.usgs.gov/circ/2004/circ1268/htdocs/text-ir.htmlexternal (accessed Jul 24, 2021).

USEPA. *Methane: Global Warming Potentials*; United States Department of Environmental Protection Agency: Washington, D.C., 2006; pp. 1–72.

Van Der Velde, M.; Tubeilo, F. N.; Vrieling, A. Bouraouj, S. Impact of Extreme Weather on Wheat and Maize in France: Evaluating Regional Crop Simulations against Observed Data. *Clim. Change* **2012,** *113*, 751–765.

Viana, D. S.; Gangosa, L.; Bouten, W.; Figuerosa, J. Overseas Seed Dispersal by Migratory Birds. *Proc. R. Soc. B. Biol. Sci.* **2016,** *283*, 211–213.

Wade, L. Microbes Survive and May be Thrive High in the Atmosphere [Online] 2013, pp 1–3. https://www.sciencemag.org/news/2013/01/microbes-survive-and-maybe-thrive-high-atmosphere/ (accessed Jul 20, 2021).

Walthall, C. L.; Hatfield, J.; Backlund, P.; Lengnick, L.; Marshall, E.; Walsh, M.; Adkins, S.; et al., Climate Change and Agriculture: Effects and Adaptation. *USDA Technical Bulletin* 1935; USDA: Washington, DC, 2012, pp 1–198.

Weather Online. Haboob [Online] 2020, pp 1–4. https://www.woespana.es/reports/wind/haboob.htm/ (accessed May, 2020).

Wehner, M. Scientists Say Earth's Atmosphere is Raining Viruses and Bacteria Down on the Entire Planet. *J. Int. Soc. Microb. Ecol.* **2018,** 1–3. https://bgr.com/2018/02/06/viruses-in-the-sky-bacteria-atmosphere-earth/ (accessed Jan 22, 2020).

Wheelock, M. J. Insect Pollinators in Corn and Soybean in Agricultural Fields [Online] 2014, pp 1–57. Iowa State University, Ames, Iowa, USA, Digital Repository. Iowa State University Capstones, These and Dissertation. https://lib.dr.iastate.edu/cgi/viewcontent.cgi?article=5201&context=etd (accessed Jul 21, 2021).

Whetter, J. Canola Growth Stalled Under A Shroud of Smoke [Online] 2018, pp 1–4. https://www.country-guide.ca/crops/canola-growth-stalled-under-a-shroud-of-smoke/ (accessed Aug 31, 2021).

Whitchraft, A. K.; Becker-Reshef, I.; Killough, B. D.; Justice, C. O. Meeting Earth Observation Requirements for Global Agricultural Monitoring: An Evaluation of the Revisit Capabilities of Current and Planned Moderate Resolution Optical Earth Observing Missions. *Remote Sens.* **2015a,** *7*, 1482–1503.

Whittaker, R. J.; Jones, S. H. The Role of Frugivorous Bats and Birds in Rebuilding Tropical Forest Ecosystem in Indonesia. *J. Biogeogr.* **1994,** *21*, 245–258.

Wikipedia, List of Unmanned Aerial Vehicles [Online] 2017. https://en.wikipedia.org/wiki/list_of_unmanned_aerial_vehicles/ (accessed Aug 3, 2021).

Wikipedia. Agriculture [Online] 2019a. https://en.wikipedia.org/wiki/agriculture (accessed Jun 2, 2021).

Wikipedia. Sky [Online] 2019b. https://en.wikipedia.org/wiki/sky (accessed Jul 4, 2021).

Wikipedia. Tornadoes in the United States America [Online] 2019c, pp 1–17. https://en.m.wikipedia.org/wiki/tornadoes_in_the_united_states/ (accessed Aug 1, 2021).

Wikipedia. Airborne Wind Turbine [Online] 2020a, pp 1–5. https://en.wikipedia.org/wiki/airborne_wind_turbine/ (accessed Jul 12, 2021).

Wikipedia. Dust Storms [Online] 2020b, pp 1–12. https://en.wikipedia.org/wiki/dust_storm/ (accessed Jul 15, 2021).

Wikipedia. Nelumba Nucifera [Online] 2020c, pp 1–5. https://en.wikipedia.org/wiki/nelumbo_nucifera/ (accessed Jul 28, 2021).

Wikipedia. Zizania Texana [Online] 2020z, pp 1–4. https://en.wikipedia.org/wiki/zizania_texana/ (accessed Jul 13, 2021).

Wilding, L. P.; Smeck, N. E.; Hall, G. F. *Pedogenesis and Soil Taxonomy 1. Concepts and Interactions*; Developments in Soil Science 11A Elsevier: Amsterdam, Netherlands, 1983a; pp 303.

Wilding, L. P.; Smeck., Hall, G. F. *Pedogenesis and Soil Taxonomy. II. The Soil Orders*; Developments in Soil Science 11B Elsevier: Amsterdam, Netherlands, 1983b; pp 410.

Wilson, A. M.; Parmentier, B.; Jetz, W. Systematic Land Cover Bias in Collection 5 MODIS Cloud Mask and Derived Products—A Global Overview. *Remote Sens. Environ.* **2014,** *141*, 149–154.

Witchcraft, A. K.; Vermote, E.; Becker-Reschof, M.; Justice, S. T. Cloud Cover throughout the Agricultural Growing Season: Impacts on Passive Optical Earth Observations. *Remote Sens. Environ.* **2014,** *156*, 438–447.

Witchraft, A. K.; Vermote, E. F.; Becker-Reshef, I.; Justice, C. O. Cloud Cover Throughout the Agricultural Growing Season: Impacts on Passive Optical Earth Observations. *Remote Sens. Environ.* **2015b**, *156*, 438–447.

Wood, S.; Sebastian, K.; Scherr, S. J. *Agroecosystems. A Pilot Analysis of Agroecosystems*; International Food Policy Research Institute: Washington, D.C. USA, 2000; pp 185.

Worland, J. How Drought and Extreme Heat are Killing World's Crops [Online] 2016, pp 1–2. https://time.com/4170029/crop-production-extreme-heat-climate-change/ (accessed Jul 14, 2021).

Wosula, E. N.; Mcmechan, A. P.; Hein, G. L. *J. Econ. Entomol.* **2015,** *108*, 1545–1552.

Xing-Guo, M.; Shu, H. U.; Zhing-Hui, N.; Su-Xia, L.; Xia, J. Impacts of Climate Change on Agricultural Water Resources and Adaptation in the North China. *Adv. Clim. Change Res.* **2017,** *8*, 92–98.

Xu, Z.; Jiang, H.; Sahu, B. B.; Kambakam, S.; Sing, P.; Wang, J.; Wang, Q.; Bhattacharya, M. K.; Dong, L. Humidity Assay for Studying Plant-Pathogen Interactions in Miniature

Controlled Discrete Humidity Environments with Good Throughput. *Biomicrofluidics* **2016,** *10*, 03410.

Yan, H.; Zhang, B.; Zhang, Y.; Chen, X.; Xiong, H.; Matsui, T.; Tian, X. High Temperature Induced Glume Closure Resulted in Lower Fertility in Hybrid Rice Seed Production. *Front. Plant Sci.* **2017,** pp 1–7. https://doi.org/10.3380/fpls.2016.01960.

Yan, W.; Ding, C.; Wang, Y.; Huo, Q. H. The Impact of Relative Humidity, Genotypes and Fertilizer Application Rates on Panicle, Leaf Temperature, Fertility and Seed Setting of Rice. *J. Agric. Sci.* **2010,** *148*, 329–330.

Yang, H. X. Acid Rain and Agriculture. *Forest Ecol. Manag.* **1989,** *126* (3), 321–329.

Yang, S.; He, H.; Lu, S.; Chen, D.; Zhu, J. Quantification of Crop Residue Burning in the Field and Its Influence on Ambient Air Quality in Suqian, China. *Atmo. Environ.* **2008,** *42*, 1961–1969.

Yanmar, Transforming Agriculture from the Sky: What is the Future Created by Remote Sensing and Unmanned Aerial Vehicle [Online] 2020, pp 1–8. Yanmar Inc.: Philippines. https://www.yanmar.com/ph/about/ymedia/a_sustainable_future/remote_sensing_1.html/ (accessed Jul 29, 2021).

Yost, R.; Doumbia, M.; Berthe, A. Systems of Diagnostics for Technology Adoption and Transfer [Online] 2002, pp 26–34. http://www.oired.vt.edu/projects/cueent/projectsynthesis.pdf.

Young, D. R.; Smith, W. K. Effect of Cloud Cover on Photosynthesis and Transpiration in the Understory Species *Arnica Latifolia*. *Ecology* **1983,** *64*, 681–687.

Yue, J.; Lei, T.; Li,. C.; Zhou, J. The Application of Unmanned Arial Vehicle Remote Sensing In Quickly Monitoring Crop Pests. *Intell. Autom. Soft Comput.* **2012,** *18*, 1043–1052.

Yue, X.; Nadine, U. Fire Air Pollution Reduces Global Terrestrial Productivity. *Nat. Commun.* **2018,** *9* (1), 5413.

Yunshan, Y.; Xu, W.; Hou, P.; Liu, G.; Liu, W.; Wang, Y.; Zhao, R.; Li, S. Improving Maize Grain Yield by Matching Maize Growth and Solar Radiation. *Nat. Sci. Rep.* **2019,** *9*, 3635.

Zell, H. Earth's Atmospheric Layers [Online] 2017, pp 1–4. National Aerospace and Space Agency-NASA/Goddard Space Centre: Washington D.C. USA. https://www.nasa.gov/mission_pages/sunearth/science/atmosphere-layers2.html (accessed Jul 21, 2021).

Zhang, J. X.; Fernando, W. G. D.; Xue, A. G. Daily and Seasonal Spore Dispersal by *Mycosphaerella Pinoides* and Development of Mycosphaerella Blight of field Pea. *Can. J. Bot.* **2005,** *83*, 302–310.

Zhang, Q. *Precision Agriculture Technology for Crop Arming*; CRC Press Inc.: Boca Raton, USA, 2015; pp 360.

Zhang, Q.; Pierce, F. J. *Agricultural Automation: Fundamentals and Practices*. CRC Publishers: Boca Raton, Florida, USA, 2013; pp 411.

Zhao, J.; Pu, F.; Li, Y.; Xu, J.; Li, N.; Zhang, Y.; Guo, J.; Pan, Z. Assessing the Combined Effect of Climatic Factor on Spring Wheat Phenophase and Grain Yield in Inner Mangolia, China. *Plosone Earts Future* [Online] 2017, pp 1–25, https://doi.org/10.1371/journal.pone.0185690 (accessed Jul 21, 2021).

Zhu, X.; Troy, T. J. Agriculturally Relevant Climate Extremes and their Trends in the World's Major Growing Regions. *Earth Future* **2018,** *6*, 656–672.

CHAPTER 2

Nutrient Dynamics in the Agricultural Sky: Its Relevance to Crop Productivity

ABSTRACT

Agricultural sky is a large source of nutrient elements essential to crops. It serves global agroecosystems. The elemental composition varies in the atmosphere with geographic location, altitude, season, weather patterns, and diurnal period. It is not easy to provide a standard nutrient distribution profile of the atmospheric layers. Among several elements traced in the atmosphere, we are concerned here with nutrients required for crop's growth and productivity. Nutrient elements are found in particulate, fluid, or gaseous/vapor forms. They are easily absorbed by crop's foliage. The geographic features, soils, crops, and the agroecosystem *per se* may yet support a unique atmosphere with specific nutrients. Nutrient dynamics too could be specific to agrarian sky above each crop belt. We need to characterize them in detail and reap advantages. The agrarian sky may induce or impose restriction on crop productivity via nutrient dynamics. We have to realize that agrarian sky is actually an important portion of global nutrient cycle.

Wind is a major atmospheric factor capable of affecting the nutrient dynamics. Wind affects nutrient distribution in the agrarian sky. Wind brings nutrients into farms via sand particles and organic particulate matter. At the same time, turbulence in wind may remove massive quantities of nutrients from soil surface. They transport soil particles via dust storms and deposit them at a relatively nearer or farther location. Farm belts often suffer loss of nutrients as topsoil. Globally, there are specific dust/nutrient-bearing storms experienced like the *Harmattan* in *Sahel, Haboob, Khamsin, Schekele* in Northeast Africa, or *Sirocco* in Southern Europe etc. Factors, such as dust

The Agricultural Sky: A Concept to Revolutionize Farming. K. R. Krishna, PhD (Author)

storms, tornadoes, droughts, and heat waves generally affect crops detrimentally. They affect the nutrient dynamics in the atmosphere. Strong local wind leads to reduction in fertility, eventually causing lower crop productivity. Wind aids movement of dust particles, microbes, and nutrients across continents. Transcontinental dispersal of sand/dust and nutrients is an important phenomenon related to agricultural sky. It is often induced by the high-altitude wind jet streams. Several giga tons of dust, sand, and nutrients transit across large distances over 15,000 km. Microbes with potential to transform nutrients too move between continents via high-altitude jet streams in the sky.

Clouds, rain, fog frost, snow, and even smog bring in nutrients into soil phase via deposition. Nutrients are found in dissolved form. Ambient temperature can affect nutrient absorption by crops. There are also physico-chemical transformations of nutrient elements that occur in the atmosphere. These are aerobic transformations.

During farming, greenhouse gas (GHG) emission is a major atmosphere-related phenomenon that farmers have to encounter. Gaseous compounds, such as NH_4, NO_2, N_2, CH_4, CO, CO_2, and volatile organic compounds (VOCs) affect atmosphere. Like soils, agricultural sky too may experience deterioration, if emissions are in excess of threshold. It seems in 2005, 10–12% of emissions were caused by anthropogenic activity. Agriculture contributed 60% of anthropogenic emissions. Yet, we may realize that based on geographic features, crops may reap benefits or loss due to GHG emissions/warming. Carbon sequestration procedures that reduce C emissions into atmosphere above cropping belts is worth exploring. Equally so, C sequestration is essential. It seems atmospheric C is increasing at the rate of 2 ppm y^{-1}.

Agricultural sky supports large-scale emissions of sulfur compounds. Sulfur is a pollutant noticed in the atmosphere. It has the potential to induce acid rains. There are other emissions in agrarian sky, such as the VOCs, ozone, acid rains. Crop residue burning induces accumulation of particulate matter in the aerospace. It leads to formation of carbonaceous smog that drifts above cropping expanses.

Phyllosphere is a crucial region on foliage/canopy. It is a region of the agrarian sky. It is comparable to rhizosphere in the soil phase, in terms of moisture and nutrient acquisition. Foliar nutrition of crops takes advantage of agricultural sky and aerobic conditions that prevail. Foliar nutrients received via atmosphere have to compulsorily transit this region—that is, phyllosphere. The current trend to apply fertilizer N as fluid makes phyllosphere important. The nutrient dynamics that ensue in this region could affect crop nutrition. Microbial load, its diversity and relevance to foliar nutrition is worth greater attention. Overall, atmosphere, its characteristics, essential nutrients, and their dynamics in different strata in the agricultural sky need greater attention. We should manage nutrients in the agrarian sky better.

2.1 INTRODUCTION

This chapter is to highlight the fact that atmosphere (agrarian sky) above crops is a huge repository of several nutrient elements required by crops. In this regard, agrarian sky is just like soil or water bodies traced on ground surface. Nutrient elements occur in gaseous or particulate form or as colloids or in dissolved state as in precipitation. The elemental composition of agrarian sky differs markedly based on distance between geographic locations, nature of agricultural terrain, land use pattern, fertilizer/nutrient supply to the crops, absorption of nutrients from the soil/crops, dust storms, and emissions. Nutrient cycling patterns and rates that occur exclusively in the agrarian sky could affect crop productivity. Within limits, we can characterize the atmosphere above the crop fields based on its physicochemical constitution. However, we are interested here on essential nutrients needed for agricultural pursuits. Nutrient distribution in the atmosphere is also influenced by the altitude above the crops. Nutrients that influence with crop's canopy, at low altitudes need greater attention. Also, portion of nutrients impinge on crop's canopy and soil, which is absorbed through aerial parts of crops, such as leaves, phyllosphere. Nutrient loss from the crop's canopy and soil profile into the atmosphere is of great concern. It supposedly occurs via wind and/ or water erosion, leaching, percolation, and greenhouse gas emission to a great extent (Heathcote, 1980; Helms, 1990; Krishna, 2008, 2014, 2015). The GHG emissions affect the cropping belts both by depleting the nutrients from the soil profile and by deteriorating the atmospheric quality. Plus, GHG increases ambient temperature that crops have to negotiate. At the bottom line, nutrient distribution in the atmosphere affects crop growth. We have to bestow equal, if not, greater importance to nutrient dynamics in the atmosphere. Nutrient dynamics takes place exclusively in the atmosphere, also in conjunction with soil, crops, and water phase of the ecosystem. We may note that nutrient dynamics in the agrarian sky is as important as that in soil, or in the physiological parts of the crops or water bodies (irrigation water). We have accrued massive amount of knowledge about several factors that affect the nutrient dynamics in the atmosphere above crops. Yet, we have not pooled the different data and inferences. We have not collated them properly prior to making apt prescription. Also, we do not identify or classify agrarian sky based on nutritional traits or their dynamics.

It is not easy to provide a typical nutrient profile of atmosphere above crops that indicates composition, distribution, and intensity, like we do for soil or even water resources. This is because of freely diffusive gaseous nature, convection processes, local wind, its turbulence, clouds, and precipitation. Despite it, we can easily recognize that nutrient element distribution and other aspects, such as particulate matter and pollutants differ. The GHG emissions may still be specific to certain agrarian regions (Engel et al., 2005; Halvorson et al., 2009, 2011, Halvorson and Del Grosso, 2012; Parton et al., 2011). For example, atmosphere above Great Plains farming regions is distinctly different from those encountered in crop belts of other continents. Similarly, atmosphere above the semi-arid/arid Sahelian belt with its wind, heat waves, dusty air, and storms is different from that encountered in the Equatorial/Tropical Africa, the Indo-Gangetic plains, or European agrarian regions (see Krishna, 2014, 2015). Overall, globally, atmosphere is a key region of any agroecosystem. Nutrient dynamics in the atmosphere needs greater attention than provided now by the agricultural researchers and farmers. Nutrient dynamics in the agrarian sky is really influenced by a wide range of physical, abiotic, biotic, and human factors. Let us try to discuss their details in the following sections.

We should aim at understanding the differences in the nature of agrarian sky in the different farming belts. Then, we have to understand as to what extent crops derive direct/indirect benefits from the agrarian sky. Later, we should learn about nutrient distribution and dynamics above crops in each region. Finally, we should try to develop appropriate prescription to maximize benefits from the agrarian sky and/or to avoid detriments, if any. At the same time, we should still maintain the quality of agrarian sky and its usefulness to farmers. Not an easy task to conduct. We may have to repeat several research steps similar to what we did with soil profile in order to maximize the soil quality and its productivity. The atmosphere above crop extends from ground surface/crop's canopy and till 14–16 km altitude. So, efforts should be taken to understand the nutrient dynamics in the profile above crops. The agrarian sky is a 3D region and variations occur in all three directions, spatially and temporally. Nutrient cycling in the agrarian region is an aspect that has drawn attention during recent times. Agrarian sky is part of the nutrient cycles of major nutrients, such as N, P, K, S, C.

Excessive N inputs have resulted in N fluxes nearing 200 tera grams N per year. Phosphorus accumulation has reached 10–10.5 tera gram per year in farming belts of developed nations (Lavelle et al., 2005). Agrarian sky could be altering the stoichiometry of nutrient transformations. Since many of the agronomic procedures adopted envisage high nutrient inputs into the fields, rapid nutrient cycling in the soil/atmospheric phase is important. In fact, they say, not just N, but P and S too have been accumulating in the ecosystem. We have to adopt procedures that induce rapid conversion and usage/transport of nutrients in the atmosphere (Lavelle et al., 2005).

A review by Hayashi and Yan (2009) suggests about N load in the farm sky. Nitrogen dynamics in the agrarian sky is influenced significantly by emissions, such as NH_3, NO_2, N_2O, N_2, nitric acid, nitrous acid, particulate NH_3. The N load in the farming regions of China, Japan, and other Far-eastern nations seems to have increased enormously. It seems despite rapid N cycling, perhaps, N load has reached its limits in the agrarian sky. It is attributable predominantly to chemical fertilizers. It has reached 3.56 Tg N $year^{-1}$ and the N load of organic manures has reached 2.04 Tg N $year^{-1}$ in China. In Japan, the figures for N load were 0.059 and 0.069 Tg N $year^{-1}$. Overuse of fertilizer N and lack of appropriate cycling seems to affect the ecosystem (FAO 2017; Hayashi and Yan, 2009). In fact, atmospheric depositions of N have been estimated at 20–60 kg N per $year^{-1}$ in these regions.

Firstly, the agrarian sky may bestow various kinds of beneficial and/or detrimental effects on crops. Many of them could be related to nutrient dynamics within the atmosphere. For example, dust bowls, tornadoes, wind, and heavy precipitation events in the Central Great Plains may bring in nutrient-rich soil/dust particles. It adds to nutrient pool of soil. At the same time, these atmospheric processes may induce soil erosion, loss of topsoil fertility, leaching of nutrients out of farming regions, etc. Frost and snow too are atmospheric aspects that affect nutrient dynamics in the wheat fields of temperate regions. Deterioration of atmosphere due to GHG emissions is another consequence attributable to excessive fertilizer N supply. These are intricately related to aerial mechanisms operative above farms. Therefore, crops have to negotiate and adapt to a set of aerospace/weather parameters within the Great Plains of North America (see Krishna, 2014, 2015; Peterson et al., 1996; Undersander et al., 1990). The atmospheric parameters of the agrarian sky in the Great Plains may actually be specific to a location within the vast stretches of crops found in the plains. Farmers in each small pocket of farm belt may need detailed analysis of aerial factors. Also, they should understand their influence on nutrient dynamics and crop productivity. For example, farmers in the dry, semi-arid regions of Texas, who grow cotton/sorghum will experience a distinct set of factors operative in the agrarian sky compared with Central or Northern plains. The agrarian sky of each subregion needs appropriate characterization. So that farmers take correct measures and accurate prescription (Heathcote, 1980; Peterson et al., 1996). *So, even within subregions of plains, analyzing the aerial factors operative above crop's canopy is a necessity.*

The same set of crop species (e.g., wheat, barley, brassicas, etc.) may have to negotiate a separate set of parameters of agrarian sky within individual regions, such as the Northeast China or European plains. The intensity with which a particular atmospheric factor affects the nutrient dynamics and crop productivity differs based on each agrarian region. In European agrarian region, it is precipitation, snow, frost, fogs, smogs, and storms that bring in nutrient-rich deposits. They affect nutrient dynamics in the farms. This is in addition to nutrient dynamics that operate in the soil profile. Emissions due to excessive soil nutrients (e.g., fertilizer N) may occur if fertilizer inputs exceed. In the case of Indo-Gangetic plains, atmospheric N depositions, precipitation, storms, etc. may add to nutrients. Nutrient dynamics may alter if incessant cropping and high fertilizer inputs occur. Several nutrient dynamics-related effects of atmosphere (agrarian sky) need greater attention, in future. Even neighboring farms may experience differential effects of nutrients.

We need to understand that crop species, agronomic procedures adopted, and fertilizer inputs in turn affect the agrarian sky. For example, paddy is cultivated in the tropical Asia using flooded fields. Paddy is also grown as upland dry crop in certain areas. The nutrient dynamics that ensue in the flooded anaerobic soils emit large quantities of CH_4, NH_3, and NO_2. The drier paddy fields with aerophilic conditions may emit less of GHG. We have to devise cropping systems that help in reducing the GHG. Higher fertilizer N inputs to flooded fields can induce soils/crops to emit NH_4, NO_2, and N_2O in proportionately larger quantities. This may affect the N dynamics in soil (Lagomarsino et al., 2016). In addition, microbial flora in the puddled paddy soils may alter as we flood and drain the fields. The microbial flora varies from anaerobic to aerobic species. Microbial transformations in soil and in the atmosphere fluctuate. Consequently, emissions from a paddy belt may fluctuate.

Factors, such as microbes in the atmosphere too need to be investigated in detail in different farming regions. Microbes relevant to dynamics of N, P, S need attention. Microbes may float through sandstorms or wind jet streams.

They carry the microbial inoculum to short distances say few km or for long distances like different continents (Acousta-Martinez et al., 2015). Aerial transport of microbes is important, particularly, when we consider microbes relevant to biological nitrogen fixation (BNF). The extent of N inputs via BNF organisms could be quantified. Also, natural aerial dispersal of BNF microbes, mycorrhizal propagules, and others (pathogens) may be important to note. Long stretches of fields could get inoculated with symbiotic microbes via atmospheric processes. Such aerial dispersal may be playing a vital role in supporting crop production. Overall, characterizing the agrarian sky and its influence on nutrient dynamics, in the ecosystem, including the aerospace, soil, water, and crops is important.

In the following sections, we discuss several of the atmospheric factors that influence the nutrient distribution and dynamics in the agroecosystem and their influence on crop productivity.

2.2 WIND IN THE AGRICULTURAL SKY: ITS INFLUENCE ON SOIL AND NUTRIENT DYNAMICS IN THE AGRARIAN REGIONS

The main purpose here is to discuss about wind as an atmospheric factor that affects soils, their structure, and nutrient dynamics in the agricultural field. Wind brings in nutrients when it blows sand particles, dust, and other organic debris into a field. It removes nutrients when turbulence is higher. Surface soils usually loose soil fertility rapidly due to gusts of wind. The nutrient carrying capacity of soil particles and organic dust portion are important. Nutrient-rich sand particles will rapidly deplete the surface fertility of soil.

The wind-driven emission, transport, and deposition of sand/dust are known as aeolian processes. The word "aeolian" is derived from Greek god's name *Aeolus*, who is the keeper of the winds. Aeolian processes occur wherever there is a supply of granular material. Also, there should be atmospheric winds of sufficient strength to move them. On Earth, this occurs mainly in deserts, on beaches, and in other sparsely vegetated areas, such as dry lake beds (Kok et al., 2012). The drift of sand/silt and dust due to wind in these regions affects the topography, vegetation, and water resources. Wind aids the formation of sand dunes and ripples. Wind induces the erosion of rocks, soil formation, and transport of soil particles. Moreover, airborne dust particles can be transported thousands of kilometers from their source region thereby affecting nutrient dynamics in the atmosphere and in soil/crop phase. It affects the ecosystem productivity. The hydrological cycle and various other components of the earth system too get affected (Acosta-Martinez et al., 2015; Kok et al., 2012; Romheld et al., 2000; Sterk et al., 1996, 1998; Stoorvogel and Smalling, 2000).

Wind gets generated in the agrarian sky due to differential pressures in the atmosphere. Wind could be light, slow in speed, and transit for only short distance above crops. Wind could appear in gusts, cause turbulence, and disturb surface soil and its nutrients. Wind could cover long distance above the agrarian terrain when the speed is higher and occurs at an altitude above crops. Wind may occur close to crop's canopy. Here, it is usually slower because of obstacles caused by the topography, vegetation, and natural features of earth. Wind speeds are uniform and faster as we move to higher altitudes into atmosphere. There are "jet streams" that occur at high altitudes in the atmosphere. They are of significance to agricultural crop production in different ways. Wind disperses microbes of utility such as the nitrogen fixing symbionts. High-altitude winds may disperse seeds/propagules of vegetation and cultivated crop species. Wind also aids intercontinental dispersal of microbes, dust, and nutrients attached to them. Wind disperses microbial species that are dreaded disease-causing pathogens on crops (e.g., rust, blotches, blasts, mosaics etc.) (Acosta-Martinez et al., 2015).

Wind is a major atmospheric factor that affects nutrient dynamics in any agroecosystem. Wind aids movement of clouds with nutrients held in them. It has potential to transport large quantities of nutrients via drift of surface soil into air. Sand, dust, particulate matter, and water/clouds also get transported via wind. Wind causes detriments, such as erosion, loss of topsoil and its fertility due to storms and tornadoes. Wind shifts nutrients from above the fields by displacing atmospheric gases and particles. Wind may bring in errors during foliar fertilization. Wind causes drifts of foliar sprays of water and fertigated nutrients.

Wind erosion is not rampant in all food production zones. It affects only a few of them. There are also matching agronomic methods, such as adopting cover crops, mulching, ridging, ploughing, and leaving clods in the field, wind breaks. Such methods reduce wind caused detriments. Wind and its impact on various food grain generating expanses on earth has been given due attention. Particularly, in the dryer, arid and sandy areas of the earth. For example, in Sahel, wind (e.g., Harmattan, Haboob) moves copious quantities of surface soil, crop debris, dust particles, moisture droplets, etc. from one region to other. It moves the soil/sand plus nutrients attached with it for long distances of several thousand kilometers. Such wind occurs at low altitudes within Sahel and it induces movement of sand/nutrients only for short distances. The intensity of wind, its turbulence, quantity of sand particles moved, and nutrient carrying capacity of the sand particles are principal factors to consider. Fields may lose topsoil plus fertility from one location due to wind. However, in regions where the wind-borne sand/dust storms settle, the same wind is a factor that improves soil fertility. Often, Sahelian farmers are advised to adopt surface mulching with crop residue. Crop residue traps the sand particles but will not allow for it to raise into atmosphere again and move away. The mulch helps in conserving the sand plus nutrients (see Beilders et al., 2002a, 2002b; Romheld et al., 2000; Sterk et al., 1996, 1998). Such wind-mediated erosion of soil and organic debris occurs in different farming belts.

The Sahelian wind, dust storms (harmattan) and heavy precipitations are important modes of transfer of soil/dust. They also transport microbes from one location to another (Acousta-Martinez, 2015; Bielders et al., 2002a, 2002b, Romheld et al., 2000; Wilke et al., 1984). The transport of nutrients occurs from short distances, after which they get deposited at a second location/farm. *The reports about increase in nutrients, such as N, P, K and micronutrient due to dust/sandstorms are actually a difference between inputs and loss of nutrients in a single location.* Mulching traps the nutrients at a single place. It means dust/sandstorms do not pick the soil from the second location. Also, in each location, the data pertain to nutrient dynamics, such as input/loss ratio of nutrient-rich dust particles. Long distance transport of dust/sand enriched with nutrients is also reported. The Sahelian nutrients get transported to several thousands of kilometers through dust/sandstorms and high-altitude wind jets streams (Krishna, 2015).

Wind erosion induces land degradation in the semi-arid tropics. It has been estimated that in the arid semi-arid zones of the world, 24% of cultivated land and 41% of pastures are affected by wind erosion. Within these climatic zones, the Sahel is considered to suffer from particularly acute wind erosion problems (UNEP, 2016).

2.3 DUST STORMS, DROUGHTS, AND HEAT WAVES IN THE AGRARIAN SKY INFLUENCE NUTRIENT DYNAMICS IN THE ECOSYSTEM

Globally, dust storms are well spread out in different continents. They last for different lengths of the period. They vary in their intensity, volume, and nutrient carrying capacity. Conspicuous dust storms have been identified.

Many of them are traced in North Africa. A few examples of dust storms are Bodele storms of Central Africa, dust storms in Mali and Mauritania, in Southern Oman/Arabia/Saudi Arabian Border, of Eastern Sahara in Libya, of Makran coast in Southwest Asia, Taklamakan Tarim basin, Etosha in South-eastern Namibia, Lake Eyre basin, Mkgadikgadi basin in Botswana, Salar de Uyuni in Bolivia, Great basin in North America, Dust bowls in USA, etc. (Goudie and Middleton, 2006). Some of these storms are called dust cycles that occur each year and affect aeolian dust content and nutrients in the agrarian sky (Shao et al., 2011). Youlin et al. (2001) have reviewed in detail the physical, chemical, and atmospheric processes that generate dusts above different agrarian regions. Their periodicity, nature of dust particles, and devastation they cause too have been delineated.

Sand and dust storms (SDS) occur when unchecked, strong or turbulent wind combines and blows over exposed loose soil and dry field surfaces. These conditions are common in semi-arid regions (UNEP, 2016, Youlin et al., 2001). Sandstorms are lower atmospheric events. They result from wind erosion. It involves liberating sediment particles from the ground surface. Dust storms occur relatively close to the ground surface. However, finer dust particles may get lifted a few kilometers high into the atmosphere. At high altitude, strong winds transport dust to long distances. Dust storms may travel thousands of kilometers from source areas, passing over land and ocean, and depositing materials far from their source areas (UNEP, 2016). The source of dust storms in Sahara is large at 260–710 million tons per year. On a global basis, atmospheric dust generation may range from 500 to 5000 million tons per year depending on the region (Goudie and Middleton, 2006). Since dust particles carry nutrients, they affect the nutrient dynamics proportionately within the region.

A compilation by Goudie and Middleton (2006) states that the dust/nutrient-bearing winds are given names. Such names depend on the local dialects in vogue in the vast agrarian regions. They are "*Irifi*" in Western Sahara, "*Ghibli*" (meaning "wind from Mecca") in Tripoli, Libya, *Guebli* ("Southern wind") in Tunisia and Algeria, *Harmattan* (to blow) in West African Sahel, *Brume seche* in West Africa, *Haboob* ("to blow") in Sudan, *Khamsin* in Egypt, *Schekele* in Tunisia, *Dchani* in Southern Sahel, "*Gobar*" in Ethiopia, "*Sirocco*" in Southern Europe, "*Levanta*" in Canary Island, etc. Each of these dust storms has potential to alter nutrient dynamics in the atmosphere and agrarian soils. They also affect the agrarian sky. The cropping belt through which the storm moves may receive deposits of particles. Saharan/Sahelian "*haboobs*" are potentially a large source of atmospheric dust from the African continent. They are usually produced by convective systems that develop into storms of 1–1.5 km height and tens or hundreds of kilometers in length. They propagate westwards from Sudan, particularly during the summer monsoon season (Roberts and Knippertz, 2012).

The Sahara/Sahelian region is perhaps the largest source of global dust storms. The aeolian dust storms that transport large quantities of nutrient-rich particles are important. They supposedly induce nutrient cycles in the atmosphere. They supposedly regulate sparse natural vegetation and subsistence farming belts of Sahel. Sahelian dust is an atmospheric phenomenon. It has large consequences to North Africa. It also affects North America. The dust trajectories show that aeolian dust from Sahel also reaches several locations in Mediterranean and European regions. The Saharan/Sahelian dust basically transports large quantities of dust. Such dust that is embedded/laden with nutrients moves from one location to another. This phenomenon maintains the natural nutrient cycles in the arid regions. Goudie and Middleton (2006) state that Sahara produces aeolian soil dust more than any other desert. Saharan dust has important impact on climatic processes, nutrient cycles, soil formation, and sediment cycles. These influences the spread of dust far beyond Africa thanks to the great distances over which Saharan dust is transported. The African dust storms have two major sources, namely, the *Harmattan* dust storm originating due to "Bodele depression" and the other regions covering Western Mali, Mauritania, and southern Algeria. They are the major sources of dust particles to Atlantic Ocean and beyond into Caribbean and North America. Of course, as stated above, there are other dust storms known to occur in different continents.

In addition to nutrient dynamics, such dust storms anywhere on earth affects the landscape, soil, its structure, fertility, nutrient distribution in the profile, ultimately, the vegetation/crops and their productivity. The agricultural sky under the spell of dust storms can be devastating on crops, at times. The hot sand particles can destroy the just germinated seedling. Hot sand affects leaf growth and productivity. Dust storms often submerge seedlings under hot sand and result in the loss of crop stand. Some such effects of dust finally affect the nutrient dynamics in the soil profile. When such phenomenon occurs in conjunction with scarce water, low precipitation trends, and low soil fertility, it can reduce crop productivity to its lowest levels. No doubt, the Sahelian agrarian sky is a potential source of dangerous dusts, droughts, and rainstorms that destroy crops. We must identify and classify the "Dust storm prone Sahelian sky." It helps farmers to adopt suitable agronomic processes that suit the subsistence farming trends. In practical farming, agronomic procedures, their frequencies, and intensity must match the potential devastation that Sahelian sky bestows.

The Sahelian sky supports transport of dust plus nutrients from North Africa to the Eastern Mediterranean. It occurs often during spring. It is associated with the eastward passage of frontal low-pressure systems. Dust from sources in the Middle East typically get transported to the Mediterranean in the autumn (Dulac et al., 1987; Ganor and Mamane, 1982; Herut and Krom, 1996).

Dust storms commonly traced in different continents affect the ambient temperature. Dust storms influence atmospheric cooling or warming, reception of solar radiation, interception of photosynthetic radiation, nutrients in the surface soil/water bodies, phytoplankton in the water bodies, nitrogen fixation trends in the atmosphere and soil, and microbial diversity (Alpert et al., 1998; Tegen and Fung, 1994, 1995). Dust storms can immensely influence nutrient transformations (Goudie and Middleton, 2006). Some of these aspects are dependent on particle size, its density, and distance to which they travel from the point of generation of storms. Oceans, such as North Atlantic, Mediterranean Sea, Black Sea, South China sea receive large quantities of dust enriched with nutrients. Of course, adjacent agrarian regions too get nutrient-rich dust. It is an important earth process that maintains natural nutrient cycles. For example, Saharan dust aerosols influence the nutrient dynamics and biogeochemical cycling of both terrestrial and oceanic ecosystems (e.g., in Atlantic Ocean).

The atmosphere above crops supports the transport of dusts/nutrients, gaseous elements, and moisture for short or long distance. High-altitude dust storms in Sahara are examples for long distance dispersal of dust particles, nutrients, and moisture. Low wind speeds that occur at low altitude usually support short distance dispersal of dust/nutrient-rich particles. Both processes are important in terms of nutrient cycling in the ecosystem. Natural vegetation may depend immensely on such phenomenon. At the same time, we may note that hanging dust storms, slow moving dust masses, and haze for longer period affect the cropping duration. It does reduce the photosynthetic radiation that impinges the crops. Carbon fixation in the crop's canopy gets reduced, proportionately. Dust deposition may also affect gas exchange and the normal activity of stomata. Phyllosphere microbes and their activity too get influenced by the deposition of dusts on the canopy of crops.

Most often, satellite imagery is utilized to first trace the dust storms. But beyond that the dust, haze, and water vapor if any obscures the imagery of ground surface and standing crops. Still, we can understand the effects of dust on biomass accumulation, by periodically studying the NDVI (Normalized difference vegetation index). This way, we can compute the influence of dust storms, haze, tornadoes, and heavy precipitation events on nutrient dynamics in the fields. The agrarian sky could also be efficiently utilized by adopting drones that fly at low altitude. Drones should be sturdy enough to withstand atmospheric turbulence and offer imagery of dust storms/tornadoes and crops (Bentley et al., 2001).

Microbes traced in the atmosphere above crops, including on the surface of canopy, leaf, and branches are versatile, particularly, with regard to their ability to utilize nutrients available on dust particles, crop's canopy, leaf, and branches. Every dust storm, precipitation event, or tornadoes potentially introduces large population of diverse aerial microbes. Microbes that are lifted from the soil surface get deposited at a different location or float in the aerospace for longer duration. Behzad et al. (2018) state that dust and sandstorms inject substantially large population of microbes into agroecosystems. Microbes survive harsher conditions in the atmosphere. Also, they conduct wide range of nutrient transformation and energy generating biochemical reactions. Sometimes, such dust storms add precious inoculum of beneficial microbes capable of fixing N from the atmosphere. They may also be adding fungi that help in P dynamics in soil/root interface. A large number of species of microbes capable of biochemical transformations of organic particles too are added into new region. Behzad et al. (2018) and Kellogg (2004) further point out that microbes that could disturb the normal nutrient dynamics in crop's canopy and soil too may get transported and deposited via the aerial dust plumes.

2.4 CLOUDS, PRECIPITATION, FOG, FROST, SNOW, AND SMOG INFLUENCE NUTRIENT DYNAMICS IN THE AGRARIAN SKY

Clouds that occur intermittently or even for longer durations at a stretch does reduce photosynthetic radiation that impinges the crop's canopy. Crops respond to fluctuating cloud cover and intensity. The photosynthetic activity alters based on crop, season, daytime, cloud cover, etc. Slattery et al. (2019) has stated that clouds can substantially reduce the incident light. Their intermittency introduces intensity changes. Atmospheric aerosols are a complex and mixture of solid and liquid particles. Aerosols can interact with the earth's radiation budget and climate. Further, they state that carbon fixation and its dynamics in the canopy (above ground) is also influenced by the crop species. The morphological features, such as foliage, leaf area, leaf angle, and shade effects of the crop species are important. So, carbon dynamics is intricately linked to cloud cover, aerosols, dust, fog, and smogs that hang above the cropping regions. Indirectly, as photosynthetic efficiency and evapotranspiration pull reduces, consequently, the nutrient acquisition into the crop's canopy from soil too decreases.

Frost is natural aerial phenomenon. It affects the physiological activity and nutrient dynamics in the cropped fields and plantations. Frost influences several crop species. Frost damage of various types occur based on crop species. For example, frost that occurs on forest plantations, fruit trees, and field crops affect foliage formation. It retards the normal photosynthetic activity and gas exchange. In the cereal belts, frost affects seedling survival and foliage formation to a great extent. Therefore, inbuilt genetic tolerance to frost damage is necessary for the cultivars grown by farmers. This way, farmers may overcome frost-related detriments imposed by the agricultural sky. For example, in the case of wheat, in Australia (Victoria and South Australia), researchers report that up to 33 million A$ are lost due to frost damage of wheat crop in the winter season. Frost events that occur close to flowering stage can cause floret sterility. Obviously, it significantly reduces the crop's yield. Subsequent frosts during grain filling can result in downgrading of grain quality. There are innumerable examples of frost-mediated alterations to normal crop growth. Frost affects carbon fixation, nutrient acquisition, flowering, and grain/fruit productivity. Frost damage is traced in regions that experience extreme cold, frost/snow during the cropping season (Eglinton, 2013).

Frost which is an aerial phenomenon varies with regard to the extent of damage it causes. Frost-induced damage varies based on geographic location, crop species, plantations, etc. For example, in Spain, during the past few years, a certain province has experienced frost that has wiped out almonds, totally. While in others damage has been reported as only 70% (Schmidt and Schmidt, 2018). Several factors operative in the agricultural sky may actually influence the severity of frost effects on crops. Agricultural sky could be characterized as source for severe, moderate, or low frost damage.

Frost is a weather-related phenomenon. Here, the agrarian sky can play havoc with crop survival and productivity. In agrarian regions, frost damage begins as ambient temperature falls below zero. Frost damage can be severe when aerial temperatures above standing crops or plantation fruit trees reach below −7°C. Physiologically, frost damage appears as temperatures cross below the critical limits for normal functioning of the tissue. Frost damage to crops does not result from cold temperature but mainly due to extracellular (i.e., not inside the cells) ice formation. It draws out water and dehydrates the cells. This process leads to injury to the cells (De Mello-Abreu, 2018; De Mello-Abreu et al., 2016). Root growth, nutrient absorption, leaf functions, buds, fruit growth, and maturity may get retarded. Biomass accumulation will be zero if frost damages the leaf photosynthetic tissue. Both, nutrient acquisition from the soil and carbon fixation from the atmosphere will get retarded. Respiratory activity too could be low due to frosty conditions.

There are passive and active methods of regulating frost damage during crop season. Most of these methods are protective. A few examples are site selection, managing cold air drainage, plant selection, pruning, plant covers and screens, soil operations for heat conductivity enhancement, planting date for annual crops, bacteria control, chemicals. Long-standing active methods of regulating frost damage include heaters, sprinklers, foggers, wind machines, helicopters, foams, and combination methods (Garcia et al., 2010; Snyder and De Mello-Abreu, 2005).

Often, plastic covers are used to warm the soil and increase protection. Transparent plastic warms the soil more than black plastic. Wetting the soil before applying the plastic further improves effectiveness. Sometimes vegetative mulches are used during dormancy of tree crops. This is to help prevent damage to roots, due to freezing and soil heaving.

Field crops, vegetables, fiber crops, cash crops, and fruit plantation and forest tree species differ with regard to their physiological responses to frosts that occur in the ambient atmosphere (Agricultural Department, 2019; FAO, 2014). Critical temperature values of most crops that encounter frosts during their growth are known. Blackberries and blueberries are hardier than grapes especially when buds are in the dormant phase. Kiwifruit and strawberries are less hardy than grapes. Citrus plantations are sensitive to frost. In case of vegetables such as tomato, it is said that even in frost-free areas, winter production of tomatoes should not be attempted where mean temperatures frequently drop below the critical margin of 12°C. Frost damage and suppression of nutrient acquisition depends on stage at which crops get exposed to severe frosts. Therefore, timing of sowing is important. It should be done, so that, frost period is avoided. Overall, freezing temperatures in combination with plant growth timing determine frost damage (Eccel et al., 2009; Papagiannaki et al., 2014; Rigby and Porporato, 2008). Warmer climate restores normal nutrient acquisition and carbon fixation trends. Frosts do affect nutrient acquisition and primary productivity of pastures and grassland (Bjerke et al., 2015). It seems, sowing legume that are frost-sensitive with oilseed crops that tolerate the frosts helps in restoring growth rate, biomass formation, and nitrogen use efficiency. Clearly, intercropping avoids detrimental effects of frost (Cadoux et al., 2015).

Foliar spray of irrigation water is said to reduce frost effects. Energy consumption of sprinklers is about 10% of that used in frost protection with heaters. The operational costs are low compared with heaters. Also, the labor requirement is less than that required for other methods. It is relatively nonpolluting. The main disadvantages are high installation cost and large amounts of water needed. In many instances, limited water availability restricts the use of sprinklers. In other cases, excessive use can lead to soil waterlogging, which could cause root problems as well as inhibit cultivation and other management. Nutrient leaching (mainly of nitrogen) is a problem where sprinkler use is frequent.

The Canadian prairies supports production of wheat, barley, lentils, brassicas, vegetable, and fruits such as apples, strawberries. These crops may often undergo frost damage. The severely low cold temperatures such as –5°C affects the crop's root formation, its functions, such as nutrient absorption, foliage development.. In the case of strawberries, the buds formed may tolerate up to –5°C, bud petals and full bloomed flowers are susceptible to even –2.2°C or –1°C. Irrigating soils using sprinklers or drip tapes prior to the onset of frost seems to restore higher temperatures. Crops may absorb normal levels of nutrients and perform photosynthesis. Frost basically affects the physiological activity of crops and nutrient accumulation in the foliage and fruits.

In the wheat belt of Australia, crops are prone to suffer damage due to cold temperature and regular frost. "Frost tolerance" or "frost escape" both may have to be adopted for proper wheat production. The degree of tolerance shown by a plant to freezing varies. It depends largely on the stage of development at which the stress occurs.

Wheat seedlings of 8–10 weeks are capable of tolerating cold/frost damage in the field. Wheat seedlings often become dormant during severe winter. Nutrient absorption and biomass accumulation is restricted by the severity of cold temperature. Spring wheat, it seems, lacks this ability of overwintering through dormancy. Yet, they may survive low temperatures of –10°C during the crop period. Often, wheat seedlings get hardened during early season. This helps them to absorb nutrients and grain formation. Frost has generally affected nutrient absorption due to dormancy. Sometimes the loss due to severe frost has reached 30–50% of normal grain formation levels (Agricultural Department, 2020). Wheat, maize, soybean, brassicas, and vegetables are crops prone to suffer damage due to frost, snow, and wind (blizzards). It is known that n USA, at 5% damage due to frost, farmers may lose 158 million bushels of maize grains. At 10% frost damage, over 316 million bushels of maize grains are lost. This is because crops tend to overwinter. Root elongation and nutrient uptake lessen. Fertilizer use efficiency gets reduced markedly due to extended frost periods (Good, 2019). Basically, cold temperature reduces nutrient removal from the soil. It reduces carbon fixation in the canopy. In the North China wheat belt, winter wheat is known to suffer frost damage. Frost affects root elongation, nutrient acquisition, and several yield components. Thus, frost affects total grain harvests (Wu et al., 2013).

2.5 GREENHOUSE GAS EMISSIONS INTO AGRARIAN SKY

It is easy to suggest that keeping agricultural sky clean and devoid of undue greenhouse gas emissions is a necessity for global farming. It is much easier to state that each farmer should maintain the farm sky (atmosphere) above their cropped fields with least emissions. *Although, legally, no farmer ever owned the aerial space above his farm.* There is no legislation in any country that lists space above crops in his field as owned by farmer. Releasing GHGs, pollutants, dust gusts, and toxic wastes into other's farms or even his own is prohibited. Trespassing into neighboring plot's airspace (i.e., agricultural sky) via drone is illegal. Unmanned aerial vehicles entering aerospace of other farmer's fields is prohibited. Yet, we may realize that crop residue burning, fossil fuel burning generate methane, CO_2, and other GHGs. They do drift into aerospace. Rules for farmers, nations, and policy makers with jurisdiction of larger agrarian regions seems hazy. Agricultural sky deteriorates due to emissions. The detriment continues at various rates of intensity depending on the characteristics of farm location. Agrarian sky that deteriorates could have devastating consequences on farmer's exchequer. *Like soils, agricultural sky may also deteriorate irreversibly.*

The GHG emissions are part of nutrient cycling that occurs in nature. They are monitored in regions with natural vegetation, agrarian regions, forest plantations, and even industrial and urban zones. Major nutrients essential to crops, such as carbon (as CO_2, CO, CH_4), nitrogen (as NH_3, NO_2, N_2O, N_2), sulfur as SO_2 are cycled between soil, crops, atmosphere, water bodies, and other natural features. Our focus here is to know the various transformations, accumulation, leaching, emissions, and percolation within the agroecosystem. Agrarian sky plays host to several of these natural phenomena during crop production. A few of the transformations are congenial to crop production. However, others could be detrimental. Foremost, we must be able to state which sky is apt and which is not for farming a particular crop species. *Selecting correct type of sky is perhaps as important as soils.*

The vast agricultural sky absorbs and acts as repository of several GHGs. Agricultural activity induces release of significant amounts of CO_2, CH_4, and N_2O into the atmosphere (Cole et al., 1997; IPCC, 2001; Massey et al., 2017; Paustian et al., 2004; Smith et al., 2007). The CO_2 is released largely from microbial decay, putrefaction, and other transformations. Burning of plant litter releases CO_2 into atmosphere (Smith, 2004; Smith et al., 2007; Janzen, 2004). Methane (CH_4) is produced when organic material decomposes under anaerobic conditions. Ruminant livestock (cattle) and rice grown under flooded conditions are important emitters of CH_4 (Mosier et al., 1998). Nitrous oxide (N_2O) is generated by microbial transformation of nitrogen in soils and manures. Nitrogen emissions become conspicuous when soil N exceeds crop's requirements (Oenema et al., 2005; Smith and Conen, 2004). Now, in the following paragraphs, we shall consider few important aspects about GHG emissions and the nature of agrarian sky that gets affected due to it.

2.5.1 METHANE (CH_4) AND CARBON DIOXIDE (CO_2) EMISSIONS INTO AGRARIAN SKY

Agricultural activity accounted for an estimated emission of 5.1–6.1 $GtCO_2$-eq/yr., in 2005. It amounts to 10–12% of total global anthropogenic emissions of greenhouse gases (GHGs). Methane contributes 3.3 $GtCO_2$-eq/yr. and N_2O 2.8 $GtCO_2$-eq/yr. Of the global anthropogenic emissions in 2005, agriculture accounted for about 60% of N_2O and about 50% of CH_4. Globally, agricultural CH_4 and N_2O emissions have increased by nearly 17% between 1990 and 2005. It is equivalent to an average annual emission increase of about 60 $MtCO_2$-eq/yr. There are options for mitigation of GHG emissions in agriculture. The cumulative GHG mitigation potential from agriculture (excluding fossil fuel offsets from biomass) by 2030 is estimated to be ~5500–6,000 $MtCO_2$-eq/yr.

Methane is an important anthropogenic gas. It affects the atmospheric energy balance also called "radiative forcing." Methane is a greenhouse gas generated frequently from agrarian regions. Its radiative potential is much greater than CO_2. It is 26 times greater than CO_2. Indirectly, methane emissions induce formation of ozone. So, it may pollute the atmosphere above crops. Methane emissions are induced by anaerobic processes caused by microbes in soil, flooded rice fields and in cattle shed. Methane is also generated when coal gas is burnt. Estimates indicate that 32–44% of methane emission could be attributed to farming activity. Emissions from flooded rice culture in the southeast Asia itself contributes about 11% of global methane emissions into atmosphere.

Methane emitted into air reacts chemically with OH radicals in the atmosphere. Methane also reacts with chloride substances. These processes reduce methane formation and emission. More important is the fact that methane reacts with OH group compounds generating ozone. Ozone is a pollutant. It suppresses crop growth. It affects CO_2 fixation rates. Methane is also removed into soil air space (micropores). It could reach 5% of that generated. They ratio of methanogenic versus methanotrophic microbes in soil decides the extent of methane generated and emitted into the atmosphere. Methane is not a cumulative factor in the sky. It breaks down in the atmosphere. But emission of CO_2 is considered as a long-term pollutant because it accumulates in the atmosphere. A few recent reports about methane emission from cattle rearing regions in Europe suggest that CH_4 emissions are actually greater than previously estimated. Cattle rearing accounts for 11% of the total methane emission in the area. So, it means we are spoiling the agrarian sky to an extent little more than previously estimated. Methods to mitigate should be increased proportionately (Agence France Presse, 2019).

Ruminant livestock (cattle, sheep, goat) and rice production are important CH_4 and CO_2 emitters. These are greenhouse gases with perceptible effects on the agrarian sky. Global warming potential (GWP) of CH_4 is relatively high. These gaseous emissions could be considered a process of polluting the agricultural sky. Isotopic studies have shown that agriculture is among key reasons for recent increases in methane emission into troposphere. Both expansion and intensification of rice production zones seem to have led to enhanced methane emission.

Extensive worldwide research during the last decade has shown that the global emissions from rice fields are more likely to be in the range of 30–80Tg/yr. China dominated the estimates of methane emissions from rice fields. This is because, it has vast expanses of flooded rice culture. It is the largest producer of rice. We may note that major increases in rice production had taken place in that country over the last several decades (Khalil et al., 1998, 2001).

In the Southeast Asia, tropical rice cultivation is the major methane emitter. It is supposedly 45 times more than several of the previous estimates (Gabbatiss, 2018). A study during 2000–2016 by Zhang et al. (2016) estimates that CH_4 emissions from global rice fields ranged from 18.3 ± 0.1 Tg CH_4/yr under intermittent irrigation. It was 38.8 ± 1.0 Tg CH_4/yr under continuous flooding during 2000–2016. Methane emissions from rice fields were dependent on flooding. It seems during past 110 years, simulated results showed that global CH_4 emissions from rice cultivation increased by 85%. *The expansion of rice fields were the dominant factor for the increasing trends of CH_4 emissions.* Rice production also induced elevated CO_2 concentration, and nitrogen fertilizer use (Khalil et al., 1998, 2001; Zhang et al., 2016).

In Indonesia, highest CH_4 emissions were obtained at the active tillering phase (35 DAP) of rice crop. It was attributed to flooded conditions. Methane is the main cause of global warming and climate change. Organic materials which easily decompose are the main sources of CH_4 production in paddy fields (Kurniawati et al., 2018). Tillage systems adopted may affect several soil physicochemical processes. A few of them may induce methane emissions greater than others. For example, generally, no tillage systems that reduce soil disturbance emit lower levels of methane. Comparisons of tillage and cropping systems adopted in the Chinese southwest has shown that conventional tillage is a major generator of methane from flooded paddy fields. About 77% of methane is emitted during flooded rice phase. Only negligible quantities are emitted during the succeeding arable rape cultivation season. Conventional tillage of turning soils and deep tillage prior to rice transplantation and flooding cause maximum efflux of methane (2.96 ± 0.04; mg × (m^2 h) (−1). No tillage with ridge planting during rape phase emitted least methane (1.83 ± 0.21 mg (m^2 × h) (−1) (Wang et al., 2017; Zhang et al., 2019). There are indeed several natural and cultivation-related factors that affect CH_4 emission from rice fields (Jia and Cai, 2003; Khalil et al., 1998, 2001, 2004; Wang et al., 2012). Lauren et al. (1994) have sown that green manuring practices during flooded rice cultivation influences CH_4 emissions. Further, interactions of temperature and redox potential of soil too influenced methane generation from rice fields.

In USA, Linquist et al. (2018) have conducted detailed studies on the influence of agronomic procedures on CH_4 and NO_2 emissions from rice fields. Water and residue management practices, such as alternate wetting and drying, stand establishment method (water vs. dry seeding), and the amount of residue from the previous crop had the largest effect on growing season CH_4 emissions. However, soil texture, sulfate additions, and cultivar selection also affected growing season CH_4 emissions. We should be able to regulate and reduce CH_4 emissions. It could be achieved just by choosing the correct agronomic procedures related to flooding and draining of the paddy fields. Emissions into agrarian sky could be kept at minimum. In USA, agriculture as such emits 9% of the total GHG emissions. Forestry and land use change is a net carbon sink. It accounts for 11.1% carbon fixation into soil ecosystem (Oertal et al., 2016).

In the above paragraphs, we have listed numerous factors that affect methane emissions. We can reduce CH_4 emissions by carefully selecting the farming procedures. Suppression of methane emissions from rice fields is important. Several measures, such as mulching with different industrial wastes, crop residues, and even inert material from diverse sources have been examined. For example, Ito (2020) reports that slag from industries, such as steel, cement, or iron ore was effective in reducing methane emission into agrarian sky.

Major countries emitting >10000 t CO_2 are China, USA, India, European Union countries, and Russia. In some agrarian regions, carbon emissions as CO or CO_2 have not reduced to the expected levels. For example, in Australia, calculations of CH_4, CO_2, and CO emissions from agricultural zones have increased to 583 m t year^{-1}. Estimations of carbon fixation due to reforestation, no-tillage, and reduction in crop burning have still not been effective in reaching the targeted decrease in C emissions (Cox, 2018; Zhang et al., 2003, 2013).

2.5.1.1 *CARBON SEQUESTRATION AND AGRARIAN SKY*

The agricultural sky is an important region that affects the global carbon cycle. The agrarian sky transfers large quantities of carbon from the atmosphere into vegetation. Photosynthesis is the key natural process that induces C sequestration into natural vegetation, forest plantations, field crops, etc. A portion of atmospheric C is also deposited on soil surface via precipitation, dust storms, etc. Carbon sequestration processes are essential to replace C compounds lost into the atmosphere. It is an important portion of global carbon cycle.

It seems, during 2019, carbon measured in the Great Plains around Iowa's wheat/maize belt reached a peak at 415 ppm. Therefore, priority should be given to reduce such emissions. Agricultural researchers are propagating the idea of farming practices that sequester atmospheric carbon back to vegetation. The aim is to sequester about one trillion-ton carbon from atmosphere back to forest plantations, crops, and shrub vegetation. This may involve preferential selection of crop species, no-tillage, stubble incorporation, and cover cropping practices (Reiley, 2019). Such a project has been named "Terraton Initiative." Essentially, we are manipulating global cycle to avoid deterioration of agrarian sky. At the same time, undue emission of organic C from soil substrate also gets subsided. Basic requirement here is to develop methods that allow regular monitoring of agricultural sky for carbon compounds. Then, adopt suitable remedies.

Lal (2008) mentions that C sequestration is essential; particularly, if the intention is to reduce ill effects of C emissions into atmosphere. It is an important aspect of global C cycle. The C sequestration rates may fluctuate but it is important to be conversant about it. Further, he states that, sequestering of atmospheric CO_2 is necessitated because of its present concentration of 385 ppm. It is increasing at the rate of 2 ppm y^{-1}. In comparison with the emission of 300 Pg C between 1850 and 2000, total emission of C during the 21st century is estimated at 950 to 2195 Pg. The annual rate of emission is 20–35 Pg C y^{-1}. We can reduce atmospheric C by sequestering it or by reducing the emission.

In some regions, carbon sink has increased. However, it is still not sufficient to mitigate industrial emissions. Terrestrial vegetation that includes agrarian crops, it seems, absorbs about 29% of carbon emissions that reach the atmosphere. This is accomplished via photosynthetic carbon fixation. Certain reports state that plants also take in atmospheric C as carbonyl sulfide. So, the quality of farm sky is dependent on emissions versus the fixation due to forests, shrubs, and field crops on the globe.

During the past 2 decades, from 1995 till 2010, there seems to be a decrease in GHG emission. It has been attributable to adoption of agronomic procedures that reduce emissions. Crop residue incorporation, avoiding crop burning and adoption of no-tillage has led to better C dynamics in soils. The capture and long-term storage of atmospheric carbon dioxide (CO_2) in soil is important. It is called carbon sequestration. Currently, C sequestration is the best solution to counterbalance the increase of greenhouse gases. Such measures include enhancing biomass production, application of low-cost plant growth regulators and bio-fertilizers, agricultural conservation practices (no till), nitrogen fixation by leguminous crops, reduced pesticide use, crop rotation, and crop-livestock mixed-farming.

2.5.2 EMISSION OF NITROGENOUS GASES INTO AGRARIAN SKY

The dynamics of N compounds in the agroecosystems is of great concern to agriculturists. Nitrogen is among the important essential plant nutrients. It is consumed by crops in higher quantities compared with all other elements. We may note here that dynamics of N compounds is stringently interlinked between the atmosphere (sky), crop biomass, and soil phase of any farming region. However, within the context of this volume, we try to emphasis the nature, physicochemical transformations and transfer of N compounds in the atmosphere. Knowledge about aerial concentrations of N compounds is important. Also, emissions of soil N into atmosphere are of relevance to agrarian environment and crop productivity. During practical farming, nitrogen traced in the soil profile is biologically or chemically transformed. Major N transformations are brought about through mineralization, immobilization in soil organic matter, nitrification, and denitrification. The basic physicochemical functions of soil get affected when the N inputs to the soils exceeds the N output. Removal of N by the crop, tree, or vegetation falls short. An excess input of N to soil affects N transformation processes. It affects soil organic matter content. High N input as fertilizers could decrease soil biodiversity. Much of the N surplus on the soil surface (on average about 50%–60% of the N surplus) is lost as the harmless gas N_2, followed by NH_3 emission, NO_3 leaching, and emissions of N_2O, and NOx.

During the past six decades, the concentration of greenhouse gases in the atmosphere has increased. As stated earlier, important greenhouse gases emitted are CO_2, CH_4, and N_2O. Most of the N_2O emissions take place from soils and are related to agricultural activities. Global N_2O emissions have reached about 17.7 Tg of N per year. About 6.7 Tg (37.8%) is from human activity. Agricultural soils share 2.8 Tg of N per year, that is, 15.3% of the total amount of emissions, or 41.8% of anthropic emissions (Denman et al., 2007). According to the second Brazilian GHG inventory, N_2O emissions were about 576 Gg in 2005, that is, 7.7% of the national emissions. It has reached 2,200 Tg of CO_2-eq (Brazil, 2010; Jantalia, 2008). The agriculture sector was responsible for about 87.2% of N_2O emissions (Cerri et al., 2009). Agricultural soils are among the major emitters of N_2O into atmosphere. Nitrous oxide is a greenhouse gas with global warming potential much greater than CO_2. Nitrous oxide emissions from soil are rampant because of fertilizer N supply that could be greater than that utilized by the crop. Soil could accumulate fertilizer N. This source of N is vulnerable for emission into atmosphere as NO_2, N_2O, NH_3, and N_2. We can assess N emissions using several different methods (Ma, 2010; Signor et al., 2013).

A few agricultural/environmental specialists are of the view that we have already surpassed the limits for NH_3, N_2O, and NO_2 emissions. Their accumulation due to fertilizer inputs has now induced definite pollution effects. For example, regions like Indo-Gangetic Plains are already under the grip of N pollution.

In this section, we discuss the aspects related specifically to emission of nitrogenous compounds, such as NH_3, N_2O, NO_2, and N_2. Emission of NH_3 into agrarian sky is traced frequently in the flooded rice agroecosystem and in regions supporting anaerobic conditions. Here, microbial transformations of N compounds release NH_3. Rice growing regions are the major emitters of NH_3. Similarly, cattle rearing too releases large quantities of NH_3. The N_2O formation in the soil occurs through nitrification and denitrification processes. These processes are influenced by soil moisture, temperature, oxygen concentration, amount of available organic carbon, nitrogen, and soil C/N ratio. Among these factors, those related to soil could be easily altered by management practices. Therefore, understanding the processes of N_2O formation in soils and the factors influencing these emissions is fundamental to developing efficient strategies to reduce N_2O emissions in agricultural soils.

Venterea et al. (2012) point out that small N_2O losses have large GHG impacts. The primary challenge for N_2O mitigation is directly associated with its high global warming potential. Its global warming potential is ~300 times greater than that of carbon dioxide (CO_2) and 12 times greater than that of methane (Forster et al., 2007). Small N_2O losses have large GHG impacts.

It seems deep placement of urea during rice cultivation is a good idea. The alternate wet and drying phases during rice cultivation affects N_2O and NO_2 emissions. Islam et al. (2018) state that deep placement urea with irrigation improves N-use efficiency and grain yield. The conflict between accelerating demand for agricultural crops and the ecological impact of their production is most evident with respect to nitrous oxide (N_2O) emissions. Such emissions are linked to applications of nitrogen (N) fertilizers and manures to soil (Davidson, 2009). Several naturally occurring soil processes can convert N into N_2O gas. which then rapidly diffuses into the atmosphere. Globally, consumption of synthetic N fertilizers is expected to increase by ~2% per year (FAO, 2017). Consequently, fertilizer derived N_2O emissions are expected to increase by those same percentages.

Nitrous oxide (N_2O) is often the largest single component of the greenhouse gas budget of individual cropping systems. Here, we highlight the factors that help in mitigating N_2O emissions from fertilized agroecosystems. Mitigating nitrous oxide emission is a difficult challenge. Therefore, we require innovative technologies and fresh ideas. Modification of the rate, source, placement, and/or timing of nitrogen fertilizer application has been an effective way to reduce N_2O emissions. However, the efficacy of existing approaches to reduce N_2O emissions, while still maintaining crop yields across locations is uncertain. It is because of the interaction of multiple factors that regulate several different N_2O-producing processes in soil. Although these processes have been studied well, our understanding of key aspects plus our ability to manage them to mitigate N_2O emissions remain limited (Signor et al., 2013).

Ammonia emissions into agrarian sky needs special attention whenever farmers apply Urea or NH_3 liquor/gaseous dissolved in water to fields. Ammonia is among the major nitrogenous gases found in the atmosphere. It is an alkaline gas of great importance to biotic component of the aerospace of a cropping ecosystem. Agricultural activity, particularly, fertilizer N supply into soils is the major source of NH_3 gas into the atmosphere. NH_3 emissions can be significant if fertilizer N supply to soils is more than sufficient for crop production (Bahera et al., 2013). Other sources of NH_3 are the industrial activity, transport vehicles, and volatilization of /emissions from burning of crop trash. Data accrued during recent years clearly suggests that NH_3 emissions from soil/crops into agricultural sky is increasing. Ammonia emission into agrarian sky can affect the quality of atmosphere. It has potential to enhance ambient temperature. It adds to global warming potential. Most agricultural/environmental researchers caution us that excessive emission of NH_3 will affect crop productivity. There is still a large gap in our knowledge about various forms of N emissions, their extent, sources that add to NH_3 and physicochemical behavior of NH_3. Ammonia may interact with various other gaseous compounds, deposits, and dust to cause smog (Bahera et al., 2013). Smog in turn affects carbon fixation into crops.

There are many agrarian regions on the earth wherein farmers tend to supply larger quantities of fertilizer N. The idea here is that crops will always have sufficient soil N reserve to absorb. Soil N deficiencies may not occur during the entire season if fertilizer supply is held high. However, soil N that accumulates is vulnerable. It could be lost as emissions. Nitrogen cycle in the specific field or region is actually primed with higher levels of N element. For example, in the USA, fertilizer N supplied is utilized for food grain, fuel, and fiber generation (Philip et al., 2018). The excess could be traced in the nearby irrigation channels, water ways, and ground water. A large quantity of soil N is also emitted to atmosphere as NH_3, if NH_4 or urea-based fertilizer formulations are utilized. Soil N gets transformed into N_2O and NO_2 due to nitrification. Particularly, when crops are grown in dry arable condition. In wetlands, like rice fields, NH_3 gets emitted. Such emission, no doubt, affects the environmental quality. Animal farms are among chief emitters of NH_3. The effluents from dairy does affect the drainage channels with NH_3. It results in the contamination of irrigation water. Farmers who supply extraneous N as urea or NH_4 salts actually induce an interaction between soils, fertilizer N and atmospheric quality. Farmers who desire not to spoil the agrarian sky may have to adopt precision farming. The aim should be to adopt correct dosages and timing of fertilizer N. This will ensure that there is no extra soil N left for emissions (Jiang, 2010; Ruser, 2006).

No doubt, minimizing emission of soil N as NH_3 is important. Several factors influence the extent of NH_3 emission farm agricultural fields. It is important to obtain a prior knowledge about loss of N from fields. It helps in supplying accurately the exact quantities of fertilizer N and in time when plants need the nutrient most during growth phase. Nitrification inhibitors are among useful adjuvants to fertilizer N. Such nitrification inhibitors (e.g., Urease inhibitors, NBPT – N-(n-butyl) thio-phosphoric tri-amide-coated urea, DCD, dicyandiamide-coated urea, etc.) thwart undue conversion of soil N to NH_3. It avoids the formation of easily volatilizable NH_3. Therefore, it reduces the loss of N from soil phase to agrarian sky. Consequently, it reduces greenhouse gas effect, and it helps in maintaining the atmospheric quality. There are indeed several reviews and research papers depicting the usefulness of nitrification inhibitors, especially in improving fertilizer N use efficiency, reducing NH_3 emissions, and avoiding loss of N to ambient atmosphere. (See , Brady, 1975; Krishna 2003; Liu et al., 2017).

Ammonia is emitted during various stages of the animal production and animal manure management. Emissions may occur from (i) animal feed (silage productions), (ii) animal manure excreted in housing systems and in pastures, (iii) animal manure in storage systems, and (iv) from animal manure applied to crop land. In addition, ammonia emissions may occur from fertilizer application (especially from urea- and ammonium-based nitrogen fertilizers). We can reduce emissions of NH_3 (Brady, 1975). It depends on the sources of NH_3 emissions. Deposition of NH_3 can cause acidification of soils (Van Breemen et al., 1983). It leads to eutrophication of natural aquatic and terrestrial ecosystems. This results in the loss of biodiversity. In addition,

NH_3 is a precursor for secondary particulate matter in the atmosphere. It has adverse effects on human health (Erisman, 1993; Erisman and Schaap, 2004; Moldanová et al., 2011; Van Damme et al., 2014). NH_3 is also a source of nitrous oxide (N_2O) which is a potent greenhouse gas. Soil injection of NH_3 (liquid) is a common practice. Reductions in emission of NH_3 from field soil is possible. Methods, such as slurry injection reduces NH_3 emission by 70–90% of control plots. Shallow soil injection reduces NH_3 emission by 60–80%, dilution reduces NH_3 emission by 30–40% of original. Solid manure incorporation reduces NH_3 emission by 30–80% depending on soil type and timing of N inputs (Sangomla, 2017; Scils, 2008).

As stated earlier, globally, NH_3 emissions are of greater importance among N gaseous compounds that are traced in the atmosphere. Of course, agricultural activities, such as application of fertilizer NH_3 gas dissolved in water, application of urea or nitrate foliar spray and soil injections liberate NH_3. Similarly, rice fields kept under flooded anaerobic conditions, wetlands, and arable soils supplied with higher quantities of fertilizer N also contaminate the atmosphere above agrarian regions with gaseous NH_3. Crop residue burning in the open fields and fossil fuel too generate NH_3 that contaminates agrarian sky (Krupa, 2003). Excessive NH_3 gas in the air above crop's canopy is deleterious. It induces leaf burn symptoms. Crop's foliage is always vulnerable to damage by extra NH_3 if farmers adopt foliar sprays of fertilizer N. Atmospheric NH_3 gets deposited on foliage in the form of fine particles. Usually, NH_3 that emanates from a field sprayed with NH_3 gas dissolved in water or urea solution may liberate NH_3 that drifts. It may drift for 3–4 km from the epicenter, that is, field that has been fertilized with N compounds. European farmers utilize foliar spray with great accuracy. Yet, they too find that NH_3 applied drifts in the agricultural sky to a distance of 5 km. Most higher plants, including the crops, such as wheat, maize, and vegetable are susceptible to higher concentrations of NH_3 in the air and also to deposits that reach crop's canopy/foliage. Sometimes, SO_2 in the atmosphere too may have its effects on crop's canopy simultaneously, particularly, if sulfur-containing fertilizers and plant protection chemicals are used in the aerial sprays (Krupa, 2003). They reported that in addition to foliage degeneration, NH_3 may have detriments on the atmospheric microbes and their activity. It may affect the symbiotic N fixation, photosynthetic and respiratory activity of leaves, etc. NH_3 toxicity is caused when plant's innate physiological ability to detoxify it is less. In such cases, the rates of detoxification of NH_3 will be slower. Rapid detoxification using aerial sprays of water is possible. There are state regulations to the use of aerial sprays of

N compounds. Depositions of N compounds on crops can be periodically monitored. Critical levels in European farms are said to be 5–10 kg N deposition per year. This includes both wet and dry deposition of atmospheric N compounds on the crop's foliage. Farm advisors often adopt concepts, such as N saturation and critical N level in air. This helps in avoiding excessive N depositions.

2.5.3 SULFUR COMPOUNDS IN THE AEROSPACE ABOVE AGRARIAN REGIONS

Knowledge about sulfur cycle in the agrarian region is important. Sulfur emissions have been progressively reduced in Europe and North America. However, sulfur emissions have not yet subsided in the emerging industrial areas of the world, such as China, India, South Africa, and the southern parts of South America. A global assessment of acid deposition threats suggests that tropical ecosystems are at high risk (Lavelle et al., 2005)

Sulfur dioxide is among important pollutants of the agrarian sky. Its dynamics in the agrarian ecosystem has far reaching effects on the quality of crop produce, water resources, and atmosphere. A major portion of sulfur emissions occur via industrial activity. Burning coal is the single largest man-made source of sulfur dioxide. Coal burning accounts for 50% of annual global emissions. Whereas oil burning accounts for a further 25–30% (Husar et al., 1978).

Sulfur dioxide occurs both in the troposphere and stratosphere in quantities that it can affect microbes, crops, insects, animals, and the general atmospheric chemistry. Sulfur dioxide that emanates in gaseous form may react with moisture to form H_2SO_4. It forms condensation nuclei for aerosols and clouds and acidifies the rain. This results in the pollution of agrarian sky. The half-life of SO_2 in troposphere is only few days. But in stratosphere, it is several weeks.

Currently, agrarian sky might be holding more of S as gas and particulate matter compared with yester years. Farms supporting both crops and domestic animals are among emitters of sulfur compounds. Sulfur is emitted from manure in the form of hydrogen sulfide (H_2S). The H_2S released gets converted in the atmosphere to sulfur dioxide. Hydrogen sulfide thereby has the same effects as sulfur dioxide. Such H_2S gets emanated from manure slurry evaporation in cattle sheds. However, major sources of sulfur in the atmosphere are mostly oil, coal, and natural gas. Sulfur emission into the atmosphere also occurs via internal combustion engines and power plants.

With the development of proton-transfer reaction mass spectrometry, it is possible to monitor emissions of sulfur compounds from livestock. Sulfur gets combined with other substances in the atmosphere. It forms particles that are hazardous to crop's health.

In natural ecosystems, which includes the atmosphere above crops, sulfur undergoes a few types of transformations. They are mineralization of organic sulfur compounds/dust particles, etc. leads to inorganic forms like H_2S. The oxidation of sulfide to elemental S and its emanation from industries, conversion of S to sulfate-containing compounds, reduction of sulfates to H_2S are other transformations. Microbial flora is involved intricately with sulfur cycle in the atmosphere. Microbial immobilization is common when sulfur containing particulate matter is richer in the air. Assimilative sulfate reduction induces the formation of sulfate. Sulfate is reduced to organic sulfhydryl compounds. Dissimilative sulfur reduction too occurs in the atmosphere. It results in the formation of H_2S. This H_2S gets emitted into atmosphere. Major S resources are S-containing ores and soil. It occurs in gypsum, pyrites, in coal, etc. Oceans, freshwater bodies too contain S in organic form. In the atmosphere, it is traced as methane sulfonic acid. It is released as H_2S and as particulate matter when crop residues are burnt.

Sulfur cycle and its reaction rates are important for crop production. If not, this element has to be replenished as S-containing fertilizers (gypsum). Sulfur dioxide may react with moisture rapidly in the clouds and cause acid rains. Sulfur may also stay in the smogs that occur above he agrarian regions. As stated earlier, atmosphere above crops receives large quantities of S via emissions from soil, crop burning, and from industrial activity. About 70 Tg S per year reaches atmosphere in the form of sulfur dioxide. It comes from fossil fuel combustion and industry. About 2.8 Tg S emanates from wildfires and 8 Tg S per year from volcanoes.

Sulfur affects crops by retarding the photosynthetic activity. Therefore, carbon fixation rate into foliage gets reduced if SO_2 pollution is rampant. Emissions of SO_2 may also affect the microbial flora, insects, and avian activity in the agrarian aerospace. Sulfur dioxide deposits that occur via clouds, precipitation or particulate matter affects the pH (acidity) of water bodies (Varshney et al., 2009). The quality of grains and fruits are utterly depleted if SO_2 deposits are severe. The SO_2 dynamics in the atmosphere and in soil are often intricately connected. Hence, SO_2 should be monitored periodically both in aerospace and soil/water resources. As stated earlier, metabolic activity of crops, respiration, and photosynthesis are altered due to SO_2 deposits on foliage (Linzon, 1979). It seems lichens and bryophytes (like mosses) are sensitive to SO_2 deposits. These organisms could be utilized by the farmers to ascertain SO_2 contamination of aerospace (Parmar, 2016).

During past decades, crop's requirements for S element were generally supplied from atmospheric deposition. Sulfur applications were done along with other nutrients using formulations, such as S-coated urea, S-coated phosphatic fertilizers. However, during recent years, S is to be applied to reach a balanced fertilization of soils. Sulfur dynamics both in soil and atmosphere seems to be heading for a change.

Sulfur inputs to farming regions have increased during the recent past. For example, oilseeds need higher amounts of fertilizer-S for S accumulation in seed. The N:S ratio preferred is 5:1. For cereals N:S ratio is 10: 1, for potatoes it is N:S ratio of 10:1 and for pastures N:S ratio is 12:1 in soil. Usually, oilseeds such as brassicas are supplied 75 kg S ha compared with only 20–25 kg S ha^{-1} for cereals. The S-fertilizer is applied to achieve correct ratios of N versus S in the soil. This is because soil-S depletion has been significant. Obviously, researchers and farmers seem to be alert to S dynamics in the soil/atmosphere. This is to ensure optimum yield of crops.

2.5.4 EMISSIONS OF VOLATILE ORGANIC COMPOUNDS (VOCS) INTO AGRICULTURAL SKY

The study of VOC emissions from plants is of great interest. It has ramifications into nutrient dynamics of the agroecosystem. The diversity of vegetation, its seasonal variations, and phenological stage affects several processes in the forest/farmland. These manifestations have direct influence on the VOCs. The gaseous composition of VOCs emitted varies depending on the phenology of vegetation. For example, foliar emissions of isoprenoids, methane, and monoterpene vary depending on the botanical diversity of the vegetation. Methanol was the major VOC emitted when crop leaves were young compared with older leaves (Bracho-Nunez et al., 2011).

Terrestrial vegetation is a source of atmospheric volatile organic compounds (VOCs) (Guenther et al., 2006). The VOC emission has significant influence on the physicochemical processes of the atmosphere (Atkinson and Arey, 2003). The degradation of VOCs influences the oxidative capacity of the atmosphere (Lelieveld, 2008; Lelieveld et al., 2002; Penueles and Staudi, 2009).

Non-methane volatile organic compounds (NMVOC) get released from cropping zones to varying extents. For example, in China, wheat farming generates about 0.34 Tg NMVOCs. If we consider similar levels of NMVOC emission for other crops cultivated in China, it amounts a total of 2.33 g NMVOCs per year (Kudo et al., 2014). Photochemical oxidation of NMVOCs plays an important role in the formation of secondary organic aerosol (SOA) as well as photochemical oxidants like tropospheric ozone (O_3) (Atkinson, 2000; Jenkin and Clemitshaw, 2000). Both SOA and O_3 affect air quality and climate on regional and global scales. Reports since past decade have clearly shown that photochemical oxidation of NMVOCs induces the formation of secondary organic aerosols (SOA) and ozone (O_3). Both SOA and O_3 reduce air quality above crops canopy. They affect agricultural sky at small local scales and global scales if left unattended. Actually, substantial amounts of VOCs are generated via biomass burning during farming (Christian et al., 2003, 2004; Yokelson, 2009, Yokelson et al., 2011). Recently, the formation of SOA from oxidation of OVOCs (Oxygenated VOCs) has been examined extensively. Chan et al. (2010) reported that SOA was produced in relatively high quantity through oxidation of unsaturated aldehydes (e.g., methacrolein) under high-NO_2 conditions. However, there are still many reports that state CO and VOCs could get transported in the agrarian sky, from one location to other for long distances (Friedli et al., 2001; Greenberg et al., 2006; Hecobian, 2011).

2.5.5 PARTICULATE MATTER EMITTED INTO AGRARIAN SKY

Emissions of particulate matter in the agricultural expanses originate from at least two different sources. First one is the suspension of already existing particles because of wind, activity of animals, or farm vehicles. These factors lift particles into the atmosphere. Particles remain in the air for longer period. It depends on the size of the particles. Such airborne material gets dominated by larger size particles. The second source of particulate matter is the combustion of farm waste or cropland burning. Particles may be formed from the gas phase during this waste burning. Particulate matter of 2.5 μm and smaller also get released to the atmosphere (Amann, 2012).

2.6 ATMOSPHERIC POLLUTION AFFECTS CROP PRODUCTIVITY

Agricultural sky is among the major regions where GHGs gets collected. The extent of GHG emissions vary markedly. The emission of GHGs may vary only marginally based on several reasons related to crop production methods. Pollution caused via atmospheric sources is clearly a phenomenon related to agricultural sky. Here, the agricultural sky plays a detrimental role by accumulating agents that are suppressive to crops. Pollutants reduce crop quality. Pollutants can be dangerous to the entire ecosystem based on the extent of toxicity and ill effects on biotic aspects, such as crops, animals, and human population. Pollution of atmosphere above cropping zones can be severe depending on the location. Pollution of agrarian sky has its immediate effect on crops, their respiration, photosynthetic activity, nutrient absorption, and productivity (Joshi and Swami, 2009; Seyyednejad et al., 2011). Polluting agents are no doubt well distributed in most of food crop generating systems of the world. Emission of gases, such as sulfur dioxide, nitrogen oxides, CO_2, CO, CH_4, VOCs, ozone, hydrogen fluoride, peroxy-acyl nitrates, and particulate matter may all induce deleterious effects on crops (Noormets et al., 2001). Clearly, the agricultural sky needs to be monitored periodically for pollutants above each field, farm, a small agrarian region, or even large agroecosystems. It has certain advantages. A farm characterized with pollutants above crops can be treated specifically for prevention or remedial methods. Farms marked as "agricultural say prone to pollutants" or "pollutant affected" could be segregated or even abandoned for some time. Pollutants usually affect the leaf morphology, their functions such as respiration and photosynthesis. Pollutants may affect normal pigmentation of foliage. Pollutants could affect fruit/grain quality. A few gaseous pollutants can reduce grain yield, bring about discoloration and reduce the market value. Pollutants that penetrate the plant's tissue can affect normal biochemical/physiological activity, such as sugar, protein, fat, and secondary metabolism. The agrarian sky, here, literally penetrates the crops to bring about devastation. In fact, while initiating farming, prospective farmers can be informed of the nature of agrarian sky in the location, particularly about pollution, its manifestation during different seasons, etc.

The main air pollutants are represented by gaseous forms, particles in suspension, different ionizing radiation, and noise. The gaseous pollutants possess carbon (CO_2, CO, CH_4), nitrogen (NO_2, NO, N_2O_4, NH_3, NH_4^+), SO_2, O_3, C_6H_6 vapors, Hg, volatile phenols, Cl_2, etc. The particulate forms are: PM10 μm and PM 2.5 μm particulate matter, heavy metals with toxic effect (Pb, Ni, Cd, As), polycyclic aromatic hydrocarbons PAHs, etc. Atmospheric pollutants have a negative effect on the plants. They can have direct toxic effects. Indirectly, they have toxic effect by changing the soil pH followed by solubilization of toxic salts of metals like aluminum (Georghhe and Ion, 2011; Mittal et al., 2009). The particulate matters have a negative effect. They cover the leaf blade reducing light penetration and blocking the opening of stomata. These impediments strongly influence the process of photosynthesis, the rate of which declines sharply. Also, the leaves of the trees have an important role in retention of the particulate matters. They are mostly affected when the wet and dry atmospheric depositions increase. The vegetation plays an important positive role in atmospheric purification and air pollutants reduction.

Reports suggest that 0.02 mg Pb kg^{-1} dry matter is derived from the soil by crops. The remainder is usually derived from the atmosphere above the crops. Therefore, soil supplies only a portion of lead (Pb), but rest is from atmosphere. It is received in the form of particulate matter/dusts (Mosbeak et al., 2014). Periodic monitoring of lead (Pb) in the agrarian sky is possible. It may be necessary in regions with industries that release lead containing fumes into the sky.

2.6.1 ACID RAIN POLLUTES AGRARIAN SKY

Air pollution has been a severe problem for the forests. Acid deposition is major problem. Commonly, it is called the "acid rain." This phenomenon occurs when emissions of sulfur dioxide (SO_2) and oxides of nitrogen (NO_x) react in the atmosphere with water, oxygen, and oxidants. It leads to the formation of various acidic compounds. These compounds then fall to the earth in either dry form (such as gas and particles) or wet form (such as rain, snow, and fog). Thus, polluted air can damage trees directly in the dry form or indirectly through its effects on the chemistry of water and soils further by making trees more vulnerable to other biological and environmental stressors. Specifically, acid rain weakens trees by damaging their leaves, limiting the nutrients available to them, or exposing them to toxic substances slowly released from the soil. Acid rain that flows into streams, lakes, and marshes also has serious ecological effects. In watersheds where soils do not have a buffering capacity, acid rain releases aluminum. Overall, acid rain is an aerial phenomenon. It materializes in the atmosphere above agrarian expanses. It requires careful monitoring. Adoption of remedial measures is important. Its ill effects can be severe on forest stands, plantation fruits, and field crops. Sulfur dioxide reacts on the surface of a variety of airborne solid particles (aerosols). It is soluble in water and can be oxidized within airborne water droplets, producing sulfuric acid. This acidic pollution can be transported by wind over many hundreds of kilometers. Later, it gets deposited as acid rain (Erisman, 1993; Erisman and Schaap, 2004; Van Damme et al., 2014).

2.6.2 PHYLLOSPHERE AND EMISSIONS

Microbes on the foliage and canopy seem to have a role in the emission of GHG, including VOC. Microbes may be able to regulate GHG emissions. The metabolic activity of microbes in the phyllosphere needs detailed investigation. Brengel and Couce (2015) make it clear that phyllosphere microbes situated at the leaf surface influences emissions, such as GHG and VOCs. Leaching that occurs from leaves could play a vital role in the nutrient dynamics within a cropped field. Microbial flora on the phyllosphere along with physicochemical transformations occurring on the leaf surface actually represents a continuum with regard to nutrient dynamics. Phyllosphere region connects the plant with atmospheric transformations. Further, they state that VOC emissions could be altered or absorbed preferentially right at the phyllosphere by the abundant microflora found in the region. Microbial communities on or around plants play a key role in plant's physiological functions, nutrient absorption, and crop vigour. During the process of regulating nutrient acquisition, emissions, and interactions with atmospheric gases, phyllosphere microbes may encounter uncongenial atmosphere, such as nutrient dearth, low water content on leaf surface, pollutants, low stomatal activity, and gaseous exchange and emissions. A few could be beneficial to crop nutrition but others detrimental. Overall, microbial activity and its relevance to nutrient dynamics on the phyllosphere, particularly GHGs including VOCs is worth investigating further. Investigations aimed at understanding phyllosphere conditions and emissions may help us to maximize crop response to foliar nutrition.

2.6.3 CROP RESIDUE BURNING AND NUTRIENT DYNAMICS

Fire is a disturbance on the global agricultural landscape. Several million hectares of vegetation are burnt every year. Land and forest fires are collectively referred to as wildland fires. They occur annually on every continent, except Antarctica. Wildland fires have many serious negative impacts. They include GHG emissions and impairment of normal nutrient dynamics observed in agrarian regions. Wildfires drift rapidly into vast regions on land. The air pollution could drift through the wind currents to vast areas. Wildfires transgress into vast areas (de Groot et al., 2006, 2007)

Global fire data are available online. The forest, natural vegetation, and crop-related fire/emissions data are collected and presented based on regions, on 3 hourlies to seasonal intervals or yearly intervals (Randerson et al.,

2015). Data about emissions are available for carbon (C), dry matter (DM), carbon dioxide (CO_2), carbon monoxide (CO), methane (CH_4), hydrogen (H_2), nitrous oxide (N_2O), nitrogen oxides (NOx), non-methane hydrocarbons (NMHC), organic carbon (OC), black carbon (BC), particulate matter 2.5 μ (PM 2.5), total particulate matter (TPM), and sulfur dioxide (SO_2). The data about CH_4 fraction of carbon emissions is also provided. Considering the context of this volume, we may suggest that it is important to collect fire/emission data based on the food grain generating regions of the world. We may have to concentrate on demarcating the cropping belts/food crop generating regions. It is worthwhile to collect data pertaining exclusively to farming zones. Of course, emissions may also drift in the atmosphere. Fire and emissions noticed in wheat producing regions definitely differ if compared with general natural vegetation or plantations. We have to identify the source of fire, first. We may then understand the relevance of fire and emissions on the nutrient dynamics of the cropping zone. There are atlases that depict fire and GHG emissions on a regional basis (Arino et al., 1999; Giglio et al., 2003, 2006; Van der Werf et al., 2004, 2010).

European farmers burn crop residue to clear up land for the next season. Burning is an inexpensive process. It causes minor pollution. Nutrient emissions do occur, but it is minimal or negligible. This activity is a minor source of several pollutants. Crop residue burning improves disease and pest control in certain crops. However, burning of stored agricultural crop wastes, dry forage/pasture, straw, or similar material is prohibited.

Crop residue burning is in vogue in most agrarian regions, particularly, those involved in the production of wheat, maize, millets, sugarcane, etc. Wheat stubble burning is a common practice in different continents. It releases nutrients into atmosphere. It causes pollution of agrarian sky (Dhammapala et al., 2006; Jenkins et al., 1992; Xinghua et al., 2007). Like other agrarian regions, in European farming zones too, burning of crop residues leads to the emission/loss of essential nutrient elements from the soil/field (Webb et al., 2019). It also emits atmospheric pollutants, such as ammonia (NH_3), oxides of nitrogen (NOx), non-methane volatile organic compounds (NMVOCs), sulfur dioxide (SO_2), carbon monoxide (CO), and particulate matter (PM). It includes black carbon (BC1). Burning these residues will also give rise to emissions of heavy metals (HM) and dioxin. It is the extent of release of pollutants that is of concern. They say, nutrient element loss can be easily replenished by computing fertilizer dosages properly. For example, the principal source of the NH_3 and NOx is plant nitrogen. Although some NH_3 is likely to originate from the soil underlying the crop. So, fertilizer calibration should consider the loss of N to atmosphere, if any.

Northern Gangetic plains and other regions in peninsular India supports large areas of cereal, sugarcane, fiber crops, and other oilseed crops. Crop residue burning is a common practice. A good share of large amounts of crop residue generated (500 million tons each year) is composted *in situ.* About 92 Mt crop residue is burnt in open fields in the plains and Deccan Plateau. It affects the nutrient dynamics of the agricultural regions. The emissions could be toxic resulting in pollution. The GHG emissions further accentuate problems, such as variable weather patterns, etc (Bhuvaneshwari et al., 2019). Among cereals, wheat in the North Indian Plains and Rice in other regions constitute 80% of crop residue burning. In the case of rice stubble burning, emissions from Indian rice belt are computed at 127265 Gg, that of Thailand at 11, 667 Gg, and Philippines 10727 Gg. Such open burning definitely alters nutrient dynamics in the atmosphere. It affects the atmospheric quality (Jain et al., 2014; Tripathi et al., 2013). An interesting report has identified rice farmers to be the major generators of crop residue. Hence, during post-monsoon, farmers tend to generate large quantities of smoke and smog with variety of gases that reach the agrarian sky. They pollute and induce global warming. About 60% of crop residue burnt in the rice-wheat belt of Indo-Gangetic Plains is contributed by rice stubbles and about 33% by wheat. This agronomic procedure does affect carbon sequestration into the Inceptisols/Mollisols. It emits CO, CO_2, and CH_4 in quantities that can be detrimental. Ultimately, carbon from crop's canopy/stubbles is lost to aerospace. As a result, the agrarian sky deteriorates.

As stated earlier, crop residue burning done in preparation for the next crop can be a well-regulated agronomic procedure. It is indeed practiced in many regions. It is adopted since burning is cost-effective and a rapid procedure. Often, in certain regions, time lapse between one crop to the next may be short. For example, rice–wheat in Gangetic plains, Wheat/soybean in China, etc. Let us consider an example. In Louisiana, wheat stubble is rapidly burnt to make way for soybean seeding. Crop burning is the mechanism adopted by farmers. In such regions, farmers are trained to avoid drifts of smoke/smog and reduce particulate matter (LSU Ag Centre, 2019).

Lederach (2009) has compiled data about forest and crop burning instances in North America. No doubt, forest/crop burning is deemed a procedure that is detrimental to agrarian sky and its quality. Yet, it is a fact that there are procedures that prescribe a certain amount of burning. In the North American forests/farm belts, burning results in emissions. It affects nutrient dynamics in the aerospace above crop's canopy. Major emissions are CO_2 constituting 57%, carbon monoxide (CO) 6%, water vapor 25%, particulate matter like soot and dry dust, etc. 1%. In addition, crop burning often affects

SO_2 cycle in the ecosystem. SO_2 emissions can be severe if the crops are burnt. Particles of soot, ash, sulfur compounds and other compounds are the cause of "global dimming." It means lower photosynthetic efficiency. Carbon dioxide conversion to vegetation too reduces. So, here, the quality of air in the agrarian sky determines the nutrient dynamics of the ecosystem to a certain extent.

Crop burning is often a well-managed procedure. An overview of vast regions that are prone to crop/forest burning is useful. Remote sensing helps in providing advance information of deterioration of aerospace above crop belts. For example, in the rice-wheat belt of Northern Indian plains, intensive farming procedures induce accumulation of large quantity of straw and stubbles. They are cleared by adopting crop burning procedures. Crop burning results in emissions of certain nutrient compounds, such as CO_2, N, N_2O, and NO_2. (Lasko et al., 2017; Punjab Remote Sensing Board, 2015; McCarty, 2011; Mittal et al., 2009). The Gangetic plain is slowly heading toward rampant accumulation of pollutants in the agrarian sky. For instance, "*dimming of agrarian sky*" can make application of high doses of nutrients ineffective. This is because the photosynthetic efficiency goes low due to fog, smog, and particulate contamination. Burning forest and field crop residues is a problem in many regions of the globe. For example, in the Indo-Gangetic plains, crop burning season-after-season results in large accumulation of GHG and particulate matter. Smogs can be devastating on both rural and urban human communities. This is in addition to alterations in the nutrient dynamics that vegetation burning results in farming zones. Further, GHG, particulate matter, and smog can drift through the wind and other atmospheric processes. Smog generated in plains are actually known to be cross-boundary problems in South Asia. Gaseous emissions, dusts, and crop burning does generate smog. Such smog may move across continents (see Chapter 1 and 6). (Nasim, 2015; Shabbir et al., 2019).

2.7 AGRONOMIC PROCEDURES AND NUTRIENT DYNAMICS IN THE AGRARIAN SKY

2.7.1 FOLIAR APPLICATION OF WATER AND NUTRIENTS

Firstly, foliar applications – can they be a substitute to soil injection? Are they more efficient? Here, we should note that, foliar fertilization is primarily a phenomenon-related agricultural sky. It is the application of dissolved nutrients on to crop's canopy. It is a phenomenon that happens in the atmosphere in an oxidized environment. Nutrient diffusion/absorption via leaves (stomata/hydathodes) occurs in an aerobic environment. No doubt, foliar fertilization affects the nutrient dynamics in the cropped field. The aerial mist may show up residual nutrients still lingering in the atmosphere. This is in addition to that found on leaf surface, branches, and bark.

Foliar feeding is basically an agronomic procedure that involves uniform spray of nutrients in liquid form or in dissolved state in water. Generally, plants absorb foliar nutrient spray via stomata, epidermis, hydathodes, and lenticels. Nutrient requirements are appreciably low if we adopt aerial route for nutrient supply. Foliar feeding of crops, particularly, in-season split dosages has both advantages and lacunae. Net advantages of foliar sprays of fertilizers depend on a variety of factors related to crops, soils, environment, and economic gains (Gazeley, 2012; Roemheld and El-Fouly, 1999)

We may have to realize that the moment farmers shift from soil application of fertilizers to aerial sprays via center-pivot aerial sprayers, they actually affect the dynamics of nutrients, particularly spatial pattern of nutrients in the cropped fields. Aerial application of nutrients like N, entirely alters its dynamics in the ecosystem, that is, in soil, in crops, water bodies, and atmosphere. Nutrient dynamics shifts proportionately from soil phase to aerial canopy of crop and atmosphere (agricultural sky). The agricultural sky is important since nutrients are channeled via aerial sprays. On a larger scale, a sizeable nutrient enters the crops via agricultural sky, if we adopt aerial sprinklers for fertigation. Nutrients are absorbed into leaf tissue instantaneously, after they pass the leaf cuticle.

Phyllosphere is a crucial region through which foliar applied nutrients move into crops, mandatorily. Practically, the focus of nutrient dynamics shifts from underground soil and roots to agricultural sky. Of course, at present, our understanding is that foliar fertilizer application is efficient, offers quick response, corrects nutrient deficiency, if any. It requires lower quantities of nutrients compared with soil application, because nutrient supply is conducted based on actual demand by the crop. Therefore, it reduces greenhouse gas (GHG) emissions. Undue accumulation of nutrients due to excessive application is minimal. Nutrient losses due to seepage, leaching, and percolation into soil too are avoided. Contamination of aquifers is also reduced.

Farmers have generally supplied the nutrients, such as N, P, and K via soil application as broadcasting on land surface, band placement, broadcasting, or soil injection or fertigation through irrigation in furrows. In the case of N, foliar sprays are efficient compared with soil application. In a cropped field, when farmers change over to foliar application of N, they avoid soil N transformations and root-mediated absorption. Soil-related impediments to nutrient absorption is totally avoided. Loss of N due to percolation, leaching, fixation into organic fraction is avoided if foliar application is adopted. Translocation of N from underground soil phase to canopy is not required. Nutrient gets directly absorbed into canopy (i.e., leaf tissue).

Aerial application of micronutrients is also practiced. Again, this procedure helps in overcoming the micronutrient deficiency rapidly, when compared with soil application. Dynamics of micronutrients too shifts from soil, roots/soil interphase, and rhizosphere to aerospace above canopy, leaves/canopy, and phyllosphere. Several of the complicated physicochemical processes that micronutrient ligands go through in the soil phase is totally avoided. This is because micronutrients are channeled directly to canopy in dissolved state or at best in chelated form.

Nitrogen and micronutrients are supplied as foliar sprays, mostly to satisfy the in-season requirements. The top dressings are done usually in 2–3 split dosages. This procedure helps in improving nutrient use efficiency. Incidentally, in future, unmanned aerial vehicles (UAVs) with sprayer bars could revolutionize the way we supply fertilizer-based nutrients to individual fields or even entire agroecoregion. *A sprayer drone is rapid, efficient, and accurate during aerial supply of plant nutrients, pesticides, hormones, and even herbicides. Such drones will revolutionize nutrient supply to crops. These sprayer drones will also markedly alter the nutrient dynamics. They are already doing so in the nations of Fareast.*

Rice is a wetland crop that often prefers flooded condition during growth period from seedling almost till final maturity of panicles/grains. Water and nutrient absorption take place mostly when the soil is held wet/inundated with water. Soil nutrients, such as N, P, K, and others are absorbed through roots spread out into flooded soils. A flooded soil provides anaerobic or microaerophilic condition for roots. Soil is held in a chemically reduced state. Nutrients undergo transformation pertinent with anaerobic medium (soil). Conventional methods of rice cultivation, no doubt, involve broadcasting seeds first then thinning to achieve rows or transplanting seedlings from nursery, in rows. It is followed by broadcasting and/or band placement of fertilizer -N, during in-season. In the study reported by Bhuyan et al. (2012), they have compared foliar application of water and fertilizer N. *Foliar application takes advantage of agricultural sky and aerobic condition that prevails in the atmosphere above the canopy.* Soil applied N undergoes physicochemical transformation. It has to experience all the different constraints to efficient absorption of water/dissolved nutrients. In comparison, foliar applied nutrient reaches the rice leaf tissue directly through the phyllosphere. Loss of nutrients due to leaching, transformation or emissions observed with soil N are absent. Total nutrient N required to satisfy crop's need at a particular stage of crop is much less if nutrients are supplied through the agricultural sky, that is, foliar nutrition. Bhuyan et al. (2012) states that foliar supply enhanced rice grain production by 9.3% compared with soil applied plots. Nutrient quantity to be applied via foliar mode is always much less when compared with soil application. It seems that about 39% of water and nutrients could be saved if foliar nutrition is adopted instead of conventional soil application. Clearly, we have to take advantage of agrarian sky better than at present. It is observed that water and nutrient use efficiency of foliar applied fertilizer N is higher than conventional soil injection. Numerous studies have shown that small amount of nutrients (nitrogen, potash, or phosphate) applied via foliar spraying increases the yield of crops (Bhuyan et al., 2012; Haq and Mallarino, 2000). In fact, foliar fertilization does not totally replace soil applied fertilizer, but it does increase the uptake and hence the efficiency of the nutrients applied to the soil. Foliar application technique is especially useful for micronutrients.

In case of submerged fields, we may note that N supplied via soil is held in anaerobic or microaerophilic conditions. Nitrogen fertilizer undergoes several transformations specific to anaerobic conditions. Roots absorb N as NH_3 more than as NO_3. The above aspects of nutrient dynamics are replaced, if in-season N is supplied as aerial (foliar) nutrients. Crops have to be versatile and accept foliar nutrients via phyllosphere, in aerobic conditions. Rice genotypes that absorb basal N (submerged, microaerophilic conditions) from soil and in-season foliar application (aerobic, diffusion via phyllosphere) are needed. Nutrient transformation, if any, are feeble. The fertilizer selected by farmer is important.

The wetland rice is a peculiarly versatile crop species. It absorbs nutrients from soil held in a chemically reduced state and in microaerophilic or anaerobic condition. Simultaneously, it absorbs nutrients applied (sprayed) as foliar sprays. Foliar absorption is conducted in an oxidized state. Farmers/plant breeders have to opt for proper ratios of foliar/root absorption of nutrients. In future, we will be utilizing agrarian sky better than now. This is because foliar application trends are getting popular for in-season fertilizer application. If we opt for foliar route, the genetic selection of crops for greater leaf surface, stomata and hydathodes is appropriate. In the case of wheat too, plant breeders may have to shift focus to selecting for improved leaf surface and great absorption of moisture/nutrients if foliar sprays are preferred. Root systems may be active in absorbing only the basal dosages of fertilizers. Accept it or not, we would have altered nutrient dynamics in a crop field just because of adoption of foliar application of nutrients.

Nutrient transformations in soil differ markedly between dryland with soil held in oxidized aerobic conditions and those of submerged paddy fields where soil is held in chemically reduced state. Rice roots are known to absorb N as NH_3 from microaerophilic soil conditions. Soil is held under reduced conditions and pH too may differ from that encountered in foliar N applied fields.

Low-land rice is grown under submerged conditions, but using foliar sprays of fertilizer N absorbs N via leaves. Nutrient absorption is conducted in oxidized state like most other dryland crops. Obviously, rice is a versatile crop in terms of nutrient absorption either via roots or through foliar tissue. There are reports that foliar applied N is absorbed efficiently by rice crop. Consequently, it yields higher than crops provided with N via soil layers (Bhuyan et al., 2012). As stated above, when rice crop is held in puddled, flooded fields, it absorbs N as NH_3. Nitrogen absorption is from soils held in anaerobic soil (i.e., chemically reduced condition). Simultaneously, when the mid-season foliar sprays of urea or NH_4NO_3 are conducted, absorption of N occurs via leaves under oxidized state (aerobic leaf surface). Nitrogen emissions from both the soil and canopy (i.e., leaf surface) may occur simultaneously. Emissions of NH_3, N_2O, NO_2, and N_2 may occur from both soil/leaf surfaces.

At present, tendency of farmers in prime cereal growing belts is to supply basal nutrients to soil prior to sowing. Soil injection or deep placement is practiced, for example, in the case of maize, wheat, and rice. Basal dosages of 50–80 kg fertilizer N is applied to soil prior to seeding. The rest 90 kg fertilizer N is supplied as foliar nutrients (sprayed on the canopy) in three to four split applications. So, we shifted the focus of nutrient dynamics from soil profile into crop's canopy and agrarian sky. The agricultural sky, crop's canopy, and leaf surface are central focus during later stages of crops. Literally, in due course, a sizeable portion of nutrient dynamics in a cropped field is influenced by the atmospheric processes (i.e., agricultural sky), not soil phase of the agricultural ecosystems (Ling and Silberbush, 2007).

Crop genotypes that withstand rapid drying of moisture on leaf surface will get preferred. Scorching is actually caused by the dry nutrients still left on the leaf surface. This situation has to be avoided. Genotypes with leaf tissue and cuticle susceptible to dry fertilizer formulation will suffer loss of tissue. We should be focusing our plant selection on the nature and physiological functions of leaf, cuticle, and stomata.

In case we decide to adopt aerial sprays more frequently and achieve better fertilizer use efficiency, then perhaps we should select cultivars with relatively thin cuticle, more of stomata, and one that withstands scorching. It is not easy, perhaps, collect all leaf traits useful foliar absorption at the same time. Making cuticle thin or permeable may affect crops resistance to leaf pathogens, such as blotches, rusts, cutworms, miners leaf eaters. We may have to strike a balance.

In future, larger fraction of essential nutrients may get supplied through aerial sprays. This happens when we adopt center-pivot aerial sprayers for water and dissolved nutrient supply. Also, when we adopt sprayer drones more frequently above the crop's canopy to spray water/nutrients. So, to that extent, we avoid circuitous soil/root phase-mediated acquisition of soil solutes. Roots will continue to be organs of anchorage. But, less involved in nutrient acquisition. This aspect holds true with field crops, such as wheat, rice, maize, tree crops, plantations. At the same time, crop's canopy, particularly leaves take over important activities of nutrient acquisition, imbibing plant chemicals (systemic chemicals), hormones, etc. If drones, aerial sprays, and split applications increase, then crop's canopy gains importance. Farmers must be alert to consequence of foliar nutrition (Rahman et al., 2014; Zain et al., 2015).

Our knowledge about leaf surface traits and its physiological aspects are important. Similarly, physicochemical nature of foliar nutrient should be known. Most importantly, the nature of agricultural sky, its climate, season, atmospheric pollutants, wind, temperature are important. The atmospheric pollution that foliar sprays may induce needs due consideration, particularly residual nutrients, pesticides, herbicides, etc. The influence of aerial sprays on nontarget microbes, insects, and birds needs to be evaluated and farmers must be informed about it. Foliar nutrition needs a permeable cuticle. Micronutrients, such as Fe, Cu, B, Zn, and Mn are amenable for application via leaves. Not all crops are amenable for foliar supply of nutrients. Also, there are series of factors, such as timing (Jones and Olson-Ruitz, 2013), formulation, crop species, season that need attention. They should be congenial for foliar supply of nutrients (Khan et al., 2009; Marschner, 2012; Reena et al., 2018; Sela, 2014; Zain et al., 2015).

Foliar nutrients are usually supplied when farmers intend to correct the deficiency quickly. Foliar supply affects nutrient dynamics in canopy of crops, but without affecting the soil nutrient status perceptibly. The chemical used to spray is selected carefully based on its solubility, absorption rate, and levels of tolerance by the crop's canopy. There are manuals to help farmers in choosing right number of sprays, chemical, stages of crops, and quantity of formulation (see Curley, 1994). For example, N is supplied as urea (0.02% solution), K as KSO_4 or KNO_3, P as HPO_3, Ca as $CaCl_2$, Mg as $MgSO_4$, Zn as $ZnSO_4$, or ZnO_2, Cu as Bordeaux mixture or $CuSO_4$, B as Borax, Mo as $MoSO_4$, Fe as $FeSO_4$. Several crop species are amenable for foliar nutrition. The tolerance for the number of foliar sprays, concentration of nutrients, and total quantity that could be routed through leaf tissue varies (see Ali et al., 2016; Curley, 1994; Krogmeier et al., 1990). Foliar sprays of certain nutrients, such as K or Ca add to vigor and resistance to the crop. This is in addition to correcting the deficiency, if any.

Success of foliar supply of nutrients depends on atmospheric conditions, crop species, and foliar nutrient. It seems we should avoid chloride and nitrate derivatives of foliar nutrients. Spray droplet, its size and time to drying on leaves should be apt. It is often preferable to conduct aerial spray when relative humidity is high. So that foliar chemical does not dry quickly be of detriment to leaf tissue. Foliar nutrition should be avoided if droughts and dryness in atmosphere are suspected. Aerial application of nutrients is done only upon seeing nutrient deficiency. Mostly, N and micronutrients are applied as foliar nitrogen. Phosphorus is supplied as basal dose, often prior to sowing or at early seedling stage (Singh, 2016; Verbree et al., 2016).

Foliar nutrient absorption could be enhanced by using chelates. Chelates induce better translocation, assimilation, and avoids leaf burn. Nutrients are needed in relatively lower amounts. Foliar formulation with chelates are compatible.

There is tendency to use more organic fertilizers on crops. In such cases, soil application is beneficial. Foliar application is apt for soluble inorganic fertilizers supplied via leaves. Whereas organic fertilizers that undergo series of transformations and are held in soil phase are better supplied via soil. Organic fertilizers have to be in liquid and easily absorbable form (Akanbi et al., 2007; Bhat et al., 2013). There are indeed several organic fluid fertilizers available in the market for farmers, to use on their crops.

When we practice soil injection of nutrients, usually entire quantity of fertilizer N is first held in soil phase. It undergoes several physicochemical transformations. A good share of nutrients gets lost from soil due to seepage, leaching, percolation, emissions. There is every reason to believe that any excess that is left unabsorbed by crop's roots gets accumulated. This soil N is vulnerable for leaching into farm's irrigation channels. Or it could be utilized as residual N if it is still in the roots zone. Often, farmers may lose out on N-use efficiency due to N percolation into soil horizons lower than root zone. Such soil-related aspects do not affect an aerial spray of urea in dissolved state or application of liquid ammonia. The crop's canopy absorbs fertilizer directly through the cuticular surface via stomata/hydathodes. As we continue to supply fertilizer N as foliar sprays, we should realize that at least half the fraction or even more of the fertilizer N stays in the crop's canopy and aerospace (atmosphere). So, foliar application means greater activity in the aerospace than in soil. Nitrogen emission due to soil physicochemical reaction is curtailed. The GHG emissions from soil phase of the agroecosystem is reduced proportionately. A larger fraction of fertilizer N may escape into atmosphere from the canopy. Usually, foliar N inputs are accurately calculated and supplied. Therefore, loss of nutrients into the aerospace may be reduced. "Agricultural sky" influences the nutrient dynamics in the field if foliar sprays are utilized.

The tolerance to foliar nutrients sprayed (number of sprays) on to canopy depends on crop species and genotype. We could aim at breeding cultivars that tolerate greater number of foliar sprays on foliage. Crop genotype selection has to be conducted with foliar methods of nutrient supply in focus. Root activity could be restricted to basal dosage of N and its absorption.

Let us consider a few examples of foliar fertilization in different farming regions. Foliar sprays of water, fertilizer-based nutrients, that is, macro- and micronutrient, pesticides with toxic moieties, hormones, stimulants are gaining popularity with farmers in developed nations. North American and European agrarian regions have adopted them briskly. They are economically more efficient, rapid, and provide excellent crop response and increased grain/fruit yield. For example, a report from Bulgaria suggests that foliar sprays are common in the region, particularly during vegetable crop production. The enhanced vegetable yield due to foliar sprays was dependent on crop species, its genotypes, season, fertilizer formulation, nutrients applied (Haytova, 2013).

Foliar fertilization is practiced on cereals, legumes, and vegetable cultivated in large commercial farms. Nutrients, such as N and trace elements, such as Zn, Fe, and B are applied using center-pivot aerial sprinklers. Fertigation that supplies both irrigation water plus nutrients is economically efficient during large-scale cereal production. Fageria et al. (2008) have mentioned that foliar supply definitely affects nutrient dynamics in the crop belt. Usually, it is preferable when leaf area of the crop species is sufficiently higher. Thickly planted cereals, broad-leaved legumes and vegetables are best suited for nutrient supply via leaves. Fertilizer formulation should be easily soluble in water. We may note that foliar fertilization is used to supplement soil application. In a large cereal farm, foliar inputs are opted when nutrient deficiency (e.g., N deficiency) is detected. This is because foliar sprays help in overcoming nutrient deficiency rapidly. Foliar nutrients reach the plant tissue directly via leaf surface. There is no time lapse, which otherwise is required when soil application is adopted. Soil-related transformation in the roots zone is avoided when foliar inputs are practiced. Overall, South American farms that adopt foliar fertilization actually are shifting the nutrient dynamics proportionately to above ground from soil profile. The in-season nutrients supplied are held mostly in canopy and aerospace. Although a certain amount of nutrient applied as foliar sprays may reach (drip) on soil surface.

There are several reports suggesting that foliar application is better in terms of amount of N absorbed by dry crops. In the case of soil application, the efficiency is about 62% of the applied fertilizer N. It means 38% of N injected into soil is lost. At the extreme, only 15% of soil applied N may reach plant roots. Rest 85% could be lost in soil profile. It may contaminate ground water. However, in the case of aerial sprinkler-aided fertigation, about 75–90% of N is absorbed on the canopy (Dixon, 2003). It is preferable to adopt soil injection for basal supply and rest via agricultural sky. Foliar supply of N reduces undue accumulation of this element in soil. It avoids series of detriments that N-accumulation in soil may cause.

There are several known factors that affect the efficiency of foliar fertilization and its impact on crop productivity. For example, Fernandez (2015) states that in southern Europe, field crops, vegetables, and fruit crops are supplied nutrients via foliar channel. Factors that affect the efficiency are grouped into environmental factors (light, relative humidity, temperature, wind speed, precipitation, fog, snow), fertilizer formulation factors (solubility, rate of retention, pH, electrical charge, molecular size, metabolic incorporation into plant system), and plant-related factors (surface topography of canopy, physicochemical nature of phylloplane/phyllosphere, circadian rhythm, stomatal activity, xylem flux, water status, and stress if any of the crop, etc.).

In the case of broad-leaved crops, such as cotton, it is said, aerial supply of nutrients can be beneficial. Foliar sprays can alter nutrient dynamics in the crop. Foliar nutrition is being increasingly preferred in the case of cotton. This is because, farmers are opting cultivars that are short-duration types. Foliar supply results in rapid absorption and response (Oosterhuis and Weir,

2002). In due course, when most farmers shift to foliar sprays, the cotton belts may experience a definite shift in nutrient dynamics in soil, in the crop's canopy and tissue as a consequence, also in the atmosphere. Emissions from cotton fields could decrease if foliar supply is timed accurately and in correct concentration/quantity.

Foliar fertilization of citrus belt in Florida has been in vogue for over 50 years now (Obreza and Morgan, 2017; Zekri et al., 2020). Therefore, this procedure has influenced the nutrient dynamics in the groove proportionately. A sizeable in-season nutrient has reached the crop via aerial route, each season. Fertilization of citrus groove is an important aspect. Tree nutrition affects health, productivity, and quality of the fruits. Nutrients have been channeled to citrus groove by adopting several different methods. A single method could be inefficient. So, Zekri et al. (2020) opine that, a combination of fertilizer supply methods is preferable. It is also true that foliar fertilization is among the most important methods of supplying nutrient. A large canopy, leaf area, and high density of stomata on the leaf surface is congenial for foliar supply. Fertilizer N is commonly supplied, using a 0.02% (water soluble) Urea-N. Higher N in solution can cause leaf burning. Micronutrients, such as Zn, Cu, Mn, Fe, and B are also supplied using foliar spray. Maintenance of optimum nutrient dynamics in the ecosystem depends, to a certain extent, on foliar sprays of nutrient solutions. The total quantity of mineral nutrients supplied to soil gets lower. So, undue accumulation of nutrients in the soil phase of citrus ecosystem is avoided if foliar inputs are practiced. Overall, we should realize that agricultural sky above citrus groove needs greater attention.

Now let us consider an example of micronutrient supplied as foliar fertilizer. In the case of copper (Cu), farmers usually supply it as $CuSO_4$ $5H_2O$ solution. It is supplied at 0.2 kg ha if it is foliar nutrition. Whereas a soil application would mean a range of 2.0, 4.0, or 5 kg Cu ha^{-1} supplied as soil injection (Karamanos et al., 2003). Preference for aerial sprays depends on crop species, its demand for Cu and deficiency experienced by the soil. Foliar nutrition is preferred when farmers observe that soil Cu distribution is highly variable. They say, soil application provides perceptible crop responses. In the case of Cu applied foliar spray, factors, such as stage of the crop, number of sprays, crop species need to be appropriate. Excessive Cu inputs via foliar route may become ineffective (see Karamanos et al., 2004; Singh et al., 1985).

A standard foliar spray of nutrients or even other agricultural chemicals may cause alterations in the physicochemical aspects of the area (atmosphere) above crop's canopy. Firstly, we should confirm the borders of such area in the atmosphere which comes under the influence of sprays (mists). In addition, it may cause upheavals in the microbial flora and population in the atmosphere. Microbial flora may include beneficial ones like BNF species or pathogens. We should have good knowledge about the effect of mists on microbes and their physiological activity. Certain microbes of benefit to farming may actually get accentuated by the hanging mist with nutrients. Even birds that occupy the sky immediately above fields may change.

We should note that not all field or plantation crops are amenable for foliar nutrition. Crop species with foliage that has thick cuticle and not so well-spread stomata turn out to be inefficient user of foliar sprays with nutrients. For example, Avacado, with thick leaves and cuticles is not a suitable candidate for foliar nutrition. Micronutrients, such as Zn, Fe, Mo, Mn are not absorbed quickly. For example, in Californian orchards, farmers tend to supply micronutrients, such as Zn via soil phase/roots because foliar methods are slow (University of California Cooperative Extension Service, 2020). In fact, Whiley and Pegg (1990) report that foliar applications of zinc have been found to be highly ineffective in Queensland orchards. Obviously, leaf area, stomatal activity, and cuticle may together affect the nutrient dynamics in the aerial parts of crops.

Since past 5 decades, variety of pastures and forage grasses too have been supplied with foliar nutrients. The grasses recover rapidly and show excellent biomass gain in response to foliar spray of nutrients (Houston and Van Der Sluus, 1972). We may make a few educated guesses about the nutrient dynamics in the aerospace above pasture in different agrarian regions. Firstly, for rapid assessment, we may adopt spectral data obtained from UAVs that occupy the agricultural sky to various periods depending on the tasks programmed. In future, aerial sprays could become common. Mists/sprays with nutrients and other chemicals may affect the nutrient dynamics. Foliar spray also affects the biotic aspects of agrarian sky. Nutrient dynamics relevant to crop's canopy and atmosphere immediately above crops may gain importance. Emissions may get curtailed since nutrient supply is generally accurate, if aerial sprays are adopted. Any excess is, of course, emitted easily and quickly, since it is held above leaf or in the canopy of crops.

As such, stomata on the leaf surface have a good influence on gaseous exchange. When we practice aerial sprays, stomatal activity influences the mineral nutrient entry into plant tissue and its translocation. Reduction or loss of foliage may have greater impact on crop's nutrient dynamics.

2.7.2 PHYLLOSPHERE: A REGION OF GREAT POTENTIAL TO GLOBAL AGRICULTURE, IN FUTURE

Phyllosphere refers to a region of 0.2 mm thickness on leaf surfaces. It is a good habitat for microorganisms, including filamentous fungi, yeasts, and bacteria (Lindow and Brandl, 2003; Lindow and Leveau, 2002). Phyllosphere represents a large area on the globe. It occurs on the foliage, branches, and barks/stems of crop plants. It is perhaps functionally too important if we extrapolate its large area. It has immediate impact on the survival of microbes, nutrient dynamics, and crop's survival.

Phyllosphere is a region of immense importance to global agricultural crop production. It will be felt more so in near future. It gains importance as we develop procedures and adopt aerial sprays using UAVs, that is, sprayer drones. While we studied soils and their potential to supply nutrients and moisture to crops and continue to do so, it is time we shifted our focus to agricultural sky, crop's canopy, and phyllosphere. This is a necessity if our aim is to reduce fertilizer inputs to cropped fields. In this manner, we may be able to regulate nutrient dynamics in the atmosphere better.

Perazzolli et al. (2014) opine that phyllosphere is an open space in the agroecosystem. It is a large area if we consider the general vegetation and agrarian belts. It has impacts on several global manifestations of agricultural sky, such as the carbon cycle, nitrogen cycle, biological nitrogen fixation, global greenhouse gas emissions, pollutant accumulation, and their transformations. In fact, Saikonen et al. (2015) have provided proof to show that phyllosphere fungi are intricately involved in global nutrient cycle.

Phylloplane is physically the surface of leaf. Phyllosphere is a small zone on the leaves exposed to agrarian sky. Like roots/rhizosphere in the soil phase, phyllosphere/phylloplane is functionally an important aspect of cultivated crops. The area plays crucial role in determining dynamics of moisture, nutrients, and biotic fraction on leaf surface. Here, we are concerned with crop's canopy along with foliage, branches, barks, and even fruits that protrude into agrarian sky (atmosphere). Phyllosphere is an area of 0.2 mm in thickness surrounding the leaves. It harbors microbes, insects/insect eggs, organic /inorganic dust particles, pollutants, nutrients, and moisture. Phyllosphere also helps in gaseous exchange between crop and atmosphere (agrarian sky).

Considering the high diversity and number of microbes encountered in the phyllosphere, Vacher et al., (2016) have rightly termed the region as an area on leaves that supports a "jungle of microbes." Phyllosphere microbes are situated at the confluence of plant foliage and ambient atmosphere. Phyllosphere microbes are involved in a variety of ecosystematic functions. Phyllosphere is central to many of the processes related to agrarian sky, crops, and nutrient dynamics in an agroecosystem. Crop's canopy and its various traits, include leaves and their physical and physiological aspects. Phyllosphere is the small area just surrounding the surface of the leaf (cuticular layer). Phylloplane is the surface of leaves. Stomatal number and its functioning are critical to normal absorption of moisture. The exchange of dissolved nutrients and gaseous compounds also depend on leaf traits. Therefore, phyllosphere is an important concept. It concerns nutrient dynamics in the crop and atmospheric phase of an ecosystem. Phyllosphere is central to nutrition of crops provided with foliar water/nutrient sprays. Sprinkler irrigation and aerial mode of fertilizer supply should consider the various traits of crop's canopy leaves and phyllosphere activity. Phyllosphere supports a posse of microbes of wide diversity. They could be beneficial like nitrogen fixers, commensals, or pathogens that try to penetrate cuticular layer and cause disease. Phyllosphere also supports deposits of dust, pesticides, herbicides, and other plant protection chemicals when aerial sprays are adopted. Phyllosphere interactions involving moisture, nutrients, microbes, organic deposits, and pesticides are aspects that needs greater attention (Ning et al., 2010). For example, just by scanning the microflora on the phyllosphere, we may be able to avoid diseases that are caused via leaves. A thick tough cuticle may resist fungal pathogens at the phyllosphere.

Let us consider a few aspects of phyllosphere which are crucial to proper functioning of foliar sprays conducted on large areas of crops. Phyllosphere on leaves protruding into agrarian sky is equivalent to rhizosphere of roots traced in soil phase of an agroecosystem. In due course, we may try to channel more of nutrients and other agrochemicals via phyllosphere. Hence, this portion of crop gains importance. We should have thorough knowledge regarding leaves, their surface traits, stomata, gaseous/solute exchange traits, etc. Foliar application literally shifts our focus to canopy. Nutrient dynamics involving foliar nutrition needs special attention. Nutrients supplied via canopy and that held in the plant tissue may increase.

"*Phylloremediation*" is the word used to describe adoption of phyllosphere plus microbes to avoid nutrient dynamics that could be detrimental to crops. In other words, phylloremediation is the reduction of "air pollution" effects. For example, emissions of CO, CH_4, N_2, NO_2, VOCs, and ozone are considered suppressive to crops, in case they occur rampantly. Phyllosphere is indeed a large surface area on earth. It occupies all of the natural vegetation and agricultural region traced on earth. Microbes on phylloplane/phyllosphere are equally important. Wei et al. (2017) have discussed about a novel approach to reduce the effects of air pollution on the large agrarian zones. Phyllosphere microbes with potential to reduce the release of GHG and thwart the pollutant effects should be identified. We should try to preferentially increase the population of phyllosphere microbes that reduce air pollution.

2.7.2.1 PHYLLOSPHERE MICROBES ARE RESIDENTS OF AGRARIAN SKY

Foremost, Vorholt (2012) points out that our knowledge about phyllosphere and its microbial ecology is lagging if we considered other aspects of ecology of crops. Rhizosphere of crop plants has been studied in greater detail. Fundamental aspects of below-ground soil, microbes, and nutrient aspects have been studied by numerous biologists.

Phyllosphere bacteria are also important. They may represent an important source of bacteria to the atmosphere (Lighthart, 1997). Phyllosphere microbes may play a key role in nutrient cycling (Freiberg, 1998; Jones, 1970). Despite their importance, we know surprisingly little about the diversity and biogeography of phyllosphere bacterial communities.

Microbes flourish on leaf surfaces despite environmental constraints. Although fungi and archaea are known to colonize leaves, bacteria are numerically dominant in the phyllosphere environment (Andrews and Harris, 2000; Lindow and Brandl, 2003; Morris, 2004). These leaf dwelling bacteria can have either neutral, negative, or positive influences on their host plants. They could be commensals or pathogens or biocontrol agents preventing leaf colonization by pathogens (Kishore et al., 2005).

Phyllosphere is a region on the surface of leaves that supports microbes of different phylogenic diversity. Although fungi, archae, and actinomyces could be traced, it is the bacterial number and diversity that has been examined more frequently, by crop researchers (Redford and Feirer, 2009; Vokou et al., 2012; Williams and Marco, 2014). Often, cereals, broad-leaved legumes, vegetables, and other species have been studied with reference to phyllosphere microbiology and its relevance to nutrient dynamics. Trees leaves too have been examined in detail for microflora. Bacterial abundances on tree leaf surface can be determined by 4,6-diamidino2-phenylindole (DAPI). For example, it ranged from 2.4 to 10^4 cells cm^{-2} on *Juniperus scopulorum* leaves to 6.0 10^3 to 10^5 cells cm^{-2} on *Cedrus libani* with an average of 2.1 to 10^5 cells cm^{-2}. In the case of *Pinus ponderosa*, they have studied microflora on the needles (Lambais et al., 2006; Osono and Mori, 2005). Redford et al. (2010) and Redford and Fierer (2009) have reported that microflora on leaves could be influenced by the geographic location of trees. The bacterial communities differ with regard to phylogenic/genetic nature and population. Over 10 different bacterial communities could be traced on the needles. Each of them may interact specifically with other bacteria. Nutrient dynamics may depend largely on the microbial activity on the needles (Osono and Mori, 2005). Such details on microbes involved in nutrient transformation on leaf surface will be of utmost value to farmers. After all, the efficiency of foliar application of nutrients may get altered based on phyllosphere microflora.

The quantification of phyllosphere microbes is important. Equally so, determining their genetic and functional diversity is important. It may actually form the basis for several inferences about the phyllosphere effects. Methods of sampling phyllosphere microbes, their enumeration and quantification have been reviewed (Jacques and Morris, 1995; Morris, 2004). Appropriate methods have to be selected from among several procedures. This is in order to derive accurate inferences. Drahos (2018) has made it clear that data accrual and its utility from experiments related to phyllosphere microflora is crucial. Sophisticated probes, equipment and computer programs are needed to get accurate measurements of microbial activity and their consequences on the crops. Methods aimed at deriving rapid data from phyllosphere, particularly about microbes is critical. At the same time, we may note that phyllosphere microbes and their interactions with the environment, including the nutrient dynamics, physicochemical changes and bacterial succession can be studied at various levels. We can focus single microbial cell and its fate in a phyllosphere environment. We may consider a few million bacteria or a colony on the leaf surface and its relevance to nutrients in the environment. We may study the role of phyllosphere in microbial changes, that is, regular succession and its relevance to disease, nutrients, and plant vigor (Remus-Emsermann and Schlecter, 2018). Muller and Ruppel (2014) have opined that phyllosphere microbes are not easily cultivable. Usual methods of isolation and assessment of their role in media is not easy. So, methods that involve molecular analysis such as 16S rRNA are being adopted (Rastogi et al., 2013).

Qian et al. (2018) have provided evidence to show that even a single fungal species residing on the phyllosphere of certain tropical trees could be monitored. For example, they have reported that fungal species on the phyllosphere varied based on the altitude at which the trees (*Mussaenda shikokiana*) were found. Aerial factors could be involved in the establishment of phyllosphere fungi.

We should note here that phyllosphere is a vast region that supports microbial life. It is traced all over in the vegetated area (Kembal and Meuller, 2014; Vorholt, 2012). Microbial flora encountered on the phylloplane and phyllosphere may have been derived from the agricultural sky (Copeland et al., 2015; Ottesen et al., 2016). While several studies show correlations between microflora found in agrarian sky with that traced on leaves, there are also reports where they do not exactly correlate. The nutritional composition of phyllosphere too may preferentially support different microbes. Whatever the source of microbes in the phyllosphere, one of their primary functions involves ecological interactions and nutrient dynamics. Several types of nutritional transformations may get mediated via phyllosphere microbes. A few of them could be crucial for proper absorption of nutrients by leaf tissue.

Phyllosphere microbial diversity itself may be dependent on nutritional status of cuticular surface and phyllosphere (Williams et al., 2013; Williams and Marco, 2014). Microbes on leaves interact for various environmental and nutritional reasons. Microbes in the phyllosphere may also be beneficial to plant. Certain phyllosphere microbes may thwart the multiplication of pathogens (Kishore et al., 2005). Nutritional status of the leaf surface may actually be central to survival, germination, and perpetuation of pathogens. Formation of appressoria may get influenced by other microbes and leaf tissue characteristics. Some of these aspects may need greater attention by the plant breeders. Geneticists may have to prefer cultivars with ability to induce phyllosphere microbes that suppress pathogens. They may have to select genotypes with cuticles and palisades that do not allow easy entry of pathogens. At the same time, they must select for better leaf surface traits, such as stomatal number, functioning, etc. Also, in simple terms, shift from soil inputs to foliar methods means we look to phyllosphere, leaf area, and aerial sprays carefully and in greater depth. Here, the agricultural sky and crop canopy are central to crop productivity. Hence, geneticists will have to optimize canopy traits to suit the agrarian sky and its factors. It calls for a shift in focus from below-ground roots to foliage and its traits. We are actually looking for aerial parts (foliage) to do more for the crops, so that productivity is greater.

2.7.2.2 PHYLLOSPHERE INTERACTIONS

Phyllosphere is a large interactive surface within a cropping zone. It supports the close interaction with plant's physiological status, microbes residing on the phylloplane and outer atmosphere. According to Brengel and Couce

(2015), phyllosphere is an excellent habitat for microbes on earth. Phyllosphere microbes colonize extreme and changing environments prevalent in the crop belts. Phyllosphere microbes and the atmosphere present a "dynamic continuum." Phyllosphere microbes interact with volatile organic compounds and atmospheric trace gasses. Phyllosphere could actually wield a certain influence on the nutrient dynamics. Phyllosphere microbes may be involved in complex interactions than previously thought (Yang et al., 2001).

Rivas-Ubach et al. (2019) have made detailed review about the role of phyllosphere in an agroecosystem. They have highlighted the relevance of interactions of aerosol, leaf surface microbes, and nutrients within phyllosphere. In fact, they have cautioned that phyllosphere studies are not conducted, but to just know, the cuticular surface microbes, surface distribution of plant pathogens, and physiology of cuticle and stomata. Since, aerosol sprays are conducted more frequently on a large scale in different cropping systems, the importance of phyllosphere is felt conspicuously. The foliar sprays that create aerosols of nutrient, microbes in the atmosphere, and moisture need to be understood in greater detail. We have to focus our attention on the role of canopy (size, shape, pattern) leaf area and phyllosphere with regard to crop nutrition and various ecosystematic functions. They have further argued that the plant phenotype and canopy pattern itself may after all be decided by the phyllosphere interactions. The aerosol spray, its droplet size, microbial load on surface, physicochemical and biological interactions may all impose a kind of restriction on the crop's phenotype, ultimately affecting the light interception and nutrient acquisition. In particular, nutrient acquisition via foliage gets regulated directly by the phyllosphere interaction (Rivas-Ubach et al., 2019).

Cucumber produces broad leaves with large leaf area and phyllosphere zone. Its phyllosphere supports a large posse of diverse bacterial, fungal, and archaea flora. Attempts have been made to know the influence of pesticide sprays on microflora found on foliage. Zhang et al. (2008) have reported that their studies indicated a reduction of fungi, pests, and insect eggs. However, bacterial flora got increased and stabilized on the leaf surface. The gram+ to gram- bacterial population ratio too decreased. They state that we should be able to identify the indicator bacterial flora that responds specifically to pesticide inputs. Also, we must decipher the result of lessening of fungal flora on the leaf surface. Also, we may realize the preferential improvement of bacteria. We should understand the microbial interactions and their relevance to crop nutrition via foliage for as many crops as possible.

Reports suggest that phyllosphere actinomycete population, such as *Actinomyces, Streptomyces, Micromonospora*, etc. are capable of producing bioactive compounds. They are detrimental to survival of bacterial leaf blight causing pathogen *Xanthomonas campestris*. Hence, phyllosphere actinomycetes should be playing a beneficial role if we consider their role as antibacterial agents (Ilsan et al., 2016). Pepper leaves support diverse microbes, such as bacteria, fungi, actinomycetes, yeasts. It seems that archae are not abundant on phyllosphere surface of vegetable crops analyzed (Knief et al., 2012).

Epiphytotic microorganisms could serve significant functional roles including (a) suppression of plant pathogens in the phyllosphere of agricultural crops (Whipps et al., 2012) and (b) nitrogen fixation (Miyamoto et al., 2004), methanol utilization, and degradation of organic pollutants (Ning et al., 2010). Whipps et al. (2012) further state that our knowledge about phyllosphere microbes drawn from culturable species is indeed based on limited data. This is because only 1–3% of phyllosphere microbes have been cultured and studied. Therefore, we should know the exact ramifications of phyllosphere microbes and their role in nutrient dynamics, and for this to occur, we should devise methods that can decipher, enumerate, and relate their functions even without being cultured in an artificial medium (Fermor et al., 1993).

According to Perazzolli et al. (2014), phyllosphere microbial communities take part intricately in the arial processes, such as carbon cycle, nitrogen fixation, and degradation of organic pollutants, pesticide residues included. Phyllosphere microbial communities are also thought to be relevant for plant development and health.

Moulas et al. (2013) have attempted to understand the influence of spraying crops, such as pepper with systemic fungicides, such as Metalaxyl on microflora, their interactions and role in nutrient transformations. Metalaxyl has induced colonization of phyllosphere by Enterobacteriaceae. They could be involved in the degradation of residual fungicide particles resting on the leaves. Most common fungal flora noted were *Erysiphe* spp., and *Cladosporium* spp. Yeasts such as *Cryptococcus* spp. were abundant in the phyllosphere.

Although not related to nutrient dynamics and productivity, insects do colonize the vast leaf area present in the agrarian regions and natural vegetation. Smee et al. (2019) found that several strains of *Psuedomonas syringae*, about 21 were colonizing the phyllosphere of crop plants. These epiphytotic microbes were actually detrimental to aphids – *Acyrthosiphon pisum* that attack the pea crops. This is one other aspect of phyllosphere microbes and their role in the dynamics of insect, namely, aphids.

Overall, we may conclude that studying agricultural sky with an aim to improve on nutrient dynamics, crop growth, and productivity is a worthwhile pursuit. It needs greater emphasis than that provided at present.

KEYWORDS

- **acid rain**
- **crop residue burning**
- **carbon sequestration**
- **foliar application**
- **greenhouse gas**
- **nutrient dynamics**
- **particulate matter**
- **phyllosphere**
- **pollution**

REFERENCES

Acosta-Martinez, V.; Van Pelt, S.; Moore-Kucera, J.; Baddock, M. C.; Zobeck, T. M. Microbiology of Wind-Eroded Sediments: Current Knowledge and Future Research Directions. *Aeolian Res.* **2015,** *18*, 99–113.

Agence France Presse. Methane Emissions from Cattle are 11% Higher than Estimated [Online] 2017, pp 1–4. https://www.theguardian.com/environment/2017/sep/29/methane-emissions-cattle-11-percent-higher-than-estimated/ (accessed July 25, 2021).

Agricultural Department. Effect of Frost on Cereal Grain Crops. Department of Agriculture, Victoria, Australia [Online] 2019, pp 1–24. http://agriculture.vic.gov.au/agriculture/grains and-other-crops/crop-production/effect-of-frost-on-cereal-grain-crops/ (accessed July 30, 2021).

Agricultural Department. Effect of Frost on Cereal Grains [Online] 2020, pp 1–5. Department of Agriculture, Victoria, Australia. http://agriculture.vic.gov.au/agriculture/grains-and-other-crops/crop-production/effect-of-frost-on-cereal-grain-crops (accessed July 30, 2021).

Akanbi, W. B.; Adebayo, O. A.; Togun, O. A.; Adeye, A. S.; Olaniran, O. A. The Use of Compost Extract as Foliar Spray, Nutrient Source and Botanical Insecticide in *Telfairia occidentalis*. *World J. Agric. Sci.* **2007,** *3* (5), 642–652.

Ali, A.; Hussain, M.; Habeeb, H. S.; Kiani, T. T.; Anes, M. A.; Rehaman, M. A. Foliar Spray Surpasses Soil Application of Potassium for Maize Production Under Rainfed Condition. *Turk. J. Field Crop.* **2016,** *21*, 36–43.

Alpert, P.; Kaufman, Y. J.; Shay-El, Y.; Tanre, D.; da Silva, A.; Schubert, S.; Joseph, J. H. Quantification of Dust-Forced Heating of the Lower Troposphere. *Nature* **1998,** *395*, 367–370.

Amann, M. *Emissions from Agriculture and their Control Potentials. International Institute of Applied Systems Analysis*; European Commission TSAP Report No. 3, Version 2.1. 2012, pp 1–47.

Andrews, J. H.; Harris, R. F. The Ecology and Biogeography of Microorganisms of Plant Surfaces. *Annu. Rev. Phytopathol.* **2000,** *38*, 145–180.

Arino, O.; Rosaz, J.-M.; Goloub, P. The ATSR World Fire Atlas. A Synergy with 'Polder' Aerosol Products. *Earth Obse. Q.* **1999,** 1–6.

Atkinson, R. Atmospheric Chemistry of VOCs and NOx. *Atmos. Environ.* **2000,** *34*, 2063–2101.

Atkinson, R.; Arey, J. Gas-Phase Tropospheric Chemistry of Biogenic Volatile Organic Compounds: A Review. *Atmos. Environ.* **2003,** *37*, 197–219.

Bahera, S. N.; Sharma, M.; Aneja, V. P.; Balasubramanian, R. Ammonia in the Atmosphere: A Review on Emission Sources, Atmospheric Chemistry and Deposition on Terrestrial Bodies. *Environ. Sci. Pollut. Res. Int.* **2013,** *20*, 8092–8131.

Behzad, H.; Mineta, K.; Gojobori, T. Global Ramifications of Dust and Sandstorm Microbiota. *Genome Biol. Evol.* **2018,** *10*, 1970–1987.

Bentley, M. L.; Mote, T. L.; Thebpanya, P. Using Landsat to Identify Thunderstorms Damage in Agricultural Regions. *Bull. Am. Meteorol. Soc.* **2001,** *164*, 363–376.

Bhat, N. R.; Albaho, M.; Suleiman, M.; Thomas, B.; George, P. Fertilizer Formulation and Methods of their Application Influence Vegetable Growth and Productivity in Organic Greenhouse Tomato. *Asian J. Agric.* **2013,** *5*, 67–70.

Bhuvaneshwari, S.; Hettiarchachi, H.; Meegoda, J. N. Crop Residue Burning in India: Policy Challenges and Potential Solutions. Int. *J. Environ. Res. Public Health.* **2019,** *16*, 1–9.

Bhuyan, M. H. M.; Ferdousi, R. F.; Iqball, M. T. Foliar Spray of Nitrogen Over Raised Beds Increases Yield of Transplanted Aman Rice over Conventional Method. *ISRN Agron.* **2012,** *2012*, 1–8.

Bielders, C. L.; Rajot, J. L.; Koala, S. Wind Erosion Research in Niger: The Experience of ICRISAT and Advanced Research Organizations [Online], 2002a. 'ICRISAT Sahelian Centre, BP 12404, Niamey, Niger 'ORSTOM, BP 11416, Niamey, Niger. http://oar.icrisat.org/6219/ (accessed July 30, 2021).

Bielders, C.; Rajot, J.; Amadou, M.; Skidmore E. On Farm Evaluation of Wind Erosion under Traditional Management Practices [Online] 2002b, p. 6. ICRISAT.Org. CCER.htm. doi:10.13031/2013.4585 (accessed July 30, 2021).

Bjerke, J. W.; Tommericke, H.; Zeilke, M.; Jorgensen, M. Impacts of Snow Season on Ground-Ice Accumulation, Soil Frost and Primary Productivity in a Grassland of Subarctic Norway. *Environ. Res. Lett.* **2015,** *10*, 9.

Bracho-Nunez, A.; Welter, S.; Staudt, M.; Kesselmeier, J. Plant-Specific Volatile Organic Compound Emission Rates from Young and Mature Leaves of Mediterranean Vegetation. *J. Geophys. Res.* **2011,** *116*, D16304.

Brady, N. C. *Nature and Properties of Soils*; Prentice Hall of India Inc.: New Delhi, India, 1975; pp 576.

Brazil. 2010 Ministry of Science and Technology. General Coordination of Global Climate Change. Brazil's second national communication to the United Nations Framework Convention on Climate Change [Online] 2010, pp 1–78. Ministry of Science and Technology:

Brasília, DF. https://antigo.mctic.gov.br/mctic/export/sites/institucional/ciencia/SEPED/clima/arquivos/Comunicacoes_Nacionais/TCN_port/TCN_Volume_2a.pdf/

Brengel, F.; Couce, S. Pivotal Roles of Phyllosphere Microorganisms at the Interface between Plant Functioning and Atmospheric Trace Gas Dynamics. *Front. Microbiol.* **2015,** *6* (468) 1–14.

Cadoux, S.; Sauzet, G.; Valantin-Morison, M.; Pontet, C.; Champolivier, L.; Robert, Lieven, J.; Flénet F.; Mangenot, O.; Fauvin, P.; Land, N. Intercropping Frost-Sensitive Legume Crops with Winter Oilseed Rape Reduces Weed Competition, Insect Damage, and Improves Nitrogen use Efficiency. *EDP Sci. OCL* **2015,** *22* (3), D302.

Cerri, C. C. Brazilian Greenhouse Gas Emissions: The Importance of Agriculture and Livestock. *Scientia Agricola* **2009,** *66* (6), pp 831–843.

Chan, A. W. H.; Chan, M. N.; Surratt, J. D.; Chhabra, P. S.; Loza, C. L.; Crounse, J. D.; Yee, L. D.; Flagan, R. C.; Wennberg, P. O.; Seinfeld, J. H. Role of Aldehyde Chemistry and NOx Concentrations in Secondary Organic Aerosol Formation. *Atmos. Chem. Phys.* **2010,** *10* 7169–7188.

Christian, T. J.; Kleiss, B.; Yokelson, R. J.; Holzinger, R.; Crutzen, P. J.; Hao, W. M.; Saharjo, B. H.; Ward D. E. Comprehensive Laboratory Measurements of Biomass-Burning Emissions: 1. Emissions from Indonesian, African, and Other Fuels. *J. Geophys. Res.* **2003,** *108* (D23), 4719.

Christian, T. J.; Kleiss, B.; Yokelson, R. J.; Holzinger, R.; Crutzen, P. J.; Hao, W. M.; Shirai, T.; Blake, D. R. Comprehensive Laboratory Measurements of Biomass-Burning Emissions: 2. First Intercomparison of Open-Path FTIR, PTR-MS, and GC-MS/FID/ECD. *J. Geophys. Res.* **2004,** *109*, D02311.

Cole, C. V.; Duxbury, J.; Freney, J.; Heinemeyer, O.; Minami, K.; Mosier, A.; Paustian, K.; Rosenberg, N.; Sampson, N.; Sauerbeck, D.; Zhao, Q. Global Estimates of Potential Mitigation of Greenhouse Gas Emissions by Agriculture. *Nutr. Cycl. Agroecosyst.* **1997,** *49*, 221–228.

Copeland J. K.; Yuan, L.; Layeghifard, M; Wang. P. W.; Guttman, D. S. Seasonal Community Succession of the Phyllosphere Microbiome. *Mol. Plant Microbe Interact.* **2015,** *28* (3), 274–285.

Cox, L. Australia's Carbon Emissions Highest on Record, Data Shows [Online] 2018. https://www.theguardian.com/australia-news/2018/dec/13/australias-carbon-emissions-highest-on-record-data-shows/ (accessed July 30, 2021).

Curley, S. *Foliar Nutrition*; Midwest Laboratories Inc: Omaha, USA, 1994; pp 1–77.

Davidson, E. A. The Contribution of Manure and Fertilizer Nitrogen to Atmospheric Nitrous Oxide Since 1860. *Nat. Geosci.* **2009,** *2*, 659–662.

de Groot, W. J.; Field, R. D.; Brady, M. A.; Roswintiarti, O.; Mohamad, M. Development of the Indonesian and Malaysian Fire Danger Rating Systems. *Mitig. Adapt. Strateg. Glob. Chang.* **2006,** *12*, 165–180.

de Groot, W. J.; Goldammer, J. G.; Keenan, T.; Brady, M. A.; Lynham, T. J.; Justice, C. O.; Csiszar, I. A.; O'Loughlin, K. In *Developing a Global Early Warning System for Wildland Fire*, Proceedings of the V International Conference on Forest Fire Research, November 24th to December 1st, 2006; Viegas, D. X., Ed.; Figueira da Foz, Portugal, Elsevier: Amsterdam, Netherlands, 2007, pp 1–12.

De Mello-Abreu, J. P. Protecting Fruit Production from Frost Damage. European commission, EIP AGRI focus group [Online] 2018, pp 1–36. https://ec.europa.eu/eip/agriculture/sites/

default/files/eip-agri_fg_frost_damage_final_report_2019_en.pdf/ (accessed July 30, 2021).

De Mello-Abreu, J. P.; Francisco, J.; Villalobos, Luciano Mateos. Frost protection. In *Principles of Agronomy for Sustainable Agriculture*; Villalobos F. J., Fereres, E., Eds.; Springer International Publishing, 2016; pp 443–457.

Denman, K. L.; Brasseur, G.; Chidthaisong, A.; Ciais, P.; Cox, P. M.; Dickinson, R. E.; Hauglustaine, D.; Heinze, C.; Holland, E.; Jacob, D.; Lohmann, U.; Ramachandran, S.; Da Silva Dias, P. L.; Wofsy, S. C.; Zhang, X.; Coupling between Changes in the Climate System and Biogeochemistry. In *Climate Change 2007: The Physical Science Basis*; Solomon, S., Qin, D., Manning, M., Chen, Z., Marquis, M., Averyt, K. B., Tignor, M., Miller B. R., Eds.; Contribution of Working Group I to the Fourth Assessment Report of the Intergovernmental Panel on Climate Change; Cambridge University Press: Cambridge, UK and New York, US, 2007; pp 500–587.

Dhammapala, R.; Claiborn, C.; Corkill, J.; Gullett, C. Particulate Emissions from Wheat and Kentucky Bluegrass Stubble Burning in Eastern Washington and Northern Idaho. *Atmos. Environ.* **2006,** *40*, 1007–1015.

Dixon, R. C. Foliar Fertilization Improves Nutrient Use Efficiency. *Fluid J.* **2003,** *129*, 1–5.

Drahos, D. J. Methods for the Detection, Identification and Enumeration of Microbes. In *Microbial Ecology of Leaves. Springer Book series in Contemporary Bioscience*, 2018; pp 135–157.

Dulac, F.; Buat-Menard, P.; Ezat, U.; Melki, S.; Bergametti, G. Atmospheric Input of Trace Metals to the Western Mediterranean Sea. 1. Factors Controlling the Variability of Atmospheric Concentrations. *J. Geophys. Res.* **1987,** *92*, 8437–845.

Eccel, E.; Rea, R.; Caffarra, A.; Crisci, A. Risk of Spring Frost to Apple Production Under Future Climate Scenarios: The Role of Phenological Acclimation. *Int. J. Biometeorol.* **2009,** *53*, 273–286.

Eglinton, J. Frost Prone Grain Producing Regions of Australia. Grain Research and Development Centre, Victoria, Australia [Online] 2013, pp 1–4. https://grdc.com.au/resources-and-publications/grdc-update-papers/tab-content/grdc-update-papers/2013/01/developing-frost-tolerance-in-wheat/ (accessed July 30, 2021).

Engel, R.; Dusenbury, M.; Miller, P.; Lemke, R. In *A First Check of Nitrous Oxide Emissions under Cropping Systems Adapted for the Northern Great Plains,* Vol. 6; Western Nutrient Management Conference: Salt Lake City, Utah, USA, 2005; pp. 25–31.

Erisman, J. W. Acid Deposition to Nature Areas in the Netherlands. Part I. Methods and Results. *Water Air Soil Pollut.* **1993,** *71*, 51–80.

Erisman, J. W.; Schaap, M. The Need for Ammonia Abatement with Respect to Secondary PM Reductions in Europe. *Environ. Pollution.* **2004,** *129* (1), 159–163.

Fageria, N. K.; Barbosa Filhho, M. P.; Moreira, A.; Guimaraes, C. M. Foliar Fertilization of Crop Plants. *J. Plant Nutr.* **2008,** *32* 1044–1064.

FAO. Frost Damage: Physiology and Critical Temperature. Food and Agricultural Organization of the United Nations: Rome, Italy, Chapter 15 [Online] 2014, pp 1–25. http://www.fao.org/3/y7223e/y7223e0a.htm/ (accessed July 30, 2021).

FAO. World Fertilizer Trends Summary Report 2017 Food and Agricultural Organization of the United Nations: Rome, Italy [Online] 2017, pp 38. http://www.fao.org/3/i6895e/i6895e.pdf/ (accessed July 30, 2021).

Fermor, T. R.; Lilley, A. K.; Lynch, J. M.; McCormack, P. J.; McQuilken, M. P.; Purdy, K. J.; Rainey, P. B.; Whipps, J. M. Quantitative and Qualitative Seasonal Changes in the

Microbial Community from the Phyllosphere of Sugar Beet (*Beta vulgaris*). *Plant Soil.* **1993,** *150*, 177–191.
Fernandez, V. Foliar Nutrition of Crops: Facts, Myths, Perspectives. New Ag.International. com. *Agron. Econo.* 2015, pp 71–78.
Forster, P.; Ramaswamy, V.; Artaxo, P. Changes in Atmospheric Constituents and in Radiative Forcing. In *Climate Change 2007: The Physical Science Basis*; Solomon, S., Qin, D., Manning, M., et al., Eds.; Cambridge University Press: New York, NY, 2007.
Freiberg, E. Microclimatic Parameters Influencing Nitrogen Fxation in the Phyllosphere in a Costa Rican Premontane Rain Forest. *Oecologia* **1998,** *117*, 9–18.
Friedli, H. R.; Atlas, E.; Stroud, V. R.; Giovanni, L.; Campos, T.; Radke, L. F. Volatile Organic Trace Gases Emitted from North American Wildfires. *Glob. Biogeochem. Cycle* **2001,** *15* (2), 435–452.
Gabbatiss, J. Rice Farming up to Twice as Bad for Climate Change as Previously Thought. *Study Reveals* [Online] 2018, pp 1–4 (accessed July 30, 2021).
Ganor, E.; Mamane, Y. Transport of Saharan Dust Across the Eastern Mediterranean. *Atmos. Environ.* **1982,** *16*, 581–587.
Garcia, R.; Andre, J. P.; de Melo-Abreu, R.; Nieto, P.V. P.; White, D. Climate and weather Risk Assessment for Agricultural Planning. In *Guide to Agricultural Meteorological Practices (GAMP)*, 3rd ed.; World Meteorological Organization: Geneva. Vol. 134, Chapter 7. 2010; pp 7.1–7.33.
Gazeley, H. Foliar Feeding Foley or Wisdom [Online] 2012, pp 1–7. https://www.growveg.com/guides/foliar-feeding-folly-or-wisdom/ (accessed February 25, 2021).
Georghhe, I. F.; Ion, B. The Effects of Air Pollutants on Vegetation and the Role of Vegetation in Reducing Atmospheric Pollutants [Online] 2011, pp 1–43. https://www.intechopen.com/chapters/18642/ (accessed July 31, 2021).
Giglio, L.; Kendall, J. D.; Mack, R. A Multi-Year Active Fire Dataset for the Tropics Derived from the TRMM VIRS. *Int. J. Remote Sens.* **2003,** *24*, 4505–4525.
Giglio, L.; van der Werf, G. R.; Randerson, J. T.; Collatz, G. J.; Kasibhatla, P. Global Estimation of Burned Area using MODIS Active Fire Observations. *Atmos. Chem. Phys.* **2006,** *6*, 957–974.
Good, K. Freezing Temperature and Snow Bring Harvest Challenges. Farm Policy New [Online] 2019, pp 1–4. https://farmpolicynews.illinois.edu/2019/10/freezing-temperatures-and-snow-bring-harvest-challenges/ (accessed Jul 30, 2021).
Goudie, S. A.; Middleton, N. J. *Desert Dust in the Global System*; Springer, 2006; pp 318
Greenberg, J. P.; Friedli, H.; Guenther, A. B.; Hanson, D.; Harley, P.; Karl, T. Volatile Organic Emissions from the Distillation and Pyrolysis of Vegetation. *Atmos. Chem. Phys.* **2006,** *6*, 81–91.
Guenther, A. B.; Karl, T.; Harley, P.; Wiedinmyer, C.; Palmer, P. I.; Geron, C. Estimates of Global Terrestrial Isoprene Emissions using MEGAN (Model of Emissions of Gases and Aerosols from Nature). *Atmos. Chem. Phys.* **2006,** *6*, 3181–3210.
Halvorson, A. D.; Del Grosso, S. J. Nitrogen Source and Placement Affect Soil Nitrous Oxide Emissions from Irrigated Corn in Colorado. *Better Crops.* **2012,** *96*, 7–9.
Halvorson, A. D.; Del Grosso, S. J.; Alluvione, F. Nitrogen Rate and Source Effects on Nitrous Oxide Emissions from Irrigated Cropping Systems in Colorado. *Better Crops.* **2009,** *93*, 16–18.

Halvorson, A. D.; Del Grosso, S. J.; Jantalia, C. P. In *Nitrous Oxide Emissions from Several Nitrogen Sources Applied to a Strip-Tilled Corn Field*, Proceedings of Fluid Form, Fluid Fertilizer Foundation. Scottsdale, Arizona. 2011; pp 1–6.

Haq, M. U.; Mallarino, A. P. Soybean Yield and Nutrient Composition as Affected by Early Season Foliar Fertilization. *Agron. J.* **2000,** *92* (1), 16–24.

Haytova, D. A Review of Foliar Fertilizer of Some Vegetable Crops. *Ann. Rev. Res. Biol.* **2013,** *3*, 455–465.

Hayashi, K.; Yan, X. Airborne Nitrogen Load in Japanese and Chinese Agroecosystems. *Soil Sci. Plant Nutr.* **2010,** *56* (1), 2–18.

Heathcote, R. L. Perception of Desertification on the Southern Great Plains-A Preliminary Enquiry. *Internal Report* [Online] 1980, pp 134. United Nations University, Tokyo, Japan. https://archive.unu.edu/unupress/unupbooks/80190e/80190E00.htm/ (accessed July 30, 2021).

Hecobian, S. Comparison of Chemical Characteristics of 495 Biomass Burning Plumes Intercepted by the NASA DC-8 Aircraft during the ARCTAS/CARB-2008 Field Campaign. *Atmos. Chem. Phys.* **2011,** *11*, 18589–18631.

Helms, D. The Soil Conservation Service in the Great Plains. *Natural Resource Conservation Service* [Online] 1990, pp 8. , United States Department of Agriculture: Beltsville, Maryland, USA. https://www.nrcs.usda.gov/wps/portal/nrcs/detail/national/about/history/?cid=nrcs143_021395 (accessed Jul 31, 2021).

Herut, B.; Krom, M. Atmospheric Input of Nutrients and Dust to the Southeast Mediterranean. *Environmental Science and Technology Book Series*. 1996.

Houston, W. R.; Van Der Sluus D. H. Foliar-Applied Urea and Ammonium Nitrate Fertilizers on Short Range Grass. *J. Range Manag.* **1972,** *26*, 358–364.

Husar, R. B.; Lodge, J. P Jr; Moore, D. J. In *Sulphur in the Atmosphere, Proceedings of the International Symposium held in Dubrovnik*, Former Yugoslavia, 1978; pp 816.

Ilsan, N. A.; Nawangsih A. Wahyndi, A. Rice Phyllosphere Actinomycetes as Biocontrol Agent of Bacterial Leaf Blight Disease on Rice. *Asian J. Plant Pathol.* **2016,** *10*, 1–8.

IPCC Climate Change. The Scientific Basis. In *Contribution of Working Group I to the Third Assessment Report of the Intergovernmental Panel on Climate Change*. Houghton, J. T., Ding, Y., Griggs, D. J., Noguer, M., van der Linden, P. J., Dai, X., Maskell, K., Johnson, C. A., Eds.; Cambridge University Press, 2001; pp 881.

Islam, S. M. M.; Gaihre, Y. K.; Biswas, J. C.; Singh, U.; Ahmed, M. N.; Sanabria, J.; Saleque, M. A. Nitrous Oxide and Nitric Oxide Emissions from Lowland Rice Cultivation with Urea Deep Placement and Alternate Wetting and Drying Irrigation. *Sci. Rep.* **2018,** *8*.

Ito, K. *Suppression of Methane Gas Emission from Paddy Fields*, Nippon Steel and Sumitamo Metal Inc. Technical Report No 109: 2020; pp 1–4.

Jacques, M. A.; Morris, C. E. A Review of Issues Related to the Quantification of Bacteria from the Phyllosphere. *FEMS Microbiol. Ecol.* **1995,** *18* (1), 1–14.

Jain, N.; Bhatia, A.; Pathak, H. Emission of Air Pollutants from Crop Residue Burning in India. *Aerosol Air Qual. Res.* **2014,** *14*, 422–430.

Jantalia, A. Fluxes of Nitrous Oxide from Soil Under Different Crop Rotations and Tillage Systems in the South of Brazil. *Nutr. Cycl. Agroecosystems.* **2008,** *82* (2), 161–173.

Janzen, H. H. Carbon Cycling in Earth Systems—A Soil Science Perspective. *Agric. Ecosyst. Environ.* **2004,** *104*, 399–417.

Jenkin, M. E.; Clemitshaw, K. C. Ozone and Other Secondary Photochemical Pollutants: Chemical Processes Governing their Formation in the Planetary Boundary Layer. *Atmos. Environ.* **2000,** *34*, 2499–2527.

Jenkins B. M.; Turn S. Q.; Williams, R. B. Atmospheric Emissions from Agricultural Burning in California: Determination of Burn Fractions, Distribution Factors, and Crop-Specific Contributions. *Agric. Ecosyst. Environ.* **1992,** *38*, 313–330.
Jia, Z.; Cai, Z. Effects of Rice Plants on Methane Emission from Paddy Fields. *J. Appl. Ecol.* **2003,** *14*, 2049–2053.
Jiang, M. Nitrous Oxide Emissions from Chinese Cropland Fertilized with a Range of Slow-Release Nitrogen compounds. *Agric. Ecosyst. Environ.* **2010,** *135* (3), 216–225.
Jones, K. Nitrogen Fixation in Phyllosphere of Douglas Fir, *Pseudotsuga douglasii*. *Ann. Bot.* **1970,** *34*, 239–244.
Jones, C.; Olson-Ruitz, K. Timing of Foliar Application. Montana State University Agricultural Extension Service [Online] 2013, pp 1–28. https://landresources.montana.edu/soilfertility/documents/PDF/pub/EEFEB0188.pdf/ (accessed Jul 30, 2021).
Joshi, P. C.; Swami, A. Air Pollution Induced Changes in the Photosynthetic Pigments of Selected Plant Species. *J. Environ. Biol.* **2009,** *30*, 295–298.
Karamanos, R. E.; Goh, T. B.; Harapiak, J. T. Determining Wheat Responses to Copper in Prairie Soils. *Can. J. Soil Sci.* **2003,** *83*, 213–221.
Karamanos, R. E.; Pomarenski, Q.; Goh, T. B.; Flose, N. A. The Effect of Copper Applied on Grain Yield and Quality of Wheat. *Can. J. Plant Sci.* **2004,** *84*, 47–56.
Kembel, S. W.; Mueller, R. C. Plant Traits and Taxonomy Drive Host Associations in Tropical Phyllosphere Fungal Communities. *Botany*. **2014,** *92*, 302–311.
Kellogg, C. Characterization of Aerosolized Bacteria and Fungi from Desert Dust Events in Mali, West Africa. *Aerobiological*. **2004,** *20* (2), 99–110.
Khalil, K.; Mary, B.; Renault, P. Nitrous Oxide Production by Nitrification and Denitrification in Soil Aggregates as Affected by O_2 Concentration. *Soil Biol. Biochem.* **2004,** *36* (4), 687–699.
Khalil, M. A. K.; Rasmussen, R. A.; Ren, L.; Dalluge, R. W.; Duan, C. Seasonal Production and Emission of Methane from Rice Fields. Final Report: DE-FG03-97ER62401/, 2001.
Khalil, M. A. K.; Rasmussen, R. A.; Shearer, M. J. Factors Affecting Methane Emissions from Rice Fields. *J. Geophys. Res.* **1998,** *103*, 25219–25231.
Khan, P.; Memon, M. Y.; Imtiaz, M.; Aslam, M. Wheat Foliar and Soil Application of Urea at Different Growth Stages. *Pak. J. Bot.* **2009,** *41* (3), 1197–1204.
Kishore, G. K.; Pande, S.; Podile, A. R. Biological Control of Late Leaf Spot of Peanut (*Arachis hypogea*) with Chitinolytic Bacteria. *Phytopathology*. **2005,** *95*, 1157–1165.
Knief, C.; Delmotte, N.; Chaffron, K. Metaproteogenomic Analysis of Microbial Communities in the Phyllosphere and Rhizosphere of Rice. *ISME J.* **2012,** *6*, 1378–1390.
Kok, J. F.; Parteli, E. J. R.; Michaels, T. I., Bou Karam, D. The Physics of Windblown Sand and Dust. *Rep. Prog. Phys.* **2012,** *75*, 106901.
Krishna, K. R. *Agrosphere: Nutrient Dynamics, Ecology and Productivity*; Science Publishers Inc. Enfield: New Hampshire, USA, 2003.
Krishna, K. R. *Peanut Agroecosystem: Nutrient Dynamics and Productivity*. Alpha Science Publishers: Oxford, England, 2008; p 298.
Krishna, K. R. *Agroecosystems: Soils, Climate, Crops, Nutrient dynamics and Crop productivity*; Apple Academic Press Inc. Waretown: New Jersey, USA, 2014; p 546.
Krishna, K. R. *Agricultural Prairies: Natural Resources and Crop Productivity*; Apple Academic Press Inc. Waretown: New Jersey, USA, 2015; pp 496.
Krogmeier, M. J.; McCarty, G. W.; Bremner, J. W. Phytotoxicity of Foliar Applied Urea. *Proc. Natl Acad. Sci. USA* **1990,** *86*, 8189–8191

Krupa, S. V. Effects of Atmospheric Ammonia (NH_3) on Terrestrial Vegetation: A Review. *Environ. Pollut.* **2003,** *124*, 179–221

Kudo, S.; Tanimoto, H.; Inomata, S.; aito, S.; Pan, X.; Kanaya, Y.; Taketani, F.; Wang, Z.; Chen, H.; Dong, H.; Yamaji, K. Emissions of Nonmethane Volatile Organic Compounds from Open Crop Residue Burning in the Yangtze River Delta Region, China. *J. Geophys. Res. Atmos.* **2014,** *119*, 7684–7698.

Kurniawati, F. D.; Setyanto, P.; Suntoro, I.; Cahyani, V. R. The Dynamics of Methane (CH4) Emissions in Organic and Conventional Paddy Fields on Alfisols and Andisols at Karanganyar Regency, Indonesia. *IOP Conf. Ser.: Earth Environ. Sci.* **2018,** *200*, 012026.

Lagomarsino, A.; Agnelli, A. E.; Pastorelli, R.; Pallara, G.; Rasse, D. P.; Silvennoinen, H. Past Water Management Affected GHG Production and Microbial Community Pattern in Italian Rice Paddy Soils. *Soil Biol. Biochem.* **2016,** *93*, 17–27.

Lambais, M. R.; Crowley, D. E.; Cury, J. C.; Bull, R. C.; Rodrigues, R. R. Bacterial Diversity in Tree Canopies of the Atlantic Forest. *Science* **2006,** *312* (5782), 1917.

Lal, R. Sequestration of Atmospheric CO_2 in Global Carbon Cycle. *Energy Environ. Sci.* **2008,** *1*, 86–100.

Lasko, K.; Krishna, P.; Vadrevu, V.; Tran, V. T.; Ellicott, E.; Nguyen, T. N. T.; Bui, H. Q.; Justice, C. Satellites May Underestimate Rice Residue and Associated Burning Emissions in Vietnam. *Environ. Res. Lett.* **2017,** *12*, 085006.

Lauren, J. G.; Pettygrove, G. S.; Duxbury, J. M. Methane Emissions Associated with a Green Manure Amendment to Flooded Rice in California. *Biogeochemistry* **1994,** *24*, 53–65.

Lavelle, P. D.; Dugdale, R.; Scholes, R. Nutrient Cycling. In *Ecosystems and Human Well-being: Current State and Trends*; Berhe, A. A., Carpenter, E., Codispoti, L., Izac, A. M., Lemoale, F. L., Scholes, M., Tiessen, F., Eds., 2005, pp 321–351.

Lederach, W. *The Effects of Fire in Agriculture and Forests Ecosystems*; International Society for Tropical Forestry: Bethesda, Maryland, USA, 2009, pp 1–18.

Lelieveld, J. Atmospheric Oxidation Capacity Sustained by a Tropical Forest. *Nature*. **2008,** *452*, 737–740.

Lelieveld, J.; Peters, W.; Dentener, F. J.; Krol, M. Stability of Tropospheric Hydroxyl Chemistry. *J. Geophys. Res.* **2002,** *107* (D23), 4715.

Lighthart, B. The Ecology of Bacteria in the Alfresco Atmosphere. *FEMS Microbial. Ecol.* **1997,** *23*, 263–274.

Lindow, S. E.; Brandl, M. T. Microbiology of the Phyllosphere. *Appl. Environ. Microbiol.* **2003,** *69*, 1875–1883.

Lindow, S. E.; Leveau, J. H. J. Phyllosphere Microbiology. *Curr. Opin. Biotechnol.* **2002,** *13*, 238–243.

Ling, F.; Silberbush, M. Response of Maize to Foliar vs Soil Application of Nitrogen, Phosphorus-Potassium Fertilizers. *J. Plant Nutr.* **2007,** *25* (11), 2333–2342.

Linquist, B. A.; Marcos, M.; Adviento-Borbe, A.; Anders, M.; Harrell, D.; Linscombe, S.; Reba, M. L.; Runkle, B.; Tarpley, L.; Thomson, A. Greenhouse Gas Emissions and Management Practices that Affect Emissions in US Rice Systems. *J. Environ. Qual.* **2018,** *47*, 395–409.

Linzon, S. N. Sulphur Concentrations in Plant Foliage and Related Effects. *J. Air Pollut. Control Assoc.* **1979,** *29*, 520–525.

Liu, S.; Wang, J. J.; Tuan, W.; Wang, X; Harrison, T. Ammonia, and Greenhouse Gas Emissions from Subtropical Wheat under Different Nitrogen Fertilization Strategies. *J. Environ. Sci.* **2017,** *57*, 196–210.

LSU Ag Centre. Louisiana smoke management Guidelines for agriculture. Louisiana Agricultural Centre Publications 3246 [Online] 2019, pp 1–8. https://www.lsuagcenter.com/NR/rdonlyres/EC7D865A-9BA3-4C5A-9BCC-F800DF664A71/90040/pup3246louisianasmokemanagementguidelinesforagricu.pdf/ (accessed Jul 31, 2021).

Ma, B. L. Nitrous Oxide Fluxes from Corn Fields: On-Farm Assessment of the Amount and timing of Nitrogen Fertilizer. *Global Change Biol.* **2010,** *16* (1), 156–170.

Marschner, P. *Mineral Nutrition of Higher Plants*, 3rd ed.; Academic Press: London, UK, 2012; pp 178–189.

Massey, R.; McClure, H.; Schnieder, R. Agriculture and Greenhouse Gas Emissions. [Onine] 2017, pp 1–8. https://extension2.missouri.edu/g310/ (accessed Mar 15, 2021).

McCarty, J. Remote Sensing-Based Estimates of Annual and Seasonal Emissions from Crop Residue Burning in the Contiguous United States. *J. Air Waste Manag. Assoc.* **2011,** *61* (1), 22–34.

Mittal, S.; Singh, N.; Agarwal, R.; Avasthi, A. Ambient Air Quality during Wheat and Rice Crop Stubble Burning Episodes in Patiala. *Atmos. Environ.* **2009,** *43*, 238–244.

Miyamoto, T.; Kawahara, M.; Minamisawa, K. Novel Endophytic Nitrogen-Fixing Clostridia from the Grass Miscanthus Sinensis as Revealed by Terminal Restriction Fragment Length Polymorphism Analysis. *Appl. Environ. Microbiol.* **2004,** *70* (11), 6580–6586.

Moldanová, J.; Grennfelt, P.; Jonsson, Å.; Simpson, D.; Spranger, T.; Aas, W.; Munthe, J.; Rabl, A. Nitrogen as a Threat to European Air Quality. In *The European Nitrogen Assessment. Sources, Effects and Policy Perspectives*, Chapter 18; Sutton, D. C., Ed.; Cambridge University Press. 2011, pp. 405–433

Morris, C. E.; Kinkel, L. L. Fifty Years of Phyllosphere Microbiology: Significant Contributions to Research in Related Fields. In *Phyllosphere Microbiology*; Lindow, S. E., Ed.; APS Press: St. Louis, MO, 2004.

Mosbeak, H.; Tjell, J. C.; Havmond, M. J. Atmospheric Lead Input to Agricultural Crops in Denmark. *Chemosphere*. **2014,** *19*, 1787–1799.

Mosier, A.; Kroeze, C.; Nevison, C.; Oenema, O.; Seitzenger, S.; Cleemput, O. V. Closing the Global N2O Budget: Nitrous Oxide Emissions through the Agricultural Nitrogen Cycle: OECD/IPCC/IEA Phase II Development of IPCC Guidelines for National Greenhouse Gas Methodology. *Nutr. Cycl. Agroecosystem.* **1998,** *52*, 225–248.

Moulas, C.; Petsoulas, C.; Rouslalaow, K.; Pershucon, C.; Karas, P.; Karpouzas, D. C. Effects of Systematic Pesticides Imidaclorpid an Metalaxyl on the Phyllosphere of Pepper Plants. *Biomed. Res. Int.* **2013,** *2013*, 8.

Muller, T.; Ruppel, S. Progress in Cultivation of Independent Phyllosphere Microbiology. *FEMS Microbiol. Ecol.* **2014,** *87*, 1–27.

Noormets, A.; McDonald, E. P.; Dickson, R. E.; Kruger, A.; Sober, J. G.; Isebrands, J. G.; Karnosky, D. F. The Effect of Elevated CO_2 and O_3 on Leaf and Branch Level Photosynthesis. *Tree Struct. Funct.* **2001,** *15*, 262–270.

Nasim, U. Paddy Fields Burning Smog Shrouds National Capital. *Down to Earth* [Onlne] 2015, pp 1–4 https://www.downtoearth.org.in/news/air/burning-paddy-fields-deteriorate-air-quality-in-delhi-51687/ (accessed Jul 29, 2021).

Ning, J.; Bai, Z.; Gang, G. Functional Assembly of Bacterial Communities with Activity for the Biodegradation of an Organophosphorus Pesticide in the Rape Phyllosphere. *FEMS Microbiol.Lett.* **2010,** *306* (2), 35–143.

Obreza, T. A.; Morgan, K. T. *Nutrition of Florida Citrus Trees*, 3rd ed.; University of Florida Institute of Food and Agricultural Sciences: Lake Alfred, Florida, USA, 2017; pp 1–101.

Oenema, O.; Wrage, N.; Velthof, G. L.; van Groenigen, J. W.; Dolfing, J.; Kuikman, P. J. Trends in Global Nitrous Oxide Emissions from Animal Production Systems. *Nutr. Cycl. Agroecosyst.* **2005,** *72*, 51–65.

Oertal, C.; Matschullat, I.; Zurba, K.; Zimmerman, F. Greenhouse as Emission from Soils-A Review. *Geochemistry* **2016,** *76*, 327–352.

Oosterhuis, D. M.; Weir, B. L. Foliar Fertilization of Cotton. *Physiol. Cotton* **2002,** 272–288.

Osono, T.; Mori, A. Seasonal and Leaf Age-Dependent Changes in Occurrence of Phyllosphere Fungi of Giant Dogwood. *Mycoscience* **2005**, *46*, 273–279,

Ottesen, A. R.; Gorham, S.; Reed, E.; Newell, M. J.; Ramachandra, P.; Canida, T.; Allard, M.; Evans, P.; Brown, E.; White, J. R. Using a Control to Better Understand Phyllosphere Microbiota. *PLoS One*. **2016,** *11* (9), e0163482.

Papagiannaki, K.; Lagouvardous, K.; Kotorni, V.; Papagiannaki, G. Agricultural Losses Related to Frost Events: use of the 850hPa Level Temperature as an Explanatory Variable of the Damage Cost. *Nat. Hazards Earth Syst. Sci.* **2014,** *14* (9) 2375–2386.

Parmar, T. K. Bioindicators: The Natural Indicator of Environmental Pollution. *Front. Life Sci.* **2016,** *9* 110–118.

Parton, W. J.; Gutmann, M. P.; Mechant, E. R.; Hartman, M. D.; Adler, P. R.; McNeal, F.; Lutz S. M. Measuring and Mitigating Agricultural Greenhouse Gas Production in the US Great Plains, 1870-2000. *PNAS* **2015,** *112* (34), E4681–E4688.

Paustian, K.; Babcock, B. A.; Hatfield, J.; Lal, R.; McCarl, B. A.; McLaughlin, S.; Mosier, A.; Rice, C.; Robertson, G. P.; Rosenberg, N. J.; Rosenzweig, C.; Schlesinger, W. H.; Zilberman, D. *Agricultural Mitigation of Greenhouse Gases: Science and Policy Options*; CAST (Council on Agricultural Science and Technology) Report, R141 2004, ISBN 1-887383-26-3, 2004; pp 120.

Penueles, J.; Staudi, M. BVOCs and Global Change. *Trends Plant Sci.* **2009,** *15*, 133–144.

Perazzolli, M.; Antonielli, L.; Storari, M.; Puopolo, G.; Pancher, M.; Giovannini, O. Resilience of the Natural Phyllosphere Microbiota of the Grapevine to Chemical and Biological Pesticides. *Appl. Environ. Microbiol.* **2014,** *80* (12), 3585–3596.

Peterson, G. A.; Schlegal, A. J.; Tanaka, D. L.; Jones, O. R. Great Plains Cropping systems Symposium. *J. Prod. Agric.* **1996,** *9*, 179–186.

Philip, R. G.; Bruulsema, T. W.; Gehl, R. J.; Kanter, D.; Mauzerall, D. L.; Rotz, C. A.; Williams, C. O. Nitrogen–Climate Interactions in US Agriculture. *Biogeochemistry* **2013,** *114*, 41–70.

Punjab Remote Sensing Board. Monitoring Residue Burning through Satellite Remote Sensing [Online] 2015. Punjab Pollution control Board, Patiala, Punjab, India. https://gis-prsc.punjab.gov.in/residue/Index.aspx/ (accessed July 31, 2021).

Qian, X.; Chen, L.; Guo, X.; He, D.; Shi, M.; Zhang, D. Shifts in Community Composition and Co-Occurrence Patterns of Phyllosphere Fungi Inhabiting *Mussaenda shikokiana* Along an Elevation Gradient. *Peer J.* **2018,** *6*, e5767.

Rahman, I. U. R.; Afzal, A.; Mana, S. Foliar Application of Plant Material Nutrition on Wheat: A Review. *J. Agric. Allied Sci.* **2014,** *3*, 19–22.

Randerson, J. T.; Van der Werf, G. R.; Giglio, L.; Collatz, G. J.; Kasibatla, P. S. Global Fire Emissions Database Version 4 (GFEDv 4). Oakridge National Laboratory, Tennessee, USA [Online] 2015. http:// dx.doi.org/10.3334/ORNLDAAC/1293/ (accessed Jul 30, 2021).

Rastogi, G.; Coaker, G. L.; Leveau, J. H. J. New Insights into the Structure and Function of Phyllosphere Microbiota through High through-put Molecular Approach. *FEMS Microbiol. Lett.* **2013,** *348* 1–10.

Redford, A. J.; Bowers, R. M.; Knight, R.; Linhart, Y.; Fierer, N. The Ecology of the Phyllosphere: Geographic and Phylogenetic Variability in the Distribution of Bacteria on Tree Leaves. *Environ. Microbiol.* **2010,** *12*, 2885–2893.

Redford, A. J.; Fierer, N. Bacterial Succession on the Leaf Surface: A Novel System for Studying Successional Dynamics. *Microbiol. Ecol.* **2009,** *58*, 189–198.

Reena, S.; Verma, O.; Datta, S.; Datta, D. Response of Field Crops to Foliar Nutrition: A Review. *Chem. Sci. Rev. Lett.* **2018,** *7*, 402–408.

Reiley, L. The New Plan to Remove a Trillion Tons of Carbon Dioxide from the Atmosphere: Bury It. Climate Centre [Online] 2019. https://theclimatecenter.org/the-plan-to-remove-a-trillion-tons-of-carbon-dioxide-from-the-atmosphere-bury-it/ (accessed Jul 30, 2021).

Remus-Emsermann, M. N. P.; Schlecter, R. O. Phyllosphere Microbiology: At the Interface between Microbial Individuals and Plant Host. *New Phytol.* **2018,** *218* (4), 1327–1333.

Rigby, J. R.; Porporato, A. Spring Frost Risk in a Changing Climate. *Geophys. Res. Lett.* **2008,** *35* (12), L12703.

Rivas-Ubach, A. R.; China, S.; Smallwood, C. R.; Handakumubura, P. P.; Sharma, N.; Pasa-Tolic, L.; Ahkami, A. H.; Guenther, A. B.; Janson, C. Aerosol-Microbe-Plant Interactions in the Phyllosphere. AGU Conference [Online] 2019. https://agu.confex.com/agu/fm19/meetingapp.cgi/Paper/622242/ (accessed July 31, 2021).

Roberts, A.; Knippertz, P. Haboobs: Convectively Generated Dust Storms in West Africa. *Weather* **2012,** *67* (12), 311–317.

Romheld, D.; Marschner, H.; Becker, K.; Schecht, E.; Hulsebusch, C.; Buerkert, A. Effects of Site Factors on the Efficiency of Soil Amendments and on Nutrient Fluxes in Subsistence-Oriented Cropping of the West African Sahel [Online] 2000, pp 1–24. http://www.troz.uni-hohenheim.de/research/sfb308/Ebcont/bcheck. (accessed July 30, 2021).

Roemheld, V.; El-Fouly, M. M. In *Foliar Fertilization. Proceedings of 2nd International Workshop on "Foliar Fertilization"*, Bangkok, Thailand, 1999; pp 1–32.

Ruser, R. Emission of N_2O, N_2 and CO_2 from Soil Fertilized with Nitrate: Effect of Compaction, Soil Moisture and Rewetting. *Soil Biol. Biochem.* **2006,** *38* (2), 263–274.

Saikonen, K.; Mikola, J.; Helander, M. Endophytic Phyllosphere Fungi and Nutrient Cycling in Terrestrial Ecosystems. *Curr. Sci.* **2015,** *19*, 120–126.

Sangomla, A. We have Breached the Planetary Limit for Nitrogen Emission. [Online] 2017, pp 1–4 https://www.downtoearth.org.in/news/agriculture/when-n-means-noxious-59279 (accessed Jul 31, 2021).

Schmidt, D.; Schmidt, C. Frosts Reduce Almond Production in Spain. The Clipper [Online] 2018, pp 1-5. https://theclippermag.com/index.php/nuts/almonds/frosts-reduce-almond-production-in-spain// (accessed Jul 21, 2021).

Scils, R. L. M. Nitrous Oxide Emissions from Multiple Combined Applications of Fertilizer and Cattle Slurry to Grassland. *Plant Soil.* **2008,** *310*, 1–2, 89–101.

Sela, G. Foliar Feeding. Smart Fertilizers. [Online] 2014 pp 1–14. https://www.smart-fertilizer.com/articles/foliar-feeding/ (accessed Jul 31, 2021).

Seyyednejad, M.; Niknejad, M.; Koochak, H. A Review of Some Different Effects of Air Pollution on Plans. *Res. J. Environ. Sci.* **2011,** *5*, 302–309.

Shabbir, M.; Jumaid, A.; Zahid, J. Smog: A Transboundary Issue and its Implications in India and Pakistan. Sustainable Development Policy Institute (SDPI) Policy Paper 67, 2019, pp 1–34.

Shao, Y.; Wyrwoll, K. H.; Chappell, A. Dust Cycle: An Emerging Core Theme in Earth System Science. *Aeolian Res.* **2011,** *2*, 181–204.

Signor, D.; Eduardo, C.; Cerri, P. Nitrous Oxide Emissions from Agricultural Soils. A Review. *Pesq. Agropec. Trop. Goiânia* **2013,** *43*, 322–338.

Singh, C. K. Foliar Application of Fertilizer for Increasing Nutrient use Efficiency and Crop Production. *Rashtriya Krishi* **2016,** *1*, 79–80.

Singh, J. P.; Karamanos, R. E.; Kachanoski, R. G. Spatial Variation of Extractable Micronutrients in a Cultivated and a Native Prairie Soil. *Can. J. Soil Sci.* **1985,** *65*, 149–156.

Slattery, R. A.; Walker, B. J.; Weber, A. P. M.; Ort, D. R. The Impact of Fluctuating Length on Crop Performance. *Plant Physiol.* **2019,** *176*, 990–1003.

Smee, M. R.; Real-Ramirez, J.; Hendry, A. Insects as Phyllosphere Microbiome Engineers: Effects of Aphids on Plant Pathogens. 2019. doi.org/10.1101/797738/.

Smith, P. Engineered Biological Sinks on Land. In *The Global Carbon Cycle: Integrating Humans, Climate, and the Natural World*; Field, C. B. Raupach M. R. Eds.; SCOPE 62, Island Press: Washington D.C., 2004; pp 479–491.

Smith, K. A.; Conen, F. Impacts of Land Management on Fluxes of Trace Greenhouse Gases. *Soil Use Manag.* **2004,** *20*, 255–263.

Smith, P.; Martino, D.; Cai, Z.; Gwary, D.; Janzen, H.; Kumar, P.; McCarl, B.; Ogle, S.; O'Mara, F.; Rice, C.; Scholes, B.; Sirotenko, O. Agriculture. In *Climate Change 2007: Mitigation. Contribution of Working Group III to the Fourth Assessment Report of the Intergovernmental Panel on Climate Change*; Metz, B.; Davidson, O. R.; Bosch, P. R.; Dave, R.; Meyer, L. A., Eds.; Cambridge University Press, Cambridge: United Kingdom and New York, NY, USA, 2007; pp 499–532.

Snyder, S.; De Melo-Abreu, *Frost Protection: Fundamentals, Practice, and Economics*; Vol. I. FAO, United Nations, Rome. This volume details the technical aspects of frost protection, 2005. http://www.fao.org/docrep/008/y7223e/y7223e00.htm/ (accessed July 31, 2021).

Sterk, G.; Herman, L.; Bationo, A. Windblown Nutrient Transport and Soil Productivity Changes in Southwest Niger. *Land Degrad. Dev.* **1996,** *7*, 325–335.

Sterk, G.; Stroosnijder, I.; Raats, P. A. C. Wind Erosion and Control Techniques in Sahelian One Niger [Online] 1998, pp 1–4. http://www.wu.ksu.edu/symposium/proceedings/sterk.pdf (accessed July 31, 2021).

Stoorvogel, J. J.; Smalling, E. M. A. *Assessment of Soil Nutrient Depletion in Sub-Saharan Africa*; 1983–2000 Volume 1 Main Report. The Win and Staring Centre, Wageningen: The Netherlands, 2000, p 342.

Tegen, I.; Fung, I.; Modelling of Mineral Dust in the Atmosphere: Sources, Transport and Optical Thickness. *J. Ž. Geophys. Res.* **1994,** *99* (D11), 22879–22914.

Tegen, I.; Fung, I. Contribution to the Atmospheric Mineral Aerosol Load from Land Surface Modification. *J. Geophys. Res.* **1995.** *100* (D9), 18707–18726.

Tripathi, S.; Singh R. N.; Shaishav, S. Emission for Crop/Biomass Residue Burning Risk to Atmosphere Quality. *Int. Res. J. Earth Sci.* **2013,** *1*, 24–33.

Undersander, D. J.; Smith, L. H.; Kaminski, A. R.; Kelling, K. A.; Doll, J. D. Sorghum-Forage. Alternative Field Crops. University of Wisconsin- Madison, USA [Online] 1990, *pp 1–7 http://www.hort.purdue.edu/newcrop/afcm/forage.html* (accessed Jul 31, 2013).

UNEP. *Global Assessment of Sand and Dust storms. United Nations Environment Program*; World Meteorological Organization and United nations Convention to Combat Desertification: Paris, 2016, pp 1–139.

University of California Cooperative Extension Service. Foliar Nutrient Sprays. Division of Agriculture and Natural Resources, [Online] 2020 pp 1–8 http://ceventura.ucanr.

edu/?blogtag=foliar%20nutrient%20sprays&blogasset=19305/ (accessed February 26, 2021).

Vacher, C.; Hampe, A.; Porte, H. J.; Saer, U.; Company, S.; Morris, C. E. The Phyllosphere: Microbial Jungle at Plant-Climate Interface. *Annu. Rev. Ecol. Evol. Syst.* **2016,** *47*, 1024.

Van Breemen, N.; Mulder, J.; Driscoll, C. T. Acidic Deposition and Internal Proton Sources in Acidification of Soils and Waters. *Nature.* **1983,** *307*, 599–604.

Van Damme, M.; Wichink Kruit, R. J.; Schaap, M.; Clarisse, L.; Clerbaux, C.; Coheur, P.-F.; Dammers, E.; Dolman, A. J.; Erisman J. W. Evaluating 4 Years of Atmospheric Ammonia (NH3) over Europe using IASI Satellite Observations and LOTOS-EUROS Model Results, *J. Geophys. Res. Atmos.* **2014,** *119*, 9549–9566.

van der Werf, G. R.; Randerson, J. T.; Giglio, L.; Collatz, G. J.; Mu, M.; Kasibhatla, P. S.; Morton, D. C.; DeFries, R. S.; Jin, Y.; van Leeuwen, T. T. Global Fire Emissions and the Contribution of Deforestation, Savanna, Forest, Agricultural, and Peat Fires (1997–2009), *Atmos. Chem. Phys.* **2010,** *10*, 11707–11735.

van der Werf, G. R.; Randerson, J. T.; Collatz, G. J.; Giglio, L.; Kasibhatla, P. S.; Avelino, A.; Olsen, S. C.; Kasischke, E. S. Continental-scale Partitioning of Fire Emissions during the 1997–2001 El Nino/La Nina Period. *Science* **2004,** *303*, 73–76.

Varshney, C. K.; Garg, J. K.; Lauenroth, W. K.; Heitschmidt, R. K. Plant Response to Sulphur Dioxide Pollution. *CRC Crit. Rev. Environ. Control.* **2009,** *9*, 27–49.

Venterea, R. T. Challenges and Opportunities for Mitigating Nitrous Oxide Emissions from Fertilized Cropping Systems. *Front. Ecol. Environ.* **2012,** *10*, 562–570.

Verbree, D.; McClure, A. T.; Leib, B. Fertilization of Row Cropping Using Overhead Irrigation [Online] 2016, pp 1–5. University of Tennessee Extension Service. https://irrigation.tennessee.edu/wp-content/uploads/sites/176/2020/08/Fertigation-of-Row-crops-using-overhead-Irrigation-W303.pdf/ (accessed Jul 31, 2021).

Vokou, D.; Vareli, K.; Zarali, E.; Karamanoli, K.; Constantinidou, H. I.; Monokrousos, N. Exploring Biodiversity in the Bacterial Community of the Mediterranean Phyllosphere and its Relationship with Airborne Bacteria. *Microb. Ecol.* **2012,** *64* (3), 714–724.

Vorholt, J. A. Microbial Life in the Phyllosphere. *Nat. Rev. Microbiol.* **2012,** *10*, 828–840.

Wang C.; Lai D. Y. F.; Sardans, J.; Wang, W.; Zeng, C.; Peñuelas, J. Factors Related with CH_4 and N_2O Emissions from a Paddy Field: Clues for Management implications. *PLoS One.* **2017,** *12* (1), e0169254.

Wang, J.; Zhang, X.; Xiong, Z.; Khalil, M. A. K.; Zhao, X.; Xie, Y.; Xing, G. Methane Emissions from a Rice Agroecosystem in South China: Effects of Water Regime, Straw Incorporation and Nitrogen Fertilizer. *Nutr. Cycl. Agroecosys.* **2012,** *93*, 103–112.

Webb, T.; Hatchings, N.; Amon, B. *Field Burning of Agricultural Residues. EMEP/EEA Pollutant Emission Inventory Guidebook-2019*; European Environmental Agency Brussels: Belgium, 2019, pp 1–14.

Wei, X.; Lyu, S.; Zonghua, W.; Liu, H.; Pan, D.; Chen, J. Phylloremediation of Air Pollutants: Exploiting the Potential of Plant Leaf Surfaces and Leaf Associated Microbes. *Front. Plant Sci. Rev.* **2017,** *8*, 1318.

Whiley, A. W.; Pegg, K. G. Correction of Micro-Nutrient Deficiencies and Control of Phytophthora Root Rot in Avocado. In *Talking Avocados*, 2nd ed.; May/June, 1990, p 11.

Whipps, J. M.; Hand, P.; Pink, D.; Bending, G. D. Phyllosphere Microbiology with Special Reference to Diversity and Plant Genotype. *J. Appl. Microbiol.* **2012,** *105* (6), 1744–1755.

Wilke, B. M.; Duke, B. J.; Jimoh, W. L. O. Mineralogy and Chemistry of Harmattan Dust in Northern Nigeria. *Catena* **1984,** *11*, 91–96.

Williams, T. R.; Marco, M. L. Phyllosphere Microbiota Composition and Microbial Community Transplantation on Lettuce Plants Grown Indoors. *mBio*. **2014,** *5* (4), e01564–14.

Williams, T. R.; Moyne, A. L.; Harris, L. J.; Marco, M. L. Season, Irrigation, Leaf Age, and Inoculation Influence the Bacterial Diversity in the Lettuce Phyllosphere. *PLoS One*. **2013,** *8* (7), e68642.

Wu, Y. F.; Zhong, X. L.; Hu, X.; Ren, D. C.; Lu, G. H.; Wei, Y. Frost Affects Grain Yield Components in Winter Wheat. *N. Z. J. Crop Hortic. Sci.* **2013,** *42*, 194–204.

Xinghua, L.; Shuxiao, W.; Lei, D.; Jiming, H.; Chao, L.; Yaosheng, C.; Liu, Y. Particulate and Trace Gas Emissions from Open Burning of Wheat Straw and Corn Stover in China. *Environ. Sci. Technol.* **2007,** *41*, 6052–6058.

Yang, C. H.; Crowley, D. E.; Bomeman, J.; Keen, N. T. Microbial Phyllosphere Populations are More Complex than Previously Realized. *Proc. Natl. Acad. Sci.* **2001,** *98*, 3889–3894.

Yokelson, R. J. Emissions from Biomass Burning in the Yucatan. *Atmos. Chem. Phys.* **2009,** *9*, 5785–5812.

Yokelson, R. J.; Burling, I. R.; Urbanski, S. P.; Atlas, E. L.; Adachi, K.; Buseckpher, P. R.; Wiedinmyer, C.; Akagi, S. K.; Toohey, D. W.; Wold, C. E. Trace Gas and Particle Emissions from Open Biomass Burning in Mexico. *Atmos. Chem. Phys.* **2011,** *11*, 6787–6808.

Youlin, Y.; Squires, V. R.; Qi, L. *Global Alarm: Dust and Sandstorms from the World's Drylands*; United Nations Convention to Combat Desertification, United Nations Building: Bangkok, Thailand, 2001; pp 345.

Zain, M.; Khan, I.; Qadri, R. W. K.; Ashraf, U.; Hussain, S.; Siddique, A.; Bashir, M. Foliar Application of Micronutrients Enhances Wheat Growth, Yield and Related Attributes. *Am. J. Plant Sci.* **2015,** *6*, 864–869.

Zhang, H.-L.; Bai, X.-L.; Xue, J.-F.; Chen, Z.-D.; Tang, H.-M.; Chen, F. Emissions of CH_4 and N_2O under Different Tillage Systems from Double-Cropped Paddy Fields in Southern China. *PLoS One*. **2003,** *8* (6), e65277.

Zhang, J. K.; C. S.; Hao, Q. J.; Tang, Q. W.; Cheng, B. H.; Li, H.; Chen, L. H. Effects of Tillage-Cropping on Methane and Nitrous Oxide Emissions from Agroecosystems in a Purple Paddy Soil. **2019,** *33* (6), 207–213.

Zhang, X. L.; Pan, X. J.; Xiang, Z. G.; Wang, J. X.; Yang, B.; Linu, X. L.; Liu, P. L. Effects of Farming Management on the Global Warming Potential of CH and N2) from a Rice-Wheat Rotation System based on the Analysis of DNDC Modelling. *J. Appl. Ecol.* **2013,** *24*, 690–696.

Zhang, B.; Tian, H.; Ren, W.; Tao, B.; Lu, C.; Yang, J.; Banger, K.; Pan, S. Methane Emissions from Global Rice Fields: Magnitude, Spatio-Temporal Patterns, and Environmental Controls. *AGU Public. Glob. Biogeochem. Cycles*. **2016,** *30* (9), 1–18.

Zhang, B.; Zhang, H.; Tangling, J.; Yang, J.; Li, B.; Zhuan, G. G.; Baiz, H. Effect of Cypermethrin Insecticide on the Microbial Community in Cucumber Phyllosphere. *J. Environ. Sci.* **2008,** *20*, 1356–1362.

Zekri, M.; Schumann, A. W.; Vashisth, T.; Kadyampakeni, D. M.; Morgan, K. T.; Boman, B.; Obreza, T. *Florida Citrus Production Guide. IFAS Extension Service*; University of Florida: Lake Alfred, Florida, USA, 2020; pp 1–78.

CHAPTER 3

Natural Biotic Factors of the Agricultural Sky Relevance to Crop Production

ABSTRACT

The agricultural sky is abode to a wide range of biotic factors that potentially affect global crop production trends. Among them, our discussions in this chapter confine to aerosol microbes, insects, and Aves. They are important to farming enterprises. The focus here is on biotic aspects of lower atmosphere close to crop's canopy. Microbes traced in the atmosphere influence crops in several ways. Microbes occur on the surface of soil/crop's canopy. Investigation of air samples in the atmosphere indicates that diverse species of microbes are distributed at different population densities, at altitudes up to 13–15 km in the troposphere. Microbes are also found in the stratosphere up to 41–74 km altitude from the crop's canopy. The atmosphere with its harsh environment is said to be home to microbes that are versatile. They withstand relatively harsh environment. Molecular methods of detection of microbes utilizing rRNA suggest that microbes are traceable even in space. Microbes may descend on earth's surface (on crops) periodically. Like soil, atmosphere, particularly, at low altitudes above crop's canopy too harbors diverse flora of microbes. We need to decipher in detail their influence on crops. We should aim at manipulating aerosol microbes to our advantage if they are beneficial. We should curb those microbes that cause diseases, epidemics/pandemics on staple crops. Of course, several microbial species could be commensals. Bioaerosol microbes could include bacteria, fungi, viruses, archaea, algal cells, lichen fragments, protozoans, etc. Clouds may be a good medium for aerosol microbes to survive, perpetuate, and get dispersed. Microbes have a role in enucleation and formation of snow, frost, and fog. Microbes found on smog/organic particulate matter may aid

The Agricultural Sky: A Concept to Revolutionize Farming. K. R. Krishna, PhD (Author)

biological transformation of nutrients. Airborne microbial species transit long distances along high-altitude winds and jet streams. Microbes are capable of intercontinental dispersal. Microbes transit via dust storms, clouds, precipitation, and wind in the atmosphere above agrarian regions. For example, aeolian dust that moves from Sahara/Sahel into North American agrarian carries microbes. The agricultural sky is an important region with regard to the initiation and spread of crop disease pandemics. Airborne microbes include several species that cause devastating crop diseases. The propagules such as spores, hyphae, and cells may get disseminated via intercontinental dust movement. The turbulence of low wind distributes pathogenic microbes only for shorter distance. Whereas high altitude jet streams carry spores of disease-causing fungi, such as *Puccinia graminis tritici* (PgT), to long distances of over 15,000 km. Often such winds are responsible for the initiation and dispersal of pandemics. Several examples pertaining to epidemic/pandemics on crops. We may note that agrarian sky has been devastating farms and distributing catastrophic disease, periodically. Perhaps agrarian sky has been more dangerous to crops than the subsoil region of cropped fields. Pathogenic microbes are distributed by vectors. Particularly, insect species like aphids, flies, and bugs that hover above crop canopy. Long-distance transport of disease microbes is possible. Climate change is an added factor that farmers have to negotiate. The climate change effects could affect crops plus the disease-causing aerosol microbe to different extents. It could enhance virulence of epidemic/pandemic causing fungal pathogen or may reduce it. Phyllosphere is a large aerial surface. It supports microbes of diverse physiological capabilities. Phyllosphere microbes aid biochemical transformations nutrients traced of foliage. Phyllosphere microbes may also interact with microbial pathogens that invade crops via the leaves. Several microbes are known to reduce disease spread by suppressing crop pathogens. Phyllosphere microbes may also include asymbiotic biological nitrogen fixers and plant growth-promoting bacteria. Studies on insect distribution in the atmosphere above crops began during the 1920s. Aerospace from 200 to 16,000 m altitude supported insect species and their eggs floating in the air. The agricultural sky supports several insect pests that attack crops. Cereal aphids such as Russian aphid (*Diuraphis noxia*), green bugs, pod borers (*Helicoverpa armigera*), leaf eaters, cut worms, boll eating bugs, etc., are commonly traced in the atmosphere above crops. Locusts are migratory pests. Locust (short-horned grasshoppers) spread is an important aerial phenomenon in certain parts of the world. Locusts move as clouds of billions of insects and initiate epidemics/pandemics. Locusts destroy large areas of natural vegetation and crops in a short period. Aerial insects may act as predators of insect pests. They are useful biological control agents. Insects like honeybees, moths, butterflies, vasps, bugs act as important pollinators and induce good seed set. Honeybees in particular produce honey. Aves (birds) are important components of agrarian sky. There are several bird species that could be pests of crops (e.g., weaver birds, red-winged blackbird, finches, sparrows, rock pigeons, parakeets, etc.). The predatory species (e.g., falcons, raptors) are helpful in reducing insect pest and other pest bird species. Birds are important biological control agents. Birds are also efficient pollinator species. A few crops like sunflower depend on bird-mediated pollination for good seed set. Bird scaring using aerial vehicles like Helikites is new to agrarian sky. Eradication of bird pests through chemicals is generally not an option. Diversion of birds by removing nesting zones near crop fields is practiced. Birds aid in dispersal of seeds. So, they could induce volunteers, weeds, and other undesirable plant species in farms. Birds that flock the aerospace in flocks also offer "bird refuse" called as "Guanos." Guanos are rich in nutrients needed by crops.

3.1 INTRODUCTION

During the past century, we have accrued a great deal of knowledge about the atmosphere above crops, particularly the seasons, weather patterns, snow, fog, frost, cold fronts, heat waves, droughts, storms, floods, and their influence on crops. Also, we have disseminated our know-how about the selection of field sites, soils, seasons, crops, their genotypes, and agronomic procedures. We have bestowed a lot of our time and intellect, to study the abiotic factors and their manifestations in the agricultural sky. For whatever reason, we overlooked biotic factors in the atmosphere above crops. Such biotic factors have not been studied to the same extent as we did with soils or crops. For example, we have not even characterized and classified the sky above crop fields, to any great extent. Usually, a farmer does get soils analyzed. He knows several aspects of about crops that he cultivates. However, it is not the same with agrarian sky. Our inquisitiveness has often confined to understanding the weather aspects of the atmosphere which are mainly abiotic factors. Among biotic factors, we know more about devastating aerial pests, such as locusts. Particularly, as they move across conspicuously. We have understood about aerial insects that attack crops, to a certain extent. A few fungal pathogens have been investigated in greater detail. However, this is not all that we should know about the agricultural sky. Particularly, if we want to regulate and exploit this region of the agroecosystem–aerospace. Believe it or not, right now our ability to manipulate agricultural sky is not as good as we can do so with soils or crops. This is because there is lacunae in our understanding about agrarian sky. There are no schools, faculties, or courses that cater, exclusively, to study aeolian nature of agrarian regions. Integrated approaches to study agrarian sky are missing, unlike soils or crops.

The theme of this chapter is to highlight the facts about biotic components of the agrarian atmosphere (i.e., agricultural sky). Here, we confine our discussions to biotic factors such as microorganisms, their distribution, and functions displayed by them. Microbes that are distributed from the soil surface (field) to an altitude of 12, 21, 28, 41, and even 70 km have been discovered. Any of them may have an important role to play in an agroecosystem. The importance of Aeolian microbes to the sustenance of crops, if any, needs mention. There are indeed innumerable microbes that attack and cause devastations to standing crops. Plant pathogens vary in their impact on crops. A few of them may cause severe pandemics, or epidemics or stay as endemic pathogens on crops. Here we highlight the importance of atmospheric factors and their influence on crop disease spread, intensity, and resultant losses to grain productivity. The diversity of airborne diseases above cropping expanses could be governed, to a certain extent, by the atmosphere. Farmers need to be alerted to changes in disease-causing potential of aerospace above their fields.

Entomological aspects of the atmosphere need greater attention than presently bestowed. In addition to major aerial insect pests, there are also beneficial aspects such as pollination, collection of honey, etc. They are related to insect activity above crops. Plant protection chemicals applied are known to contaminate the agricultural sky. Residual effects of pesticides in the aerosols above crops need attention.

Aves (birds) that operate incessantly in the sky can reach pest proportions, if they are voracious grain eaters/destroyers. They could be immensely deleterious to farmer's exchequer since they destroy the yield. Aves also operate as biological control agents, by feeding on various stages of insect pests. Birds keep the pest population and their destructive activity under the threshold. Aves are supposedly good pollinators of commercial crops such as sunflower. The seed set of this oilseed crop depends on Aves in agrarian sky, as much as to soil fertility and phloem activity—do we realize it? Aves also offers fertilizer (e.g., guanos) in some agrarian regions. For example, in Sahelian farming region and more importantly in the South American farming zones of Peru, Chile, and Argentina. Some of these aspects of the agricultural sky and its components are highlighted in this chapter.

3.2 MICROORGANISMS IN THE AGRICULTURAL SKY

Historically, microbes were first detected in the air by Antoni van Leeuwenhoek. He suggested that microbes were traced in the aerial habitats (Gregory 1971, 1973; van Leeuwenhoek, 1941). It seems that Darwin (1846) detected organic life forms in the aerial dusts aboard the ship used for navigation. Pasteur (1861) made a rudimentary effort to study aerial microbes. During the mid-1900s, air samples were collected using airplanes, and microbes were studied (Fierer and Lindbergh, 1935). In the 1970s, microbial culture medium for in-vitro culture of aerial microbes was designed. Artificial media was utilized to study metabolic abilities and reproduction trends of microbes. Aerospace microbes were isolated on an artificial medium and studied (see Womack et al., 2010, Imshenetsky et al., 1978). Molecular methods of identification of microbes were applied to prove the existence and density of microbes in atmospheric air samples. Usually, such samples were obtained from different altitudes, including stratosphere (Bowers et al., 2009; Smith, 2013; Smith et al., 2012, 2013; Spring et al., 2018; Yamaguchi et al., 2012).

Space microbiology needs commensurate attention regarding the way it might influence cropping expanses, if any. Microbes may descend onto earth's surface, periodically. During the past decade, that is, 2010–2020, air samples from stratosphere were collected using space vehicles, shuttles, and International Space Station. They traced microbes that could tolerate the harsh habitats in the upper troposphere and lower stratosphere. We have no knowledge about their biochemical/physiological abilities, particularly, to withstand UV radiation, extremely cold temperatures, and low oxygen tension (ISRO, 2009, Smith et al., 2009; Amato 2012; Amato et al., 2017 a. b; Horneck et al., 2010; Fierer et al., 2008). A few trials using helium balloons have shown that bacteria such as haloarcheae which are salt-loving could survive in the lower stratosphere. It is worth noting here that if one wants, we could seed the stratosphere with microbes of desired traits. However, previous reports by NASA suggest that *Bacillus pumillus* is a known extreme weather tolerant microbial strain. Still, it could not survive or flourish in the stratosphere under conditions that prevailed at 31 km altitude from the ground. About 99.9% of over 40 million *B. pumillus* cells seeded into stratosphere perished within 8 h. Overall, stratosphere is believed to be the upper limit for the survival of many microbial species. The existence of microbes even for a short duration requires certain specialized traits such as tolerance to desiccation, UV radiation, dearth for moisture, etc. Stratosphere seems to help in long-distance journey of microbes. Upon landing on agrarian land surface, stratosphere microbes may proliferate and influence the microbial composition of soils, root zone, leaf surface flora, etc. (DasSarma and DasSarma, 2018).

Atmosphere, that is, aerospace above the crops in the present context is a repository of microbes of diverse genera. Their physiological abilities are worth noting considering that atmosphere is a "feeble medium" for microbes to flourish. Womack et al. (2010) in fact, aptly point out that we have not pursued studies on aerial microbes, to any great extent. We have focused more on knowing microbial diversity and microbe related biochemical/physiological transformation and diverse eco-systematic functions attributable in soil and water. She further cautions that the atmosphere is not just a conduit to microbes that transit from one terrestrial ecosystem (soil surface) to other, etc. Instead, microbes utilize the atmosphere for survival in the aerospace above crops and natural vegetation. They are known to mediate several biochemical transformations involving sugars, organic acids, and complex proteins traced on particulate matter in the atmosphere (Womack et al., 2010; Pearce et al., 2009). A few photoautotrophs have also been detected in the air above vegetation (Brodie et al., 2007; Herlihy et al.1987). Therefore, Womack et al. (2010) argued that aerial microbes are metabolically active. Perhaps in due course, we could exploit their physiological/metabolic prowess to our advantage.

Womack et al. (2010) further listed and showed us that atmosphere is not as harsh a habitat as others where microbes survive or actively flourish. The moderate or low level of organic sources, moisture, and gaseous components together are actually fairly congenial for microbes. Particularly, to grow and reproduce for several generations. Regarding ambient conditions for microbial survival and growth, Womack et al. (2010) stated that pH, temperature, moisture, air, and UV radiation are not as harsh as in soil, oceans, or other habitats. Microbes may negotiate much harsher climatic conditions and proliferate. For example, microbes that tolerate pH 3.0 (extreme acidity) and pH 9 (alkalinity) are known. Microbes that tolerate −18°C and survive are known. There are microbes that tolerate UV radiation via appropriate DNA repair mechanism. Therefore, compared to such extreme habitats, aerospace above crops at different altitudes are not too harsh or alien for microbes. The atmosphere above natural vegetation is known to contain several types of organic acids, carboxylic, acids, alcohols, complex proteins, and other not easily degradable organic complexes. However, arial microbes that utilize such compounds in the atmosphere are known (Womack and McAlpine, 2015; Kawamura and Kaplan, 1990). A few of the microbial species traced in the atmosphere at different altitudes are *Bacillus sp., Staphylococcus aureus, Xanthomonas, Penicillium notatus, Micrococcus albus, Mycobacterium luteum, Aspergillus niger, Papulospora anamola, Circinella muscae*, etc.

Despite the above observations, the atmosphere is still considered an extreme environment where organisms are subject to low temperatures and high radiation. Many of the microorganisms detected appear in resistant forms. Otherwise, they show mechanisms of adaptation designed to withstand these extreme conditions. Airborne microorganisms may play an important role in the global climate system, biogeochemical cycling, nutrient supply, and crop's health. Dust storms are the atmospheric phenomenon that moves more topsoil through the earth's atmosphere (Aguiliera et al., 2018). Gat et al. (2017) have shown that the composition of bioaerosol or dust storm microbes depend on the source. For example, they say, it is possible to trace differences in dust storm microbes based on the dust storms that got generated from locations, such as North Africa, Syria, and Saudi Arabia (Gat et al., 2017; Toepfer et al., 2011). The impact of microbes from different origins on agrarian regions needs some thought and experimentation. Locations that infuse plant pathogens into dust storms need to be identified carefully. If possible, remedial measures to thwart the pathogens selectively need to be devised.

The atmosphere is a region suitable for "extremophile microbes" since the conditions could be really harsh for any organism (Smets et al., 2016; Madigan and Marrs, 1997; DasSarma and DasSarma, 2018). It is known that airborne cells can maintain viability during their atmospheric residence. They can exist in the air as spores or as vegetative cells. It is dependent on diverse molecular mechanisms of resistance and adaptation (Smets et al., 2016, DasSarma and DasSarma, 2018). The big question is whether some of them can be metabolically active, divide, and multiply. Also, how well they could survive and land on agricultural expanses, to affect the soils/crops.

Microbial flora of atmosphere above agrarian regions has not yet been studied extensively. There is a general opinion that we have not explored the microbial diversity or biochemical transformations related to several airborne microbes. This is particularly true if we compared the way we accrued knowledge, about microbes in agrarian soils, water resources or crop surfaces, etc. Studies about functional aspects of atmospheric microbes are meagre (Polymenakou, 2012). In the atmosphere, beginning from the ground surface until the frontier in the lower stratosphere, microbes may get attached to dust particles, or stay in raindrops, fog, snow, or hail or particulate matter in general (Bauer et al., 2002; Fuzzi et al., 2006; Sattler et al., 2001; De Guillame et al., 2008). These microbe-laden particulate matter could return to earth, say, on soils in agricultural fields. There is much to understand about the impact of bioaerosols on agrarian ecosystems. We only have a few partly detailed studies about bioaerosols emanating from agrarian regions. For example, they say, precipitation plays a vital role in the dissemination of certain plant pathogens, such as *Pseudomonas syringae* and *Botrytis cinerea* (Monteil et al., 2014).

Atmosphere above any region on earth does support bioaerosols. Dust particles and microbes are ubiquitous in the atmosphere. Microbial composition of bioaerosols may vary depending on source, altitude, environmental parameters, etc. The particulate matter may vary in size. Microbes in bioaerosols are distributed on particles of size < 2.5 µm. It may constitute 3–11% of total microflora in the atmosphere. About 25% of particulate matter could be occupied by microbes (Boreson et al., 2004; Peccia and Hernandez. 2006; Jones and Harrison 2004; Polymenakou 2012).

Bioaerosol particles can include viruses, bacteria, fungi and their spores, lichen fragments, protists (e.g., protozoa, algae, and diatoms), spores and fragments of plants, pollen, small seeds, invertebrates (e.g., nematodes, mites, spiders, and insects), and their fragments as well as faecal material (Polymenakou, 2012).

The concentration of microbes (bacteria and fungi) detected in atmosphere above ground surface at various levels could be on an average 10^3 microbial cells per cm^3. In fact, many of the microbes detected in the atmosphere often are similar to the one traced in the agrarian soils/crop canopy, if we considered the agricultural belts. However, a few fungi like *P. notatus* has been traced even at 77 km altitude. There are few other bacteria and fungi reported from altitudes of 12, 43, and 51 km above farming belts.

Atmospheric microbes traced above agrarian regions could be a source of human diseases such as meningitis (*Neisseria meningitis*), staphylococcal infections (*S. aureus*), viral influenza, etc. In this way, the airborne microbes may be affecting farmers' health (Griffin et al., 2007; Sultan et al., 2007). Microbes in the bioaerosol above farms may also carry infectious agent that attacks domestic farm animals, birds, and useful insects. Considering the theme of this book, our focus is restricted to microbes that are pathogenic/beneficial to crop production.

It is interesting to note that microbes above agrarian regions may be capable of several types of biochemical transformations relevant to us (Polymenakou, 2012). For example, metabolically active bioaerosols may have bacteria that degrade pesticides/herbicides. They may also harbor a wide range of enzymes that tolerate the harsh atmospheric conditions and still accomplish several types of biochemical transformations. We need to isolate them and utilize them efficiently. The metabolic potential of bioaerosols above agrarian regions is still unexplored. We may trace innumerable microbial species capable of rapidly degrading pesticide residues, fungicides, and herbicides (Durand et al., 2006; Ariya and Amyot, 2004). Aerial microbes may also degrade other harmful organic material lifted into the agricultural sky. Amato et al. (2007) reported that microflora from clouds was able to utilize carboxylic compounds, such as farmates, acetates, lactates, succinate, etc. A few had the ability for degradation formaldehyde and methanol derivatives. Airborne strains of *Pseudomonas, Sphingomonas,* and yeasts traced in cloud water were able to degrade and transform several carbon compounds.

Now, let us define the term "Aeromicrobiology." Aeromicrobiology is the study of microbes and their activity in the aerosols. Microbial population found in the air is collectively termed "bioaerosols" (Al-Dagal and Fung, 1990; Despres et al., 2012; Klein et al., 2016; Pepper and Dowd, 2009; Pepper and Gerba, 2015). Aerially, the physical environment that microbes in bioaerosol negotiate, survive, and perpetuate is important. Several physical factors may affect the launching (levitation), transport, and redeposition of aerial microbes at a different locations. Wind induces microbes to launch from soil/ground surface into the atmosphere. Wind turbulence is good force to introduce microbes into air. Human activity too induces launching of soil microbes into air (e.g., ploughing of soil). Bioaerosols get pulled due to gravity resulting in deposition of microbes, sometimes in large scale when drifts subside. So, the environmental parameters above the crop's canopy decide, to a good extent as to what happens with microflora in the air. No doubt, during this period, atmospheric factors such as UV radiation, moisture, light, temperature, clouds, dust particles may affect the survival of microbes in the air (DeGuillame et al., 2008).

Clouds are conducive to the survival and proliferation of bioaerosols. Cloud's moisture suspended dust particles and nutrients could help microbes, to survive longer and transit for longer distances. Of course, appropriate breeze must be available. Clouds avoid desiccation of microbes. They avoid ill-effects of UV radiation on aerial microbes. Clouds offer congenial pH in the range of 3–7. So, acidophilus microbes may proliferate better. Clouds offer surfaces on which microbes can transit relatively longer distances. Later, as precipitation occurs, the aerial microbes may descend via water drops on agrarian soils (Amato 2012; Vaitilingom et al., 2010). Overall, in an agrarian atmosphere, clouds are important medium through which microbes disseminate. In the current context, we should note that in addition to natural factors, abiotic man-made factors such as drones and airplanes too may disperse aerial microbes. This aspect needs attention in due course, particularly, when we consider spread of plant pathogenic diseases.

Generally, desiccation is the major factor that affects survival of microbes in the atmosphere above crops. Warmer temperatures too can affect survival of aerial microbes. They say, certain Gram-positive microbes tolerate higher relative humidity. Hence, such microbes may find a niche in the clouds. Extremes of low or high temperatures could denature proteins. Therefore, enzyme activity relevant to survival and proliferation in atmosphere get affected. Agrarian sky that supports lots of pesticides, herbicides, and other foliar applied chemicals may affect bioaerosols. Microbial communities get retarded to an appropriate extent. The quantity of toxic material released its efficacy and tolerance by microbes need attention. Crop residue burning too affects bioaerosols immensely.

Genetically, widely diverse microorganisms can be traced in aerosol form in the atmosphere. They include viruses, bacteria, actinomycetes, fungi, archaea, yeasts, and protozoans. The aeolian microbes must invariably get adapted to harsh climatic characteristics of the agricultural sky. They include extreme temperature, gases, and humidity. Certain species of microbes can survive harsh conditions, by forming resting spores (also called endospores in the case of *Bacillus sp.*). Fungi too form such fruiting bodies and spores that help in overcoming harsh climate. These resting structures can withstand harsher atmospheric conditions (Al-Dagal and Fung, 1990; Khodadad et al., 2017; Reche et al., 2018). A cruise height of 12 km above the soil surface allows for longer transit of stratosphere microbes. Air sample studies by Smith et al. (2018) indicate that microbes have been detected in the stratosphere up to 38 km from earth's surface. Previous reports indicate that microbes do survive in the stratosphere. They could trace a few cocci and rods in the samples at 41 km altitude from the ground surface. However, recently Wainwright et al. (2003) reported about attempts to culture organisms traced in the stratosphere air sample. Microbes such as *Bacillus simplex*, *Staphylococcus pasteurii*, and *Engydontium album* were cultured. Although extraneous contamination cannot be ruled out. Previous studies have shown that stratosphere organisms do occur at 18, 31, and 42 km from earth's surface. We have no idea about the role played by them in the ice nucleation, biochemical transformations, or other eco-systematic functions, in the atmosphere.

Airborne microbes are distributed at different altitudes from the soil surface. In fact, microbes closer to soil surface may transit only small distance before landing. Particularly, if the wind is feeble or due to common obstructions. Microbes at higher altitudes in atmosphere, say, at upper troposphere or lower stratosphere may float for longer distances. They utilize the wind currents and dust particles. It is imperative that low-altitude microbes do affect the microbial composition of soil. Microbial population, their diversity, and physiological/biochemical abilities of stratosphere microbes need attention (Deleon-Rodriguez et al., 2013). Smith et al. (2018) have observed that microbes are homogeneously distributed in the stratosphere, up to 12 km altitude. They point out that uniformity in microbial population is dependent on atmospheric conditions, aerosols, and trajectory of microbes that transit in the troposphere/stratosphere (Smith et al., 2009, 2011). Smith et al. (2018) have also point out that it is worthwhile knowing the change in the population and diversity as microbes move from stratosphere toward tropopause and then onto the ground surface. Further, they point out that tera-grams of microbe-laden dust moves across different agrarian regions (Acosta-Martinez et al., 2015; Mahaffee and Stoll, 2016) and a good share may land on the soil surface. So, it is necessary to know the consequences of airborne microbes on the eco-systematic functions (particularly, soil and water bodies) on the ground.

Amato (2012) and Amato et al. (2017 a, b) believe that clouds are a good medium for aerospace microbes, that is, bioaerosols, in terms of survival and transit. Their significance to agrarian regions in terms of influence on crops, soil nutrient transformation, and general biotic activities of the soil profile is worth knowing, in detail. The major microbial genera traced in the upper troposphere are *Lachnospiraceae, Ruminococcaceae,* and *Erysipelotrichaceae.* Also, *Bacillus sp., Micrococcus so, Arthrobacter sp., Staphylococcus sp., Clostridium, Mogibacterium, Corynebacterium, Bacteroides, Prevotella, Pseudomonas,* and *Parabacterioides* are commonly traced, both, in upper troposphere and stratosphere (see Smith et al., 2009, 2011; Yamaguchi et al., 2012; De Guillaume et al., 2008). As stated earlier, in Chapter 1, there are at least three microbial species detected from stratosphere. The cryo-sampling was done from 21 and 42 km from surface of soils/crops, above a Southern Indian location. Microbiologists have identified the stratospheric species as *Janibacter hoylei, Bacillus aryabahatta*, and *Bacillus isronensis.*

Their ability to tolerate variations in temperature, freezing, dehydration, UV radiation, or their role in soil/atmospheric functions is yet to be deciphered.

3.2.1 MICROBIAL FLORA IN THE ATMOSPHERE ABOVE AGRARIAN REGIONS

We are concerned here with the microbiology of the agrarian atmosphere. In other words, microbes in the agricultural sky and its relevance to crop growth and productivity. Further, we should have a semblance of understanding about the impact of aerospace microbes that may exist either in low and/or high altitudes above crops on the agrarian expanses. Aeromicrobiological aspects include those that positively improve crops in the field (e.g., N-fixing microbes, plant growth promoting bacteria, mycorrhizal propagules), those causing deleterious effects (e.g., pathogens), or those whose saprophytic activities may not have immediate effects of any scale, on agrarian crops. Generally, genetic diversity and intensity (population) of aerial microbes, say, specifically above agricultural fields may be relatively lower. Particularly, when compared with that encountered in soils of farming zones. Agricultural soils, in comparison, are richly endowed with microflora. Soil microbes can bring about several types of nutrient transformations and other processes of greater relevance to crops. The root zone is of significance to agricultural crops, since essential nutrients, gaseous material, and microbes reach plant tissue via rhizosphere. Aerosols (aerial microbes) too are equally important although microbial diversity and population could be of lower order. Phyllosphere in the aerospace above crops is comparable to what is rhizosphere in soils. Phyllosphere is the region through which nutrients applied as foliar nutrients and moisture gets absorbed into plant tissue. Gaseous exchange vital for CO_2 fixation and respiration also require phyllosphere. Microbial flora resident on leaves too needs greater attention. Phyllosphere microbes are important as far as nutrient transformations are concerned. Knowledge about pathogenic activity of aerosols is important. Basically, the shoot system, canopy, foliage, leaf surface characteristics, and leaf area attract greater attention while we study agricultural sky.

Clouds and raindrops that fall on the agrarian belts are a major source of plant pathogens, in addition to other microbes. For example, Monteil et al. (2014) have shown that raindrops may support large inoculum of fungi such as *Botrytis sp* and bacteria like *P. syringae*. So, in general, the quality of atmosphere above the crop's canopy evaluated in terms of pathogen's population is a necessity. A "sickly atmosphere," in other words, a pathogen-filled-air above crops can always be troublesome when it rains. Precipitation is a repository of microbes. It also disseminates microbes to a certain distance, including plant pathogens. Precipitation also induces variability in microbes traced in atmosphere and in soil. It is due to uneven precipitation patterns of events.

A study dealing with bacterial species above tomato crops has indicated that population of pathogenic bacteria such as *Xanthomonas* and *Pseudomonas* may fluctuate, depending on the diurnal time. Population of bacteria is highest in the bioaerosols during late afternoon. Natural precipitation and wind did affect the bacterial population in the air samples (McInnes et al., 1988). Certain agronomic procedures too reduced the bacterial population in the aerosols. We should give due attention to sampling time, location, and stage of the crops, if we aim at accurate forecasts about disease incidence due to a "sickly agrarian sky."

Generally, terrestrial regions are better sources of airborne microbes compared with oceans, lakes, or riverine regions. Actually, here we are concerned with bioaerosols generated from soils and crop's canopy of agroecosystems. We should note that in agrarian regions and expanses of natural vegetation or dry land with shrubs and scant vegetation, it is the soil that generates airborne microbes. Plants cover soils of agrarian regions to different extends based on seasons. Airborne microbes (e.g., bacteria, fungi, and viruses) are greater during crop season. It might be low during winter. Plant canopies are good sources of microbes that are traced in the bioaerosols. For example, during a crop season, plant pathogens that exist on plant surfaces like leaves, stem, and fruits may launch into the air and be part of bioaerosols. However, most of microbial diversity and population traced in the bioaerosol originally generated from soil, as dust storms. Soil surface and dusts that launch into the air are the best sources of airborne microbes (Lindmann et al., 1982).

It is interesting to note that bacteria involved in "Ice Nucleation" originated from regions with good plant cover or crop canopies (Lindow et al., 1978a, b). Microbes in bioaerosol may be influenced by the atmospheric processes. Microbes in the aerosol (bacteria and fungi) are known to cause changes in the atmosphere through ice nucleation. Proteobacterial species such as Burkholdreales and Moraxellacea were dominant in the atmosphere. However, these same bacterial/fungal species did not have a direct impact on ice nucleation (Bowers et al., 2019). Reports from Montana in the USA suggest that *P. syringae* traced in the atmosphere above wheat fields could induce nucleation, frost, and snow formation above cropland (Tsang, 2019). This is an example of how bioaerosol microbe can affect the physical property of the atmosphere via biological ice nucleation and snow formation.

We have previously stated in clear terms that bioaerosols possess both beneficial and deleterious microbes. A study by McInnes et al. (1988) suggests that viable bacteria detected in the aerosol above fields and agricultural experimental station in Georgia included *Xanthomonas campestris* and *P. syringae*. Aerial samples collected immediately after rain or foliar irrigation of tomato crops showed a lower density of these plant pathogens. They also found certain agronomic procedures such as clipping of tomato plants induced a higher population the two pathogens in the bioaerosol. Aerial sprays of Cu fungicides did not alleviate the situation as far as pathogenesis is concerned. Population of bacterial pathogens in bioaerosol did not alter significantly, despite Cu fungicide sprays.

Microbes in the crop's canopy are important. They could be deleterious and infect the crop upon landing on the foliage or other parts. Microbes may thrive on leaf surfaces using the exudates or organic dust particles as energy source. They get dispersed as the wind blows or insects/birds pick them while foraging on crops. Soil microflora immediately below the canopy is also important. Soil disturbance may launch microbes on to crop's canopy and leaf surface. Let us consider an example. Perennial grass species were analyzed for phyllosphere microflora, using 16s rRNA. It was found that the microbial taxa got altered based on seasons. Most of phyllosphere microbes were originally traced in the soils under the plants canopy (Grady et al., 2019).

Microbes in the phylloplane/phyllosphere could also be regular pathogens, like rust, blotch, and rot causing fungi. They could be blight or blast causing bacterial agents or even mosaic causing viruses. Such aerial microbial pathogens could enter plants via stomata, hydathodes, or open injured tissue. Sometimes they penetrate plant tissue actively via production of cellulases and penetrating cuticle. Leaf surface microbes enter plants passively via insect bites or wounds, etc. Whatever be the mode of entry, microbes on the crops' surface (canopy) can be dangerous to crops. Therefore, prophylactic measures have to be commissioned, in time. The microclimate in the crop's canopy and under it are factors that need consideration. Particularly, while judging the airborne microbes above the canopy.

At this juncture, we should note that phyllosphere microbes are important. Although, they are not usually ascertained or studied in detail about their possible role in crop productivity. Phyllosphere surface supports several different microbes of diverse physiological/biochemical abilities. Phyllosphere in the agrarian sky could be compared to rhizosphere in the root zone. All nutrients and gaseous exchanges occur in the phyllosphere region. Often the soluble nutrients that move into crops do encounter phyllosphere microbes on the cuticular surface. Leaf exudations into phyllosphere need to be deciphered in detail to know how it might affect phyllosphere microbes and their physiological activity. We have no detailed knowledge that if nutrients do get transformed before they transit into leaf paranchyma via stomata. Foliar sprays applied do encounter phyllosphere microbes. So, each foliar spray containing bactericides, fungicides, herbicides, or nutrients disturbs the phyllosphere microbial diversity and population density. Specific ecosystematic functions exhibited by phyllosphere microbes could get disturbed or reduced/accentuated due to foliar sprays. Pathogens resting on leaf surface are of course destroyed upon foliar spray of plant protection chemicals. Overall, we may realize that microbes coexisting with crop's canopy could influence plant growth through several mechanisms.

3.2.2 AEOLIAN DUST AND TRANSPORT OF MICROBES ABOVE AGRARIAN REGIONS

Microbiologists point out that once airborne, microbes of small dimensions such as those of 0.009 mm drifted long distances crossing continents. Generally, microbes below 0.02 mm can travel thousands of km before landing in a new crop belt. Larger microbes travel only for short distances. These microbes could influence crops or other biotic aspects in the new location/soil. However, first, they have to acclimatize and multiply. For example, a pathogen may have to find a leaf surface or a grain, to proliferate and affect crops. A symbiotic microbe such as *Bradyrhizobium* that falls on soil may have to proliferate and colonize the roots. Similarly, symbiotic fungi such as mycorrhiza may need a root system of a live plant, to proliferate and sporulate. There might be large number of commensalistic microbes in the air that do not affect crops to any extent.

According to Yamaguchi et al. (2012), it is clear that microbes are resident on aeolian dust particles. Also, bacterial cells move across continents via dust storms. The potential for global dispersal of aeolian bacterial species is greater. It depends on the characteristics of aerial dusts. Frequency and location where dust storms get generated are important. There are several reports proving that airborne microbes move across continents and large expanses of cropping zones (Prospero, 1999; Prospero et al., 2005; Bowers et al., 2013; Brown and Hovmoller, 2002; Maki et al., 2013, 2017; Weil et al., 2017; Tang et al., 2016; Stein et al., 2015). Let us consider Asian dusts and their transport across continents. They say, bacterial species from Asian locations transit over 3000–5000 km via Asian dust storm generated in the Gobi deserts. They may land at any place along the route. For example, in Japan, they were traced using the 16S rRNA sequencing (Yamaguchi et al., 2012; Griffin 2001, 2003, 2004; Griffin et al., 2010, 2017). There is also clear evidence to prove that microbes from farm locations in Africa such as Sahel and Sahara deserts transit on the dusts, to regions in other continents, such as Caribbean and North American plains region (Smith et al., 2013). They may travel to locations as far as 15,000 km away and land on agricultural soils. Reports prove that Saharan/Sahelian dust from desert/crop fields does reach farming expanses of Europe, lakes and mountains with temperate vegetation (Hervas, 2009; Smith et al., 2012). In fact, a few reports make it clear that dust from locations in Asian agrarian soils can make a circuitous route around the globe (Uno, 2009). If dust moves, then bacterial/fungal propagules too transit the same locations and reach the original spot, from where they first levitated into the sky. The above evidence perhaps makes it clear that global microbial flora is distributed rapidly all over, into every agrarian region. Crops normally encounter various genera of microbes from different continents. The circulating winds at high and low altitude facilitate distribution of aerosol microbes through long distances. The microflora that transits long distance on the globe and lands on crops could be plant pathogens/symbionts or just commensals (see Bouvallius et al., 2006; Mahaffee and Stoll, 2016; Toepfer et al., 2011; Stein et al 2015).

Smith and Griffin (2013) opine that despite great strides in our ability to pick air samples and analyze them for microbes, there is still dearth of easier methods that offer authentic data. In fact, studies on Aeromicrobiology are hampered, to certain extent, due to inadequacy of methods that allow us, to study bioaerosols. In the present context, we may note that methods to study bioaerosols above agrarian regions are few and their ability to offer accurate data need improvisation. A few basic steps/methods involved in studying the aeolian bacteria need mention. To detect microbes on the aeolian dust particles, it is feasible to pass the dust material through the Millipore filter. Then, grow the adhering microbes/bacteria on a selective medium such as nutrient-rich PYG medium (broth). We can also visualize the bacteria on the dust particles, using microscope. For checking the abundance of microbes on the aeolian dust, we can adopt rRNA analysis (Yamaguchi et al., 2012). It is useful to measure the size of the aeolian dust particles. Aeolian dust particle and bacterial cells could be distinguished, using fluorescence microscopy. Samples of aeolian dust are usually collected using samplers located under the airplane or small UAVs, or tethered aerostats (see Bryan et al., 2014). Elemental composition of dust particles is deciphered, using the scanning electron microscope. Here, we may realize that large quantities of aeolian dust that transits removes/adds different elements at a spot (crop fields) in agrarian regions. There are physical difficulties during the collection of authentic aeolian samples particularly from high altitudes above crop's canopy and then analyze them rapidly. Yet, there are methods that allow sampling of air above agrarian regions, so that, they could be utilized for detecting microbes via 16S rRNA procedures (Spring et al., 2018).

The diversity and intensity of microbes attached to dust particles traced in the sky depend on environmental parameters. Normally, bacteria in aeolian dust survive despite harsh climatic conditions that they have to negotiate while floating in the sky, also in the locations (soils/water) wherever they land. The bio-aerosol bacterial species will have to get acclimatized, to the new ecosystem that they reach after the transit, by attaching themselves to the dust particles. For example, Yamaguchi et al. (2012) have reported that bacterial diversity and population vary depending on the source and the region from where atmospheric dust particles are sampled. Deinococci was traced in all samples from Asian dust. Other species found but at lower population density are *Caldilineae, Cyanobacteria, Acidiobacteria, Planctomycetaceae, Vericomicrobiaea, Fusobacteria, Actinobacteria Clostridia, Erysiphalotricia, Flavobacteria*, etc. The population and diversity are definitely affected by the season, environmental parameters, and altitude of the sampling. A few bacterial genera such as *Betaproteobacteria, Deltaproteriobaceria,* and *Gammaproteriobacteria* were feeble (Bowers et al., 2009, 2013; Finlay, 2002; Finchel and Finlay, 2004).

We may note here that bacterial community above Asian regions could show up good similarity despite sampling from different locations. This has been confirmed by Nishimura (2010). The rRNA gene-targeted studies were adopted on microbes of Asian dust samples. No doubt, microbial abundance on dust particles (aeolian dust), its biodiversity, and role in atmosphere, if any, should be known in detail. The ability of bacteria in the bioaerosol to transfer genes through cocultivation is an important trait to note. Bacteria including important plant pathogens are liable to be mutated or genetically altered. This happens due to gene transfer from bacterial species that might be transited long distance while resting (sticking) above the dust articles.

There are chances for native aerial bacterial species to become virulent upon genetic hybridization via gene transfer from bacterial species on aeolian dust. In addition, the aeolian bacterial species may itself be a virulent plant pathogen. It could even be a useful biological N-fixing organism. Symbiotic bacteria too may transit long distance each season. Later, land in regions growing legumes/cereals. In fact, Schmale and Ross (2015) have opined that there are highways of microbial distribution and transport all over the globe, in the troposphere. Such highways are of concern when it comes to transport of plant pathogen that induces pandemics (e.g., routes of aerial distribution of *P. graminis tritici* routes). Similar aerial highways can be of great consequence to agrarian regions which are seeded with symbiotic microbes such as N-fixers, mycorrhizal propagules, nutrient transformers, etc.

Regarding aeolian dust, it is said bacteria that migrate should be able to survive, multiply, and propagate, at any place that dust particles descend. In this regard, Beijerinck (1913) has commented that microbes could be everywhere in the nature, including atmosphere, agrarian regions, sea, soil, and other structures. However, it is the environment that selects and decides the survival and proliferation of the bacteria. So, it means each agrarian region may harbor bacterial genera of diverse characteristics and ability. Yet, its successful establishment in soil or as pathogen on crops is dependent on the nutritional source and environmental conditions prevailing in the cropped field.

No doubt, bacterial species are transported along long distances (Duce, 1980; Kellogg and Griffin, 2006; Bouvallius et al., 2006). As they alight on land and mix with indigenous microflora, there is the possibility that they may introduce new genes into indigenous bacterial flora and generate hybrids. For example, Hua (2007) stated that Asian dust particles arise from regions covering China, Mongolia, Korea, and Japan. They introduce genes to indigenous bacterial species resident in agrarian soils, at each place that they land after transit. Evidence indicates that Asian dust may transit for over 13–150,000 km distance and land in a new soil ecosystem (see Duce, 1980; Kellogg and Griffin, 2006; Griffin 2001, 2003). They could introduce new genes. Such dust storms can induce greater bacterial diversity (Yamaguchi et al., 2012; Burrows et al., 2009; Behzad et al., 2018). Some of the consequences of transport of microbes via aeolian dust are worth exploring further. Clearly, airborne microbes may transfer genes to native soil microflora upon landing and establishment in the new location. This aspect has great relevance to spread of pathogenesis and virulence (via plasmids).

Periodic sampling of atmospheric dust may show up a large diversity of bacterial species over the agricultural sky. We may have to also characterize the physiological and biochemical nature of the aeolian bacteria, particularly, their ability to mediate nutrient transformation. In case, some of them are pathogens on crops, then we may have to check for virulence. This way, perhaps, we can forecast the spread of disease.

The airborne microbiome is a topic of high interest in ecology, biogeochemistry, and environmental sciences. Large amounts of microorganisms are continuously exchanged among ecosystems and continents through the atmosphere. The consequences of such shifts in microbes for ecosystem services and biogeochemical cycling are not easily predictable (Caliz et al., 2018; Tragard, 1977). Smith et al. (2012) believe that arid zones with sparse vegetation and well-exposed topsoil are a source of large quantities of aeolian dust. Such aeolian dust is created by winds that blow at low altitudes. Aeolian dust from such zones with natural vegetation and crops may contain sparse microbes. Although sampling procedures, detection, identification, and classification are difficult, a few microbial species from aeolian dust from arid regions have been characterized. For similar reasons, understanding the long-distance transport of microbes was difficult. Yet, using 16S rRNA-based microarray analysis has allowed detection of innumerable microbial species. In fact, thousands of bacterial taxa could be detected and identified. The aeolian plumes showed *Proteobacteria, Actinobacter, Fermicutes*, etc. (Smith et al., 2012). The bacterial species detected depended on natural characteristics of the region/place from where the aeolian dust originated. Therefore, further investigation to characterize the microbial flora of aeolian dust from say, high altitudes that originated from thick or sparse cropping zones, mountains, coastal areas, and marine regions is needed. Seasonal variations in airborne microbes with regard to diversity, population, and distance that they travel are yet to be known in detail. However, recently, Caliz et al. (2018) have shown that airborne microbial species detected are strongly correlated to seasonal changes in wind currents and atmospheric conditions. This is true, particularly, in the areas such as mountains, rich cropland, arid regions, etc.

Wade (2013) explained that in addition to traveling through long distances, dust particles do inoculate atmosphere with aerial microbes. Such inoculation could occur even at high altitude above the ground surface. A study in the troposphere of Caribbean farming regions has shown that atmosphere carries more active bacterial cells (5×10^3 cells per cm^3 of air) compared to fungal propagules. Over 314 families of bacteria were represented in the samples drawn from 10 km above the Caribbean region. While from other studies, it is clear that troposphere samples could contain several important fungal species, including a few pathogens/symbionts.

Recent investigations by microbiologists at Kellogg's experimental station in Michigan show that fog offers better niche for microbes in bioaerosols (Michigan State University, 2020). They travel for longer distance on fog than on dry aerosols (like sands or dusts). Fogs offer relatively better nutritional support for microbes, to multiply. Fog is a good vector for microbial transport across long distances. Microbial samples from fog, clear sky, and raindrops were compared. It was found that fog supported a higher population of microbes in the atmosphere. Fogs above agricultural sky could transport several different genera of microbes, from one location to another. Once transported microbes that alight on farmland may induce changes in microbial diversity and activity. Fog may be a good source of bacterial and fungal pathogens that spread into farmland during the winter season. Fog has a role to play in transmitting plant pathogens. In addition, fog may carry bacteria and fungi that eventually help in "Ice Nucleation" in the winter months.

Onetoa et al. (2020) state that fungi move within and outside cropping expanses via spore liberation into atmosphere. Fungal spore dispersal takes different patterns. A few fungal species release spores intermittently; others release spores at specific times of the day. The trajectory of fungal spores in the atmosphere is important because it decides the distance to which the propagules may disperse. Generally, individual spores may follow unpredictable trajectories due to turbulence. While aggregates of spores or fruiting bodies may take definite path and distance. Fungal spores released during mid-day may transit for several days in the atmosphere. It depends on fungal spore buoyancy. On the contrary, spores released at night may settle to ground surface within a short time.

3.3 MICROBE CAUSED CROP DISEASES AND THE AGRICULTURAL SKY

3.3.1 AIRBORNE MICROBES AND CROP DISEASES

Historically, agricultural sky has played havoc with crops and farmers and continues to do so even at present. Agricultural sky has served as a major source and dispersal medium of disease propagules. Among the factors related to agricultural sky, it is the air, that is, wind that has been identified as an excellent conduit for disease propagules. The drift of particulate matter that carries with it the pathogen can be traced by placing traps/nets of various sized pores. Microbes that actually cause the disease needs to be identified by cultural methods or microscopic examination. Wind, as a source of crop disease, has been realized by farmers since several millennia. Mahaffee and Stoll (2016) have compiled historical evidence concerning aerial spread of microbial diseases on crops. They state that during ancient period, the spread of diseases into small zones or even pandemics was traced to air (agricultural sky). The specific microbe was not detectible. However, the disease agent and its source as "agrarian sky" were clearly identified, by farmers since the days of Sumerian civilization. Historically, Egyptians too identified crop disease as being spread via wind. Diseases such as blights and rusts on crops were known to ancient Chinese and Indians. Romans associated crop disease to aerial factors such as wind. Roman farmers detected "diseaseborne wind" in the fields. During recent history, there are several reports about microbe-caused epidemics and pandemics that attack crops. Sampling, detecting, and identifying pathogens in the air samples are indeed a crucial step. It helps us in devising remedial measures. Avoiding aerial spores from spreading and landing on new locations where susceptible crops may exist is the crux.

Monitoring the dispersal of airborne spores/propagules and spread of disease is important, in any agrarian region. We can do so by actually trapping and identifying the existence of disease-causing microbe. We can also develop sick plots/trap plots by growing highly susceptible variety of crops, then identifying lesions or other disease symptoms. Currently, there are several rapid tests based on nucleic acid assay (Capote et al., 2012; Summers et al., 2015; DeBoer and Lopez, 2012, Mahaffee and Stoll, 2016). Here, we have to note that identifying the airborne inoculum of pathogens depends on the intricacies of biochemical extraction, amplification, and estimation of nucleic acids. The nucleic acid signatures have to tally as accurately as possible. More importantly, our ability to accurately relate the airborne inoculum to disease causation, particularly the severity is also questionable, in some cases. Whatever, the complications of the detection of disease, we can still take precautions, by just knowing that a particular disease is disseminated via agricultural sky. Just like, we do with soilborne or waterborne crop disease, we should be able to integrate our approaches and reduce aerial spread of disease. At this juncture, we should be thinking of "sanitation and hygiene" in the aerospace above crops. For short it could be termed "aerial hygiene" in farms. Like periodic monitoring of soils, irrigation water, crop residues, etc., for disease propagules, we may have to send drones up above the crop's canopy and collect air samples and monitor for the notorious rust, blotch, blight spores, bacterial cells and spores, actinomycetes, etc. Prophylactic sprays of aerospace up to several 100 ft above cops are after all feasible.

Knowledge about the infectious microbe, its infective propagule and ability to spread/transit via different media such as surface (leaf surface, stem, or grain) or crop canopy or longer distances via atmosphere is important. In other words, scale of dispersal of an airborne disease agent should be known in order to devise control measures. Usually, we identify dispersal within a leaf/plant canopy, a field, farm, agroecosystem, region, and at continental scales when a pathogen is considered. We should realize that the longer the distance of dispersion the higher is the requirement of viability period of pathogen or its propagule. Generally, airborne pathogen's propagules differ with regard to life span. Microbial pathogens that transit in the air could be viable only for few seconds, minutes, hours, day, a month, or even more. Longer the distance of dispersal, the basic requirement is longer period of viability and ability to infect (Lang and Evans, 2011).

The agricultural sky is the region where the propagules transit. The wind, precipitation, clouds, fog, and smog are some factors that induce dispersal. There are different scales of wind that decide the distance to which the pathogen may travel. Minor turbulence in the air may allow a shorter distance of dispersal. Higher speed winds, tornadoes, and thunderstorms may allow spores to travel for 10^3–10^4 m. Strong sea breezes, hurricanes, and tropical storms may disperse pathogens to longer distances, say 10^6 m. High-altitude jet streams really allow longer distance of dispersal, say, to different continents (Mahaffee and Stoll, 2016; Griffin et al., 2017). Agrarian sky and various factors really decide the transit of plant pathogens. Nature of spores/propagules is also important factors. They decide the extent to which a pandemic may spread all over the globe.

Nicolaisen et al. (2017) opine that there is insufficient knowledge about aerial fungal spores in European farms. Farmers need to be more conversant with the disease-causing potential of aerospace above the crops. It is worthwhile for farmers to know just how sickly or healthy is their sky. Knowing about soils and crops in detail does not suffice. At any time, agricultural sky may harbor diverse sets of fungal spores belonging to crop pathogens. Let us consider an example, Nicolaisen et al. (2017) have studied the distribution and diversity spores/propagules of fungal diseases on brassica and other crops. They have attempted to compare the fungal spore population above both crops and urban settings. They report that often spores on urban locations were derived from above the cropland. Fungal spores detected in the atmosphere immediately above the copped fields often belonged to diseases that have attacked the standing crop. Fungal spores of a few different disease were detected frequently above the crops. Fungal spore diversity was similar, if the agricultural expanses studied were uniformly cropped with similar crop, for long stretch of time.

It is generally agreed that to forecast epidemic of rusts or powdery mildew on wheat, we need detailed aerobiological information about the pathogens, their strains, distribution in the agrarian sky above the fields, and its vicinity. This is in addition to knowledge about the disease resistance traits of the host-, that is, wheat cultivar. Detailed data about airborne fungal distribution in the aerospace above cropping expanse are important. Airborne spores of rust fungi could be distributed in the canopy, and air above it for up to 6–10 m altitude. Some reports suggest that urediospore concentration could be uniform right up to 1 km altitude from the crop. Others say, it is uniform only up to 500 m above the canopy. Studies conducted some decades ago, using Hirst traps, have shown that urediospores of rust fungi are distributed at different levels in the atmosphere above crops (Eversmeyer et al., 1972). The uredospore distribution close to crop canopy is vital to know if epidemics will occur or not. Eversmeyer et al. (1972) reported that *P. graminis* and *Puccinia recondita* urediospores density fluctuated hourly, based on time of the day. At mid-morning and afternoon, spore release is greater. Spore release peaks may also occur during the early evening and night. It depended on weather conditions. Regarding crop stage, it is said maximum spore release occurs during panicle development. Many of the details listed above also apply to wheat powdery mildew—*Erysiphe graminis* (Eversmeyer et al., 1972). The conidia release and infection development are aided by congenial conditions surrounding wheat canopy. Usually, warm temperature above 25°C, relative humidity above 70%, and windy conditions help the spread of airborne conidia (Thomas and Jayasena, 2020).

3.3.1.1 DISPERSAL OF AIRBORNE MICROBIAL PLANT PATHOGENS

Agricultural sky and its factors play a vital role in the dispersal of plant pathogens, symbiotic microbes, and several commensals. Here we are concerned more with plant pathogenic fungi that cause epidemics/pandemics on crops. Microbes are commensal or symbiont in the atmosphere or a potential plant pathogen that could get transported across long distances, by hitch-hiking on the aeolian dust particles (Yamaguchi et al., 2012, Griffin 2001, 2003). The bacteria, for example, are known to hitch a ride over aeolian dusts generated in the deserts of Asia [e.g., Taklamakan and Gobi Desert or African dust (Sahel and Saharan dust), Australian dust (Australian Desert), North American dust (Great Plains), etc.]. The transit of bacterial cells on dust has been shown by several researchers (Yamaguchi et al., 2012).

Dust storms, clouds, and wind together transport sizeable quantities of microbes from one continent to another. Clearly, agricultural sky plays a vital role in the dispersal of microbes. New microbes may impinge into agroecosystems. Such agroecosystems may not have been previously exposed to them. Many of the microbes could be pathogens and a few of them symbionts too. Usually, microbes that disperse through to long distances resist certain stress conditions related to the environment (Behzad et al., 2018). The microbes that disperse long distance could be deadly plant pathogens. Therefore, it is a general caution that they may induce epidemics in the locations where they land. The susceptible host (substrate) is a requirement. Clearly, atmosphere and its factors play vital role in either inducing an epidemic or not on crops. For example, rust fungi that transit long distances may induce pandemics on crops.

Canopy level interactions between crop, pathogen, and atmospheric conditions are important. It is called the "disease triangle" (Pangga et al., 2011). In nature, canopy level interactions among the crop, pathogen, and environment, in this context the "agricultural sky" or "aerospace above crops" is important. While developing models and forecasts about disease progression and severity, we have to consider the above three factors. There are reports wherein the crop, its growth pattern, and dynamics of bacterial/fungal pathogen have been considered. In addition, we have to pay due consideration to ambient weather, air quality, wind, radiation, precipitation pattern, and their influence on pathogen. Recent changes in the climate, that is, prominent climate change factors too need consideration while developing models about disease spread within a crop canopy, a field, or even large expanses (Pangga et al., 2011). A field study using *Styllosanthes scabra* and anthracnose organism, namely, *Colletotrichum gloeosporioides* suggests that ambient factors are important. For example, elevated CO_2, moisture, temperature, diurnal changes, etc., may affect the disease causation, its impact on crop growth and productivity. They say, periodic infection cycles too depend on agricultural sky to a certain extent. Crop-pathogen interactions may get enhanced due to elevated CO_2 emissions. Further, an elevation in temperature in the aerospace, it seems, increases population, aggressiveness, and infectivity of certain fungal pathogens that affect wheat. So, warmer climate may support potentially higher density of pathogen propagules in the air. Warmer sky may actually induce better spore release/dehiscence.

Plant pathogens may reside on various locations within the crop canopy. The distribution of propagules of the pathogen within the canopy needs careful attention. Localization of pathogen could be easily affected by atmospheric factors like wind, precipitation or irradiance, etc. Upon release from the infection focus, the propagules may land within a few meters. The distribution pattern within a crop decides the fate of pathogen, its spread and severity of disease (Legg, 1983). The nature of causal agent and its ability to transit within a field is important. Factors like wind, physical nature of spores, timing of release, and air turbulence may affect the rapid spread of a plant disease within a field.

Dispersal through whatever means is central to spread of an epidemic/pandemic causing plant pathogen. The wind is among chief factors of dispersal of plant pathogens across short or long distances, including through different continents. Researchers have shown a greater interest in deciphering the modes of dispersal of microbes mainly fungi, viruses, and bacteria in that order. Many of the microbes studied are fungi. They have the potential to cause epidemics/pandemics. Especially, if they encounter congenial substrate and conditions for spread of their propagules (Golan and Pringle, 2017; Roper et al., 2010). Aylor (1990) points out that rapid dispersal at both temporal and time scales is key to the development of epidemics/pandemics of crop disease. Repeated cycles of infection are a requirement for disease to become epidemic.

Fire may also play a role in initiating long distance dispersal (LDD). Because it can rapidly heat air and cause high-velocity updrafts (Golan and Pringle, 2017; Roper et al., 2010). For example, trajectory modeling of air masses suggests that viable spores captured from smoke over the Gulf Coast of Texas originated, 1500 km away in forest fires in the Yucatán Peninsula, Mexico. Sugarcane agriculture provides an excellent model for exploring how fire promotes LDD. This is because fields (stubbles) are burnt after harvest. They are often plagued by one of the most widely referenced putative long-distance dispersers. *Puccinia melanocephala* is among the first fungal species described that has undergone trans-Atlantic dispersal. Fire-borne updrafts, perhaps often caused by humans, may facilitate the spread of *P. melanocephala*. It challenges the hypotheses of the unassisted dispersal of spores across oceans.

Del Ponte et al. (2003) examined wheat fields in New York state to trace the source of inoculum of the Fusarium head blight (FHB). They have shown that debris from previous cereal crops does initiate new infection. Ascospores of the pathogen could be traced for up to small area from the inoculum (debris). In the wheat fields, both limited distance dispersal from infected crop debris and long-range through airborne ascospores occur. Indeed, airborne ascospores are traced not just above crop canopy or low latitudes but even in the boundary layers of atmosphere. It means ascospores do escape from the debris through turbulent air. Ascospores reach to different altitudes in the atmosphere. Viable conidia are easily traceable in the surrounding fields and atmosphere above the zone. Ascospores are known to travel long distances in the agrarian sky and perhaps initiate new infections elsewhere. Aerial spores found clustering into dense inoculum could definitely induce new infections, after traveling longer distances say over a few km. Based on spatial pattern of FHB found in wheat fields in New York state, Del Ponte et al. (2003) opine that airborne inoculum that falls on spikes is important, in terms of long-range dispersal and rapid spread of FHB.

Let us consider a few more examples. Spores of powdery mildew fungi are released at mid-day. A few other fungal pathogens release spores at night or in the early morning like *Mycosphaerella spp.* (Leaf blotch fungus), *Gibberella zeae* (neck rot in cereals), and *Venturia inequalis* (apple scab fungus) (Aylor, 1992, 2003). Their further dispersal in the atmosphere depends immensely on ambient factors, such as wind, temperature, humidity, UV radiation. The airborne plant pathogen and its adaptability to resist harsh atmospheric conditions are important. Several morphological, biophysical, and physiological characters are usually considered while assessing the potential of epidemic/pandemic causing microbe.

In an agroecosystem, the genetic diversity of crops cultivated is important. The genetic variation of pathogenic bacteria, fungus, or virus is equally important when we think of spread of a crop disease. The distance to which the propagules disperse quickly is crucial information that farmers have to know, prior to considering prophylactic procedures. The dispersal kernel is supposedly a trait that could be indicative of ability of fungus or bacteria to spread. Let us consider an example, Rieux et al. (2014) conducted an extensive experimental trial dealing with a fungus. Particularly, the spores that spread through the wind in the banana plantations. They have found that spore type, wind velocity, season, and genetic diversity of crops sown can influence the spread of the disease. Wind that lifts and makes fungal spore (*Mycospearella fijiensis*) float through the fields is a major determinant of virulence of the pathogen. Ascospores that are sexual spores were found to be dispersed closely to the plant (inoculum focus). Rarely, the ascospores were traced beyond 1000 m from the center of infection. Asexual spores showed a gradient in density up to 12.5 m. Some long-distance dispersal of over 1000 m occurs due to wind. So, we should consider both the wind and airborne spore types while forecasting spread of fungal disease, such as *Mycophhaerella fujiensis* rot.

Agricultural sky plays a vital role in the dispersal of microbial diseases. Fungal spore and its morphological traits are also important. This is in addition to their physiological ability to overcome harsh conditions during dispersal. Re-establishment on a substrate in a different location needs due consideration. Basically, it is the fungal spore, medium of transport, and substrate that determine the spread of a microbe. Ordinarily, wind and water currents are two main media through which the plant pathogens spread. This is in addition to animals, insect vectors, birds, etc., that aid fungal spore dispersal.

Calhim et al. (2018) opine that sexual spores of pathogenic fungus show remarkable adaptability to environmental vagaries, during dispersal. Actually, they identify three main aspects, namely, the versatility of spores, medium of transport, and substrate on which they land. They studied over 780 fungal species for variations in spore morphological traits, such as size, density, shape, ornamentation, etc. They concluded that morphological traits of fungus are important for wind/waterborne dispersal (Buller, 1909; Pringle et al., 2015; Norros, 2014). More important is the fact that spores should land on a congenial substrate. According to them, spore dispersal trends actually decide the population dynamics of the pathogenic fungi. This aspect has direct relevance to spread of disease.

Now, let us consider water as a dispersal agent of plant pathogens. Water in the form of clouds, precipitation, storms, or dust plus water storms does induce dispersal of pathogens. These are partly aerial mechanisms. Dispersal of pathogens across small or long distance in agrarian regions both are important. Like wind/air, water is an important factor of atmosphere. Water-aided dispersal of plant pathogens deserves attention since it can induce epidemics. Rainwater is said to be an important dispersal agent in nature. Spores of fungi, bacteria, and viral clumps may move across fields. Splashing rainwater can disperse spores to short distance around a focus of inoculum.

Water as medium for dispersal of pathogens of crops was identified first in 1880 A.D. Rainwater was shown to collect spores of fungi and bacterial cells. The number of propagules in the rainwater increased depending on the location of fucus of inoculum. Earliest evidence for rainwater dispersal came from observation on *X. campestris var malvacearum*, the rot bacteria that affected cotton in the USA. Xanthomonas was found in the rainwater collected from dust and rainstorm (See Fitt et al., 1989).

Black rot of cabbage is predominantly a seedborne disease. Yet, the secondary spread from a field to another is caused by rainwater or irrigation channels contaminated with bacteria *X. campestris var campestris* (Agriculture Canada, 2020). Rain splashed bacteria from contaminated plant residue left on the soil or from neighboring diseased plants is the primary method of disease spread throughout a field. No doubt, short distance spread from one plant to another is also important.

Several aspects of atmosphere (i.e., agricultural sky) affect the development of bacterial blight of soybeans caused by *Pseudomonas savastanoi.* This disease develops into an epidemic. Firstly, it depends on season. Wind and precipitation that splashes can disseminate the pathogen from infected crop residues found on the soil surface. Wind gusts can spread the bacterial pathogen from one plant to another through the rubbing of plants in the rows. Stomata and ruptured tissue due to injury are the entry points for the bacterial pathogen (Giesler 2011).

3.3.1.2 CLIMATE CHANGE EFFECT ON AERIAL PATHOGENS

Climate change effects impinge on both crop host and the pathogen. Each abiotic factor may have differential effects on pathogen/crop to different extents in space and time. So, the net result is difficult to forecast, prior to its occurrence. Jones (2012) has reviewed the entire gamut of climate change effects on plant disease and epidemics in different parts of world. It has been pointed out that in the case of bacterial and viral diseases, climate change influences crop, pathogen, and vectors simultaneously. Data relevant to forecasting the final effects on crop growth and biomass/grain yield are not easy to secure. Yet, we can strive at tangible forecasts if detailed data for several years are available. Environment immediately above the crop canopy has its influence, both, on crop and the pathogen. Weather parameters such as temperature, humidity, and CO_2 emissions may influence the physiological aspects of pathogen. They also influence pathogen's ability to cause devastation (Velasquez et al., 2018; Zhou et al., 2019). Environment is known to modulate sporulation and virulence of fungal pathogens. A few observations made about ambient air, weather parameters, and crop disease incidence are worth noting. In Asia, bacterial and sheath blight of rice could be observed due to higher temperature in the air. In Europe, increased temperature may enhance incidence of leaf spot of sugar beet. In Brazil, enhanced temperature may show greater incidence of *Colletotrichum falcatum*. Elevated temperatures in North America and Europe may induce higher incidence of rust fungi, etc. In Africa, a few viral diseases may show aggressive spread due to active vector population. Clearly, temperature alteration in the agrarian sky could affect pathogen and its ability to spread.

Climate change, particularly, enhanced levels of CO_2 in the ambient atmosphere that may affect crop's resistance to pathogens. They say crops defense is related to salicylic acid and/or jasmonic acid. They are specific chemicals elaborated by crops. Molecular analysis (genes) too suggests that the expression of certain genes may regulate the defense mechanism of host plant against the pathogen. Especially, under elevated CO_2 conditions. For example, it has been found that crops show enhanced resistance against certain necrotrophic diseases such as *Botrytis cinera*, under high CO_2 conditions. Whereas resistance to *P. syringae* by tomato was reduced due to CO_2 elevation. While a few diseases such as those caused by *Rhizoctonia solani* or *Fusarium oxysporum* may not get affected by changes in ambient CO_2 concentrations. Therefore, susceptibility/resistance of crops to pathogens may have to be understood with climate change effects in view.

As stated earlier in Chapter 2, soil is a medium that allows roots to penetrate then to derive anchorage, water, and nutrients. Above the ground surface, it is atmosphere (aerospace) which is a medium for crop's canopy to grow protruding into agrarian sky. The crop's canopy conducts gaseous exchange and draws moisture plus dissolved nutrients, in addition to photosynthesis. In the present context, we are concerned more with aerial canopy and how it can be modified to reduce plant pathogens from attacking crops. McDonald et al. (2013), for example, have stated that crop canopies can be modified to reduce crop disease expression, in field. Lateral trimming of carrot canopy by 30–40% at seedling stage, it seems, reduces *Sclerotinia sclerotiorum* disease, to almost zero. This is even without application of fungicides. Firstly, they say, trimming reduces relative humidity within the crop canopy but increases the temperature. This change in canopy microclimate affects the formation of apothecia. Trimming along with fungicide application similarly reduced fungal disease. Overall, agricultural sky has its influence on crop genetic expression with regard to susceptible/resistance to pathogens.

3.3.2 PHYLLOSPHERE MICROBES AND CROP DISEASE

Basically, phyllosphere is a part of agrarian sky. It is in constant interaction with ambient atmosphere and its factors. Phyllosphere is an important habitat for aerial microbes. Leaf exudates are good source of nourishment to phyllosphere microbes. Phyllosphere microbes may include members of diverse genera of bacteria, fungi, viruses, and Archaea. Phyllosphere microorganisms play a crucial role in helping their host against pathogens (Lacava et al., 2006; Lindow and Leveau, 2002; Lindow and Brandl, 2003; Mejía et al., 2008; Rajendran et al., 2011; Zhang et al., 2018; Schiro et al., 2018). Airborne microbes may invade host plant tissue via phyllosphere. They may actually lodge themselves as phyllosphere microbes for a stretch of time before causing pathogenesis. Later, they develop into pustules, blotches or necrotic spots, etc. Phyllosphere microbiome involves interactions of plant pathogens with other microbes. This aspect needs attention from farm researchers. The phyllosphere microbes are known to impart a certain degree of resistance against pathogens traced on leaf surface. Host genotype is also an important factor affecting phyllosphere exudation, microbial interactions, and pathogenesis, if any (Sapkota et al., 2015).

Let us consider a few plant diseases that involve foliage and their surface. Powdery mildew fungi that land on the phyllosphere may encounter a certain degree of hostile environment. Particularly, if the phyllosphere microbes are deterrent to it. There are several studies conducted in recent years, using rRNA methods. Such studies try to first ascertain the genetic diversity, then the quantity of pathogens on the leaf surface. Plus, they try to enumerate other microbes in phyllosphere microbiome, such as antagonists, diazotrophs, plant growth promoting bacteria (PGPR), pathogens, etc. Later, make appropriate decisions based on antagonistic interaction, if any, between pathogens and other microbes. It has been pointed out that microbial dynamics in the root-rhizosphere/rhizoplane have been investigated in greater detail. While phyllosphere microbial interactions were bestowed relatively meagre interest. During recent years, the availability of rRNA-based molecular methods to detect microbes has allowed us to study the phyllosphere microbiome too, in detail (Zhang et al., 2018).

Further, Zhang et al. (2018) believe that studies on deciphering benefits of phyllosphere microbes in terms of control of leaf-affecting pathogens are still rudimentary. It needs to be explored. Powdery mildew, for example, affects cucurbits. *Golovinomyces cichoracearum* (syn. *Erysiphe cichoracearum*) and *Podosphaera xanthii* (syn. *Sphaerotheca fuliginea*) cause powdery mildew.

Powdery mildew affects leaves. Leaves turn powdery-white due to fungal spores. Infected leaves show reduced photosynthesis. They later turn yellow, brittle, and may senesce (Lebeda et al., 2010). Phyllosphere microbes may reduce pathogenesis caused by powdery mildew fungus. However, impacts of pathogens on the phyllosphere microbiome are largely underexplored. Microbial flora on leaf surface may play an important role in the resistance to *B. cinerea* (Ritpitakphong et al., 2016; Leveau and Tech, 2011).

3.3.3 TECHNIQUES TO STUDY AIRBORNE MICROBIAL PATHOGENS

Agricultural sky is a repository of diverse species of microbes that attack crops. At any time, the atmosphere may carry a sizeable density of pathogens depending on season and stage of the crop. We should be able to monitor the pathogen, at the earliest, using various aerobiological techniques.

Microbes that may be airborne and can be carried over long distances through air currents need attention. A few other plant pathogens disperse through shorter distances utilizing rains and raindrop splashing. Dispersal of plant pathogens in the air (wind) is common mode for spread of crop diseases. Hence, air sampling methods are important. Lacey and West (2006) and West and Kimber (2014) have reviewed the topic of air sampling and detection of pathogens in the atmosphere. They state that since the beginning of "Hirst trap" in the 1950s to detect aerial microbes there has been improvement, in terms of efficiency of trapping plant pathogens found in the ambient atmosphere. Hirst trap has been utilized most frequently to obtain samples useful in aerobiological studies. During recent years, unmanned aerial vehicles (UAVs) fitted with GPS connectivity and facility to trap aerial spores of plant pathogens have been adopted. These UAVs could be adopted to detect pathogens that disperse to longer distances away from the field of origin. Also, spores could be detected at different altitudes from crop's canopy. Agricultural drones could get popularized based on how economical and easy it is to conduct air sampling. The situation is similar with aerostats and Helikites that are being experimented in farms (Krishna, 2020a, b).

Fang and Ramaraja (2015) list several methods that are suitable for the detection of plant pathogens. A few are difficult to apply. A few are inadequate on accuracies. Yet, they are good enough when no other quick methods are available. A few are costly. Also, need elaborate laboratory processing and trained technicians. For example, methods that are based on enzyme-linked immunosorbent assays, immune-fluorescence methods, flow cytometer methods, and gas chromatographic techniques are costly. Methods that rely on DNA/PCR that could be applied on small sample size are gaining in acceptances among crop researchers. Aerial microbiology in general and our knowledge about airborne plant pathogens may depend, to a large extent, on how accurate are the air samples. This has to be followed by identification and quantification of the infective propagule. There are few molecular methods that allow rapid identification of microbial pathogens of crops (Capote et al., 2012; Lopez et al., 2003; Martinelli et al., 2015; Dean et al., 2012; Khiyamia et al., 2014; Sinclair and Dhingra, 1995; Kaczmarek et al., 2019). However, there are also hyperspectral analysis-based techniques that study the crop's reflectance. The spectral signatures are utilized to identify the expression of disease right in the field (in situ). Such in situ methods may be helpful in rapid detection of pathogen in the early stages of spread in an agrarian region. Such methods have clear advantages because they detect the pathogen based on symptoms. There could be airborne spores but may not be infective due to seasonal variations and lack of virulence. Plus, weather patterns may not be congenial. Otherwise, crops could be resistant. Such UAV-based hyperspectral analysis is useful since it shows only established disease patterns on crops. We should note that UAV-aided aerial analysis of crops is quick and absolutely accurate. Particularly, regarding the incidence of pathogenesis. It can be applied repeatedly on short notice. The UAV-based hyperspectral analysis is suited best to screen disease-resistant germplasm lines, and also to monitor the crops on large fields. UAVs can be utilized to map the progression of the disease in the fields.

Williams et al. (2016) have attempted to standardize a method of air sampling and PCR analysis to identify airborne fungi traced. This method is aimed at analyzing microbes on crops or in the aerospace of a field location, like cropped field. Detection of spores of pathogens of crops is not easy. Conventional methods that rely on sampling the air followed by microscopic analysis have its inaccuracies. Spores of certain disease-causing fungi or even bacterial cells may not be easily identifiable. Microscopic analysis is tedious. Similarly, cultural methods using selective medium or a generalized nutrient medium or sticky petri plates with pads may not be easy to adopt. For example, slow-growing fungi may get overlooked on certain nutrient media. A few pathogens may never appear on nutrient medium (e.g., obligate pathogens).

The PCR methods are advantageous because they are accurate regarding DNA/rRNA sequences. Sample size could be small and may not require long time of incubation. Immunological methods were adopted but currently, they seem costly to operate. The antigenicity of fungal spore is a requirement. Also, raising antibodies is again tedious and time consuming. Its accuracy is questionable when cross-reacting antigens appear. Particularly, closely related species or pathotypes are to be detected. Under these conditions, West and Kimber (2014, 2015) and several other plant pathologists opine that PCR-DNA-based methods are suitable to be adopted and to study plant-pathogen dynamics in the aerospace above crops. We can develop reasonably good forecasts, if we conduct elaborate air sampling. Then, decipher the pattern of spread of pathogens (bacteria, fungi, or even virus). For example, McCartney et al. (2003) adopted the PCR method, to detect microbial pathogens affecting oilseed rape in Great Britain.

As stated earlier, gas chromatographic methods are also applied to detect airborne plant pathogens. They could be used when crops emit certain volatile compounds specific to disease condition. Iozzin (2014) suggests the detection of plant pathogens using pattern of evolution of volatile organic compounds (VOC). The assumption here is that emission pattern of VOC and quantity are signatures of specific crop diseases. Robotic miniaturized gas chromatographs that detect VOC are being recommended, to trace the gases rapidly. Analysis is done *in situ* in the aerospace above crops. Iozzin (2014), in fact, states that a one-inch sensor mounted on robotic platform attached to farm vehicles such as tractors or pick-up vans is good enough. It can detect a disease, using gas samples of aerospace above crops.

Nanophytopathology is a cutting-edge science that uses nanotechnology for detecting, diagnosing, and controlling plant disease. It aims at regulating the pathogens, at an early stage. It helps in the protection of crops from epidemic diseases. Traditional methods of sampling leaves, stem or other tissues, and culturing the causal agent (microbes) are tedious. It could also be costly and time-consuming. It may not be accurate in certain cases. If the causal microbe is not amenable for in vitro culture, using the culture media Identification using traditional methods is difficult to apply. Microscopic methods are used to detect microbes within leaf tissue or sample from canopy. At present, methods to detect the microbes involved in crop disease are ordinarily based on quantitative polymerase chain reaction (Q-PCR) (Falacy et al., 2007). They use rRNA sequences specific to disease-causing microbe to detect its presence. Rapid diagnostic kits are available. These seem to need a relatively larger-sized sample to process accurately, then identify and trace the pathogen in the canopy/leaf. Nanotechnology is yet to make a mark in detection and monitoring of airborne plant pathogens. Nanotechnology involves tracing nanoparticles, quantum dots, biosensors, nanoimaging, and nanopore DNA sequencing tools to detect plant pathogens. These nanotechniques are supposedly quick. Nanotechniques detect pathogens with greater accuracy. Samples could be smaller. Often, they serve as good rapid methods to monitor disease. Once, rapid diagnostic kits are developed for any of the severe diseases attacking the crops, they can be made routine. We can adopt through-put methods, to detect airborne inoculum of epidemics or pandemic causing bacteria, fungal spores or viruses (Jones 2020), etc.

3.3.4 ROLE OF AGRARIAN SKY IN THE EPIDEMICS AND PANDEMICS OF CROP DISEASES

The emphasis in the following paragraphs is on establishing the fact that "agricultural sky" and its factors play vital role, in the initiation and spread of crop disease, that is, epidemics/pandemics. Also, the effects of epidemics are devastating on biomass/grain accumulation. There are indeed several ways microbe-induced epidemics can affect food generation (Godfrey et al., 2016; Dean et al., 2012; Jones, 2020). Let us first consider definitions of various levels of pathogenic situations on crops. The disease is termed "endemic" if it occurs season-after-season or year-after-year affecting the host species. Also, it is caused by the same pathogen. The causal agent is said to be endemic to that cropping region. Pathogen actually develops a niche with crop host (species) when the crop is grown repeatedly in a location. Most often we encounter "sporadic infection." The crop disease is localized in small area and occurs irregularly, due to a microbe or its variants. "Epidemic" is a situation when the disease occurs frequently. Disease expresses itself rapidly, spreading quickly, and devastating the crops to greater levels. Pandemic is a situation when an epidemic disease spreads fast into different parts of a continent or it crosses into different continents. It affects crop productivity. Pandemics result in depreciated grain harvests in several parts of the world, simultaneously.

There are several known epidemics/pandemics caused by viruses, bacteria, fungi, nematodes, etc. "Late Blight of Potato" caused by the phycomycetous fungus *Phytophthora infestans* is an example of epidemic caused by microbe. The blight epidemic affected the staple carbohydrate source of rural communities in Europe. It is an airborne fungus. In the 1840s, it even caused starvation in the USA and certain European nations. In the 1990s, too several farms in the USA lost their entire potato crop, due to late blight.

Proper sanitation of farms was required to avoid spores from spreading the fungus. Appropriate crop rotation and fungicides (Metalyxyl) were then applied to potato crop. Cultivars resistant to this airborne fungus was needed to avoid the disease. Similarly, an epidemic on grapes called the "Downy Mildew of Grapes" affected the European grapevine regions. The secondary spread of spores via atmosphere was to be controlled. Application of Cu fungicides helped in reducing the infection cycles (Roberts and Ryan, 2016; Schumann and D'Arcy, 2005; Aylor et al., 2011; also see Dean et al., 2012).

Let us list a few epidemics caused by airborne fungi, bacteria, and viruses on major food and cash crops. Usually, a large area of the crop is affected (see 2Blades Foundation, 2020). The infection often develops into intense rust pustules, mildews, spots, blotches, wilts, cankers, etc., depending on the pathogen. For example, coffee rust is caused by *Hemilia vastetrix*. It is a basidiomycetous fungus with ability for release of airborne teliospores and urediospores. It caused devastation of coffee plantations in Srilanka, in 1868. Once again coffee rust fungus was deadly. This time, in 1975, the rust fungus devastated coffee gardens in Central America. Coffee rust fungus reappeared as an epidemic in 2015 in the Caribbean region.

Next, Grape Downy Mildew devastated grapevines in France from 1878 to 1885. This epidemic occurred due to *Plasmopara viticola*, an airborne fungus. In the 1930s, rice blight caused by *Cochliobolus miyabeans* is supposed to have devastated rice culture in Northeast Gangetic Plains. The epidemic was so severe that over 2 million humans starved and perished. Since humans in this region depended entirely on rice, for their carbohydrate nutrition. Again, this genus of fungus, *Cochliobolus heterostrophus* destroyed maize crops in North America. Maize grain productivity reduced to the tune of 20% of original harvests (Strange and Scott, 2005).

Historically, wheat rust spores (urediniospores) were excavated in Israel. The rust spores were dated to 1300 B.C. Writing by Italians provides support to the fact that wheat rust was a severe disease around 43 B.C. to 17 A.D. The first detailed report about cereal stem rust caused by *P. graminis* was provided by Italians, Fontana, and Tozzetti, in 1797. Later, in 1815, stem and leaf rust diseases of wheat were distinguished as caused by two different pathogens (USDA Agricultural Research Service, 2019). In 1941, black stem rust of wheat became a serious epidemic covering entire North American wheat growing regions. *P. graminis* is an airborne fungus that has frequently caused epidemic. American wheat has suffered major epidemics caused by these rust fungi. The pathogenic fungus releases lots of urediospores that are airborne. Such urediodpores spread rapidly into the wheat farming belts.

During the 1930s till the 1960s, North American wheat crop experienced severe bouts of stem rust. It reduced grain yield of wheat by 50% of original levels. In 1999, wheat grown in Middle East Asia suffered due to a new stain of rust fungus known as "Ug99." It resulted in 30–40% loss in yield in some regions. An epidemic lasted for up to 5 years in the Great Plains region. This epidemic was caused by wheat stripe rust fungus (*Puccinia striiformis*). It threatened wheat yield in North America, in 2010 (Lucas et al., 1992). In 2016, *P. striiformis* caused an epidemic in East Africa (Wandale et al., 2016; Plate 3.1). Wheat blast caused an epidemic in Northeast India and Bangladesh during 2016–2018.

During the 1920s, Panama disease of bananas reached epidemic proportions. It is caused by *Fusarium fujiense*. This fungus induced epidemics in the 1990s in Southeast Asia. The spores are airborne during spread of the inoculum. On rice crop, brown spot occurred as an epidemic in 1942, in India. Again, in the 1960s, the rice belt in eastern India suffered an epidemic caused by *Pyricularia oryzae*. It causes bacterial blight of rice. Maize necrosis disease and streak viruses became epidemic in several regions of world, during 2015–2017. In 1975, Rossette virus spread into groundnut fields in Northern Nigeria. Airborne aphids are vectors. The epidemic almost reduced groundnut cropping to negligible levels. It continues so even now in 2020. Agricultural sky-imposed biotic stresses do affect groundnut in West Africa (Yayock et al., 1976; Waliyar et al., 2007; Musa et al., 2020). From the above examples, it is clear that "agricultural sky" can mediate devastating diseases. It needs to be monitored periodically for such detrimental biotic factors.

During the whole decade of the 1990s till 2000, citrus crop in Florida suffered the loss of yield and fruit quality. It is due to citrus canker caused by *Xanthomonas citri*. More recently, citrus plantations in Central Florida suffered epidemic due to "Hualongbing disease" also called "Citrus Greening." It is caused by bacteria known as *Candidatus liberibacter*. It is spread by airborne insect vectors. It affected citrus yield during 2013–2019. Olives in Italy and Spain suffered an epidemic caused by *Xylella fastidosa* a severe bacterial pathogen, in 2017 (Fasano, 2020; see Almeida, 2018). It seems over a million trees were destroyed due to this organism. It got spread by aerial insect vectors. Lack of sanitary conditions in the farm regarding plant pathogens and water splashing also induced its spread. Overall, global agriculture has suffered epidemics/pandemics due to a "hostile agricultural sky" and its factors, both biotic and abiotic. The crop losses suffered have been severe. So severe that it has even induced famines in past centuries. Particularly, when it concerned staple cereal sources. Wheat pandemics have almost become routine worldwide due to a rapid spread of rust spores into all continents.

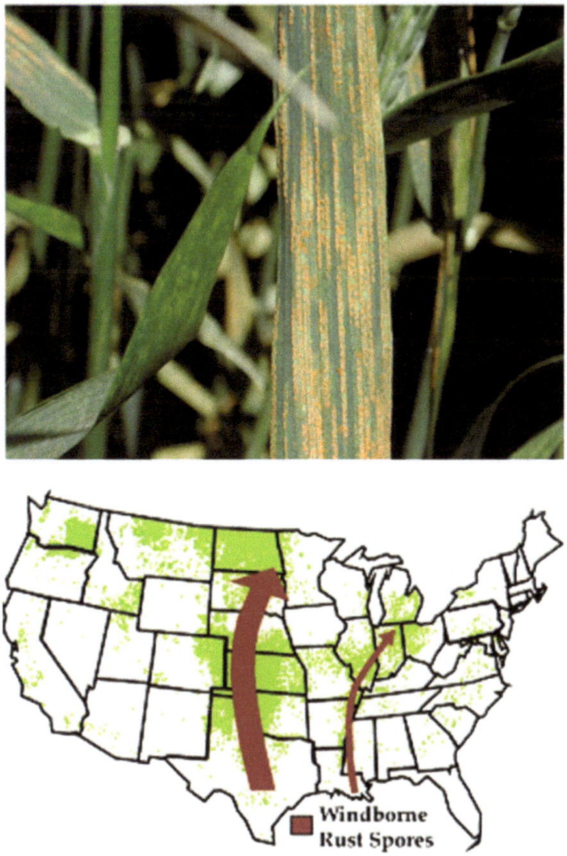

PLATE 3.1 Top: Yellow stripe rust of wheat caused by *Puccinia striformis*.
Bottom: General pathway of different species of Puccinia such as the graminis tritici, striformis, etc.

Source: Cereal Disease Laboratory, United States Department of Agriculture, St. Paul, Minnesota, USA

Source: Dr Shahriar Kianian, Wheat Specialist, Research Staff, Cereal Disease Laboratory, United States Department of Agriculture, St. Paul, MN, USA. https://www.ars.usda.gov/midwest-area/stpaul/cereal-disease-lab/docs/cereal-rusts/wheat-stripe-rust/
Note: Rust fungi could potentially cause epidemics or pandemics based on atmospheric traits such as wind its direction, speed, altitude, relative humidity, temperature, etc. The viability of fungal propagules is important.

Agricultural sky, meaning the wind/water and airborne vectors like insects and birds, in almost all continents have played havoc with crop productivity. Generally, they say, weeds are the most severe yield reducers in a cropped field. Weeds account for 30–35% biomass/grain yield loss. It is followed by diseases that constitute 16% of crop yield loss. Out of this 16% grain yield loss, fungi are said to be the most dreaded aerial factors in a farm. They can also affect vast stretches because their propagules are often highly infectious. Such propagules spread via atmosphere (wind and water). Fungi are the most dreaded disease-causing organisms on crops. They account for 70% for the loss of food grain/fruit productivity. No, doubt monitoring the atmosphere for plant pathogens is the most important task for farmers. They need to adopt integrated procedures that reduce spread of pathogen via air/water in the atmosphere. Planting resistant genotypes of crops is important. Since it avoids spores from striking a congenial substrate after transit via wind currents. Crop rotations with nonhosts during pandemic season are useful.

Cassava is a major carbohydrate source for low-income population in Central and East African nations. During the 1980s and 1990s, this crop was affected severely by a few different strains of Cassava Mosaic Virus. It resulted in reduction of harvests by over 20–30% of original harvest levels, in the region. Several farmers abandoned growing cassava due to viral attacks (Strange and Scott, 2005; see Jones, 2020).

Among the various abiotic factors, it is the wind, its speed, currents, and jet streams that aid long-distance journey of propagules of microbial plant pathogens (Hau and Vallavieille-Pope, 2006, Schmale and Ross, 2015). LDD and establishment of infection focus are perhaps the basic requirements, if the microbe (fungus, bacteria, or virus) has to become a pandemic. Aylor (1990) has pointed out that intermittent wind gusts that keep the propagules afloat is also important. Aspects like efficient dispersal, infectivity, virulence, and rapid multiplication are a few essentials of a pandemic microbe. Let us consider an example. Nagarajan and Singh (1990) express that rust fungi which cause pandemics across most of the wheat-growing regions of the world, actually elaborate enormous quantities of sexual spores (urediospores). The urediospores are transported to long distances via wind (jet streams). It allows fungal spores to transit and land in different continents. Initially, the spores are scattered by the rust fungus forcefully in all directions. After dispersing in air, they land on either congenial wheat belt again or stray uncongenial vegetation. Usually, the spore transit in the agrarian sky is regulated by the wind speed, its currents and high-altitude jet streams. These are essential for rapid dispersal and initiation of infections.

This leads to an epidemic/pandemic depending on how far the spores travel into different agroecosystems or continents in the world. There are indeed several fungal pathogens that are capable of developing into epidemics/pandemics. They are aided by wind during transport (see Table 3.1). There are bacterial pathogens that disperse due to precipitation, clouds, fog, and initiate epidemics (e.g., black rot caused by *X. campestris*). The bacterial pathogen gets disseminated rapidly through rain/water droplets.

It is worthwhile studying the windborne dispersal of both ecto- and arbuscular mycorrhizas across different continents. Each year, perhaps, such wind-mediated dispersal may introduce several new mycorrhizal fungi/genotypes. Vesicular arbuscular mycorrhizas are highly promiscuous with regard to crop species. They may strike a host definitely and rapidly once they land in a location after high-altitude transit. Hence, their spread in any vegetation located at short or long distance can be definite/quick, in comparison to disease-causing fungi that are prone to high host specificity.

As stated above, many of the plant pathogens are severe on crops. They devastate large patches of crops situated in different continents. These are actually transported for long distances by the wind currents in the atmosphere. The ability to track the spores of such microbes is important (Schmale and Ross, 2015). The dispersal routes and density of the microbial pathogen could be monitored. We can study the spread of epidemics at the farm level by monitoring the pathogen. We can do so even at continental scale by intercepting the spores/bacteria in the atmosphere. At present, we have UAVs that could be flown in the sky as many times at different altitudes, to trap spores of pathogens. Such information on fungal spores or bacteria will be useful while forecasting the appearance of epidemics/pandemics. We can accordingly develop prophylactic measures (sprays). This is to thwart the further transit and spread of the pathogen.

Fusarium blight of wheat that occurs in North America is caused by airborne inoculum of *G. zeae* (anamorph *Fusarium graminearum*). It is a severe disease that results in loss of grain yield. Also, it depreciates value of grains because of mycotoxin, Deoxynivelinol (DON), in the wheat grain. This fungus over-winters in the crop debris as a saprophyte. It also affects a few different cereals and grasses. Ascospores and macroconidia produced on over-wintered residues act as inoculum source. The spores infect spikelets, at anthesis to kernel stage. The fungus spreads all through the spikelet in 3–5 days and induces premature senescence. Humid conditions favor sporulation and secondary spread of the pathogen in the wheat field. Del Ponte et al. (2003) examined wheat fields in New York state to trace the source of

TABLE 3.1 Long Distance Dispersal of Fungal Plant Pathogens and Symbionts: A Few Examples.

Crop disease	Causal agent	Spore type	Size	Shape	References
Fusarium head blight	*Gibrella zeae/Fusarium graminarium*	Ascospore	13–28 × 10–13	Long ellipsoid	Leslie and Summerell (2006)
Black leaf streak	*Mycosphaerella Fijians*	Ascospore	11–16.5 × 2.5–5	Fusiform	Parnell et al. (1998)
Black sigatoka	*Mycosphaerella graminicola/Septoria tritici*	Ascospore	8–10 × 2.5–3	Fusiform	Parnell et al. (1998), Linde et al. (2002)
Sigatoka leaf spot	*Mycosphaerella musicola*	Ascospore	14.7–17 × 4.6	Fusiform	Parnell et al. (1998)
Cottony rot, stem rot	*Sclerotina sclerotiarum*	Ascospore	12.5 × 4–6	Ellipsoid	Qandoh and del Rio-Mendoza (2011)
Ectomycorrhiza	*Laccaria laccata*	Basidiospore	6.16 × 8.47	Globose	Vincenot et al. (2012)
Ectomycorrhiza	*Laccaria amethystina*	Basidiospore	4.42 × 5.17	Globose	Vincenot et al. (2012)
Loose smut	*Ustilago nuda*	Teliospore	6.5 × 5.8	Subglobose	Simmonds (1994)
Leaf rust of coffee	*Hemilia vastetrix*	Urediospore	29.7–34.5 × 5–8	Globose	Isard et al. (2005)
Soybean leaf rust	*Pakospora pachyriza*	Urediospore	18–34 × 15–24	Globose	Carvalho et al. (2011)
Wheat stem rust	*Puccinia graminis tritici (Pgt)*	Urediospore	28–30 × 17.5	Globose	Anikster et al. (2005)
Sugarcane rust	*Puccinia melanoocephala*	Urediospore	28–33 × 18–23	Obovoid	Anikster et al. (2005)
Wheat yellow rust	*Puccinia striformis*	Urediospore	14–36 × 13–21	Ellipsoid	Anikster et al. (2005)
Sugar beet rust	*Uromyces betae*	Urediospore	25–35 × 12–40	Spiny, dotted/ Globose	Pammel (2017)

Excerpted from Golan and Pringle (2017), Lucas et al. (1992), 2Blades Foundation (2020).

Note: *Laccaria laccata* and *Laccaria amithystina* are ectomycorrhizal fungi. The above fungi may get dispersed from a few m to >1000 m or even regions within an agroecosystem or even to different continents. It seems spore size, type, shape, surface ornamentation may all contribute to long-distance floatation over wind/jet streams that occur at high altitudes. A few fungi like *Pgt* can get dispersed globally and initiate infection in different locations (Uredospore = Urediniospore = Urediospore).

inoculum of the FHB. Debris from previous cereal crops do initiate new infection. Ascospores are traced for up to small area from the inoculum (debris). Both, limited distance dispersal from infected crop debris and long-range through airborne ascospores occur. Indeed, airborne ascospores are traced even in the boundary layers of atmosphere. The ascospores do escape from the debris through turbulent air. Spores reach to different altitudes in the atmosphere. Viable conidia are easily traceable in the surrounding fields and atmosphere above the zone. Aerial spores found clustering into dense inoculum could definitely induce new infections after traveling longer distances over a few km. Based on spatial pattern of FHB found in wheat fields in New York state, Del Ponte et al. (2003) opine that airborne inoculum that falls on spikes is important.

In many farming regions, the pathogen that causes epidemic could be actually a seedborne microbe. Initiation of infection begins with seed germination and develops into severe disease on a plant. However, for the disease to progress in a field, in other wards to disperse and became epidemic (i.e., secondary infection), it is the weather parameters that are important. Atmospheric parameters such as rain, temperature, relative humidity, wind, also sprinkler irrigation affect dispersal. Lack of culling of disease-affected seedlings, lack of prophylactic sprays, and plus farm vehicle movement may also cause rapid dissemination of propagules. For example, in rice, the blight which is seedborne is actually disseminated better by the congenial ambient weather parameters. Blight is commonly observed when strong winds and continuous heavy rains occur. Here, agricultural sky and its parameters induce the epidemic from a simple seedborne infection (Diekmann and Bogyo, 1992). During the 1970s, blight had devastated rice farming areas of Southeast Asia (IRRI, 2020)

There is greater emphasis on monoculture and intensive production practices. Just a few cultivars are getting preferred for cultivation by farmers. Large commercial farms in developed nations have been preferring monoculture. These trends could easily induce epidemics. Further, we should realize that fungal species that cause epidemics have short life cycles. So, they may mutate and evolve into virulent strains rapidly (Owings, 2020). Despite breeding cereals for resistance, the fungal pathogens may overcome the host resistance and induce epidemics.

At any time, in any agroecosystem, the sky (atmosphere) holds diverse microbes. They could be severe pathogens, mild pathogens, symbionts (e.g., *Bradyrhizobia*, ecto- or arbuscular mycorrhizal spores, plant growth promoting rhizobacteria, etc.) or commensals of no tangible consequence to crops. Overall, we should aim at characterizing and classifying the agricultural sky (atmosphere) above the region. We have to consider microbial load and diversity. Sky above a farm or a field could be characterized based on its ability, to induce disease on crops. We have to base it also on specific crops/cropping systems followed. Agricultural sky with dense population of virulent strain of pathogen naturally induces disease. Such a sky has to be classified as "most sickly sky." One with the mild pathogen as "prone to mild disease pressure." Sky with sizeable population of symbionts as "friendly sky" prone to induce *N* fixation or *P* mobilization, etc. A sky termed as "deficient in symbionts" is a clear possibility. One with commensal may have to be termed sky with commensals of no conspicuous advantage or trouble to standing crops. Sky prone to endemics/pandemics is the ones that need to be watched regularly. Farmers may encounter a "sick sky" with multiple pathogens hanging in the air above canopy. Perhaps, it is worthwhile identifying all of the pathogens and potential levels of devastations.

3.4 ENTOMOLOGICAL ASPECTS OF THE AGRICULTURAL SKY

3.4.1 AGRARIAN SKY IS A REPOSITORY OF HARMFUL PESTS ON FIELD CROPS AND PLANTATIONS

Anywhere in the sky above the different agrarian regions of the world, we encounter a posse of diverse insects. They thrive on the natural vegetation/crops on the ground. Their distribution in the atmosphere may get affected by the altitude they localize, ambient weather parameters, greenery available, etc. We usually encounter pests and nonpests that thrive on vegetation, etc. Also, predators and parasitoids thrive on insects. Knowing the pest diversity, migratory trends, their population, and damage potential are of utmost importance to agriculturists. Knowledge about pests in the *agrarian sky* can provide us with an idea about the control measures that should be adopted. Prior knowledge about the distribution of aerial pests provides us with sufficient reaction time. It helps us in adopting the most appropriate methods.

Earliest of the intentions to study the distribution of insects from the ground surface to higher altitudes began in the 1920s, in the USA. Prior to it, insect distributions were estimated by installing traps on roofs of tall buildings, mountains, treetops, forest observation posts, and lighthouse. Let us narrate one example of investigation related to airborne insects. It was conducted in North American farming zones.

The Entomological Bureau of USDA began studying the insect distribution in the atmosphere in 1926–1931. They used airplanes flown at altitudes from 200 m to 16,000–20,000 ft altitude in the sky. Airplanes were fitted with insect traps smeared with sticky glue to trap them. Several genera of insects were recorded both during day and night at altitudes from 200 to 16,000 m above the fields. Reports by Glick (1939) stated that during aerial analysis using airplanes in 1926–1931, researchers collected 22,850 insect specimens at altitudes from 200 to 15,000 m. The distribution of insects differed based on altitudes. Their efforts showed that in all 28,739 specimens occurred during day and night in Louisiana and about 1294 in the atmosphere in Mexico. Insects flying in the air differed during day and night. There were more insects in the atmosphere during day than in night. Precipitation had a definite effect on the numbers and species of insects found in the upper air. Atmospheric parameters such as season of the year, month, ambient temperature, relative humidity, light also affected the insect distribution (Felt, 1928a, b; Felt and Chamberlain, 1935). In the ambient atmosphere above Louisiana farms (Talaulah, LA), they traced 18 orders of insects, 216 families, 824 genera, and 700 species. They found 24 new species yet to be described. Some of the insect orders that were predominant in the collections were Homoptera, Hymenoptera, Lepidoptera, Thysanoptera, Orthoptera, Neuroptera, Coleoptera, Heteroptera, etc. It seems size, weight, and buoyancy of an insect contribute directly to the height to which it may be carried by air currents. Of the Homoptera, numbers of the small species of Cicadellidae, Fulgoridae, Psyllidae, and Aphididae were traced up to 14,000 ft. (Glick 1939). Details about proportion of crop pests and their distribution at different altitudes in the agrarian sky above the fields were not available. Also, how many of the insect species traced in the sky were actually crop pests also were not mentioned. We may, however, assume that several of them could invariably be pest on vegetation/crops on the ground. There are several other studies depicting insect distribution above crop's canopy. They describe insect diversity and population from different agrarian regions of the world.

Discussing all aerial pests that attack the large number of different crops grown in agrarian regions may be difficult considering the theme and context of the book. There are several treatises, journals, reports, manuals, etc., about pests and their control. Our aim here is to substantiate that agrarian sky affects the crop-pest interaction. It affects crops' tolerance to pests. A pest could be severe or mild. The predators/parasitoids that thrive on pests too play an important role in the aerospace above cropped fields. Examples drawn here pertain mostly to wheat in Great Plains plus a few others only. A few of the severe pests that routinely affect wheat crop grown in the Great Plains are as follows: Armyworms such as *Helicoverpa* affects the foliage and panicles. They are migratory causing severe damage to wheat fields as they move from one field to the next. Greenbug (aphid) and Russian Aphids are dreaded aerial pests. They suck the sap. Simultaneously, they inject toxins into plant tissue. Russian aphid is known to serve as vector for leaf streak viruses and mosaic viruses. Hessian fly can cause large-scale devastation of wheat crop. Their larvae feed on stems at the leaf base and leaf sheathe. Farmers have to sow the wheat crop after October 15, that is, during Hessian-free dates, if they want to avoid hessian fly attacks. Cereal leaf beetles are also common aerial pests on wheat, oats, barley, and millet. Beetles feed on succulent aerial tissues. There stink bugs that destroy foliage and panicles at ripening sage. We may note here that there are indeed innumerable aerial pests that attack field crops, plantations, and forests. No doubt, pests are among most important aerial phenomenon. It is worth studying in greater detail. So that, we can regulate the agrarian sky to support pests only up to below economic threshold or even lower levels. We have to monitor agrarian sky consistently and frequently, to overcome aerial pests. As stated in Chapter 1 (Section 1.3.2), insect pests usually traced in the agrarian sky could fall into one of these descriptions. Namely, the leaf hoppers (aphids, jassids, and thrips), leaf eaters and cutworms, armyworms (*Helicoverpa*, *Spodoptera*), leaf rollers, shoot flies, stem borers, head miners and head eaters, blister beetles, head bugs, greenbugs, grasshoppers, locusts, and other species that migrate long distances to devour crops, etc. Wheat midge (*Sitdiplosis mosellana*) is traced in most of the agrarian regions of the world. It affects wheat crop gown in the Canadian Prairies and several states of the Northern Unites States like Dakotas, Minnesota, Montana, and Idaho (Agriculture Knowledge Centre, 2020). Midge also affects other cereal crops such as barley, oats, couch grass, and brassicas in the plains. In addition, there are several insect species that are vectors of microbial disease (e.g., Aphids transmit viruses) (see Prescott et al., 1986; Whitworth et al., 2021; Agriculture Knowledge Centre, 2020; CESAR, 2019)

3.4.2 AERIAL PESTS OF MAJOR CROPS: A FEW EXAMPLES

Globally, perhaps, there are no field or plantation crops that are not attacked by "aerial insects." Agrarian sky is no doubt detrimental to crops to a certain extent. A few crop species suffer aerial attacks by multiple pest species, simultaneously or in sequence. Such pests bring down the biomass/grain productivity. A few pests occur periodically in a location and are entrenched to the area (endemic). Few others are migratory insects. They cause pandemics by distributing themselves to different agrarian regions. Many are sporadic and occur feebly too.

Cereal aphids such as the Russian aphid (*D. noxia*) and green bug (*Schiziphis graminum*) are among dreaded aerial pests. They attack the foliage, stem, and even grains of major crops such as wheat, barley, and others. Cereal aphids, it seems, have got accentuated. They become almost endemic to semiarid plains. They appear year-after-year in the agrarian sky above the wheat. Aphids occur in the entire wheat belt in the Great Plains from Canadian Prairies in Alberta to Rolling Plains in Texas (Giles et al., 2008). Monoculture of wheat continuously for several years, covering long stretches of fields, also lack of natural enemies increase pest pressure on wheat. However, Giles et al. (2008) suggested that ecologically the absence of habitats that support natural enemies in these monoculture agricultural systems is considered a primary reason for aphid build-up. That is why populations of aphids increase above economic threshold levels (Elliott et al., 1998b, 2002; French and Elliott, 1999a; Brewer et al., 2001; French et al. 2001; Giles et al., 2008; Brewer and Elliott, 2004). Economic losses associated with both Greenbug and Russian wheat aphid could reach a few hundred million US$, annually, across the Great Plains of the USA (Webster, 1995). Aphid attack is avoided or reduced using resistant wheat varieties. Diversifying cropping systems seems to reduce aphid attack. Application of higher dosages of pesticides could induce insecticide resistance in the crop pests. Therefore, one of the suggestions is to include crops such as canola, millet, sorghum, clover, lucerne, cotton, and sunflower. These crops are known to support multiplication of natural predators and parasites of aphids (Parajulee and Slosser, 1999; Elliott et al., 1998a, b 2002; Brewer et al., 2001; French et al., 2001; Brewer and Elliott, 2004; French and Elliott, 1999a, b; CESAR, 2019; Whitworth et al., 2021). Conservation of natural enemies needs priority, especially, in monoculture zones.

Crop pests that spend part of their cycle in the aerospace above crop's canopy could also be migratory. Migratory trends are often guided by availability of moisture and greenery. Insect movement may also be controlled by wind, its speed, and direction at various altitudes above the canopy of crops. Several of pests that are known to affect crops are also transported by wind.

The outbreaks of pests depend on the number of individuals that migrate or carried by wind to other locations. The density of airborne pest population is an important factor that initiates a pest attack, in a location (Joyce, 1983).

Globally, there are several aerial pests that attack the food grain crops like maize, wheat, millets, pulses, oil seeds, etc. It is worth noting that these crops, wherever cultivated are attacked by aerial insects the moment they emerge from seeds. Specific stages of crops could be more susceptible. A few species of insects may be virulent. So, they may gain upper hand over the crop. Yet, we can classify the aerial pests that attack crops into those affecting at the early seedling stage, mid-season growth stages, milking/seed fill stage, and panicle/seed maturity stage. Similarly, we can also classify the aerial pests based on feeding habit or destruction pattern. They are leaf hoppers, sucking insects, cutworms, chewing insects, and panicle pests. Whatever the crop stage or feeding habit, it is important to trace the pest in the atmosphere, at the earliest, then adopt appropriate remedies. Many of the remedies too could be mediated aerially. It means that we are applying aerial techniques (e.g., spraying) to control a pest in the agrarian sky. At the bottom line, we need to keep the agrarian sky clean and with low intensity of insect pests, if we aim at higher harvests.

Let us mention a few examples of aerial insect pests of various crops grown in different continents. The following examples pertain primarily to crops grown in Queensland, Australia (Department of Agriculture and Fisheries, 2020a). Maize is an important fodder and grain crop. Aerial insect pests that attack the crop can be grouped as sucking pests, leaf cutter, chewing insects, and armyworm. The armyworm *H. armigera* is the most devastating pest. It affects the seedlings. It devours leaves, therefore, reduce photosynthetic surface. It attacks cobs at silking stage and eats the succulent seeds. Maize leaf hoppers, maize thrips, and cutworms are aerial pests that attack the early seedling stages. Locusts, beetles, greenbugs, aphids, and armyworm devastate the crop, during growths stages. At silking, armyworm attacks the cobs. At seed-fill stage yellow moths, bugs, and mites reduce seed yield and its quality. Clearly, agrarian sky above the maize crop can harbor several detrimental pests. We need appropriate prophylactic measures (Department of Agriculture and Fisheries, 2020a).

Sorghum is attacked by armyworm (*H. armigera*) and sorghum midge (*Stenodiplosis sorghicola*). These are serious aerial pests in the Queensland region. Minor aerial pests are aphids, head caterpillars, and panicle bugs. Again, reports suggest that sorghum cultivated for fodder/grains is exposed to insect attack from emergence till grain-fill stage. Agrarian sky is again a detrimental aspect of sorghum cultivation, if we consider the aerial insects (Department of Agriculture and Fisheries, 2020b). Biological control agents traced in the aerospace above the sorghum crops could be utilized. A few examples of parasites are *Trichogramma spp* and *Microplitis sp*. These are wasps that affect caterpillars. There are aerial predatory bugs, beetles, spiders, and lacewings. They may keep the pests at threshold.

Airborne pests are not a major problem on the winter cereals such as wheat, barley, oats, and triticale. Yet, they do hover around the canopy and affect the crop species. We can detect armyworms, aphids, armyworms, and mites (Department of Agriculture and Fisheries, 2020c; GRDC, 2016).

The agrarian sky above sunflower fields in Queensland harbors several aerial insect pests. They can be detrimental to crop productivity. Sunflower is attacked by aerial insects such as armyworms (*Helicoverpa spp*), cut worms, thrips, beetles, whiteflies, loopers, and green vegetable bug (Department of Agriculture and Fisheries, 2020d).

Major aerial pests that attack peanuts grown in Queensland are the *Helicoverpa sp*, mites, and web moth. They affect foliage during growth stages of the crop (Department of Agriculture and Fisheries, 2020e). Major pest on cotton cultivated in Queensland is *Helicoverpa spp*, spider mites, aphids, whiteflies, thrips, and beetles. They affect the cotton crop from the moment the seeds germinate and last till the boll ripens. So, cotton is among the crops most affected by aerial insects. Commensurately, it is said that farmers apply cotton with maximum amount of plant protection chemicals anywhere in the world (Department of Agriculture and Fisheries 2020f).

The above discussions pertain to crops and their aerial insect pests encountered in a specific agrarian region. However, we may note that almost all crop production zones are invariably infested by aerial pests. They could play a definite detrimental role.

Climate change and its influence on crop pest species in the Agrarian sky is perhaps an important topic. It needs immediate and greater attention, than, bestowed presently by the scientists. Foremost, we should be alert to the influence of climate change on the crops that pests attack. Many ecological patterns depend strongly on phenology of crops (Harrington et al., 1999). While increased CO_2 may result in greater photosynthate, it may also reduce foliage quality as plant defensive compounds increase in concentration. The carbon–nitrogen ratio will also increase, affecting C_3 plants (soybeans, rice, and wheat) more than C_4 plants (corn, sorghum, and sugarcane). Changes in foliage quality affect the insects feeding on them.

It also affects the incidence of plant diseases and higher-order interactions of predation and parasitism (Zvereva and Kozlov, 2006; Velasquez et al., 2018). Climate change can create trophic dislocation, that is, it affects the synchrony between crop and pest species. It could be beneficial or utterly deleterious, if crop growth gets depreciated simultaneously increasing pest's destructive activity. For example, they say synchrony between wheat crop and winter moth (*Operophtera brumata*) both, get affected, due to climate change. If we consider pests in isolation, it is said climate change can affect traits such as migratory trends and resident activity if the pest is stationary in a location. It can affect pest's preferences to attack different crops. Climate change may affect pest's fecundity and multiplication rates. Climate change affects pest's damage potential in a location. Several physiological traits such as days for eggs to hatch, degree days required by the pests to reach adult stage, generation time, foraging pattern may all alter due to climate change (Taylor et al., 2018). Climate change affects several pests that attack wheat and other crops in the Great Plains. They are as follows Coleoptera: bean leaf beetle (*Ceratoma trifurcata*); Mexican bean beetle (*Epilachna varivestis*); Lepidoptera: Armyworm (*Pseudaletia unipuncta*); black cutworm (*Agrotis ipsilon*); corn earworm (*Heliothis zea*); European corn borer (*Ostrinia nubilalis*); stalk borer (*Papaipama nebris*); velvetbean caterpillar (*Anticarsia gemmatalis*); Homoptera: Potato leafhopper (*Empoasca fabae*). Overall, climate change affects both crops and pests.

Climate change effects on ecological aspects of crops, pests, and predators/parasitoids that attack pest can be complex. It is worthwhile knowing at least the basics about how climate affects the synchrony and trophic aspects of the three organisms (crop, pest, and parasitoid) as they operate in nature (agrarian sky). Climate change affects the tritrophic aspects in nature. First it affects the phenology of crops, next affects the insect pest's life cycle events and thirdly its natural predators/parasitoids in a field. For example, Castex et al. (2017) have shown that climate change factors such as temperature, solar radiation, diurnal pattern, availability of sap/photosynthates, nesting surfaces, egg laying locations, etc., humidity, and CO_2 evolution are important. These changes affect the availability of green vegetation in a crop field. Succulence of leaves may increase pest population and cause greater damage to crop. In other situations, pest life cycle may get altered and population explosion may be totally absent leaving crops to grow better. Simultaneously, population of biological control agents (predators/parasitoids) may get affected first due to climate change factors operating directly, their life cycle events, and their population build-up. It may also affect the availability of pest on which they predate. For example, synchrony between grapevine growth pattern (*Vitis vinifera*), pests such as *Lobesia borana,* and its parasitoid *Trichogramma spp* may get altered, due to climate change effects. Sometimes, adopting integrated pest management practices (IPM) may become difficult due to changes in trophic mechanisms. Add to it, if the insect pest is migratory, then synchrony between its migration and biological control agent may get altered further. Overall, climate change is expected to alter the timing, intensity, duration, population, and extent of devastation, if any, by the pest. Simultaneously, predator/parasiticide's activity is also altered.

Agrarian sky could be utilized efficiently to trace and map the insect pest attacked zones, in a farm. Satellites, small UAVs (also called agricultural drones), Helikites, parafoils, blimps, even kites that fly in the aerial space above crops could be utilized, to obtain spectral data (Krishna, 2020a, b). Aerial imagery using sensors is the centerpiece of the farm procedures. Precision methods involving spectral analysis of crops, to detect pest-attacked zone is in vogue. Perhaps, only in some farms. It is yet to make a mark all over the agrarian regions of the world. Several experimental stations spread across North America, Europe, and Fareast are evaluating the procedures. It involves the use of pesticides most appropriately, that is, only at locations that are attacked, not on the entire farm. The arial imagery of fields and digital signatures of pest attacks are utilized, to guide the variable-rate nozzles located on the sprayer. Let us consider an example. The UAVs (drones) with facility to obtain multispectral imagery have been adopted, to distinguish healthy and European corn borer (*Ostrijnia mubilis*) affected maize crops. They have targeted carotenoid and anthocyanin pigments on the healthy/infested crops. However, targeting chlorophyll pigments of healthy and infested crops was more accurate in depicting the actual situation on the ground about the intensity of corn borer infestation. Corn borer infestation was easily detected and analyzed using multispectral sensors (Caroll et al., 2008; Yue and Lei, 2012; Barbedo, 2019). Aphid-affected crops were detected accurately using sensors on drones (Kim et al., 2020; Marston et al., 2020). Occurrence of locust swarms in the aerospace above agrarian regions too could be detected, by flying the UAVs and obtaining the aerial imagery.

Agricultural drones with sprayer bar and variable-rate nozzle are flown over the agrarian sky, to spray the crop. In places not adopting precision farming methods, crops are being sprayed, using aerial drones with sprayer facility. Entire field is sprayed rapidly by a drone with preplanned flight path. Often sprayer tanks hold 5–10 L of pesticide formulation. Here, we are adopting an aerial method to control an aerial insect pest (See Yue and Lei, 2012; Krishna, 2018, 2020a; RMAX, 2015; Iost and Kong, 2019). In some cases, such as locusts, the sprayer drone, it seems can be flown into the locust cloud to distribute the pesticide. This is an example of abiotic man-made aerial factor (drone aircrafts and other UAVs) being utilized, to control an aerial pest of severe consequence to crops.

3.4.3 LOCUSTS IN AGRARIAN SKY: INSECT PANDEMICS ON CROPS

Locusts in the agrarian sky is among the best examples that drive us to study the aerospace above crops with greater intensity, seriousness, and alertness. We should not overlook the agrarian sky or just investigate vagaries related to climate (i.e., radiation, wind, clouds/precipitation, temperature, and air quality). Instead, we have to also assess the aerospace region for various biotic detrimental factors. Then, apply remedial measures. Just like, we do with soils and crops. It is better to halt the locust pandemic right in the sky than to perceive it at a greater intensity on the green vegetation or food crops.

Desert locusts are said to be most devastating on standing crops (Arthur, 2008; Plate 3.2). Food and Agricultural Organization of the United Nations (2020) lists, at least a few more migratory locusts (short-horned grasshoppers) that operate in the agrarian sky of different continents. They are African Migratory Locust (*Locusta migratoria migratorioides*)—Africa; Oriental Migratory Locust (*Locusta migratoria manilensis*)—South-East Asia; Red Locust (*Nomadacris septemfasciata*)—Eastern Africa; Brown Locust (*Locustana pardalina)*—Southern Africa; Italian Locust (*Calliptamus italicus*), from western Europe to Central Asia; Moroccan Locust (*Dociostaurus maroccanus*)—North-West Africa to Asia; Bombay Locust (*Nomadacris succincta)*—South-West to South-East Asia; Australian Plague Locust (*Chortoicetes terminifera*)—Australia; tree locusts (*Anacridium sp.)*—Africa, Mediterranean, Near East.

Let us consider the role of desert locusts that cause widespread attacks and pandemics on food crops of South Asia, Arabian Peninsula, Eastern Africa, parts of West Asia, and extensive farming zones of sub-Sahara, especially Sahelian belts. They are deleterious to crops. They spread rapidly from one continent to the other. They often move in swarms of large populations. They create food grain insecurity to populations residing in their path if they alight at a location. Desert locusts are indeed devastating (Plate 3.2). So deleterious that greenery of entire stretches of several km of natural vegetation and food grain crops get wiped out in a matter of few hours (See Plate 3.2). During the years 2003–2005, it seems the food security of over 8 million people residing in Sahel was at risk, due to locust-driven pandemics. Food grain yield depressions were severe. So, it resulted in large quantities food grain import and food aids in the West African Sahel (Food and Agricultural Organization of the United Nations, 2017). Clearly, agricultural sky harboring short-horned grasshoppers in swarms plays detrimental role on food generation. Here, agricultural sky supports insects that diminish food grain production.

Life span of short-horned grasshoppers is about 3 months. Eggs take 2 weeks to hatch. Hoppers become adults in 6 weeks. Adults need 1 month to mature and initiate the next cycle of egg laying. Grasshoppers breed rapidly. It results in 20 times increase in 3 months, and 400 times in 6 months, and 8000 times in 9–12 months. It is this trait of the locusts that cause devastation. Locust swarms need large quantity of greenery to feed themselves. They take to agrarian sky in search of foliage and grains. They fly in swarms containing millions of individual grasshoppers. Desert locusts consume 2 g of green vegetation each day per individual. They consume crop's foliage and grains. Crop species attacked by locusts include pearl millet, sorghum, maize, wheat, rice, barley rice, date palms, banana plants, etc. Several noncrop plant species found in natural vegetation are also consumed voraciously, by the grasshoppers.

Most recently, in 2020, locust monitoring agencies have reported that the global desert locust situation deteriorated (Food and Agricultural Organization of the United Nations, 2002; Benaim, 2020; Dunne, 2020). It was attributed to favorable climatic conditions that allowed widespread breeding of the pest in East Africa, Middle East, South Asia, and the area around the Red Sea. Every increase in locust population proportionately deteriorates the quality of the agrarian sky, particularly, in terms of aerial pests and their impact on food grain generation.

Managing locust invasion is crucial. Prior warning with longer reaction time is a necessity. They say, technique followed by International Agricultural agencies and local farm organizations relied predominantly on aerial observations via air crafts, local folk information, and scouting farming zones for grasshoppers. These methods offered very less reaction time (Joffe, 1995). Especially, considering that these flying swarms invade crops swiftly and move to next alighting location. Later, came the use of satellite imagery of clouds of grasshoppers. We could monitor the swarms periodically within a few hours interval and forecast their migratory routes and inform the farmers accordingly. This method meant that locusts were identified as they group and develop into swarms. More recent satellite methods involve the use of imagery showing greenery. In other words, images showing green natural vegetation and crops during rainy season and later were identified as possible locations for locust attack. So, a standing crop was required to be imaged. It allowed about less than a month reaction time for preparing prophylactic measures. Recently, we have special satellites that detect soil moisture (Soil Moisture and Ocean Salinity Mission-SMOS Satellite of European Space Agency). Soil moisture distribution and its levels in upper layers are fairly good indicators of green vegetation. Hence, such maps can be prepared well ahead of actual cropping season and utilized. Longer reaction time seems crucial while managing crops against locusts or even severe spells of droughts. They say, European Space Agency's satellite SMOS is capable of it. Routine observations all through the year, using "Copernicus Sentinel Satellites" has also been recommended, to locate locust swarms above Sahel/Sahara region. This method allows up to 2 months reaction time for farm agencies, to supply control measures like pesticides, and adopt diversionary tactics. The imagery and digital data offered by this satellite have been evaluated in sub-Saharan nations such as Algeria, Mali, Mauritania, and Morocco. They have provided about 70 days of reaction time to farmers. Signs of moisture and vegetation have accurately imaged. So, forecasts that are useful to farm agencies dealing with the control of locusts could be made available (Food and Agricultural Organization of the United Nations, 2017). There are computer-based models that help us forecast locust swarm movements (Ariel and Ayali, 2015). Locust preparedness is important because it saves foliage and food grains (Editor, 2020).

More recent reports suggest that agricultural drones may have a better chance to be used against locusts. Drones may monitor the sky regularly and bring in appropriate information (see Krishna, 2018, 2020a). There are sprayer drones that could directly enter the grasshopper clouds and apply the pesticides or even biological control agents, like formulations of *Metarrhizium acridum,* an entomophagus fungus.

PLATE 3.2 A desert locust resting on cereal stem.
Source: Food and Agricultural Organization of the United Nations, Rome, Italy; http://www.fao.org/locusts/en// pp 1–7 (December 28, 2020).
Note: A swarm of 1.0 km^2 area of grasshoppers that are ordinarily observed in the agrarian sky above South Asia, East Africa, or Sahel can encompass about 40 million individual short-horned grasshoppers. Once they alight at a location, they are known to devour green vegetation and crops very rapidly. Cereal food grains good enough to feed 35,000 people are consumed by locusts in 1 day (see Food and Agriculture Organization of the United Nations, 2020). Desert locusts are usually restricted to drier areas of Africa, the Near East, and South-West Asia that receive less than 200 mm of rain annually. This is an area of about 16 million square kilometers, consisting of about 30 countries (see Food and Agricultural Organization of the United Nations, 2020).

3.4.4 POLLINATION BY AIRBORNE INSECTS IS VITAL FOR OPTIMUM CROP PRODUCTIVITY: A BENEFICIAL ROLE OF INSECT SPECIES IN AGRARIAN SKY

Discussions in the above paragraphs were confined to insects that are active as aerial pests on valuable food grain and fruit crops. They are examples of deleterious effects of insect pests that attack parts of crop's canopy. Many of them ruin the crops in situ causing localized damages. Several others such as armyworms move longer distances from the focus of infection. There are pests like locusts that travel long distances and damage crops. We must realize that the same "agricultural sky" also harbors several insect species that are beneficial to crops and humans. They too localize above crops and in their canopies. Most important beneficial function among them, in terms of agrarian aspects, is "pollination." There are honey-producing bees, wax producers, lac insects, etc., In the following paragraphs, let us consider the beneficial effects of agrarian sky bestowed to crops/humans.

Globally, pollination via insects is attracting greater attention in recent years. Particularly in light of deficiency of pollinators caused due to rampant use of pesticide, herbicides, and fungicides.

Pollination is a natural phenomenon operative in the agrarian sky. The process is vital for propagation of natural vegetation and crops. Pollination has a greater impact on grain yield and its stability, in the case of intensively grown cross-pollinated crops (Bartomeous et al., 2014; Kunast et al., 2013; Pearce, 1998; Blanche and Cunningham, 2005; Blanche et al., 2006; Roubik, 1995). Seed-set in such crops is dependent on optimum pollination levels. Here, we are basically concerned with pollination mediated by insects dwelling in the agrarian regions.

Among naturally wild flowering plant species, about 87% of them depend on insects, like honeybees, wasps, moths, butterflies, beetles, bugs, and ants for their pollination. Of course, in addition, birds and animals too mediate pollination. The extent of the dependence of natural vegetation and crops on insect-mediated pollination may vary. Such variations are attributable to several factors such as the crop species, geographical location, ambient weather conditions, characteristics of the agrarian sky, insect diversity, diversity of vegetation/crops, etc. Globally, tropical crop species are more dependent on insect-aided pollination. They say, in some tropical locations, mainly in Central Africa, Asia, and Latin America, predominately cross-pollinated corps are cultivated. About 98% of pollination is dependent on insect species traced in the agrarian sky. Among domesticated crops consumed routinely by humans about 60% of crops do not entirely depend on insect pollination for seed sets. About 35% need support from insects for seed sets. The remaining crop species are yet to be characterized (Intergovernmental Science Policy Platform on Biodiversity and Ecosystem Services 2018; University of California at Berkeley, 2006).

Among major food grain crops, such as cereals, millets, pulses, oil seeds, and fruits about 13 depend entirely on pollination by insects, about 30 depend to a great extent on insects for seed-set. About 27 crop species are moderately dependent, 21 crop species are slightly dependent, and few of them show no definite trend regarding dependence on insect-mediated pollination for seed production. Consequently, about 70% of food grain generation across different agrarian regions of the world depends on insects and other aerial biotic species, for regular pollination and seed set (Klein et al., 2003a, b; Bartomeous et al., 2014; Pearce, 1998).

In a natural vegetation, pollination has certain extra importance in terms of genetic diversity and geneflow in a plant species. It is important to note that pollen and seed dispersal are the critical stages for gene flow. Pollen-mediated gene movement is often important for genetic diversity within and among plant populations (Barleunga et al., 2011; Imbert and Lefevre, 2003; Heuertz et al., 2003; Fijen, 2019). Here is an example, where in, agrarian sky has its impact on gene-flow in a plant species. Crops that are pollinated may be under the influence of pollinators for a proper gene-flow and good progeny.

Since past couple of decades, agricultural researchers have expressed concern about maintaining optimum or high levels of population of honeybees and other pollinators in cropped fields. The activity of honeybees is crucial for optimum crop production. It is observed that certain species of honeybees including the managed ones such as *Apis mellifera* has been showing a decline. A few countries use excessive dosages of pesticides, particularly, in large farms. They have suffered a conspicuous decline of honeybee population. Agrarian regions in the USA, Brazil, Great Britain, Spain, and Germany have shown a decline of *Apis species* (European Honeybee). There are only a few reports about reduction in pollinator activity, in farming regions of Africa (Intergovernmental Science Policy Platform on Biodiversity and Ecosystem Services, 2018). A few genera of honeybees such as *Bombus sp.* have shown a steep decline. Some species and their biotypes of honeybees/pollinators have even gone extinct. Most of the ecologists blame it on excessive use of pesticides, removal of natural vegetation surrounding field, climate change effects on both pollinator and crop (Potts et al., 2014; Sustainable Agriculture Research and Education-SARE, 2020).

Klein et al. (2003a, b) have listed several species of bees and other insects that are active in the agrarian sky. They contribute to pollination of crops to varying extents. Following are a few examples: managed honeybees are useful when the crop is not frequently visited, by the naturally available wild bees. A few species of domesticated "managed bees" are *Apis cerana, A. dorsata, A. florea*, and *A. mellifera*. There are several "stingless bees" that operate in the aerospace above crops and pollinate the flowers. They are *Melipona subnitida, Melipona quadrifasciata, Nanotrigona perilampoides, Nanotrigona testaceicornis, Trigona cupira, Trigona iridipennis, Trigona terminata, Trigona Minangkabau*, and *Scaptotrigona depilis*. "Bumble bees" such as *Bombus affinis, Bombus californicus, Bombus hortrum, Bombus lapidaries, Bombus terrestrius* are a few other examples of wild bees pollinating crops. There are several solitary bees that operate swiftly in the agrarian sky and conduct effective pollination of crop plants. There are "solitary bees" *Amegilla chlorocyanea, Amegilla holmesi, Andrena ilerda, Anthophora pilipes, Centris tarsata, Habropoda laboriosa, Halictus tripartitus, Megachile addenda., Osmia aglaia Sandhouse, Osmia cornifrons, Osmia cornuta, Osmia lignaria lignaria, Peponapis limitaris, Peponapis pruinosa, Pithitis smaragdula, Xylocopa dejeanii, Xylocopa frontalis,* and *Xylocopa suspecta.* Next, we have wasps such as *Blastophaga psenes*; "hover flies" such as *Eristalis cerealis, Eristalis tenax* and *Trichometallea pollinosa*; beetles such as *Carpophilus hemipterus* and *Carpophilus mutilates*; thrips such as *Thrips hawaiiensis* and *Haplothrips tenuipennis* that are able to conduct pollination in the crop's canopy.

Klein et al. (2003 a, b) have discussed pollination-related constraints that might affect a few crops of importance. For example, sunflower (*Helianthus annuus*) is a crop that depends immensely on insect-aided pollination for seed set. The ratio of the crop to natural vegetation area is important. Availability of honey species diversity seems vital for a good crop of sunflower (Greenleaf and Kremen, 2006b). The number of visits by the wild bees and managed ones affect the seed set. For crops such as tomato, distance to natural vegetation from field, and population of *Bombus sp, Apis melliferae,* and other species affected pollination and fruit yield (Greenleaf and Kremen, 2006a, b). In the case of pineapple (*Anona squamosa*), it is the honeybee species diversity that could affect pollination. Low population of honeybees can affect pollination and seed set in *Brassica napus, Brassica rapae,* etc. (Lindstrom, 2017; Adamidis et al., 2019). In the case of watermelon (*Citrullus sp*), it is the pollen density that affects the pollination. In the case of *Coffee arabica,* in addition to honeybee species, it is pollen grains, stigma, fruit set, and seed mass that are of concern (Ricketts, 2004 and Ricketts et al., 2004). In the case of citrus plantations, it is the distance between trees and flowers that bees have to transit that may affect pollination and fruit productivity. The above examples make it immensely clear that insect-aided pollination is important in agrarian regions. There could be a few specific reasons related to honeybee activity, crop/natural vegetation ratio, and weather parameters that affect pollination. Whatever the reason, whichever the species of honeybees, certain crop species do depend on insects for pollination and seed set. So, farmers need to monitor honeybees above crop's canopy. They should have an idea about how frequently they are visiting flowers. If not, they may have to harvest chaffy panicles and/or low grain yield. As stated earlier, 35–39% of global crop species are dependent on honeybees and other insects in the atmosphere, for pollination.

There are insect pollinators like honeybees and syrphid flies that collect pollen from naturally wind-pollinated crop species. They conduct the usual transfer of pollen from one flower to another (Saunders, 2017). Jha and Dick (2010) stated that it is the distance from the hive that bees cover that matters during pollination. In a farm, crops are found grown along with natural vegetation. Such tree species, shrubs, and bush species are visited by native bees. These native bees are capable of traveling relatively longer distances compared to domesticated European bees. For example, a single individual may forage an area with radius of 1800 m from hive. This allows better gathering. It primarily induces long-distance pollination in a field or farm. No doubt, agrarian sky supports both, short and long-distance pollination though different honeybee species.

Pesticide application may ruin the farm atmosphere. The agricultural sky may deteriorate due to residual activity of toxic plant protection chemicals. Farm chemicals applied during flowering and pollination can be deleterious to hone bee activity. Pesticide drifts due to wind can affect pollinator population even in neighboring fields too. And their activity added to pesticides during spray has also proved to be toxic. Several researchers have reported that the availability and application of harmful pesticides is key to declining population of honeybees. Countries with high per capita use of pesticides in agrarian regions suffer greater decline in pollinators.

The degree of dependency of yield on pollinators varies among crops. However, we should note that pollinators are responsible, in a direct way for a relatively small share (5–8%) of total agricultural production. Pollinators are also responsible for many indirect contributions, such as the production of many crop seeds for sowing, but not for consumption. It means seed production plots should not experience dearth of pollinators. A deficiency in pollinators in the agricultural sky may lead to chaffy or low-quality seeds meant for sowing.

Honeybees are said to be the most prolific pollinators. There are several species of plants within a natural vegetation patch that offer nectar to insects other than honeybees. Moths, butterflies, wasps, bugs, and beetles too pollinate flowers. Many of these insect species could be hovering the cropped fields in search of nectar flower. In the process, they induce pollination leading to better seed set. A few examples to quote are the moths (Hawk Moths). They were found to collect nectar, using long proboscis but simultaneously pollinating the flowers.

There are pollinators other than honeybees in almost all agrarian regions. Natural dispersal or deliberate introduction of honeybees from one agrarian zone to the other has at times, offered great advantages. Agrarian sky and crops accommodate such intrusions and even perform better than usual. For example, one wild pollinator from West Africa has made a major economic contribution to the world. It is oil palm weevil (*Elaeidobius kamerunicus*).

This is a West African species that was introduced into Malaysia. It is also known to pollinate several different global food crops. Interestingly, West African oil palm beetles seems to have performed well on many of the Southeast Asian crops. Seed set of crops like sunflower, oil seed rape, and few vegetables have been better in fields visited by beetles.

Honeybees hovering over crops do provide a few other ecosystem services in addition to the pollination of food/fruit crops. The large number of beehives in farm belts also offer wax. Globally, this aerial process is said to generate 65,000 t of beeswax and 518,000 tonnes of honey, annually (Intergovernmental Science Policy Platform on Biodiversity and Ecosystem Services, 2018). Honey and wax production require low initial investment. Hence, they have been preferred activities by even subsistence farming households. We could utilize these beneficial insects better by improving native vegetation/crops. Here, agricultural sky offers benefits to farmers.

Let us consider aerial insect pollinator species and their role in farming regions of Africa. The African continent has large zones of dense natural vegetation with plant species that support insect pollinators and derive benefits from them. There are also regions in Sub-Sahara that supports scattered or often scanty vegetation. Sahelian vegetation too supports the large number of beneficial insects like pollinators, honey, and wax producers. African farming regions are said to use relatively lower levels of pesticides. In many regions, the subsistence farming trends do not allow the use of pesticides. So, deleterious effects of farm chemicals on pollinators (honeybees) are expected to be low.

No doubt, African farming and forestry zones support diverse and large posse of pollinators. They are of great value for the propagation of natural vegetation/crops in the continent. Major crops that derive benefits of optimum pollination levels through honeybees are food crops like cereals, pulses, oilseeds, vegetables, cash crops, forest plantations, firewood species, etc. Pollination services through agrarian sky affects both quality and quantity of food grain/fruit generation in the African farming zones. For example, it said fruit quality and productivity are affected by pollinator activity (Asiko, 2012; Vaissiere et al., 2012). Several horticultural species depend on *A. melliferae* activity, in the sky. Honeybee species such as *Hypotrigona grobdoi* is said to improve fruit set of *Capsicum annum*. In East Africa, honeybee population and its activity are vital for good harvests of sweet pepper (Intergovernmental Science Policy Platform on Biodiversity and Ecosystem Services 2018; Potts et al., 2018).

Diversity of plant species like trees, shrubs, bushes, small hedge plants, surface creepers all of them which ever flowers is useful for honeybees. It helps honeybees to operate in the aerospace above the vegetation. We need to ascertain the diversity of both the crops and honeybee species and their preferences. This suggestion by farm researchers applies to all agrarian regions that support cross-pollinated species, in greater number. Reports suggest that several African bee species have been moved around all over the world's agrarian regions. Like *A. melliferae*, the European domesticated honeybees, a few African species are found as well-managed beehive series in farms across the world. They say managed bees are prone to suffer diseases. A few reports about effectiveness honeybees in African farms are interesting. Wild bees have performed more efficiently in producing better seed-set of several crop species, coffee plantations, onions, tomato, oilseed rape, etc. They were better compared to domesticated bees. Agricultural researchers point out that there are only a few studies on pollinators in Southern African Farming zones. Aspects like pollinator and crop diversity, flower visitation, honey production, and seed set need greater attention. Overall, there are clear suggestions that farming zones in Africa need to be studied in greater detail, to improve further the role of pollinators on crop productivity.

In summary, we need to improve the symbiotic activity of crops and insect pollinators. Honeybees have prime place among pollinators. Their upkeep in terms of population, genetic diversity, and activity in the agroecosystems need due consideration. We should adopt cropping systems that induce better pollinator activity. This way we would be utilizing aerial pollinator insect in the agrarian sky, better.

3.5 AVES IN THE AGRICULTURAL SKY

3.5.1 AVES AS PESTS OF AGRICULTURAL CROPS

Bird pests on crops are worldwide in distribution. Aves are a class of vertebrates that comprises birds. There are several bird species that affect quite a wide range of economically useful crops. Yet, they say, accurate bird damage estimation on a worldwide basis is not available easily, particularly, if we consider it for each bird pest species (De Grazio, 1978).

Sunflower crop is a good example that thrives on aerial pollinators like honeybees and birds. Agricultural sky provides a beneficial aspect to pollination and seed set of this oil seed crop. At the same time, farmers find that sunflower is also the crop that suffers loss of seeds, due to bird pests. Bird pests are a major detriment posed from the agricultural sky.

Sunflower seeds sown in the fields are often at risk of being eaten by birds. It is easy to identify because birds would have dug wholes in soil at intervals where sunflower seeds were sown. Therefore, seedling establishment and crop stand get affected. Gap filling will be needed. Bird damage to sunflower crop occurs as early as four leaf stage and in different forms. Bird pests like pigeons, sparrows, crows, etc., damage sunflower grains. Birds leave chaffy seed coats on the head after consuming the kernel (Peer et al., 2003).

Bird control using baits and pesticides is not recommended. It results in loss of bird fauna and diversity. Clearing nesting sites and encouraging shift is possible. Scaring birds away and avoiding them from entering ripe sunflower farms is the procedure adopted, in most farmers. It could be done by bird scaring guns, bangers, balloons, shiny plastic ribbons that reflect sun rays or Helikites fitted with sounds of raptors and predatory birds. Such methods scare away the bird pests. Bird scaring using Helikites seems effective on several crops. So, it could be tried on sunflower fields too (Allsopp et al., 2013; Allsopp Helikites Ltd., 2019; Hagy et al., 2005).

Here is an example of how agricultural sky with a high density of bird population can affect small grain cereal crops. Again, aerospace above millets can be beneficial when it encourages bird-mediated pollination, insect pest control (i.e., biological control), or even seed dispersal. However, birds in sky can be severe detriments to millet grain yield. Several different millet species are cultivated in the Indian subcontinent. They are sorghum (*Sorghum bicolor*), pearl millet (*Pennisetum glaucum*), finger millet (*Eleusine coracana*), foxtail millet (*Seteria italica*), Kodo millet (*Paspalum scrobiculatum*), little millet (*Panicum sumatrense*), proso millet (*Panicum miliaceum*), and barnyard millet (*Echinochloa fromentanceae*). Most, if not all of these millets suffer bird damage right from the seeding stage. Birds pull out seeds from soil and reduce seedling density. Birds are severe on millets at panicle ripening stage. They can leave chaffy seed coats by removing the soft mild grains. Birds that disperse the millet grains may also cause volunteer problems in the fields in the vicinity. Usual bird control measures adopted in the aerospace above fields are to put bird scaring bangers, sound guns, reflectors, balloons, scaring sound of falcons and other predators, use of Helikites, kites, etc. Bird damage can range from less than threshold (2–3%) to severe (28–35%). Sparrows, parakeets, pigeons, crows, muniyas, bayas, and mynas are severe on millets cultivated in South Asian drylands (Patel, 2011, 2016). Maize is an important cereal grain and forage crop in the Indian subcontinent. This crop is affected by a few bird pests. They are blue rock pigeons (*Columbia livia*), rose-ringed parakeet (*Psittacula krameria*), and house crow (*Corvus splendens*). Bird control measures during cob formation and maturation is essential (Ikisan, 2021). Major crop species of the Deccan region of the Indian subcontinent such as sorghum, pearl millet, pigeon peas, chickpeas, and brassicas were affected by bird pests. Bird species such as sparrows (*Passer domesticus*), weaver birds—bayas (*Ploceus philippinus*), and rock pigeons (*Columba livia*) are most common pests in the region. Reduction in grain yield ranged from 42 to 52% compared to unaffected fields. Conventional bird repelling methods were effective if applied in time (Kale et al., 2014).

Millets grown anywhere in the agrarian regions of different continents seem to be vulnerable to birds. Particularly, during panicle ripening and seed maturation stages. Perhaps there are no exceptions unless farmers have sown bird pest tolerant cultivars. In the sahelian region, major millets such as sorghum and millet are of course susceptible to bird damage. Bird scaring methods are needed to safeguard the harvest (panicles). There are several methods of detecting onset of bird damage (Manikowski and Camara-Smeets, 1979). Damage control measures include scaring.

Bird pests can be a challenge in millet growing regions of Eastern Africa. Quelea is serious bird pest on sorghum and pearl millet. Control procedures involve positioning of bird detractors, reflectors, scaring gadgets, etc. In the Nigerian tropics and Sahelian zones, crops are severely affected by bird pests. Several species of birds have been identified as pests. A few bird pests noted on cereals and other grain crops are doves (red-eyed turtle dove, red-eyed wood dove, laughing dove), weaver birds (*Ploceus cuclatus, Ploceus nigerrum*), red-headed quelea (*Q. erythropus, Q. quelea*), bush fowls and crakes (Oluwadare, 1980).

Passerines, pigeons, rooks, blackbirds, parakeets, starlings, and gulls are among the most notorious pests that throng agrarian sky of Great Plains and Midwest regions of North America. Farmers have to manage the birds and their population in the sky. They are a threat to optimum harvest, despite good crop stand, growth, optimum seed set and maturity. This happens because these bird pests devour soft and mature grains of cereal, legume also fruits in orchards. Scaring birds from nesting zones near farms is a requirement. Migratory birds in large flocks are also common pests on crops.

The red-winged blackbird (*Agelaius phoeniceus*) is one of the most abundant birds in North American agrarian regions. Their population exceeds 300 million birds. The black male with striking red and yellow shoulder patches is slightly larger than the female (Bent,1965).

Dolbeer and Linz (2016) have pointed out that the term "blackbird" refers to a set of 10 species. They inhabit North American plains and Midwest region. Blackbirds belong to the avian family Icteridae. Examples of a few species are as follows: red-winged blackbird (*A. phoeniceus*), common grackle (*Quiscalus quiscula*), great-tailed grackle (*Quiscalus mexicanus*), brown-headed cowbird (*Molothrus ater*), yellow-headed blackbird (*Xanthocephalus xanthocephalus*), brewer's blackbird (*Euphagus cyanocephalus*), and rusty blackbird (*Euphagus carolinus*)

Blackbird roosting sites are important indication about the extent of damage to crops. For example, studies of blackbird damage to ripening corn in Ohio and North Dakota have revealed that almost all losses exceeding 5% of the crop have occurred in fields, within 8 km of marshes that contain large blackbird roosts, in late summer (Dolbeer and Linz, 2016).

Blackbirds provide some benefits by feeding on harmful insects, such as beetles (*Diabrotica spp.*) and corn earworms (*Helicoverpa zea*), and on weed seeds, such as Johnson grass. Further, because of their abundance, redwings are a food source for a variety of avian and mammalian predators.

Managing agrarian sky prone to blackbirds is the requirement. Usually, control measures may involve manipulation of agrarian sky. Exclusion of bird pest using a net above the crop is possible. Provided the plots are small and easily manageable. Cultural methods include delayed planting, modification of crop rotations (Wilson et al., 1989), deep planting of cereals, and planting crop species not preferred by blackbirds in the roosting areas. Planting hybrids tolerant/resistant to blackbirds. Scaring birds using sounds of predators of blackbirds.

In a given agrarian region, crops are often exposed to both native birds and nonindigenous species. Native bird species have often evolved to a niche. They encounter vagaries of foraging grounds, diseases, natural predators, biological control agents, etc. Let us quote an example. Crows (rooks), it seems are not native to Victoria in Australia. Crow is a major pest of agriculture and is known to devastate crops such as wheat, maize, and sunflower. It can cause severe damage to vegetables and fruit crops, including mango, guava, pawpaw, fig, apple, pear, grape, and stone fruits. The nonindigenous birds could be new to the agrarian zone. They may not experience stress of predators, otherwise felt by native bird species. Sometimes, nonindigenous bird species reach higher populations. So, it is generally suggested that the agrarian sky needs to be regulated regarding bird pest species that it harbors. Several legislations and restrictions to bird species movement have been suggested in Victoria. Such efforts will limit migratory species and nonindigenous birds from damaging crops. It is the sky that needs to be monitored continuously, throughout the crop period. The bottom line is that remedies should effectively reduce population of bird pests.

3.5.2 AVES AS BIOLOGICAL CONTROL AGENTS OF INSECT PESTS

The intention in this section is to provide examples of bird-mediated biological control of insect pests attacking vegetation. Birds do get active above the crop's canopy for different purposes. It could be for perching, to collect nectar in flowers, eat grains/fruits, consume stray edible items on land, small rodents and other organisms in soil, insects on crop's canopy, etc. Birds are swift and agile in the air above fields. Birds prey on insect pests that appear in large numbers on fields attacked by pests. Experimental observations have shown that predatory birds or even generally, several avian species reduce the population of insect pests. They consume insects, particularly ones found in large numbers. This aerial phenomenon, therefore, results in reduction of pest population to levels known as below threshold. In a way, bird species that hover in crop fields regulate the population and resultant destruction by insects. Firstly, we should reduce the use of harmful chemicals in farms to control insects. We may adopt biological control methods involving birds that are predatory and devour insect pests. In such a case, requirement for pesticides reduces. We may even avoid the use of plant protection chemicals. This is because birds keep pest population below threshold levels (Kirk et al., 1996)

Birds in the aerospace above crops are indeed beneficial too. They are not just pests on grain crops/fruit plantations (Heath and Long, 2019). Birds devour insect pests attacking crops such as apples, coffee, cacao, oil palm, corn, cabbage, cereals, legumes, oilseeds, cotton, etc. A few large raptors, owls, and hawks feed on even small rodents that otherwise affect crop produce in the fields. No doubt, birds are good biological control agents in the open cropped fields. Yet, farmers in food grain production belts often perceive birds as pest on cereal panicles. In other words, a detriment to good quantity grains. We need a location-based rational judgment about birds.

Firstly, detection of crops infested by insect pests is a requirement if bird-mediated biological control is to be successful. Field foraging birds particularly those preying on insect pests are known to be sensitive to plant volatiles. They say plant volatiles released from crops/trees that harbor or are attacked by insect pests are detected by birds. Such birds could be of use during biological control of insect pests (Rubene et al., 2019).

Birds differ in their foraging habits. Natural vegetation and cropping systems followed may affect the predatory activity of birds. Aves are traced frequenting in woody regions (patches) found in an otherwise predominantly food grain farming zone belts. Here is an example from cereal farming zone in Nebraska. Puckett et al. (2009) found that predatory birds occur in the woody regions. They move out to fields to forage for grains and insect pests. Observations were made regarding movement and foraging habits. Birds seem to fly out of woods until 20 to 30 m into cereal fields to feed on insect pests. Clearly, woody natural habitats interspersed with wheat fields seem a good idea. So, avian predators can keep insect pests at threshold population. Crops such as soybean on the fringes and outside the woods were efficiently foraged for insects (Puckett et al., 2009). In a farmland, birds generally forage for grains, fruits, and insects. It is the predominantly insect eaters that are used as biological control agents. A list of bird species useful in the control of insect pests is provided in Table 3.2 (see Laliberte, 2019).

TABLE 3.2 Bird Species Involved in the Biological Control of Insects—A Few Examples.

Predatory birds	Insects foraged and controlled by predatory birds
Bluebirds (*Motacilla sialis; Sialia sialis*)	Grasshoppers, crickets, beetles, larvae, moths
Cardinals (*Cardinalis cardinalis*)	Beetles, grasshoppers, leafhoppers, stinkbugs, snails
Chickadees (*Poecile atricapillus*)	Aphids, whitefly, scale, caterpillars, ants, earwigs
Falcons (*Falco eleonirae*)	Grasshoppers, larvae, beetles, bugs
Grosbeaks (*Mycerobus melanozanthus; M. icterioides*)	Larvae, caterpillars, beetles
Nuthatches (*Sitta carolinensis; S.c. mexicana*)	Tree and shrub insects like borers, caterpillars, ants
Sparrows (*Passer domesticus*):	Beetles, caterpillars, cutworms
Swallows (*Hirundo rustica*)	Moths, beetles, grasshoppers
Tits (*Periparus sp*)	Caterpillar, bugs, beetles
Warblers (*Setophaga petachia*)	Caterpillars, aphids, whitefly
Woodpeckers (*Chrysocolaptes lucidus*)	Larvae, beetles, weevils, borers

Sources: Benayas et al. (2017), Cela and Ramos (2016), Laliberte (2019), and Mols and Visser (2002).

Note: This list is not exhaustive. Only small-sized insectivorous birds have been mentioned. There are several falcons, raptors, hawks, vultures, etc., that consume insects, rodents, even small mammals, etc. They are useful as biological control agents in farms.

There are several examples of insectivorous birds involved in the control of crop pests. Small birds such as passers, small falcons, and raptors are a few bird species that have been examined for use as biological control agents. Experimental evaluations of predatory birds have been generally encouraging. For example, Great Tits (*Parus major*) were evaluated in German Apple orchards, for their efficacy in controlling insect pests. Mols and Visser (2002) have reported that Great Tits reduced insect larvae by 19%. As a result, fruit yield was 66% higher in orchards where Tits were introduced, using nest boxes. Armyworms (*Spodaptera exigua, Spodaptera litoralis*, etc.) are common in the Californian vegetable and grapevine growing regions (Jedlicka et al., 2011). The placement of nest boxes for predatory birds was effective in improving their population. This step increased control of larvae of armyworms. An example from Central American coffee regions shows that predatory birds reduced insect pest, on plantations. It seems placing nest boxes enhanced bird breeding in farms. This resulted in effective reductions of insect pests (Karp et al., 2013).

Bats that operate efficiently in the night have also been evaluated in agroforestry tree species and fruit orchards. These flying mammals keep insect pests under threshold levels (Maas et al., 2013). There are several instances in farms and plantations located in different continents which prove that providing next boxes improves bird population. Consequently, it resulted in the reduction of insect pests.

Macadamia nut is a tree crop originally found grown in Australia. It was later planted in different regions of the world. Currently, South Africa is a major producer and supplier of macadamia nuts. It seems insect species, like stink bugs are becoming severe on macademia nut trees. Macadamia nut Grower's Association in South Africa reports that every year, 60–80% of nuts are damaged by the stink bugs. If stinkbugs attack the kernels early in the season, the nut carries a dark brown mark that optical sorters can pick up automatically. This reduces the commercial value of nuts. Reports suggest that there are several bird and bat species that feed on stink bugs. They regulate their population in the orchard to below threshold. A few forecasts point out that stink bug population and its damage get doubled if bat population gets reduced. Therefore, macadamia nut farmers are applying lowered levels of pesticides. Farmers are trying to manage bat population at higher levels by building homes (nest boxes) for them in farms (Sishuba, 2017).

This is an example depicting insect pest control within natural vegetation with cork (*Quercus suber*) and holm oak (*Quercus rotundifolia*) trees. It seems that there was a decline in oak biomass accumulation due to infestation by insect pests, particularly, in the Iberian Peninsula region. Insect pests such as wood borers, defoliators, and leaf miners were severe on oak trees. Cela and Ramos (2016) found that several predatory birds were active in devouring insect pests. Bird foragers were eating overwintering pupae of defoliator and wood eating insect pests. Among the birds, improving population of passerine insectivorous species were useful. Placing nest boxes further improved passerines population and activity. Improving population of Great Tit, Blue Tit, Nuthatch, and Short-toed Treecreeper were also effective in reducing pests on oak trees.

There is a need to emphasize the use of predatory birds as biological control agents of insect pests in farms, all over the world. Predatory birds may reduce the use of harmful plant protection chemicals. We may have to regularly monitor predator bird activity in farms and take advantage of this useful aerial phenomenon.

3.5.3 POLLINATION OF AGRICULTURAL CROPS BY AVES IS MEDIATED IN THE AGRARIAN SKY

Pollination of flowers by Aves is a beneficial activity in the aerospace above crops. Pollination of flowers of plant species occurring in natural vegetation within agrarian regions plus those of field/plantation crops is important. Aves-mediated pollination of natural vegetation and agricultural crops is a global phenomenon. Pollination helps in perpetuation of a botanical species. Pollination discussed here is a good example of natural phenomenon mediated within "agricultural sky." The aerial biotic agents such as honeybees, several other insect species, birds and bats, also animals conduct pollination in the agrarian ecosystems. Here, in this section, we are concerned with bird species that have direct impact in cross-pollination and seed set of crops. Birds operate predominately within the crop's canopy and above it. Pollination of flowers by Aves in the cropped fields is an example of beneficial role. There are reports that several plant species around the world are endangered because they are exclusively dependent on birds and bats for their reproduction, that is, pollination/seed set. Japanese primrose, agaves in the wild, and several wild ancestors of cultivated crops are becoming endangered due to the lack of pollinator birds (Buchmann and Nabhan, 1996; Gallai et al., 2009).

The productivity of crops in farmland too has reduced in many regions of the world. It is because of rampant use of pesticides. Pesticides deter birds from operating above the crops. There are also diseases of birds that have reduced the pollinator number and their activity in the aerospace. Natural vegetation patches within the cropland are important. Because plant species in them helps to sustain bird population in the absence of crops and say during the fallow period. Such natural vegetation corridors with nectar-bearing flowers are getting reduced. This has led to depreciated population of bird/ bat. Naturalists have cautioned that such loss of habitat and lack of crops can be disastrous to bird population in an ecosystem. Several intricately linked ecosystem functions get affected if bird population is scanty.

Pollination as a natural phenomenon does not cost directly to farmers. It is an important aerial phenomenon. It should not be overlooked. Pollination aided by bees, birds, and animals seems to affect the fertilization of flowers and productivity of several crop species. About 35% of global crops are dependent on pollinators. In many locations, honeybees and birds can be critical to optimum grain yield. For example, yields of brassicas, cross-pollinated cereals, and plantations such as almonds, apples, etc., are affected by pollinators. Every cross-pollinated crop species has its requirement for insects/birds for pollination. For example, researchers at the University of California at Berkeley (2006) found that for an almond orchard of 550,000 ha, farmers need 1.4 million honeybee colonies plus several species of birds and animals.

Out of the 115 crops studied, 87 depended to some degree upon animal pollination, accounting for one-third of crop production, globally. Of those crops, 13 are entirely reliant upon animal pollinators, 30 are greatly dependent, and 27 are moderately dependent. In some farming zones, domesticated or managed honeybees have been introduced in great numbers to overcome deficits of natural honeybee population. Such an effort needs to be replicated with bird pollinators. We should be able to introduce bird species that are prolific pollinators. At Harvard University Robotics research section, a few experts are trying to mimic bees and create small bumblebee-sized autonomous bees (Albright. 2014; Shaw, 2017; Wood et al., 2000). They are being tested for ability to accurately pollinate crops. A hummingbird-sized autonomous pollinator bird (robotic machine) seems a good idea. Such an effort may help crops in areas experiencing dearth for bird pollinators. Pesticide application will not matter if we adopt robotics. Robots are not influenced by seasons, breeding efficiency, or population. It seems that declining bird and bat population have reduced fruit production in apple, pears, and orchards in North America. Several farms in Central America have reported reduction of plantation productivity. So, small robot bees, can they replace birds/bees.

There is a syndrome called "ornithophily." It is a phenomenon that makes flower-bearing plants to evolve and produce such flowers that attract bird pollinators in greater number. Ornithophilous crops, it seems often produce red-colored flowers. When we consider bird-mediated pollination, it is said scent emanating from the flowers, if any is not of great concern. Bright and deep colors actually attract birds. Red colors on petals it seems deters honey-bees but simultaneously encourages birds that pollinate the flowers. One other character is the production of abundant nectar that helps in accentuating the bird-mediated pollination of flowers. Abundant nectar and secondary perches are traits of crops that help in improving bird-mediated pollination (Cronk and Ojeda, 2008). Report from Australian natural vegetation zones states that flowers visited by birds are often red, though yellow (Adenanthos) and green (e.g., Amyema, Correa) are common. Hairs in tubular flowers and lack of attractive smell may deter insects without affecting birds (e.g., Astroloma) (Ford et al., 1979).

No doubt, we may utilize trends in flowering and petal colors to match the crops versus birds traced in an agroecosystem (Ley et al., 2006). The evolution of crops toward ornithophily may need appropriate natural conditions. It could be a slow process taking several generations of a plant species. However, we can always improve ornithophily in an agro-ecoregion by deliberately selecting crops with bright-colored flowers and hoe with red-colored petals. So, apt cropping systems can enhance bird-mediated pollination. Perhaps, it will keep bird population levels in optimum levels, without dwindling.

Diversity of flowering plants (herbs, shrubs, and trees) and cultivated crop species found in an agro-ecoregion may differ seasonally (temporal) and spatially too. Birds that inhabit or those that migrate into a given farmland too may periodically change. The dominant crop versus bird relationship is worth noting prior to planting seeds of a particular crop. We can take advantage of bird pollination by suitably selecting time of sowing, crop species, and birds that hover in a given farming location. We have to achieve the best combination of these natural factors. Let us consider farming regions/forest plantations of Australia, as an example. Ford et al. (1979) stated that about 100 species of birds with the ability to pollinate plantation trees, shrubs, and crops have been noticed. They are to inhabit regions with forest belts and farms. Flowers of about 250 plant species are known to get pollinated by these bird pollinators. Honeyeaters and lorikeets are the most persistent flower feeders. Some bird species depend immensely on nectar as a source of energy. Silvereyes, parrots, wood swallows, pardalotes, thornbills, and a few other species of passerines visit flowers. Many plant genera get pollinated by birds. They are Eucalyptus, Callistemon, Banksia, Grevillea, Adenanthos, Dryandra, Epacris, Astroloma, Amyema, Correa, Xanthorrhoea, Anigozanthos, and Eremophila. No doubt, we have to note that such genera and opt for best farming/forest practices regarding bird pollination. About a third of the 300 families of flowering plants have at least some members with ornithophilous ("bird-loving") flowers, that is, flowers attractive to birds. Conversely, about 2000 species of birds, belonging to 50 or more families, visit flowers more or less regularly to feed on nectar (The State of Victoria, 2019).

PLATE 3.3 Left: Hummingbird; right: scarlet honeycreeper.
Source: United States Department of Geological Service and United States Department of Agriculture; Beltsville, MD, USA. https://www.fs.fed.us/wildflowers/pollinators/animals/birds.shtml
Note: Hummingbirds are among prolific pollinators traced in natural vegetation and cropped fields. They easily get attracted to flowers with bright or similar hue.

Naturalists have also cautioned that declining bird/bat population can be a problem, in due course, in the fruit plantations. Bats operate in the sky during night. They are important agents of cross-pollination of fruit trees. In Central America, they say, decline in hummingbirds has reduced crop yields perceptibly (Plate 3.3). Incidentally, primates such as lemurs and macaques are important in the wild vegetation. Since they too aid cross-pollination of flowers in the natural vegetation.

No doubt, bird-mediated pollination is an aerial phenomenon. Crops that bear flowers get help from such pollinators in sizeable number, to cross-pollinate the flowers. Such cross-pollination has important influence on seed set and grain productivity. It also has relevance to generation of genetically diverse germplasm. Pollinator activity can determine to a good extent the genetic nature of the succeeding generation, in case of regularly cross-pollinated crops. Clearly, agricultural sky wields a certain degree of influence on the genetic nature of natural vegetation and/or crops grown in a region, via the bird pollinators. Birds have the ability to preserve and induce wider genetic base of composites (cultivars). No doubt, birds can influence crop grain yield.

As stated earlier, birds play a vital role in pollinating flowersborne by natural vegetation and agricultural crops. There are at least over 2000 bird species that conduct flower pollination. In the North American region, hummingbirds have been identified as important pollinators of crops (Plate 3.3). They conduct pollination while they forage for nectar flowers. Similarly, honeycreepers too are prolific pollinators (Plate 3.3). In the Australian forests and annual vegetation zones, honey eaters are said to be good pollinators of flowers. There are parrots, sunbirds, and a few other species that pollinate flowers in the oriental regions (United States Forest Service, 2020). Three important bird families involved in pollination are Trochilidae (hummingbirds), Nectarinidae (sunbirds), and Meligphagidae (honeyeaters) (Southwick and Southwick, 1999; National Research Council of the National Academies, 2006).

In the arid and drylands of southwest Great Plains, it seems white-winged doves are important migratory birds. Their migratory trend matches the flowering of certain plant species (e.g., saguaros). Several plant species are pollinated efficiently by the doves. In an agroecosystem, teaming population of birds that hover in the sky has the potential to influence the crop productivity. It also has the potential to affect the genetic constitution of the cross-pollinated crops. It can influence genetic composition of F1 seeds. In the neighboring natural vegetation patches, birds have the potential to affect perpetuation of different native species. Perpetuation in greater number may also be influenced by the bird and/or bee pollinators.

There are regular manuals released by farm agencies and forest service personnel that depict various procedures to be followed in cropped fields, gardens, forest plantations, and expanses having natural vegetation. This is to improve habitat and aerial activity of hummingbirds and other pollinators (Plate 3.3). There are few preferences exhibited by the birds engaged in pollination of crops/natural vegetation. Bird pollinators usually go to flowers that are tubular. Flowers are borne on stronger branches so that they can perch. Birds, it seems, do not search flowers with good scent. Flowers open during the daytime are required. Flowers that produce good quantity of nectar are preferred (United States Forest Service, 2020).

3.5.4 AVES MEDIATED DISPERSAL OF CROPS, FRUIT/FOREST TREES, AND WEEDS

Agrarian sky is an abode to birds involved in dispersal of seeds and other propagules of plants. Seed dispersal is movement, spread or transport of seeds, grains/panicles, bulbs, tubers, corms, stolons, fruits/seeds, spores, etc., away from the parent plant. Plants are stationary. Hence, they rely upon a dispersal agent to transport their propagules. Dispersal agent could be abiotic (wind, water) or biotic (insects or birds or animals).

Birds may disperse seeds that get attached to their bodies, feet, and feathers. Birds could disperse crop seeds/weed seeds to shorter distance from the focus or to longer distance. It depends on bird's flight characteristics, seed viability, and probability of hitting congenial substrate once the seeds have been dropped by the birds. LDD means actual spread of seeds to locations situated at farther distance, say above 1 km from mother plant. LDD is crucial for a few species. It decides the survival and fitness of a plant/crop species. LDD may help the spread of plant species into areas hitherto not colonized area (Gutierrez, 2014). Certain migratory birds may transport seeds from one agro-ecoregion to other situated several 100 s of km away. Seed dispersal between continents is also possible through birds migrating for long distances. Perhaps, shrewdly devised experimental evaluations and authentic data are needed to prove LDD of seeds to farther continents.

The decline in range and density of frugivorous birds worldwide could have consequences for the functioning of ecosystem processes, such as seed dispersal. The low density of frugivorous birds may mean requirement of other animal species to help in dispersal. Fruit traits such as pulp, size, and sugar content too may affect dispersal of seeds. The distance to which birds disperse seeds from the mother plant and the area covered is important. The consequence of lack of frugivorous birds or lower density is generally reduced spread of seeds around the focus (Wyman, 2013).

LDD of seeds by birds, no doubt, offers new soils, water resources, and habitats to plant species. It is also an important survival mechanism. Particularly, in the light of climate change effects. Frugivorous and granivorous birds are said to be well suited for LDD. Water birds that pick sediments laden with seeds are another example of LDD of seeds. There could be several thousand bird species capable of and regularly conducting LDD of seeds. Bird-mediated LDD of seeds has been proved in some cases. For example, Duarte et al. (2016) stated that birds that migrate from Canary Islands to African mainland carried seeds in their guts and dropped them on soils in the continental region. This was detected by examining the guts of birds predated by falcons. Duarte et al. (2016) mentioned that at least 21 species of birds could be acting as LDD agents in the Spanish ecosystems. A few examples are European flycatcher (*Ficedula hypoleuca*), common red stark (*Phoenicurus phoenicurus*), and common quail (*Coturnix coturnix*). These birds are known to conduct LDD in the aerospace. Some observations indicate that landscape corridors are useful of migratory birds involved in LDD (Levey et al., 1999).

Bird-mediated dispersal of seeds is important for perpetuation, survival, and productivity of crops in an agroecosystem. Birds usually disperse seeds through bodies, beaks, claws, body/feathers, while eating fruits/panicles/grains/through droppings. They say, a few plant species including agricultural crops have coevolved with birds. They may obligately depend on birds for the completion of their life cycle (Moon 2017). In some cases, both pollination and dispersal of seeds are immensely dependent on bird species (e.g., sunflower, brassicas).

Bird-mediated dispersal of seeds is concerned with the variety of plant species, including several agricultural crops. Aves-aided dispersal of seeds of food grain crops, fruits, nuts, creepers, and other forest bush/tree species is an aerial phenomenon. Agricultural sky has immense influence on the processes such as flowering, pollination, seed formation, its maturity, and dispersal. Ambient weather patterns also affect the activity of birds, including dispersal of seeds. Seed dispersal by birds is an important ecosystem service connected to birds. Birds that are active during the grain maturity period are often prolific in picking panicles/grains and dispersing them to different distances. Similarly, fruits in plantation and natural tree stands are picked seasonally when they are ripened. Dispersal of seeds of forest tree/shrub species is an important aerial biological process. Human activity related to clearing forests, lumbering, residue burning, sapling planting, etc., may affect bird aided dispersal of seeds. Let us consider an example, depicting natural changes and their influence on bird-aided dispersal of forest species.

A report by Arnold Arboretum at Harvard University states that birds and their activity in an ecosystem (or natural vegetation patch) are related to seasons, fruit set, and maturity patterns. Birds begin nesting during summer, particularly the migratory birds. Bird species may congregate and then forage on tree species that produce fleshy fruits. Birds roam frequently in the area picking several types of fruits, not just a specific type. Trees producing large number of fruits at high density are preferred. The attractive colors of ripening fruits are another factor that affects bird activity during dispersal

(Arnold Arboretum, 1967). Fleshy colorful fruits such as cherries may get preferred, for feeding and dispersal seeds. Usually, fruits are stripped of flesh and seeds may pass through their guts or even discarded. A few examples of preference exhibited by bird fauna in the New England Forest area (Harvard Forest, Petersham, USA) during autumn season are Asiatic sweet leaf (*Symplocos paniculate*), sassafras (*Sassafas albidum*), dogwoods (*Cornus species),* blue berries (*Gaccinium species*), spice brush (*Lindera benzoin*), etc. In addition, there are indeed several fruit-bearing tree species whose seeds get dispersed by the birds that flock the forests. They are *Acer, Aescuelus, Betual, Carguana, Chaenonceles, Chamaecyparis, Chionanthus, Cornus, Cotoneater, Elacagnus, Fraxinus, Kaloponax, Lonera, Magnolia, Morus, Phellodendron, Prunus, Quercus, Sophora, Tsuga, vaccinium, Viburnum,* etc.

Let us consider an example of closed forest vegetation and neighboring cropland ecosystem. It is an island in the Central African region (Sao Tome). It is located in the Gulf of Guinea. It is an example of biodiversity hotspot, wherein the number of endemic species (flora and fauna) is high. However, anthropogenic activity has increased enormously during recent years. It is threatening several aspects of ecosystem function, for example, bird fauna and their seed dispersal activity. Alterations in tree composition and habitat in general are imminent, if both disperser (birds) and dispersed (fruit trees) depreciate in number or altogether disappear. Such bird-mediated dispersal is key to biodiversity expressed within the island of Sao Tome (Coelho, 2018). The report suggests that seed dispersal by birds, which is an aerial process, is essential. It affects the forest dynamics, including the tree species that get spread faster and dominates the landscape. Studying the seed dispersal mechanism by birds in different seasons is useful. Researchers have adopted methods that involve collecting bird droppings and tacking seeds of trees and understory species in the droppings. Birds were captured using mist nets. They were described and tagged.

Seed dispersal is a function that is essential to forest ecosystem. In the tropics, most seeds are dispersed through mutualist relationships between plants and animals. The effectiveness of seed dispersal depends on the disperser itself, the treatment they provide to the seeds, and the number of seeds dispersed. The loss of a key disperser in an ecosystem can compromise the viability of plant populations and alter vegetation dynamics.

As stated above, the island of São Tomé, in the Gulf of Guinea, is a biodiversity hotspot with a remarkable number of endemic species and unique forest ecosystems. Much of its biodiversity is currently threatened by the increasing human population and associated habitat change. Seed dispersal was assessed by detecting intact seeds in the droppings of mist-netted birds. Birds were captured using mist nets operated at ground level, opened before sunrise, and left open while climate conditions were favorable (i.e., no heavy rain, wind, or fog).

Bird-aided dispersal was often conducted by a single species of bird. Common bird species that dispersed at least over 16 species of forest trees are as follows: *Oriole crassirostrus* (Oriole), *Zosterops feae* (White Eye), *Turdus oleviascus* (Thrush), *Speirups lugubris* (Speirops), *Serinus rufunneus* (Seed eater), and *Columba larvata* (Lemon dove). These and other birds were responsible for dispersing about 49 tree species. Out of which 14 were native, 6 were introduced, and 23 from unknown regions. Most frequently dispersed tree species by birds was *Psydrax subcordata*. It was followed by introduced species *Rubus rosifolius*. Again, preference by birds depended on several factors such as fruit bearing season, bird species, and its activities. We may realize that biodiversity of tree species in the island is definitely affected by the bird fauna and its dispersal activity.

Now let us consider an example from Central American tropical forests. It substantiates specificity between forager and fruit tree in an agrarian/forest aerospace. Birds may prefer a single tree species and its fruits for foraging and dispersal. This happens despite the presence of several other trees with ripe fruits. For example, in Costa Rica, Howe (1977) found that Masked Tityra (*Tityra semifasciata*) is an effective seed dispersal agent. Special attraction to a certain species such as *Casearia corymobasa* has been observed. *Casaeria corymbosa* is known to fruit when others tree species are not in a position to do so. Specificity between bird and tree species is also attributable to a few other reasons. For example, Howe (1977) stated that *Tityra* picks fruits and regurgitates viable seeds. *Tityra* is a frequent visitor to Casearia trees too frequently in all seasons. The frequency of not just visits but feeding on fruits is high. *Tityra* birds also pick fruits that shed/fallen on ground in the vicinity. *Tityra* species is known to depend to a greater degree on *Casearia* and disperse seeds efficiently. This occurs when 21 other bird species traced in the same set of trees were not efficient dispersers of seeds. Seed dispersal by birds other than ones that have specificity occurs simultaneously. For example, in the same region, parakeets too were active in dispersal of seeds of *Casearia corymba*. There were 14 other bird species that were casual visitors of the same trees. Clearly, in a forest or field crop production zone, we have to first identify bird species that forage and disperse seeds efficiently. Such birds could be encouraged to multiply. We should also trace and encourage other species that disperse seeds. Each set of bird species has its degree of importance in an agro-ecosystem.

We may note that efficiency of seed dispersal by birds needs attention. For example, in a Costa Rican Forest, Jordan (1983) found that several bird species foraged and dispersed the *Ficus continifolia* seeds. They forage fruits regularly. Birds such as orioles, parakeets, tanagers, trogans, and flycatchers were common. Parakeets were dominant dispersers of Ficus seeds. Yet, a sizeable section of fruits was not dispersed but lost due to activity of other vertebrates. Dispersal of seeds as an aerial mechanism has to be efficient. We may note that birds and bats occur in natural vegetation zones or in farms. Both do disperse seeds. Bats are particularly active during the night (Medellin and Gaona, 1999).

3.5.4.1 WEEDS SEED DISPERSAL BY BIRDS

In nature, birds do disperse seeds of weeds and other unwanted invasive species into farmland. It can be a detrimental activity. Particularly, if weed population exceeds limits. There are indeed several instances of weed seed dispersal by birds. Let us consider an example from polish plains that support wheat crop production. Here, amaranth species is a weed whose seeds are dispersed by frugivorous birds. Orlowski and Czarnecka (2009) reported that droppings of the grey partridge *Perdix perdix L.* wintering on a field were found to contain 99.3% of *Amaranthus retroflexus* and 0.7% of *Chenopodium album* seed coat fragments. A bird consumed on average 3008 (± 95% CL = 2699–3317) weed seeds per 1 g of droppings. Weed seeds form the staple diet of numerous seed-eating birds (Wilson et al., 1999, Holland et al., 2006; Holmes and Froud-Williams, 2005). Further, they say, availability and high density of weed seeds is a factor allowing the survival of many birds. Especially in the autumn–winter period. During this period, diet of most of the farmland species wintering in the temperate northern hemisphere is exclusively granivorous (Moorcroft et al., 2002).

Let us consider another example of weed seed dispersal. It seems fleshy fruits of weeds are dispersed in large quantities by several bird species. In some farming regions of New Zealand, eight birds and two mammals were found to be the main dispersers of weed seeds. They are blackbirds (*Turdus merula*), silvereyes (*Zosterops lateralis*), starlings (*Sturnus vulgaris*), kereru (New Zealand pigeons, *Hemiphaga novaeseelandiae*), song thrushes (*Turdus philomelos*), tui (*Prosthemadera novaeseelandiae*), bellbirds (*Anthornis melanura*), mynas (*Acridotheres tristis),* brushtail possums (*Trichosurus vulpecula*), and feral pigs (Sus scrofa). Fruiting phenology differs between weeds and native plants, with many weed species fruiting from late autumn until early spring (May–September) when native fruits are scarce (Wotton and McAlpine, 2015). Weed fruiting duration does not differ from natives. So frugivorous species do thrive on the weed population in the absence of regular vegetation. There are also secondary dispersers of seeds of weeds. For example, raptors that feed on small frugivorous birds/mammals/rodents may disperse seeds of certain weeds (Lopez-Darias and Nogales, 2016).

3.5.5 AVES AS SOURCE OF ORGANIC MANURE-GUANOS

Birds are part of aerial factors operative in the agroecosystem. Guanos organic refuse is derived from bird defecations. Guanos have been adopted as fertilizer for crops by farmers, since the past 1500 years. They are particularly popular as organic fertilizer source in South America. Guanos and similar nutrient sources are in vogue even in other continents such as in the West African Sahel. Guanos are derived from bird droppings.

The term "Guano" applies to natural mineral deposits consisting of excrements, eggshells, and carcasses of dead seabirds found in almost rainless, hot–dry climatic regions. Guanos are classified according to age, genesis, geographical origin, and chemical composition (Schnug et al., 2018; Riechmann, 2003). White guano indicates a recent formation. It is produced daily by animal excrements by seabirds. It consists of 10–12% nitrogen, 10–12% phosphoric acid (P_2O_5), and 3% potash (K_2O). There are products derived from sea birds, owls, and parakeets that are of use as organic fertilizers to crops grown in South America. Seabird guanos are said to be rich in major nutrients like N, P, K, and Ca. The quality of guanos depends of course on the diets of birds. Sometimes the nutrient quality of guanos fluctuates as the foraging pattern and availability of usual food sources of birds change, depending on the season. Fruit-bearing trees after all are season in any case. Guanos may contain micronutrients, useful microbes, and other growth factors useful to crops. A few nutrients such as N and C may be lost to atmosphere if the droppings are not utilized timely and left to dry in the location itself. To overcome the loss of nutrients, a few products are made by dispersing the guanos in water and applying to crop fields. Guanos are an excellent method of recycling nutrients within the agrarian ecosystem (Totten, 2016).

KEY WORDS

- **aeolian dust**
- **airborne insects**
- **biological control**
- **biotic factors**
- **guanos**
- **locusts**
- **microbial flora**
- **pandemics**
- **phyllosphere**
- **seed dispersal**

REFERENCES

Acosta-Martinez, V.; Van Pelt, S.; Moore-Kucera, J.; Baddock, M. C.; Zobeck, T. M. Microbiology of Wind-Eroded Sediments: Current Knowledge and Future Research Directions. *Aeolian Res.* **2015,** *18*, 99–113. DOI: 10.1016/j.aeolia.2015.06.001 (accessed July 28th, 2021).

Adamidis, G. C.; Carter, R. V.; Melathopoulus, A. P.; Pernal, S. F.; Hoover, S. E. Pollinators Enhance Crop Yield and Shorten the Growing Season by Modulating Plant Functional Characteristics: A Comparison of 23 Canola Varieties. *Sci. Rep. Nat. Res.* **2019,** *9*, 14208. https://doi.org/10.1038/s41598-019-50811-y/ (accessed July 26th, 2021).

Agriculture Canada. Black Rot of Crucifers, Ministry of Agriculture, Ontario, Canada, 2020, pp 1–5. http://www.omafra.gov.on.ca/english/crops/facts/02–025.htm (accessed July 28th, 2021).

Agriculture Knowledge Centre. Wheat Midge. Saskatchewan, Canada, 2020, pp 1–16. https://www.saskatchewan.ca/business/agriculture-natural-resources-and-industry/agribusiness-farmers-and-ranchers/crops-and-irrigation/insects/wheat-midge/ (accessed July 31st, 2021).

Aguiliera, A.; Diego-Cstella, G.; Osuna, S.; Bordera, R.; Mendi, S. S.; Blanco, Y.; Gozalex-Toril, E. Microbial Ecology in the Atmosphere: The Last Extreme Environment, 2018, pp 1–23. DOI: http://dx.doi.org/10.5772/intechopen.81650 (accessed July 12th, 2021).

Albright M. B. Could Robot Bees Help Save Our Crops. National Geographic, Washington, D. C., 2014, pp 1–5 https://www.nationalgeographic.com/culture/food/the-plate/2014/08/21/could-robot-bees-help-save-crops// (accessed January 11th, 2021).

Al-Dagal, M.; Fung, D. Y. Aeromicrobiology—A Review. *Crit. Rev. Food Sci. Nutr.* **1990,** *29* (5), 333–340.

Almeida, R. P. P. Emerging Plant Disease Epidemics: Biological Research Is Key But Not Enough. *PLoS Biol.* **2018,** *16* (8), e2007020. https:// doi.org/10.1371/journal.pbio.2007020 (accessed July 22nd, 2021).

Allsopp, S.; Reynaud, L.; Mohorcic, M. Integrated Project: ABSOLUTE- Aerial Base Stations with Opportunistic Links for Unexpected and Temporary Events, 2013, pp 21–25. https://cordis.europa.eu/docs/projects/cnect/2/318632/080/deliverables/001-FP7ICT20118318632ABSOLUTED23v10isa.pdf (accessed March 10th, 2019).

Allsopp Helikites Ltd. Aerial Photography. Allsopp Helikite Ltd, Hampshire, England, 2019, pp 1–7. http://www.allsopphelikites.com/index.php?mod=page&id_pag=33 (accessed July 29th, 2021).

Amato, P. Clouds Provide Atmospheric Oases for Microbes. Microbe Magazine: American Society for Microbiology. Environmental Science Microbe Magazine, 2012. DOI: 10.11128/MICROBE.7.119.1/ (accessed July 28th, 2021).

Amato, P.; Brisebois, E.; Draghi, M.; Duchaine, C.; Fröhlich-Nowoisky, J.; Huffman, J. Main Biological Aerosols, Specificities, Abundance, and Diversity. In *Microbiology of Aerosols*; Delort, A. M., Amato, P., Eds.; John Wiley & Sons, Inc.: Hoboken, NJ, 2017a; pp 1–21.

Amato, P.; Joly, M.; Besaury, L.; Oudart, A.; Taib, N.; Moné Active Microorganisms Thrive Among Extremely Diverse Communities in Cloud Water. *PLoS ONE* **2017b,** *12*, 1–12, e0182869. DOI: 10.1371/journal.pone.0182869.

Amato, P.; Demeer, F.; Melaouhi, A.; Fontanella, S.; Martin-Biesse, A.-S.; Sancelme, M.; Laj, P.; Delort, A.-M. A Fate for Organic Acids, Formaldehyde and Methanol in Cloud Water: Their Biotransformation By Microorganisms. *Atmos. Chem. Phys.* **2007,** *7*, 4159–4169.

Anikster, Y.; Eilam, T.; Bushnell, W. R.; Kosman, E. Spore Dimensions of *Puccinia* Species of Cereal Hosts as Determined by Image Analysis. *Mycologia* **2005,** *97* (2), 474–484. DOI: 10.3852/mycologia.97.2.474 (accessed July 28th, 2021).

Ariel, G.; Ayali, A. Locust Collective Motion and Its Modelling. *PLoS Comput. Biol.* **2015,** *11* (12), e1004522. DOI: 10.1371/journal.pcbi.1004522/ pp 1–25 (accessed July 28th, 2021).

Ariya, P. A.; Amyot, M. New Directions: The Role of Bioaerosols in Atmospheric Chemistry and Physics. *Atmos. Environ.* **2004,** *38*, 1231–1232.

Arnold Arboretum. Seed Dispersal by Birds and Animals in the Arnold Arboretum. *Arnoldia* (Harvard University) **1967,** *27*, 73–85.

Arthur, S. Grasshoppers and Locusts as Agricultural Pests. In *Encyclopaedia of Entomology*; Capinera, J. L., Ed.; Springer, Dordrecht: The Netherlands, 2008. **DOI:** https://doi.org/10.1007/978–1-4020–6359–6_1167 (accessed July 28th, 2021).

Aylor, D. E. The Role of Intermittent Wind in the Dispersal of Fungal Pathogens. *Annu. Rev. Phytopathol.* **1990,** *28*, 73–92.

Aylor, D. E. Release of *Venturia inaequalis* Ascospores During Unsteady Rain: Relationship to Spore Transport and Deposition. *Phytopathology* **1992,** *82*, 532–540. http://dx.doi.org/10.1094/Phyto-82-532 (accessed July 28th, 2021).

Aylor, D. E. Spread of Plant Disease on a Continental Scale. Role of Aerial Dispersal of Pathogens. *Ecology* **2003,** *84*, 1989–1997.

Aylor, D. E.; Schmale, D. G.; Shields, E. J.; Newcombe, and Nappo, C. J. Tracking the Potato Late Blight Pathogen, in the Atmosphere Using Unmanned Aerial Vehicles and Lagrangian Modelling. *Agric. Forest Meteorol.* **2011,** *151*, 251–260.

Barbedo, J. G. A. A Review on the Use of Unmanned Aerial Vehicles and Imagery Sensors for Monitoring and Assessing Plant Stress. *Drones* **2019,** *3* (2), 40. DOI: 10.3390/drones3020040/ (accessed July 28th, 2021).

Barleunga, M.; Austerlitz, F.; Elzinga, J. A.; Teixeira, S.; Goudet, J.; G Bernasconi, G. Fine-Scale Spatial Genetic Structure and Gene Dispersal in *Silene latifolia*. Latifolia **2011,** *106*, 13–24.

Bartomeous, I.; Potts, S. G.; Stefen-Dewenter, I.; Vaissiere, B. E.; Woyciechowski, M.; Krewenka, K. M.; Tcheulin, T.; Roberts, T. A. M.; Westphal, C.; Bommorco, R. Contribution of Insect Pollinators to Crop Yield and Quality Varies with Agricultural Intensification. Peer J. **2014,** 2, 1–20, e328. https://doi.org/10.7717/peerj.328/ (accessed December 26th, 2020).

Bauer, H.; Kasper-Giebl, A.; Löflund, M.; Giebl, H.; Hitzenberger, R.; Zibuschka, F.; Puxbaum, H. The Contribution of Bacterial and Fungal Spores to the Organic Carbon Content of Cloud Water, Precipitation and Aerosols. *Atmos. Res.* **2002,** *64*, 109–119.

Behzad, H. Mineta, K.; Gojobori, T. Global Ramifications of Dust and Sandstorm Microbiota. *Genome Biol. Evol.* **2018,** *10*, 1970–1987.

Beijerinck, M. W. De infusies en de ontdekking der backterie¨n, Jaarboek van de Koninklijke Akademiev. Wetenschappen. Muller, Amsterdam, The Netherlands, 1913, p 28.

Benaim, R. D. Unbelievably Large Swarm of Locusts Threatens Middle East, 2020, pp 1–8. https://weather.com/science/environment/news/2020–03–18-swarm-of-locusts-threatens-middle-east (accessed December 28th, 2020).

Benayas J. M.; Meltzer J.; de las HerasBravo, D.; Cayuela, L. Potential of Pest Regulation by Insectivorous Birds in Mediterranean Woody Crops. *PLoS ONE* **2017,** *12* (9), e0180702. https://doi.org/10.1371/journal.pone.0180702/ (accessed July 28th, 2021).

Bent, A. C. *Life Histories of North American Blackbirds, Orioles, Tanagers, and Allies*; Dover Publications, Inc.: New York, 1965; p 54.

2Blades Foundation. Major Crop Disease Outbreaks: Disease Epidemics Are an On-Going Condition, 2020, pp 1–8. https://2blades.org/major-crop-disease-outbreaks/ (accessed December 20th, 2020).

Blanche, R.; Cunningham, S. A. Rainforest Provides Pollinating Beetles for Atemoya Crops. *J. Econ. Entomol.* **2005,** *98*, 1193–1201.

Blanche, K. R.; Ludwig, J. A.; Cunningham, S. A. Proximity to Rainforest Enhances Pollination and Fruit Set in Macadamia and Longan Orchards in North Queensland, Australia. *J. Appl. Ecol.* **2006**. DOI: 10.1111/j.1365–2664.2006. 01230.x (accessed July 28th, 2021).

Boreson, J.; Dillner, A. M.; Peccia, J. Correlation Bioaerosol Load with PM2.5 and PM10cf Concentrations: A Comparison Between Natural Desert and Urban Fringe Aerosols. *Atmos. Environ.* **2004,** *38*, 6029–6041.

Bouvallius, Å.; Roffey, R.; Henningson, E. Long-Range Transmission of Bacteria. *Ann. NY Acad. Sci.* **2006,** *353*, 186–200. DOI: 10.1111/j.1749–6632.1980.tb18922.x / (accessed July 28th, 2021).

Bowers, R. M.; Clements, N.; Emerson, J. B.; Wiedinmyer, C.; Hannigan, M. P.; Fierer, N. Seasonal Variability in Bacterial and Fungal Diversity of the Near-Surface Atmosphere. *Environ. Sci. Technol.* **2013,** *47*, 12097–12106. DOI: 10.1021/es402970s/ (accessed July 28th, 2021).

Bowers, R. M.; Lauber, C. L.; Wiedinmyer, C.; Hamady, M.; Hallar, A. G.; Fall, R. Knight, R.; Fierer, N. Characterization of Airborne Microbial Communities at a High-Elevation Site and Their Potential to Act as Atmospheric Ice Nuclei. *Appl. Environ. Microbiol.* **2009,** *75*, 5121–5130.

Brewer, M.; Elliott, N. C. Biological Control of Cereal Aphids in North America and Mediating Effects of Host Plant and Habitat Manipulations. *Annu. Rev. Entomol.* **2004,** *49*, 219–242.

Brewer, M. J.; Nelson, D. J.; Ahem, R. G.; Donahue, J. D.; Prokrym, D. R. Recovery and Range Expansion of Parasitoids (Hymenoptera: Aphelinidae and Braconidae) Released for Biological Control of Diuraphis noxia (Homoptera: Aphididae) in Wyoming. *Environ. Entomol.* **2001,** *30*, 578–588.

Brodie, E. L.; DeSantiz, T. Z.; Parker, J. P. M.; Zubietta, L. X.; Piceno, Y.M and Anderson, G. L. Urban Aerosols Harbour Diverse and Dynamic Aerosol Population. *Proc. Natl. Acad. Sci. USA* **2007,** *104*, 299–304. DOI: 10.1073/pnas.0608255104 (accessed July 26th, 2021).

Brown, J. K.; Hovmoller, M. S. Aerial Dispersal of Pathogens on the Global and Continental Scales and Its Impact on Plant Disease. *Science* **2002,** *297*, 537–541. DOI: 10.1126/science.1072678 (accessed July 28th, 2021).

Bryan, N.; Stewart, M.; Granger, D.; Guzik, T.; Christner, B. A Method for Sampling Microbial Aerosols Using High Altitude Balloons. *J. Microbiol. Methods* **2014,** *107*, 161–168. DOI: 10.1016/j.mimet.2014.10.007/ (accessed July 28th, 2021).

Buchmann, S.; Nabhan, G. P. *The Forgotten Pollinators*; Island Press: New York, 1996, p 312. https://islandpress.org/books/forgotten-pollinators/ (accessed July 28th, 2021).

Buller, A. H. R. Research on Fungi. An Account of the Production, Liberation and Dispersion of the Spores of Hymenomycetes Treated Botanically and Physically. Also, Some Observations Upon the Discharge and Dispersion of the Spores of Ascomycetes and *Pilobolus*. Longmans, Green and Co.: London, New York, Bombay and Calcutta, 1909. p 318. https://archive.org/details/researchesonfung01bull/page/n15/mode/2up? (accessed July 28th, 2021).

Burrows, S.; Elbert, W.; Lawrence, M.; Pöschl, U. Bacteria in the Global Atmosphere–Part 1: Review and Synthesis of Literature Data for Different Ecosystems. *Atmos. Chem. Phys.* **2009,** *9*, 9263–9280.

Calhim, S.; Halme, P.; Petersen, J. H.; Laesoe, T.; Bassler, C.; Heilman, J. Fungal Spore Diversity at Any Time, Agrarian Sky May Hold Diverse Microbial Pathogens That Can Potentially Cause Disease on Crops Reflects Substrate Specific Deposition Challenges. *Sci. Rep.* **2018,** *8*, 53–56. DOI:1038/s41598-018-23292-8 (accessed July 28th, 2021).

Caliz, J.; Triardo-Margaritz, X.; Camerero, L.; Casameyer, E. O. A Long Term Survey Unveils a Strong Seasonal Pattern in the Airborne Microbiomes Coupled to General and Regional Atmospheric Circulations. *Proc. Natl. Acad. Sci.* **2018,** *115*, 12229–12234.

Capote, N.; Pastrana, A. M.; Agauda, A. Sanchez, -Torres, P. Molecular Tools for Detection of Plant Pathogenic Fungi and Fungicide Resistance. In *Plant Pathology Cumagan*; C. J. R., Ed.; InTech Rijrcka: Croatia, 2012; pp 151–202. https://www.intechopen.com/chapters/34844 (accessed June 29th, 2021).

Caroll, M. W.; Glaser, J. A.; Hellmich, R. L.; Hunt, T. E.; Sappington, W.; Celvin, D.; Copenhaver, K.; Fridgen, J. F. Use of Spectral Vegetation Indices Derived from Air Borne Hyper-Spectral Imagery for Detection of European Corn Borer Infestation in Iowa Corn Plots. *J. Econ. Entomol.* **2008,** *101*, 1614–1623. https://doi.org/10.1603/0022–0493(2008)101(accessed 1614:UOSVID)2.0CO:2/ (accessed July 28th, 2021).

Carvalho, C. R.; Fernandes, R. C.; Carvalho, G. M. A.; Barreto, R. W.; Evans, H. C. Crypto-Sexuality and the Genetic Diversity Paradox in Coffee Rust, *Hemileia vastatrix. PLoS One* **2011,** *6*, 1–12, e26387 (accessed July 28th, 2021).

Castex, V.; Beniston, M.; Calanca, P.; Fleury, D.; Moreau, J. Pest Management Under Climate Change: The Importance of Understanding Tri-Trophic Relations, 2017. https://doi.org/10.1016/j.scitotenv.2017.11.027/ (accessed July 28th, 2021).

Cela, R. S.; Ramos, J. R. Birds as Predators of Cork and Oak Pests. *Agroforestry Syst.* **2016,** *90*, 159–176. DOI: 10.1007/s10457–014–9749–7/ (accessed July 28th, 2021).

CESAR. Russian Wheat Aphid, 2019, pp 1–6. http://cesaraustralia.com/sustainable-agriculture/pestnotes/insect/Russian-wheat-aphid/ (accessed July 31st, 2021).

Coelho, A. Seed Dispersal by Birds: Implications for Forest Conservation. Report for African Bird Club. Unpublished Report, University of Lisbon, Portugal, 2018, pp 1–14. https://www.africanbirdclub.org/sites/default/files/2015_Seed_dispersion_Sao_Tome.pdf/ (accessed July14th, 2021).

Cronk, Q.; Ojeda, I. Bird Pollinated Flowers in an Evolutionary and Molecular Context. *J. Exp. Bot.* **2008,** *59*, 715–727. doi:10.1093/jxb/ern009/ (accessed July 12th, 2021).

Darwin, C. An Account of the Fine Dust Which Often Falls on Vessels in the Atlantic Ocean. *Quart. J. Geol. Soc.* **1846,** *2*, 26–30.

DasSarma, P.; DasSarma, S. Survival of Microbes in Earth's Stratosphere. *Curr. Opin. Microbiol.* **2018,** *43*, 24–30. DOI: 10.1016/J. MIB.2017.11.002 (accessed July 28th, 2021).

Dean, R.; Van Kan, J. A. L.; Kim, P. E.; Antonio de Piotra, H.; Spanu, P. D.; Rudd, S. J.; Dickman, M.; Kahmann, R.; Ellis, J.; Foster, G. D. The Top 10 Fungal Pathogens in Molecular Plant Pathology. *Mol. Plant Pathol.* **2012,** *13*, 414–430.

DeBoer, S. H.; Lopez, M. M. New Grower Friendly Methods for Plant Pathogen Monitoring. *Annu. Rev. Phytopathol.* **2012,** *50*, 197–218.

De Guillaume, L.; Leriche, M.; Amato, P.; Ariya, P.; Delort, A.-M.; Pöschl, U. Microbiology and Atmospheric Processes: Chemical Interactions of Primary Biological Aerosols. *Biogeosci. Discussions* 2008, *5*, 841–870. DOI: 10.5194/bgd-5–841–2008 / (accessed July 28th, 2021).

DeLeon-Rodriguez, N.; Lathem, T. L.; Rodriguez, R. L.; Barazesh, J. M.;erson, B. E.; Beyersdorf, A. J. Microbiome of the Upper Troposphere: Species Composition and Prevalence, Effects of Tropical Storms, and Atmospheric Implications. *Proc. Natl. Acad. Sci. USA* **2013,** *110*, 2575–2580. DOI: 10.1073/pnas.1212089110 (accessed July 29th, 2021).

Del Ponte, E. M.; Shah, D. A.; Bergstrom, G. C. Spatial Patterns of Fusarium Head Blight in New York Wheat Fields Suggest Role of Airborne Inoculum. *Online. Plant Health Progress* 2003. DOI: 10.1094/PHP-2003–0418–01-RS (accessed July 31st, 2021).

Department of Agriculture and Fisheries. Insect Pest Management in Maize. Queensland Government, Australia, 2020a, pp 1–6. https://www.daf.qld.gov.au/business-priorities/agriculture/plants/crops-pastures/broadacre-field-crops/insect-pest-management-specific-crops/insect-pest-management-maize/ (accessed July 31st, 2021).

Department of Agriculture and Fisheries. Insect Pest Management in Sorghum. Queensland Government, Australia, 2020b, pp 1–4. https://www.daf.qld.gov.au/business-priorities/agriculture/plants/crops-pastures/broadacre-field-crops/insect-pest-management-specific-crops/insect-pest-management-sorghum (accessed July 31st, 2021).

Department of Agriculture and Fisheries. Insect Pest Management in Winter cereals. Queensland Government, Australia, 2020c, pp 1–6. https://www.daf.qld.gov.au/business-priorities/agriculture/plants/crops-pastures/broadacre-field-crops/insect-pest-management-specific-crops/insect-pest-management-winter-cereals/ (accessed July 31st, 2021).

Department of Agriculture and Fisheries. Insect Pest Management in Sunflower. Queensland Government, Australia, 2020d, pp 1–8. https://www.daf.qld.gov.au/business-priorities/agriculture/plants/crops-pastures/broadacre-field-crops/insect-pest-management-specific-crops/insect-pest-management-sunflowers/ (accessed July 31st, 2021).

Department of Agriculture and Fisheries. Insect Pest Management in Peanuts. Queensland, Australia, 2020e, pp 1–4. https://www.daf.qld.gov.au/business-priorities/agriculture/plants/crops-pastures/broadacre-field-crops/insect-pest-management-specific-crops/insect-pest-management-peanuts/ (accessed January 1st, 2021).

Department of Agriculture and Fisheries. Insect Pest Management in Cotton. Queensland Government, Australia, 2020f, pp 1–8. https://www.daf.qld.gov.au/business-priorities/agriculture/plants/crops-pastures/broadacre-field-crops/insect-pest-management-specific-crops/insect-pest-management-cotton/ (accessed January 1st, 2021).

Despres, V.; Huffman, J. A.; Burrows, S. M.; Hoose, C.; Safatov, A.; Buryak, G.; et al., Primary Biological Aerosol Particles in the Atmosphere: A Review. *Tellus Ser. B Chem. Phys. Meteorol.* **2012,** *64,* 15598. DOI: 10.3402/tellusb.v64i0. 15598.

De Grazio, S.; John W. World Bird Damage Problems. *Proc. 8th Vertebrate Pest Conf.* **1978,** 13. https://digitalcommons.unl.edu/vpc8/13/ (accessed January 18, 2021).

Diekmann, M.; Bogyo, T. P. Distribution of Bacterial Blight of Rice (*Xanthomonas campestris var. oryzae*) Depending on Climatological Factors. *J. Plant Dis. Protection* **1992,** *99,* 127–136.

Dolbeer, R. A.; Linz, G. M. Blackbirds. United States Department of Agriculture, Animal and Plant Health Inspection Service. Wildlife Damage Technical Series, Beltsville Maryland, 2016, pp 1–17. http://digitalcommons.unl.edu/nwrcwdmts/1/ (accessed January 18th, 2021).

Duarte, S. V. Gangoso, L.; Bouten, W.; Figuerola, J. Overseas Seed Dispersal by Migratory Birds. *Proc. R. Soc. B Biol. Sci* **2016,** *283* (1822), 20152406. DOI: 10.1098/rspb.2015.2406/ (accessed January 16th, 2021).

Duce, R. A. Long-Range Atmospheric Transport of Soil Dust from Asia to the Tropical North Pacific: Temporal Variability. *Science* **1980,** *209,* 1522–1524.

Dunne, D. Are the 2019–2020 Locust Swarms Linked to Climate Change. Carbon Brief: Clear on Climate, 2020, pp 1–9. https://www.carbonbrief.org/qa-are-the-2019–20-locust-swarms-linked-to-climate-change/ (accessed December 28th, 2021).

Durand, S.; Amato, P.; Sancelme, M.; Delort, A.-M.; Combourieu, B.; Besse-Hoggan, P. First Isolation and Characterization of a Bacterial Strain That Bio Transforms the Herbicide Mesotrione. *Lett. Appl. Microbiol.* **2006,** *43,* 222–228.

Editor. A Lack of Locust Preparedness Costs Lives. *Nature* **2020,** *174,* 1–3. DOI: 10.1038/d41586–020–00692–3/ (accessed December 28th, 2021).

Elliott, N. C.; Hein, G. L.; Carter, M. C.; Burd, J. D.; Holtzer, T. J. Armstrong, J. S.; Waits, D. A. Russian Wheat Aphid (Homoptera: Aphididae) Ecology and Modelling in Great Plains Agricultural Landscapes. In *Response Model for an Introduced Pest—The Russian Wheat Aphid*; Quisenberry, S. S., Peairs, F. B. Eds.; Thomas Say Publications, Entomological Society of America: Lanham, MD, 1998a; pp 31–64.

Elliott, N. C.; Kieckhefer, R. W.; Lee, J. H.; French, B. W. Influence of Within-Field and Landscape Factors on Aphid Predator Populations in Wheat. *Landscape Ecol.* **1998b,** *14,* 239–252.

Elliott, N. C.; Kieckhefer, R. W.; Michels, J. D. Jr.; Giles, K. L. Predator Abundance in Alfalfa Fields in Relation to Aphids, Within-Field Vegetation, and Landscape Matrix. *Environ. Entomol.* **2002,** *31,* 253–260.

Eversmeyer, M. G.; Kramer, C. L.; Burleigh, J. R. Vertical Spore Concentration of Three Wheat Pathogens Above a Wheat Field. *Phytopathology* **1972,** *63,* 211–218.

Falacy, J. S.; Grove, G. G.; Mahaffy, W. F.; Galloway, H.; Glawe D. A.; Larsen, R. C.; Vandemark, G. J. Detection of *Erysiphe nector* in Air Sampling Using the Polymerase Chain Reaction and Species-Specific Primers. *Phytopathology* **2007,** *97,* 1290–1297.

Fang, Y.; Ramaraja, P. R. Current and Prospective Methods for Plant Disease Detection Methods. *Biosensors* **2015,** *4*, 537–561. DOI: 10.3390/bios5030537 (accessed January 20th, 2021).

Fasano, S. Olive Tree Pandemic Wired, 2020, pp 1–4. https://www.wired.co.uk/article/italy-olive-trees-epidemic.

Felt, E. P. Dispersal of Insects by Air Currents. *NY State MuS. Bull.* **1928a,** *274*, 59–129.

Felt, E. P. Insect Inhabitants of Upper Air. *Fourth Int. Congress Entomol. Trans.* **1928b,** *2*, 869–872.

Felt, E. P.; Chamberlain, K. F. The Occurrence of Insects at Some Height in the Air. *NY State Mus. Circ.* **1935,** *17*, 70.

Fierer, F. C.; Lindbergh, C. A. Collecting Microorganisms from the Arctic Atmosphere: With Field Notes and Material. *M Sci. Monthly* **1935,** *40*, 5–20.

Fierer, N.; Liu, Z.; Rodriguez-Hernandez, M.; Knight, R.; Henn, M.; Hernandez, M. T. Short Term Temporal Variability in Airborne Bacterial and Fungal Populations. *Appl. Environ. Microbiol.* **2008,** *74*, 200–207. DOI: 10.1128/AEM.01467–07 (accessed July 29th, 2021).

Fijen. T. P. M. Towards Ecological Intensification: The Relative Importance of Wild Pollinators as an Agricultural Input in Seed Production. PhD thesis, Wageningen University, Wageningen, the Netherlands, 2019, p 174.

Finchel, T.; Finlay, B. J. The Ubiquity of Small Species: Patterns of Local and Global Diversity. *Bioscience* **2004,** *54*, 777–784.

Finlay, B. J. Global Dispersal of Free-Living Microbial Eukaryote Species. *Science* **2002,** *296*, 1061–1063.

Fitt, B. D. L.; McCartney, H. A.; Walklate, P. J. The Role of Rain in Dispersal of Pathogen Inoculum. *Annu. Rev. Phytopathol.* **1989,** *27*, 241–270. https://doi.org/10.1146/annurev.pv.27.090189.001325 (accessed July 29th, 2021).

Food and Agricultural Organization of the United Nations. New Satellite Imaging to Better Forecasting Locust Plagues, 2017, pp 1–3. https://www.preventionweb.net/go/53911/ (accessed July 27th, 2021).

Food and Agricultural Organization of the United Nations. Desert Locust, 2020, pp 1–7. http://www.fao.org/locusts/en// (accessed July 28th, 2021).

Ford, H. A.; Paton, D. C.; Forde, N. Birds as Pollinators of Australian Plants. *NZ J. Bot* **1979,** *17* (4), 509–519. DOI: 10.1080/0028825X.1979.10432566 / (accessed July 2021).

French, B. W.; Elliott, N. C. Temporal and Spatial Distribution of Ground Beetle (Coleoptera: Carabidae) Assemblages in Grassland and Adjacent Wheat Fields. *Pedobiologia* **1999a,** *43*, 73–84.

French, B. W.; Elliott, N. C. Spatial and Temporal Distribution of Ground Beetle (Coleoptera: Carabidae) Assemblages in Riparian Strips and Adjacent Wheat Fields. *Environ. Entomol.* **1999b,** *28*, 597–607.

French, B. W.; Elliott N. C.; Kindler, S. D.; Arnold, D. C. Seasonal Occurrence of Aphids and Natural Enemies in Wheat and Associated Crops. *Southwestern Entomol.* **2001,** *26*, 49–61.

Fuzzi, S.;reae, M. O.; Huebert, B. J.; Kulmala, M.; Bond, T. C.; Boy, M.; Doherty, S. J.; Guenther, A.; Kanakidou, M.; Kawamura, K. Critical Assessment of the Current State of Scientific Knowledge, Terminology, and Research Needs Concerning the Role of Organic Aerosols in the Atmosphere, Climate, and Global Change. *Atmos. Chem. Phys.* **2006,** *6*, 2017–2038.

Gallai, N.; Salles, J. M.; Settele, J.; Vaissiere, B. E. Economic Valuation of the Vulnerability of World Agriculture Confronted with Pollinator Decline. *Ecol. Econ* **2009,** *68*, 810–821.

Gat, D.; Mazar, Y.; Cytryn, E.; Rendich, Y. Origin-Dependent Variations in the Atmospheric Microbiome Community in Eastern Mediterranean Dust Storms. *Environ. Sci. Technol.* **2017,** *51*, 6709–6718.

Giesler, L. J. Bacterial Blight. Crop Watch, Institute of Food and Natural Resources, University of Nebraska, Lincoln, 2011, pp 1–3. https://www.soybeanresearchinfo.com/pdf_docs/bacterialdiseases_G2058_NE.pdf (accessed July 21st, 2021).

Giles, K; Hein, G. L.; Peairs, F. 2008 Areawide Pest Management of Cereal Aphids in Dryland Wheat Systems of the Great Plains, USA Panhandle Research and Extension Centre 33, 2008. https://digitalcommons.unl.edu/panhandleresext/33/ (accessed July 29th, 2021).

Glick, P. A. The Distribution of Insects, Spiders and Mites in the Air. Bureau of Entomology and Plant Quarantine. United States Department of Agriculture, Washington, DC, Technical Report No.673, 1939, pp 1–157.

Godfrey, C. J.; Mason-DeCruz, D.; Robinson, S. Food System Consequences of a Fungal Diseases Epidemic in a Major Crop. Royal Society Publishing Philosophical Transactions B, 201, pp 1–10. http:/dx.doi.org/10.1098/rstb.2015.0467/ (accessed July 21st, 2021).

Golan, J. J.; Pringle A. Long-Distance Dispersal of Fungi. *Microbial Spectrum* **2017,** *5* (4):FUNK-0047–2016. doi:10.1128 /microbiolspec.FUNK-0047-2016 (accessed July 29th, 2021).

Grady, K. L.; Sorensen; Stopnisek, N.; Guittar, J.; Shade, A. Assembly and Seasonality of Core Phyllosphere Microbiota on Perennial Biofuel Crops. *Nat. Commun.* 201**9,** *10* (4135), 1–8 https://www.nature.com/articles/s41467-019-11974-4 (accessed July 28th, 2021).

GRDC. Whet: Insect and Other Pest Control. Grain Research and Development Centre, Australia, 2016, pp 1–38.

Greenleaf, S. A.; Kremen, C. Wild Bees Enhance Honeybees' Pollination of Hybrid Sunflower. *Proc. Natl. Acad. Sci. USA* **2006a,** *103*, 13890–13895. DOI: 10.1073/pnas. 0600929103/ (accessed July 29th, 2021).

Greenleaf, S. A.; Kremen, C. Wild Bee Species Increase Tomato Production and Respond Differently to Surrounding Land Use in Northern California. *Biol. Conserv*. **2006b**. DOI: 10.1016/j.biocon.2006.05.025/ (accessed July 29th, 2021).

Gregory, P. H. The Leeuwenhoek Lecture 1970. Airborne Microbes: Their Significance and Distribution. *Proc. R. Soc. Lond. B* **1971,** *177*, 469–483.

Gregory, P. H. *The Microbiology of the Atmosphere*; John Wiley and Sons Inc.: New York get pages, 1973.

Griffin, D. W. African Desert Dust in the Caribbean Atmosphere: Microbiology and Public Health. *Aerobiologia* **2001,** *17*, 203–213.

Griffin, D. W. Atmospheric Microbiology in the Northern Caribbean During African Dust Events. *Aerobiologia* **2003,** *19*, 143–157.

Griffin, D. W. Terrestrial Microorganisms at an Altitude of 20,000m in Earth's Atmosphere. *Aerobiologia* **2004,** *20*, 135–140. DOI: 10.1023/B:AERO.0000032948.84077.12 (accessed July 29th, 2021).

Griffin, D. W.; Gonzalez, C.; Teigell, N.; Petrosky, T.; Northup, D. E.; Lyles, M. Observations on the Use of Membrane filtration and Liquid Impingement to Collect Airborne Microorganisms in Various Atmospheric Environments. *Aerobiologia* **2010,** *27*, 25–35. DOI: 10.1007/s10453–0109173-z (accessed July 29th, 2021).

Griffin, D.; Gonzalez-Martin, C.; Hoose, C.; Smith, D. Global Scale Atmospheric Dispersion of Microorganisms. In *Microbiology of Aerosols*; Delort, A. M., Amato, P., Eds.; John Wiley and Sons, Inc.: Hoboken, NJ, 2017; pp 155–194.

Griffin, D. W.; Kubilay, N.; Kocak, M.; Gray, M. A.; Borden, T. C.; Shinn, E.A Airborne Desert Dust and Aeromicrobiology Over the Turkish Mediterranean Coastline. *Atmos. Environ.* **2007,** *41*, 4050–4062.

Gutierrez, D. Birds Do Not Just Disperse Seeds When They Eat Fruits—They Can Also Spread Seeds Stuck to the Mud of Their Feet. *Natural News*. **2014,** pp 1–3. https://www.naturalnews.com/044514_birds_seed_dispersal_plant_propagation.html/ (accessed July 14th, 2021).

Hagy, H.; Raetzman, J.; Linz, G.; Bleier, W. Decoy Cropping Methods for Luring Blackbirds Away from Commercial Sunflower: USDA Wildlife Conservation Sunflower Plots. *Proc. Wildlife Damage Manage. Conf.* **2005,** *11*, 304–310.

Harrington, R.; Woiwod, I.; Sparks, T. H. Climate Change and Trophic Interactions. *Trends Ecol. Evol* **1999,** *14*, 146–150.

Hau, B.; Vallavieille-Pope, S. Wind Dispersal of Disease. In *The Epidemiology of Plant Disease*; Springer, 2006; pp 387–416.

Heath, S.; Long, R. Birds Are Beneficial Too, 2019, pp 1–4. https://ucanr.edu/blogs/Green/index.cfm?tagname=Alfalfahttps://ucanr.edu/blogs/Green/index.cfm?tagname=Alfalfa/ (accessed January 12th, 2021).

Herlihy, L. J.; Galloway, J. N.; Mills, A. L. Bacterial Utilization of Formic and Acetic Acid in the Rainwater. *Atmos. Environ.* **1987,** *21*, 2397–2402.

Hervas, A. Viability and Potential for Immigration of Airborne Bacteria from Africa That Reach High Mountain Lakes in Europe. *Environ. Microbiol.* **2009,** *11*, 1612–1623.

Heuertz, M.; Vekemans, X.; Hausman, J-F.; Paladas, M.; Hardy, O. J. Estimating Seed vs. Pollen Dispersal from Spatial Genetic Structure in the Common Ash. *Mol. Ecol.* **2003,** *12*, 2483–2495.

Holland, J. M.; Hutchinson, M. A. S.; Smith, B.; Aebischer, N. J. A Review of Invertebrates and Seed-Bearing Plants as Food for Farmland Birds in Europe. *Ann. Appl. Biol.* **2006,** *148*, 49–71.

Holmes, R. J.; Froud-Williams, R. J. Post-Dispersal Weed Seed Predation by Avian and Non-Avian Predators. *Agric. Ecosyst. Environ.* **2005,** *105*, 23–27.

Horneck, G.; Klus, D. M.; Mancinelli, R. L. Space Microbiology Reviews, 2010, pp 121–156. DOI: 10.1128/MMBR.00016–0 (accessed July 28ht, 2020).

Howe, H. F. Bird Activity and Seed Dispersal of a Tropical Wet Forest Tree. *Ecology* **1977,** *58*, 539–545. https://doi.org/10.2307/1939003/ (accessed July 28th, 2021).

Hua, N. P. Detailed Identification of Desert-Originated Bacteria Carried by Asian Dust Storms to Japan. *Aerobiologia* **2007,** *23*, 291–298.

Ikisan. Maize. Ikisan AgriInformatics and Services, 2021, pp 1–14. http://www.ikisan.com/tg-maize-Bird-Damage.html/ (accessed January 28th, 2021).

Imbert, E.; Lefevre, F. Dispersal and Gene Flow of *Populus nigra* (Saliaceae) Along a Dynamic River. *J. Ecol.* **2003,** *91*, 447–456.

Imshenetsky, A. A.; Lysenko, S. V.; Kazakoov, G. A. Upper Boundary of the Biosphere. *Appl. Environ. Microbiol.* **1978,** *35*, 1–5.

Intergovernmental Science Policy Platform on Biodiversity and Ecosystem Services. Regional Report for Africa on Pollinators and Pollination and Food Production. Proceedings of SBSTTA (Subsidiary Body on Scientific, Technical and Technological Advice), Kaulalumpur, Malaysia, 2018, pp 1–32.

Iost, F. H.; Kong, Z. Drones: Innovative Technology for Use in Precision Pest Management. Research Gate, 2019, pp 1–26. https://www.researchgate.net/publication/337826660/ (accessed Jan 2nd, 2021).

Iozzin C. Robots That Can Sniff Out Crop Disease, 2014, pp 1–4. Smithsonian.com (accessed December 10th, 2020).

IRRI. Disease and Pest Resistant Rice. International Rice Research Institute, Manila, Philippines, 2020, pp 1–8. https://www.irri.org/disease-and-pest-resistant-rice (accessed July 29th, 2021).

Isard, S. A.; Gage, S. H.; Comtois, P.; Russo, J. M. Principles of the Atmospheric Pathway for Invasive Species Applied to Soybean Rust. *Bioscience* **2005,** *55*, 851–861. http://dx.doi.org/10.1641/0006–3568(accessed 2005)055 [0851:POTAPF]2.0.CO;2. / (accessed July 209th, 2021).

ISRO. New Microbes Discovered in Earth's Stratosphere. Indian Space Research Organization, Ahmedabad, India, 2009, pp 1–3. https://www.sciencedaily.com/releases/2009/03/090318094642.htm/ (accessed July 28th, 2021).

Jedlicka, J. A.; Greenberg, R.; Letourneau, D. K. Avian Conservation Practices Strengthen Ecosystem Services in California Vineyards. *PLoS ONE* **2011,** *6* (11), e27347. https://doi.org/10.1371/journal.pone. 0027347 PMID: 22096555 (accessed July 29th, 2021).

Jha, S.; Dick, C. W. Native Bees Mediate Long-Distance Pollination in Coffee Landscape. *Proc. Natl. Acad. Sci.* **2010,** *107*, 13760–13764. https://doi.org/10.1073/pnas.1002490107/ (accessed July 29th, 2021).

Joffe, S. R. Desert Locust Management: A Time for Change. International Bank for reconstruction and Development, Washington, DC, World Bank Discussion Papers 284, 1995, pp 1–76.

Jones, R. A. C. Influence of Climate Change on Plant Disease Infection and Epidemics Caused by Viruses and Bacteria. CAB Reviews and Perspectives, 2012, pp 1–32. DOI:10.1079/PAVSNNR20127022/ (accessed July 29th, 2021).

Jones, R. A. C. Disease Pandemics and Major Epidemics Arising from New Encounters Between Indigenous Viruses and Introduced Crops. *Viruses* **2020,** *12*, 1388. DOI:10.3390/v12121388/ (accessed July 22nd, 2021).

Jones, A. M.; Harrison, R. M. The Effects of Meteorological Factors on Atmospheric Bioaerosol Concentrations—A Review. *Sci. Total Environ.* **2004,** *326*, 151–180.

Jordan, P. Fig Seed Predation and Dispersal by Birds. *Biotropica* **1983,** *15*, 38–41.

Joyce, R. J. V. Aerial Transport of Pests and Pest Outbreaks. Paper presented at the EPPO/WMO Symposium on Meteorology for Plant Protection, Geneva (Switzerland), 1983. https://doi.org/10.1111/j.1365–2338.1983.tb01585.x (accessed July 29th, 2021).

Kaczmarek, A. M.; King, K. M.; West, J. S.; Stevens, M.; Sparkes, D.; Dickinson, M. A Loop-Mediated Isothermal Amplification (LAMP) Assay for Rapid and Specific Detection of Airborne Inoculum of Uromyces betae (Sugar Beet Rust). *Plant Dis.* **2019,** *103*, 417–421. https://doi.org/10.1094/PDIS-02–18–0337-RE/ (accessed July 29th, 2021).

Kale, M. A.; Dudhe, N.; Kasambe, R.; Bhattacharya, P. Crop Depredation by Birds in Deccan Plateau. *Int. J. Biodiversity* **2014,** Article ID 947683, pp 1–8. http://dx.doi.org/10.1155/2014/947683/ (accessed July 29th, 2021).

Karp, D. S.; Mendenhall, C. D.; Figueroa, R. S.; Chaumont, N.; Ehrlich, P. R.; Hadly, E. A. Forest Bolsters Bird Abundance, Pest Control and Coffee Yield. *Ecol. Lett.* **2013,** *16*, 1339–1347.

Kawamura, K.; Kaplan, I. R. Stabilities of Carboxylic Acids and Phenols in Los Angeles Rainwaters During Storage. *Water Res.* **1990,** *24*, 1419–1423.

Kellogg, C. A.; Griffin, D. W. Aerobiology and the Global Transport of Desert Dust. *Trends Ecol. Evol.* **2006,** *21*, 638–644.

Khiyamia, M. A.; Almoammara, H.; Awadh, Y. M.; Alghuthaymic, M. A.; Abd-Elsalamd, K. A. Plant Pathogen Detection Techniques: Forthcoming Changes. *Biotechnol. Biotechnol. Equip.* **2014,** *28*, 775–785. http://dx.doi.org/10.1080/13102818.2014.960739 (accessed July 29th, 2021).

Khodadad, C. L.; Wong, G. M.; James, L. M.; Thakrar, P. J.; Lane Micheal, C. J. A.; Smith, D. J. Stratosphere Conditions Inactivate Bacterial Endospores from a Mars Spacecraft Assembly Facility. *Astrobiology* **2017,** *17* (337), 350. DOI: 10.1089/ast.2016.1549 (accessed July 28th, 2021).

Kim, H. G.; Park, J. S.; Lee, D. H. Potential of Unmanned Aerial Sampling for Monitoring Insect Populations in Rice Fields. *Florida Entomol.* **2020,** *101* (2), 330–334. https://doi.org/10.1653/024.101.0229/ (accessed January 2nd, 2021).

Kirk, D. A.; Evenden, M. D.; Minaeu, P. Past and Current Attempts to Evaluate the Role of Birds as Predators of Insect Pests in Temperate Agriculture, 1996. DOI:10.1007/978–1-4615–5881–1_5 (accessed July 29th, 2021).

Klein, A. M.; Bohannan, B. J.; Jaffe, D. A.; Levin, D. A.; Green, J. L. Molecular Evidence for Metabolically Active Bacteria in the Atmosphere. *Front. Microbiol.* **2016,** *7*, 772. DOI: 10.3389/fmicb.2016.00772 (accessed July 29th, 2021).

Klein, A. M.; Steffan-Dewenter, I.; Tscharntke, T. Fruit Set of Highland Coffee Increases with the Diversity of Pollinating Bees. *Proc. R. Soc. B Biol. Sci* **2003a,** *270*, 955–961. DOI: 10. 1098/rspb.2002.2306 (accessed July 29th, 2021).

Klein, A. M.; Steffan-Dewenter, I.; Tscharntke, T. Pollination of *Coffea cenephora* in Relation to Local and Regional Agroforestry Management. *J. Appl. Ecol.* **2003b,** *40*, 837–845. DOI: 10.1046/j.1365-2664.2003.00847.x (accessed July 29th, 2021).

Krishna, K. R. *Agricultural Drones: A Peaceful Pursuit*; Apple Academic Press Inc.: Waretown, NJ, 2018, p 376.

Krishna, K. R. *Unmanned Aerial Vehicles in Crop Production*; Apple Academic Press Inc.: Palm Bay, FL, 2020a, p 683.

Krishna, K. R. *Aerial Robotics in Agriculture: Parafoils, Blimps, Aerostats and Kites*; Apple Academic Press Inc.: Palm Bay, FL, 2020b, p 394.

Kunast, C.; Riffel, M.; de Graeff, S.; Whitmore, G. Pollinators and Agriculture. European Initiative for Sustainable Development of Agriculture, Brussels, Belgium A Report, 2013, pp 1–48.

Lacava, P. T.; Araújo, W. L.; Azevedo, J. L.; Hartung, J. S. Rapid, Specific and Quantitative Assays for the Detection of the Endophytic Bacterium *Methylobacterium mesophilicum* in Plants. *J. Microbiol. Methods* **2006,** *65* (3), 535–541. DOI: 10.1016/j.mimet.2005.09.015/ (accessed July 29th, 2021).

Lacey, M.; West, J. *The Air Spora: A Manual for Catching and Identifying Airborne Biological Particles*; Springer: Dordrecht, The Netherlands, 2006. http://dx.doi.org/10.1007/978–0-387–30253–9 (accessed July 29th, 2021).

Laliberte, K. How to Attract Bug-Eating Birds, 2019, pp 1–8. https://www.gardeners.com/how-to/attracting-bug-eating-birds/8103.html/ (accessed July 29th, 2021).

Lang, A.; Evans, J. L. Introduction to Tropical Meteorology. University Corporation for Atmospheric Research, Boulder, CO, Version 2.0, 2011, p 344 (accessed July 29th, 2021).

Lebeda, A.; Magrath, M. T.; Sedlakova, B. Fungicide Resistance in Cucurbit Powdery. *Mildew Fungi. Fungicides* **2010,** *11*, 221–246. DOI: 10.5772/1408 (accessed July 29th, 2021).

Lecoq, M. Desert Locust Management: From Ecology to Anthropology. *J. Orthoptera Res.* **2005,** *14*, 179–186. https://bioone.org/journals/Journal-of-Orthoptera-Research/ (accessed July 29th, 2021).

Legg, B. J. Movement of Plant Pathogens in the Crop Canopy. *Phil. Trans. R. Soc. Lond. Ser. B Biol. Sci.* **1983,** *302*, 559–573.

Leslie, J. F.; Summerell, B. A. *The Fusarium Laboratory Manual*; Blackwell: Ames, IA, 2006; pp 1–388. https://onlinelibrary.wiley.com/doi/book/10.1002/9780470278376/ (accessed July 29th, 2021).

Leveau, J. H. J.; Tech, J. J. Grapevine Microbiomics: Bacterial Diversity on Grape Leaves and Berries Revealed by High-Throughput Sequence Analysis of 16S rRNA Amplicons. *Acta Hortic.* **2011,** *905*, 31–42. DOI: 10.17660/ActaHortic.2011.905.2/ (accessed July 29th, 2021).

Levey, D. J.; Bolker, B. M.; Tewksbury, J. J.; Sargent, S.; Haddad, W. M. Effects of Landscape Corridors on Seed Dispersal by Birds. *Science* **1999,** *309*, 145–148.

Ley, E. L.; Buchmann, S.; Stritch, L. Selecting Plants for pollinators: A Regional Guide for Farmers, Land Managers and Gardeners in the Great Plains Steppe and Shrub Province Including Oklahoma and Texas. North American Pollinator Protection Campaign-NAPPC, San Francisco, California, 2006, pp 1–24.

Linde, C. C.; Zhan, J.; McDonald, B. A. Population Structure of *Mycosphaerella graminicola*: From Lesions to Continents. *Phytopathology* **2002,** *92*, 946–955. http://dx.doi.org/10.1094/PHYTO.2002.92.9.946 (accessed July 2021).

Lindmann, J.; Constantinidou, H. A.; Brachet, W.; Upperi, C. D. Plant Sources of Airborne Bacteria, Including Ice Nucleation Active Bacteria. *Appl. Environ. Microbiol.* **1982,** *44* (5), 1059–1063.

Lindow, S. E.; Brandl, M. T. Microbiology of the Phyllosphere. *Appl. Environ. Microbiol.* **2003,** *69* (4), 1875–1883.

Lindow, S. E.; Leveau, J. H. Phyllosphere Microbiology. *Curr. Opin. Biotechnol.* **2002,** *13* (3), 238–243. DOI: 10.1016/S0958–1669(02)00313–0.? (accessed July 29th, 2021).

Lindow, S. E.; Arny, D. C., Upper, C. D. Distribution of Ice Nucleation-Active Bacteria on Plants in Nature. *Appl. Environ. Microbiol.* **1978a,** *36*, 831–838.

Lindow, S. E.; Arny, D. C.; Upper, C. D. *Erwinia herbicola*: A Bacterial Ice Nucleus Active in Increasing Frost Injury to Corn. *Phytopathology* **1978b,** *68*, 523–527.

Lindstrom, S. A. M. Insect Pollination of Oilseed Rape. Doctoral Thesis, Swedish University of Agricultural Sciences Uppsala, Sweden, 2017, pp 1–57.

Lopez, M. M.; Bertolini, E.; Olmos, A.; Caruso, P.; Gorris, M. T.; Llop, P.; Penyalver, R.; Cambra, M. Innovative Tools for Detection of Plant Pathogenic Viruses and Bacteria. *Int. Microbiol.* **2003,** *6*, DOI: 10.1111/ibi.12360 233–243/ (accessed July 29th, 2021).

Lopez-Darias, M.; Nogales, M. Raptors as Legitimate Secondary Seed Dispersers of Weed Seeds. *Ibis* **2016,** *158* (2). https://www.researchgate.net/publication/293824770 (accessed July 29th, 2021).

Lucas, G. B.; Campbell, C. L.; Lucas, L. T. Disease Caused by Airborne Fungi. In *Introduction to Plant Disease*, 1992, pp 192–242. https://link.springer.com/chapter/10.1007/978–1-4615–7294–7_13/ (accessed July 29th, 2021).

Maas, B.; Clough, Y.; Tscharntke, T. Bats and Birds Increase Crop Yield in Tropical Agroforestry Landscapes. *Ecol. Lett.* **2013,** *16*, 1480–1487.

Madigan, M. T.; Marrs, B. L. Extremophiles. *Sci. Am.* **1997,** *276* (4), 82–87.

Mahaffee, W. F.; Stoll, R. The EBB and Flow of Airborne Pathogens: Monitoring and Use in Disease Management Decisions. *Phytopathology* **2016,** *106*, 420–431. DOI: 10.1094/PHYTO-02–16–0060-RVW.

Maki, T.; Hara, K.; Iwata, A.; Lee, K. C.; Kawai, K.; Kai, K. Variations in Airborne Bacterial Communities at High Altitudes Over the Noto Peninsula (Japan) in Response to Asian Dust Events. *Atmos. Chem. Phys.* **2017,** *17*, 11877–11897. DOI: 10.5194/acp-17–11877–2017 (accessed July 29th, 2021).

Maki, T.; Kakikawa, M.; Kobayashi, F.; Yamada, M.; Matsuki, A.; Hasegawa, H. 2013 Assessment of Composition and Origin of Airborne Bacteria in the Free Troposphere Over Japan. *Atmos. Environ.* **2013,** *74*, 73–82. DOI: 10.1016/j.atmosenv.2013.03.029/ (accessed July 29th, 2021).

Manikowski, S.; Camara-Smeets. Estimating Bird Damage to Sorghum and Millet in Chad. *J Wildlife Manage.* **1979,** *43*, 540–544. https://www.jstor.org/stable/3800369/ (accessed July 29th, 2021).

Marston, Z. P. D.; Cira, T. M.; Hodgson, E.; Knight, J. F.; Macrae, I. W.; Koch, R. L. Detection of Stress-Induced by Soybean Aphid (Hemiptera: Aphididae) Using Multispectral Imagery from Unmanned Aerial Vehicles. *J. Econ. Entomol.* **2020,** *113*, 779–786. https://doi.org/10.1093/jee/toz306/ (accessed July 29th, 2021).

Martinelli, F.; Scalenghe, R.; Davino, S.; Scuderi, G.; Ruisi, P.; Villa, P.; Stroppiana, D.; Boschetti, M. Goulart, L. R.; Dvis, C. E.; Dandekar, A. M. Advanced Methods of Plant Disease Detection: A Review. *Agron. Sustain. Dev.* **2015,** *35*, 1–25.

McCartney, H. A.; Freeman, J.; Calderon, C.; Foster, S. J.; Ward, R. Detecting Airborne Inoculum of Fungal Pathogen of Oilseed Rape, Using Polymerase Cain Reaction (PCR) Assays, 2003, pp 1–3. https://www.gcirc.org/fileadmin/documents/Proceedings/IRC2003Copen/AGRONOMICS%20MANUSCRIPTS%20FINAL%2018.06.03/AO8.pdf/ (accessed June 29th, 2021).

McDonald, R. M.; Gossen, B. D.; Kora, C.; Parker, M.; Boland, G. Using Crop Canopy Modifications to Manage Plant Disease. *Eur. J. Plant Pathol.* **2013,** *135*, 581–593.

McInnes, T. B.; Gitaitis, R. D.; McCarter, S. M.; Jaworski, C. A.; Pathak, S. C. Airborne Dispersal of Bacteria in Tomato and Pepper Fields. *Plant Disease* **1988,** *72*, 575–579 https://www.cabdirect.org/cabdirect/abstract/19891121322/ (accessed July 29th, 2021).

Medellin, R. A.; Gaona, O. Seed Dispersal by Bats and Birds in Forest and Disturbed Habitats of Chiapas, Mexico Biotropica, 1999. DOI: 10.1111/j.1744–7429.1999.tb00390.x (accessed July 29th, 2021).

Mejía, L. C.; Rojas, E. I.; Maynard, Z.; Bael, V.; Arnold, A. E.; Hebbar, P. Endophytic Fungi as Biocontrol Agents of *Theobroma cacao* Pathogens. *Biological Control* **2008,** *46* (1), 4–14. DOI: 10.1016/j.biocontrol.2008.01.012 (accessed July 29th, 2021).

Michigan State University. MSU Biologist Sheds Light on Fog's Role in Microbial Transport. MSU Today, 2020, pp 1–4 https://msutoday.msu.edu/news/2018/msu-biologist-sheds-light-on-fogs-role-in-microbial-transport-/ (accessed July 28th, 2021).

Mols C, M.; M.; Visser, M. E. Great Tits Can Reduce Caterpillar Damage in Apple Orchards. *J. Appl. Ecol.* **2002,** *39*, 888–899.

Moon, A. How Do Birds Disperse Seeds, 2017, pp 1–3. https://sciencing.com/how-do-birds-disperse-seeds-12517955.html/ (accessed July 29th, 2021).

Monteil, C. L.; Borddia, M.; Morrris, L. E. Features of Airmasses Associated with the Deposition of *Pseudomonas syringae* and *Botrytis cinerea* by Rain and Snowfall. *Int. Soc. Microb. Ecol. J* **2014,** *8*, 2290–2304.

Moorcroft D.; Whittingham M. J.; Bradbury R. B.; Wilson J. D. The Selection of Stubble Fields by Wintering Granivorous Birds Reflects Vegetation Cover and Food Abundance. *J. Appl. Ecol.* **2002,** *39*, 535–547.

Musa, V. H.; Akogu, S. E.; Agbaji, F.; Adah, H. Genetic Associations Under Aphids and Rosette Stress in Groundnuts (Arachis hypogea). *World J. Agric.* **2020,** *8*, 89–96. DOI: 10.12691/wjar-8–3-4/ (accessed July 29th, 2021).

Nagarajan, S.; Singh, D. V. Long Distance Dispersal of Rust Fungi. *Annu. Rev. Phytopathol.* **1990,** *28*, 139–153.

National Research Council of the National Academies. *Status of Pollinators in North America.* National Academy Press, Washington, DC, 2006, p 326. https://www.nap.edu/catalog/11761/status-of-pollinators-in-north-america/ (accessed July 29th, 2021).

Nicolaisen, M.; West, J. W.; Sapkota, R.; Canning, G. M.; Shoen, C.; Justesen, A. F. Fungal Communities Including Plant Pathogens in Near Surface Air Are Similar across North-Western Europe. *Front. Microbiol.* **2017,** *8*, 1–14. Article 1729. DOI:10.3389/fmicb.2017.01729 (accessed July 29th, 2020).

Nishimura, Y. Similarity of Bacterial Community Structure Between Asian Dust and Its Sources Determined by rRNA Gene-Targeted Approaches. *Microbes Environ.* **2010,** *25*, 22–27.

Norros, V. Do Small Spores Disperse Further Than Large Spores? *Ecology* **2014,** *95*, 1612–1621.

Oluwadare, F. Mammals and Birds Affecting Food Production and Storage in Nigeria. *Proceedings of the 9th Vertebrate Pest Conference*, 1980, p 13. https://digitalcommons.unl.edu/vpc9/13/ (accessed July 29th, 2021).

Onetoa, D. L.; Golan, J. J.; Mazzinoe, A.; Pringle, F.; Seminara, A. Timing of Fungal Spore Release Dictates Survival During Atmospheric Transport. *Proc. Natl. Acad. Sci.* **2020,** *117*, 5134–5143.

Orlowski, G.; Czarnecka, J. Granivory of Birds and Seed Dispersal: Viable Seeds of *Amaranthus retroflexus* L. Recovered from the Droppings of the Grey Partridge *Perdix perdix* L. *Polish J. Ecol.* **2009,** *57*, 191–196 (accessed January 15th, 2021).

Owings, L. Crop Disease Pandemics Coming Sooner Than Later. *Sci. Dev. Net.* 2020, pp 1–4. https://www.scidev.net/global/features/crop-disease-pandemic-coming-sooner-rather-than-later/ (accessed July 29th, 2021).

Pammel, L. H. *Fungus Diseases of Sugar Beet*; Iowa State University: Ames, Iowa. *Bulletin* **2017,** *2* (15), Article 4. http://lib.dr.iastate.edu/bulletin/vol2/iss15/4/ (accessed July 28th, 2021).

Pangga, I. B.; Hanan, J. Chakraborty, S. Pathogen Dynamics in a Crop Canopy and Their Evolution Under Changing Climate. *Plant Pathol.* **2011,** *60*, 71–80. https://doi.org/10.1111/j.1365–3059.2010.02408.x.

Parajulee, M. N.; Slosser, J. E. 1999 Evaluation of Potential Relay Strip Crops for Predator Enhancement in Texas Cotton. *Int. J. Pest Manage.* **1999,** *45*, 275–286.

Parnell, M.; Burt, P. J. A.; Wilson, K. The Influence of Exposure to Ultraviolet Radiation in Simulated Sunlight on Ascospores Causing Black Sigatoka Disease of Banana and Plantain. *Int. J. Biometeorol.* **1998,** *42*, 22–27. http://dx.doi.org/10.1007/s004840050079/ (accessed July 29th, 2021).

Patel K. B. Damage in Seedling Stage on Pearl Millet Due to Pigeon and Doves. *Life Sci. Leaf* **2011,** *15* (1), 524–527.

Patel, K. B. Studies on Control of Bird Pests Damage to Millet Crops in Semi-Arid Zone of Northern Gujrat. *J. Environ. Res. Dev.* **2016,** *11*, 132–138.

Peer, B. D.; Homan, H. J.; Linz, G. M.; Bleier, W. J. Impact of Blackbird Damage to Sunflower: Bioenergetic and Economic Models. *Ecol. App.* **2003,** *13*, 248–256.

Pearce, D. A.; Bridge, P. D.; Hughes, K. A.; Sattler, B.; Psenner, R.; Russel, N. J. Microorganisms in the Atmosphere over Antarctica. *FEMS Microbiol. Ecol.* **2009,** *69*, 143–157.

Pearce, F. Birds Do It, Bees Do It, -Pollinators Are Vital for Crop Yields. But They Are Dying Out. *New Scientist*, 1998, pp 1–3. https://www.newscientist.com/article/mg15721219–600-birds-do-it-bees-do-it-pollinators-are-vital-for-crop-yields-but-they-are-dying-out/#ixzz66HS47fco/ (accessed July 30th, 2021).

Pasteur, L. Memoir sur les corpuscules organises qui existent dans l'atmosphere: exami de la doctrine des generations spontanees. *Annals de Science Nationale* **1861,** *16*, 5–98.

Peccia, J.; Hernandez, M. Incorporating Polymerase Chain Reaction-Based Identification, Population Characterization, and Quantification of Microorganisms Into Aerosol Science: A Review. *Atmos. Environ.* **2006,** *40*, 3941–3961.

Pepper, I. L.; Dowd, S. E. *Aeromicrobiology. Environmental Microbiology*; Academic Press: San Diego, 2009, pp 83–101.

Pepper, I. L.; Gerba, C. P. Aeromicrobiology. In *Environmental Microbiology*; Gerba, C. P.; Gentry, T. J., Eds., 3rd ed.; Elsevier Inc.: Pepper, IL, 2015, pp 89–110. http://dx.doi.org/10.1016/B978–0-394626–3.00005–3.

Polymenakou, P. N. Atmosphere: A Source of Pathogenic or Beneficial Microbes. *Atmosphere* **2012,** *3*, 87–102. DOI: 10.3390/atmos3010087 (accessed July 29th, 2021).

Potts, S.; Garratt, M.; Senapathi, D.; Breeze, T. Policy and Practices. Insect Pollinators Initiative Note No 13. Living with Environmental Change: Polaris House, London, 2014, pp 1–4.

Prescott, J. M.; Burnett, P. A.; Saari, E. E.; Rauson, J.; Bowman, J.; deMilliano, J.; Singh, R. P.; Bekele, G. Wheat Diseases and Pests: A Guide for Field Identification. International Maize and Wheat Centre, Mexico, 1986, pp 1–73. https://wheat.pw.usda.gov/ggpages/wheatpests.html/ (accessed June 29th, 2021).

Pringle, A.; Vellinga, E.; Peay, K. The Shape of Fungal Ecology: Does Spore Morphology Give Clues to a Species' Niche? *Fungal Ecol.* **2015,** *17*, 213–216.

Prospero, J. M. Long-Range Transport of Mineral Dust in the Global Atmosphere: Impact of African Dust on the Environment of the South-Eastern United States. *Proc. Natl. Acad. Sci. USA* **1999,** *96*, 3396–3403. DOI: 10.1073/pnas.96.7.3396 / (accessed July 29th, 2021).

Prospero, J. M.; Blades, E.; Mathison, G.; Naidu, R. Interhemispheric Transport of Viable Fungi and Bacteria from Africa to the Caribbean with Soil Dust. *Aerobiologia* **2005,** *21*, 1–19. DOI: 10.1007/s10453–004–5872–7 (accessed July 29th, 2021).

Puckett, H. L.; Brandle, J. R.; Johnson, R. J.; Blankenship, E. E. Avian Foraging Patterns in Crop Field Edges Adjacent to Woody Habitat. *Agric. Ecosyst. Environ.* **2009,** *131*, 9–15. DOI: 10.1016/j.agee.2008.08.015/ (accessed July 29th, 2021).

Qandah, I. S.; del Río Mendoza, L. E. Temporal Dispersal Patterns of *Sclerotinia sclerotiorum* Ascospores During Canola flowering. *Can. J. Plant Pathol* **2011,** *33*, 159–167 http://dx.doi.org/10.1080/07060661.2011.554878 (accessed July 29th, 2021).

Rajendran, L.; Ramanathan, A.; Durairaj, C.; Samiyappan, R. Endophytic *Bacillus subtilis* Enriched with Chitin Offer Induced Systemic Resistance in Cotton Against Aphid Infestation. *Arch. Phytopathol. Plant Protect.* **2011,** *44* (14), 1375–1389. DOI: https://doi.org/10.1080/03235408.2010.499719/ (accessed July 29th, 2021).

Reche, I.; D'Orta, G.; Mladenov, N.; Winget, D. M.; Suttle, C. A. Deposition Rates of Viruses and Bacteria Above the Atmospheric Boundary Layer. *ISME J.* **2018,** *12*, 1154–1162. DOI: 10.1038/s41396–017–0042–4 (accessed July 29th, 2021).

Riechmann, D. Guano—Das weiße Gold Perus, 2003. http://www.scinexx.de/dossierdetail-153–11.html/ (accessed July 29th, 20201).

Ricketts, T. Tropical Forest Fragments Enhance Pollinator Activity in Nearby Coffee Crops. *Conserv. Biol.* **2004,** *18*, 1262–1271. DOI: 10.1111/j.1523–1739.2004.00227.x

Ricketts, T.; Daily, G. C.; Ehrlich, P. R.; Michener, C. D. Economic Value of Tropical Forest to Coffee Production. *Proc. Natl. Acad. Sci. USA* **2004,** *101*, 12579–12582. DOI: 10.1073/pnas.0405147101 (accessed July 29th, 2021).

Rieux, A.; Soubeyrand, S.; Bonnot, F.; Klein, E. K.; Ngando, J. E.; Mehl, A.; Ravigne, V.; Carlier, J. Long-Distance Wind Dispersal of Spores in a Fungal Pathogen: Estimation of Anisotropic Dispersal Kernels from an Extensive Field Experiment. Plos.org, 2014, pp 1–22. https://journals.plos.org/plosone/article?id=10.1371/journal.pone.0103225/ (accessed July 29th, 2021).

Ritpitakphong, U.; Falquet, L.; Vimoltust, A.; Berger, A.; Métraux, J. P.; L'Haridon, F. The Microbiome of the Leaf Surface of Arabidopsis Protects Against a Fungal Pathogen. *New Phytologist* **2016,** *210* (3), 1033–1043. DOI: 10.1111/nph.13808/ (accessed July 29th, 2021).

Roberts, P.; Ryan, D. R. Late Blight of Potato and Tomato. Get full ref., 2016.

Roper, M. A.; Seminara, A.; Bandi, M. M.; Cobb, A.; Dillard, H. R.; Pringle, A. Dispersal of Fungal Spores on a Cooperatively Generated Wid. *Proc. Natl. Acad. Sci.* (PNAS) **2010,** *41*, 17474–17479.

Roubik, D. W. Pollination of Cultivated Plants in the Tropics. FAO Agricultural Services Bulletin No. 118: 1–259, Food and Agricultural Organization of the United Nations, Rome, Italy, 1995.

RMAX. RMAX Specifications. Yamaha Motor Company, Japan, 2015, pp 1–4. http://www.max.yamaha-motor.Drone.au/specifications (accessed September 8th, 2015).

Rubene, D. Leidefors, M.; Ninkovic, V.; Eggers, s. Low, M. Disentangling Olfactory and Visual Information Used by the Field Foraging Birds. *Ecol. Evol.* **2019,** *11*, 545–552. DOI: 10.1002/ece3, 4773.eCollection2019 Jan (accessed June 19th, 2020).

Sapkota, R.; Knorr, K.; Jørgensen, L. N.; O'Hanlon, K. A.; Nicolaisen, M. Host Genotype Is an Important Determinant of the Cereal Phyllosphere Mycobiome. *New Phytologist* **2015,** *207* (4), 1134–1144. DOI: 10.1111/nph.13418.

Sattler, B.; Puxbaum, H.; Psenner, R. Bacterial Growth in Supercooled Cloud Droplets. *Geophys. Res. Lett.* **2001,** *28*, 239–242.

Saunders, M. E. Insect Pollinators Collect Pollen from Wind-Pollinated Plants: Implications for Pollination Ecology and Sustainability. Insect Conservation and Diversity, 2017, pp 1–41. Wiley online library. https://doi.org/10.1111/icad.12243.

Schiro, G.; Verch, G.; Grimm, V.; Müller, M. E. H. Alternaria and Fusarium Fungi: Differences in Distribution and Spore Deposition in a Topographically Heterogeneous Wheat Field. *J. Fungi* **2018,** *4*, 63. DOI: 10.3390/jof4020063/ (accessed December 23rd, 2020).

Schmale, D. G.; Ross, S. D. Highways in the Sky: Scales of Atmospheric Transport of Plant Pathogens. *Annu. Rev. Phytopathol.* **2015,** *53*, 591–611. DOI: 10.1146/annurev-phyto-080614–115942 (accessed January 22nd, 2021).

Schnug, E.; Jacobs, F.; Stronue, K. Guano: The White Gold of the Seabirds, 2018. https://www.researchgate.net/publication/327449209/ (accessed September 22nd, 2021).

Schumann, G. L.; D'Arcy, C. J. The Plant Health Instructor, 2005, pp 1–8. DOI: 10.1094/PHI-I-2000–0724–01/ (accessed December 22nd, 2020).

Shaw, J. Robobee Collective: Exploring Engineering's Limits to Build a Mobile Microbot Colony, 2017, pp 1–6. https://harvardmagazine.com/2017/11/robobee-harvard/ (accessed January 20th, 2021).

Simmonds, N. W. Some Speculative Calculations of the Dispersal of Sugarcane Smut Disease. *Sugar Cane* **1994,** *1*, 2–5.

Sinclair, J. B.; Dhingra, O. D. *Basic Plant Pathology Methods*; CRC Press Inc.: Boca Raton, FL, 1995; p 448.

Sishuba, S. Using Bats and Birds to Control Macadamia Crop Pests, 2017, pp 1–5. https://www.farmersweekly.co.za/agri-technology/farming-for-tomorrow/using-bats-birds-control-macadamia-crop-pests/ (accessed January 12th, 2021).

Smets, W.; Moretti, S.; Denys, S.; Lebeer, S. Airborne Bacteria in the Atmosphere: Presence, Purpose, and Potential. *Atmos. Environ.* **2016,** *139*, 214–221. DOI: 10.1016/j.atmosenv.2016.05.038 (accessed January 3rd, 2021).

Smith, D. J. Microbes in the Upper Atmosphere and Unique Opportunities for Astrobiology Research. *Astrobiology* **2013,** *13*, 981–990. DOI: 10.1089/ast.2013.1074 (accessed January 25th, 2021).

Smith, D. J.; Griffin, D. W. Inadequate Methods and Questionable Conclusions in Atmospheric Life Study. *Proc. Natl. Acad. Sci. USA*. **2013,** *110*, E2084. DOI: 10.1073/pnas.1302612110 (accessed July 30th, 2021).

Smith, D. J.; Griffin, D. W.; McPeters, R. D.; Ward, P. D.; Schuerger, A. C. Microbial Survival in the Stratosphere and Implications for Global Dispersal. *Aerobiologia* **2011,** *27*, 319–332.

Smith, D. J.; Griffin, D. W.; Schuerger, A. C. Stratospheric Microbiology at 20km Over the Pacific Ocean. *Aerobiologia* **2009,** *26*, 35–46. DOI: 10.1007/s10453–009–9141–7 / (accessed July 29th, 2021).

Smith, D. J.; Jaffe, D. A.; Birmele, M. N.; Griffin, D. W.; Schuerger, A. C.; Hee, J. Free Tropospheric Transport of Microorganisms from Asia to North America. *Microb. Ecol.* **2012,** *64*, 973–985 (accessed January 1st, 2021).

Smith, D. J.; Ravichandar, J. D.; Jain. S.; Griffin, D. W.; Yu, H.; Tan, Q.; Thissen, J.; Lusby, T.; Nicoll, P.; Shedler, S.; Martinez, P.; Osorio, A.; Lechniak, J.; Choi, S.; Sabino, K.; Iverson, K.; Chan, L.; Jaing, C.; McGrath, J. Airborne Bacteria in Earth's Lower Stratosphere Resemble Taxa Detected in the Troposphere: Results from a New NASA Aircraft Bioaerosol Collector (ABC). *Front. Microbiol.* **2018,** *9*, 1752. DOI: 10.3389/fmicb.2018.01752/ (accessed July 28th, 2021).

Smith, D. J.; Timonen, H. J., Jaffe, D. A.; Griffin, D. W.; Birmele, M. N.; Perry, K. D. Intercontinental Dispersal of Bacteria and Archaea by Transpacific Winds. *Appl. Environ. Microbiol.* **2013,** *79*, 1134–1139.

Southwick, E. E.; Southwick, L. Estimating the Economic Value of Honeybees (Hymenoptera: Apidae) as Agricultural Pollinators in the United States. *J. Econ. Entomol.* **1999,** *85* (3), 13.

Spring, A. M.; Docherty, K. M.; Domingue, K. D.; Kerber, T. V.; Mooney, K. M.; Lemmer, K. M. A Method for Collecting Atmospheric Microbial Samples for Set Altitudes for Use with Next-Generation Sequencing Techniques to Characterize Communities. *Air Soil Water Res.* **2018,** *11*, 1–12.

Stein, A.; Draxler, R. R.; Rolph, G. D.; Stunder, B. J.; Cohen, M.; Ngan, F. NOAA's HYSPLIT Atmospheric Transport and Dispersion Modelling System. *Bull. Am. Meteorol. Soc.* **2015,** *96*, 2059–2077.

Strange, R. N.; Scott, P. Plant Disease: A Threat to Global Food Security. *Annu. Rev. Phytopathol.* **2005,** *43*, 83–316. DOI: 10.1146/annurev.phyto.43.113004.133839/ (accessed July 22nd, 2021).

Sultan, B.; Labadi, K.; Guegan, J. F.; Janicot, S. Climate Drives the Meningitis Epidemics Onset in West Africa. *PLoS Med.* **2007,** *2005*, 2. DOI: 10.1371/journal.pmed.0020006. (accessed July 28th, 2021).

Summers, C. F.; Adair, Gent, D. H.; McGrath, M. T.; Smart, C. D. *Pseudoperenospora cubensis* and *P. humili* Detection Using Species Specific Probes and High-Definition Melt Curve Analysis. *Can. J. Plant Pathol.* **2015,** *37*, 315–330.

Sustainable Agriculture Research and Education-SARE. Cover Cropping for Pollinators and Beneficial Insects. United States Department of Agriculture, Beltsville, MD, 2020, pp 1–16.

Tang, K.; Huang, Z.; Huang, J.; Maki, T.; Zhang, S.; Ma, X. Characterization of Atmospheric Bioaerosols Along the Transport Pathway of Asian Dust During the Dust-Bioaerosol 2016 Campaign. *Atmos. Chem. Phys. Discuss.* **2016,** *18*, 7131–7148.

Taylor, R. A. J.; Herms, D. A.; Cardina, J.; Moore, H. Climate Change and Pest Management: Unanticipated Consequences of Trophic Dislocation. *Agronomy* **2018,** *8*, 1–7. DOI: 10.3390/agronomy8010007/ (accessed July 29th, 2021).

The State of Victoria. Non- Indigenous Bird Management Policy. State of Victoria, Australia, 2019, pp 1–17. http://agriculture.vic.gov.au/agriculture/pests-diseases-and-weeds/protecting-victoria/legislation-policy-and-permits/policies-and-strategies/non-indigenous-bird-management-policy/ (accessed January 18, 2021).

Thomas, G.; Jayasena, K. Managing Powdery Mildews in Wheat. Department of Primary Industries, Agriculture and Food, Australia, 2020, pp 1–8. https://www.agric.wa.gov.au/spring/managing-powdery-mildew-wheat?nopaging=1 (accessed December 14th, 2020).

Toepfer, I.; Favet, J.; Schulte, A.; Schmölling, M.; Butte, W.; Triplett, E. W. Pathogens as Potential Hitchhikers on Intercontinental Dust. *Aerobiologia* **2011,** *28*, 221–231. DOI: 10.1007/s10453–0119230–2.

Totten K. What a Load of Guano: 5 Facts You Did Not Know About Bird Poop. National Museum of American History- Behring Center, 2016, pp 1–8. https://americanhistory.si.edu/blog/what-load-guano-5-facts-you-didnt-know-about-bird-poop/ August 20th, 2021).

Tragard, C. Sampling of Aerobiological Material from a Small Aircraft. *Grana* **1977,** *16*, 139–143. DOI: 10.1080/00173134.1977.118 64649.

Tsang, J. *Snow Is Coming-What's That Have to Do with Microbes*; American Society of Microbiology: Washington, DC, 2019; pp 1–4. https://asm.org/Articles/2019/January/Snow-Is-Coming-Whats-That-Have-to-Do-with-Microbe/ (accessed July 29th, 2021).

United States Forest Service. Bird Pollination. US Department of Forests, 2020. https://www.fs.fed.us/wildflowers/pollinators/animals/birds.shtml/ (accessed July 29th, 2021).

University of California-Berkeley. Pollinators Help One-Third of the World's Food Crop Production. Science Daily, 2006, pp 1–4. https://www.sciencedaily.com/releases/2006/10/061025165904.htm/ (accessed January 11th, 2021).

Uno, I. Asian Dust Transported One Full Circuit Around the Globe. *Nat. Geosci.* **2009,** *2*, 557–560.

USDA Agricultural Research Service. Cereal Rust Laboratory Report. United States Department of Agriculture, St Paul, MN, 2019, pp 1–17.

Vaitilingom, M.; Amato, P.; Sancelme, M.; Laj, P.; Leriche, M.; Delort, A. M. Contribution of Microbial Activity to Carbon Chemistry in Clouds. *Appl. Environ. Microbiol.* **2010,** *76*, 23–29.

van Leeuwenhoek, A. *The Collected Letters of Antoni van Leeuwenhoek*, Vol. 2; Amsterdam, Netherlands: Swets and Zeitlinger, 1941, pp 1–187.

Velasquez, A. C.; Castro Verde, C. D. M.; He, S. Y. Plant and Pathogen Warfare Under Changing Climate. *Curr. Biol.* **2018,** *28*, 619–634.

Vincenot L, Nara K, Sthultz C, Labbé J, Dubois M-P, Tedersoo L, Martin F, Selosse M-A. Extensive Gene flow Over Europe and Possible Speciation Over Eurasia in the Ectomycorrhizal Basidiomycete *Laccaria amethystina* complex. *Molecular Ecology* **2012,** *21*, 281–299. https://doi.org/10.1111/j.1365–294X.2011.05392.x. (accessed July 29th, 2021).

Wade, L. Microbes Survive and May Be Thrive High in the Atmosphere, 2013, pp 1–3. https://www.sciencemag.org/news/2013/01/microbes-survive-and-maybe-thrive-high-atmosphere/ (accessed July 29th, 2021).

Wainwright, M.; Wickremasinghe, N.; Narlikar, J.; Rajaratnam, P. Microorganisms Cultured from Stratospheric Air Samples Obtained at 41km. *FEMS Microbiol. Lett.* **2003,** *218*, 161–165. DOI: 10.1111/j.1574–6968.2003.tb11513.x/ (accessed July 29th, 2021).

Waliyar, F.; Kumar, P. L.; Monyo, E.; Nigam, S. N.; Reddy, A. S.; Osiru, M.; Diallo, A. T. *A Century of Research on Groundnut Rosette Disease and Its Management*. Technical Report. International Crops Research Institute for Semi-Arid Tropics (accessed ICRISAT), Patancheru, Andhra Pradesh, India. **2007,** *75*, 1–44.

Wandale, L.; Ayalew, H.; Woldeab, G. Mulugeta, G. Yellow Rust (*Puccinia striiformis*) Epidemics and Yield Loss Assessment on Wheat and Triticale Crops in Amhara Region, Ethiopia. *Afr. J. Crop Sci.* **2016,** *4* (2), 280–285.

Webster, J. A. Economic Impact of the Greenbug in the Western United States: 1992–1993. Volume Publication No. 155, Great Plains Agricultural Council, Stillwater, Oklahoma, USA, 1995. https://agris.fao.org/agris-search/search.do?recordID=US9557787/ (accessed July 29th, 2021).

Weil, T.; De Filippo, C.; Albanese, D.; Donati, C.; Pindo, M.; Pavarini, L. Legal Immigrants: Invasion of Alien Microbial Communities During Winter Occurring Desert Dust Storms. *Microbiome* **2017,** *5*, 32. DOI: 10.1186/s40168–017–0249–7/ (accessed July 29th, 2021).

West, J. S.; Kimber, R. B. E. Innovations in Air Sampling to Detect Plant Pathogens. *Ann. Appl. Biol.*, 2015.

West, J. S.; Kimber, R. B. E. Innovations in Air Sampling to Detect Plant Pathogens. *Ann. Appl. Biol.* **2014,** *166*, 4–17. doi:10.1111/aab.12191/ (accessed July 28th, 2021).

Willocquet, L.; Berud, F.; Raoux, L. A. Clerjeau, M. Effects of Winds Relative Humidity, Leaf Movement and Colony Age on Dispersal of Conidia of *Uncinula necator* Causal Agent of Grape Powdery Mildew. *Plant Pathol.* **1998,** *47*, 234–242

Williams, R.; Ward, E.; McCartney, H. A. Methods for Integrated Air Sampling and DNA Analysis for Detection of Airborne Fungal Spores. *Appl. Environ. Microbiol.* **2016,** *67*, 2453–2459.

Wilson, E. A.; LeBoeuf, E. A.; Weaver, K. M.; LeBlanc, D. J. Delayed Seeding for Reducing Blackbird Damage to Sprouting Rice in Southwestern Louisiana. *Wildlife Soc. Bull.* **1989,** *17*, 165–171.

Wilson J. D.; Morris A. J.; Arroyo B.; Clark S.; Bradbury, R. A Review of the Abundance and Diversity of Invertebrate and Plant Foods of Granivorous Birds in Northern Europe in Relation to Agricultural Change. *Agric. Ecosyst. Environ* **1999,** *75*, 13–30.

Whitworth, J.; Michaud, J. P.; McCornack, S. Wheat Insect Pest Management. Kansas State University Experiment Station and Cooperative Extension Service, 2021, pp 1–8. https://bookstore.ksre.ksu.edu/pubs/mf745.pdf/ (accessed July 29th, 2021).

Wood, S.; Sebastian, K.; Sherr, S. J. *Agroecosystems: A Pilot Analysis of Agroecosystems*; International Food Policy Research Institute: Washington, DC, 2000, p 185.

Womack, A. M.; Bohannan, B. J. M.; Green, J. L. *Phil. Trans. R. Soc. London B* **2010,** *365*, 3645–3653. DOI: 10.1098/rstb.2010.0283 (accessed July 29th, 2021).

Wotton, D. M.; McAlpine, K. G. Seed Dispersal of Fleshy-Fruited Environmental Weeds in New Zealand. *NZ J Ecol.* **2015,** *39* (2), 155–160. http://www.newzealandecology.org/nzje/ (accessed January 29th, 2021).

Wyman, T. E. Consequences of Reduced Bird Densities for Seed Dispersal. University of Canterbury, Kent, United Kingdom, PhD thesis, 2013, p 150.

Yamaguchi, N.; Ichijo, T.; Sakoyani, A.; Baba, T.; Nasu, M. Global Dispersion of Bacterial Cells on Asian Dust. *Sci. Rep.* **2012,** *2*, 525, 1–6. DOI: 10.1038/srep00525 (accessed July 28th, 2029).

Yayock, J. Y. Rossel, H. W.; Harkness, C. A. Review of the 1975 Rosette Epidemic in Nigeria. Paper presented at African Groundnut Council Symposium on pest of Groundnut and Millet in the field. Kaolock-Senegal, 1976, 12.

Yue, J.; Lei, T. The Application of Unmanned Aerial Vehicle Remote Sensing in Quickly Monitoring Crop Pests. *Intell. Autom. Soft Comput.* **2012,** *18*, 1043–1052.

Zhang, Z.; Luo, L.; Tan, X.; Kong, X.; Yang, J.; Wang, D.; Zhang, D.; Jin, D.; Liu, Y. Pumpkin Powdery Mildew Disease Severity Influences the Fungal Diversity of the Phyllosphere. *Peer J* **2018,** *6*, e4559. https://doi.org/10.7717/peerj.4559/ (accessed July 28th, 2021).

Zhou, Y.; Van Leeuwen, S. K.; Pieterse, C. M. J.; Bakker, P. A. J.; Van Wees, S. C. M. Effect of Atmospheric CO_2 on Plant Defence Against Leaf and Root Pathogens of *Arabidopsis. Eur. J. Plant Pathol.* **2019,** *154*, 31–42. https://doi.org/10.1007/s10658–019–01706–1 / (accessed July 31st, 2021).

Zvereva, E. L.; Kozlov, M. V. Consequences of Simultaneous Elevation of Carbon Dioxide and Temperature for Plant-Herbivore Interactions: A Meta-Analysis. *Global Change Biol.* **2006,** *72*, 27–41.

CHAPTER 4

Man-Made Abiotic Factors in the Agricultural Sky

ABSTRACT

We investigated soil profile intensely in order to prescribe suitable procedures and amendments. Land preparation, seeding, interculture, application of soil amendments, fertilizer inputs, and harvesting have involved adoption of series of different ground vehicles. During recent years, GPS-guided robotic ground vehicles have been invading the agrarian regions. We have mended soils intensely using such ground vehicles. However, we have not mended or managed the agricultural sky to the same extent. Farmers have overlooked most aspects of agrarian sky. Aerial vehicles are the only recent introduction into agrarian sky.

The aim of this chapter is to show that man-made aerial vehicles are gaining acceptance in agrarian regions. They are termed Unmanned Aerial Vehicles (UAVs or drones). There are several agronomic procedures that utilize aerospace above crops. For example, foliar supply of irrigation water and fertilizers via sprinklers. More recently there are tethered aerostats with turbines lofted above the farms to produce wind energy.

The primary aim of this chapter is to enlist the different kinds of UAVs that have invaded the agrarian sky. They are capable of utilizing the vantage points above crops and provide useful digital data. Such digital data can then be utilized to conduct various agronomic procedures more accurately than previous years. There are indeed several types of UAVs (drones) being adopted by farmers. Yet there is no single drone called agricultural drone. They are multipurpose aerial vehicles suitable for variety of tasks. Recently, certain farms are adopting "sprayer drones" with ability to get aerial imagery, digital data, store plant protection chemicals/fertilizer solution, and spray

The Agricultural Sky: A Concept to Revolutionize Farming. K. R. Krishna, PhD (Author)

the crops. There are sprayer drones capable of rapid transit and application of chemicals. These are perhaps the best candidates to be called "Agricultural Drones." There are multicopters, fixed-winged aircrafts like drones, autonomous single roto helicopters, and copters with facility for detection of pollutants in the atmosphere. They are provided with probes for gaseous contaminants etc.

There are several other types of aerial vehicles that take to aerospace above crops. They are semiautonomous and robotic parafoils, microlights, aerostats, balloons, and even kites have been utilized to sample study the agricultural sky. These aerial robots are capable of obtaining digital imagery, surveillance, and even spray chemicals. Autonomous aerial vehicles (UAVs) could be lightweight (<5 kg), medium (50–200 kg), heavy 1–2 t). Fixed-winged lightweight versions are popular with farmers. They are small and swift. There are fixed-winged drones with fuselage fused with wings, single rotor helicopters, multicopters, vertical take-off and lift (VTOL) hybrid drones, etc. These have been discussed in greater detail.

Aerial spray above crop canopy affects the tranquil and chemicals composition of agrarian sky. Foliar sprays are conducted using either ground vehicles or aerial vehicles. Aerial sprays using center-pivot is popular in developed countries. Small UAVs with fluid tanks and spray vars getting common in the in North America and Far-eastern nations. (e.g., RMAX; Hercules; AGRAS MG-5B). Foliar sprays, that is, chemigation could alter chemical composition of atmosphere crops, although for a short duration until the original conditions are regained.

There are several satellites transiting the space above agrarian regions. They are helpful in offering data, monitoring cropping zones, aerospace above agroecosystems, and provide GPS data that offers greater accuracy.

Farm power generation using aerial vehicles like tethered aerostats is gaining acceptance. It could be a multipurpose vehicle if utilized for general surveillance farm activity, obtaining digital data of crop's status, disease/pest attack if any and to scare birds using tapped sounds of predatory birds.

We should note that we intensely researched, invented, devised, and utilized vehicles, implements and amendments to improve agricultural soil. We have to perhaps bestow greater interest to study the "agricultural sky" by adopting aerial vehicles, mend agrarian sky to reduce disease/pest attacks, supply inputs, and reap better harvests.

These aerial robots that take to sky are man-made. They could revolutionize farm activities. They can potentially reduce farm drudgery by rapidly obtaining data and conducting agronomic procedures like spraying farm chemicals.

4.1 INTRODUCTION

In this chapter, we deal with abiotic aerial factors that manifest in the agricultural sky. Such abiotic factors could be natural. For example, the weather parameters such as precipitation, storms, floods, scanty rainfall, droughts, arid conditions, high ambient temperature, cold front, snow, frost, fog, smog, etc. These are natural factors. These aspects have been dealt in Chapter 1. However, here, we are concerned entirely with "man-made abiotic factors" that throng the agricultural sky, like aerial vehicles and their utility in farming. There are indeed several types of aerial vehicles that could fly over the crop's canopy. In agrarian regions, most of them are being utilized for aerial survey, surveillance of farms, and for obtaining aerial imagery. Aerial imagery has often involved procurement of still or video pictures of farms, crops, and other natural resources. The imagery is generally obtained, using the visual, multispectral, infra-red, and lidar sensors.

Aerial robots, particularly, the small unmanned aerial vehicles (UAVs) are also called drones. These UAVs are being continuously modified and upgraded with sophisticated electronics, remote control systems, and computer software. Such superior electronics allow greater flexibility, while fixing flight path, flying, and conducting aerial imagery (Krishna, 2018, 2020a; Banjo and Ajayi, 2020). Such aerial robots do affect the agricultural sky. Particularly, the way we utilize the aerospace above crops. A few types of small autonomous UAVs and their models have been improvised with "spray bars" and "variable-rate nozzles." They serve the farmer by distributing water, fertilizers, and plant protection chemicals from the aerospace above crop's canopy. Such aerial vehicles with spray bars are termed "Sprayer drones" or "Sprayer Microlights." There are also aerial vehicles used to obtain samples from atmosphere or water bodies in the agricultural areas (MacDonell et al., 2013; Scentroid Inc., 2017). There are several models of aerial robots that collect imagery of storms, tornadoes, volcanoes, terrain, etc. Each of these aerial robots have definite impact on the agricultural aerospace. They affect the way we negotiate and overcome problems related to farming.

Then, there are ground vehicles operated by farmers such as planters, soil inoculators, fluid fertilizer, pesticides, herbicides, and fungicide applicators that cause disturbances in the aerospace immediately above the crop's canopy. Currently, large center-pivot irrigation machines are utilized to spray or to channel farm chemicals, on to crops via agricultural sky. The atomized irrigation water does affect various parameters of the sky above crops, in addition to the intended effects on crops. These machines are remotely controlled. They are semiautonomous or entirely robotic too. They do not fly above fields but transit on the ground in the fields. Yet, they affect atmosphere immediately above crop's canopy.

There are satellites placed way away from the crop's canopy in the space, at 300 km altitude. They help in assessing the large patches of crops. Their resolution is of much lower order at 2–5 m. So, the imagery is good enough to be utilized only, when it concerns large patches, regions, or agroecosystems. A small farmer's field is not easily amenable for aerial imagery via satellite. Satellite imagery could be a basis for several decisions regarding agricultural regions. They may influence crop species selected and cropping systems adopted. Fertilizer, pesticides, and irrigation schedules in large areas could be guided, by the imagery that satellites offer to farm agencies (e.g., SPOT, LANDSAT 8 satellites).

There are a few other farming activities that relate to the agricultural sky. For example, spraying foliar nutrients, pesticides/fungicides, hoisting aerostats at low or high altitudes to get wind energy generated, using lightweight helium-filled turbines (Altaeros Energies Inc., 2019), etc. They would soon be a part of agrarian sky in almost all major food-generating regions.

4.2 UAVs IN THE AGRICULTURAL SKY

A few remarks about where we started-off with UAVs even prior to introducing them into the agricultural sky. Aerial robots, namely, UAVs or colloquially known as "drones" are in use, since several decades. They are flown in the sky at different altitudes, speeds and for a wide range of applications. During past few decades, UAVs were predominantly utilized for military purposes. They were efficiently used for surveillance of international borders and battle zones. They are used for reconnaissance of enemy aircrafts, drones, and military barracks. They were perhaps most efficiently used during the Middle East war, in recent times, particularly, for bombing small targets. Drones (autonomous UAVs) are capable of stealth, by flying at high altitudes and descending only while conducting bombing sorties. Then, they may fly back to the place of origin. They are robotic and do not involve military personnel to be physically present in the battle zone. These drones are called military drones (e.g., Global Hawk, Predator). Some of the military UAVs are known to fly several thousands of km before accurately bombing the location and returning home. They fly at high altitudes in the intercontinental air space at 30,000 ft (e.g., Global Hawk) (Krishna, 2018, 2020a; Newcome, 2004; Nicole, 2015; Keane and Carr, 2013; Tetrault, 2014). Such drones do not disturb the agricultural sky, in any way. They serve military purposes exclusively. While UAVs such as Global Hawk or Predator are larger-sized heavy drones (>2000 kg). There are several medium-sized drones used by the military establishments. They fly closer to ground, but still do not affect agricultural sky. They confine to higher altitudes than agricultural drones (e.g., Insight-20, also called ScanEagle) (Litchman, 2015). Krishna (2018, 2020a) points out that, during the past few years, say since 2000 A.D., there is a clear trend to adopt relatively low-altitude small and medium-sized military UAVs in the farming sector. Some of the models have been transformed to suit the agricultural sky. They are utilized for aerial surveillance and imagery of natural resources, crops, and pest/diseases. Small UAVs, in particular, have made an excellent transition from military sky to agricultural sky. Such a transition has been quick, smooth, and accurate. Currently, the trend is to adopt UAVs in agricultural sky much more than in urban or military locations (Garland, 2014; Green, 2013; Elmquist, 2015; King, 2013; Spence, 2013; Yintong Aviation Supplies Ltd., 2012, 2014).

So far, the UAV machines that fly either in the military space or those that may throng the agricultural sky, do not seem to cause perceptible disturbance to atmosphere. The drone machine (petrol engine) emits CO_2 into air at low levels. The noise pollution that agricultural drones cause is often bearable. Military UAVs are flown in low number and feebly. They are flown only when absolutely required. They may not obstruct or distract the general conditions in the sky. In case of agricultural UAVs, again, they are flown in the crop season. Drones are utilized only when required, to conduct aerial imagery or during spraying. They could be used in swarms. They lift off into sky and cause noise or apply harmful chemicals in intervals, during the season. Agricultural drones (small air crafts) are not a permanent fixture in the atmosphere, that is, agrarian sky. The blimps and aerostats may hold on to locations in the sky for longer duration or for the entire season. However, they are used singly or in few numbers over large areas of farm. Clearly, military drones were disturbing the peaceful conditions prevalent in the ground, but not the aerospace *per se*. However, we may soon realize that agricultural drones, particularly, the "sprayer drones" are destined to spill a large quantity of farm chemicals into the air space above crops, pollute and contaminate. So, they take part in activities that reduce the air quality, to various extents. We ought to realize that agricultural drones are trifle more troublesome than military counterparts when we consider a few aspects. A few precautions and environmental regulatory aspects are to be considered while dealing with drones in the agriculture sky.

Krishna (2020a) states that, historically, drones were utilized in the sky above urban locations. They transited long distance, perhaps, above peaceful agricultural terrain, before entering a battle zone. Drones were first adopted in military some eight decades ago. However, there is no set of characters to call a UAV as "agricultural drone." Agricultural drones are a recent initiative. They occupy relatively low altitudes above cropping belts.

The relatively larger UAVs that are classified as medium, heavy, or super heavy UAVs are mostly adopted, by agencies involved in aerial imagery of natural resources such as geological features, soil types, large agrarian regions, river valleys, coastal line, etc. They often fly at higher altitude of >3000–30,000 ft above ground. Sensors are relatively far away from ground surface. Therefore, images are of low resolution. However, agricultural drones are light-weighted small drones. Agricultural drones are invariably flown close to crop's canopy. This helps to derive high-resolution images and digital data. While spraying, the copter drones are held at levels as low as possible, above crop's canopy.

The agricultural sky itself extends to several km above a ground location. Yet, we can stratify the sky in general. The UAVs used in farm sector gets stratified at low altitudes of 10–300 m above crop. Agricultural drones may stray into airports and defense installations or neighbors plot if flight path programmed goes astray.

There are several types of drones that throng the sky above mines, civilian, regions, industries, mines, traffic line, railway lines, policing, transport of small packets, etc. Here, we are concerned only with agricultural sky and the drones adopted in this zone. Agricultural drones are flown above crop fields/plantations, water bodies, irrigation channels, dams, farm vehicles, and farm installations.

No doubt, in a few more years, agricultural UAVs are expected to be major abiotic aerial factors above the farming enterprises. Their influence may stretch into vast agrarian regions of the world. They are actually recent introductions into agricultural sky. Their greatest advantage is that they allow aerial imagery and spectral analysis, from vantage location above farms. Such aerial imagery was never possible for several millennia. Hence, they are supposed to offer useful insights about crops/soils to farmers. Forecasts are that drones could revolutionize the way we conduct agronomic procedures.

4.2.1 TYPES OF UAVS UTILIZED IN THE AGRICULTURAL SKY

Agricultural sky is invaded by specific types of UAVs. Not all UAV models are amenable for use in the agricultural aerospace. The preference to certain types of UAVs is based mainly on the farmer's purpose, ease with which the aerial flights and spectral analysis could be conducted, economic efficiency, and regulatory aspects. Clearly, not all models produced by the aerospace companies are drafted for use by the farmers. Agricultural sky is actually exploited better by the UAVs flying at lower altitudes, that is, close to the crop's canopy. Small fixed-winged UAVs are predominantly utilized for aerial imagery.

In recent decades, there have been significant efforts to increase flight duration (endurance), flight altitude, the payload, and tolerance to various weather conditions. This has resulted in UAV configurations with different sizes, duration of autonomy, and competencies (Simelli and Tsgaris, 2015). A key criterion currently used to distinguish among UAVs is the size and flight duration: High altitude long endurance UAV, such as, Northrop-Grumman Ryan's Global Hawks (65.000 feet altitude, flight time 35 h, payload 1900 lbs) are larger drones. These are not agricultural drones. They are adopted for high-altitude reconnaissance and bombing expeditions, by the military.

Medium Altitude Long Endurance UAV too are preferred mainly by non-agricultural agencies. They may be used however to provide aerial survey and imagery of large agricultural zones. They may offer low-resolution imagery compared with UAVs that fly at low altitudes (e.g., Predator) (Gogarty and Robinson, 2012, See Krishna, 2018). A predator drone, which is often used for military purposes reaches 27.000 feet from ground surface. Its endurance is 30–40 h per flight and payload is 200–250 kg. The above two types of drone aircrafts are not meant commonly for agricultural sky.

Small and portable UAVs with short period of endurance are being produced in large numbers. These are among the most preferred small UAVs flown in the agricultural sky. They can be launched frequently, using hands and holding them, at shoulder height (e.g., *eBee*, UX5, AgEagle). Small UAVs with relatively longer endurance are preferred when fields are large. Also, when few km of farmland has to be aerially surveyed and photographed. They can be launched at shorter intervals during the crop season. Agricultural sky is more efficiently used by the small drones while flying at low altitudes over the crops. Yet, we may note that, relatively higher altitudes are utilized by drone aircrafts, for preparing maps. Also, while assessing crop stand and natural vegetation of vast regions. For example, in order to study the agroecosystems in greater detail, drones capable of reaching relatively higher altitudes are preferred (Krishna, 2018).

UAVs have been classified based on several different manufacturing characteristics and usage. For example, Krishna (2020a) has listed characteristics of the UAVs utilized to classify them. Primarily, the small UAV (autonomous aircrafts) can be classified as:

(a) Flat Fixed-winged drones (e.g., SenseFly's *eBee*, Precision Hawk's Lancaster, Aeromao's AeroMapper; Trimble's Gatewing x-100, etc.). This group also includes fused-flat fixed-winged drones (Sense Fly's; E-Bee, Trimble's UX5; Bramox; Wingtra One; Delta-M, Bramor ppX, etc.).
(b) Fixed-winged Hybrids that show traits of both fixed-winged and copter drones (Quantix; Thresher-03 etc.).
(c) Single Rotor Helicopters (Yamaha's RMAX; HSE's Hercules; Yintong-5).
(d) Multicopters: There are quad-, hexa-, octa-, and dodeca-copters adopted by farmers. Quadcopters such as DJI's Phantom, Micro Drone's md-3000 are popular.

Based on lift-off, UAVs are grouped as those needing short/long runway, or vertical lift-off. Here, all fixed-winged models irrespective of their size or payload need short runway. Otherwise, they could be released in hand-held state, from shoulder height into air (e.g., *e*-Bee). Helicopters and multicopter drones are vertical lift off and landing UAVs. There are hybrids such as AeroVironment's Quantix; Thresher-03; Terrahawk; and ALTI-Transition that have vertical lift-off. But forward flight in the air is like fixed-winged drones. Most microlights need short runway of 100–1000 m. They could be launched from cliffs too. Autonomous parafoils also need a short runway for lift-off. Blimps and Aerostats are filled with lighter-than-air (LTA) gas such as helium. They lift from ground in a vertical fashion.

A few other important characteristics (specifications) are utilized to classify the agricultural UAVs. They are weight, payload size/weight, endurance in the sky, altitude attained by the drone during flight and while spraying, wing loading, engine type that could be energized using petrol or lithium batteries, and sensors fitted on the drone's fuselage. For example, they could be surveillance drones with just visual sensors. They could be aerial imagery and spectral data collectors. Otherwise, they could be "sprayer drone," if fitted with payload tank and spray bar. They could be multifunctional drones. Multifunctional drones are capable of aerial imagery and spraying pesticides in the agricultural sky. A few drones could be adopted for transport of small payloads from one location to another. This is in addition to other agricultural functions.

We can consider various applications of UAVs and classify them accordingly. Obviously, there are several uses for drones that fly in the sky. For example, Banjo and Ajayi (2020) prefer to classify drones based on their functions in the sky. They say, classification could be based on purposes such as aerial photography, surveillance, cinematography, traffic monitoring, filming public events, etc. A 2018 census of companies/institutions specializing the production of UAVs provided a list of 1200–1400 models of them. Most of the companies were manufacturing UAVs for civilian purposes. They were new initiatives and start-ups in developed nations. At present, these UAVs have found their niche in the civilian and agricultural skies of North America, China, Japan, and Fareast. Some companies were churning out drones meant exclusively for military (see Krishna, 2018, 2020a; Wikipedia, 2020). We may have to match the drone after considering its specifications and costs with the purpose. Later, release the drone models for specific use in the agricultural sky. There are now agricultural drones that are specifically meant for aerial imagery and spectral analysis. There are also drones meant exclusively, for spraying farm chemicals. These groups again need some kind of further classification (see Krishna, 2020a).

4.2.2 AUTONOMOUS FIXED-WINGED AIRCRAFTS

At present, "Fixed-winged UAVs" are among the popular aerial robots in the sky. They were first utilized by the military establishments for reconnaissance, aerial photography of enemy terrain, fix locations to bomb, and in detection of other aircrafts in the airspace. They were usually fitted with high-resolution multispectral sensors. They were controlled, using a ground control station (GCS) with facilities for image processing. Drones, relay vital information to military headquarters. However, during past decade, such small flat-fixed-winged drones have been sought by the farm sector. The farm sector is opting for fixed-winged UAVs with visual, multispectral, near infra-red, and infra-red cameras. Many of the UAVs carry Lidar cameras too. Around 2012–2015, there were several start-up factories that aimed at producing small UAVs for agricultural sector. Several models of different specifications are now being churned by these companies. They vary in size, swiftness (speed), cameras, endurance, energy sources, etc. They are now occupying the agricultural sky. The UAV market worldwide is flooded with innumerable models of small drone aircrafts (platforms). Farmers are to pick the right model for tasks envisaged.

Krishna (2020a) has utilized the weight of the fixed-winged drones, to group them for use in farms. Let us consider a few examples of fixed-winged drones. We should know the purposes for which they are used in agricultural sky (see Table 4.1). The fixed-winged UAVs are smaller version of a large transport airplane. They possess small wings with ailerons, a tail piece for providing stability and direction, to the drone. They fly past the crops/soil at low altitudes of 20–50 m above ground, while conducting aerial imagery. However, their ceiling altitude could be much higher at 1000 m above ground surface. They are launched from shoulder height using hand (e.g., *eBee*, Precision Hawk). They fly at a fast pace of 20–50 mph. Hence, they cover larger areas of crop fields at a rapid rate. They need a runway if they have undercarriage with wheels.

The fixed-winged drones have been classified based on weight. Major weight-based classes of fixed winged UAVs are Microdrones (<5 kg); Lightweight UAVs (5–50 kg); Medium-weight (50–200 kg) and heavy-weight (200–2000 kg) and Super heavy weight (> 2000 kg).

4.2.3 MICROWEIGHT UAVS (<5 Kg)

Microweight and lightweight UAVs are the most frequently adopted drones above the agricultural terrain. Medium and heavy-weight drones are feebly utilized in the agricultural sky. It is because of various reasons such as basic cost of drone aircraft, need for high-resolution imagery, cost of launching accessories, monitoring facilities in the GCS and lower efficiency, in terms of benefit: cost ratio. Aerial imagery too could be of low resolution. Super heavy-weight UAVs that hold large quantities of fuel and facility, for carrying bombs are often flown at high altitudes, by the military establishments. They are not seen in the agricultural sky. Perhaps, rarely they are used to get aerial imagery of farms/crops. This is because, they are flown at higher altitude in the sky. They are not apt for assessing small patches of crops. Imaging individual plants for disease/pest is difficult. Small fixed-winged drones are currently utilized by the agricultural consultancy agencies dealing with land holdings. They are utilized for demarcating farm boundaries. They also provide status reports on soil and crop, to those involved in providing periodic advisories to farmers. Agricultural experimental stations are utilizing the small fixed-winged drones. Mainly, for aerial surveillance of farms, aerial photography of research material, in collecting data such as NDVI, CSWI, chlorophyll content, disease/pest incidence, etc. The aerospace above experimental stations is utilized efficiently compared with previous years.

The agrarian regions have been introduced with implements (e.g., hoes, plows, tined harrows, weeders), ground vehicles such as tractors for deep/shallow tillage of soil, soil clod crushers, leveling hoes, and furrow formers. Planters have been adopted in the fields, to seed the fields at a rapid pace. Interculture and spraying pesticides/fungicides are done, using ground vehicles. *Most recent among the machines to invade the agrarian regions is the "UAVs." However, these UAVs invade the agricultural sky, unlike the ground vehicles that transit on the terrain.* They offer excellent vantage spots in the aerospace for us to conduct spectral analysis and store digital data about crops. We may note here that, farmers have not been utilizing the aerospace very efficiently, since long. Introduction of linear and center-pivot sprinkler irrigation is relatively recent. They too affect the agricultural aerospace.

Farmers had not utilized imagery and spectral data collected from an altitude above the crops, to any great extent. Hand-held leaf meters that assess chlorophyll pigmentation helped to decide the fertilizer-N requirements and dosages. These leaf meters operated above the crop's canopy, that is, aerospace. Otherwise, usually, farm technicians toured the large farms and conducted the scouting. Scouting on ground is a difficult and time-consuming task. It involves higher costs. Human fatigue and errors do show up.

Arial imagery and analysis of crops requirements are recent. It is now becoming common in many parts of the world. Fixed-winged drones are routinely utilized for aerial imagery and collection of spectral data of farms/crops. Fixed-winged drones are in vogue in many farms of North America, Europe, and Fareast. However, fixed-winged drones utilized are small. They are not adopted to spray farm chemicals. Normally, copters with flexible flight pattern, hovering, and low altitude capability are preferred, for spraying farm chemicals. Fixed-winged drones are adopted for rapid transit above farms, to collect spectral data of large farms. They complete aerial imagery at a rapid pace, covering 50–100 ha in a matter of 30–45 min. This is because, they fly in the sky above crops, at 45–60 kmph.

Fixed-winged drones carry only small quantity of fuel source/lithium batteries. Fuel quantity determines the endurance in flight. Their endurance is short at 15–60 min, per flight. However, they can be launched repeatedly, in short intervals, into the farm sky. Fixed-winged drones used above farms can be controlled, using a remote controller with radio or internet connectivity. They offer images tagged with GPS coordinates. The flight path and imaging schedule can be pre-programmed. We use computer software such as *e-Motion*, Pixhawk Atlas flight path, etc. to fix the path.

Agricultural experts have already started using UAVs above their experimental stations. They have adopted them, to collect vital data about crops. Krishna (2020a) lists a series examples of tasks conducted by UAVs, when flown above experimental farms. This list is expected to increase exponentially as UAVs become sophisticated, versatile in collecting data, and efficient (Table 4.1).

No doubt, in future, agrarian aerospace, that is, sky above experimental station will be utilized most efficiently. Micro- and lightweight drone models are best to assess the crops, using sensors. Agricultural experts may find it easier, to provide accurate spectral data about crops growth, nutrient (plant N), and water status. These aerial robots avoid errors that creep in if farm technicians are adopted. They can repeat the tasks as frequently as the researchers require them to do so. Agricultural drones are agronomically efficient. They complete the task quickly and are less costly, , that is, economically efficient. Regular farmers may use the UAVs during entire crop season. They can be launched and utilized, at various stages of growth and maturity. However, in experimental farms, they will have to adopt UAVs more frequently.

Micro-UAVs with gimbal that supports a set of visual, multispectral, hyperspectral, infra-red, and lidar sensors will be most useful, above the agricultural experimentation. This is because, crop researchers may have to evaluate their crops and collect data from vantage aerial spots, frequently. They may have to launch micro- or even lightweight drones repeatedly in a day, to collect data required for growth analysis of various genotypes of crops. UAVs may have to collect spectral data frequently while developing Crop Surface Models, Again, during ripening stage both farmers and experts may fly the UAVs frequently to assess the panicles/grains.

Agricultural sky above farms may differ because of natural resource, farm inputs, crops, and cropping systems adopted. Agronomic procedures adopted such as fertilizer sprays, pesticide/fungicide also affect agrarian sky. Small fixed-winged drones are expected to be adopted frequently. Heavy and Super heavy drones may not find use in the agricultural aerospace. The heavier drones used for military may get stratified in the higher altitudes.

4.2.4 LIGHTWEIGHT FIXED-WINGED DRONES (5–50 Kg)

Like microdrones, lightweight UAVs are also small. They are preferred because of relatively smaller weight, handiness, transportability within farm, and ability for repeated launching. A light-weight UAV once launched has relatively longer endurance. They can occupy the aerospace and assess crops, for a stretch of 40 min to even 3 h. It depends on the model and power, that is, fuel source. There are models that are energized, using lithium batteries and others fitted with petrol engine. Those with petrol engine release carbon particles and CO_2 gas through exhaust into the agricultural sky. Light-weight drones can conduct aerial survey and relay imagery or digital data, to GCS. Light-weight UAVs also offer data about crop stand, NDVI, leaf chlorophyll content (i.e., leaf-N status), and crop's water status. The imagery also helps in identifying the pest/disease-affected patches of crops or even weed infestation. These lightweight drone aircrafts are however not amenable for aerial sprays. Their small payload ability does not allow fixing of a pesticide tank or spray bar. However, they can collect air samples above the agricultural farms. Such air sampling can be done for extended time. They can monitor larger areas, assess crop vigor, disease, and water status. It depends on the flight path and way points chosen. In Germany, models such as Carolo 360, have been adopted, to collect greenhouse gas samples for assessment. This is in addition to usual functions in the agricultural sky, such as aerial photography, assessment of NDVI, CWSI (Crop Water Stress Index), plant-N status, etc (Altstadler et al., 2015; Weheren et al., 2016; Shi et al., 2016, Cornell, 2015).

In this section, let us quote a few prominent examples wherein micro-weight and lightweight fixed-winged UAVs have been adopted. They are utilized to monitor farms and crop's vigor (NDVI, leaf N), water stress index (CWSI), disease/pest attack if any, weed infestation in the fields/plantations, etc. Such examples provide us with evidence to show that, UAVs are indeed new aerial factors, in the agricultural sky. The UAVs would definitely play a vital role during crop production. In fact, introduction of drones may significantly alter the crop production tactics. They make agronomic procedures easier and efficient (see Krishna, 2018, 2020a, 2020b). At the International Centre for Maize and Wheat Mexico (CIMMYT), experts have adopted fixed-winged UAVs such as *eBee*, UX5, Precision Hawk. This is to assess maize/wheat crop's vigor related parameters, disease tolerance, water stress index and drought tolerance, weed infestation, and grain yield (panicle traits) (Tropnevad, 2013; Shi et al., 2016; Trimble, 2015; Zhenkun et al., 2017; Rice, 2015; Pena et al., 2013, 2015; Swain, 2014). Similarly, there are several experimental stations in USA, Great Britain, and other nations, where in, fixed-winged drones are employed. This is to periodically assess experimental research material (e.g., cultivars of a crop). These aerial robots are known to reduce farm drudgery otherwise required by technicians/experts. Krishna (2020a) provides a list of crops assessed for growth-related traits such as NDVI, GNDVI, canopy chlorophyll content, spectral reflectance signatures for disease/pest attack, water stress index, etc. These abiotic aerial factors introduced into the sky i.e. fixed-winged and copter UAVs have been adopted, on crops such as wheat (CIMMYT, 2012; Song and Wang, 2016; Kansas State University, 2015); on maize (Tremblay et al., 2014; Agape-Palilo, 2014; Dronologista, 2015); sorghum (Shi et al., 2016); potato (Hunt et al., 2015; Allen, 2016), rice (Swain et al., 2010; Shi et al., 2016); sunflower (Aguera et al., 2011); grapes (Sky Squirrel, 2017; Zarco-Tejada et al., 2013); banana (Machovina et al., 2016); citrus (Ehsani and Sankaran, 2010; Sankaran et al., 2010, 2012, 2015), and few other crop species.

UAVs have a few definite roles to play, in future, in the sky above fruit tree plantations and forests. Firstly, they are suited to take an inventory of trees along with their growth/health status. They could be utilized to obtain infra-red spectral data, tree's water stress index, and then consider quantum of irrigation. Nutrients, especially fertilizer-N requirements could be assessed carefully and accurately, using leaf chlorophyll data. Farmers may also obtain data about fruit yield and make suitable forecasts, at an earlier stage. (Shi et al., 2016; Fiske, 2020). It should also be possible to assess tree biomass accumulation pattern.

4.2.5 MEDIUM (50–200 Kg), HEAVY (200–2000 KG), AND SUPER HEAVY WEIGHT (>2 T DRONES)

There are several models of medium, heavy, and super heavy weight drones available for use by military, civilian, and even agricultural farmers. They are larger in size when compared with small/light UAVs, utilized predominantly in the agrarian regions. The medium-weight UAVs (e.g., Arcturus by Arcturus Inc.; Gamma by AvaSys LLC; Sky Robot X-450 by Robot Aviation Ltd.) are utilized for aerial survey of natural vegetation, forest plantations, and cropped field. They weigh about 50–80 kg, and their payload is 30–35 kg. A medium-weight UAV may collect aerial imagery from 200 to 300 ha of cropland per day of 6–8 h. Their endurance ranges from 12 to 15 h, depending on fuel tank size.

The heavier drones stratify themselves at higher altitude in the sky. They offer aerial imagery and reconnaissance data. They are adopted in situations where long-distance transit and proportionate long endurance are needed. There are also large 2 t drones that transit in the sky from one continent to other. They transit for over 1500–2000 km in a day (e.g., Global Hawk, Predator, etc.). Generally, these heavier drones do not intrude into agricultural aerospace. They fly at 3000–5000 m or even 10,000 m above ground surface during transit. They even launch guided missiles. These military drones carry highly sophisticated electronics.

The utility of heavy drones in peaceful civilian tasks is restricted to aerial imagery from high altitudes. They are adopted for studying natural resources, geological features, like rivers, mountainous tracts, vast forest, and agrarian regions. They offer low-resolution imagery because sensors are located at farther distance from the geographical features to be studied. The payload of a medium/heavy drone ranges from 30 to 50 kg. Its altitude of flight ranges from 100 m to 3000 m above ground surface. Endurance is greater than 15–24 h per flight. They carry all the requisite sensors such as visual, multispectral, infrared, and LIDAR. Plus, they may carry bombs, missiles, and reconnaissance instruments such as RADAR, heat seekers, etc. The heavy drones need runway and fuel that lasts for couple of days. They are monitored and guided in the airspace, by regular satellite-connected tracking facilities. These heavy drones are not of great utility in farming. Right now, their discussion in the present volume is curtailed.

4.2.6 SMALL UAVS WITH FIXED WINGS FUSED TO FUSELAGE

Aerospace engineers have been striving to produce more efficient versions of small UAVs. They are also aiming at placing as many sensors as possible and useful computer software. Such a payload enhances utility and accuracy of UAVs meant for use in farm world. A few models produced by companies possess wings that are fused with fuselage. Such a design is supposed to increase flight speed and efficacy. They may also allow better control during flight and imagery. For example, there are models such as *eBee* by SenseFly Inc. Switzerland (SenseFly, 2015, 2017), Atlas UAV, Bramor gEO, and Dramor ppX by C-Astrol Inc, Slovenia, and MH850 by MAVT, Italy that have fused wings. These drones are swift and relay images of cropped fields, rapidly. Such small, fused-winged drones are among most frequently used UAVs, in farming sector (CIMMYT, 2012; Guglieri and Ristorto, 2016). They are supposedly better in aerodynamics than other types of small UAVs.

The agrarian sky is a free space above cropland. It is a region with dynamic weather conditions. Ambient parameters such as wind, moisture, temperate, relative humidity, radiation may fluctuate in space, and time. The region above crops is indeed highly dynamic based on season, time of the day/night, and actual location in the sky, that is, altitude. A good GPS tagging is necessary to demarcate the various weather parameters and fluctuations above crops. Now, above the crops' canopy, the windy turbulent conditions may not allow certain types of UAVs, to perform equally well and still collect useful data. The performance is evaluated in terms of transit speed, stability during aerial imagery, and collection of air samples, if any. Performance may differ based on type of UAV adopted. Now, there is a unique type of UAV with special "Flexible Membrane Wing." These are not the usual fixed wings made of plastic or light carbon fiber or aluminum material. Such flexible wings are made of toughened membrane material. They are foldable during storage. Flexible wings allow better control and maneuverability, in windy conditions. Such harsh climate with windiness may be encountered frequently in temperate countries, or during light stormy period or dusty conditions that prevail, prior to monsoon in other regions. Flexible-winged UAVs are micro- or light-weight vehicles. They are designed to fly specifically in harsh and windy conditions with moderate weight/size payloads.

4.2.7 FIXED-WINGED VTOL UAVS

These models combine features of the fixed-wing and the multirotor UAVs. Usually, it is the multirotor copters that possess VTOL features. The fixed-winged VTOL types discussed here are often fitted with propellers that are foldable/flexible, with reference to direction. Propellers can face up words while conducting vertical take-off. They can stay straight, like any normal fixed-wing drone while transiting at rapid speed, in a horizontal path above the crop's canopy. They do possess the usual set of sensors, namely visual (red, green, and blue) wavelength band, multispectral, infrared, red edge, near infra-red, and LIDAR. Since they can hover, they may be utilized above crop's canopy to conduct sampling of air and particulate matter. They are able to perform VTOL as well as hovering above crop's canopy, at low altitudes just like the multirotor and single rotor copters. These versatile fixed-winged VTOLs are amenable for aerial imagery plus precision spraying of plant protection chemicals, if needed. A few good examples of hybrid fixed-winged VTOL UAVs utilized with excellent results, in the agrarian sky are ALT Transition by ALTI UAS Inc. of South Africa, FT2 VTOL by GTF Aviation Tech Beijing, China, and Quantix produced by AeroVironment Inc. California, USA. Quantix is utilized above farms, to conduct aerial survey and collect data good enough, for precision farming procedures (see Krishna, 2020a; AeroVironment, 2017). A few other models of fixed-winged VTOL UAVs are WingtraS125H produced by Wingtra One- ETH, Zurich, Switzerland, Thresher03 by TechnoSys Systems,

New Delhi, India, Terrahawk VTOL by Phoenix Aerial Systems California, USA. All of these above-mentioned UAV models are amenable for use above crop's canopy, at low altitudes.

4.2.8 AUTONOMOUS SINGLE-ROTOR HELICOPTERS

Autonomous single rotor helicopters are in vogue within the agrarian sky, since past two decades. They are among versatile abiotic factors introduced into farm sky. Farmers utilize them as aerial vehicles capable of flights close to crops. Helicopter UAVs utilize agrarian sky efficiently, by accomplishing foliar sprays of fertilizer solutions and pesticides/fungicides. They are fitted with pesticide tanks and spray bars, if adopted to spray plant protection chemicals. There are several models such as RMAX, Aintong, Autocopter, Hercules, etc. Helicopter sprayers such as RMAX by Yamaha Inc. Japan have successfully spread into agrarian regions of China, Japan, and other Fareast nations (Yamaha, 2014; RMAX, 2015). They reach an altitude of 5–30 m above cropland while spraying plant protection chemicals. Helicopter drones have a higher ceiling altitude (300 m above ground), but in the agrarian regions they localize at low altitudes. Localizing at low altitudes lessens drift of pestcides/fungicides to neighboring farms.

Helicopter UAVs have significantly longer flying time than their multirotor counterpart, as they are often powered by gas engines. These UAVs are also highly maneuverable. They are more efficient than the multirotor types. Like multirotor UAVs, they are capable of hovering at a location, Hence, they are useful for aerial photography and precision spraying. Despite these beneficial attributes, they come with higher operational risks as the large-sized rotor blades usually pose a risk (Banjo and Ajayi, 2020).

During aerial survey and imagery, helicopter UAVs are versatile in flight and spectral analysis of crops. They hover, or even standstill at low altitude above the crop's canopy. They have been utilized even to study a single plant and its branches. Also, we can study the spread of disease/or drought effects on a plant or a small experimental plot (see Krishna, 2017, 2018, 2020a; RMAX, 2015; Banjo and Ajayi, 2020; HSE, 2015a, b, 2017).

4.2.9 AUTONOMOUS MULTIROTOR COPTERS

Multirotor copters are versatile aerial vehicles introduced in the aerospace above crops. At present, they are among popular aerial robots with civilian agencies and farmers. Multirotor UAVs are further classified based on the number of rotors fitted on the platform. Models with three rotors are called tricopter. Those with four rotors are called quadcopter, with six rotors are called hexacopters, and those with eight rotors are called octocopter (see Banjo and Ajayi, 2020; Krishna, 2020a). Farmers tend to use them frequently for both aerial imagery and spraying, depending on the model. Multicopters are usually fitted with visual (RGB), high resolution multispectral, infrared, red-edge, and LIDAR sensors. Sprayer drones possess pesticide tanks and sprayers. The sprayer nozzles are controlled by computer instructions. There are multicopters that can obtain air samples at different altitudes. Also, a few models obtain underwater samples, and bring them on to surface. These are versatile multicopters indeed (e.g., Scentroid). The ability for hovering and close-up shots of crops from aerospace is a noteworthy characteristic of multicopter drones. They can also be used to carry a payload and transport it. Farm agencies dealing with regular surveillance can monitor the crops for water requirements, drought effects/wilting symptoms if any, even in small patches (Barbedo, 2019). This is because of high-resolution cameras, plus ability to hover at low altitude. Similarly, if data banks have requisite spectral signatures, these multicopters in the aerospace can provide useful data about disease/pest attack of crops. In fact, pest/disease-affected regions could be mapped, and digital data used on precision sprayers. Image processing and detection of disease signatures are important.

The multicopters are classified based on size, weight, propellers, payload, endurance, and altitude that are attributable to them. Right now, certain quadcopters such as "DJIs Phantom" are popular in the civilian and farm aerospace. Similarly, some models of quadcopters are popular with farmers who practice aerial spraying and precision farming procedures. For example, AGRAS MG-1 (Real Agriculture, 2017) or md-3000 (Microdrone, 2020). There are companies that produce hexacopters and octocopters that are useful for monitoring crops for longer duration. They carry larger payload tanks. Hence, they distribute (spray) larger quantities of pesticides/fluids onto crops, at a time. Multicopters are excellent for use above agricultural experimental stations, for general surveillance of farms, crops, irrigation channels, farm vehicles, and intrusions, if any. They also offer detailed spectral data about agricultural experimental material.

At present, multicopters are popular with farmers. They are easily noticeable with farmers in North America, Europe, and Fareast. In some developing countries, multicopter sprayers are gaining acceptance. The power source for multicopters is often lithium batteries. Such lithium polymer batteries allow 15–40 min endurance, in the air above crops. The petrol engines do emit CO_2 into the sky. Recent census suggested that, globally, there are at least 500–1000 small industries and several start-ups that manufacture multicopters. They are useful in farming. In future, multicopters may become more frequent in the agricultural sky. We should note that aerial sprays will replace the ground vehicle-supported sprayers. Therefore, aerospace above crops needs careful monitoring for pollution, contamination, and emissions once, the multicopter has been adopted, to conduct various tasks. Pesticide drift, residual foamy chemicals, mist, smog, and particulate matter still afloat in the air above crops need to be examined, carefully. However, it is to be noted that, multicopters keep farmers away from harmful chemicals, during the spraying exercise. Aerospace could also be utilized efficiently by using multicopters to distribute seeds of forest trees, preparing maps of weed infestation, studying the botanical diversity of natural vegetation, etc.

Banjo and Ajayi (2020) state that, multicopters are among common type of UAV in the agrarian sky. They are not difficult to manufacture. They are relatively low-cost type of UAVs. Flying a multirotor UAV does not require exceptional skill unlike the other types of UAVs. Multirotor UAVs though cheap and easy to manufacture have a few drawbacks. Such lacunae include limited flying time, endurance and speed. They can only sustain an average flying time of between 20 and 30 min. This is because a large percentage of their energy is expended in fighting gravity and wind, to remain stable in the air.

Let us consider a few examples wherein multicopters have been utilized, to study the agricultural crops by obtaining spectral data. At this juncture, we should note that, multicopters are versatile regarding flight path, maneuvering, and providing vantage spots in the aerospace for aerial photography. *Such vantage points were never possible for farmers in the previous centuries.* Tall perches or mountainous locations never provided good vantage locations directly above the crops, for close or long-distance photography.

The multicopter drones with a full complement of sensors such as visual, high resolution multispectral, infrared, red-edge, and LIDAR are among aerial vehicles utilized, by grapevine farmers. In California, grapevines are being assessed for NDVI, leaf chlorophyll, diseases, and pest, if any, using low-flying copter drones. They transit through the agrarian sky, at a rapid rate of 10–50 ha a day (Sky Squirrel, 2017). Similarly, in Germany, France, and Italy, grape vines are being monitored and assessed periodically, using multicopters. Mainly, these farmers are using the drone's imagery to know about crop vigor, growth rate, and diseases. In Australia, grape vine pests are identified at an early stage, by using aerial imagery obtained through multicopters. In addition to above-mentioned advantages, Gago et al. (2015a, b, and c) suggest that crop's water stress index too could be known by flying the multicopters above the grapevines. Similarly, it is clear that UAVs, particularly, multicopter are destined to be *new aerial factors* introduced above the plantations, like citrus (CREC, 2015; Sankaran and Ehsani, 2012; Sankaran et al., 2010, 2015; She et al., 2017; Barbedo, 2019). Here, we should note that, this multicopter could be flown at higher altitudes when farmers or farm agencies require imagery and spectral data of large patches of plantation. Unlike, the satellite imagery, these could be flown at just 100–100 m above the plantation. So that, images of better resolution could be obtained. At low altitudes, even a single plant could be assessed, in detail. Such versatile imagery is not possible with other aerial vehicles. Of course, multicopters with payload tank, spray bars, and variable-rate nozzles could be a good idea, if one wants to use agricultural sky above plantations, more efferently. They could use them to spray plant protection chemicals or even foliar nutrients like N, Zn, and other micronutrients.

Single-rotor helicopter sprayer UAVs and multicopter sprayer UAVs are useful while applying herbicides, to eradicate weeds. The spectral data and maps showing weed infestation in the inter-row area are introduced into computer processors of the drone. The sprayers provided with nozzles operate based on directions from commuter processor. Hence, the precision spraying conducted will be efficient, since only patches that need weedicide will only be sprayed (Cornett, 2014; RMAX, 2015; Huang et al.,2009; HSE, 2015b).

4.2.10 SMALL UAVS FOR SPECIAL TASKS IN THE ATMOSPHERE ABOVE CROPS

We may note that UAVs (e.g., Scentroid DR300 Sampling Quadcopter) with ability for versatile flight and hovering ability could become common in the aerospace above farms. They are suitable to sample atmospheric gases above cropland. Agrarian regions are known to emit GHG and particulate matter, at a higher rate. This happens when crop burning is conducted. Stubble burning may cause pollution even in the neighboring regions, located away from fields that were burnt. Atmospheric drift spreads pollutants at a rapid pace. Farmers and environmental agencies need accurate information about pollutants and gaseous composition, at different altitudes above the crops. There are special types of UAVs fitted with accessories that aid in accurate sampling of air, at various altitudes. The gaseous samples could be assessed in situ or at the GCS. According to MacDonell et al. (2013), we can assess gaseous components of agrarian sky, such as carbon dioxide (CO_2), carbon monoxide (CO), methane (CH_4), ammonia (NH_3), nitrogen dioxide (NO_2), nitrous oxide (N_2O), nitric oxide (NOx), sulfur dioxide (SO_2), oxygen (O_2), Volatile Organic Compounds, and particulate matter, using appropriate chemical probes placed on the UAVs. These sensors offer data about GHG with acceptable resolution and range. Analysis of atmosphere above farmland may be required frequently during certain periods of the year. Also, when certain agronomic procedures such as pesticide application, N application as ammonia gas, crop burning, etc. are conducted. It seems chemical constituents could be sampled from altitudes up to 125 m above crop's canopy. Then, analyzed, using such versatile processors.

4.3 PARAFOILS, BLIMPS, AEROSTATS, AND KITES IN THE AGRARIAN SKY

In this section, we deal with a set of aerial vehicles that could be piloted or remote controlled. They could also be entirely autonomous while they transit in the agrarian sky. They are different from the several types of small UAV aircrafts discussed in the previous sections. The autonomous parafoils, blimps, aerostats, and kites too have found utility in the agrarian sky. They seem to be most efficient in certain regions. They are apt for accomplishing specific tasks in the farmland. Autonomous parafoils that fly low over the cropped fields are gaining in popularity, in European farms (Thamm, 2011; Thamm and Judex, 2006; Thamm et al., 2013; Pudelko et al., 2012). Blimps and aerostats have a role in the agricultural sky. They are suited for aerial imagery, and in obtaining digital data. They are good for continuous surveillance of farm installations and monitoring research material (e.g., crop genotypes) in experimental farms. Free-floating blimps may carry payload and help in transport of agricultural goods. Kites may find utility in regions where farmers cannot afford highly sophisticated and costly aerial vehicles. Kites are cheaper than small fixed-winged or copter drones with set of sensors and computer programs for fixing flightpath, way points, and processing the images.

4.3.1 PARAFOILS IN THE FARM SKY

Parafoils are recent introductions into the agrarian sky. *Parafoils could become common in the farm sky because of several advantages that they offer to farmers*. They allow farmers to obtain greater details through spectral data. Autonomous parafoils tend to gain popularity with farmers adopting precision farming (Thamm et al., 2013; Pudelko et al., 2012; Fig. 4.1). Aerial photography (visual band width) and digital data about crops are important, while practicing precision farming methods. Parafoils may be utilized above the large expanses of agrarian regions. As such, parafoils were developed in 1970s by Domina Jalbert, for use in recreation. They were adopted in the military as aerial transport vehicles. Also, to drop (land) and to land paratroopers. Paratroopers were provided with extra advantage of maneuverability, gliding, descending, and accurate landing. This was possible because parafoils could be guided by the paratrooper. An autonomous parafoil is fitted with small engine. It is powered by fuel or lithium batteries. It carries a set of sensors and appropriate computer software.

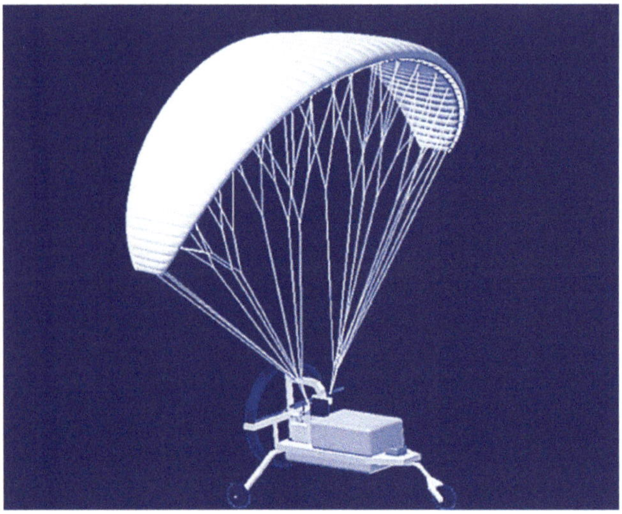

FIGURE 4.1 Autonomous Parafoil SUSUI 62.
Source: Thamm, H.P., Geo-Techniques GMBH, Linz Am Rhine, Germany.
Note: SUSI 62 is a versatile parafoil. It is autonomous. It transits in the air utilizing pre-programmed flight path and way points. The aerial photography and spectral data are captured from sensors such as multispectral, infra-red, and lidar. This autonomous parafoil (SUSI 62) transits at low altitude above crops. So, it offers high-resolution clear images of crops. It has relatively short endurance of 30–45 min, depending on the engine and fuel tank. It does not crash since parafoils float once the fuel is exhausted, or engine fails in the mid-air. Perhaps, in due course, they could be modified to carry a payload of pesticide tank and sprayer bar. Such modification will make them useful in spraying farm chemicals. Autonomous parafoils seem to have clear edge over other aerial robots in the farm sky. Its acceptability depends on pilot demonstration, propaganda, and usage.

In addition to autonomous parafoils, we should realize that there are other aerial vehicles. They are equally useful and equally efficient in terms of their performance in the farm sky. For example, piloted microlights (Triches), piloted parafoils, semiautonomous, and entirely autonomous blimps/free-floating aerostats. Microlights have been utilized to monitor the field crops and plantations. They have provided clear aerial imagery of vegetation, farms, and cropping expanses.

4.3.2 BLIMPS, AEROSTATS, BALLOONS, HELIKITES, AND KITES IN THE AGRICULTURAL SKY

We may realize that, in addition to small UAVs, there are a few other useful aerial vehicles. They can be of great help to farmers thriving in different agrarian regions. Several of them are efficient in terms of manufacture, use, and in obtaining useful spectral data about crops to farmers. Clearly, there is a competition among several types of aerial vehicles, that are entirely autonomous or semiautonomous types. Each region may have its influence on the type of aerial vehicle preferred by farmers, in their sky. Right now, small UAVs seem to have attracted greater attention, imagination, and investment from farmers/companies. Soon, aerial vehicles may fly in the agrarian sky, more intensely.

There are several models of blimps, aerostats and even balloons that have found utility in farming zones (see Krishna, 2020a, b). The blimps produced by different aviation industries occupy the agricultural sky, Blimps fitted with sensors provide excellent aerial images, spectral data and help in transit of payload. Certain models are also utilized in the aerospace above farms for transport. Aerostats are small, tethered, LTA stationary platforms. They can support different kinds of payloads such as multispectral sensors. Sensors provide data for use in the precision agriculture (Primicerio et al., 2012; Tetracam Inc., 2015; Institute for Engineers-India, 2016). Aerostats, particularly, tethered aerostats have been evaluated above farms/fields. They are assessed for their efficiency in providing data about crops, experimental stations, and movement of agricultural vehicles. They are often lofted for long stretch of time. Tethered aerostats with sensors placed at 100–330 m offer high-resolution images, on a continuous basis. They are apt aerial vehicles to keep a vigil of agricultural experimental farms that evaluate plant breeders' material. They are well suited to monitor crop genetic stock and assess them- all through the season. Blimps and aerostats filled with LTA gas are efficient in terms of economics. Blimps and aerostats are utilized in military for aerial surveys, surveillance, and aerial photography. However, we are still in the early stages of their adoption in agricultural fields. Among regions, North America, Europe, and China are major utilizers of blimps/ aerostats in the sky. There are forecasts that almost each farm may exhibit an aerostat (tethered) or small blimps, in future. This is to obtain spectral data at a rapid rate. Networks of blimps/aerostats are a clear possibility. Particularly, when tracking pandemic disease of crops such as wheat, maize, sugarcane, etc.

At present, aerostat usage is confined to the military, border surveillance, and advertisements industry. It's spread into agricultural domain is feeble, if any. However, it is supposed to increase exponentially when farmers start using them for obtaining aerial imagery of their crops, on a continuous basis. Tethered blimp is said to be preferred in farms. Tethered helium-filled small aerostats, (e.g., helikite), that is, a combination of tethered aerostat and kite are said to be popular with farmers in Europe and in North America (Allsopp Helikites Ltd., 2017; Cousins, 2018; see Krishna,2020b; see Plate 4.1). It acts as bird scaring and surveillance device in the agrarian sky. Bird scaring aerostats/helikites have been used against several bird species. For example, they scare away crows, rooks, wood pigeons, seagulls, large birds, small birds, starlings, pheasants, finches, etc. They say, a single helikite can be effective in an area of 10–25 ac of cereals, oil seed brassicas, or peas. They are effective on 3–5 ac of grapes, bush crops, strawberries, etc. (see Krishna, 2020b; Bajoria et al., 2018).

Perhaps, production and marketing figures for aerostats would change as farmers realize the utility of helikites. They would prefer them in greater number. So, in the near future, aerostats may have field day in the agrarian sky. According to marketing research agencies, United States of America with 16.4% of global share of sales of aerostats, is said to be top user. European nations such as Germany, France, and Britain, also China and Japan adopt aerostats, in their skies. Globally, the aerostat systems market is forecasted to grow, at 15.4% each year and reach 12 billion US$. Agricultural applications of aerostats may hold the key for their production, dissemination and market size, in future (Laura, 2020).

In North America and Europe, tethered aerostat, elliptical blimps, and balloons networks have been successfully evaluated. They have served in tactical surveillance, monitoring borders, transport vehicles, natural vegetation, and even cropping expanses. Tethered aerostats, in particular, are being utilized in Brazil for telecommunication. Also, to spread mobile services into rural and remote farms. These advantages from aerial vehicles are being replicated in Brazil, by companies such as ALTAVE (Brazil) and Airstar Aerospatiale (France) (Altave and Airstar Aerospace, 2017).

In Australia, tethered aerostats lofted to different altitudes have served as mobile phone towers. Particularly, in the rural areas and in remote agricultural locations (BAL, 2016; UAS Vision, 2012; Glass, 2018). We may note that, most of the farm vehicles are currently started and guided, using computer programs available on mobile phones. Regulation of farm vehicles and their turn out needs use of GPS-connected mobiles. In this regard, tethered aerostat placed on mobile vans or fastened to ground locations are useful. The agricultural sky could also be used efficiently in making "aerostats towers." This is to disseminate agricultural information to farms. This way, aerostats in the agrarian sky will induce a revolution in the use of mobile-controlled ground vehicles/robots.

PLATE 4.1 A helikite used as mobile 4G tower to transmit signals to farm community in a European rural location.
Source: Allsopp Helikites Ltd. Fording Bridge, Hampshire, Great Britain.
Note: Helikites could be utilized for continuous aerial photography of farms. They are also used for bird scaring.

Kites with gimbals containing the usual visual, multispectral, and infrared sensors have been tried as useful platforms. For example, in the drylands of Sahel, agricultural researchers have utilized kites with sensors to study the millet and cowpea. The spectral data showing spread of viral disease has been mapped carefully and utilized while taking precautionary measures in the field (Gerard, 2016). Kites could be useful in regions where obtaining the services of electronically sophisticated aerial vehicles is difficult (Krishna, 2020b).

Regarding adoption of several types of autonomous aerial robots and those guided/piloted, Krishna (2020a, 2020b) states that, a kind of competition among the type and model that dominates the agrarian sky has already ensued. The model that allows us to utilize agricultural sky efficiently for diverse functions and most helpful in farming will win the race. There could be some models suitable for specific function based on location, economics, and ease of operation. A few such as tethered aerostats may stay always in the sky with least requirement of technical attention. They may get opted above experimental stations that need day/night surveillance and continuous data collection. A few farms that are large may opt for small fixed-winged drones. This is because, small fixed-winged UAVs transit at high speed and conduct aerial imagery, rapidly. No doubt, there is a race among aerial robots to dominate the agrarian sky. A few models have already become more popular than others. Therefore, aerospace above crops is set for marked changes. *Perhaps, a revolution in farming is imminent through better utilization of agricultural sky.* Of course, it could be mediated by drone aircrafts, aerostats, blimps, helikites, kites, etc. A tethered aerostat/or a helikite like other aerial vehicles needs to be watched. They may rule the agrarian sky, in future. Some of the features associated with it, such as low cost, low technology, efficiency in terms of economics, long endurance of few days to months, continuous surveillance, and multiple uses like wind power generation, aerial surveillance, bird scaring, mobile towers, etc. make them most useful in the agrarian sky.

As such, autonomous ground vehicles appearing on the agricultural farms could induce a drastic change, in the way we conduct farm operations. Added to it, aerial vehicles such as tethered aerostats/helikites and small drone aircrafts could offer the needed GPS guidance, mobile connectivity and digital data for autonomous function of all of the ground vehicles in the farm (Glass, 2018). *So, like it or not, farms are getting autonomous (or robotized). A robotized farm is a boon to all, such as farmers, farm technicians, farm laborers, investors, and agricultural marketing agencies. Robotics definitely leads to reduction in human drudgery. Robotics offer greater agronomic and economic efficiency. Perhaps, a concerted and coordinated function of aerial and ground robots is the requirement of the hour.*

4.4 SATELLITES IN THE AGRICULTURAL SKY

Satellites are among the earliest of the several types of aerial vehicles adopted in the agrarian regions. They were adopted during the past century. Satellites have played a great role in the evolution and modernization of agrarian regions. It refers, particularly, to gathering data about weather patterns in the agrarian sky, aerial imagery of large acreages of farm belts, surveillance of farms/crops, monitoring climatic disasters, if any and helping farmers with requisite information prior to adoption of various agronomic operation in the fields.

During past five to six decades, satellites have also played an important role in offering aerial images of natural resources relevant to farming sector. Such aerial imagery has generally pertained to natural vegetation, its biodiversity, gross maps of natural vegetation, water resources, soil types, cropping expanses, cropping systems followed, etc. The satellites have offered excellent data regarding the agroclimatic conditions prevailing, over farming belts. Advanced information about tornadoes, hurricanes, heavy rainfalls, dust storms, droughts, soil erosion even disease and insect attacks (e.g., locusts, armyworms, helicoverpa), etc. have been most useful to global agriculture. For example, in the agrarian regions of Great Plains of USA, they say, annually at least 20,000 storms of different intensities occur. Such storms vary in space and time regarding occurrence. Most of the storms occur during peak crop season. The damage to crops it seems reaches around US$ 2 billion annually. Such disastrous effects mediated via agrarian sky could be reduced, by using satellites such as Landsat system. Satellites provide early warning to farmers. This helps in securing the farm property, installations, and even protecting crops in open fields. Also, in reducing soil loss due to erosion. The Landsat data has also been utilized to assess the crop damage, using the satellite imagery. Insurance companies could use the aerial imagery of damage, obtained from Landsat, to pay the compensation (Bentley et al., 2002; NASA, 1999; Schaefer and Brooks, 2000). Satellite imagery is useful in detecting the ozone and particulate matter pollution and its effects on the crops. Large tracts of crops could be assessed, using satellite image for crop loss due to ozone/particulate matter pollution. The agricultural sky, in USA alone, it seems mediates a loss of 3.1 billion US$ annually via ozone plus particulate matter pollution. It is expected to increase to US$ 8.3 billion, by 2030. Satellites in the sky could help in reducing the impact to a certain extent (NASA, 2020).

No doubt, satellites are now an important part of agricultural sky. During recent years, satellites stationed in the space have actually aided the rapid spread of precision agriculture. *Soon, satellites are supposed to bolster adoption of robotics and make them invade farm world.* That means, satellites would reduce the use of farm human labor. Satellites will make food production easier. Let us consider a few examples. Satellites that transit in the space way away at high altitude of >300 km have a few definite roles attributable to them, within the realm of present agriculture. Recent concepts such as precision farming, use of GPS-guided semiautonomous or entirely robotic vehicles, during the conduct of several different agronomic procedures depends on satellite guidance. Satellite guidance of farm vehicles is among the most important aspect of "Push Button Agriculture." A concept that allows for totally autonomous conduct of all agronomic procedures (Krishna, 2018).

Satellites are placed several kilometers away from the crop fields and canopy. Hence, sensors placed on them do not offer high-resolution close-up images. Instead drones that fly close to crop's canopy has to be employed. Drones provide high-resolution imagery of crop plants/soil features. The revisit time for satellites is long. It could be 15 days. During this period, certain changes such as insect attack or floods, crop loss due to highspeed winds may have occurred without being noticed by the said satellite. However, satellites are good bet when studying large areas of farming. Policy decisions that need prior knowledge about topography, soil types, water resources, crop suitable, and their status during the season could be studied better, using satellites (e.g., NASA's LANDSAT 8 system, IRS, mention US soil water satellite name). Satellites have large swath of 200 km. The best resolutions possible are about 20 cm (e.g., SPOT satellites of France).

During recent years, several agronomic procedures beginning from land selection based on topographic features, soils, and water resources are decided, based on satellite imagery. Monitoring large expanses of crops and individual large field, monitoring insects/diseases, harvest procedures have all required GPS/GLONASS connectivity aided by satellites. More recent developments such as soil sampling, soil fertility mapping, autonomous seed planters, estimation of plant density, identification of gaps within crop rows, application of in-season fertilizers using robotic ground vehicles, combine harvesting using autonomous GPS-guided vehicles, and preparing yield maps have required satellite guidance. Such satellites are stationed above the agrarian regions. Knowledge about variations in soil fertility, nutrient availability, and grain productivity are vital for adopting precision farming methods, in large farms. A few examples are autonomous GPS connected planters manufactured by Kinze Autonomous Systems, Iowa, USA, New Holland, John Deer, weeders (Vitirover, 2014; Klose et al.,2012), in-season ground and aerial sprayer drones (RMAX, AGRAS MG-5B, Hercules), fertilizer inoculators (Deutz Fahr-Agrosky, 2020; see Plate 4.2), Combine Harvesters (John Deer, Massey Ferguson, and Chase Inc.). In addition, we should note that the recent trend in North American farms is to adopt a complete set of satellite-guided autonomous (robotic) vehicles for harvesting, processing, transport, and filling-up the grain storage bins. For example, Kinze Autonomous Systems Inc. Iowa, USA offers autonomous harvesting and grain transport system. It removes physical drudgery by farm technicians and laborers during harvest, processing, and storing of grains. (See Krishna, 2018; Kinze Manufacturers Inc., 2013, 2014; McMahon, 2012; Garber, 2014). Agrarian revolution based on robotics, i.e., autonomous ground and/or aerial vehicles will immensely depend on satellite in the space! (see Plate 4.2).

At present, satellites are no doubt part of farm world. Satellites with ability for offering useful geospatial information, GPS guidance, and forecasting atmospheric conditions are perhaps indispensable to global agriculture. Satellites and their guidance are almost mandatory for robotic vehicles in the farms. Several of the autonomous farm vehicles and even machines are interlinked, using GPS connectivity/satellite signals (Garber, 2014).

4.5 AERIAL SPRAYS ABOVE CROPPED FIELDS AFFECT AGRICULTURAL SKY

4.5.1 AERIAL SPRAY OF IRRIGATION WATER AND OTHER FLUIDS USING GROUND VEHICLES AFFECTS THE QUALITY OF AGRICULTURAL SKY

Archaeological evidence suggests that irrigation, that is, supplemental supply of water to crops as an agricultural practice is believed to have begun around 6000 BC in both Egypt and Jordan (Hillel, 1994). At present farms adopt a few different types of irrigation. There are three types of irrigation systems, namely surface flood irrigation, sprinkler irrigation (aerial), and drip irrigation. The second type of irrigation is sprinkler irrigation. It is a method of applying water through a system of pipes and spraying it into the air (atmosphere) through sprinklers. This system mimics rainfall (Brouwer et al., 1990).

PLATE 4.2 Satellite-guided farm vehicles in a European farm.

Top left: Electronic Instrumentation in the cabin; Top right: A GPS-guided tractor discing and crushing clods in the soil;

Middle left: An autonomous GPS-guided tractor conducting interculture, using tined hoe. Middle right: Tractor conducting leveling and marking.

Bottom: Semiautonomous/autonomous ground vehicles preparing land (ridging) in a farm.

Note: Satellite guidance from agricultural sky is an important aspect of such semiautonomous or totally autonomous (robotic) farm vehicles. Agronomic operations such as land preparation, soil tillage, planting, interculture, in-season fertilizer application, harvesting, and grain storage depend on satellite guidance (GPS/GLONASS). Such satellite guidance is based on accurate geographical coordinates. In this case, satellites in the agricultural sky offer greater accuracy and electronic sophistication to farm operations.

Also see: https://www.youtube.com/watch?v=nj_EYZeSkhM; and Jay bridge Robotics/Kinze Grain Cart system: https://www.youtube.com/watch?v=k0Lj_5MBu8w/ (October 28th, 2020).

Source: Kinze Autonomous Systems, Iowa, USA; Landoll Inc., Kansas, USA; Deutr Fahr-Agrosky, Germany;

First "Self-Propelled Sprinkling Irrigating Apparatus" was invented in 1948 by Frank Zybach (Evans, 1998; Christiansen, 1942; Allen et al., 1998). The early designs were improved to develop the current self-propelled center-pivot and linear move irrigation systems. They have been sought worldwide. High irrigation efficiency, uniformity of application of fluids, large area coverage, low labor needs, and ability to supply fertilizers/chemicals are the major reasons, for their acceptability with farmers in USA, Europe, and other nations. About 29% of crops in USA are irrigated using center-pivot self-propelled machines (Evans, 1998). Crops such as sugar cane, maize, small grain, millets, potatoes, vegetable crops, vines, and orchard fruit crops are amenable for irrigation with center-pivot water application systems (Evans, 1998; Waller and Yitayew, 2019).

The two most commonly used sprinkler systems are center-pivot and linear systems. A center-pivot irrigation system is a movable pipe structure that rotates around a central pivot point connected to a water supply. Center-pivot irrigation systems are the most popular sprinkler irrigation systems in the world because of their high efficiency, high uniformity, ability to irrigate uneven terrain, low capital requirement, low maintenance, and management costs. The history of center-pivot irrigation systems began in Nebraska, in the 1950s. There are now hundreds of thousands of center-pivot irrigation systems in the world. Center-pivots are "perhaps the most significant mechanical innovation in agriculture since the replacement of draft animals by the tractor." The systems move through the field by electrically powered tractor wheels. Sprinkler flow rates increase toward the outer end of the pivot because the end of the pivot travels faster.

Aerial irrigation involves use of space above the crop's canopy. This is true with irrigation method that involve sprinklers. Sprinklers are usually placed on ground vehicles, like tractors, center-pivot travellers, etc. They do affect the agricultural sky above the soil surface to different extents in terms of ambient temperature, canopy temperature, relative humidity, and chemical contents, if fertigation or pesticide application is practiced. During recent years, center-pivot irrigation has gained popularity with most agricultural farms because of a few reasons. Center-pivots that cover large areas in shorter span of time were first developed in 1950s, in Nebraska. They were preferred because of low labor needs, low maintenance cost and greater irrigation efficiency of applicators (nozzles). It is a convenient system for large farms. It is flexible and works at a stretch for long duration (New and Fipps, 2019).

Let us consider a few aspects about fertigation, using center-pivot or agricultural UAVs with spray bars. The application of fertilizers with irrigation water, or fertigation, is often referred to as "spoon-feeding" the crop. Fertigation is common and has many benefits. Most fertigation uses soluble or liquid formulations of nitrogen, phosphorus, potassium, magnesium, calcium, sulfur, and boron. Nitrogen is most commonly applied because crops need large amounts of it. Nitrogen is highly soluble and has the potential to leach; it needs to be carefully managed. There are a few more characteristics attributable to aerial sprays, particularly fertigation. Fertilizer-based nutrients can be channeled at any time during the crop season. Nutrient supply is often uniform, but it depends on the uniformity attained by the sprinklers. If the chemicals to be applied are compatible, then, fertilizer, pesticides, and herbicide combinations can be supplied simultaneously. Certain chemicals like liquid ammonia cannot be sprayed because it is highly volatile. It leads to excessive loss of ammonia. There are also corrosive chemicals that should be avoided, during aerial sprays using center-pivot systems. They say, fertigation using aerial sprinklers avoids severe percolation and groundwater contamination. Aerial sprayers may not damage standing crops.

Let us discuss some comparative aspects of different types of irrigation. Flood irrigation, sheet irrigation, furrow irrigation, or spot applications are among the most commonly practiced irrigation systems. They are in use since ancient times. They have offered certain key advantages in terms of crop's survival and productivity. These methods may not affect the agricultural sky in any way, perceptibly. Evapotranspiration process occurs to affect the ambient relative humidity and temperature. Often it is a mere transitory effect persistent for short period (FAO, 2002). Here, in this chapter, we are concerned more with facts, relative advantages, and consequences to ambient air (i.e., agricultural sky).

Farmers find a few good reasons to adopt aerial sprinkler irrigation, using linear or center-pivot systems. Foremost, center-pivot sprinklers are efficient since they require relatively lower quantities of irrigation water to achieve the same effect (soil wetting). Particularly, if we compare it to other systems such as the furrow or sheet /flooding systems. Flooding does not soak the fields uniformly, Hence, over-watering is done at the ends of the field. Laser leveling is to be adopted to get soil profile to same moisture saturation levels. It could be costly. Center-pivots can be adopted even on undulated fields (Waller and Yitayew, 2019).

Center-pivot sprinklers need 30% less irrigation water than furrow or sheet irrigation. Flood irrigation affects soil nutrients and pre-emergent plant protection chemical applied to soil. Soil nitrogen percolates to lower horizons. Soil-N distribution could be affected leading to leaching or accumulation at various levels in the soil profile if flooding is adopted. Sprinklers offer greater uniformity if fertigation is adopted. However, farm chemicals applied through center-pivot does affect the aerospace and atmosphere above the crop's canopy. Again, it could be transitory. Reports suggest that crop productivity is higher if aerial sprinklers are adopted. In addition, aerial sprinklers consume less water, less nutrient formulations if fertigation or chemigation is adopted. Labor needs and cost on energy to pump water are lower if sprinklers are adopted. *In toto* it is clear that aerial sprinklers offer greater advantages, but they could pollute the agricultural sky, if chemigation/fertigation is adopted. Crops also affect ambient atmosphere. We have to also understand the effect of aerial sprays on biotic factors such as microbes, insects, birds and even dust/particulate matter in the region just above the crop's canopy.

Aerial application of irrigation water and dissolved chemicals is efficient compared to ground application. Soil application of irrigation water using sheet, furrow drip irrigation involves loss of moisture to soil layers. While aerial spray reaches leaf stomata, lenticels, hydathodes. Water is imbibed directly into plant tissue.

Adoption of center-pivot sprinkler irrigation induces certain types of loss and decrease in irrigation efficiency. For example, Rogers et al. (1997, 2017) state that, reduction in irrigation efficiency could be severe. Types of losses encountered are air loss (drift, droplet evaporation), canopy loss (canopy evaporation, foliage interception), surface loss (surface water evaporation, surface runoff, soil evaporation), and deep loss (percolation). These processes related to center-pivot method of irrigation could affect agricultural sky and its normal manifestations.

Adoption of irrigation apparatus to channel fertilizer in dissolved state is now getting common in large farms of developed world. This procedure is called fertigation. Fertigation could also be achieved using underground drip irrigation systems. This aspect is out of context since in this volume we are concerned with aerospace above the canopy. When irrigation systems are adopted to channel not only fertilizers, but also the fungicides, pesticides, herbicides, the general term is "Chemigation." Chemigation is a very efficient system if used shrewdly and accurately, during crop production. Chemigation basically involves water irrigation system involving sprinkler mounted on center-pivot vehicles or other linear sprayers. Adoption of chemigation can be detrimental to the environment just above the crops. The agricultural soil/

water may show transitory increase in the harmful chemicals. Also, when chemicals get into air along with atomized water droplets. The mist, along with dissolved or miscible harmful pesticides, herbicides, or even fertilizer can affect the agricultural sky.

Chemigation involves use of irrigation water supply system along with dissolved chemicals such as fertilizers, herbicides, pesticides, etc. Such a spray can affect agricultural sky above the crops, detrimentally. The extent of deterioration of aerospace above crops depends on nature of the chemicals, spray quantity, intensity, and duration. Usually, fertilizer/plant protection chemicals marked as amenable for aerial spray are only used. Precautions to be followed too are mentioned on the product.

Such use of chemigation may affect several other aspects of agricultural sky. For example, it may affect the biotic components such as microbes in the atmosphere above the crops. Usually, labels on the plant protection chemical show up weather the contents could be adopted for fertigation or aerial spray via irrigation sprinklers. In each country, environmental protection agencies have certain conditions and regulations to follow before aerial chemigation is practiced. This is to protect the atmosphere above the cropping belt. Water supply systems and air should not get contaminated with harmful plant protection chemicals. Hence, farmers are asked to adopt highly precise and accurate chemicals, their concentrations, (active ingredient), nozzles, and spray control during chemigation. Chemigation involving arial sprays is definitely a procedure that needs greater accuracy and precaution. This is because it has direct effects on agricultural sky, its biotic factors and even soil and its profile when the chemicals start percolating into soil.

There are certain advantages of using chemigation. They are, chemigation offers greater uniformity of application of plant protection chemicals. Chemicals can be applied at precise quantities, using precision methods (i.e., variable-rate nozzles). Chemigation is economically efficient, since lower quantities of chemicals are required. We can apply chemicals at right times. It avoids soil compaction that otherwise occurs when heavy tractors are adopted. Operator safety is possible. It is excellent when aerial copters are used to spray plant protection chemicals in dissolved state (see Krishna, 2017, 2018, 2020a, b).

There are a few disadvantages attached with chemigation using aerial sprinklers. Farm technicians with excellent training are needed to handle autonomous center-pivot systems or drone (copters with spray bar); knowledge about irrigation and chemigation is needed; knowledge about fertilizers, both major nutrients and microelements is necessary if fertigation is adopted;

farmers should be highly conversant with plant protection chemicals and their dispersal in the atmosphere; Injection/sprinklers should be accurate, and without malfunctions. Let us quote an example. Aerospace above vast stretches of pastures could be utilized efficiently to irrigate them in times of drought/stress. Sprinkler methods that adopt foliar feeding are indeed utilized, to supply water plus chemicals. For example, in the arid regions of West Asia, it seems pastures are amenable for aerial sprays of water and proline. This is in order to overcome water and salinity stress to pastures. It is less costly and easier to handle (Bowman, 2015; Kandil and Shareif, 2016; Siddique et al., 2015; Talat et al., 2013).

The misty atmosphere created by sprinkler's droplets or atomized chemicals affects interception of sunlight. Radiation does not reach the leaf surface directly. Leaves that receive the sprays may hold droplets of water or fluid chemicals for a while until drying. These features reduce photosynthesis. These effects are definitely transitory and occur for a negligible length of period if the whole crop is considered. Mists and droplets above leaf surface may also affect gas exchange functions of stomata. We need to assess this aspect before adopting aerial sprays using linear or center-pivot aerial sprinklers. We should also consider the physiological and physical limits for foliar irrigation and chemigation sprays. Each individual chemical may have specific effects on physiological activities in leaves or entire plant. The above points are related to adoption of agricultural sky, to spray irrigation water/chemicals. Such aerial spray is done instead of practicing soil injection or band application during interculture. Yet, forecasts suggest fertigation/chemigation, using center-pivot or sprayer drone could revolutionize farming.

4.5.2 PILOTED AIRCRAFT CAMPAIGNS TO CONDUCT AERIAL SPRAY OF IRRIGATION WATER AND OTHER FLUIDS

Piloted aircrafts have been adopted in agrarian sky to make aerial observations of crops. Airplanes have helped in identifying and locating spots with destructive effects of floods, tornadoes, rain, pests, and diseases. Aircraft surveillance is costly. Aircraft campaigns have been adopted for the past five to six decades by geologists, natural resource mappers, and agriculturalists.

Airplane campaigns are feasible only when farms are large. Such campaigns have been utilized, in different geographic regions to study geographic features, terrain, natural vegetation, and water resources.

Agrarian regions too were studied in greater detail using aerial imagery from airplane campaigns. The quality of aerial imagery depends on the altitude and speed adopted during the flight.

Piloted aircrafts were being used in some farms since many years. However, airplane campaigns are costly. Fuel costs could make it not so profitable. Plus, farmers have to hire well trained pilot capable of low-altitude flying and uniform distribution of water/chemicals. Repairs are costly too. Disasters are possible while flying at low altitude in airplanes capable of spraying agricultural chemicals.

4.5.3 FOLIAR SPRAY OF IRRIGATION WATER AND FERTIGATION/CHEMIGATION, USING ROBOTIC/SEMIAUTONOMOUS UAVS (SPRAYER DRONES): A RECENT TREND IN FARMING

Historically, the "Sprayer UAVs" are a very recent introduction into the agrarian regions. Sprayer helicopters were first utilized in countries such as China, Japan, and South Korea to distribute pesticides, to rice crop. Rice crop received two to three sprays of pesticide, using such drones. Early models were heavy helicopter models such as RMAX and its predecessor. Later, during 2005–2015, there was a trend in the Fareast farming regions, to convert military UAVs into farm drones (Krishna, 2017; Zhu, 2014; Zhang, 2016a, b, Chen et al., 2016; Banjo and Ajayi, 2020). These sprayer UAVs are extremely efficient in covering larger areas of crops, at a stretch. They can also keep the human farm labor and farmers away from pesticide/fungicide spraying zones. So, they avoid exposure of farmers to harmful chemicals. Yet, they induce environmental pollution as they spray the harmful chemicals high up in the agricultural sky. They can potentially contaminate sky above the cropping expanses. When used in large number, that is, simultaneously as swarms, they can be a serious detriment for biotic aspects of aerospace above the crops. The area contaminated is proportional to the height from which the crops are sprayed with agricultural chemicals.

During recent years say past 5–8 years, UAV companies, agricultural researchers, and farmers alike have begun using the agricultural sky efficiently, for spraying fluid/granular formulations. An agricultural sprayer UAV (drone), no doubt, is a useful aerial vehicle in the hands of farmers. It is safer for farmers to use aerial sprayers. A robotic sprayer drone does the spraying rapidly. It completes the agronomic processes like spraying pesticides, liquid fertilizer, herbicides, or fungicides in comparatively shorter duration. Particularly, if compared with hand-held sprayers operated by human farm laborers or even tractor-mounted spray equipment. Sprayer drones allow farmers to be away standing on a bund in the farm, while it sprays the pesticides efficiently. A robotic UAV utilizes an accurate flight path program (e.g., e*Motion*, Atlas Flight Plan, Pixhawk Autopilot, C-Astral Pilot 3C) (SenseFly Inc., 2017; Krishna, 2020a). It sprays fluids based on spectral (digital) data collected about areas infested with pests/fungal diseases. A sprayer drone's accuracy and efficiency in the aerospace increases, if precision techniques are adopted (see Krishna, 2018, 2020a; Mogili and Deepak, 2018). An autonomous sprayer UAV could be energized, using electric batteries or petroleum fuel. In case, it is powered using petrol, then, it leaves polluting carbonaceous material through the exhaust. Sprayer UAVs are not utilized too frequently in the atmosphere above crops. Some three to four sprays of pesticides are conducted in a season of 3–6 months. However, pollution of agricultural sky due to sprayer UAV can be severe, due to pesticides/herbicide or copper fungicides sprayed into atmosphere. This happens whether sprayed, using farm laborers, ground vehicles, or using UAVs. The agricultural sky, particularly the area polluted by sprayer UAV could be controlled. We could lessen the space between crop's canopy or soil surface and UAV's spray bar, to least. The space between the sprayer drone and crop gets polluted in any case. There are instances when single rotor helicopters have been flown high in the agricultural sky at 20–50 m above crop's canopy. This is to spray fertilizer formulations or organophosphorus pesticides. These are deleterious to biotic aspects of atmosphere, like microbes, insects (pollinators, bees, etc.), also aves many of which are nontargeted. Pesticide drift due to a high-flying sprayer UAV needs careful attention. It is best to fly the helicopter/multicopters at lowest altitude above the soil/crops. There are suggestions to adopt UAV swarms to complete a spraying task rapidly. In that case, we will be releasing large quantities of harmful agricultural chemicals at a rapid are into agricultural sky. We need more information about its consequences on biotic aspects of atmosphere.

Regarding agricultural sprayer UAVs, they say, aspects such as endurance, payload tank, and its size, spray width (swath), altitude of the drone, speed in the aerospace above crops, computer programs used to fix flight path are important. This is in addition to other usual specifications quoted by the manufacturers and mentioned in their manuals. There are indeed innumerable UAV manufacturing companies that have been commissioned all over the world. Sprayer UAVs are popular in North America (e.g., HSE's Hercules 50, Autocopter Corporation's Autocopter, DJI's Agras MJ-1, etc.).

TABLE 4.1 Examples of Agricultural Sprayer Unmanned Aerial Vehicles (UAVs) and Their Specifications Relevant to Their Ability to Apply Water or Agricultural Chemicals and to Disturb the Atmosphere Above the Crop's Canopy and Ambient Atmosphere.

UAV model	Payload lt.	Spray bar width ft.	Flying altitude m. above ground	Endurance Min. per flight	Spray area ha h^{-1}	References
Single Rotor Helicopter						
RMAX	28–31	4–5	5–8 (50–1000)	60	8–40	RMAX (2015), Cornett (2014), Krishna (2020a)
ZHNY	15	3–6	1–3	25	8–22	Krishna (2020a)
Hercules 50	17	3–4	5–200	12	10–16	Krishna (2020a)
AG-RHCD-80-1	15	5	3–4 (0–100)	15	8–10	Krishna (2020a)
Quadcopter						
HSE-AG 10	10	12–18	2–3	9–12	2.5–12	Krishna (2020a); HSE (2015a, b, 2017)
AGStar X8	4–6	3–4	5–20	30	7–15	Krishna (2020a)
Hercules 20	20	5	1–3	12	10–12	Krishna (2020a)
Hexacopter						
HSE-AG 6A	7–12	16	5–10	15	50	Krishna (2020a), HSE (2015a, b), Bowman (2015)
HSE AG-MRCD-6	6–8	9–12	1–3	< 10	3–35	
Octocopter						
HSE-AG-MRCD 18	5	9–15	50–100	10–15	21–43	Bowman (2015), HSE (2015a, b), Krishna (2020a)
AGRAS MG-1	10	5–10	5	45	5–8	Caturegli et al., (2016), DJI (2016, 2017), Krishna (2020a)

Note: Small fixed-winged autonomous drones are not feasible for spraying a payload with pesticide formulation.

Sprayer UAVs are adopted in China and Far-eastern nations (e.g., Yamaha's RMAX, DJI's MJ-AGRAS-1, Hercules 20, etc. (Wolf, 2019; Real Agriculture, 2017). Sprayer UAVs could be helicopters, quadcopters, hexacopters or octocopters. There are even deca-copters that have been modified and accessories for spraying added on to them (see RMAX, 2015; DJI, 2016, 2017; Krishna, 2020a).

Let us consider an example of a sprayer drone and its specifications. Usually, these sprayer drones are made of carbon fiber, plastic, and light aluminum. They withstand high impacts in case of disasters. Most models produced are single rotor helicopters, quadcopters, hexacopters, and a few are octocopters. They possess water/pesticide tanks made of plastic. Pesticide tanks are of different sizes and hold 2–30 lts of water/pesticide formulations. The spray bar attached to tank is an important aspect. Since it decides the swath and the time required to cover a defined cropped area. It often ranges from 2 to 8 ft in length with 4–12 nozzles (see Table 4.1; RMAX, 2015; DJI, 2016, 2017; FoxTech, 2020).

There are companies spread across all nations of the world that manufacture drones. Several of them produce agricultural drones i.e., sprayer drones that are useful during spraying water, pesticide, herbicide, fungicide formulations, or foliar nutrients. These agriculture drones are powered by electric batteries or petroleum/diesel fuel that can pollute atmosphere. Agricultural chemicals sprayed by drones may pollute the atmosphere. Agricultural chemicals sprayed by drone affects normal microflora, insects, and even birds. The space between crops' canopy and drones flying height allows for wind-aided drift. Harmful agricultural chemicals may in fact drift to neighboring areas. Foliar nutrients sprayed may affect the atmosphere by altering gaseous/particulate composition.

In all the above cases of copter UAVs, computer programs adopted for fight plan, digital data about variable release of fluids and intensity of pesticide spraying will govern the extent to which the "agricultural sky" gets deteriorated (contaminated), due to spray schedules over farms.

The spray bars could be fitted with simple atomizers (nozzles) or those that are able to deliver variable rates, depending on the computer programs. Nozzles with variable-rate facility are required if the copter drone is utilized, to conduct precision sprays of fertilizers (fertigation), pesticides, herbicides, or even water. Flight path can be controlled using remote controller or using preprogrammed paths. Computer programs such as *E-Motion-2* or *E-Motion-3*, Pixhawk flight plan could be adopted. Flight duration and capacity of the pesticide tank are important aspects of a sprayer copter drone (See Krishna, 2020a).

These sprayer drones, quite a few of them, have short endurance ranging from 15–40 min. A helicopter UAV such as RMAX may have endurance of 60 min. Otherwise, most of them, especially copters have endurance of 15–40 min per flight (see Krishna, 2020a; Table 4.1). It depends on the size of the pesticide tank and rate at which liquid formulation or granules area sprayed on to crops. The speed at which the sprayer drones cover the crop field varies from 2 to 6 m s^{-1}. Flow meters can be utilized to regulate rate of dispensation. The height attained and stability of copter drone are crucial. The agricultural drones have to hover at low altitudes above crop canopy. This will ensure that harmful chemicals do not drift much. Also, it helps in lessening disturbance to agricultural sky in terms of mist/haze. Accuracy and efficiency of spray depends to a great extent on the constant low altitude attained by the sprayer drone over field crops or plantations. They cover about 2–15 ha h^{-1}. Payload and drone's weight without load needs due consideration. (DJI, 2017; HSE, 2015a; Krishna, 2018, 2020a).

Reports suggest that about 30% of the agricultural cropping zones in South Korea and 70% in Japan are sprayed using "Sprayer UAVs." They are sprayer drones such as RMAX, Frazer, AGRAS MG-1, or few new brands of multicopters (see Plate 4.3). Sprayers are gaining in popularity in USA, at a rather brisk pace in the farm belts of Great Plains, California, and in the plantations. They say, major drone companies from the Fareast such as DJI, Yamaha, and others have induced a kind of revolution in pesticide spraying and foliar fertilizer supply. The in-season nutrient supply is being conducted using sprayer drones, in USA (Wolf, 2019; see Krishna, 2018, 2020a). The situation with drone sprayer usage is similar in Canadian Great Plains. The wheat and brassicas are getting pesticide sprayers from sky through the rapidly flying drones (Wolf, 2019). In future, we may expect several specific regulations regarding the height at which the drones can fly while spraying pesticides or foliar fertilizers. The type and size of nozzles, droplet size, that is, fine sprays or coarse droplets, speed of the drone, etc. This is in order to restrain undue contamination of agricultural sky. Right now, drift control and safety of areas not targeted seems important. They say, regulations regarding spray tank volume (2–24 lts size), spray bar size, nozzle types, droplet size, fine/coarse sprays (Gao et al., 2013; Qin et al., 2014; Qiu et al., 2013; Qin et al., 2016; Zhang et al., 2016a, b), and height of the spraying drone requires standardizations (Zhang et al., 2012, 2014). This is true irrespective of drone type and model.

PLATE 4.3 A multicopter sprayer drone applying pesticides on to a rice crop.
Source: Vinveli Unmanned Systems Inc. Cedar Rapids, Iowa, USA
Note: Firstly, it utilizes the aerospace above the rice crop efficiently. Spraying is done rapidly and uniformly, without human-related errors due to fatigue and mistakes. It can cover 20–50 ha of cropped area in an hour, which is faster than ground-based farm technicians operating sprayer vehicles. It keeps farmers/technician away from harmful chemicals. At the same time note that it may contaminate aerospace above crop's canopy, up to 3–5 m hight. Its influence on biotic aspects of agroecosystems is yet unknown.

Sprayer UAVs are being tested for use on citrus trees to spray pesticides (CREC, 2015; Zhang et al., 2014, 2016a, b). It seems tree size, its shape, and foliage distribution are important, to decide about the height from which the sprayer drone releases the pesticide (Zhang et al., 2016b) Several aspects of pesticide drift, droplet size, rate of dispensation of chemicals need to be standardized. If not, they may affect the agricultural sky above the plantations (Yang et al., 2018). Drones should perform excellently in hill tracts of Western ghats of Indian where coffee plantations are intensely cultivated.

Olson (2020) states that sprayer UAVs are set to totally replace airplane campaigns to apply pesticides/herbicides. Agricultural sprayer UAVs cause much less drift in the atmosphere compared to airplanes. Rice farmers in China and Fareast are among the pioneers who have popularized use of agricultural sprayer drones (Olson, 2020; Zhu, 2014; Krishna, 2020a). In

South Asia, sprayer drones are being evaluated for use by farmers engaged in production of dryland crops (Yellappa et al., 2017; Krishna, 2018, 2020a).

Agricultural drones are being used increasingly to spray plant protection chemicals in the rough terrain. Also, in areas that are not accessible easily. Like the rice terrace farming zones in Philippines or in mountainous tracts of Andes etc. This is in addition to preference for drone sprayers in plains zone (Yang et al., 2018).

Generally, farmers obtain data about wind direction, its speed, drift patterns, etc. Then, they match the drone's spray system. Electrostatic nozzles seem to help in reducing the drift. They allow better focus of spray direction and swath (Olson, 2020).

Lv et al. (2019) opine that, for farmers in the South-eastern Asian nations, crop spraying is usually a difficult task. Farmers and companies with large land holdings expend laborers "time and money" at higher rates. Sprayer UAVs that can reach the sky and distribute pesticide are being accepted by the farmers. These sprayers are autonomous. They adjust their flight altitude based on the quantity and speed of the vehicles. These sprayer drones are provided with computer programs that allow them to adjust spray altitude, speed, and droplet size, based on the nature of the terrain. The droplet size and quantity dispensed by sprayers depends on the speed of the vehicle (Lv et al., 2019). This aspect is useful when we use sprayer drone in the sky above a mountainous tract or terraces or undulated fields. Higher accuracy in the release of pesticides makes sprayer drones, efficient in terms of economics. It seems wastage of pesticide formulation due to excessive sprays or uneven distributions are avoided.

Farmers prefer sprayer-UAVs because they are much smaller and not elaborate like a ground sprayer. A ground sprayer vehicle with its sprayer attachments is much costlier. A sprayer drone costs between 15,000 and 50,000 US $ for a piece with all attachments (2020 price levels). On an average, a ground sprayer vehicle holds 1000–1500 gallons of plant protection chemical. It distributes about 10 gallons formulation h^{-1} (Wolf, 2019). Drones too, spray similar levels of pesticides.

In addition to spraying pesticides, drones are also adopted to conduct aerial seeding of rice in countries such as China and Japan. There are also attempts to spray biological control agents onto crops, to control pests (Li et al., 2013, 2016; See Krishna, 2020a). In such cases, too agricultural sprayer drones cause certain disturbance to agriculture sky, proportionately. They may induce disturbance to biotic aspects of aerospace above the crops.

4.5.4 PRODUCTION OF UAVS AND THEIR SPREAD IN DIFFERENT AGRARIAN REGIONS

Small drone aircrafts are gaining acceptance into aerospace above crops, at a rapid rate, in the Americas. They say, at present, in USA, about 19% of total UAV market relates to their use in the agriculture. Similarly, in Brazil, drones are being actively evaluated by their national farm agencies. In Brazil, UAVs might be most helpful in the aerospace above large expanses of maize, wheat, soybean, as well as in coffee plantations.

There are reports about drone usage in the European agricultural sky. It says, there is a marked increase, firstly in the production by the drone industries. Later, in their adoption by farmers. From a mere 7 million units (fixed-winged/copters) supplied to farmers in the European region, it has increased to nearly 22 million in 2019. It is expected to be 29 million by 2021 (Digital Transformation Monitor, 2020). The primary intention is to obtain spectral data about crops. They say drone usage in the European agrarian sky is a bit more prominent. It is done mainly to assess soil, fields, crop's water stress index, aerial examination of irrigation equipment and progress of agronomic procedures. Crop spraying too is gaining in acceptance. Foliar application of in-season fertilizer-N (liquid) is being evaluated.

No doubt, European agrarian sky is an important region where drones could be tested stringently, particularly, during spraying chemicals. Now, let us consider German drone industry and its relevance to efficient utilization of German Agrarian sky. Several of the drone production and economic indices show that *per capita* utilization of drones, especially, in southern region is increasing. The German Aviation Association predicts drones will fly all over the German agrarian regions. That should make agronomic procedures easier for the famers (Bdl.aero, 2020).

In France, for example, use of plant protection chemical is increasing each year. Cost of farm labor to spray the chemicals is also increasing proportionately. Experimental evaluation of sprayer UAV-aided aerial application of plant protection chemicals, herbicides, and even foliar nutrients has been initiated. Although farmers are allowed to use the UAVs in sky for aerial photography and spectral analysis, its use in crop protection needs further evaluation (Barniere, 2020).

Experimental evaluation of fixed-winged UAVs in the Chinese agrarian sky has shown that such aerial vehicles have great future, in the farm belt. The fixed-winged UAV (e.g., YUYAN-09 by Zero Tech Inc. China) were utilized, to assess wheat crop's vigor. Also, to assess panicle maturity and forecast grain yield. They also help farmers in identifying wheat powdery mildew fungus (*Blumeria graminis var tritici*) infestation, if any, through the spectral data (Liu, et al., 2018).

No doubt, future of drones in agricultural regions also depends on aerial methods of analysis of crops. It also depends on ability of drones to make certain agronomic procedures easier (e.g., spraying). In Italy, they estimate that most farming regions would soon be covered, by drones (Stein, 2017). There are indeed several ways, for drones to revolutionize farming methods in European nations or even in other locations (Agritech Tomorrow, 2019).

4.6 LOW AND HIGH-ALTITUDE AERIAL TURBINES IN THE AGRICULTURAL SKY: A METHOD TO PRODUCE ENERGY REQUIRED FOR FARMS

Aerial turbines are known as good source of energy. Especially, to run farms and machines related to crop production. Ground surface-based low altitude towers are useful in generating electricity (e.g., windmills of 5–10 m height). Aerial power generation needs that turbine be lifted into air, using towers. Such towers could be permanent fixtures or temporary installations that could be dismantled. In a few cases, they could be shifted to new locations as required. Permanent towers are costly to erect. There are also ground-based windmills that generate energy, using ground-level wind energy. High altitude wind energy is a better proposition if one aims at continuous generation of power for farm. High-altitude wind turbines harness the wind streams that are steady and high-speed winds. A recent method is to adopt tethered aerial vehicles such as aerostats, to generate power, using a lightweight plastic turbine. Tethered aerostats could be platforms for low or high-altitude wind turbines. The high-altitude turbines have been touted to revolutionize the supply of power to individual farms or a group within a village (Altaeros Inc., 2018). The high-altitude wind energy grids too are proposed to be utilized, in rural areas. Reports suggest that a single turbine lofted, using a tethered aerostat with strings to transmit electricity can suffice to serve seven to eight farms and their households (See Krishna, 2020b). Tethered aerostats with turbines could be portable. They could be lifted from a farm vehicle and shifted to different locations. Overall, we should realize that automation and aerial power generation is vital for functioning of farms. Together, aerial robotics and power generation facilities may revolutionize farm operation, to a great extent. Farm productivity too increases, if, in situ power generation gets accepted by farmers. Greater details about farm power generation, using aerial vehicles such as tethered aerostats with turbines are discussed in Chapter 5. Discussions here about aerial vehicles for power generation are restricted to the context.

In addition to tethered aerostats that act as platforms for wind turbines that generate power, there are a few other installations and/or farm procedures. They too may involve aerospace above farms. They are not entirely aerial gadgets. Often, they are ground surface installations but do affect the atmosphere above the soil or crop's canopy. They offer certain advantages. For example, heating devices placed in the intercrop/interrow space. They help in raising temperatures of plantations/fields through generation of smoke and/or heat. They warm the atmosphere in the crop's canopy (e.g., in cirrus plantation in Florida). Selective crop residue burning in the intercrop region too has similar function. It alters the temperature in the aerospace above crop's canopy.

Wind brakes made of tall trees, small trees or shrubs, or other material may all help in reducing ill effect of high-speed wind that blows over dry loose soil or crops. Wind speeds noted in many agrarian locations are good enough, to induce soil erosion. It can cause loss of soil nutrients and affect soil structure too. Ind brakes may also thwart unhindered dispersal of windborne microbial pathogens of crops.

KEYWORDS

- **aeolian dust**
- **airborne insects**
- **biological control**
- **biotic factors**
- **guanos**
- **locusts**
- **microbial flora**
- **pandemics**
- **phyllosphere**
- **seed dispersal**

REFERENCES

AeroVironment Inc. Quantix, 2017, pp 1–2. http://www.avinc.com (accessed July 26th, 2021).

Agape-Palilo. Monitoring and Management of Maize Rust (*Puccinia sorghi*) by a Drone Prototype in Southern Highlands, Tanzania. Sokoine University of Agriculture, Morogoro, Tanzania, 2014, pp 1–15. http://www.academia.edu/8063999/MONITORING_AND_MANAGEMENT OF MAIZE-RUST DISEASE (accessed July 30th, 2021).

Agritech Tomorrow. Nine (9) Ways Your Agricultural Drones Revolutionize Your Farm or Ranch, 2019, pp 1–10. https://www.agritechtomorrow.com/article/2017/10/9-ways-agricultural-drones-revolutionize-your-farm-or-ranch/10311/ (accessed July 30th, 2021).

Aguera, F.; Carvajal, F.; Perez, M. Measuring Sunflower Nitrogen Status from an Unmanned Aerial Vehicle-Based System and an on the Ground Device. *Int. Arch. Photogrammetry Remote Sens. Spatial Inform. Sci.* **2011,** *38*, 1–5.

Allen, R. G.; Keller, J.; Martin, D. *Centre-Pivot System Design*; The Irrigation Association: Fairfax, VA, 1998; 176 pp.

Allen, W. Drones Detect Crop Stresses More Effectively. *ICT Update* **2016,** *82*, 10–11. http://ictupdate.cta.Int (accessed July 32nd, 2021).

Allsopp Helikites Ltd. GIS, Geomatics, Surveying and Inspection Helikites Balloons, 2017, pp 1–15. http://www.allsopp.co.uk/index.php?mod=page&id_pag=63 (accessed July 21st, 2021).

Altaeros Energies Inc. *Clean Energy*; Altaeros Energies: Somerville, MA, 2019; pp 1–3 (accessed February 18th, 2019).

Altaeros Inc. *The Super Tower ST 20*; Altaeros Inc: Somerville, MA, 2018, pp 1–12. http://www.altaeros.com/technology.html (accessed July 28th, 2021).

Altave and Airstar Aerospatiale. Altave and Airstar Aerospace Join Forces to Lead Global Market of Tethered Aerostats, 2017, pp 1–4. .https://en.prnasia.com/releases/apac/ALTAVE_and_Airstar_Aerospace_join_forces_to_lead_global_market_of_tethered_aerostats-180335.shtml/ (accessed July 27th, 2021).

Altstadler, B.; Platis, A.; Wehner, B.; Schultz, A.; Wildmann, N.; Hermann, M.; Kachner, K.; Baars, B.; Bange, J.; Lampert, A. ALADINA- an Unmanned Research Aircraft for Observing Vertical and Horizontal Distributions of Ultrafine Particles Within the Atmospheric Boundary Layer. *Atmos. Measure. Tech.* **2015,** *8*, 1627–1639. DOI: 10.5194/amt-8-1627-2015 (accessed July 24th, 2021).

Bajoria, A.; Mahto, N. K.; Boppana, N.; Pant, R. S. Design of a Tethered Aerostat System for Animal and Bird Hazard Mitigation. First International Conference on Recent Advances in Aerospace Engineering (ICRAAE). Conference Paper, 2018; pp 1–16. DOI: 10.1109/ICRAAE.2017.8297244/ (accessed July 27th, 2020).

BAL. Aerostats all Australia AAA Mobile Overage, 2016, pp 72. https://www.pc.gov.au/__data/assets/pdf_file/0006/205827/sub054-telecommunications-attachment.pdf/ (accessed July 26th, 2021).

Banjo, C. Y.; Ajayi, O. Sky-Farmers: Applications of Unmanned Aerial Vehicles (UAV) in Agriculture. IntechOpen, 2020, pp 1–17. http://dx.doi.org/10.5772/intechopen.89488/ (accessed July 26th, 2021).

Barbedo, J. G. A. A Review on the Use of Unmanned Aerial Vehicles and Imaging Sensors for Monitoring and Assessing Plant Stress. *Drones* **2019,** *40*, 1–27. DOI: 10.3390/drones3020040 (accessed July 22nd, 2021).

Barniere, C. Agricultural Drone Are Gradually Taking-Off in France. Euroactive, 2020, pp 1–8. https://www.euractiv.com/section/agriculture-food/news/agricultural-drones-are-gradually-taking-off-in-france/ (accessed July 26th, 2021).

Bdl.aero. The Analysis of the German Drone Market. Bundesverband der Deutschen Luftverkehrswirtschaft (German Aviation Association), Berlin, Germany, 2020, pp 1–34. https://www.bdl.aero/en/publication/analysis-of-the-german-drone-market/ (accessed July 22nd, 2020).

Bentley, M. C.; Mole, T. L.; Thebpanya, P. Using Landsat to Identify Thunderstorm Damage in Agricultural Regions. *Am. Meteorol. Soc.* **2002,** *122*, 363–376.

Bowman, L. UAVs for Farmers and Ranchers; Homeland Surveillance and Electronics LLC, 2015, pp 1–3. http://www.griculturuavs.com/uav_UAVs_for_farmers_rancher.htm (accessed July 23rd, 2015).

Brouwer, C.; Prins, K.; Kay, M.; Heibloem, M. *Irrigation Water Management: Irrigation Methods: Training Manual No. 5: Irrigation Methods*; Food and Agriculture Organization of the United Nations: Rome, Italy, 1990, pp 1–24. ftp://ftp.fao.org/agl/aglw/fwm/Manual5.pdf/ (accessed July 26th, 2021).

Caturegli, L.; Comeglia, M.; Gaetani, M.; Grossi, N.; Magni, M.; Miglizzi, M.; Angelini, L.; Mazzoncini, M.; Silvestri, N.; Fontanelli, M.; Raffaelli, P.; Volterrani, M. Unmanned Aerial Vehicle to Estimate Nitrogen Status of Turf Grass. *PLOS One* **2016,** 1–9. https://doi.org/10.1371/journal.pone0158268 (accessed July 27th, 2021).

Chen, S.; Lan, Y.; Li, J.; Zhou, Z.; Jin, J.; Liu, A. Effect of Spray Parameters of Small, Unmanned Helicopter on Distribution Regularity of Droplet Deposition in Hybrid Rice Canopy. *Trans. Chinese Soc. Agric. Eng.* 2016, *32* (17), 40–46 (In Chinese with English abstract).

Christiansen, J. E. *Irrigation by Sprinkling*; California Agricultural Experiment Station: Sacramento, CA, Bulletin No. 670, 1942; p 124.

CIMMYT. Obregon Blimp Airborne and Eyeing Plots; International Maize and Wheat Centre: Mexico, 2012, pp 1–4. https://www.cimmyt.org/news/obregon-blimp-airborne-and-eyeing-plots/ (accessed July 24th, 2021).

Cornell. C. Farmers Use Drones and Data to Boost Crop Yields. *The Globe and Mail*, 2015. http://www.theglobend mail.com/report-on-business/sb-growth/farm/ (accessed July 26th, 2021).

Cornett, R. *Drones and Pesticide Spraying—A Promising Partnership*; Western Plant Health Association, Western Farm Press, 2013, pp 1–3. http://westernfarmpress.com//grapes/drones-and-pesticides-spraying-partnership (accessed July 30th, 2014).

Cousins, D. A Guide to Effective Bird Scaring Kits for Farmers, 2018, pp 1–5. https://www.fwi.co.uk/machinery/guide-to-effective-bird-scaring-kits-for-farmers/ (accessed October 27t, 2020).

CREC. UAV Application in Agriculture. Citrus Research and Education Centre, Lake Alfred, Florida, 2015, pp 1–3. http://www.crec.ufl.ifas.edu/publications/news/PDF/UAVwsflyer3-pdf.pdf (accessed July 30th, 2021).

Deutz Fahr-Agrosky. Autonomous Systems: German Engineering at Its Best, 2020, pp 1–4. https://www.deutztractors.co.nz/products/precision-farming/deutz-fahr-agrosky-gps-system/ (accessed July 28th, 2020).

Digital Transformation Monitor. Drones in Agriculture. European Commission, 2020, pp 1–7. https://ati.ec.europa.eu/sites/default/files/2020-06/Drones%20in%20agriculture%20%28vf%29.pdf/ (accessed July 26th, 2021).

DJI. Above the World. UAS Magazine, 2016, p. 238. http://www.uasmagazine.com/articles/1591/dji-explains-new-book-of-UAV-captured-images (accessed July 20th, 2021).

DJI. DJ1 MG-1S Agricultural Wonder Drone, 2017. https://www.youtube.com/watch?v=P2YPG8PO9JU (accessed July 28th, 2021).

Dronologista. Drones for Pest Control, 2015, pp 1–3. http://dronologista.wordpress.com/2014/04/09 (accessed October 9th, 2015).

Ehsani, R.; Sankaran, S. Sensors and Sensing Technologies for Disease Detection. Citrus Industry, 2010, pp 1–5. http://www.crec.ifas.ufl.edu/../2010junesensoringtechnology.pdf (accessed October 14th, 2015).

Elmquist, S. Fighter Pilot Turned Farmer to Ply Drones Over Crop Land. Stars and stripes, 2015, pp 1–12. http://www.stripes.com/news/veterans/fighter-pilot-turned-farmer-to-ply-drones-over-cropland. Htm (accessed December 27th, 2016).

Evans, R. G. *Center-Pivot Irrigation Systems*; United States Department of Agriculture: Beltsville, MD, 1998, pp 1–16. https://www.ars.usda.gov/ARSUserFiles/30320500/IrrigationInfo/general%20irrigation%20systems-mondak.pdf/ (accessed July 27th, 2021).

FAO. *Crops and Drops: Making the Best Use of Water for Agriculture*; Food and Agriculture Organization of the United Nations: Rome, 2002, viewed 17 February 2017. http://www.fao.org/docrep/005/y3918e/y3918e10.htm/ (accessed July 27th, 2021).

Fiske, S. V. *Drones Effective Tools for Fruit Farmers*; American Society of Agronomy: Madison, Wisconsin, 2020. https://www.agronomy.org/news/science-news/drones-effective-tools-fruit-farmers-0/ (accessed July 27th, 2021).

FoxTech. GAIA 160 AG-Agricultural Sprayer, 2020, pp 1–3. https://www.foxtechfpv.com/gaia-160-ag-hexacopter-arf-combo.html/ (accessed July 26th, 2020).

Gago, J.; Douthe, C.; Coopman, R. F. Gallego, P.P, Ribas-Carbo, M.; Flexas, J.; Esclona, J.; Medrano, H. UAVs Challenge to Assess Water Stress for Sustainable Agriculture. *Agric. Water Manage.* **2015a,** *153*, 1–14. DOI: 10.1016/j.agwat.2015.01.020 (accessed July 27th, 2017).

Gago, J.; Martorell, S.; Douthe, C.; Fuentes, S.; Toms, M.; Hernandez, E.; Mir, Li.; Carriqui, M.; Escalona, J. M.; Gallego, P. P.; Medrano, H. Upscaling Levels for Drought Assessment in Agriculture: From Leaf to the Whole Vineyard. 1 Journadas del Grupo de Viticultura y Enologia de la SECH- Reos Actuales de I +D en Viticultura. *Agric. Water Manage.* **2015b,** *153* (issue C), 9–19.

Gago, J.; Martorell, S.; Tomas, M.; M.; Pou, A.; Millan, B.; Ramon, J.; Ruiz, M.; Sanchez, R.; Galmes, J.; Conesa, M. High Resolution Aerial Thermal Imagery for Plant Water Status Assessment in Vineyards, Using Multi-Copter-RPAS, 2015c, pp 1–2144 file:///C:/Users/Dr%20krishna/Downloads/Libro_de_actas_completo_dic_2014_final.pdf/ (accessed July 27th, 2021).

Gao, Y.; Zhang, Y.; Zhang, N.; Niu, L.; Zheng, W.; Yuan, H. Primary Studies on Spray Droplets Distribution and Control Effects of Aerial Spraying Using Unmanned Aerial Vehicle (UAV) Against Wheat Midge. *Crops* **2013,** *2*, 139–142 (In Chinese with English abstract).

Garber, H. Robotic Harvesting System. In Service Robots in Agriculture-Automatica. Sensors, 2014, pp 1–4 https://www.sensorsmag.comelectronics-computer/news/automatic-2014-service-robots-agriculture-1316.htm (accessed July 29th, 2014).

Garland, C. Drones May Provide Big Lift to Agriculture When FAA Allows Their Use. *Los Angeles Times*, 2014, pp 1–5. http://www.latimes.com/business/la-fi-drones-agriculture-20140913-story.html#page=1 (accessed July 25th, 2021).

Gerard, B. It's a Bird, It's a Plane No, It's a Super Scientist. SAT Trends, ICRISAT Newsletter, 2016, p 14. http://www.icrisat.org/what-we-do/satrends/01dec/1.htm (accessed July 23rd, 2021).

Glass, B. In Soaring 'SuperTowers' Aim to Bring Mobile Broadband to Rural Areas (By Steadler, T.). ITU News, 2018, pp 1–7. https://news.itu.int/soaring-supertowers-aim-to-bring-mobile-broadband-to-rural-areas/. (accessed July 15th, 2021).

Gogarty, B.; Robinson, I. Unmanned Vehicles: A (Rebooted) History, Background and Current State of the Art. *J. Law Inform. Sci.* **2012,** *21*, 1–18.

Green, M. Unmanned Drones may have their greatest impact on Agriculture, 2013, pp 1–4. https://www.thedailybeast.com/unmanned-drones-may-have-their-greatest-impact-on-agriculture/ (accessed July 27th, 2021).

Guglieri, G.; Ristorto, G. Safety Assessment for Light Remotely Piloted Aircraft Systems. International Conference on Air Transport- INAIR, 2016, pp 1–7. https://core.ac.uk/download/pdf/76533631.pdf/ (accessed July 27th, 2021).

Hillel, D. *Rivers of Eden: The Struggle for Water and the Quest for Peace in the Middle East*; Oxford University Press: New York, 1994; p 335.

HSE. Intelligent Imaging, 2015a, pp 1–3. http://www.uavcropdustersprayers.com/agriculture_delta_fw_70_fixed_wing_uav.htm. (accessed July 29th, 2021).

HSE. Invasive Species Detection: Canadian Thistle, 2015b, pp 1–5. http://www.uavcropdusterssprayers.com/agriculture_delta-fw70_fixed_wing_uav.htm (accessed July 29th, 2021).

HSE. Agricultural and Conservation UAS. Homeland Surveillance and Electronics LLC, Illinois, 2017, pp 1–4. https://hse-uav.com/products-htm/ (accessed July 30th, 2021).

Huang, Y.; Hoffman, W. C.; Lan, Y.; Wu, W. Fritz, B. K. Development of a Spray System for an Unmanned Aerial Vehicle Platform. *Am. Soc. Agric. Biol. Eng.* **2009,** *25*, 803–809.

Hunt, E. R.; Horneck, D. A.; Gadler, D. J.; Bruce, A. F.; Turner, R. W.; Spinelli, C. B.; Brungardt, J. J. Detection of Nitrogen Deficiency in Potatoes Using Small, Unmanned Aircraft Systems, 2015, pp 1–4. https://www.ars.usda.gov/research/publications/publication/?seqNo115=301000 (accessed July 27th, 2021).

Institute of Engineers-India. Tethered Aerostat Systems for Agricultural Application in India, Annual Technical Volume of Aerospace Engineering Division Board, 2016, pp 82–88.

Kandil, A. A.; Shareif, A. E. Productivity of Some Forage Grasses and Alfalfa Under Foliar Sprinkler Irrigation with Foliar Application of Chemical Substances Under Water Stress. *Int. J. Contemporary Appl. Sci.* **2016,** *3*, 289–327.

Kansas State University. Project Using Drones to Detect Emerging Pest Insects, Disease in crops. Department of Entomology, Kansas State University at Salina and Manhattan, USA, 2015, p 104. http://www.agprofessional.com/news/project-using-drones-detect-emerging-pest-insects-diseases-in-crops/ (accessed July 25th, 2021).

Keane, J. F.; Carr, S. S. A Brief History of Early Unmanned Aircraft. *Johns Hopkins Appl. Techn. Digest* **2013,** *32*, 558–593.

King, R. Farmers Experiment with Drones. The CIO Report, 2013. https://www.wsj.com/articles/SB10001424127887324763404578431031698188710/ (accessed July 28th, 2021).

Kinze Manufacturers Inc. Kinze Manufacturing Continues Progress on Kinze Autonomous Vehicles, 2013. https://www.kinze.com/kinze-manufacturing-continues-progress-on-kinze-autonomy-2/ (accessed July 28th, 2021).

Kinze Manufacterers Inc. Kinze Autonomous Grain Cart System, 2014, pp 1–8. https://www.prweb.com/releases/2012/9/prweb9936899.htm (accessed July 28th, 2014).

Klose, R.; Marquering, J.; Theil, M.; Ruckelshausen, A.; Marquering, J. Weedy-A Sensor Fusion Based Autonomous Field Robot for Selective Weed Control. *Agric. Eng.: Land Technik* **2012,** 167–172.

Krishna K. R. *Push Button Agriculture Drones, Robotics and Satellite guided Crop Production*; Apple Academic Press Inc.: Waretown, NJ, 2017; p 476.

Krishna, K. R. *Agricultural Drones: A Peaceful Pursuit*; Apple Academic Press Inc.: Waretown, NJ, 2018; p 425.

Krishna, K. R. *Unmanned Aerial Vehicle Systems in Crop Production: A Compendium*; Apple Academic Press Inc.: Palm Bay, FL, 2020a; pp 697.

Krishna, K. R. *Aerial Robotics in Agriculture: Parafoils, Blimps, Aerostats and Kites*; Apple Academic Press Inc.: Palm Bay. FL, 2020b; p 397.

Laura, W. Global Aerostat Systems Market Analysis, Trends and Forecasts 2019–2025, 2020, pp 1–5. https://www.businesswire.com/news/home/20200101005069/en/Global-Aerostat-Systems-Market-Analysis-Trends-and-Forecasts-2019-2025---ResearchAndMarket (accessed October 23rd, 2020).

Li, J.; Lan, Y.; Zhou, Z.; Zeng, S.; Huang, C.; Yao, W. Design and Test of Operation Parameters for Rice Air Broadcasting by Unmanned Aerial Vehicle. *Int. J. Agric. Biol. Eng.* **2016,** *9* (5), 24–32.

Li, D.; Yuan, X.; Zhang, B.; Zhao, Y.; Song, Z.; Zuo, C. Report of Using Unmanned Aerial Vehicle to Release Trichogramma. *Chinese J. Biol. Control* **2013,** *29* (3), 455–458.

Litchman, A. Humanitarian Uses of Drone and Satellite Imagery Analysis. The Promises and Perils. *Am. Med. Assoc. J. Ethics* **2015,** *17*, 931–937.

Liu, W. L.; Cao, X.; Fan, J.; Wang, Z.; Yun, Z. Detecting Wheat Powdery Mildew and Predicting Grain Yield Using Unmanned Aerial Photography. *Plant Disease* **2018,** *102*, 1981–1988. https://www.doi.org/10-1094/PDIS-12-17-1893RE/ (accessed 22nd, 2020).

Lv. M, Xiao S, Tang Y, He Y. 2019 Influence of UAV flight speed on droplet deposition characteristics with the application of infrared thermal imaging. International Journal of Agricultural and Biological Engineering. 12(accessed 3):10-17.

MacDonell, M.; Raymond, M.; Wyker, D.; Finster, M.; Chang, Y.; Raymond, T.; Temples, D.; Scofield, M. Mobile Sensors and Applications for Detection of Air Pollution. Argonne National laboratory, and Environmental Protection Agency of USA, 2013, p 278.

Machovina, B. L.; Feeley, K. J.; Machovina, B. J. UAV Remote Sensing of Spatial Variation in Banana Production. *Crop Pasture Sci.* **2016,** *67*, 1281–1287. https//doi.org/10.1071/CP16135 (accessed August 3rd, 2021).

McMahon, K. Kinze Autonomous Tractor System Tested in Field by Farmers. Farm Industry News, 2012, pp 1–8. https://www.farmprogress.com/precision-guidance/kinze-s-autonomous-tractor-system-tested-field-farmers (accessed July 27th, 2021).

Microdrone. Work Smarter with Microdrones Integrated System. Microdrones MBH, 2020, p 18. https://www.microdrones.com/en/integrated-systems/ (accessed July 22nd, 2021).

Mogili, U. M.; Deepak, B. V. L. Review on Application of Drone Systems in Precision Agriculture. International Conference on Robotics and Smart Manufacture. *Procedia Comput. Sci.* **2018,** *133*, 502–509.

NASA. The Landsat Satellites: Unique National Assets. NASA Public Affairs Fact Sheet, 1999 (accessed July 24th, 2021).

NASA. Food Security from Space: Monitoring Indica, 2020, pp 1–8. https://landsat.gsfc.nasa.gov/outreach tors of Air Quality on Crop Yields. National Aeronautics and Space

Agency, Washington, DC. https://science.gsfc.nasa.gov/610/applied-sciences/images/AirQuality_8-5-2019.pdf/ (accessed July 2021).

New, L.; Fipps, G. Centre Pivot Irrigation; Texas Agriculture Extension Service, Texas A and M University, College Station, 2019; pp 1–30.

Newcome, L. R. *Unmanned Aviation: A Brief History of Unmanned Aerial Vehicles*; American Institute of Astronautics Inc. Reston: Virginia, 2004; pp 1–29.

Nicole, P. Sustainable Technology- Drone use in Agriculture, 2015, pp 1–7. https://wiki.usask.ca/display/~pdp177/Sustainable+Technology+Drone+Use+griculture (accessed July 27th, 2015).

Olson, B. Agricultural Sprayer Drones. Liberty University, 2020, pp 11–18. https://www.researchgate.net/publication/339181546_Agricultural_Sprayer_Drones/ (accessed July 27th, 2021).

Pena, J. M.; Torres-Sanchez, J.; deCastro, A.; Kelly, M.; Granados, F. Weed Mapping in Early-Season Maize Fields Using Object-Based Image Analysis of Unmanned Aerial Vehicle Images. 2013, pp 1–18.

https://journals.plos.org/plosone/article?id=10.1371/journal.pone.0077151/ (accessed July 24th, 2021).

Pena, J. M.; Torres-Sanchez, J.; Serrano-Perez, A.; deCastro, A.; Lopez-Granados, F. Quantifying Efficacy and Limits of Unmanned Aerial Vehicle (UAV) Technology for Weed Seedling Detection as Affected By Sensor Resolution. *Sensors* **2015,** *15*, 5609–5626.

Primicerio, J.; Di Gennaro, S. F.; Fiorillo, E.; Genesio, L.; Lugato, E. A Flexible Unmanned Aerial Vehicle for Precision Agriculture. *Precision Agric.* **2012,** *13* (4), 517–523.

Pudelko, R.; Stuczynski, T.; Borzecka-Walker. The Suitability of an Unmanned Arial Vehicle (UAV) for the Evaluation of Experimental Fields and Crops. *Zemdirbyste=Agriculture* **2012,** *99*, 431–436. UDK 631.5.001.4:629.734/ (accessed July 24th, 2021).

Qin, W.; Xue, X.; Zhou, L.; Zhang, S.; Sun, Z.; Kong, W. Effects of Spraying Parameters of Unmanned Aerial Vehicle on Droplets Deposition Distribution of Maize Canopies. *Trans. Chinese Soc. Agric. Eng.* **2014,** *30* (5), 50–56 (In Chinese with English abstract).

Qin, W.; Qiu, B.; Xue, X.; Chen, C.; Xu, Z.; Zhou, Q. Droplet Deposition and Control Effect of Insecticides Sprayed with an Unmanned Aerial Vehicle Against Plant Hoppers. *Crop Protection* **2016,** *85* (6), 79–88.

Qiu, B.; Wang, L.; Cai, D.; Wu, J.; Ding, G.; Guan, X. Effect of Flight Altitude and Speed of Unmanned Helicopter on Spray Deposition Uniform. *Trans. Chinese Soc. Agric. Eng.* **2013,** *29* (24), 25–32.

Real Agriculture, DJI Spraying Drone MG-1 Agras, 2017, pp 1–6. https://www.dji.com/mg-1 (accessed July 28th, 2021).

Rice, C. Weed Detection from Drones… Finally Here. Spectrabotics, 2015, pp 1–4. https://medium.com/@spectrabotics/weed-detection-from-drones-3b49a30d3017/ (accessed July 27th, 2021).

RMAX. RMAX Specifications. Yamaha Motor Company, Japan, 2015, pp 1–4. http://www.max.yamaha-motor.Drone .au/specifications (accessed July 28th, 2021).

Rogers, D. H.; Kisekka, I.; Aguilar, J.; Lamm, F. Center-Pivot Irrigation System: Losses and Efficiency. In *Proceedings of the 29th Annual Central Plains Irrigation Conference*, Burlington, Colorado, Feb. 21–22, 2017. CPIA, 760 N. Thompson, Colby, Kansas, USA, 2017; pp 19–34.

Rogers, D. H.; Lamm, F. R.; Alam, M.; Trooien, T. P.; Clark, G. A.; Barnes, P. L.; Mankin, K. L. Efficiencies and Water Losses of Irrigation Systems. Kansas State University Research and Extension. Irrigation Management Series. MF-2243, 1997, pp 1–8. https://bookstore.ksre.ksu.edu/pubs/MF2243.pdf/ (accessed July 2021).

Sankaran, S.; Ehsani, R. A Detection of Huanglongbing Disease in Citrus Using Fluorescence Spectroscopy. *Trans. ASABE* **2012,** *55*, 313–320.

Sankaran, S.; Khot, L. R.; Espinoza, C. Z.; Jarolmasjed, S.; Pavek, M. J. Low Altitude, High Resolution Aerial Imagery System for Row and Field Crop Phenotyping. *Eur. J. Agron.* **2015,** *70*, 112–123.

Sankaran, S.; Mishra, A.; Ehsani, R.; Davis, C. A Review of Advance Techniques for Detecting Plant Disease. *Comput. Electron. Agric.* **2010,** *72*, 1–13. https://doi.org/10.1016/j.compag.2010.02.007Get rights and content/ (accessed July 27th, 2021).

Scentroid Inc. Scentroid: The Future of Sensory Technology, 2017, pp 1–12. http://scentroid.com/scentroid-sampling-drone/ (accessed July 21st, 2021).

Schaefer, J. T.; Brooks, H. E. Convective Storms and Their Impact. Preprints, Second Symposium on Environmental Applications, Long Beach, CA. *Am. Meteorol. Soc.* **2000,** *118*, 152–159.

SenseFly Inc. eBee by Sensefly, 2015, pp 1–4. http://www.sensefly.com (accessed July 27th, 2021).

SenseFly Inc. eMotion 3: Flight Plan and Control System. SenseFly-A Parrot Company, 2017. https://www.sensefly.com/software/emotion-2. Html (accessed July 27th, 2021).

She, Y.; Ehsani, R.; Robbins, J.; Leiva, J. N.; Owen J. Application of Small UAV Systems for Tree and Nursery Inventory Management. In *Proceedings of 14th, International Conference on Precision Agriculture Montreal*, Quebec, Canada, 2017, pp 1–7. https://www.researchgate.net/publication/284311667_Applications_of_small_UAV_systems_for_tree_and_nursery_inventory_management (accessed July 30th, 2021).

Shi, Y.; Thomasson.; Murray, S. C.; Puch, N. A.; Rooney, W. L.; Shafian, S.; Rajan, N.; Rouze, G.; Morgan, C. L. S.; Neely, H. L.; Rana, A.; Bagvthiannan, M. V.; Herrickson, J.; Bowden, E.; Vilsack, J.; Olsenholler, J.; Bishop, M. P.; Sheridan, R.; Putman, E. B.; Popescu, S.; Burks, T.; Cope, D.; Ibrahim, A.; McCutchen, B. F.; Baltensperger, D. D.; Vent, R.; Vidrine, M.; Yang, C. Unmanned Aerial Vehicles for High Throughput Phenotyping and Agronomic Research. *PLOS One*, 2016, pp 1–15. http://dx.doi.org/10.1371/journal.pone.0159781 (accessed July 29th, 2021).

Siddique, A. B.; Rafiqul Islam, M. D.; Anamul Hoque, M. D.; Mahmudul Hasan, M. D.; Rahman, M. T.; Uddin, M. M. Mitigation of Salt Stress by Foliar Application of Proline in Rice. *Univ. J. Agric. Res.* **2015,** *3* (3), 81–88.

Simelli, J.; Tsgaris, A. The Use of Unmanned Aerial Systems (UAS) in Agriculture, University of Macedonia, Greece, 2015, pp 1–7. http://ceur-ws.org/Vol-1498/HAICTA_2015_paper83.pdf (accessed July 26th, 2021).

Sky Squirrel. Vineyard Drone Research: Applications, 2017, pp 1–3 (accessed July 28th, 2021).

Song, Y.; Wang, J. Evaluation of the UAV-Based Multispectral Imagery and Its Application for Crop Intra-Field Nitrogen Monitoring and Yield Prediction. The University of Western Ontario Electronic thesis and Dissertation Repository No 4085, 2016 p 112 (accessed July 24th, 2021).

Spence, K. North Grumman Wants to Sell Unmanned Drones to Farmers. The Motley Fool, 2013, pp 1–3. https://www.suasnews.com/2013/12/northrop-grumman-wants-to-sell-unmanned-drones-to-farmers/ (accessed July 24th, 2021).

Stein, N. The Future of Drones in the Modern Farm Industry **2017,** *21* (5). https://mediageo.it/ojs/index.php/GEOmedia/article/view/1493 (accessed July 26th, 2020).

Swain, K. C.; Thomson, S. J.; Jayasuriya, H. P. Adoption of Unmanned Helicopter for Low Altitude Remote Sensing to Estimate Yield and Total Biomass of a Rice Crop. *Trans. Am. Soc. Agric. Eng.* **2010,** *53*, 21–27.

Swain, S. Canberra Rangers to Use Helicopter Drones to Fight Weeds, 2014, pp 1–2. http://ww.smh.com.au/it-pro/government-it/canberra-rangers-to-fight-weeds (accessed September 29th, 2015).

Talat, A.; Khalid, N.; Khalid, H.; Khizar, H. B.; Ejaz, H. S.; Aneela, K.; Sehrish A.; Sharif M. U. Foliar Application of Proline for Salt Tolerance of Two Wheat *Triticum aestivum* L. Cultivars. *World Appl. Sci.* **2013,** (4), 547–554.

Tetracam Inc. Aerostats in Agriculture, 2015, pp. 1–5. http://aeroscraft.com/precision-agriculture/4584020239/ (accessed October 27th, 2020).

Tetrault, C. A Short History of Unmanned Aerial Vehicles (UAVs). Dragon Fly Innovations Inc., 2014, pp 1–2. http://texasagrilife extension.edu/ (accessed July 24th, 2014).

Thamm, H. P. SUSI A Robust and Safe Parachute UAV with Long Flight Time and Good Pay Load. Int. Arch. photogrammetry, *Remote Sens. Spatial Inform. Serv.* **2011,** *38*, 1–6.

Thamm, H. P.; Judex, M. The Low-Cost Drone. An Interesting Tool for Process Monitoring in a High Spatial and Temporal Resolution. International Archives of Photogrammetry, Remote Sensing, Spatial information Science. ISPs Commission 7th Mid-Term Symposium. Remote Sensing: From Pixels to Process. Enchede. The Netherlands **2006,** *36*, 140–144.

Thamm, H. P.; Menz, G.; Becker, M.; Kuria, D. N.; Misana, S.; Kohn, D. The Use of UAS for Assessing Agricultural Systems in AN Wetland in Tanzania, in the Wet Season, for Sustainable Agriculture and Providing Ground Truth for Terra Sar X data. ISPRS International Archives of the Photogrammetry, Remote Sensing and Spatial Information Sciences. XL-1/w2, 2013, pp 401–406. DOI: 10.5194/isprsarchives-XL-1-W2-401-2013 (accessed July 21st, 2021).

Tremblay, N.; Vigneault, P.; Belec, C.; Fallon, E.; Bouroubi, M. Y. A Comparison of Performance Between UAV and Satellite Imagery for N Status Assessments in Corn. In *Proceedings of 12th International Conference on Precision Agriculture*, Sacramento, California, USA, 2014, p 19.

Trimble. Trimble UX5 Aerial Imaging Solution for Agriculture, 2015, pp 1–3. http://www.trimble.com/Agriculture/UX5.aspx (accessed May 20th, 2015).

Tropnevad. Object-Based Weed Analysis of Unmanned Aerial Vehicle (UAV) Images, 2013, pp 1–4. http://www.dronetrest.com/t/object-based-weed-analysis-of-unmanned-aerial-vehicle-images/ (accessed July 20th, 2013).

UAS Vision. Helikite Aerostats to Lift the World's First 4G Airborne Base Station Get Web Address, 2012.

Vitirover. Vitirover: Replace the Herbicide Chemicals with Autonomous Robots, 2014, pp 1–12. https://www.vitirover.fr/en-faq (accessed July 25th, 2021).

Waller, P.; Yitayew, M. Center Pivot Irrigation Systems. Irrigation and Drainage Engineering, Springer, Cham, 2019, pp 209–208. https://doi.org/10.1007/978-3-319-05699-9_12/.

Weheren, M.; Raunker, P.; Sommer, M. UAB-Based Estimation of Carbon Exports from Heterogenous Soil Landscapes—A Case Study from the CarboZalf Experimental Area. *Sensors* **2016,** *16*, 255–276. doi:10.3390/s16020255/ (accessed July 21st, 2021).

Wikipedia. List of Unmanned Aerial Vehicles, 2020, pp 1–18. https://en.wikipedia.org.wiki/2017/ (accessed October 13th, 2020).

Wolf, T. The Challenges of Spraying by Drone. *Sprayers* **2019,** *101*, 1–3. https://sprayers101.com/challenges-drone/ (accessed July 21st, 2021).

Yang, S.; Yang, X.; Mo, J. The Application of Unmanned Aircraft Systems to Plant Protection in China. *Precision Agric.* **2018,** *19*, 278–292.

Yellappa, D.; Veerangouda, M.; Maski, D.; Palled, V.; Bheemanna, M. Development of and Evaluation of Drone Mounted Sprayer for Pesticide Application to Crop, 2017, pp 1–9. https://www.researchgate.net/publication/32206031 2/ (accessed July 22nd, 2021).

Yamaha. RMAX-History, 2014, pp 1–4. http://www.rmax.yamaha-motor.com.all/history (accessed July 20th, 2021).

Yintong Aviation Supplies Ltd. A Precision Agriculture UAV, 2012, pp 1–3. http://www.china-yintong.com/en/productshow.asp?sortid=7&id=57 (accessed July31st).

Yintong Aviation Supplies Ltd. Yintong UAV YT-P5, 2014. https://en.wikipedia.org/wiki/Yintong_UAV? (accessed July 27th, 2021).

Zarco-Tejada, P. J.; Guillen-Climent, M. L.; Hernandez-Clement, R.; Catalina, A.; Gonzalez, M. R.; Martin, P. Estimating Leaf Carotenoid Content in Vineyards Using High Resolution Imagery Acquired from an Unmanned Aerial Vehicle (UAV). *Agricultural and Forest Meteorology* **2013,** *171*, 281–294.

Zhang, P.; Deng, L.; Lyu, Q.; He, S.; Yi, S.; Liu, Y.; et al. Effects of Citrus Tree-Shape and Spraying Height of Small Unmanned Aerial Vehicle on Droplet Distribution. *Int. J. Agric. Biol. Eng.* **2016a,** *9* (4), 45–52.

Zhang, J.; He, X.; Song, J.; Zeng, A.; Zeng, A.; Liu, Y. Influence of Spraying Parameters of Unmanned Aircraft on Droplets Deposition. *Trans. Chinese Soc. Agric. Machinery* **2012,** *43*(12), 94–96.

Zhang, D.; Lan, Y.; Chen, L.; Wang, X.; Liang, D. Current Status and Future Trends of Agricultural Aerial Spraying Technology in China. *Trans. Chinese Soc. Agric. Machinery* **2014,** *45* (10), 53–59.

Zhang, P.; Lv, Q.; Yi, S.; Liu, Y.; He, S.; Xie, R. Evaluation of Spraying Effect Using Small Unmanned Aerial Vehicle (UAV) in Citrus Orchard. *J. Fruit Sci.* **2016b,** *33* (1), 34–42.

Zhenkun, T.; Yingying, F.; Suhong, L.; Liufeng, N. Rapid Crops Classification Based on UAV Low-Altitude Remote Sensing. *Trans. Chinese Soc. Agric. Eng.* **2017,** *29*, 109–116.

Zhu, X. An Exploration on Agricultural Unmanned Aerial Vehicle for Crop Protection. *Agric. Machinery Technol. Ext.* **2014,** *5*, 31–32.

CHAPTER 5

Wind and Solar Power Generation in the Agrarian Sky

ABSTRACT

This chapter deals with the role of agricultural sky in farm power generation and related aspects. Wind and solar energy generation are two important aspects of agrarian regions.

Wind above cropping expanses is a vast source of power. It could serve the entire world's need for electricity if we adopt suitable methods. Wind is an integral part of any agrarian sky. Wind gets generated based on differential atmospheric pressure in the sky. Wind speed is an important determinant of power generation. Wind could be erratic or steady. Wind that blows at low altitude and possesses a certain degree of turbulence can affect ground surface. Loss of topsoil from the fields due to it is common. Low altitude wind encounters obstructions to its free flow. Slower wind produces low amounts of farm power. There are a few distinct types of windmills adopted in farming regions. Simplest of technology was already in vogue in 12th century in European farms. They were originally adopted to grind grains and irrigate farms. In North America windmills were again small and were in operation by 1820s. They were called "windpumps" because of their utility in supply of water to crops. By the 1940s wind power was utilized to light rural homes in USA. Most ground surface windmills generate only small amounts of power enough to irrigate, supply power to farmhouse-holds, grinding, etc. High altitude wind power generation using tethered aerostats may offer greater amount of farm power. Wind speed is crucial factor that determines electricity generation. Wind speed of 9 km h^{-1} generates only 0.01 kwh^{-1} electricity. At 54 km h^{-1} wind speed about 2.2 kw h^{-1} power can be generated. Ground surface windmills utilize wind that blows at 100–200 ft altitude.

The Agricultural Sky: A Concept to Revolutionize Farming. K. R. Krishna, PhD (Author)

Wind resources and speed varies with geographical location, topography and features of agricultural sky. Accordingly, potential to generate wind power too varies. On a Beaufort scale, at force below 3, power does not get generated. At force 7 windmills may get damaged. At high altitudes of 300–2000 m above fields, wind is uniform without intermittency. It blows at relatively higher speed. Hence, wind power from high-altitude turbine is a consistent source. Usually, a high-altitude aerostat-supported turbine generates power good enough for 7–8 farmhouse-holds and supply a small share to electricity grid. There are kite models that help in wind power generation.

Solar photovoltaic cell technology was first standardized in 1950s. At present, United States of America and China are major solar power-producing nations. Together, they generate 30% of total solar energy produced worldwide. During recent years, several countries have developed "Solar Parks." Solar energy is gaining in acceptance in the farming regions. The agrarian sky mediates the reception of solar radiation. Solar power conversion to electric power is achieved through photovoltaic cells. Solar panels of many sizes and specifications to suit each location are available. Solar energy generation may involve small or large area. Individual farms with small holdings may utilize solar panels and generate power. Often large-scale electric power generation is conducted by adopting solar panels installed in vast areas (say, few km^2). There are several countries that generate solar power using such "solar park." Farmers may install solar PV cells (panels) simultaneously in the field and reap crop as well as electric power. In Germany, such farms with both crops and solar PV panels are common. The system is called "Agro-Photovoltaics." Sometimes solar power and crops are alternated. There are several applications for the solar-PV power generated in the sky above farms. It ranges from energizing water pumps, grain grinding machines, recharging electric farm vehicles, farmhouse lighting, etc.

5.1 WIND ENERGY GENERATION IN THE AGRARIAN SKY: INTRODUCTION

Windmills have a long history of usage by rural folk of different agricultural regions. Earliest of the records suggests that farmers in Afghanistan were conversant with windmill technology in the 7th century A.D. In otherward, they had begun exploiting wind available in the agrarian sky to their advantage. They used it to grind food grains and in lift irrigation. Windmills were in vogue by the 12th century A.D. in the European agrarian regions.

Historically, natives and farmers in the Netherlands are supposed to have utilized windmills or wind pumps, to conduct a few distinct functions such as lifting water, draining the polders of excess water, irrigating farms, even grinding grains, etc. Around the 13th century, windmills adopted by Dutch natives were smaller versions of modern windmill. Similar windmills and wind pumps were adopted by farmers in Spain, Portugal, France, and a few other regions (FAO 2012). Europeans produced small windmills in large numbers for pumping water.

By 1900 A.D., many European nations had been using windmills. It seems Holland and Denmark, that are well endowed with sea breeze adopted windmills. For example, during the late 19th century in Denmark, about 2500 windmills were installed in a small area. Windmills were popular by the early 20th century among the farmers in Europe. Not just with farm belt that utilized windmills. Regions needing electrical energy too adopted windmills. Interestingly, during WWII, windmills were utilized to recharge the batteries of military vehicles. Early windmills developed in China, that is, in 19th century were utilized to pump water to rice fields.

Wind above cropped expanses is not always a detriment. Although, wind may carry propagules of pathogenic microbes that affects crop's health. It is not always an agent of soil erosion, or a destroyer of crop stand and farm installations. Wind above the farms is cleanest source of power. It needs to be utilized advantageously wherever feasible. For example, wind above the agrarian regions/plains in Europe were exploited to generate mechanical power lift and/or generate electric energy. During 19th century, wind pumps were modernized to suit the farmers.

During the mid-1800s several types of wind pumps were adopted by Americans who practiced cattle farming and food grain production. There was strong demand for irrigation of crops. Hence, wind pumps called "American all steel water pump/electricity generators" became common in the Northern Great Plains. Wind pumps and power generating towers were also utilized in Australia and Argentina. In Southeast Asia, several designs of wind pumps were adopted. For example, irrigation and water supply wind pumps like "Cretan type" are utilized in Ethiopia. Windpump types made in Kenya were called Kijito. In Pakistan, they were called Tawana. The Thai adopted vertical/horizontal wind pumps. Examples for electricity generation are Dunlite wind electricity generator. Primarily, windmills are of two types, namely, those with blades that revolve around a vertical axis and others that rotate on a horizontal axis. Chinese and Thai farmers utilized a few vertical wind pumps too. Several hundreds of them are still in vogue in some parts of Thailand (FAO, 2012). Some of the Chinese wind pumps were low-cost installations. They were made using bamboos and wires.

The American farm's windpump evolved during the period between 1860 and 1900. It coincides with the period when many millions of cattle were being introduced into the North American Great Plains region. In fact, limited surface water created a vast demand for water lifting machinery. So, windpumps rapidly became the main general-purpose power source for the cattle farms. American Farm Windpump is rarely used today for irrigation. Most of them are used for the purpose they were originally developed for, namely watering livestock and, to a lesser extent, for farm or community water supplies. Typically, wind pumps are used in the 10–100 m depth range on boreholes. Large windpumps are in regular use on boreholes of over 200 m depth (Fig. 5.1).

European Energy Commission is encouraging use of renewable resources such as wind and solar energy. In March 2007, European leaders agreed that a binding target of 20% of all energy must be from alternative source (e.g. wind). It seems agrarian regions in Denmark, Netherlands, and Germany depend on wind power to the extent of 15%–20% of their normal total consumption. The aim is to uniformly deploy wind turbines in most of the agrarian locations of European plains.

Let us consider windpump adoption in USA. Historically, the golden period for introduction of windmills in USA extended from 1870s till 1930s. American colonists used windmills to grind wheat and corn, pump water, and to cut wood. As early as the 1920s, Americans used small windmills to generate electricity. Particularly, in rural areas not provided with electric service. When power lines began to transport electricity to rural areas in the 1930s, local windmills were used less frequently. Although they can still be seen on some Western ranches. Currently, wind power is utilized to generate electricity in 40 states of USA. The Texas state leads the nation This state produces over 25% of the wind-generated electricity (Wikipedia, 2021).

Earliest of the wind power generators utilized for street lighting and other functions were developed in North Dakota. In 1940, they had five large turbine systems at a place called Pettibone in 1940. In the New England zone, New Hampshire built the earliest commercial wind farm in 1980s. The wind farm generated 30 KW electric energy for use by farms and urban households. At present, USA has large wind power farms. Their installed capacity ranges from 662 MW to 1580 MW. They are operating such wind power installations in Texas, California, Indiana, North Dakota, Oregon, and Colorado (Wikipedia, 2021).

During recent period, since 1980s, farmers from the plains and other geographical locations have installed substantial number of short and taller windmills. They were usually 50 m or less in height. So, they harness low-speed surface winds that had passed through several obstacles situated above the agrarian region. The second period of greater interest in clean energy sources has begun in 1998. The "Wind Powering USA" is a national program in that country. It envisages use of clean wind power in rural and farming regions (NREL, 2018a; Union of Concerned Scientists, 2021). Wind turbines worth 60 billion US $ were installed. They were situated mostly in the farming zones of Great Plains. Such a program may last a few more years until it plateaus and stabilizes as a routine system. The wind farming programs it seems offer energy to about 800,000 households and farms, in USA. The wind energy farms that began to appear initially in the Californian coasts have now become more common in the Great Plains and Midwestern region. Wind power helps farmers in irrigation and lighting their homes. There are also farms that sublet the facilities and sell the generated power to the grid. It adds to farmer's exchequer. Here, farmers exploit agrarian sky to their economic advantage.

During 1930s and 1940s, farmers in USA adopted only small windmills. They generated power equivalent to 400 watts to 40 kilowatts. Ranchers used windmills to pump water to crops, to supply water to households and cattle. The total wind power potential of USA, it seems, is five times more than that is required annually by all sectors. Therefore, in the Great Plains, wind power has to be exploited. One of the incentives since 1800s is that cost of installing wind turbines has reduced by several folds. Further, there are reports that high altitude wind power generation above the farm belt of Great Plains is indeed a good proposition (Jiang et al., 2006). During summer months from April till end of September, the Great Plains agrarian sky is supposed to experience low-level jet streams. This is a good source of wind power. It needs to be harvested using high-altitude turbines.

The economics of owning a turbine depends on many factors. They include wind speeds, the size and cost of the wind turbine, interest rates, taxes, and electricity prices. One key issue is how much of the power generated the farm uses in situ and how much is sold back to the utility (Union of Concerned Scientists, 2021; NREL 2018b).

Wind resources above the farm belt varies based on several factors related to geographic location, terrain, topography, natural features, trees, plantations, field crops, water bodies, etc. The kinetic energy in wind is converted to mechanical energy, using the blades. Certain windpumps utilize drag to convert the wind energy. A few other use blades that are slightly inclined so that a little lift is created to the moving wind. The power generated by a windmill is directly dependent on the wind speed and direction. Also, altitude of the wind pump affects energy generated and the intermittency, if any. Wind speed of 9 km h^{-1} generates 0.01 kW m^2; 18 km h^{-1} generates 0.08 kW m^2; 27 km h^{-1} generates 0.27 kW m^2; 36 km h^{-1} generates 0.64 kW m^2; 54 km h^{-1} generates 2.2 kW m^2; and 144 km h^{-1} generates 41 kW m^{-2} (FAO 2012). Most farms utilize wind resources harvestable up to 100–200 ft. A few towers may reach taller locations of 300–300 ft (Marvel et al. 2012). Globally, wind power generated varies depending on geographic location and features related to wind speed, direction, topographic features, etc (see Fig. 5.2).

Windmills have been generally utilized for two purposes in the rural and agrarian regions. They are water supply for human households and cattle and the other is for irrigation. For household use, the wind pump has to be steady in generating energy and lift water. For irrigation even intermittent lift of water and distribution is possible. Irrigation duties on the other hand are seasonal. So, the windmill may only be useful for a limited fraction of the year. They involve pumping much larger volumes of water through a low head, and the intrinsic value of the water is low. Therefore, any windpump developed for irrigation has to be low in cost. Such a prerequisite tends to over-ride most other considerations. Most of the wind pumps even many of those still in operation in different continent were made of heavy steel and wooden blades. However, in recent times, windpumps are lightweight. They are made of lightweight plastic/graphite and simple in design.

Water pumping windmills are economically competitive in areas that do not already have electricity for powering irrigation pumps. In these circumstances, the alternatives are small engine driven pumps that are expensive. However, they require costly fuel. Their maintenance may also be costly. Low-lift irrigation applications for high-value vegetable farming may be economically competitive in many parts of the world. The economic appeal of locally built windmills is even greater. Particularly, when the savings of scarce foreign exchange is considered. Windmills are preferable for remotely located villages and farmland. Other advantages of locally built windmills include the creation of village capital using local labor and materials, much lower initial cost, and avoidance of maintenance problems associated with portable engine-driven pumps.

FIGURE 5.1 Windmill for generation of electricity and pumping irrigation water.
Left: A typical American all steel Windmill adopted for water pumping and electricity generation. Right: Typical farm wind pump configurations
Top: borehole to raised storage tank, middle: well to surface storage tank bottom: surface suction pump.
Source: FAO, 2012; Food and Agricultural Organization of the United Nations, Rome, Italy.

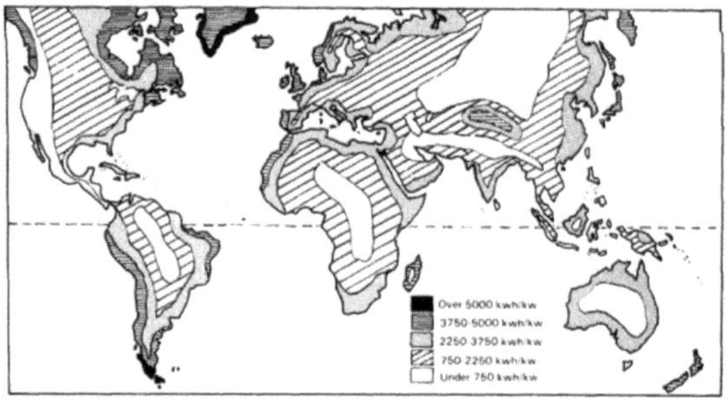

FIGURE 5.2 Annual mean wind power generated in various parts of the world.
Source: FAO, 2012; Food and Agricultural Organization of the United Nations, Rome, Italy.
Note: Wind power generation depends on wind speed and several other factors. Wind is simply air in motion. Wind is generated by the uneven heating of the Earth's surface by energy from the sun. Uneven atmospheric pressure is also a cause of wind in the atmosphere. An average wind speed of 8–16 miles per hour (mph) is needed to convert wind energy into electricity, economically. However, wind speed is just <5 mph in several agrarian locations.

5.2 LOW ALTITUDE GROUND SURFACE WINDMILLS

Bielllloo (2012) has pointed out that, wind turbines are among best choices for power generation both on land (agrarian regions) and on seas (offshore).

Wind turbines can supply energy equivalent to four times the present need for power by the entire world. If we adopt robotic aerostats with turbines, kites, and similar methods we can perpetually generate wind power for farm world as well as for others. Literally, it is said "sky is the limit for wind power generation." "Agrarian sky" that accounts for a 3rd of region above land mass. Agricultural sky is a big source of wind power. Globally, there is a trend to reduce fossil fuel and greenhouse gas emission. Hence, in countries such as USA, European, and Chinese farm zones, they are adopting wind power to a greater extent. They are supposed to be generating 239 gigawatts of energy for their use adopting wind turbines (Bielloo 2012). Four million turbines are spread around the globe. They could easily and sustainably produce 7.5 terawatts of power. It is nearly half of all power used today (Beilloo, 2012).

Wind energy is gaining in popularity due to few reasons. Firstly, it is the need for alternative sources of energy that are clean without pollution and GHG emissions. A few others are need for cost effective in situ electrical energy generation for farmers. Next, it is availability of facilities and models of windmills that are efficient in power generation. Finally, it is the easier access to established grids. The potential for wind energy, particularly, high-altitude wind turbines is high. Wind energy seems to re-emerge as popular mode of power generation.

The working principle is that when the wind passes through the blades. They experience a lift due to the aerodynamic shape of the airfoil. The blades move and start rotating due to the lift generated. The yaw unit aligns it towards the incoming wind direction. The rotation of the blades is transmitted through the gear train and couplings to the generator. Windmill models manufactured at present are aptly called Wind turbines, because most if not all are capable of generating electric power. This is in addition to functions like water lifting and/or grinding food grains. There are two types of wind turbines identifiable. They are vertical axis and horizontal axis types. A few examples of horizontal axis windmills are Single bladed, Double bladed, Three-bladed, Multibladed US farm windmill. A few other examples are Bicycle type bladed, Upwind, Sail wing, Multirotor, Counter rotating blades, Crosswind Savonius, Crosswind paddles, Diffusers, Concentrators, Unconfined vortex. There are several types of vertical axis windmills. A few examples are Savonious, Multibladed savonius, Plates, Cupped, Darrieus, Giromill, Turbine, Magnus, Airfoil, Deflectors, Sunflight, Venturi, and Confined vortex types. Several of the above models could be eventually of great use in exploiting wind power through the agrarian sky.

A simple typical wind power generator installed on ground consists of components. They are

(a) The rotors consisting of the blades and the hub;
(b) The drive train, gears, and mechanical brakes. The brakes are used in the maintenance work and whenever a storm is coming;
(c) The generator that generates electricity;
(d) The yaw system that rotates the housing toward the direction of the wind.
(e) Tower and foundation.
(f) Battery and the electrical system to transmit to the grid.

As stated earlier, the basic mechanism involved in a typical windmill involves, firstly the ambient agrarian wind blows through the blades. As a consequence, blades experience a lift due to aerodynamic foil-like structure. The blades start moving and rotating over a horizontal axis. The yaw aligns the blades toward the incoming wind based on its direction. The mechanical energy generated via rotation is transmitted via gears and couplings the electricity generator. The electricity, so generated is transmitted and stored in batteries that get charged. The electric power generated is often supplied directly to grids. Farmers may derive exchequer for selling power to the grid or use it directly on machines utilized in the farm. Households in a village often depend on windmill generated power if it is easily available.

The power output from the windmills is often mentioned, using "Beurfort scale." The power generation values consider steady state wind without gusts and intermittency. A force of 6 on the Beauford scale means it is a strong wind above crops. It is good enough to generate electricity and lift irrigation water.

Small wind turbines generally operate between force 3 and force 7 on the Beaufort scale. The rated capacity is commonly defined at force 6 with a wind speed of 12 m/s. Below force 3 the wind turbine will not generate significant power. At force 3, wind speeds range from 3.6–5.8 m/s (8–13 mph). Wind conditions are described as "light," that is, when leaves flutter. At force 7, wind speeds range from 14–17 m/s (32–39 mph). Wind conditions are described as "strong," that is, whole trees are in motion. With winds above force 7 small, domestic wind turbines should be shut down, to prevent damage. Large turbines used in the electricity grid are designed to work with wind speeds of up to 25 m/s (90 km/h, 56 mph). It corresponds to between force 9 (severe gale, 23 m/s) and force 10 (storm, 27 m/s) on the Beaufort Scale.

Basically, wind power generated by using turbine is calculated, using the following formula:

The power **P** available in the wind impinging on a generator is given by

$$\mathbf{P = \tfrac{1}{2} C A \rho v^3}$$

where in;

C is an efficiency factor known as the "Power Coefficient" which depends on the machine design,

A is the area of the wind front intercepted by the rotor blades (the swept area),

ρ (rho) is the density of the air (averaging 1.225 kg/m^3 at sea level), and

v is the wind velocity (Electropaedia, 2021).

Low altitude wind power generators are most common in farming belts. They are among best examples of efficient exploitation of farm sky. Agricultural farmers, particularly, those producing crops in regions bestowed with good amount of wind power have been using the sky above crops. Since several centuries, farmers have been adopting windmills to reap wind power and to conduct lift irrigation in the fields or to grind food grains. More recently the ground surface wind turbines have been supplying power to farm gadgets and lighting farm households. Agrarian sky has been exploited at low altitudes of 3–100 m high. A few farmers have been lending the windmills and power generated, if any, to others in the region. Leasing windmills both permanent ones and others temporarily installed ones have been an excellent idea, particularly, in years when crops have failed or yielded low. In the general course, in addition to crops, farmers with windmills can earn a small exchequer by lending windmills. It adds to stable income.

A few clear advantages stated for low altitude wind turbines operational in farms are as follows. Farmers can install the low-altitude towers on a permanent basis. It does not occupy large area in cropped fields. Each turbine needs about 0.25 ac area. Even a good wind farm mixed with cropping zones is a possibility. Usually, farmers allocate 2–3% of land for windmills. Most sowing zone in farm is left undisturbed. Irrigation lines, tiles drainage systems and other installations and pathways in the field are usually left undisturbed. Mass crop dusting sorties by airplane is possible. At present, farmers are adopting sprayer drones efficiently without being disturbed by any number of wind turbines in a farm. The above information regarding windmills and low-altitude turbines makes it clear that, "agrarian sky" is a highly remunerative proposition when we consider wind power generation. Almost each farmer can adopt the wind power generation system. Farmers could exploit agrarian sky using low-altitude wind power.

Let us consider an example of a region or state that has been exploiting wind power, in its predominantly, prairie vegetation and cereal belt. Reports from US Department of Energy suggest that wind energy in the northern prairie states like North Dakota is getting exploited. There is still a large potential for windmills still unutilized in this vastly agrarian belt. At present, only 3000 MW of energy is produced using wind turbines. It is about half the potential of this wheat belt. In North Dakota, Prairie Wind farms have reached a capacity of 43% of the total wind power feasible in the region. Clearly, there is a large potential to be exploited, both, by using low latitude ground tower turbines and high-altitude turbines (NREL, 2018b). Multi-purpose farms with crop production and wind farm facilities are indeed efficient. It involves allocating a small area of fields to wind power (towers).

How would the wind power generation scenario get affected by climate change? What would be the consequences on wind power generation in future over European farming regions? The climate change experts are forecasting global warming and rise of 1°C–1.5°C. Hosking et al. (2018) have tried to ascertain various related facts plus authentic forecasts and trends in farming practices in relation to wind energy potential, in future. They say, there could be a north word shift of "wind jets" in the Atlantic and European agrarian regions. This may increase surface winds over European plains and British islands. There could be a decrease in surface winds in the southern European farming zones encompassing nations such as Italy, Greece, Turkey Romania, Hungary, etc. But, they say, the decrease in wind energy generation potential of southern Europe is negligible. Even if a warmer climate persists. Global installed capacity for electricity is 487 GW and European share is 154 GW (GWEC, 2017; Wind Europe 2017; Mitchell et al. 2016). In Europe, about 18% of energy is contributed by wind power. Since wind power is cleaner, there is impetus to improve installed wind power in the European agrarian regions.

5.2.1 AIRBORNE TURBINES ABOVE CROP'S CANOPY

One of the definitions for airborne energy states that, it is generation of wind energy using aerodynamic or aerostatic lift devices (i.e. platforms for turbines). These methods harvest high-altitude wind power. This contrasts with ground-based turbines situated on towers that utilize surface winds. High-altitude wind energy could be harvested by keeping the turbines on various kinds of platforms. A few examples are kites, kytoons, aerostats, gliders, airfoils, etc. The first step is to get kinetic energy of wind converted to mechanical energy. This is later converted to electricity. Such power could be channeled directly to traction, lift energy using pump sets to irrigate crops, to light households in a village, etc. The mechanical energy could be converted to heat, sound, electricity, light, tension, lasers, microwave, chemical changes, etc.

Several types of airborne energy harvesters have been developed and evaluated in previous years. A few not all of them are relevant here considering the context of this book. A few of them are related to airborne wind power harvesting above the agricultural regions. Also, a few of them have been successfully commercialized and utilized, although not with high degree of acceptance. A few of the airborne wind power systems could directly utilized to move the ground vehicles on the field. Helicopters with wind turbines have also been evaluated. The copter stays at an altitude of 15,000 ft above ground while its turbines generate wind energy.

A few examples are Dutch researchers at Delf University have demonstrated an airborne system called "ladder mill." It consists of kites. It is used to run a generator. The "Kitegen Inc." is an airborne turbine manufacturing company. It produces vertical axis wind turbines. It employs high-altitude kites to place turbines. Rotokite is a model developed by Italian researchers. It employs a tethered kite lofted to a height with turbine. A more recent development in high-altitude wind energy generation systems is the placement of turbines on aerostats that are lofted to low and high altitudes, using light weigh helium aerostats. (Glass, 2015; Wikipedia, 2020).

5.3 HIGH ALTITUDE WIND TURBINES TO GENERATE ENERGY ABOVE CROPS AND FARMS

Agrarian sky is potentially a large source of wind energy. Wind energy could be generated using not just low-altitude wind gusts, but we can effectively adopt turbines that churn out steady electrical energy from high-altitude winds. High altitude winds, say, 1000 ft above cropping belts can be utilized effectively. Wind blowing at high altitudes, as stated earlier is constant, steady, and almost perpetual, if we compare it with low-level winds. There are currently few aerostat-producing companies. They actually produce light-weighted aerostats with turbines mounted and lofted at altitudes from 300–1500 ft., above natural vegetation or cropped fields.

The Airborne Wind Turbine (AWT) uses a helium-filled inflatable shell to ascend to higher altitudes. Wind is more consistent at higher altitude. They are over five times stronger than those received by traditional tower-mounted turbines. Strong tethers hold the AWT steady and send electricity down to the ground (see Rein, 2021). There are recently devised airborne turbines that are called "Buoyant Airborne Turbines (BAT)." These are light-weighted helium balloons fitted with generators that stay afloat at high altitudes. Usually, above 2000 m from ground surface/crop fields. These are known to harness the steady highspeed wind to generate electrical energy (Altaeros Energies Inc. 2021). The "BAT" can float at much greater heights than the traditional turbines attached to aerostats. They can also withstand wind gusts, hurricanes, and other turbulence better than other wind turbines.

Basically, an AWT consists of the following components. They are shell, turbine, tethers, and ground station. A shell is filled with helium for buoyancy. It is made from industrial fabric that stabilizes in the air. Turbine is a lightweight three-blade horizontal axis wind turbine. It is fixed within the helium-filled aerostat shell. Ground station could be a permanent structure or portable vehicle-based facility. It collects energy and channels it to grids where they are further sent to farms and villages. Fully automated robotic ground stations that have control switch and distribution programs are also available.

Among the wind turbines adopted by farmers, so far, traditional windmills on ground surface (30 m above ground), and slightly taller wind turbines (100–150 m above ground surface) located on permanent towers, generate only small quantities of electrical energy. They utilize the low-speed erratic wind to generate electrical energy equivalent to 50–350 W m^{-2}. They say world's tallest permanent tower holding a wind turbine is just 240 m high from ground surface. In comparison, a simple aerostat-mounted lightweight AWT reaches 1000 m and the presently tested BAT reaches above 20,000 m. These improved wind turbines generate over 1200 W m^{-2}.

It seems BAT offers electrical energy to farmers at lower cost. It is efficient in terms of economics. It serves a few villages if a ground station is connected to a network or grid. Forecasts suggest that high altitude wind energy potential is at least 100 times more than other sources, cumulatively. If we harness it, we may be able to supply the human inhabitation and their agricultural farms over 100 times than they actually need (Altaeros Energies Inc. 2021)

Let us quote an example. Altaeros Energies Inc. at Salem, in Massachusetts, USA produces balloon-like lightweight helium-filled aerostats. They are shaped like wind turbines and fixed with blades made of light-weighted plastic. Such high-altitude wind turbines are versatile. They can be moved into areas where wind speeds are constant and congenial in terms of speeds. The energy generated is usually channeled to a ground control station and connected to a grid. (Altaeros Energies Inc. 2021; Glass, 2015).

High altitude wind energy generation above vast expanses of natural vegetation and/or agricultural regions has been drawing attention of policy-makers, entrepreneurs, and farmers. Almost a decade ago, several start-up companies were provided the funds, impetus, and personnel. Aviation engineers were enthused to develop facilities for wind energy. Levitan (2012) states that, while engineering aspects made strong progress, particularly, in utilizing light-weight aerostats to generate high-altitude wind power, regulatory aspects are still to be decided. Farmers who utilize the agrarian sky above their farms may have to be clear about who can utilize such high-altitude aerostat-supported wind turbines. We may note that, wind power generation over farms is a possibility for almost each and every farmer with a moderate-sized farm holding. He may get electric power at low cost for use in his own farm. Plus, the wind power that farmers generate will be compliant and free of GHG emissions. During the past decade, most high-altitude power generators placed above agrarian regions held at around 300–1000 ft above the ground/crop surface. This is because winds are faster at higher altitude. Higher placement of aerostat in the jet stream is a possibility. This is because wind is steady and fast in the jet streams. Conventional wind power facilities in the Northeast of USA suffer from erratic wind speeds, obstacles, intermittent gusts, and space, if it is near urban locations (Levitan, 2012).

There are Dutch-based companies that have designed efficient high-altitude tethered aerostats. The energy is transported directly to a ground station via tethers (electrical wires). There are wind power generators, such as Magenn Inc. of California, USA. Their prototypes have aerostats that suit the wind direction and speed. There are also several other models of aerostat-aided wind turbines. We may have to choose the best model and try to adopt them above farms. In some regions of Europe, there are trials conducted with high-altitude tethered aerostats with wind turbines. Such turbines have been operated continuously for over a year. Obviously, they provide uniform flow of electric power.

Euan (2016) mentions several salient features about high-altitude wind energy systems. Such features are applicable to power generation turbines placed atop platforms above the agrarian sky. A few of them are as follows: The equipment, models of lightweight turbines, platforms such as aerostats, tether types, and ground station equipment need further refinement. Preliminary trials of equipment were done a decade ago. There are several items that still need improvement in terms of design, usage, energy efficiency, cost, ease of operation, flexibility, and versatility for use above crops.

In most regions of the world, the ground surface wind speed is relatively feeble at 5 ms^{-1}. This is well below the threshold for power generation, using ground-based concrete towers or wooden windmills. Hence, conventional ground surface wooden or steel or concrete tower-mounted turbines cannot be used everywhere. This is unlike high-altitude energy generation which is possible in most locations of the agrarian sky. Wind speed at 250 m above crops is most often well-suited for power generation, using high-altitude wind turbines. Several companies have developed aerostats and kites that suit wind speeds available at 200–2000 m above crops.

Tethers and transmission lines too are available for electric power to be transmitted to ground station. It could then be converted into mechanical energy. They could be utilized for electric transport vehicle, to charge their batteries. Let us consider a few advantages of adopting high-altitude wind energy systems. Firstly, high altitude wind energy is apt when wind source, speed, and continuity at the ground surface is erratic and below threshold. Often, intermittency is feeble at high altitude. Therefore, wind energy production could be a continuous system. Power generation is high. It increases by the cube root of wind speed. It means lifting aerostat or kite a bit more in the atmosphere enhances power generation, perceptibly. Hence, agrarian sky is best suited for high-altitude power generation, using tethered aerostats or kites.

Power increases by the cube of wind speed. Therefore, going a little higher in altitude may produce a lot more power. High altitude devices are significantly less massive than traditional turbines of equivalent rating. Hence, high-altitude devices could be significantly cheaper to make. The wind front of a ground-based turbine is defined by the area swept by the turbine's blades. This limits the amount of kinetic energy that a turbine can extract from the wind. Such lacunae are avoided by high-altitude aerostat-based turbines.

5.4 AIRBORNE KITE POWER

There are several models of Kite-based wind energy generators. For example, in Germany, kite wind power is utilized for traction. This is an example of direct transfer of power to the spot where it is to be utilized. There are also tests conducted to pull farm vehicles using kite power (see Krishna, 2020). There are Dutch producers who are manufacturing autonomous glider planes tethered to ground generators (e.g., Ampyx Power), Italian high altitude kite manufacturers too are adopting them to develop wind power (e.g., Kitegen Inc.). Altaeros Inc. located in New England zone of USA too are designing aerostats with tethers. They support turbines and transmit energy to ground stations.

Kite power generation technology if applied at higher altitudes seems to be efficient. It is a low-cost method (Aziz, 2012). Further, it is possible to produce more power at less initial cost, at the same site with same wind conditions when compared to conventional wind turbines. High altitude wind energy generation systems also adopt tethered kites to lift a turbine and generate wind energy. The kite often reaches 200–800 m depending on the wind currents, tethers, their length and quality. A report from Italian wind power generation companies suggests that it is possible to generate electricity, using kitegen or kite mounted turbines. Italian government has also instituted a separate "KITEGen Project" that explores use of kites above agrarian regions, to generate in situ power.

Near-surface wind power is a mature technology. It is a fast-growing renewable power source. It substantially reduces greenhouse gas emissions. However, here we are concerned with high-altitude energy production. In the Middle East, like many other regions on earth, surface wind resources could be low and intermittent (Yip et al., 2017). However, high-altitude wind resources are abundant, continuous, and persistent. Wind speeds are usually congenial. They are readily available at altitudes of 200–2000 m above ground. We may exploit this source to serve farm purposes. They may provide alternative energy resources although the Middle eastern region is basically a fossil-fuel-dependent region (e.g. petrol). Further, Yip et al. (2017) state that, airborne methods that adopt turbines mounted on aerostats, kites, or even tall concrete towers are steady and abundantly possible. The high-altitude wind-borne turbines adopt tethers with facility to transmit electrical energy instantaneously, to ground station. It is continuous and often dependable as a source of energy for small set of villages/farmers.

5.5 WIND POWER GENERATION AND ECOLOGICAL CONSEQUENCES TO AGRARIAN REGIONS

Wind energy is an aerial process. It is useful to human activity. At the same time, wind energy has certain impacts on aerial ecological aspects such as fauna (birds, insects, perhaps microbes), and ground vegetation. Wind energy-related installations and activity of turbines have a few ecological costs (National Research Council, 2007). We need to ascertain the detrimental effects on the ecological aspects. Wind energy generation is generally free from pollution and GHG emissions. It does not destroy habitats. Yet, we may realize that construction of towers of different sizes, and other permanent structures for transmission lines, ground control stations in natural reserves, also in farming zones does cause a little disturbance to natural functions in an ecoregion. The wind energy generation facilities can actually affect the ecosystem in two ways. First, by directly affecting the natural vegetation, grasses, pastures, crop fields, water ways, etc. Secondly, it can modify the habitats suited to fauna and flora. For example, several species of birds are detrimentally affected by the wind turbines. Bird fatalities due to wind turbine are negligible, if we compare the same with effects of pollutants, pesticides etc. Raptors, songbirds, passerines are among the bird species that show vulnerability to wind turbine blades. Yet, if we compare the total bird fatalities due to anthropogenic causes, it is said wind turbines in USA have been attributed as responsible for only 0.003% bird deaths. Even this, sometimes is due to complex nature of habitat, bird demography, and population density. It is difficult to decipher the cause of bird damage, accurately (National Research Council, 2007).

Wind turbines are surprisingly lethal because they kill insects. It is not just a few aerial insects. It has been observed that each wind turbine kills about 12,000 insects per day. It equates to some 1200 tonnes of dead insects per year. Further, for each bug that is killed, perhaps ten bugs are injured or dazed (Eschenbech, 2020). Such aspects need further detailed investigation.

5.6 SOLAR ENERGY IN AGRARIAN REGIONS: INTRODUCTION

In the present context of this book, solar energy generation is a phenomenon related radiation capture via agrarian sky. Agrarian belts in tropics are usually endowed with sumptuous quantities of sun's radiation. Solar photovoltaic (PV) cells were first developed in the early 1950s. Solar PV cells were invented by researchers at the Bell Laboratories in the United States of America, in 1954. Solar cells have found extensive use in space science applications, particularly in satellites, space stations, etc (Torshizhi and Mighani, 2017). PV cells can be manufactured and arranged to concentrate energy. Solar cells could be arranged as a cone or as flat panel to collect and transmit electrical energy. Flat panels are common in many parts of the world. Since 1992, interest in solar voltaic cells increased massively.

Globally, solar PV cell technology improved from negligible level in 1992 to about 512 GW power generating facility in 2018.

Globally, there is a trend to utilize renewable energy sources. This is in preference to those that utilize fossil fuel. The purpose is to reduce carbon emissions perceptibly during power generation. Agrarian sky mediates transit of sun's radiant energy on the crops. Radiation impinges on the solar panels (cells) that generate electricity. During the past five years, the solar energy capacity has increased by 60%. It has reached 485.82 GW annually in 2018 (Power Technology, 2021). Together, United States of America and China account for 30% of total global solar energy generation. Solar generation units are also called "Solar farm" or "Solar park." The ten largest solar energy farms are Tengger Desert Solar Park, China—1547 MW; Sweihan PV Independent Power Project, UAE—1177 MW; Yanchi Ningxia Solar Park, China—1000 MW; Datong Solar Power Top Runner Base, China—1070 MW; Kurnool Ultra Mega Solar Park, India—1000 MW; Longyangxia Dam Solar Park, China—850 MW; Enel Villanueva PV Plant, Mexico—828 MW; Kamuthi Solar Power Station, India—648 MW; Solar Star Projects, US—579 MW; and Topaz Solar Farm/Desert Sunlight Solar Farm, USA—550 MW. These large solar parks harvest solar energy that impinges on the vast ground surface, using solar panels. They are located in regions endowed with enormous quantities of solar radiant energy that impinges the ground surface.

The energy generated from these large solar parks are utilized for different purposes. For example, the "Topaz Solar Farm" at San Luis in California supplies electricity to 180,000 households, throughout the year. Yet another example is Solar Star Projects. It includes Solar Star 1 and Solar Star 2 projects. They are situated in the Kern and Los Angeles counties of California, USA. They generate about 579 MW electric power. Together, the farms have 1.7 million solar modules installed on 3200 acres of land. Solar Star project supplies electric power to 255,000 households daily. They are equipped with technology that positions the panels to track the sun, during the day. This technique improves energy capture by up to 25%. In states like Oklahoma, hybrid power generation using both wind and solar energy installations are being preferred. The rate of installation of new solar parks has seen fluctuations each year, during past few years (Motyka, 2021). In Central America, largest penetration of solar power in electricity production is found in Honduras (14%).

Since past 3 decades, Germany is manufacturing wind turbines and equipment necessary for solar power technology (Wehmann, 2020). This is to thwart environmental deterioration caused by fossil fuels and depend more on clean solar energy production. Reports suggest that Spain too has embarked on installing solar panels, in agrarian regions. According to Spanish Photovoltaic Cells Organization, it seems, they have to install solar panels that produce 3000 MW each year to meet the requirement of 32% of total energy consumption from solar panels. This leads to reduction in use of fossil fuels and GHG emissions.

Earlier this year, the African Development Bank announced plans to facilitate production of 10 GW of solar energy across the semi-arid Sahel region of Sub-Saharan Africa (Kenning, 2018). The aerospace above the ground surface and vegetation including sparse crops that are cultivated in the Sahel needs to be exploited better. The Sahara and its fringes the Sahel are said to be potentially rich sources of wind and solar power. The demand for power from Europe and Middle east could be satisfied, to a certain extent if we exploit aerial space above Sahara/Sahel properly. In the Sahel, farmers are expected to improve both crop productivity and their own economic status, if they took to wind/solar energy more seriously than at present (Armstrong, 2018; Al-Habaibeh, 2018; Freitas, 2019). We may note that, wind and solar installations have local effects on heat, humidity, and other climatic factors. Such effects may be beneficial or detrimental to the region. An investigation finds that wind and solar installation in the Sahara Desert and Sahel would increase local temperature, also precipitation and vegetation. Overall, the researchers state that effects would benefit the region.

One of the interesting forecasts states that, if solar panels are spread all through the dry Sahara, it results in enhanced temperature, rains, and wet tropical climate (Yan et al., 2018; United Nations Environment Program, 2020; University of Illinois at Urbana Champaign, 2018). This trend may convert vegetational pattern. Trees and prolific annuals may thrive in the Sahara. Large-scale adoption of wind/solar panels could increase ambient temperature by 2 °C and precipitation by 0.2 mm per day. Whatever be the forecast about climate and vegetation in Sahara/Sahel due to solar panels, one aspect that is sure is that it is a vast region that has potential to churn really large amount of power. It seems the power generated in this area could be 40 times that generated in UK, at present (Armstrong, 2018).

China is among the major solar energy producers in the world. It exploits the sky to generate power. It still has a large potential to install solar panels across vast expanses of waste land, natural vegetation zones, and crop belts but only during off-season. China manufactures about 60% of the total solar panels produced in the world. At present, China generates 130 GW of electricity through solar energy farms. China has the world's largest solar farm at Tennger Desert. It generates 15,000 MW electric power, annually. A few other large solar farms too add to solar power in the nation. For example, Longiaxia Dam facility on the Tibetan Plateau has 4 million solar panels. This is to generate electric power for households and farms. Solar energy has also been used to heat up the frost and reclaim vegetation zones in parts of Tibet. It is an attempt to make the area more appealing, to Chinese settlers (Baraniuk, 2018). The trend in many locations is to adopt crop production and solar panels together at staggered heights above ground surface.

There are two ways to convert solar energy into electrical energy; a system using PV technology and another that uses solar capture heating systems (Torshizhi and Mighani, 2017). In the PV system, the sun's rays are converted directly to electricity by semiconductors. Solar capture involves conversion of radiant energy to heat via concentrating the energy. It also involves thermodynamic processes, heat exchange equipment. Such solar energy can be further converted to mechanical energy. It seems wind and solar power share similarities as power generation sources. Wind and solar projects rely on freely available and renewable resources of the wind and sun (Moore, 2017). Combining solar PV power with wind power, in a hybrid system, is useful. It has the potential to reduce variability in renewable electrical energy production. Such a combination could aid power grid operators in planning and to help power companies with critical investment decisions (Brewster and Degelia, 2016; Blumsack and Richardson 2012; Fagiano et al. 2010).

5.7 SOLAR POWER PRODUCTION ON FERTILE AND WASTE LAND IN FARMS

Dual production of crops and solar power could be remunerative to farmers in regions bestowed with good sunshine hours. For example, in Wisconsin and Massachusetts farmers are growing cranberries. Simultaneously, they have installed solar panels to generate power and sell it to the nearest power grid for obtaining exchequer. Interestingly, farmers here say that, in years when cranberry prices fall, it is the generation of solar power that offers extra exchequer (Marcelo, 2019). They have built vast number of solar panels on bogs in a farm. Farmers state that, it is a novel way of blending renewable solar energy and cranberry production in a single field and adjacent areas. Farmers are also adopting dual production system. First, they plant crops and then placing solar panels at a height not to overshade the foliage. This way they can avoid any reduction to photosynthetic efficiency due to shade from solar panels. Clearly, these are excellent propositions to overcome any low or bad harvests of major crops, due to disease, pests, natural disasters or economic recession, and low-price levels in the market.

German farmers refer to the dual use of land for crops and solar energy production as system of "Agrophotovoltaics" or "APV." The APV has the potential to open up new space that is urgently needed for the PV expansion in Germany. One of the experimental demonstrations in the German plains has involved 720 bi-facial solar panels. It covered about 0.33 hectare of agricultural land. Solar panels were mounted high enough to allow the crops planted below to receive sunshine at normal levels. Further, sufficient space was also allocated for farm machinery to operate beneath them. After a year of trials, the research showed that the dual use system increased the total productivity of the land, by 60%. The first crops evaluated were winter wheat, potatoes, celery, and clover plus grass mixture. "The crop yield of clover/grass under the PV array was only 5.3% less than the reference plot." The yield for potatoes, wheat, and celery were about 19% less. It is believed that "Agrophotovoltaics" could improve land use efficiency.

Solar energy generation is increasing in many parts of Indian subcontinent. During recent years, farmers in agrarian regions of India are preferring to install a certain number of solar panels. Along with it even windmills are located to utilize the power in situ. There are several variations of farm models, both small and large being experimented in the Indian agrarian regions. Solar panels have been installed sometimes temporarily in agricultural fields, during off-season. In certain cases, the height of the solar panel is increased to the extent that its shadow will not obstruct solar radiation falling on cropped land. In certain cases, they have even examined the effect of solar panel created shadows on crops grown in the neighboring regions. Harinarayana and Vasavi (2014) state that, if dual system of crop and solar energy production is practiced, then, it results in greater land use efficiency. Farmers will achieve better economic profits from their fields (Gujarat Energy Research and Management 2014; Das et al., 2018). Further, there are several solar panel parks situated in waste land stretches in South India (Plate 5.1)

PLATE 5.1 Solar panels at farms and industrial locations in Karnataka, India.

Source: Amplus Solar Energy Solutions Ltd, Bengaluru, https://amplussolar.com

Top: Veera Project, Gadag District, Karnataka state, India; Middle: Veera Project Gadag District, Karnataka state India; Bottom: Nayaka Project, Chitradurga District, Karnataka state, India. These solar parks are situated in the vertosols plains of South India.

Note: Such practices during off-season improves efficiency with which farmers utilize agricultural sky, particularly, in agrarian regions endowed with ample sunlight hours per day, all through the year.

Japanese farmers are generating electricity simultaneously while growing crops. The practice of food and energy double-generation is known as "solar sharing," in Japan (Wan-Ho, 2013). Solar-shared farming may be appropriate for other regions such as in Africa, Latin America, and the Middle East. Here, the incident sunlight is intense plus the water is scarce. Farmers are required to report their annual crop cultivation. If the amount cultivated on the solar-shared farmland falls below 80%, then, they will be required to dismantle the PV system. This is to reduce shade created by solar cell panels. The structures are made of pipes and Solar PV panels are mounted above-ground and arranged at certain intervals. This is to allow enough sunlight for photosynthesis. These structures can be easily dismantled. The system allows adequate sunlight for crops and space for agricultural machinery to be used. Overall, installation of solar panels in the agrarian regions, in fields, in areas with natural vegetation or wasteland with scanty vegetation has its impact on biomass production trends and general greenery. We could study the impact of solar panels or even simulate it on computers and then decide on the extent of solar panels to be distributed in a farm location (Hernandeza et al., 2015).

For example, farmers in Japan installed 348 PV panels in a small 750 square meter farm. They mounted solar cell panels on pipes that are 3 m high from the ground. The PV system produces 35,000 kWh a year. Under the solar panels, they have cultivated peanuts, yams, egg plants, cucumber, tomatoes, taros, and cabbages, during the winter. The vegetables are sold at a nearby street market and consumed by the neighbors. Japan requires about 2.5 million acres of land to supply the entire country's electricity, using PV cells. Under the solar-share scheme, it will take about 7 million acres of farmland to supply the same amount of electricity. At present, Japan has more than 11.3 million acres of available farmland. A sizeable share of it could support solar panels in the future.

5.8 APPLICATIONS OF SOLAR ENERGY IN AGRARIAN REGIONS

Solar energy is a natural gift channeled via the agrarian sky to the farmers. There are limits to its applications during sunshine hours. However, it could be harnessed and stored for use later as required. Solar energy could become a key resource of energy for global farming. It could hold the key to efficient use of farm vehicles, gadgets, and other devices of immediate requirement to farming. Solar energy promises to be cost efficient to farmers of different regions and different economic disposition. Since capital needs are within limits, solar panels could just spread all over the developing regions of the world and farms. There are clear suggestions that, in future, solar energy reduce production costs of each individual farm/farm household, all over the world. Solar energy therefore needs greater attention in near future. Farmers may soon opt for using solar energy over the crop fields, particularly, during off-season. They could adopt them over the dairy sheds and cover the roofs with solar panels. Vineyards may opt to possess solar panels in between the rows. Orchards may have well spread out set of solar panels placed in the open space where tree shades are absent. Farmers may actually put sun's energy to work for them better than at present, by planting crops and installing solar panels.

According to Union of Concerned Scientists (2008), amount of energy from the sun that reaches earth each day is enormous. All the energy stored in earth's reserves of coal, oil, and natural gas is equal to the energy from only 20 days of sunshine. The amount of sunshine received of course differs based on location within a geographic region. For example, in USA, arid areas such as Arizona and Nevada get more sun than other parts of the United States. It is accepted that most areas receive enough sunshine to make solar energy practical.

Further, Union of Concerned Scientists (2008) state that, solar energy can be used in agriculture in a number of different ways. Solar energy conversion is an efficient process. Solar energy has several applications relevant to farming. It saves money, increases self-reliance, reduces GHG emissions and pollution. Solar energy can cut a farm's electricity and heating bills. Solar heat collectors can be used to dry crops and warm homes, livestock buildings, and greenhouses. Solar water heaters can provide hot water for dairy operations, pen cleaning, and homes. PVs (solar electric panels) can power farm operations and remote water pumps, lights, and electric fences. Buildings and barns can be renovated to capture natural daylight, instead of using electric lights. Solar power is often less expensive than extending power lines (Union of Concerned Scientists, 2008; Agriculture Energy LLC., 2021; Hanley 2017; Harris, 2019).

A few applications most useful to farmers immediately on the farm are as follows:

Solar Powered Water Pump Generator, Solar energy powered center-pivot irrigation system, Solar Warehouse Units, etc. (Agriculture Energies LLC., 2021). Torshizhi and Mighani (2017) mention other uses of solar energy in a farm setting. Such uses pertain mostly to farm households or farm-related industries. For example, in drying, preserving, and transport. Preservation of food through drying is one of the oldest and the most widespread methods. It can be used to enhance the strength of the food. Solar cookers are common in rural areas of developing nations. They utilize clean energy that is drawn from sun's radiant energy. Water heaters are one of the useful applications of solar energy for farms and industrial applications. For example, solar dryers of crop residues, grains, and other farm products. They are being supplied rapidly in areas that intend to reduce use of fossil fuel and consequent GHG emissions. Water heating systems are also common in both rural and urban households worldwide.

At present, there are innumerable options available for adopting PV systems in agriculture and horticulture. Farm equipment and gadgets that exploit solar energy are being designed at a rapid rate, for adoption in fields and farm houses. According to Xiarchos and Hay (2019), designing or updating building material for solar energy generation periodically is a good idea. It helps to take advantage of the sunlight better. It often costs relatively less and is efficient. They could enhance farm's productivity. The PV panel systems can harness the sun's energy to pump water for irrigating crops and operating farm machinery. Solar technologies can also be effectively used to provide light, heat, and ventilation to the farm-house and livestock buildings. There are indeed several ways available to exploit solar energy in agrarian regions (Xiarchos and Hay, 2019). Solar technology offers farmers an opportunity to reduce their energy costs. Some solar strategies involve an initial investment, but once those costs are recovered, fuel is free.

The greenhouse is a structure that is utilized in agriculture to grow plants with the best quality fruits and vegetables. Recently, solar energy has been used for heating greenhouses. So that, such greenhouses are known as "solar green houses." Solar energy can be used to provide light to the greenhouse (Torshizhi and Mighani, 2017).

There are clear reports that energy requirements of crop production using greenhouse is high. Particularly, if we are using fossil fuel, petroleum, or similar sources that are costlier. Adoption of greenhouse for mass production vegetable particularly in temperate regions such as northern Europe, Canada, and Northern Great Plains involves prohibitive cost. Hence, one of the suggestions is to adopt solar PV to generate electricity above the roofs of greenhouses or even above the crops inside the glasshouses. This is to generate power that could be utilized in situ. Several of the technical hurdles have been overcome. Agricultural technologists are aiming at efficient solar energy system. Further, this way we would be utilizing the agrarian sky better than ever before (Hassanien et al., 2016).

KEYWORDS

- **agricultural sky**
- **ecology**
- **high altitude**
- **kite power**
- **solar power**
- **windmills**
- **wind power**

REFERENCES

Agriculture Energy LLC. Agriculture Solar and Wind Energy. [Online] 2021, pp 1–8. http://www.agriculturesolar.com/3b_wind_&_solar_energy_in_agriculture.html#.Xb_0qpIzapo/ (July 29, 2021)

Al-Habaibeh, A. Should We Turn the Sahara Desert Into a Huge Solar Farm. *The Conservation*, 2018, pp 3–7. https://theconversation.com/solar-panels-all-over-the-sahara-desert-imagine-newsletter-2-116320 (July 21, 2021)

Altaeros Energies Inc. Buoyant Airborne Turbines: What's New in the Wind Fair World. The Altaeros Buoyant Airborne With Turbines; Altaeros Energies Inc.: Salem M, USA, 2021, pp 1–4. https://w3.windfair.net/wind-energy/news/17116-what-s-new-in-the-windfair-world-the-altaeros-buoyant-airborne-wind-turbine/ (July 25, 2021)

Armstrong, A. Massive Solar and Wind Farms Could Bring Vegetation Back to the Sahara. *The Conservation*, 2018, pp 1–15. https://qz.com/africa/1384503/solar-and-wind-power-farms-could-turn-sahara-green-again/ (July 21, 2021)

Aziz, I. Design of a High-Altitude Wind Power Generation System. Master's Dissertation, Linkoping University, Institute of Technology, Linkoping, Sweden, 2012, p 63. http://liu.diva-portal.org/smash/get/diva2:697691/FULLTEXT01.pdf/ (July 25, 2021)

Baraniuk, C. China's Giant Solar Energy Projects. *BBC New Letter*, 2018, pp 1–8. https://www.bbc.com/future/article/20180822-why-china-is-transforming-the-worlds-solar-energy/ (July 28, 2021).

Bielloo, D. The Sky is the Limit for Wind Power. *Sci. Am.* **2012,** 1–2.
https://www.scientificamerican.com/article/no-limit-for-wind-power/ (July 24, 2021)

Blumsack, S.; Richardson, K. Cost and Emissions Implications of Coupling Wind and Solar Power. *Smart Grid Renew. Energy* **2012,** *3*, 308–315.

Brewster, K. A.; Degelia, D. K. In *Availability and Variability of Potential PV Solar and Wind Power Production in Oklahoma*. 7th Conference on Weather, Climate and the New Energy Economy, New Orleans, LA; American Meteorological Society: Massachusetts, 2016, Paper No. 3.3, pp 1–9.

Das, S. K.; Naidu, B.; Manikanta, G. In *Application of High-Altitude Wind Power Generation as Distributed Generation and its Optimal Location and Sizing.* 2nd International Seminar On "Utilization of Non-Conventional Energy Sources for Sustainable Development of Rural

Areas ISNCESR', Bhillai, India, 2016. https://www.ijsr.net/conf/PARAS16/paras2016.php/ (July 24, 2021)

Electropaedia 2021 Wind Power. Battery and Energy Technologies. pp 1–18 https://mpoweruk.com/wind_power.htm/ (July 24, 2021)

Eschenbech, W. Explaining Wind Turbine Lethality. [Online] 2020, pp 1–3.
https://wattsupwiththat.com/2019/08/21/explaining-wind-turbine-lethality/ (July 27, 2021)

Euan, M. High Altitude Wind Power Reviewed. [Online] 2016, pp 1–12.
http://euanmearns.com/high-altitude-wind-power-reviewed/ (July 22, 2021)

Fagiano, L.; Milanese, M.; Pigo, D. *High Altitude Wind Power Generation.* IEEE Transactions on Wind Power Technical Report No 27082009, 2010, pp 1–25. DOI: 10.1109/TEC.2009.2032582/ (July 25, 2021)

FAO 2012 Wind Power. FAO Repository Article. Food and Agricultural Organization of the United Nations, Rome, Italy, 2012, pp 1–40. http://www.fao.org/3/ah810e/AH810E10.htm/ (July 18, 2021)

Freitas, W. 2019 Solar panels all over the Sahara Desert. The Conversation. pp 2–3
https://theconversation.com/solar-panels-all-over-the-sahara-desert-imagine-newsletter-2-116320/ (July 25th, 2021)

Glass, B. Floating Wind Turbines Bring Electricity to Where it is Needed. National Science Foundation: Washington, D.C., 2015. https://phys.org/news/2015-02-turbines-electricity.html pp 1-3 (February 14, 2019)

Gujrat Energy Research and Management. Dual-Use Farms-Installing Solar Over Crops. [Online] 2014.
https://www.energymatters.com.au/renewable-news/em4172/ (July 25, 2021)

GWEC. *Global Wind Energy Council.* Global Wind Report: 2016 Annual Market Update. [Online] 2017, pp 1-1-3 http://gwec.net/publications/ global-wind-report-2/global-wind-report-2016/ (July 25, 2021)

Hanley, S. Combining Solar Panels With Agriculture Makes Land More Productive. Clean Technica. 2017, pp 1–12.
https://cleantechnica.com/2017/11/24/combining-solar-panels-agriculture-makes-land-productive/ (August 28, 2021)

Harinarayana, T.; Vasavi, K. S. V. Solar Energy Generation Using Agriculture Cultivated Lands. *Smart Grid Renew. Energy* **2014,** *5*, 31–42.

Harris, M. Higher, Cheaper, Sleeker Wind Turbine of the Future- in Pictures. [Online] 2019, pp 1–3. https://www.theguardian.com/sustainable-business/gallery/2017/may/08/renewables-wind-energy-turbines-tech-kites-drones-in-pictures (July 15, 2021)

Hassanien, R. M. E.; Li, M.; Lin, W. D. Advanced Application of Solar Energy in Agriculture Green Hours. *Renew. Sustain. Energy Rev.* **2016,** *54*, 989–1001.

Hernandeza, R. R.; Hoffackere, M. K.; Murphy-Mariscale, M. L.; Wind, G. C.; Allencef, M. F. Solar Energy Development Impacts on Land Cover Change and Protected Areas. *Proc. Natl Acad. Sci.* **2015,** *112*, 13379–13589. www.pnas.org/cgi/doi/10.1073/pnas.1517656112 / (July 26, 2021)

Hosking, S.; Macleod, D.; Phillips, T.; Holmes, C. R.; Watson, P.; Shuckburgh, E. F.; Mitchell D. Changes in European Wind Energy Generation Potential Within a 1.5°C Warmer World. Environmental Research Letters 13: 054032 [Online] 2018, pp 1–10. https://doi.org/10.1088/1748-9326/aabf78/ (July 25, 2021)

Jiang, X.; Ngar-Cheung, L; Issac, M. H.; Ployshay, J. J. Mechanism of the Great Plains Low-Level Jet Steams as Simulated in an AGCM. *J. Atmos. Sci.* **2006,** *64*, 4–12.

Kenning, T. Halting Desertification in the Sahel With Solar Power. [Online] 2018, pp 1–8. https://www.pv-tech.org/long-reads/long-read-halting-desertification-in-the-sahel-with-solar-power/ (July 29, 2021)

Krishna, K. R. *Aerial Robotics in Agriculture: Parafoils, Blimps, Aerostats and Kites*; Apple Academic Press Inc.: Palm Bay, Florida, 2020, p 396.

Levitan D. High Altitude Wind Energy: Huge Potential and Hurdles. Yale School of Forestry Studies: Connecticut, 2012, pp 1–8. https://e360.yale.edu/features/high_altitude_wind_energy_huge_potential_and_hurdles/ (July 26, 2021)

Li, Y.; Kalnay, E.; Motesharrei, S.; Rivas, J.; Kucharski, F.; Kirk-Davidoff, D.; Bach, E.; Zeng, N. Climate Model Shows Large-Scale Wind and Solar Farms in the Sahara Increase Rain and Vegetation. *Science* **2018,** *361*(6406), 1019–1022. DOI: 10.1126/science.aar5629 (July 31, 2021)

Marcelo, P. Cranberry Farmers Want to Build Solar Panels Over Their Bogs. [Online] 2019. https://techxplore.com/news/2019-10-cranberry-farmers-solar-panels-bogs.html/ (July 28, 2021)

Marvel, K.; Kravitz, B.; Caldeira, K. Geophysical Limits to Global Wind Power. Nature Climate Change, 2021, pp. 22–23. DOI: 10.1038/nclimate1683 (July 14, 2021)

Mitchell, D.; James, R.; Forster, P.; Betts, R.; Shiogama, H.; Allen, M. Realizing the Impacts of a 1.5°C Warmer World. *National Climate Change* **2016,** *6*, 735–737.

Moore, J. Powering into the Future With Renewable, Reliable Power. United States Climate and Clean Energy Program. 2017, p 21. https://www.nrdc.org/experts/john-moore/powering-future-renewable-reliable-power/ (July 29, 2021)

Motyka, M. The 2020 Renewable Energy Industry Poised to Entre New Growth Phase. *Delloitte Trans Bus. Analy LLP.* **2021,** 1–8. https://www2.deloitte.com/us/en/pages/energy-and-resources/articles/renewable-energy-outlook.html/ (July 26, 2021)

National Research Council. Ecological Effects of Wind-Energy Projects. Environmental Impacts. National Academies Press, 2007. DOI: 10.17226/11935/ (July 21, 2021)

NREL *Wind Resource Potential*. National Renewable Energy Laboratory, 2018a, pp 1–3. www.eere.energy/gov/windpoweringamerica/wind_resources.html. (July 19, 2021)

NREL. *North Dakota Energy Resources: Wind*; Great Plains Energy Corridor Annual Report: United States Department of Energy, 2018b, pp 1–2. https://www.energynd.com/resources/wind/#:~:text=North%20Dakota%20has%20more%20than,its%20power%20from%20wind%20resources/ (July 25, 2021)

Power Technology. The World's Biggest Solar Power Plants. Power Technology, 2021, pp 1–12.

Rein, A. *Altaeros Energies Achieves Breakthrough in High Altitude Wind Power*; Altaeros Energies Inc.: Salem, Mass, 2021, pp 1–5. https://www.newswire.com/altaeros-energies-achieves-breakthrough/165882/ (July 25, 2021)

Torshizhi, M. V.; Mighani, A. H. The Application of Solar Energy in Agriculture. *J. Renew. Energy Sustain. Dev.* **2017,** *3*, 234–242.

Union of Concerned Scientists. *Renewable Energy and Agriculture: A natural Fit*; Reports and Multimedia Explainer. 2008, pp 1–17. https://www.ucsusa.org/resources/renewable-energy-and-agriculture/ (July 29, 2021)

Union of Concerned Scientist. *Farming the Wind: Wind Power and Agriculture*. Clean Energy Program of Union of Concerned Scientists: Fact Sheet. [Online] 2021, pp 1–5. https://www.ucsusa.org/sites/default/files/2019-09/agfs_wind_2003.pdf (July 20, 2021)

United Nations Environment Program. Wind and Solar Sahara Could Increase Rainfall in Sahel. *News and Reports of UNEP* [Online] 2020, pp 1–3. https://www.unenvironment.org/news-and-stories/story/more-just-clean-energy-wind-and-solar-sahara-could-increase-rainfall-sahel/ (July 21, 2021)

University of Illinois at Urbana Champaign. Large Wind and Solar Farms in the Sahara Would Increase Heat, Rain and Vegetation. *Science Daily*, 2018, pp 1–8. https://www.sciencedaily.com/releases/2018/09/180906141611.htm/ (August 1, 2021)

Wan-Ho, M. Japanese Farmers Producing Energy and Crops Simultaneously.

Permaculture Research Institute. *Permaculture News*, 2013, pp 1–7.

https://www.permaculturenews.org/2013/10/16/japanese-farmers-producing-crops-solar-energy-simultaneously/ (July 28, 2021).

Wehmann, B. *Solar Power in Germany- Output, Business and Perspectives*. Clean Energy Wire, 2020, pp 1–18.

https://www.cleanenergywire.org/factsheets/solar-power-germany-output-business-perspectives/ (July 1, 2021)

Wikipedia. Airborne Wind Turbines. [Online] 2020, pp 1–8. https://en.wikipedia.org/wiki/Airborne_wind_turbine/ (July 20, 2021)

Wikipedia. Wind Power in the United States. [Online] 2021, pp 1–17. https://en.wikipedia.org/wiki/Wind_power_in_the_United_States/ (July 26, 2021)

Wind Europe. Wind Power 2017—Annual Combined Onshore and Offshore Wind Energy Statistics. [Online] 2017, pp 1–26. https:// windeurope.org/about-wind/statistics/european/windin-power-2017/ (July 26, 2021)

Xiarchos, I.; Hay, F. *Solar Energy in Agriculture-Resources*; United States Department of Agriculture: Beltsville, MD, 2019; pp 1–9.

https://farm-energy.extension.org/solar-energy-in-agriculture-resources/ (April 2021)

Yip, C. M. A.; Gunturu, C. B.; Stenschikov, G. L. High Altitude Wind Resources in the Middle East. *Sci. Rep.* **2017**, *7*(1), 29. DOI: 10.1038/s41598-017-10130-6/ (July 26, 2021)

CHAPTER 6

Aeroponics to Revolutionize Crop Production

ABSTRACT

Aeroponics involves the use of aerospace within a chamber for crop production. Aeroponic systems utilize nutrient rich aerosols. The aerosols are held within a closed chamber. The nutrient medium prepared in water is atomized on to roots held in a chamber or larger enclosures. It ensures that nutrients reach root surface. Aeroponic systems are not plausible in the open fields, yet. A point to note here is that in natural open field conditions, roots try to reach moisture and nutrients held in soil phase. However, in aeroponics, nutrients are supplied carefully right on to the surface of roots. It improves efficiency of water and nutrient supply and their acquisition.

Aeroponics have been commercialized. Crops like spinach, tomatoes, and potatoes have been produced in large-scale aeroponic systems held in closed garages. Basic requirement of a modest-sized aeroponic installation useful in vegetable production has been mentioned. Advantages and disadvantages of adopting aeroponics have been discussed. Indeed, several different types of aeroponic assemblies are available all over the world. There are aeroponic systems that spray nutrient aerosols on to both roots and shoots, that is, foliage. Foliar mode of nutrition involves absorption of moisture and dissolved nutrients via stomata, hydathode, and lenticel. Leaf characteristics are important when nutrients are supplied as aerosols to foliage.

Whatever the suggestion or arguments, aeroponics is an efficient system of crop production, particularly, vegetables, and few other species. Several nations in Europe utilize hydroponics/aeroponics to produce clean vegetable for human consumption. Aeroponics helps avoid diseases, pests, and loss

The Agricultural Sky: A Concept to Revolutionize Farming. K. R. Krishna, PhD (Author)

of crops due to natural disasters. This is because it does not involve open agrarian fields vulnerable to natural forces. Aeroponics has been utilized to produce roots for sensitive experimental purposes, for example, to study root physiology/biochemical manifestation. Roots generated under aeroponics are clean. So, biochemical analyses are not confounded by the soil phase. Crops grown in aeroponics are useful in studying pathogens and symbiotic organisms without the interference of atmosphere or soil factors.

6.1 INTRODUCTION TO AEROPONICS

Aeroponics is a topic pertinent with the central theme of this treatise, that is, "Agricultural sky." Aeroponics, as a concept, utilizes free aerospace to supply moisture, nutrient, hormones, and pesticides through aerosol, to grow crops. Perhaps, we may ultimately develop vast stretches of open fields supplied (sprayed) with aerosols, to feed crops with nutrients dissolved in water. Aeroponics is a novel system of crop production. It utilizes aerosols made of nutrients and other essential factors, if any. Aerosols are created in aerospace within a chamber to confine the fog or mist. It is the aerosol confined in the aerospace that provides nourishment and water-to-root systems. In some cases, an entire plant is held in a mist chamber. Then, gas, moisture, and nutrient exchange occur via both the root and shoot systems that are held within the aeroponic boxes or greenhouse. Therefore, in some cases, aeroponic boxes may contain mist that feeds the plants both via roots and foliage. Whatever the variations of aeroponic systems and its engineering aspects, the basic fact is that it is a soil-less system. Aeroponics avoids full set of problems related to soil that may affect crop production. This system does not occupy soil profile, nor utilizes soil-based nutrients, water, or other factors. It solely confines to aerospace above ground whether you place the aeroponic system in open field, greenhouses, or enclosed chambers.

Let us consider a few historical facts about the evolution of the idea, design, standardization of equipment, its variations, and crops amenable for production under aeroponic conditions. During second half of 1880s and till early 1900s, there were a few soil-less plant culture methods devised (Li et al., 2018; Gabreil, 2021; Gericke, 1940). For example, hydroponics or water culture was invented in 1930s and publicly demonstrated for use by gardeners and farmers by Gericke (1930). Here, the root systems do not have soil for anchorage and spread. A rooting medium too is absent. Instead, roots are grown in water that is constantly aerated to provide dissolved oxygen (Jones, 1982; Jones, 2014). Next came the aeroponics that is in vogue since mid-1900s. Aeroponics involved formation of nutrient-rich well-oxygenated aerosols. Aerosols were impinged on to hanging root surface, using atomizers. Several species of vegetables and other species were successfully cultivated, using simple aeroponic instrumentation (Zobel, 1976; Kratsch, 2006). Crops grew rapidly under aeroponics because of highly oxygenated environment on the root surface and easy access to minerals. Lighting in chamber induces rapid growth of shoot. Further, aeroponics allowed good control over the rooting zone. Also, we could add or decrease or delete nutrient or hormonal factors, as required. There would be no interference from soil-related chemical processes in an aeroponic chamber.

A soil-less technique of growing plants known as "aeroponics" was also used by plant scientists/botanists, during 1920s.Their aim was to investigate root systems in detail about their histology, physiological functions, and gross morphology (Barker, 1922). They suggested that aeroponics could be a good system to grow plants rapidly. Such soil-less systems cause least damage to roots, during cultivation. In 1944, Klotz developed a mist chamber meant to grow citrus plants. He reported that his soil-less mist chambers were good to study the root diseases of crops like citrus and avocado (Klotz, 1944). Later, in early 1950s, apple seedlings were cultivated in foggy mist chambers by Vyvyan and Travell (1953). Went (1957) reported aeroponic cultivation of coffee seedlings and tomatoes. They were actually watertight containers having fog/nutrient rich aerosol. Atomizers were adopted to release nutrient solutions into the fog chambers. It seems during 1960s, the aeroponic systems were mostly laboratory experimental systems. They provided good and clean research material (specimens), to plant researchers. As stated earlier, aeroponics avoided soil-related contamination (Stoner and Schorr, 1982). During early 1980s, aeroponics received attention from commercial nursery and crop production specialists. They tried to adopt it for commercial production of high-value vegetables.

The Epcot Centre at Disneyland (Florida, USA) had a permanent exhibit of aeroponics showing several species of vegetable and garden crops. The crop species were being grown in mist chambers, fogs, and aeroponic conditions in green houses (Peterson and Kreuger, 1988). During past 5–6 decades, hydroponics and aeroponics are being adopted in plant physiological research (Li et al., 2018; Wagner and Wilkinson, 1992). During the same period, aeroponics was also adopted to grow obligate root pathogens and symbiont like arbuscular mycorrhizas (Sylvia and Hubbell, 1986; Sylvia and Jarstfer, 1992). Aeroponics was utilized to grown soil-free spores of arbuscular mycorrhizas. Such fungal cultures allowed detailed studies of morphogenetics and inoculum production (personal observations). Such microbial specimens were suited for electron microscopic studies and detailed DNA analysis and serological studies. Since the microbes were produced in soil-less conditions, interference by soil factors during detailed biochemical and electron microscopic studies were avoided. During 1980s, soil-less aeroponics was also adopted to study legume plant-bradyrhizobium interactions at cellular and subcellular levels (Hung and and Sylvia, 1988; Sylvia and Hubbell, 1986; Sylvia and Jarstfer, 1992; Tibbitts et al., 1994).

Vinoo (2018) points out that, for several millennia, farmers have adopted soil tillage, ploughs, and harrows to loosen soil, and then, plant seeds. It is only in recent decades that hydroponics and aeroponics have been considered for mass production of crops, especially vegetable crops. Knowledge about how well we can channel nutrients in a soil-less system is crucial to develop aeroponics.

Once a proper understanding of what elements constituted plant nutrition was gained, it became easier to develop appropriate nutrient solutions. Further, it became possible to grow plants in closed systems. Water, with fertilizer dissolved in it, was circulated across plant roots—a system which we call "hydroponics." Currently, aeroponics is in vogue in many nations.

According to some agricultural entrepreneurs, aeroponics represents a quantum jump in water-use technology. *Aeroponics consumes 90% less water to produce the same amount of biomass compared to traditional field grown vegetables. Similarly, about 20% less water is used in aeroponics than the hydroponic systems* (Editorial Staff, 2021). The soil profile and its effects on roots are nil. It is a soil-less production of crops. Roots are allowed to hang freely absorbing moisture and nutrient elements from the mist. It seems in Newark, in USA, agricultural specialists are building world's largest aeroponic facility. Such an installation may perceptibly reduce water usage. In addition to adopting aeroponics, it is interesting to note that, the entire aeroponic system is powered by electricity generated by using solar cells/ panels above greenhouses. This installation in North America is supposed to be used for producing 2 million kg of vegetables like tomato, potato, capsicum, cucumber, etc. (Editorial Staff, 2021).

Aeroponic systems are being evaluated in Mexico for large-scale production of vegetables and other related edible crops. This project aims at standardizing aeroponic equipment, that is, aerosol-producing machines, atomizers, mist chambers, watering systems, and nutrient-solution recycling apparatus. Several ways of planting, that is, seedling placement methods and densities are also being standardized. They should allow proper impingement and circulation of aerosol of nutrients on roots (Reyes et al., 2012). In an aeroponic system used to produce vegetables, we can actually control humidity, temperature, pH, electrical conductivity of nutrient solution (i.e., salt concentration), and usage of water. As stated earlier, we minimize use of water and nutrients (fertilizers) immensely, by adopting aeroponics. Such large-scale aeroponic systems in the Central American Nation of Mexico could reduce the burden of using excessive fertilizers, water, and plant protection chemicals. Such inputs are necessary for a decent crop of vegetable, if grown in the open field. It is often suggested that higher capital needs of an aeroponic system easily gets paid up, since aeroponics are feebly affected by disease and pests. Marketing is easier. Harvesting can be made to synchronize with market demand. It is a soil-less system. So, soil preparation, management, and other problems related to soil phase is entirely avoided. Economic costs on soil preparation and interculture are avoided.

Baudoin et al. (2013) reported that the artificial plant growing method (e.g., greenhouse and factory farms) is one of the fundamental types of precision agriculture. Since factors such as water, nutrients, and gases like O_2, CO, N_2, NO_x are uniformly available to crops. The photosynthetic radiation is also intercepted uniformly by each similar sized/aged seedling. So, it could be called a good example of precision farming. This is not related to the precision farming that we usually find in the open fields/farms.

Aeroponics could be explained as a method to grow crops in air, by generating suitable aerosols in any part of the earth. Root system grows hanging freely in the mist. There is no need for mediums such as soil, vermiculite, peat moss, or any other rooting medium. Gopinath et al. (2017) mention that aeroponics are well suited to produce short-duration vegetable of diverse species. Aeroponics may be necessary in some regions to avoid growing on low fertility and deteriorated soils. In areas not usually cultivating vegetables but those importing them from long distance, aeroponics provides a better option for in situ production of vegetables. They say we can even try to modify the nutritional status of produce to a certain extent by manipulating the nutrient solution.

At present, aeroponic system for production of several different species of crops is in vogue. Tunio et al. (2020) opine that after realising the advantages of aeroponic systems, several nations have adopted aeroponics. A few countries regularly adopt aeroponics to support production of vegetables and seed material under controlled conditions. Aeroponics is practised in North America (Canada and USA); also, Brazil, Colombia, Bolivia, Peru,

Chile, and Argentina in South America. European nations such as France, Spain, Russia, Malaysia, Germany, Greece, Poland, Italy, and Slovakia have adopted aeroponics for vegetable production. Australia, New Zealand, Asian nations like Egypt, South Korea, Indonesia, Thailand, Japan, Mongolia, Peru, Uzbekistan, Kenya, Sri Lanka, Iran, Italy, Korea, Vietnam, Bhutan, Nigeria, India, Philippine, and Abu Dhabi have adopted aeroponics. African countries like Uganda, Tanzania, Mozambique, Malawi, Ghana, Ethiopia, and Kenya too practice aeroponics for vegetable production (see Lakhiar et al., 2018a, 2018b, 2019).

Artificial intelligence (AI), also called autonomous systems or robotics, was developed in 1956, as "the science and engineering of making intelligent machines." Lakhiar et al. (2018a, 2018b) states that to enhance the quality and productivity of the cultivated crop, we are moving toward the modern plant cultivation technologies in agriculture. Indeed, artificial intelligence (AI) is among the recent additions to control the aeroponic systems. The autonomous aeroponic systems that reduce cost on labor are being introduced. Sensors are key to the introduction of robotics to aeroponics. Several types of sensors have been introduced into aeroponic chamber maintenance during the past decades, that is, since 2000 (Lakhiar et al., 2018a, 2018b).

6.1.1 DEFINITIONS AND EXPLANATIONS FOR AEROPONIC CULTURE OF CROPS

Let us consider a few variant definitions and explanations about aeroponics offered by plant production specialists. As stated earlier, basically, aeroponics is soil-less production of crops. It adopts aerospace above the ground to spray roots with nutrient rich aerosols, using atomizers/nozzles. Aeroponics is producing crops using mist chambers. When the mist chambers engulf both roots and shoots, it means we are feeding crops both via roots and leaves (stomata, hydathodes, and lenticels). Aerial feeding supplements root-mediated absorption of water and nutrients.

Aeroponics has been called as "white collar agriculture" because farmers need not touch soil with their hands. Products are basically clean and easy to harvest. Aeroponics involves minimum use of water, fertilizers, and energy. It is being adopted to produce vegetables (lettuce), basil, parsley, celery, gherkins, zucchini, and cherry tomatoes, at rapid rate (Sharma et al., 2018; Sharma, 2017; Plate 6.1).

Aeroponics has been defined as a compact system of growing plants in aerial environment without soil or a similar medium prepared artificially, for example, nursery potting mix or rooting mixtures for tubers, grafts, seeds, etc. Aeroponics involves spraying nutrient-rich water on roots growing in a well-aerated closed chamber. The environment for crop growth is controlled to suit the species. Aeroponics is often coupled with hydroponics (Olson, 2018). Aeroponics is said to serve as a good alternative food producing system. It seems soil is getting deteriorated in some key food grain generating regions. Also, the traditional systems are emitting larger quantities of GHG. Hence, aeroponics may find acceptance.

Lakhiar et al. (2018a, 2018b) define aeroponics as a method of plant cultivation without soil being the anchor or a substrate of growth. Plants grow in air with an artificial support that holds them. The crops, basically, utilize nutrients, and water to grow and produce grains/fruits. Roots are left to hang freely in an aeroponic chamber. Incidentally, the root and shoot systems are kept separated using an artificial partition made of perspex, fibre boards, glass or wood or metal board.

There is a clear possibility to keep both root and shoot exposed to aerosol or mist, simultaneously. Then they are called mist chambers filled with nutrient-rich aerosol. Here, both roots and leafy shoots do absorb nutrient and water. This system perhaps will become more common in future since nutrient absorption efficiency is higher. We need to know the limits to which roots and shoot system can absorb nutrients. The ratio of water and nutrients absorbed by roots versus shoot may depend on crop species, the mist chamber set up, age of the plants, and nutrient concentrations maintained in the nutrient aerosol.

In aeroponic growing systems, the roots of plants are suspended in a volume where emitters continually spray the roots with nutrient solution. The relative humidity around the root is kept at 100%. Also, nutrient and oxygen concentrations are kept optimum for the crop species. However, there is a lack of a buffer around the roots for water, nutrients, and heat (van Os, 2019). This is unlike hydroponics.

Aeroponics is a farming technique involving an enclosed chamber for aeration of roots through micro-spray of aqueous nutrient mist. The word aeroponic is derived from Greek language—"aero" means air and "ponics" means labor. Sharma et al. (2018) opine that although aeroponics was initially adopted by physiologists to study roots, at present, its utility in commercial production of vegetable crop seems important.

6.1.2 ADVANTAGES AND DISADVANTAGES OF AEROPONIC SYSTEMS UTILIZED FOR CROP PRODUCTION

There are several generalized advantages and disadvantages attributable to aeroponic culture of crops. Aeroponically produced crops grow rapidly and with uniformity in terms of morphogenetics. This is because crops do not experience dearth or variation in moisture, nutrients, and photosynthetic radiation. Pathogen, insect pests, and disastrous ambient atmosphere are thoroughly avoided. Therefore, yield of fruits is higher if grown using aeroponics. There is wide choice regarding nutrient solutions and expected growth rates. Since, soil is not used, cleaning the plants/fruits off the adhering soil or dirt or pesticides is not required. Aeroponic systems can be shifted or replaced with advanced versions easily, unlike soil/land. Aeroponic systems are often compact, space saving, and low-cost fixtures compared with open fields or hydroponic systems. Aeroponics is well suited to raise seed crops. It offers good uniform seeds. There is saving in terms of water used per crop; also, since nutrients are channeled to crops at almost 80% less quantities. Soil-less system allows farmers to spray nutrients directly on root surface (rhizoplane) without any of the interfering physico-chemical transformations connected with soil-based open-field situations.

A few of the disadvantages listed are that aeroponics involves initial costs for equipment. It involves a certain degree of engineering sophistication and knowledge. It takes time to standardize the system for each location. Specific crops may need specialized nutrient solutions. The aeroponic technology is being improvised; hence, newer methods may involve extra investments.

The most relevant advantages of aeroponics compared to traditional agriculture depend on specific crops selected for aeroponic production (see Tunio et al., 2020). Let us consider an example. Firstly, aeroponic production of small potato seed material requires low quantity of water. They are suited even in deserts like in Israel, Saudi Arabia, or Arizona etc. Aeroponic systems need 1/10th–1/13th of water that is otherwise utilized in outfield (Putraa and Yuliando, 2015). The quality and productivity of potato tubers is independent of soil quality. After all, it is a soil-less system. Aeroponics offer high yield per unit area and per year. Sometimes aeroponics could be practiced in greenhouses situated anywhere in the farm, in urban areas, or even industrial zones. Greenhouses with aeroponics could be placed right near the point of consumption if it allows. Large open fields are not required. Production of crops is not affected by seasonal changes since the crops are contained in closed uniform climate inside greenhouses. Soil type, soil fertility, and nutrient composition and related aspects are irrelevant, and no extra costs are spent on them. We can develop, install, and operate aeroponics in short time. Land/soil preparation that is cumbersome is not required. Planting, weeding, thinning, interculture, and pest/disease control operations are almost nil. We can decide to harvest potatoes when they are still small tubers or medium-sized ones. Even larger tubers can be produced with due care. Fertilizer costs are least since low concentrations of nutrient solutions are sprayed directly on to root surface. Fertilizer use efficiency is high in aeroponics.

Aeroponics is a good bet in situations with paucity of space. National Aeronautics and Space Agency of USA, it seems, plans to adopt aeroponics with recycling facility for their missions to Mars, in future (Sharma, 2017). Aeroponics is suited best for urban farming conditions. Rooftops of buildings may support clean vegetable crops (Sharma, 2017). In the vertical system, about 60 plants m^2 could be grown. Mere 8–10 plants m^2 is possible in traditional hydroponics. Aeroponics technique consumes just 10% of water compared to field grown crops.

A few disadvantages attached specifically to vegetable production under aeroponics are that initial cost of equipment could be high. Longer stretches of aeroponic production may need special gadgets. It is costly to engage experts in aeroponic production, for longer times. Farmers have to be trained in order to maintain aeroponic chambers and their growing conditions, meticulously. A few accessories may be costly if they have to be replaced, for example, sensors for relative humidity, temperature, light intensity, tube maturity, size, round-the-clock closed-circuit cameras, etc. Robotic systems and computer programs, if adopted, may cost extra. Installation of robotic harvester may be costly at present, although feasible. A standby power back up is always necessary.

The aeroponic system has some substantial vulnerability like a failure of water supply pumps, improper nutrient distribution and preparation, and atomization nozzle getting clogged. These problems require special knowledge and attention to avoid damage, rapid plant death, and failure of the system. Uniformly healthy seedlings are a necessity.

6.2 BASIC REQUIREMENTS OF AN AEROPONIC SYSTEM

The aeroponic system utilizes relatively lower quantities of water and nutrients compared to open fields that support crop production. The aeroponic solution is among important factors or items that go to make a system of crop production, inside a greenhouse or similar indoor spaces. The aeroponic system often uses less water because aerosol is generated intermittently with intervals. Nutrient is sprayed precisely on to root systems that often are suspended from a board. The crops are nourished directly when a nutrient prepared in water is sprayed on to roots. We may note here that nutrient solutions prepared and used in aeroponics are of relatively low concentration compared to hydroponics. They are adopted at 10–15 folds dilution of the original nutrient solutions used to cultivate potted plants on sand or in a hydroponic system. The aeroponic solution at the minimum has to be congenial for roots. The hydrogen ion concentration (i.e. pH) is usually maintained normal, at pH 7.0. However, optimum pH in the aeroponic solution for rapid plant growth ranges from 5.5 to 6.3. Most readymade aeroponic solutions possess 6.0–6.5 pH. In case, aeroponic solution is prepared fresh, then it should be done using water at pH 7.0. So that pH decreases as aeroponic solution gets used up in the system and recycled. Low solute concentrations, that is, low electrical conductivity and disease-free water sources are required mandatorily. Crop roots do not prefer hard water. The preference for certain value of electrical conductivity, that is, mineral concentration in the aeroponic solution is often crop specific. So, distilled water low in calcium carbonate may be required while preparing aeroponic solution. The crop roots absorb minerals as ions from the aeroponic solution. For example, nitrogen is absorbed as NH_4^+ (i.e. cation) or as NO_3 (i.e. anion). The aeroponic chambers and root surfaces need to be kept sterile to the extent possible, by washing the zone periodically.

Crops require right concentration of nutrients, that is, all the mineral nutrients generally available from soil if it were to be cultivated on field soils. The ratios of mineral nutrients too need to be ascertained since crops prefer certain ratios specifically at various stages. At this juncture, we may note that, aerospace above crops or ground surface and water are good sources of carbon, hydrogen, and oxygen. These elements are needed for photosynthesis, respiration, and biomass accumulation processes. Along with the aforementioned elements, essential minerals such as major nutrients namely N, P, K, secondary minerals such as S, Ca, and Mg, and micronutrients such as Fe, Zn, Mo, Mn, B, Cu, Co, and Cl are mandatory, for optimum growth of crops in an aeroponic system. The concentrations, chemical forms, and ratios of nutrient elements need proper attention while preparing a good aeroponic solution.

Commonly used nutrient composition in aeroponic systems according to review by Gopinath et al. (2012) is as follows: nutrient concentration (g/L)

N-NH_4 0.54; N-NO_3 0.35; P 0.40; K 0.35; Ca 0.17; Mg 0.08; Na 0.04; Fe 0.09; Zn 0.03; B 0.03; and Cu 0.04. Minor variations in specific nutrients are to be expected. It is based on various factors like crop species, its genotypes, chemicals used, and other ambient factors like relative humidity, temperature, and yield goals too affect the nutrient concentrations used in the solution.

"Not all aeroponic systems are created equal. There are lots of different ways to construct an aeroponic system. But the most critical task is to ensure regular delivery of fine mist. This can be harder than it sounds."

The aeroponic systems require continuous monitoring, corrections, and adjustments regarding the various plant growth parameters. Often, the management of continuous supply of nutrient aerosol of unform concentration, droplet size, continuously circulation of aerosol/solution, maintaining pH, solute concentration, that is, electrical conductivity (EC), oxygenation of roots, etc., are cumbersome. They need attention by technicians almost round-the-clock. Several sensors are used. Their readings/data needs to be recorded all through the season. Hence, Lakhiar et al. (2018a, 2018b) believe that it is worthwhile to introduce robotic systems. Artificial intelligence could be introduced to conduct several jobs relevant to maintenance of aeroponic systems. Sensors are perhaps the most important components if autonomous systems that maintain an aeroponic system are to be commissioned.

Recently, Lakhiar et al. (2018a, 2018b) have reviewed the topic of artificial intelligence and sensors relevant to aeroponics. They say artificial intelligence provides an opportunity to adopt aeroponics in larger scale. Particularly, to cultivate crops under controlled greenhouse conditions. In fact, several of the sensors and devices that regulate aeroponic conditions can actually be controlled remotely using mobile links. Indeed, this is rapid advancement in aeroponic procedures.

Application of artificial intelligence techniques is relatively a recent trend in agriculture Since past decade, autonomous systems are being introduced to field crop production in a big way. Several improvements in farm tractors, planters, sprayers, combine harvesters, aerial drones, etc., have taken place. Such aspects are being persued with great zeal. This is true even with aeroponic production of crops. Sensors and robotic systems have been included wherever feasible in the aeroponic systems. The aim is to reduce cost of labor, need for well-trained technicians, and improve economic gains. In addition, we may note that robotics adds a certain degree of accuracy to aeroponic cultivation of crops.

Lakhiar et al. (2018a, 2018b) state that they reviewed the literature and found that implementation of advanced monitoring technology is important in aeroponics. It could provide an opportunity for the farmer to monitor and control several paraments without using laboratory instruments. The farmer can control the entire system remotely. Sensor-based methods have drawn attention from farmers (Qiu et al., 2018; Lakhiar et al. (2018a, 2018b). Sensors are currently popular during hydroponic/aeroponic production of vegetables in the greenhouses. Sensors are also used liberally during storage, grading, and marketing of produce derived from an aeroponics/hydroponics system.

According to Lakhiar et al. (2018a, 2018b), nutrient mist is sprayed through atomizers at intervals, to keep the roots moist. Several parameters of an aeroponic system like temperature, humidity, light intensity, water nutrient solution level, pH and EC value, CO_2 concentration, atomization time, and atomization need round the clock attention. Both, in situ monitoring and that via wireless remote sensing are feasible. Data could be recorded at definite intervals. Consequently, inputs could be regulated accordingly. Autonomous methods are currently in vogue in many greenhouses to relay the most accurate information about happenings inside the aeroponic boxes (chambers). Farmer can monitor several paraments without using laboratory instruments. They could control the entire system remotely. Moreover, the technique also provides a wide range of information which could be essential for plant researchers. It provides a greater understanding of how the key parameters of aeroponics correlate with plant growth in the system. It offers full control of the system, not by constant manual attention from the operator but to a large extent by wireless sensors. Furthermore, the adoption of the artificial intelligent techniques in the aeroponic system could reduce costs (Lakhiar et al., 2018a, 2018b, 2019; Pala et al., 2014; Kerns and Lee, 2017)

6.3 APPLICATION OF AEROPONIC SYSTEMS TO PRODUCE CROPS

Li et al. (2018) have listed several food crops, medicinal plants, and other botanical species that have been grown successfully, using simple aeroponic systems in the Southeast nations such as Vietnam, Korea, China, Japan, etc. (see Plate 6.1). There are private companies that adopt aeroponics to produce crops in South America and Africa. A few examples relate to aeroponic production of mini- and medium-sized tubers of potato. Li et al. (2018) mention that vegetables like tomato, capsicum, lettuce, and yam (*Diascora rotundata* and *D. alata*) have been grown luxuriantly using aeroponics. Aeroponic systems are supposedly good alternative systems for the production of root-derived herbal and phytopharmaceutical products (Hayden et al., 2004). Medicinal plants such as *Urtica dioica*, *Anemopsis californica*; herbs like valerian (*Valeriana officinales*); rhizomes like ginger and great burdock, (*Arctium lappa*) are amenable for production using aeroponics. Saffron (*Crocus sativus*) is a speciality plant grown using aeroponics. Perhaps it is worth commercializing since the product is highly priced. Tree saplings of *Acacia mangium* and woody herbaceous species like *Chrysanthemum* has also been grown under aeroponic conditions. For special experimental reasons and cultivating obligate pathogens or other speciality microbes that need pure hosts, aeroponics could be adopted (Wagner and Wilkinson, 1992). For example, to produce soil-free pure spores of arbuscular mycorrhizas (e.g., *Glomus aggregatum; G. occultum; G. constrictum*); field crops like sorghum (*Sorghum bicolor*); and perennial grass species such as Cenchrus (*Cenchrus ciliaris*) were cultivated using simple aeroponics. Such pure spores are useful for serotyping, morphological studies, electron microscopy, and for developing seed inoculum (personal observation).

6.3.1 AEROPONICS TO PRODUCE GREEN LEAFY VEGETABLES

Aeroponic systems for large-scale production of lettuce and other salad vegetables seems efficient in terms of capital, inputs, rate of vegetable leaves production, harvesting, and most importantly the quality of vegetable. Nutrient-rich misting and LED lamps allow rapid growth and biomass accumulation. Since clean water that is purified to remove solids and any chemical contaminants are used to develop the aerosol, lettuce harvested is clean and hygienically good. For example, there are companies that have over 6000 m^2 of lettuce under aeroponics placed in a controlled condition. They harvest about 3 million heads per growing cycle of 20–25 days. Harvesting and processing costs are indeed least if aeroponics is adopted to culture lettuce. No soil, no insects, disease, or water borne *E. coli* is detected in the produce (Anthony, 2020). Such aeroponics allows several cycles of fast-growing crops such leafy vegetables. No doubt, in this system of crop production, we are exploiting the aerial aspects of farming better. There are indeed several types of aeroponic systems adopted in different parts of the world. Most often they are selected considering crops, facilities available, and economic considerations (see Plate 6.1).

6.3.2 *AEROPONIC SYSTEM FOR POTATO PRODUCTION*

At present, aeroponic production of mini tubers of potato meant especially for seeding is perhaps among the most successful enterprises; particularly, if we considered the sophisticated greenhouse based soil-less production of different crops (Li et al., 2018). Let us consider an example. Aeroponic production of seed potatoes that are small and easy to handle while sowing in the outfield is a good idea. Potato tubers that are small at <20 mm are utilized as seeds in the main field. We may note that such small tubers are usually kept disease-free, using controlled greenhouse conditions. Aeroponics allows for variation in plant density during cultivation. It is said in some cases, plant density and nutrient supply could affect the quality of the tubers generated using aeroponic settings. Aeroponic set-up in this particular case consisted of a Styrofoam board with holes to fix individual potato seedlings. Seedlings were grown for 2 months under an aeroponic system. Reports by Tunio et al. (2020) suggest that mini tuber yield ranged from 11 to 13 kg per m^2. However, spacing/plant density on the aeroponics did not seem to affect productivity of tubers. Clearly, in this case, aerospace within an aeroponic chamber (in a greenhouse) is utilized efficiently to harvest better quality seeds (small tubers) (Abdullateef et al., 2012; Al-Shrouf, 2017; Hussain et al., 2014).

Thatch (2016) has reported excellent harvest of medium-sized potatoes, using the aeroponic methods. At the Institute for Agrobiology, Vietnam, he could cultivate about 7480–8470 plants per month on the aeroponics system constructed in the greenhouse. It yielded on an average 835–1160 tubers per m^2. This is supposedly higher than the expected 1000 tubers per m^2. Importantly, tubers were totally devoid of any disease or insect infestation marks. Harvesting such larger tubers was easy compared to outfield, where in, digging the soil is necessary. Such a soil-less system is indeed useful during vegetable production. The aeroponic system described by Thatch (2016) could also be used to grow strawberries, carnations, tomatoes, bell pepper, etc. Such clean and nutritious tubers were generally valued at a high price in the market.

In vitro tissue culture, hydroponics, and aeroponics are among the techniques adopted to produce potato seeds. Regarding, small tuber production, it is said aeroponics are apt. Aeroponics consume less energy and potato crop production is 45% faster than in traditional soil-based open field farming (Saveer Biotech Ltd, 2021). Plus, aeroponics provides cleaner, safer, tastier, and good quality tubers compared with other systems. Potato tuber production increased by six times if we compare it with a field-grown crop. It seems there are a set of advantages related to potato culture and farming. Aeroponics ensures that potato seed material is available regularly to farmers, at any time of the year. Aeroponic products are clean. So, it reduces cost on harvesting, cleaning, and transport. It avoids interference from climatic factors on seed quality. Farmers producing potato seeds/small tubers opine that aeroponics improves their exchequer. An important advantage attributed to aeroponics is that virus-free potato seeds could be produced, using aeroponics.

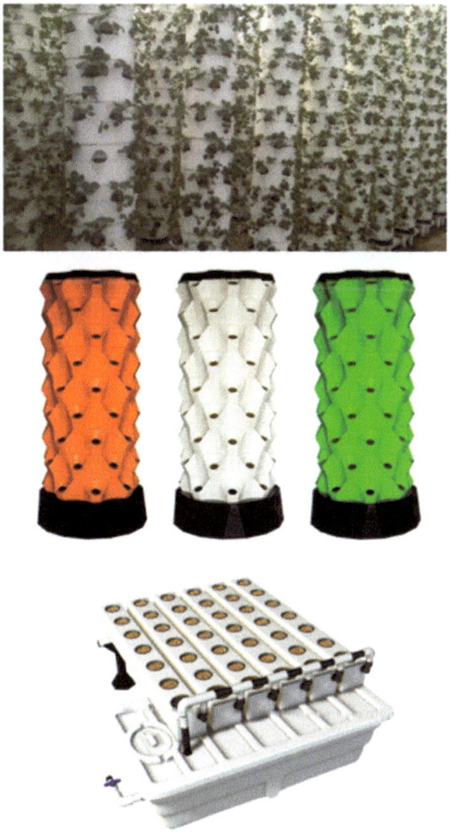

PLATE 6.1 Aeroponic systems.

Source: https://dir.indiamart.com/impcat/aeroponics-system.html

Note: Several types of aeroponic systems are available in the market. Top: An example of small seedlings of leafy vegetable growing aeroponically. Middle: Simple and efficient system of aeroponics that accommodates a few seedlings. Bottom: An automatic aeroponic system that circulates the nutrient solution and creates aerosol in the rooting chamber. Also, see http://www. https://www.aerofarms.com/.

Tunio et al. (2020) have described in detail an aeroponic system that is suited to produce potatoes of different sizes. Firstly, they say, the aeroponic technique is an alternative method for soil-less cultivation in controlled environments, such as greenhouses. The system includes the supply of a nutrient solution to an enclosed root system with an atomizing device in a darkroom. Aeroponics systems primarily consist of electrical units, light proof (dark) growth chambers, nutrient chambers, high-pressure pumps, filters, timers, and nozzles.

In general, greenhouses for aeroponics of potato crop should provide safe pest and disease-free conditions. It should provide temperature control. Usually inside the roof of greenhouse the temperature is higher. So, some methods to lessen the radiant heat could be adopted. Meshes and seals using double doors are useful. Power source without interruptions is a necessity (Buckseth et al., 2018; Riggio et al., 2019; Lakhiar et al., 2018a, 2019; Chiipanthenga et al., 2012).

Water source should have minimal levels of Na^+ and Cl^-. Water with low EC of 1–1.5 mS cm^{-1} is preferred for potato production. Water source with pH above 8.0 should be avoided (Otazu et al., 2010; Calori et al., 2017). Electricity is the basic power source to operate the greenhouse and aeroponic equipment. A good automatic generator is required in case of interruption in power supply.

Plant material, that is, potato seedlings should be disease/pest free. Preferably, it should be grown in controlled conditions. In vitro grown seedlings of solanum are excellent sources. They should be transplanted quickly and placed in the greenhouse without long gaps once taken out from test tubes. Aeroponic boxes made of metal or wood are kept insulated to avoid dissipation of or absorption of heat. They should not experience stress in terms of water or nutrients or photosynthetic sources. In case of potato usually 20 seedlings m^{-2} is optimum for transplantation into an aeroponic setup. Farmers practice a range of 12–20 seedlings m^{-2}.

Spray equipment (misters) is perhaps most crucial to a successful aeroponic system. Mist chambers are fitted with variety of atomizers/sprayers to suit the crop's requirement. The atomization is often categorized as high, medium, and low. The droplet size may range from < 10 µm to larger 1000 µm. Often droplets of 10–100 µm are used in the aeroponic boxes.

Some farmers adopt ultrasonic atomization foggers. This method involves liquid in a thin film on a vibrating surface (vibration at > 20 khz). The film brakes into fine droplets. The fog that develops provides optimum humidity in the air. Such foggers also improve oxygenation of roots. Atomizing nozzles of appropriate specificity should be adopted (Avvaru et al., 2006). Tunio et al. (2020) state that, pH of water should never exceed 7.3. Electrical conductivity (EC) values of 1.5–2.5 are optimum. The nutrient solution used for mist or spray should be monitored for EC and pH, regularly.

For optimum tuberization aeroponic chambers should be kept at 20°C during daytime and 10°C–15°C in the night. Light energy is an important environmental factor that needs careful management in the greenhouses. Morphogenesis of crops depends to a certain extent on light energy impinging on the crops. Supplemental light using LED lamps are possible (Cio'c et al., 2019). Similarly, humidity too has to be regulated at optimum levels inside the aeroponic chambers. Nutrient solutions made of fertilizers that are previously well analyzed for availability of different nutrients could adopted. There are several commercial nutrient solutions available for use in aeroponics.

Aeroponically produced potato tubers should be harvested with due care. They are usually ready in 2–3 weeks after transplanting. Tubers reach about 8 g or 0.5 cm diameter in size. Subsequent harvesting can be done in every 2-week intervals (Otazu, 2010; Mateus-Rodriguez et al., 2014). Harvesting could be continued for 6 months until the crop senesces. About 800 tubers m^{-2} could be harvested from the aeroponic system. At present, there are illustrated manuals for setting up aeroponics to produce seed potato tubers. They are usually small-sized tubers grown rapidly under aeroponics. They are later distributed to regular farmers who specialize in potato production (Otazu, 2010). The procedure if adopted offers 10 times higher tuber yield than conventional methods, in shorter time. It costs less than conventional methods.

6.3.3 AEROPONICS FOR TOMATO PRODUCTION

At present, production of crops, particularly high-value vegetables and fruits are being evaluated for their performance in controlled greenhouse conditions. Soil-less aeroponic and hydroponic systems are opted to grow such crops, like tomatoes, strawberries, tomatoes, capsicum, potatoes, etc. (Torrellas et al., 2010). Super–short-season lettuce is among the preferred crops to be grown in aeroponic systems. Such aeroponic systems allow controlled release of nutrients, water, and growth hormones. Since exposure of crops to plant pathogens, insects, and environmental vagaries are almost nil, cost on items is totally avoided. Mist chambers actually allow absorption of nutrients and water both through roots and leaves. Let us consider an example where a hybrid-systems involving aeroponics and nutrient film technique is adopted, to produce tomatoes. Dannehl et al. (2007) state that maintaining mist/aerosol without interruption is important. Since such interruptions could affect growth and productivity of aeroponically grown tomatoes. Similarly, root-zone temperature needs regular monitoring. Such soil-less aeroponic systems seems to be relatively cost and space efficient. Aeroponically produced tomatoes accumulated relatively more Na+, P, and micronutrients compared to field grown ones. They were also compact in phenotype and fruits could be harvested easily.

In case of tomato production, seedlings are first raised on nursery bed. It is held for at least 28–30 day in the nursery. Then, the seedlings are transplanted to the main field. However, if farmers adopt soil-less aeroponics, seedlings ready for transplantation can be grown within 10 days. Harvesting seedlings is easier and often kept disease-free in aeroponic chambers.

Researchers at Cornell University Cooperative Extension have found that aeroponics is suited to grow leafy vegetables, like Amaranth, in a clean disease-free condition. Aeroponics actually avoids the usual soil-related contamination of leaves with *E. coli* and other soil microbes of possible detriment. Lettuce grows efficiently utilizing less water. Constant recycling of nutrients makes the aeroponic system efficient.

Herbs like mints, nettles, ginger, and yerba mansa also have been successfully tested in aeroponics systems. Aeroponics results in rapid growth, higher yields, and multiple harvests. It is less labor-intensive than open-field farming short-season crops.

We may have to note that in future, plant breeders may have to initiate projects that develop crop cultivars apt for aeroponics, specifically. Roots that tolerate dryness for a certain length of period would be preferred. Limited but rapid growth of roots is preferable in an aeroponic set up. In comparison, in field crops situation selecting crop genotypes efficient in nutrient absorption is important. So, we selected genotypes with efficient roots systems. Nutrient absorption occurred via well spread-out extensive root system. *"In an open agricultural field, roots went for moisture, and nutrients found in soil. Tertiary roots and root hairs scavenged as much moisture from soil pores and from the surface of soil particles. However, in an Aeroponic system it is vice versa. We spray moisture and nutrients on to where the roots exist."* Excessive root formation may not be necessary in an aeroponic system. So, plant breeders have to embark on programs appropriate for aeroponics. Roots not so long or excessive but ones that are efficient in absorption of moisture and all essential nutrients are preferred. In many aeroponic systems, shoot system and its traits related to moisture and nutrient absorption are important. So, shoot and root traits specific to aeroponics need to be selected.

KEYWORDS

- **aeroponics**
- **aerosol**
- **atomizer**
- **green leafy vegetables**
- **mist chamber**
- **nutrient medium**
- **potato**
- **soil-less**

REFERENCES

Abdullateef, S.; Bohme, M. H.; Pinker, I. In Potato Mini-Tuber Production at Different Plant Densities Using Aeroponic System, XXVIII International Horticultural Congress on Science and Horticulture for People (IHC2010): International Symposium on Greenhouse 2010 and Soilless Cultivation, July 23, 2021, pp 1–8, 2012. 10.17660/ActaHortic.2012.927.53

Al-Shrouf, S. Hydroponics, Aeroponic and Aquaponic as Compared With Conventional Farming. *Am. Sci. Res. J. Eng. Technol. Sci.* **2017,** *27*(1), 247–255.

Anthony, N. S. Living Greens Unveiling Large Aeroponic Vegetable Farm. Star Tribune [Online] 2020. http://www.startribune.com/living-greens-unveiling-large-aeroponic-vegetable-farm/495216011/

Anton, A.; Torrellas, M.; Montero, J. I. Environmental Impact Assessment of Dutch Tomato Crop Production in a Venlo Glasshouse. *Acta Hortic.* **2010,** 92, 781–792.

Avvaru, B.; Patil, M. N.; Gogate, P. R.; Pandit, A. B. Ultrasonic Atomization: Effect of Liquid Phase Properties. *Ultrasonics* **2006,** *44*(2),146–158.

Barker, B. Studies on Root Development. Annual Report for Long Ashton Research Station: 1921, 9–57

Baudoin, W; Nono-Womdim, R.; Lutaladio, S. Good Agricultural Practices for Greenhouse Vegetable Crops: Principles for Mediterranean Climate Areas; Food and Agriculture Organization of The United Nations: Rome, Italy, 2013; Paper No 217, pp 22–28.

Buckseth, T.; Sharma, A. K.; Pandey, K. K.; Singh, B. P.; Muthuraj, R. Methods of Pre-Basic Seed Potato Production With Special Reference to Aeroponics. A Review. *Sci. Hortic.* **2016,** 204, 79–87.

Calori, A. H.; Factor, T. L.; et al. Electrical Conductivity of the Nutrient Solution and Plant Density in Aeroponic Production of Seed Potato Under Tropical Conditions (Winter/Spring). *Bragantia* **2017,** *76*(1), 23–32.

Chiipanthenga, M.; Maliro, M.; Demo, P.; Njoloma, J. Potential of Aeroponics System in the Production of Quality Potato (Solanum tuberosum L.) Seed in Developing Countries. *African J. Biotechnol.* **2012,** 11, 3993–3999.

Cio'c, M.; Kalisz, A.; Zupnik, M.; Pawlowska, B. Different LED Light Intensities and 6-Benzyladenine Concentrations in Relation to Shoot Development, Leaf Architecture, and Photosynthetic Pigments of Gerbera jamesonii Bolus in vitro. *Agronomy* **2019,** *9*(7), 16.

Dannehl, D.; Taylor, Z.; Suhl, J.; Miranda, L.; Ulrichs, C., Salazar, M. R.; Fitz-Rodriguez, E.; Lopez-Cruz, L.; Rojano-Aguilar, A.; Navas-Gomez, G.; Schmidt, U. Sustainable Cities: Viabilities of Hybrid Aeroponic/Nutrient Film Technique System for Cultivation of Tomatoes. *World Acad. Sci. Eng. Technol. Pub.* **2007,** *11*(6) 1–8. DOI: org/10.5281/zenodo.1131259/ (May 3, 2021).

Editorial Staff. Aeroponics Use 90% Less Water. In a bottle Italiano. [Online] 2021, pp 1–7.

https://www.inabottle.it/en/environment/aeroponics-use-90-less-water/ (July 4, 2021)

Gabreil, E. The History of Hydroponics. The Natural Farmer. Rain.org. [Online] 2021, pp 1–12.

https://thenaturalfarmer.org/article/the-history-of-hydroponics/ (July 26, 2021)

Gericke, W. Soilless Culture of Plants. Different Soilless Culture Systems and Their Management. Tuze, Y., Gul, A., Tuzel, I.H., Oztekia, G, B., Eds. *J. Agric. Food Environ. Sci.* **1930,** 73, 7–12.

Gericke, W. F. The Complete Guide to Soilless Gardening; Prentice Hall Inc.: New York, USA, 1940; p 187.

Gopinath, P.; Vethamoni, I; Gomathi, M. Aeroponic Soilless Cultivation System for Vegetable Crops. *Chem. Sci. Rev. Lett.* **2017,** 6, 838–849.

Hayden, A. L.; Yokelson, T. N.; Giacomelli, G. A.; Hoffmann, S. Aeroponics: An Alternative Production System for High Value Root Crops. *Acta Hortic.* **2004,** 629, 207–213.

DOI: 10.17660/ActaHortic.2004.629.27/ (July 29, 2021).

Hung, L. L.; Sylvia, D. M. Production of Vesicular Arbuscular Mycorrhizal Fungus Inoculum in Aeroponic Culture. *Appl. Envir. Microbiol.* **1988,** 54, 353–357.

Hussain, A.; Iqbal, K.; Aziem, S.; Mahato, P.; Negi, A. K. A Review on the Science of Growing Crops Without Soil (Soilless Culture) – A Novel Alternative for Growing Crops. *Int. J. Agric. Crop Sci.* **2014,** 7, 833–842.

Jones, J. B. Hydroponics: Its History and Use in Plant Nutrition Studies. *J. Plant Nutr.* **1982,** *5*(8), 1003–1030.

Jones Jr, J. B. Complete Guide for Growing Plants Hydroponically; CRC Press Inc.: Boca Raton, Florida, USA, 2014; p 206. file:///C:/Users/Dr%20krishna/Downloads/1%20(1).pdf/ (July 25, 2021).

Kerns, S. C.; Lee, J. L. Automated Aeroponics System Using IoT for Smart Farming, *Euro. Sci. J.* **2017,** *13* (10). https://doi.org/10.19044/esj.2017.v13n10p%25p/ (July 26, 2021).

Klotz, L. A Simplified Method of Growing Plants With Roots in Nutrient Vapours. *Phytopathology* **1944,** 34, 507–508.

Kratsch, H. A.; Graves, W. R.; Gladon, R. J. Aeroponic System for Control of Root-Zone Atmosphere. *Environ. Exp. Bot.* **2006,** *55*(1), 70–76.

Lakhiar, I. A.; Gao Jianmin, G.; Syed, T. N.; Chandio, F. A.; Buttar, N. A.; Qureshi, W. A. Monitoring and Control Systems in Agriculture Using Intelligent Sensor Techniques: A

Review of the Aeroponic System. J. Sens. 2018a, 2018, Article ID 8672769, 18. https://doi.org/10.1155/2018/8672769

Lakhiar, I. A.; Gao, J.; Syed, T. N.; Chandio, F. A.; Buttar, N. A. Modern Plant Cultivation Technologies in Agriculture Under Controlled Environment: A Review on Aeroponics. *J. Plant Interact.* **2018b,** *13*(1), 338–352.

Lakhiar, I. A.; Gao, J.; Xu, X.; Syed, T. N.; Chandio, F. A.; Jing, Z. Effects of Various Aeroponic Atomizers (Droplet Sizes) on Growth, Polyphenol Content, and Antioxidant Activity of Leaf Lettuce (Lactuca sativa L.). *Trans. Am. Soc. Agric. Bio. Eng.* **2019,** *62*(6), 1475–1487. DOI: 10.13031/trans.13168. / (July 26, 2021).

Li, Q.; Li, X.; Tang, B.; Gu, M. Growth Response and Root Characteristics of Lettuce Grown in Aeroponics, Hydroponics and Substrate Culture. *Horticulturae* **2018,** *4*, 35–39. DOI: 10.3390/horticulturae4040035/ (July 2021).

Mateus-Rodriguez, J.; de Haan, S.; Rodríguez-Delfín, A. Genotype by Environment Effects on Potato Mini-Tuber Seed Production in an Aeroponics System. *Agronomy* **2014,** *4*(4), 514–528.

Olson, J. Aeroponics: A Complement to Hydroponics and the Food System of Space. *Global Food Health Soc.* **2018,** 1–6. http://web.colby.edu/st297-global18/2018/11/26/aeroponics-a-compliment-to-hydroponics-and-the-food-system-of-space/ (July 26, 2021).

Otazú, V. Manual on Quality Seed Potato Production Using Aeroponics. p. 44. International Potato Centre (CIP), Lima, Peru [Online]. http://cippotato.org.research/publication/manaul-on-quality-seed-potatoproduction-using-aeroponics./ (July 26, 2021)

Pala, M.; Mizenko, L.; Mach, M.; Reed, T. Aeroponic Greenhouse as an Autonomous System Using Intelligent Space for Agriculture Robotics. In Robot Intelligence Technology and Applications 2. Advances in Intelligent Systems and Computing; Kim, J. H., Matson, E., Myung, H., Xu, P., Karray, F., Eds.; Springer: New York, **2014;** 274, pp 83–93.

Peterson, L. A.; Krueger, A. R. An Intermittent Aeroponics System. *Crop Sci.* **1988,** *28*(4), 712–713.

Putra, P. A.; Yuliando, H. In Soilless Culture System to Support Water Use Efficiency and Product Quality: A Review, International Conference on Agro-Industry, vol. 3, Guritno, A. D., Schlich, E., Pawelzik, E., Eds.; Elsevier: Amsterdam, pp 283–288, 2015.

Qiu, R.; Wei, S.; Zhang M. Sensors for Measuring Plant Phenotyping: A Review. *Int. J. Agric. Biol. Eng.* **2018,** *11*(2), 1–17.

Reyes, J. L.; Montoya, R.; Ramirez, R. Development of an Aeroponic System for Vegetable Production. Food and Agricultural Organization of the United Nations, Rome, Italy. [Online] 2012, pp 1–2. https://agris.fao.org/agris-search/search.do?recordID=US201301958188/ (July 26, 2021).

Riggio, G.; Jones, S. L.; Gibson, K. E. Risk of Human Pathogen Internalization in Leafy Vegetables During Lab-Scale Hydroponic Cultivation. Horticulturae 2019, 5, 25–29. DOI: 10.3390/horticulturae5010025.

Saveer Biotech Ltd, Aeroponics Facility. [Online] 2021, pp 1–6. https://www.saveer.com/eroponics-facility-for- potato.html/ (July 26, 2021)

Sharma, N. K. Future of Farming, NASA Style Method to Grow Veggies Takes Root in Lonavla. *Times of India,* **2017,** pp 1–8. https://timesofindia.indiatimes.com/city/pune/future-of-farming-nasa-style-method-to-grow-veggies-takes-root-in-lonavla/articleshow/58885841.cms/ (July 26, 2021).

Sharma, U.; Barapal, M.; Shekhawat, N. S.; Hataria, V. Aeroponics for Propagation of Horticultural Crops: Approach for Vertical Farming. *Hortic. Int. J.* **2018,** *2,* 443–444.

Stoner, R.; Schorr, S. Aeroponics Versus Bed and Hydroponic Propagation [The Process of Propagating and Growing Plants in Air]. Florists' Review (USA). *Florists Review* **1983,** *173*(4477), 1–2.

Sylvia, D. M.; Jarstfer, A. G. Sheared-Root Inocula of Vesicular-Arbuscular Mycorrhizal Fungi. *Appl. Envir. Microbiol.* **1992,** 58, 229–232.

Sylvia, D. M.; Hubbell, D. H. Growth and Sporulation of Vesicular-Arbuscular Mycorrhizal Fungi in Aeroponic and Membrane Systems. *Symbiosis* **1985,** 1, 259–267.

Thatch, N. Q. Mastering Aeroponic and Hydroponic Technologies in Production of Potato and Other Vegetables: A Good Sign for Horticulture Development. [Online] 2016. https://eng.vnua.edu.vn/slide/mastering-aeroponic-and-hydroponic-technologies-in-production-of-potato-and-other-vegetables-a-good-sign-for-horticulture-development-10426.html/ (July 26, 2021)

Tibbitts, T.; Cao, W.; Wheeler, R. Growth of Potatoes for CELSS; NASA Contractor Report 177646 for Ames Research Centre: Moffett Field, CA, 1994.

Torrellas, M.; Anton, A.; Hernandez, J. C. L.; Baeza, E. LCA of Tomato Crop in a Multi-Tunnel Greenhouse in Almeria. *Int. J. Life Cycle Assess.* **2012,** 17, 1–12.

DOI: 10.1007/s11367-012-0409-8/ (July 26, 2021)

Tunio, M. H.; Gao, J.; Shaik, S. A.; Lekhiar, I. A.; Qureshi, W. A.; Solangi, K. A.; Chandio, F. A. Potato Production in Aeroponic: An Emerging Food Growing System in Sustainable Agriculture. For Food Security. *Chilean J. Agric. Res.* **2020,** 80, 1–21. http://dx.doi.org/10.4067/S0718-58392020000100118/ (July 24, 2021).

Van OS, E. A.; Gieling, TH. H; Lieth, J. H. Technical Equipment in Soil Less Production Systems. In Soilless Culture. 2019, pp 157–207. https://lieth.ucdavis.edu/pub/Pub074_RavivLieth_SoillessCulture_CH05.pdf/ (July 29, 2021).

Vinoo, A. Aeroponics—The Future of Crop Production. [Online] 2018, pp 1–15. https://www.researchgate.net/publication/329023722_Aeroponics-the_future_of_crop_production (July 30, 2021).

Vyvyan, M. C.; Travell, G. F. A Method of Growing Trees With Their Roots in a Nutrient Mist; Annual Report for East Malling Research Station: East Malling, Kent, UK, 1953, pp. 95–98.

Wagner, R. E.; Wilkinson, T. H. An Aeroponics System for Investigating Disease Development on Soybean Taproots Infected With Phytophthora sojae. *Plant Dis.* **1992,** 76, 610–614.

Went, F.W. In The Experimental Control of Plant Growth. The Experimental Control of Plant Growth; Chronoica Botanica Company: Wlatham, MA and Ronald Press: New York, 1957; p 343.

Zobel, R. W.; del Tredici, P.; Torrey, J. G. Method for Growing Plants Aeroponically. *Plant Physiol.* **1976,** 57, 344–346.

CHAPTER 7

The Agricultural Sky Above the Major Food-Crops-Generating Regions of the World: A Few Examples

ABSTRACT

This chapter is about the nature of agricultural sky above each of the many large agroecosystems found in different continents. Such massive food grain generating systems experience different soil and environmental conditions. We are concerned here with the specific natural and man-made characteristics that operate in the aerospace above them. The aim is to convince through detailed data that agrarian sky above each large food crop generating region is unique. There are indeed several such large food crop generating regions on earth. Five distinct agroecosystems have been selected as examples to prove that the agrarian sky is specific, distinct with regard to aerospace traits and that sky imposes stringent influence on crop species and productivity. They are The Great Plains of North America, The Pampas cropping stretch in Argentina, The Sahelian agrarian zone in West Africa, The European agrarian plains extending from French Plains till Russian Steppes, and the Indo-Gangetic Plains. Of course, there are few other cropping belts with unique aerospace characteristics and located in other continents. A comparative study of these food grain generating regions with regard to agricultural sky is worthwhile.

The natural characteristics of the agrarian sky, such as the climatic parameters, namely, photosynthetic radiation, temperature, relative humidity, annual precipitation, seasons, droughts, dust storms, snow, floods have been discussed for each region. The physical features, such as dust storms, heat

The Agricultural Sky: A Concept to Revolutionize Farming. K. R. Krishna, PhD (Author)

waves, scanty rainfall, temperature make it more than clear that the food generating systems differ enormously. Yet, our efforts to understand the aerospace characteristics of such food generating regions and match agronomic procedures to tide over difficulties or to take advantage of congenial conditions are meagre. The biotic nature of aerospace above each food generating system is no doubt different. This aspect has been delineated in previous chapters, already. The agricultural sky mediates innumerable microbial diseases (epidemics/pandemics) above each food generating region. These need special attention. Similarly, there are insect pests. For example, locusts are dangerous aspect of agrarian sky in Sahel but not in several other regions. The semiarid Indo-Gangetic Plains produce crops in all seasons. It supports major cereals like rice, wheat, maize, and legumes. Whereas temperate food production zones in Europe, China, and Russian steppes supports wheat, barley, millets, and short season vegetables. The aerospace here stipulates the tropical/subtropical crops that are possible. Agrarian sky perhaps has been more devastating apparently due to harsh physical conditions, dangerous biotic fraction filled with disease/pest and birds. There are also beneficial aspects of physical nature of agricultural sky and biotic fraction that need to be recognized and accentuated, if possible. This is to achieve better crop productivity. For example, practicing foliar sprays, fertigation, improving pollinator population. Soils are being amended suitably, but similar measures to improve the agrarian sky seem to be lacking. Perhaps, at present, it is not an easy proposition. We should still attempt to amend aerospace above crops where feasible.

Overall, this chapter tries to convince that the agrarian sky above the food generating regions is unique. Let us study them in greater detail and try to enhance crop productivity. At the same time, we need to be alert to aspects like excessive fertilizers, pesticide, and loss of soil fertility. We should enhance productivity yet maintain natural conditions intact at levels congeal to flora fauna and humans too.

This is the last chapter of the treatise. Perhaps, it is now clear that we have to bestow greater interest, funds, and expertise in understanding the agrarian sky, in detail. It is time we understood the agricultural sky and its factors better than at present!! The future of global farming rests to a great extent on the upkeep of agrarian sky. Agrarian sky can potentially induce a revolution that enhances food grain productivity, at the same time, can reduce farm drudgery.

7.1 INTRODUCTION

The intention of this chapter in the volume is to highlight our knowledge about agrarian regions and factors that operate in their aerospace, that is, above the crop's canopy. The classification of agrarian regions based on natural resources, such as soil, climate, and crops are common. Agricultural sky per se too is an important natural resource. We should identify the various salient traits of agricultural sky above each farming location or agrarian region on a wider scale. This will help us to understand the role of agricultural sky in global crop production better. We have not done this in earlier years satisfactorily. Understanding mere weather patterns may not suffice. There are several other aspects, such as topsoil drifts, nutrient shifts in sky, disease propagules and pandemics, insects, their migratory patterns, beneficial microbes and their dispersal, birds, their diversity and foraging behavior, man-made aerial vehicles, foliar spraying too are important.

Globally, there are few major agrarian regions that support mass production different food crops. Some of them are relatively large expanses of cereals, legumes, oil seeds, fruit crops. Krishna (2015) and Pearson (1992) identify and mention the salient features of at least eight major agrarian belts (or agroecosystems). These agricultural prairies are responsible for serving the humanity with cereal food grains, legumes, and other products relevant to their daily life. No doubt, detailed knowledge about each of the major agricultural regions and their classification needs attention (Whittelesey, 2009; Helbaum, 2010).

A survey of literature about crops and cropping systems, in general, shows us that soils, water resources, nutrient, and water requirements plus a few special agronomic procedures needed for the crop/locations are discussed. Soil diversity and soil fertility levels are usually stressed to a good extent. The crops species compatible with the topography, soils, agroclimate, and economic aspect are again highlighted. The agricultural sky, as we realize now is different in each of the major agrarian regions. *The features of agricultural sky and their impacts may in fact be distinct within each small location or even a set of few crop fields around a village. No doubt, agricultural sky is an indispensable part of any agroecosystem. Yet, it is less clearly understood than other aspects of an agrarian region.*

Basically, we should realize that sky above the urban regions, industrial setting and agricultural expanses differ markedly. Although not mentioned frequently, farmers tend to know a few facts about the agro-climate, that is, weather patterns prevalent above crops. They try to match the crops accordingly. A step further, we may have to identify, at least in future, the salient features of the agricultural sky prevalent above each small cropping zone. Then, design our agronomic procedures, accordingly. At the bottom-line, agricultural policy makers, researchers, and even large farming companies have to understand that the major food generating systems are actually served by agricultural sky. Agricultural sky differs based on geographic location, crops preferred, also agroclimatic features. It is now a glaring fact that agricultural sky, which is crucial to farming is distinct for each agrarian region. For example, the sky above Great Plains cropping zones, Pampa's farming zones, Cerrados in Southern Brazil, European Plains, Sahelian cropping belts, Indo-Gangetic Plains, the Ukrainian steppes, Chinese Northern Plains, and Australian wheat growing regions differ. We should not overlook it. We know a lot about other aspects like soil, crops, fertilizer, and irrigation requirements. Agricultural sky, its dominant features and methods to manage or overcome detriments caused by the aerial factors are not at all understood properly.

Hence, it was felt that a chapter that discuses and highlights agricultural sky and related factors that affect the world's major food generating systems is a necessity. We may realize that each of these vast agrarian regions actually experience a wide range of detrimental factors. Several of them are mediated through the sky. Farmers for some reason have neglected to understand the details of agricultural sky. However, each farmer has to be conversant with all the various positive factors that affect crops positively/usefully or harmfully. Most if not all of the major crop disease are mediated through aerial factors. The environmental disasters, such as floods, dust storms, drought, differ based on the agrarian region. They need greater attention by researchers.

7.2 AGRICULTURAL SKY ABOVE GREAT PLAINS OF NORTH AMERICA: ITS NATURE, MANIFESTATIONS, AND INFLUENCE ON FOOD CROPS

The Great Plains of North America extends for 800 km from east to west and 3200 km from north to south. The plains occupy an area of 1,300,000 km^2 (Wishart, 2004). It is a vast plains region situated at approximately 320 m.a.s.l. The major soil types traced in the plains are characterized as Mollisols (Black Chernozems, Kastanozems), Udolls and Ustalfs (Deckers et al., 1998; FAO, 1997; Lothar Beyer, 2002; Soil Survey Staff, 1992, 1995, 1996a, b, Soil Survey Staff, 1998). They support wheat, maize, sorghum, soybean, brassicas, and few other food crops. The cropping regions within the Northern Great Plains differ markedly depending on natural resources and agricultural sky. The Northern Great Plains, Intermediate Plains, Central Plains, and Southern Plains can be easily discerned as different regions. They are markedly different with regard to agricultural sky. The subregion could be further divided and identified using distinctive atmospheric characteristics (i.e., agricultural sky). For example, Hudson (2012) subdivides just Northern Great Plains as nine (9) different agroecoregions. Their subdivision is based on several aerial characteristics plus others. Major crops are wheat, maize, sorghum, sunflower, lentils, soybeans brassicas, cotton (USDA, 2019; Putnam et al., 2012). Obviously, the specific traits of agricultural sky differ for each agroecoregion. They need to be identified, marked, and appropriately named for farmers and other experts to utilize. The cropping system in the Northern Great Plains is wheat-based or maize-based. Usually, it is wheat–brassica, wheat–soybean, wheat–lentil, etc. In the Central Plains (Nebraska, Iowa, Kansas parts of Colorado) it is wheat–maize, Wheat–wheat, Wheat–soybean, Wheat–vegetables. The Southern Plains is relatively drier. The cropping systems include wheat–fallow, wheat–sorghum, wheat–fodder, wheat–soybean, wheat–cotton–fallow (Peterson and Westfall, 2005). The agricultural sky varies between the subregions and even within each subregion. The available data of the Great Plains and its subregions must be pooled. However, agricultural sky needs to be characterized further by including data about several other aspects in detail. Then, utilized during decision-making. In fact, agricultural sky has to be characterized for each major and subregions of the Great Plains. Such an effort will help farmers, in forecasting seasonal/annual fluctuations, in the agricultural sky.

7.2.1 AGROCLIMATE ABOVE THE GREAT PLAINS OF NORTH AMERICA

The focus of this section is about climate-related factors that operate predominantly via aerospace above the crops. Soils, their characteristics and influence on crops and their productivity are not the priority here. However, we should note that soil and its influence on crop production in the Great Plains has been studied since several decades. Innumerable treatises, manuals, and handouts are available on soils of Great Plains. Basic idea is to match and adopt agronomic procedures. Particularly, to suit the climate change trends and vagaries, if any, and revising them at short notice (Easterling (1996).

The agroclimate above the Great Plains region is no doubt predominantly a natural phenomenon generated and manifested in the sky. Our focus here is climatic conditions above soil surface and crop's canopy. Several physicochemical processes interact to produce the final weather condition. The agricultural weather in the Great Plains region that supports large-scale production of cereals and other food grains varies between subregions. Weather parameters alter even within the subregions, such as the Northern, Western, Central or Southern Great Plains. For convenience, agricultural geographists have demarcated the subregions further and studied them. Generally, the aerial characteristic of Northern Great Plains is cooler and that of Southern Great Plains is relatively warmer. On an average, aerial temperatures of the cropping region range from 15°C to 18°C in summer to −12°C to −15°C in winter (January). The coldest temperature is felt in Wyoming, Montana and Canadian Plains regions, which is about 30°F (Kunkel et al., 2013). Southern Plains region may experience heat wave conditions during summer. The region experiences snowfall and frost in the sky. The snow-free days range from 120 to 150 days per year. The annual precipitation from the atmosphere ranges from 375 to 500 mm, but it differs markedly based on the location. In parts of Central Plains (Oklahoma region), the precipitation may reach even 900 mm annually (Baumhardt and Salinas-Garcia, 2005). The southern Great Plains experiences a summer peak precipitation between July and September. It is caused by a continental shift in wind flow. It is known as the "North American Monsoon" (Kunkel et al., 2013). Further, they state that precipitation may alter due to climate change and *El Nino* effects. It seems that agricultural sky above the Northern Plains experienced a rise of 1.0°C and in the Southern Plains, it has increased by 5–7°F during the past century due to climate change effects! (USGCRP, 2009a, b; USGCRP, 2012, USEPA, 2012: Kunkel et al., 2013). In the Southern Plains, since precipitation from the sky is insufficient, it experiences water deficits frequently. Extraneous sprinklers or furrow irrigation is a requirement. If the precipitation received is 900 mm annually, the evapotranspiration per year is above 1500 mm. The Great Plains receive snow/frost from the sky. Generally, the snow/frost-free days in the Great Plains region range from 120–300 days per year, depending on the location. Frost-free days decrease as we traverse from warmer south to north. Northern Plains experienced very cold sky temperatures. Diurnal period is short during winter but longer during summer. It is said that longer period of sunlight during the summer helps crops to photosynthesize and fix carbon, better. Wind erosion and loss of topsoil is conspicuous in the Central and Southern Plains region. Such a natural malady is related to events in sky.

It is overcome by deep ploughing (discing) and leaving the field with large clods. Such cloddy fields do not loose topsoil to a great extent. Fog above the crop fields is common in the winter. This phenomenon in the sky reduces the photosynthetic radiation reaching the foliage. Snowy, fogy, and cloudy days, if they are more frequently observed, then there are clear chances of reduced carbon fixation by crops in the vast plain's region (Baumhardt, 2003).

Precipitation trends in the Northern Plains is affected by seasonal changes. Northern Plains receive water from snow/frost melt during the crop season. Rivers are fed with snow melt. Temperature is generally low.

Southern Plains is relatively a warmer region. It suffers from low levels of precipitation. Precipitation may not match the need for moisture at different critical stages of the crop. Southern Plains also suffers from droughts and heat wave. Loss of topsoil due to erosion is attributable to high velocity wind. In 1920s and 1930s, dust bowl conditions were conspicuously an agricultural sky-related natural phenomenon. Normally, cropping zones in Southern Plains is supported by relatively higher levels of precipitation in the east. Rainfall lowers as we transit to west. It ranges from 450 to 250 mm annually as we transit from east to west within this region. Cereals, such as wheat require more water than that received in the region. Even, if farmers try to cultivate drought tolerant crops, such as sorghum or cotton, the need for greater precipitation use efficiency is mandatory. Appropriate planting period, mulching, cover crops, practicing proper crop-fallow systems are said to overcome the paucity of rains. Soil erosion due to high intensity rainfall and control of erosion using contour planting, crop residue application/mulching, and bunding are required This is an example of how farmers match crop species and agronomic procedures to overcome vagaries of agricultural sky.

Droughts and desertification processes are created by aerial processes, such as lack of precipitation. Often, high wind speeds induce loss of topsoil. For example, Heathcoat (1980) states that dust bowls of 1930s, a very conspicuous aerial phenomenon was created by dearth of precipitation, high velocity wind, excessive farming activity (e.g., higher tillage intensity, absence of contouring, lack of cover crops, etc.). Wind-induced soil erosion is an important phenomenon related to agricultural sky in the Central Plains. Soil erosion could be rampant if wind speed reaches beyond threshold. The threshold is 6 mph above 1 ft from the field soil surface. Obviously, atmospheric phenomenon close to the surface of soil and crop's canopy is the cause of soil fertility loss (Lyons and Smith, 2010; Borade, 2012; Fryerear et al., 2001; Popper and Popper, 2004). A few procedures are meant specifically to reduce wind caused losses of soil fertility. Farmers practice surface application of crop residues, different types of mulching, cover crops, or fallows with clods. The average wind speed experienced in the sky above the agricultural fields is 5.5 S^{-1}. Plains region also experiences high wind speeds, tornadoes, and dust bowl conditions that affect the general conditions above the crops in the sky. Soil particles with nutrients attached are often transported due to high winds and dust storms during fallow seasons. Such movement of topsoil, dust and other particles could also cause drift of plant pathogens, such as bacteria, actinomycete propagules, fungal spores, and viral particles. Insect eggs and adults too could be disseminated by high-speed winds.

7.2.2 GREENHOUSE GAS EMISSIONS FROM AGRARIAN REGIONS WITHIN GREAT PLAINS OF NORTH AMERICA

The greenhouse gas (GHG) is an important agricultural sky-related phenomenon. It affects the agroenvironment of the Great plains. Consequently, crop productivity too is affected. It is said that agricultural operations and crops are responsible for 23% of global GHG emission. Obviously, major food grain generating zones of the world, such as the Great Plains is among the major contributor to global GHG. GHG emissions vary with subregions of the Great Plains. Parton et al. (2011) says GHG emission from the Great Plains is high. It is dependent on the way the region is utilized for cropping. It seems about 30–50% of soil carbon stored through the ages were lost from the soil to agricultural sky, in just last 30 years. Mineralization phenomenon was rampant leading to loss of N as NO_2, N_2O, and NH_4. Incidentally, most commonly used fertilizer N formulations, such as NH_4NO_3, KNO_3, and liquid NH_4 applied as foliar or even soil application is highly susceptible to emission (Halvorson and Del Grosso, 2012). Soil N loss could be altered using appropriate cropping system (Engel et al., 2005). Cattle farming in the Great Plains has contributed to GHG in the aerospace. Cattle farming increased the loss of C as CH_4. Factors, such as ambient temperature, precipitation pattern, cropping systems, agronomic practices, fertilizer application either as foliar or soil application affect the emissions in time and space. For example, adoption of No-tillage practices may reduce loss of soil N as emissions. Soil oxidation-related emission of CO_2 too reduces if No-till is adopted (Halvorson et al., 2009, 2011).

Forest canopies are known to emit volatile organic compounds into atmosphere. This is in addition to GHG such as N_2O, NO_2, CO_2, CH4. Forest fires too cause deterioration of agricultural sky in the Great Plains. Forest fires emit particulate matter, SO_2, and VOC. They affect the ambient temperature, moisture, relative humidity, and precipitation pattern. Forest fires affect microbial diversity and population in the aerospace. Insect pests, pollinators, and aves too are affected, if forest stands, understorey species and crops are burnt periodically. The aerospace above the crop is an important aspect of farm world. It that gets affected by forest/crop burning episodes.

Pollution of atmosphere above crops in Great Plains is also a matter of concern. There are several industrial locations in the Great Plains that generate pollutants and emit them into sky. Smog, fog, and particulate dust that collect in the atmosphere too affect the crops and their photosynthetic potential. Severe pollution and acid rains may affect crop growth and grain yield. In the Northern Plains, smoke/smog affects crops in many ways. Firstly, it affects the penetration of sunlight and photosynthesis is impeded. Smog has deleterious effects on stomatal activity. Reports suggest that wheat, canola, soybean were all stunted, if exposed to smog, during the in-season (Whetter, 2018). In Alberta, canola matured late by 10 days compared with normal. It was attributed to occurrence of smog during growing season.

7.2.3 MICROBES IN THE SKY ABOVE CROPS

Microbes are traceable in all portions of an agrarian region, such as soil profile, roots, shoot systems, and aerospace above the canopy. They mediate several different types of nutrient transformations. Several of them are beneficial, for example, in nitrogen fixation. Beneficial microbes could be traced in the soil, on the crop surface, and even in the atmosphere. (Aerosol with microbes). Simultaneously, there are several microbial species that are pathogenic and cause severe disease on crops. A few microbes traced in the aerial portions of Great Plains are potentially capable of spreading the crop disease rapidly. Long distance transmission of microbial pathogens (e.g., *Puccinia*, *Helminthosporium*, *Erisyphae*, Maize streak virus) from one region to other within the Great Plains is also possible. It is aided usually by continental wind systems.

Let us consider soil microbes found in the fields. But only those that have relevance to conversion of atmospheric nitrogen into soil organic nitrogen. They are called commonly as biological N fixers. They could be symbiotic, such as with legumes or free-living in soil, such as *Azospirillum, Azotobacter*. These beneficial soil microbes are prone to get distributed via aerial mechanisms such as high surface speed wind, dust storms, floods. The microbes are carried into various other locations in the plains via atmospheric dust during a storm or along with a rain droplet or flowing water on the surface of soil. Precipitation that percolates may distribute beneficial microbes (also others) to lower layers of soil profile. Nitrogen fixation, that is, conversion of atmospheric N to soil organic N may actually depend on several factors related to soil and atmosphere.

Leaves in crop's canopy is a repository of several species of aerial microbes. They could be deposited due to mild or high-speed winds, breezes, morning dew, mist, or precipitation droplets. The phyllosphere is a region above crop's leaf. It harbors a wide range of microbes. The microbial population and diversity encountered in the canopy/phyllosphere is dynamic. It may depend on several aerial mechanisms. Phyllosphere microbes mediate several processes, such as N fixation, pathogenesis, commensalistic processes. Here, we should note that soil nutrients must pass through the rhizosphere/rhizoplane before reaching the internal root portions (phloem). In the agricultural sky, it is phyllosphere that has to be encountered by all aerial microbes, nutrients, and moisture that reaches the internal tissue of leaves.

Nutrients received through atmospheric deposits and precipitation in the Great Plains depends on the location. It may not suffice to support a good crop of cereal or legume, till final harvest. Hodges (1991) has reported that atmospheric deposits provided only about 2–3 kg N ha^{-1} $year^{-1}$. Yet, legume crops, such as groundnut may not respond to inorganic fertilizer N because they receive N via biological N fixation (BNF). A good crop of groundnut grown on mollisols in Southern Great Plains may fix 75–145 kg of atmospheric N ha^{-1} in a season (Burkart and Jones, 1999). Regions not practicing legume intercrops or rotations and those having poorly nodulated crops fix only feeble amounts of atmospheric N via BNF. They must be replenished using inorganic fertilizer N. They reported that in Texas, 20–25% of peanut belt has poorly nodulated crop stand (Trostle, 2002). Therefore, extraneous inoculation of *Bradyrhizobium* is mandatory. It enhances the conversion of atmospheric N into soil N. On the other hand, in many locations of North Carolina, peanut crop may not respond to inorganic fertilizer inputs from 20 to 120 kg N ha^{-1}, because it receives large portion of N through BNF. Overall, peanut occupies large tracts of mollisol in the Southern Great Plains and other regions such as North Carolina, Virginia, Georgia, and Florida. It could be utilized to convert atmospheric N to soil biomass N effectively. It lessens the need for extraneous inorganic fertilizer N inputs to the soil.

Microbial plant pathogens may occur in the Great Plains sky endemically or erupt as epidemics and affect crops at various ages. Agricultural sky above the cropping belts plays vital role in the dissemination, deposition of fungal spores, bacteria, viral particles, and insect eggs. The population/diversity of microbial plant pathogens found on the canopy and aerospace may vary, depending on several factors related to atmosphere, crops, microbe species. They have shown that microbial diversity and population of plant pathogen in the atmosphere above may differ within each closely placed crop fields. The pathogenicity of aerial microbes too varies. Hence, devastation via agricultural sky too differs based on microbes and environment, in each small field.

As stated in earlier chapters, our understanding about aerobiology of plant pathogens that may affect the crop productivity is important (see Stackman and Christensen, 1946; Yates et al., 2015). Air sampling procedures are vital. Mainly because, they decide the accuracy of data. Beginning with "Hirst spore traps," we have now devised several methods using which we can collect microbial samples. Then, study the spores and other types of propagules of fungi, bacterial and viral particles that float above the crops (West, 2012; West et al., 2008; Techy et al., 2010). Methods that are dependent on traps placed on UAVs, balloons, airplanes, even spacecraft etc. are adopted. Aerobiological samples can be tracked accurately using the GPS coordinates. Microbial densities and species could be mapped in different layers of the atmosphere. There are indeed several plant pathogens that affect growth and productivity of crops that are cultivated in the Great Plains. Crops are affected by rusts, smuts, blasts, blotches, streak mosaics, and leaf curl.

The agricultural sky above the specific fields or a set of them or a larger agrarian region is repository of Fusarial propagules. Atmosphere may actually harbor several strains (pathotypes) of *Fusarium graminearum* . Each may occur in different proportions in the agricultural sky. Therefore, during breeding for resistance, experts should expose the crops (wheat, maize etc.) to different population levels of as many fusarial strains possible. We have to understand the behavior of *F. graminearum* spores in the sky, its transport and spread. Firstly, we have to trace the effects of wind, precipitation, temperature, RH, dust bowl conditions, and tornadoes that may occur frequently in some regions of the Great Plains. Some of these suggestions about developing crop's resistance to a single or cluster of pathotypes also applies to a few other fungal pathogens, such as rusts, streak mosaics. In due course, as we develop greater insights, we should be able to easily categorize and classify the agricultural sky. At least, in certain parts of the Great Plains, agricultural sky could be classified as susceptible to fusarial disease, highly prone to epidemic, or low in propagules of pathogens. A farmer may then select a suitable crop and its genotype. No doubt we can alert the large farms of the Great Plains with sufficient knowledge about the fungal spore dynamics and its possible destructive power.

In the Great Plains, knowledge about the aerobiology of wheat rust fungi is essential. Epidemiology of disease-causing agent and expression of resistance of wheat genotypes to different strains of leaf/stem rust fungi (*Puccinia graminis, Puccinia striformis, Puccinia triticana*) needs attention. Rust epidemic depends to a sizeable extent on the aerial behavior of spores/propagules (Eversemeyer and Kramer, 2000). Long distance dissemination of uredospore of wheat rusts were shown in 1930s and 1940s. The fungal spores are said to travel about 2500 km from south to north of the Great Plains. The dissemination of spores depends on the physical nature, morphological features, spore density, and wind currents. The density and number of strains/pathotypes of these rust fungi may vary. Classification of agricultural sky based on suspected pathogenic fungal propagule density in the sky (atmosphere) is worthwhile. It helps farmers to adopt suitable prophylactic measures. It has been pointed out that chemical control of wheat leaf rust has not been effective. It is attributed to complex interactions among atmospheric conditions (agricultural sky), propagule density, virulence, and crop growth stages. In fact, Eversemeyer and Burleigh (1970) have stated that it is preferable to adopt computer models and forecast the spread of *P. graminis* and its trends, particularly its spore density, disease severity, and probable reduction in wheat grain yield. Based on such forecasts, farmers may accordingly invest on fungicides and other control measures.

Aylor et al. (2008) have shown that tracking the sporangia and other propagules of potato late blight disease in the atmosphere is useful. In fact, they have adopted unmanned aerial vehicles (UAV) to collect aerial samples. Fungal spores could be trapped and monitored. The distribution and intensity of spores in the sky could be deciphered later. They reported that fungal spores were traceable from 90 m to even 500 m above the crop's canopy in the air. Air turbulence could affect the distribution of fungal spores. Aeromicrobiological studies helps in classifying the "agricultural sky" into "disease-prone" or "less prone" or "totally clear" as far as fungal pathogen's propagules are concerned.

7.2.4 INSECT PESTS AND POLLINATORS IN THE AGRICULTURAL SKY

Historically, agricultural sky above the Great Plains has harbored certain beneficial insects (e.g., pollinators). Of course, there are large posse of insects (pests) that have brought disasters to farmers. In this section, we are concerned primarily with aerial insects, not the ones that dwell in soil. Great Plains sky has supported rapid transit and dispersal of detrimental insects. In the recent times, forecasting aerial pests and releasing digital maps on computers is a serious pre-occupation of agricultural experts. Large-scale epidemics caused by insect pests have been recorded periodically. Such methods help in preempting the spread of insect pest epidemics.

In the Canadian Plains, again, agricultural sky aids the dissemination of a posse of insect pests. They attack crops, such as wheat, maize, barley, brassicas and lentil. Agricultural sky could become uncongenial when populations of insect pests, such as the armyworm, diamondback moth, grasshoppers, midges, and bugs increase (Saskatchewan Natural Resources Department, 2020). Crops like sunflower which are native to the Great Plains do support a range of coevolved aerial insect pests. For example, Knodel et al. (2015) and Charlet and Gavloski (2011) list several aerial insects that thrive on sunflower (e.g., moths, midges, weevils, beetles, bugs). There are several species of honeybees that thrive in the Great Plains sky. They are beneficial to crops and have coevolved with crops, since ages.

Further, Knodel and Shrestha (2018) have reported that leguminous crops in the Great Plains of USA and Canada are affected severely by the aerial insects, such as the cut worms, wireworms, and moths. No doubt, sky above pulse (legume) crops too is harsh with insect pests ready to attack their canopies. They say, integrated pest management (IPM) principles must be adopted to reduce the population of insects in the ambient space above crop's canopy. One of the procedures suggested is to apply biological control agents (microbes, insects) into air/canopy to reduce the pest damage.

Cereal aphids (e.g., Russian Wheat Aphid) are yet another major detriment to optimum growth and productivity of wheat and other cereals in the Great Plains. The aphids also play a vital role in the transmission of Yellow Dwarf Virus (e.g., BLDV). Hence, agricultural sky above the large stretches of wheat in the plains needs appropriate management. Otherwise, the crop must tolerate both insect and virus caused damages (MyFields, 2020; Bynum et al., 2019; Brewer and Elliot, 2004; Giles et al., 2008). During recent years, researchers have tried to assess and adopt drones (small UAVs). They are new potent aerial vehicles useable in agrarian regions, particularly to collect spectral data, mark/map aphid and/ or armyworm attacked zones, and even conduct prophylactic sprays. Satellite imagery from the sky too has been adopted to assess insect damage and its spread (See Kansas State University, 2015; Zhang et al., 2015; Krishna, 2020a). Aphids also attack soybean crop in the Great Plains. They utilize the aerial space to disseminate swiftly into large patches of soybean (Varenhost, 2019). This aerial pest may reduce soybean grain productivity by 40% in severe cases.

Locusts are an example of how detrimental an insect intruder from the agricultural sky could be to crops, consequently to farmer's exchequer. They are generally termed grasshopper pest on cereals and other vegetables when they occur in sparse number. However, locusts are a kind of aerial phenomenon wherein billions of grasshoppers migrate from one location to other, in large numbers (clouds). In the Great Plains Sky, they do thrive in good intensity to be called locusts. Although, not felt frequently in the recent past. In the mid-19th century, locusts were among detriments for a brief period. Such locusts were mediated through agricultural sky! Locusts were most severe in the year 1874. It caused loss of over US$ 200 billion of the exchequer derived from the entire cereal belt of the Great Plains states (Lyons, 2018). They were most severe in the Central Great Plains states of Kansas, Nebraska, and Oklahoma. Locusts almost devoured crops in the entire North Texas, it seems. Locusts did affect the Dakotas where wheat is a main crop. Since then, extremely rich harvest of crops has also been possible. Particularly, in years when agricultural sky has not mediated migrations and attacks by various pests. There are several control procedures in place to overcome such pests. We need a clear map of agricultural sky over the Great Plains marked with the insect pests. Especially, maps depicting possible epidemics, such as locusts, cut worms, army worms, helicoverpas, bugs for each season. Forecasts are generally made available to farmers about possible locust attack. In fact, regions within the agricultural sky over Great Plains should be classified, using set of maps as insect pest prone or locust prone regions, so that cropping systems could be appropriately modified. At least, it helps farmers to be prepared to negotiate a detrimental factor pertaining to agricultural sky (not soil or water-mediated detriment!).

The climate change effects can be felt conspicuously on the insect pests of the Great Plains. Alterations in the weather parameters, particularly, temperature and humidity could severely affect the pest behavior and its life cycle. Warmer temperatures are known to induce emergence of adults from pupa, at an earlier date, in just 5–7 days. Climate change effects noticed in the Great Plains could be detrimental or beneficial to insect species. Its consequences on crop productivity too will alter accordingly. For example, a few cut worms deleterious on maize are known to have caused greater damage to maize plus another cereal—sorghum. This is attributed due to climate change effects to its life cycle events. Not just the insect pests, climate change is known to influence the life cycle and behavior of biological control agents (predators and parasites) traced in the Great Plains cropping regions. (Thomson, 2010). Further, we may note that, climate change affects aerial insects that are beneficial like bees, their lifecycle, and pollinator activity.

In the Great Plains, climate change has even modified the spread of viruses, particularly, those for which insects are vectors. Such manifestations have potential to cause epidemics of virus through widespread distribution of insects. For example, climate change effects on insects traced in the Central Plains has induced firstly insect migration in the sky. Then, it has induced viral epidemics as far as Minnesota in the Northern Great Plains (Tomas, 2008).

Food crops depend on pollinators. It includes honeybees (*Apis mellifera*). There are more than 3500 species of native bees (USDA-NRCS, 2016). Farmers growing crops that depend on pollination can rely on local pollinators or pay beekeepers to provide honeybees. Honeybees are beneficial to sunflower farming. Bees transport pollen from plant to plant in a field. It results in cross-pollination. Some sunflower varieties will not produce higher yield unless pollinators are present. All varieties will produce some sterile seed. However, varieties differ in their degree of dependence on insect pollinators. Autogamous sunflower hybrids do not require bees for maximum yield. They will yield the same when covered by bags or uncovered. In non-autogamous sunflower varieties, pericarp (bull) development is normal, but no ovules or meat develop. Wind is relatively unimportant in cross-pollination of sunflower. Some of the older open-pollinated varieties, such as Peredovick set only 15–20% of seed without pollinators. Hybrid sunflower sets 85–100% seeds without pollinators

There are also bee species that are beneficial in a couple of ways. In addition to aiding pollination by hovering in the airspace above crops, they also collect honey. Honeybees (*A. mellifera*) and their activity above crops of the Great Plains are an important aspect of biotic fraction of agricultural sky. Otto et al. (2016) state that insect pollinators, such as bees are needed for maintaining global food production and ecosystem function, not just the Great Plains. They also state that land-use changes occurring in the US Northern Great Plains is influencing the habitats of both wild and managed honeybee colonies. In their experimental region, about > 40% of the bee colonies were commercial and well-managed stock. They found that agricultural sky above the crops/natural vegetation has changed to a certain extent. This is attributable to corresponding change in the land use and vegetation composition, in the recent years. Here, bee farmers have often avoided corn and soybeans. Yet, farms with cereals/soybeans/corn have higher density of honeybee colonies that are productive. About 20–40 kg honey could be harvested per site. Based on the observations about honeybee colonies and their activity, they conclude that bees are changing in their activity both regarding pollen transfer and more importantly honey productivity (see Otto et al., 2016; Smart et al., 2016). It seems both climate change and land-use alterations could be affecting honeybee colonies in the Great Plains. Honeybee activity in the sky above crops and their productivity with regard to honey collection is important. So, it is necessary to keep a close watch on honeybee population and changes, if any. Perhaps, we can classify agricultural sky as "bee rich" or "bee feeble" based or pollinator population/activity.

7.2.5 AVES IN THE GREAT PLAIN'S AGRICULTURAL SKY

The natural vegetation, forests, and vast farmlands found in the Great Plains regions is indeed a big repository of diverse species of birds. The diversity ranges from predatory birds, scavengers, pollinators, biological control agents, seed dispersing species, pests on crops. (see Powell, 2019; Wilson and Henshaw, 1913). Here, we are concerned with bird pests, those acting as biological control agents to reduce insect pests, pollinators, and seed dispersal agents. These bird species are integral part of agricultural sky of the Great Plains.

Several species of birds could be playing a vital role in controlling insect pests that occur on common crops, such as wheat, maize, and soybean. Repeated measurements of bird foraging behavior for insect pests in soybean/wheat fields of Nebraska has shown that birds could provide a good biological control of insect pests. They reported that about 13 different species of birds were responsible for reduction of insect pest in the soybean fields (Pukett et al., 2009). Pest population reduced by a 56% compared with plots not allowed with bird foragers. Many of the bird species in crop fields are usually insectivorous. A few of them are pollinators and general foragers in the Northern Great Plains (Piesley et al., 2015; see Robbins et al., 1983).

The agricultural sky above the crops cultivated in the Great Plains has indeed diverse manifestations. Such manifestations relate to many of the biotic factors, such as the pests, predators, biological control agents (birds/insects), pollinators, and disease disseminators. There are a few interesting studies conducted in the Great Plains region (Nebraska, Kansas, North Texas) about the occurrence and role of birds in the farming belts. For example, Avery (2002) states that birds have been moving into farmland in search of foraging grounds, mainly for the food grains. This happens because crop land has expanded into erstwhile natural reserves. Large flocks of birds need vast food grain producing regions to feed themselves. They report that in the Great Plains, birds have been foraging on vineyards, fruit trees (apples), corn, wheat, and soybean in good number.

Dolper and Linz (2016) list at least few major bird pests that occur in the Great Plains region. These bird pests are detrimental to cropping zones. They are red-winged blackbird (*Agelaius phoeniceus*), common grackle (*Quiscalus quiscula*), great-tailed grackle (*Quiscalus mexicanus*), brown-headed cowbird (*Molothrus ater*), yellow-headed blackbird (*Xanthocephalus xanthocephalus*), brewer's blackbird (*Euphagus cyanocephalus*), and rusty blackbird (*Euphagus carolinus).*

Red-winged blackbirds can cause considerable damage to ripening corn, sunflower, sorghum, wheat, and oats, particularly in the milking and maturation stages. Grackles too cause similar damage. However, grackles feed on mature field corn, particularly at the dent stage. Grackles remove entire kernels from the cob. Common grackles pull sprouting corn and cause gaps in the crop stand. Some types of grackles, such as great-tailed grackles damage various fruits and melons. Brown-headed cowbirds can cause damage to ripening wheat, sorghum, and millet. Yellow-headed blackbirds cause localized but generally minor damage to ripening corn and oats. They are traced often in association with redwings (see Dolper and Linz, 2016; Sorghum and Millet Innovation Laboratory, 2002).

Bird damage on Great Plains crops have to be assessed authentically using appropriate methods. Perhaps, drones that are recently introduced in many regions of Great Plains could be effective in conducting aerial surveys and detecting bird populations/clouds. Usually, radars and multispectral sensors are used for detecting bird flocks. Accurate identification of bird species is indeed necessary while adopting prophylactic procedures like scaring. Bird densities need to be assessed as accurately as possible. Often, agricultural sky may harbor large flocks of birds that could be damaging. However, birds could be just leading a commensalistic life without causing perceptible detriment. Sometimes, animal-caused damage could be mistaken as bird caused one (Dolper and Linz, 2016; Linz et al., 2001, Linz and Homan, 2011; Linz et al., 2011).

In the Great Plains, there are bird species that are distinctly pests on sunflowers (e.g., red-winged blackbird). Notable bird pests are blackbird, goldfinch, dove, grosbeak, and sparrow. Actually, several methods that disrupt feeding by bird pests have been tried. A few examples are scarecrows, fright owls, aluminum strips that flutter in the wind, and carbide exploders. No techniques are 100% effective, as birds will adapt to many of these techniques. Currently, no chemicals are approved for bird control in sunflower fields. During recent years, helikites with recorded voices of predatory birds (falcons, eagles, raptors, and owls) are being evaluated for their efficacy in bird scaring.

As stated earlier, bird damage estimation is possible using UAVs (drones) and tethered aerostats. Some of the constraints mentioned to the control of bird damage of crops are financial returns. Economic advantage from a regular or elaborate control measure is insufficient. Often the birds damage is at low levels of threshold. Avery (2002) quotes examples wherein bird damage is less than 1% in most fields. Severe damages are not common since the produce is harvest timely. Also, grains are removed out of reach of the birds. A few exceptions are possible. For example, blackbird damage to sunflower heads is often very severe, wherever it occurs in the Great Plains. The agricultural sky is really detrimental (Linz et al., 2001). We can control the bird pest efficiently by using helikite scarers in the aerospace. Chemical controls are not appreciated by bird lovers. We can preferentially increase birds of prey and reduce crop/grain feeding activity of bird pests.

Avery (2002) reports that in the Great Plains, there are also clear examples of birds playing a beneficial role through their predation of insect pests. For example, certain cut worms that affect forest foliage could be controlled using birds that feed on such insects. Avian predation of insects, such as European maize borers (Straub, 1989), *Helicoverpa* on soybean, budworms, such as *Choristoneura fumiferana* could be induced by attracting birds into farms. Sometimes, a few insects that attract birds may eventually lead to further loss because birds shift to foraging the food grains as the crop matures (Crawford and Jennings, 1989). So, in all cases, we have to opt best aerial methods. We have to shrewdly regulate birds and their beneficial/detrimental activity in farms.

McGlashen (2018) states that birds, such as Cedar waxwings, American Robins, and others devour cherries and other fruit crops. For example, in northern states, they reduced fruit yield to the tune of 3.4 million US$ annually. Therefore, farmers have embarked on using a composite of control methods, such as using propane cannons to frighten the flocks, bird scaring tethered aerostats, and predatory birds. Native predatory bird population has reduced. Hence, they have tried to introduce falcons into the orchards.

Pollinator birds and bats have a great role to play in stabilizing crop yield. There are several bird species that regularly conduct aerial sorties and help in the pollination of major crops in the Great Plains. Among them, they say Hummingbirds that get attracted to panicles and flowers of natural vegetation and crop fields are major pollinator bird species (Ley et al., 2018). Generally, they carry pollen both through their beaks and feathers. Regions with warmer climates possess largest number of hummingbird species and the greatest number of native plants to support the bird's need for food. White-winged doves (*Zenaida asiatica)* are also pollinators in the Southcentral United States. Bright colored tubular flowers attract hummingbirds to gardens throughout the United States of America. American grapevine growing region in the Great Plains attracts Black-chinned hummingbirds (Ley et al., 2018). Bats are not important pollinators of crop plants. However, they do aid pollination of certain wild plant species, such as agave, cacti, hardy shrubs. (Ley et al., 2018). The above examples substantiate the role of aves in the agrarian sky.

7.2.6 AERIAL APPLICATION OF FLUID FERTILIZER USING SPRINKLER

Great Plains support innumerable farms with large land holdings. Further, cropped area that has to be irrigated is vast. As a result, usage of "Aerial Sprinkler" is common in large farms. Aerial sprinklers are a phenomenon related to above ground portion of the agrarian regions. Foliar spray of nutrients dissolved in irrigation water is also practiced. It is commonly called "fertigation." Here, the crop's canopy, foliage, the stomata, and hydathodes are active in deriving both water and nutrients. So, agricultural sky has a role in aiding crop's water and nutrient acquisition functions. Usually, in-season split doses of nutrients, particularly, N is applied as foliar spray. Urea or liquid NH_3 is also sprayed aerially on to the crops. This approach has been found successful in large-sized fields.

Aerial sprinklers need low quantities of nutrients. Foliar application of N is done at 0.02% urea. It is a glaring fact that foliar applied fertilizer N is not affected by soil factors that otherwise reduce fertilizer N use efficiency. Fertilizer N required to achieve a good response from cereals is low if aerial sprinklers are used. Foliar nutrients, as stated earlier, have to transit into plant via phyllosphere. The stomata, lenticels, and hydathodes help in the absorption of dissolved nutrients. During this process, phyllosphere microbes may get altered. We may have to understand the limits to fertilizer N that can be channeled to crops using foliar methods. Also, residual nutrient traceable on the leaf surface needs to be measured.

7.2.7 WIND AND WIND POWER GENERATION OVER THE GREAT PLAINS

As stated earlier in the section on agroclimate of Great Plains, farming in this zone has been a battle of sorts, due to weather-related vagries bestowed by the agricultural sky. Reinhardt and Ginzel (2003), while recounting the historical aspects of farming in the Great Plains state that agricultural sky has been lashed with tornadoes, hails storms, and blizards. The Great Plains also experience cold spell in winter, floods in the rainy season, drought and heat waves in summer, etc. Dust bowls in 1920s–1930s that have occurred are a trifle more severe than the generally known detrimental phenomenon inducing loss of top soil (Heathcoat, 1980; Loring, 2004). Yet, farmers in the Great Plains have shown remarkable agricultural accumen and ingenuity, to overcome a harsh agricultural sky and still offer large quantities of food grains to the populace.

The Central Plains region is endowed with winds of high velocity that carry water and dust particles. This wind system that operates during winter in the agricultural sky is called "Chinook." Tornadoes that are funnel-shaped winds are also common in the Southern and Central Plains. It is an important phenomenon in the agricultural sky. It causes damage to farms, crops, domestic animals, etc. In addition, the Great Plains experience blizzards, cold breeze, frost, and snow fall. Hence, agricultural cropping experts need to select crops that could be planted and cultivated, escaping, the fury of tornadoes (United States Department of Geology, 2013).

At present, there are several satellites (e.g., NASA's Aqua satellite) that capture the aerial images of Great Plains, continuously. They depict the developing dust bowls caused due to extremely powerful wind gusts, heat waves, and dearth for water. For example, MODIS (Moderate Resolution Imaging Spectrophotometer on Aqua satellite) rapid response team helps in informing the farming belts in the Great Plains about the dust bowls. They also monitor other harsh phenomenon related to agricultural sky. In Kansas and Oklahoma, high winds remove dust, dirt, and topsoil to about 3000–4000 m altitude into atmosphere. Topsoil (particles) could float to long distances. Actually, jet streams of winds are known to cause the weather patterns encountered in the Great Plains (Mersereau, 2017: Jiang et al., 2007; Badner, 1979; PNNL, 2016; Berg et al., 2015).

Wind is of course the moving force behind a sandstorm. When north winds blow into Texas, they displace rising hot air and snatch up any loose soil in their path. The stronger and faster the winds, the more sand they push before them. An equally strong factor is drought. The drier the land, the looser the topsoil and the more easily the upper layer is swept away. So, here, agricultural sky induces loss of soil fertility.

Wind-related disasters are frequently perceived in the Great Plains region of the USA. They say between 1950 and 2000 about 34,438 tornadoes or tornado-like activity with severe consequences to humans, crops, and infrastructure were experienced in the USA (Marchigiani et al., 2013). Tornadoes usually develop during intense thunderstorms classified as "supercell thunderstorms." The loss in crop stand, panicles, and grains in the field can be severe.

Wind encountered in various locations within the Great Plains is not all detriment (crop disease propagule carrier), disaster (e.g., Dust storms), or destruction (e.g., tornadoes). Winds aid pollination of several cross-pollinated agricultural crops. Wind also helps in the dispersal of seeds of several plant species that are native to the region. It also carries the seeds (propagules) to long distances. Wind also helps in the dissemination of beneficial microbes from topsoil and cropped fields to different places. Wind power created by the use of parafoils has been utilized to pull carts in farms. This technology has been adopted feebly to transport farm material.

Firstly, wind and wind power generation are phenomena intricately connected to the atmosphere (i.e., agricultural sky) above Great Plains' cropping zones. The potential for global wind power is large. It is important to identify locations that are most congenial for wind power generation. In this regard, the Great Plains with interruption-free constant breeze at low level and high wind intensity at higher altitude make it congenial. Usually, they identify locations with jet-like wind profiles in the lower troposphere to install wind turbines. Turbines are placed either on ground or high in the atmosphere using tethered aerostats or kites (Archer et al., 2013; Glass, 2018; Krishna, 2020b). Unlike wind turbines mounted on towers, turbines on tethered aerostats can be automatically raised and lowered to the height of maximum wind speeds. Therefore, it allows a consistent power production.

Wind turbines (windmills) placed on land can supply plenty of electrical energy (Bello, 2012). Added to it, the tethered aerostats and kites fitted with wind turbines, the total power generated is roughly 100 times more than that presently consumed in the Great Plains (see Duncan, 2016). Reports by United States Environmental Protection Agency, Washington D.C. states that wind energy generated using low-altitude wind turbines could suffice to meet 25% of the electricity requirements of the farming belt, in the Great Plains. However, selecting appropriate light-weight wind turbine that could be hoisted on a ground-based tower or tethered aerostat is important (Martin, 2019). Further, they report that due to preferential use of high-altitude light-weight wind turbines, the prices of electric power generated using wind has reduced markedly. The price of wind power has decreased from US$ 70 per MW-h in 2009 to US$ 20 per MW-h in 2018 (Timmer, 2019). Reports suggest that wind power is gaining in acceptance by the farmers in the Great Plains. Major wind power generating Great Plains states with higher installed capacity are Texas (28,843 MW yr^{-1}), Iowa (10,201 yr^{-1}), Oklahoma (8172 yr^{-1}), Kansas (6182 yr^{-1}), and North Dakota 4188 yr^{-1}) (Wikipedia, 2019).

Further, high-altitude wind power generation can be lucrative. It offers continuous power supply to farms. The present global demand for electricity is 18 tera watts (one terawatt = 1.0 trillion watts) annually. Great Plains needs only small fraction of it to support agricultural food grain generation, industrial production, and urban activity. It seems that in the Great Plains region, there is a trend to adopt wind power generation. This is to reduce GHG eEmission. If we consider merely the low-altitude wind power generation, then, possibly we will reach a threshold where wind-based generation of power could be less profitable. However, if we consider the high-altitude wind power generation using the aerostats, then, such limits to profitability may not be reached quickly. About 4% of the total electricity generated in the US is derived from airborne wind turbines lofted into the atmosphere. Wind power is attracting interest due to search for new sources of energy in the plains (Perkins, 2014). It seems large farms with wind turbines are contributing to wind power via airborne turbine. In comparison, earthbound wind turbines (windmills) are cumbersome. They are noisy and less efficient in electricity generation per unit time (Perkins, 2014). Wind speeds are erratic and feeble at low altitudes because of interferences from vegetation and other topographic aspects. The earthbound windmill blades could be deleterious to birds in the sky.

North Dakota is a major cereal grain producer in the plains. It is endowed with plenty of both low and high-altitude winds. Hence, it has the potential to generate wind power in plenty. Reports suggest that it has 3000 MW of wind power capacity which is generated using 1500 wind turbines. Wind power constitutes about 20% of total electrical energy produced by this state. Most of the Great Plains states are in a position to enhance their capacity to produce wind power. Agricultural farms could be an excellent source to hoist the light-weight wind turbines, using tethered aerostats (NREL, 2018; Glass, 2018).

Airborne wind power generation is a recent concept. It involves designing of turbines that are airborne at different altitudes, using tethers and helium-filled aerostats or kites, etc. There are indeed several designs of kites lofted into air with a light-weighted turbine, to generate electricity (Cherubuni et al., 2015; Loyd, 1980; Levitan, 2012; Krishna, 2020b; Arnolds, 2019; Mortenson, 2016; Wikipedia, 2020a). Generally, tethers are used to transmit electrical power generated at low or high altitudes in the atmosphere. Wind is consistent, less turbulent, and uniform. Therefore, it is safer to adopt high altitude wind power generation above cropland. Here, we should note thatthe Southern and Central Great Plains are characterized by rough wind systems, sometimes even at high altitudes. There are tornadoes and storms for airborne turbines to encounter.

Generation of wind power using light-weight tethered aerostats is a recent idea. It is entirely a process that is operated within the agricultural sky, at different altitudes. Low latitude wind power generation is operative since a couple of centuries, that is, using the windmills. At present, tethered aerostat with wind turbine is being touted. It could be an efficient way to harness high-altitude wind energy available above the crop fields. The agricultural sky above the vast expanses of the Great Plains is perhaps among best bets to adopt low- and high-altitude wind energy aerostat systems. High-altitude wind energy generation is a clear possibility in all the subregions of the Great Plains. The possibility of in situ power generation may induce the farmers in this belt to adopt them. A single aerostat placed at 2000–20,000 m above crop fields serves to supply electricity to seven to eight farms (Glass, 2018).

Forecasts suggest that the Great Plains with its uninterrupted winds at different altitudes is potentially a large wind power generating region. Its potential for wind power can be compared with the oil-based resources of some of the West Asian nations such as Iran or Saudi Arabia. Several states in the Northern Great Plains have series of wind turbines installed. In the Midwest USA, wind power generation could become a major source of electricity. Among the Great Plains states, North Dakota, Oklahoma, and Kansas represent 20% of wind power generation in USA. Texas that produces 10,000 MW of wind power is the major electricity producer through aerial turbines. While the above discussion is confined to the Great Plains only, we may note that globally there are many farming regions that are highly congenial for wind power generation (Jossi, 2017).

7.2.8 SOLAR POWER GENERATION IN GREAT PLAINS REGION

State energy agencies and farmers in the Great Plains area are evincing greater interest in the renewable and clean sources of energy. Wind and Photovoltaic solar sources are being preferred (Brewster and Degelia, 2016; SEIA, 2019). Major problems attached with wind power is that it could be variable and erratic. The wind strength may also be feeble. In addition, installations (solar panels) meant for capturing sun's energy and windmills both need relatively more space (Gross, 2020). Therefore, wind power alone is difficult to attach to grids. Variable power generation often causes problems of grids and their functioning. In addition, Brewster and Degelia (2016) point out that in the Great Plains, every place is congenial for wind power generation or even PV solar power. The power generation point is located in the west and away from population/farms that need them. Therefore, transmission costs increase. Regarding PV solar power, it is said that states such as Oklahoma, Kansas, and Texas are well endowed with sunshine. So, they can support consistent generation of solar energy. The Agricultural sky in the Central and Southern Great Plains is congenial for PV solar energy generation. Oklahoma state has abundant solar energy potential. It generates 2.4 MWh/m^2 at Kenton and 1.6 MWh/m^2 at in the Eastern border region with Arkansas (Brewster and Degelia, 2016; Degelia et al., 2014; SPP Market Monitoring Unit, 2015). There is pressure to reduce GHG emissions. Also, there are incentives available with PV solar makes farming communities in the Great Plains to opt for solar energy. There is a suggestion to combine wind power that is variable with PV solar installations and then channel power generated to the grids. It clearly overcomes the problems connected with wind power (Blumsack and Richardson, 2012). If we consider the energy generation using a mix of wind and solar resources, then Kansas, Oklahoma, and Iowa are the major producers at 31–33% of their total energy generation (More, 2019). Renewable energy sources, such as wind and solar sources are a natural fit with agricultural enterprises, in the Great Plains (UCSUSA, 2008). In the Great Plains, at least two different technologies are adopted to generate energy from solar resources. The photovoltaic -PV solar energy is common. It utilizes solar cells. The concentrated solar power (CSP) is a type of solar energy generation that involves focusing and concentration of solar rays at a single point (North Dakota Energy Center, 2018).

The cost per unit of solar energy is steadily decreasing because of popularity and production of accessories (SEIA, 2019). Some of the best locations to install solar and wind power facility are those where one intends to decrease CO_2 emissions caused by fossil fuel, and also, areas where pollution of atmosphere is to be reduced. Areas with a preexisting power grid are congenial (Plumer, 2013). By this way, we would be utilizing the agricultural sky in the Great Plains to our advantage and in most efficient way.

Reports suggest that in the South Dakota, some of the tribal areas and farming belts are among the best areas to set up a wind/solar power project. It reduces GHG. It trains tribal population in solar power generation systems. It is also economically advantageous to the area (Balaskovitz, 2020). Declining cost of solar power generation has induced greater interest (Motyka, 2020). Solar energy does not pollute air, water, or cause GHG emissions. So, it will not affect tribal land's environment. However, it requires space, and a few solar panels could be made of plastics. Solar power does not work, if the area is predominantly cloudy or excessively cold or frozen.

7.2.9 USAGE OF AERIAL VEHICLES FOR COLLECTING SPECTRAL DATA OF CROPS AND TO SPRAY FORMULATIONS ON FOLIAGE

This agrarian region is among the top users of piloted airplanes, unmanned aerial vehicles (UAVs), such as the fixed-winged drones, autonomous copters, aerostats, blimps, parafoils, microlights, and helikites (see Krishna, 2020a, b). They fly these aerial vehicles to conduct surveillance of fields, soils, and crops. They are used to collect digital data about crops, particularly, the nutrient, and water status of crops, biomass fixation trends. The same aerial vehicles are also used to survey, for occurrence of disease caused by plant pathogens. Also, to map insect pest attack, natural disasters, such as floods, soil erosion. In many places, within Central and Southern Plains multicopter UAVs have been adopted to spray plant protection chemicals and herbicides. The farming companies and farmers with large holdings have begun to utilize the UAVs in a big way in the agricultural sky. These UAVs are often relatively small, efficient and cover large areas of agricultural sky above the cropped fields. They are quick and cost-efficient compared with piloted aircraft campaigns. In the Northern Great Plains, small fixed-winged agricultural drones seem to promise great advantages to the farmer. The cereal farmers in the Dakotas region may really reap better harvest at low cost and low levels of farm drudgery. Reports suggest that there are at least nine different ways by which the small drone aircrafts can revolutionize the crop scouting, agronomic procedures, and monitoring the crops in the Great Plains region of North America. They are high-definition Maps for field surveys; faster and cheaper crop scouting; accurate crop health assessments with false-cColor mapping; precision agriculture; faster detection of weeds, disease, and pests; monitoring slopes, drainage, and irrigation; livestock monitoring; asset inspection; crop damage assessment and insurance (Agri-tech Tomorrow, 2019). It seems that small drones, helikites and aerostats could be most efficient in collecting useful data about experimental crops (Krishna, 2020a, b; See Table 7.1).

The agricultural sky in the Great Plains could also be utilized efficiently to transport cargo using the small UAVs, autonomous parafoils, blimps, and piloted microlights (Krishna, 2020a, b). During the recent past, there is a trend to evaluate the tethered aerial vehicles, that is, aerostats with facility for visual photography, spectral analysis and even chemical probes to judge the atmosphere above the crops.

TABLE 7.1 The Aerial Observation and Monitoring of Crops Using Autonomous Aerial Vehicles at Agricultural Experimental Stations Located Within the Great Plains of North America: A Few Examples for Efficient Usage of Agricultural Sky to Monitor Experiments.

Location	Purpose	References
Oklahoma State University, Stillwater, OK, USA	To assess performance of wheat germplasm and plant breeder's lines for tolerance to low soil fertility and specific disease. To trace genetic variation for in-season chlorophyll index, and canopy temperature, biomass accumulation and grain formation	Reynolds (2012), Babar (2006), Prasad (2007)
Texas A and M University, Brazos county TX, USA	Evaluation of elite wheat genotypes and cultivars for growth and yield performance of wheat genotypes	Shi et al. (2016), DMZ Aerial Inc. (2013)
University of Tennessee Agricultural Experimental Station, Knoxville TN, USA	Monitor cotton crop sown in experimental plots for crop's water stress index (CWSI), using thermal (Infra-red) cameras. Select cotton genotypes with better ability to tolerate drought and low soil moisture tension.	Sullivan et al. (2007), Wooten (2007)

Source: Krishna (2020a).

7.2.10 USAGE OF SATELLITES IN SPACE TO STUDY GREAT PLAINS AGRICULTURE

Let us consider an era when satellite imagery was not available. Earlier in the past century (1930s), U.S. Southern Great Plains suffered drought, dust bowls, and heat waves. It caused some 200,000 farm bankruptcies. Wheat and corn yield were reduced by as much as 50% (Rosenweig et al., 2011; Warrick, 1984). On a severe year, the situation is no better even at present. We have to understand the "agricultural sky" much better. Then, take precautionary measures, such as fallows, wind breaks, etc. A low-cost cover crop helps to reduce the atmospheric ferocity. The bottom-line suggestion is that we must be prepared in advance. For this, we have to monitor agricultural sky in detail. Then adopt suitable remedial measures swiftly.

Satellites that are placed in the space are being utilized by the agricultural experts of the Great Plains, since several decades. Perhaps, it started with the Landsat in 1970s or even prior to it. Satellite imagery has helped farmers in recognizing the weather phenomenon relevant to their area/fields. It has helped in recognizing the disease/pest infestation in larger areas and the impending danger. Satellite imagery has also helped researchers in forecasting weather patterns in the agricultural sky. It has aided in forecasting grain/biomass yield in large zones. The clouds and haze that are traced in the lower layers of agricultural sky affects the clarity and resolution of satellite imagery. Also, the re-visit time to verify the crop's status is generally not congenial. Yet, satellites are useful source of data about agrarian regions. The Great Plains agricultural agencies use the satellite imagery. The GPS tag adds to accuracy of the agronomic procedures. This is a good example of influence of agricultural sky-related man-made phenomenon on the crop production. The precision farming that is being practiced in many regions of the Great Plains depends immensely on GPS tagged data. Digital data derived from satellites could be utilized to guide a few types of farm vehicles.

Climate variability and its trends affect global crop yields. Generally, global crop yields are characterized as highly dependent on location, soil type, crop species, and irrigation. Kukal and Irmak (2018) state that U.S. Great Plains is a large food grain generator. Based on climate data from 1968 to 2013, it is clear that Great Plains experience climate variability. Therefore, it is an ideal region of investigation if we intend to study the impacts of climate on food production. Incidentally, agroclimate considered here is mostly aerial phenomenon. This paper evaluates impacts of climate on maize, sorghum, and soybean yields. About 25% of grain yield could be attributed to aerial phenomenon, such as precipitation, temperature, and humidity. During past decade, satellites that adopt high resolution infra-red sensors have offered soil moisture maps. The spectral maps provide data for only upper 6 inches of soil layer. Such data are useful in organizing large-scale planting programs and in detecting regions that need immediate supply of water. Satellite imagery of natural vegetation, its botanical diversity, and biomass trends have been useful. Such aerial images show areas prone to forest/shrub fires, crop residue burning, accumulation of pollutants, smog, and dust. At present, farming in Great Plains is about adapting agricultural production strategies to climate change effects, revising crops, inputs and yield expectations, based mostly on weather and climate change. Satellites may help farmers in accomplishing several functions on farms in a better way.

7.3 THE AGRICULTURAL SKY ABOVE THE PAMPAS OF SOUTH AMERICA

Pampas of Argentina and adjoining regions of South America is a large expanse of natural vegetation. It is composed of predominantly a grassy vegetation and agricultural cropping belt. These semiarid plains support small and tall grasses and broad-leaved shrubs. Trees are sparsely distributed. Pampas that extend into 750,000 km^2 miles is indeed a vast fertile lowland plain (Taboada, 2006). It occurs between 28°S and 39°S latitude and 50°W to 65 °W longitude. The Pampas agricultural belt also extends to parts of southern Brazil and Uruguay. This is in addition to Central and Northeastern parts of Argentina. The Pampas are classified as Rolling Pampas, Subhumid Pampas, Southern Pampas, Semiarid Pampas, Flooding Pampas, and Mesopotamian Pampas (Viglizzo, 2011; Hall et al., 1992, also see Krishna, 2015; Editors, 2020). Overall, the cropping belt within Pampas has spread considerably. In 1960, the cropping belts occupied only 33%. The rest 65% was natural prairies. In 2005, however, cropping zones had expanded to 45% and natural prairies had shrunk to 55% of the total vegetational area. This is attributable to mechanization and better economic returns due to export of grains (Viglizzo, 2005 also see Krishna, 2015; Baldi and Paruelo, 2008; Solbrig, 1997, 2005; Solbrig and Viglizzo, 2011).

Pampas is actually among the major food grain generation systems of the world. Its produce serves human population and farm animals both *in situ* and those in other regions of the world. Primarily, the agroecosystem supports production of wheat, sorghum, legumes, and oil seed. The topography and soils have been classified in detail. The Rolling Pampas supports large-scale production of wheat. Wheat and other field crops are produced by large private companies. Hence, individually owned small farms are few. Many of the agricultural operations are suitable only with large companies. Like other agroecosystems, agricultural sky above the Pampas has played its role in influencing the soils, their fertility, and crop production trends. Major soil types encountered in the Pampas of Argentina are the Hapludalfs and Haplic phaeozems. They occur in the Rolling Pampas. In the East, Pampas possess soil types classified as Arguidolls or Luvic phaeozems (Glatzle, 1999; Hall et al., 1992; Mostacelli and Pazos, 2000; Deckers et al., 1998; FAO,1997). Overall, Haplustolls are common in the cropping zones of Pampas. These soils are not high on water retention capacity. Pampa's plains possess gentle inclination. Yet, the soils suffer from loss of topsoil. The atmosphere immediately above the natural vegetation and cropping zones of Pampas region has been studied. However, comprehensive knowledge about aerial and ground characteristics are necessary. Agronomic decisions depend on careful compilation of all relevant data and their analysis. Pampas agrarian sky might have influenced the evolution of crops and their specific genetic stocks. Also, specific cropping systems that are practiced by farmers. *There is a need to characterize the "Pampa's sky" in detail and take appropriate measures to maximize agricultural productivity.*

Pampas too has been influenced immensely by the natural and man-made factors related to agricultural sky. However, so far, we have not bestowed good interest to study the details, intricacies, and interactive effects of the Pampas sky. We need to understand the "Pampa's sky" and see how best we can coordinate our efforts, to maximize crop productivity. No doubt, Pampas sky has its unique features. We have to study them in detail and adopt farming practices, accordingly. We could then improve productivity from the present level, perhaps. On a global basis, we should also be able to assess the peculiarities of different agroecosystems and food generating regions and compare them. It might help us to adopt farming practices that are both distinct for Pampas region.

7.3.1 AGROCLIMATE ABOVE PAMPAS REGION OF SOUTH AMERICA

Agricultural sky above the Pampas cropping zones supports a subtropical climate in the North and temperate climate in the South (Hall et al., 1992).

This region is classified as semiarid based on annual precipitation and its distribution. The Koppen's classification for Pampas is "Dry Mid-latitude climate (Bs) with dry summers." Wind from Atlantic is a major factor that influences the agricultural sky and weather patterns of the Pampas. ENSO (El Nino southern Oscillation) is a major factor that influences precipitation and seasonal patterns (Podesta et al., 2009; Bert et al., 2006; Wikipedia, 2013a, b). Hence, cropping systems and their productivity are immensely influenced by these ENSO factors and Atlantic wind. ENSO, which is an agricultural sky-related phenomenon influences seasonal changes and annual climate pattern. ENSO has marked influence on precipitation pattern (Podesta et al., 2009; Bert et al., 2006).

Diurnal variations and receipt of photosynthetic radiation spans 10.2–14 h in Northern Pampas and 9.5–15 h in the South. The influx of radiation on crop's canopy is high during December (26 MJ m^{-2}) but recedes to 8.9 MJ m^{-2} in June (Otegui et al., 1995). Agricultural weather stipulates that seeding be done in September to harness the moisture (precipitation), photosynthetic radiation, and heat units efficiently.

The Pampas sky bestows about 750 mm to 1200 mm annual rainfall. Precipitation varies within limits depending on the subregion. Rainfall distribution varies within a year based on location and season. Precipitation begins in October. It peaks during December till March when it reaches 100 mm per month. Precipitation is low during May till September, averaging only 40 mm per month. The ambient temperature in Pampas is moderately warm at 25°C–29°C in the day and 14°C–12°C in the night. Heat waves do occur resulting in dusty storms. Pampas aerospace experiences a period of heat wave followed by a "Pampero wind." Pampero wind is a cool breeze that induces a drop in temperature to 14°C. The Pampero wind occurs for 2–3 days at a stretch. Frosty periods occur during April–July. Crops do suffer due to frosts. Snow fall is rear but if it occurs it is sporadic, lasting for a day or two. Thunderstorms do occur. They bring about destruction of standing crops. Incidentally, long stretches of dry zero precipitation periods also occur in a year. It affects crops due to paucity of soil moisture. Precipitation pattern is affected by ENSO and El Ninos. El Ninos is connected to higher precipitation (Podesta et al 2009; Bert et al., 2006). Droughts caused by the La Nina are generally attributed to low levels of precipitation and delayed rainfall events (NASA, 2009).

Several studies involving simulation of climate change and its effects on major crops, such as wheat, barley, soybean, and sunflower in the Pampas region has been conducted, since past three decades. One of the earliest efforts has been reported by Baethgen and Magrin (1995). They say climate change does affect atmospheric characters, such as temperature and precipitation pattern. Temperature changes of 0°C–2°C were forecasted while precipitation pattern could alter significantly. Such climate change effects, particularly, higher temperature has lower impact on barley compared with wheat grown in the same area. Wheat yield was affected significantly due to alterations in precipitation and temperature. Pampas sky has its share of climate change impacts on soil characteristics, crop growth, and productivity. Perhaps, we should analyze the agricultural sky and its various factors in isolation. Later, to assess the impact on crops cultivated in Pampas.

In addition to natural precipitation, Pampas plains are endowed with several other water resources, such as riverine, lakes, dams, and canals. Irrigation using furrows is feeble (< 6%). Hence, large farm companies depend predominantly on foliar supply of water, through large linear or center-pivot sprinklers. Aerial sprinklers are automated and fitted with GPS connectivity, so they can be used during precision irrigation procedures. Sprinkler irrigation is an important aspect related to agricultural sky. Sprinkler irrigation also reaches soil and its layers. Water can be effectively absorbed by crop roots.

7.3.2 GREENHOUSE GAS EMISSION IN PAMPAS PLAINS

In the vast plains of Pampas, GHG emissions may originally get generated in subsoil or surface layers or due to crop residue burning. Such GHG may also be emanated through the crop canopy. Whatever be the source, the ultimate effect is on agricultural sky and its gaseous composition. Excessive GHG can influence the normal weather patterns. It can enhance the ambient temperature during the growing season (Paruelo and Sala, 1993).

Viglizzo (2002, 2005) has stated that land conversion and clearing of natural vegetation got accelerated since 1890s. This has resulted in a vast prairie-based agrarian region. This transformation of Pampas natural vegetation diversity to monotonous crop expanses with just few species has taken place rapidly, since 1970. Adoption of conventional tillage, reduction in pastures, heavy grazing, and neglect of no-tillage option has led to excessive emission of CO_2 from soil into the sky (Garcia-Prechac et al., 2004). Crop residue burning has caused loss of C as both CH_4 and CO_2. Fertilizer supply to the crops has increased appreciably. Particularly, fertilizer N has been channeled through soil and also via sky above the canopy using sprinkler-mediated fertigation. This has induced loss of a certain fraction of N into atmosphere as NO_2, N_2O, and NH_3 (Portela et al., 2009; Palma et al., 1998). Such a land-use change has indeed affected the agrarian sky too. Farmers may have to readjust their agronomic procedures with due alertness to changes in the agrarian sky.

7.3.3 *THE WIND FACTOR IN THE PAMPAS SKY*

Wind and water-mediated soil erosion is a common phenomenon in the plains. The Loess soil in the La Pampa region is particularly severely affected by the agricultural sky-related phenomena. For example, high wind and heavy precipitation events have induced soil erosion. Frequent intensive winds do cause loss of topsoil fertility (Michelina and Irrutia, 1995; Buschiazzo et al., 1998et al., 1999). In the La Pampa, reports suggest loss of 9.4–27.1 t ha^{-1} topsoil, annually. Severe topsoil loss amounting to 51–54 t ha^{-1} occurs in Southwest La Pampa. It is attributable solely to the wind. Topsoil loss due to wind reaches 1.82 t ha^{-1} on Ustipsamments and 0.29 t ha^{-1} on Haplustolls. Here, wind speeds of 21–25 kmph is common. The high-speed winds (>20 kmph) cause dust bowl-like conditions. It induces dust/particulate deposits. About 370–770 kg ha^{-1} dust/particulate deposits have been attributed to wind (Ramsperger et al., 1998). In comparison, topsoil loss in the Central Great Plains, that is, in Kansas USA) is around 200–300 kg ha^{-1}. In Israel, cropping regions suffer 200–400 kg dust/particulate matter per year. In the Sahelian West Africa, it is 1400–1560 kg ha^{-1} yr^{-1} loss of dust/particulate from the soil. Incidentally, wind-mediated loss of topsoil also induces loss of nutrients. For example, reports from La Pampa states that in a year about 29 kg N, 2.5 kg P, 33 kg K, 43 kg Na, 55.8 kg Ca, and 13.5 kg Mg are lost from agricultural fields. Typically, in La Pampa, each kg of dust created by wind carries 566 mg Na, 273 kg mg K, 664 mg Ca 145 mg P, and 125 mg. Obviously, wind is a major agricultural sky-related phenomenon. Efforts to lessen wind speed using wind breaks are needed. Soil loss should be reduced by contouring, mulching, and using cover crops during off-season. Droughts and dust bowl condition attributable to agricultural sky-related events are experienced frequently, in the Pampas (NASA, 2009).

Overall, wind plays a vital role in farming belts of Pampas. Wind that blows at low levels above the soil surface and high velocity could be detrimental. High-speed wind causes soil erosion by blowing of surface soil. The soil fertility depreciation depends on the N and P enrichment traced in the soil particles that are lifted and displaced from the Haplustolls found in the Pampas. Windstorms of different strengths do affect Pampas agricultural fields. Mulching is often the best bet to reduce wind caused erosion. No-tillage systems too are adopted to reduce erosion (Buschiazzo et al., 2006).

7.3.4 PHOTOSYNTHESIS AND CARBON FIXATION RATES

In the Pampas, wheat/soybean cropping system is an important terrestrial component. It sequesters massive amounts of atmospheric CO_2. Since soybean is a legume, it also fixes large quantities of atmospheric N through BNF. Accurate estimates of biomass, grain yield, and crop residue recycling are available. Natural vegetation like Pampas grass (*G. argenteum*) plus other native species crops, such as wheat, soybean, and sunflower regulate the carbon dynamics. The Pampas agroecosystem regulates C dynamics in the sky and underground portion of the Mollisols (Civeira, 2011; Alvarez et al., 2009, Alvarez, 2012; Peiretti, 1998; Trigo et al., 2009).

Atmospheric N is captured into soil microbial biomass through biological N fixation process. As such, legume/cereal rotations are adopted frequently in the Pampas. Zotarelli et al. (2002) have screened a few different legume species to determine the extent of atmospheric N converted and fixed into soil microbial biomass. They report that certain species like groundnut could convert up to 200 kg N season^{-1} from the ambient atmosphere and fix it into soil microbial biomass. Obviously, BNF is an important natural phenomenon. In fact, we can harness atmospheric N via the large tracts of Pampas with cereal/legume intercrop and rotations. For example, maize or wheat/soybean rotations and/or intercrops were designed to improve fixation of atmospheric N. The area under legumes, such as soybean or groundnut or legume pastures must be reducing the need for industrial conversion of atmospheric N. Otherwise, we have to adopt Haber's process to produce fertilizer N.

The CO_2 emissions occur due to soil oxidative processes and soil microbial respiration. No tillage methods are adopted frequently, say, on alternate years or once in 3 years. Such practices reduce loss of CO_2 to atmosphere. Garcia-Prechac et al. (2004) states that no-tillage systems began in Pampas region around 1960s. Since then, it has spread to several parts of the vast 22 m ha of cropping belt of the Pampas. The wheat/soybean, wheat/sunflower, sorghum/fallow, and wheat pasture crop sequences followed are best suited to reduce the return (loss) of CO_2 to atmosphere. Crop production in

Pampas occurs on a few different types of soils, such as Mollisols, Ultisols, Aridisols, and Hapludolls. The SOC ranges from 1 to 3% on dry weight in most regions. Since 1980s, farming companies in Pampas are adopting zero-tillage systems. Such a system avoids frequent ploughing that may result in the oxidation of soils. Zero-tillage avoids excessive loss of SOC as $CO_{2,}$ to atmosphere through soil respiration. Reports suggest that about 40% of agrarian region is under zero-tillage system. Therefore, it allows for better C sequestration than regions not pursuing zero-tillage systems. A few reports from peanut growing regions suggest that zero-tillage favored preservation of SOM. In comparison, conventional tillage induced loss of both CO_2 and soil N as oxides (Fabrizzi, 2003; Bayer et al., 2001; Sa et al., 2001).

Conservation of SOM and C sequestration is important in the Pampas. Repeated farming has reduced as much as 15 t C stocks in the soil, per season (Farage et al., 2005). Some of the reports from Pampas state zero-tillage sequesters about 0.1–0.25 t SOC ha^{-1}. Az-Zorita et al. (1999) suggest that designing agronomic procedure that enhances SOC sequestration is important. Reducing loss of SOC as CO_2 is essential to preserve the quality of soil. Fields under direct seeding with no-tillage conserve 6.9 kg ha^{-1} SOM more in soil than the conventional tillage plus seeding.

7.3.5 MICROBES IN THE AGRICULTURAL SKY

Pampas agricultural sky, like that of other agrarian belts is the main medium for dissemination of innumerable microbes, including the plant pathogens. Several bacterial and fungal diseases are spread via the atmosphere. Rust fungi and their uredospore are most frequently traced in the agricultural sky.

The cereal belt in South America is often exposed to infestation by stem rust fungi and other similar fungal species. These rust fungi are disseminated aerially based on wind direction and speed. The aerial propagules (uredospore's) are blown across the cereal belt. They say, in South America, we can identify two epidemiological regions. The Andean region (Columbia, Peru, Western Bolivia) is one of them. The other one is Pampas region (Argentina, Uruguay and southern Brazil). The Pampas region is a vast wheat belt that has *P. graminis* and other rust fungi endemically. Here, the pathogen encounters congenial atmospheric (agricultural sky) conditions. It gets spread rapidly through the air. Generally, wheat crop does not suffer high loss in grain productivity because the severity of disease is not great. However, it does get disseminated quickly through the wind.

Prior to intensive farming, Pampas supported a biologically diverse vegetation. We have no idea about the microbial component in the aerospace above such natural vegetation (grasses, shrubs and sparse trees) stretches. We have no idea therefore about the changes and upheavals of microflora in the agrarian sky that occur due to adoption of large-scale farming.

7.3.6 AGRICULTURALLY IMPORTANT INSECTS IN THE PAMPAS

The Pampas agroecosystem encompasses subtropical grasslands in the provinces of Entre Rios, La Pampa, Santa Fe, Cardoba, and San Luis. Natural grasslands have encountered prominent changes in the vegetation due to large-scale crop production (Sanchez and DeWysiecki, 2008). Yet, the agricultural sky above these areas has remained constant in some cases, for example, insect pests. Insects, such as aphids have got accentuated. However, grasshoppers, for example, remained serious detriment to crops, such as wheat, maize, sorghum in some years (Sanchez and DeWysiecki, 2008).

The cereal crops of Pampas, namely, wheat, maize, and sorghum are all attacked by aerial insect pests. Wheat in particular is infested by a series of aphids, leaf-eating cut worms, etc. The extent of loss due to aerial insect pests could vary depending on crop genotype, the prevailing stage of the crop, weather, and density of pest population. Cereal aphids are almost native and endemic pests on wheat. They have been a detrimental factor since 1960s (see Krishna, 2015). Pampas farm belt lacked the natural enemies for the many species of aphids that affected the cereals. Major aphid pests affecting crops in Pampas are rose-grass aphid (*Metopolophium dirhodum*), English aphid (*Sitobion avenae*), Bird cherry aphid (*R. padi)*, and Russian Wheat aphid (*D. noxia*). Often, a combination of aphid species and other aerial insect pests attack the cereal crops. This leads to significant decrease in grain harvests (Stary et al., 1993; Zuniga, 1990; Reed and Pike, 1990). In the later years, there was an effort made to select, accentuate and use biological control agents on these aphids. Both, endemic predators and even microbial pathogens were adopted to control aphids (Van den Bosch, 1976).

Lepidopteran pests that inhabit the canopy of sunflower fields can play havoc on crop growth and yield formation. For example, McRae (2011) has pointed out that a Lepidopteran defoliator, such as *Rachiplusia nu* can reduce foliage and seed yield of sunflower grown in Pampas. Fuente et al. (2006) believe that Pampas farming enterprises have got their farming procedures intensified. This has led to build up of insects. The diversity of species attacking the wheat/coriander rotations and their populations have both increased. Such build up has been dependent on crop species and rotations followed. Intensification has also affected the crop–weed associations in addition to pests. The diversity and population of insects in the canopy and above the crop also depended on the growth stage. Further, they state that insect population could be monitored above the canopies of the intercrop, which depended on the growth state and emissions. Clearly, aerial insect population that could be detrimental to crops has increased in the Pampas sky.

Pampas agricultural sky also harbors a range of beneficial insects. Perhaps, the crop's canopy holds the key. Honeybees and other pollinating insects are traced in the Pampas sky. Honeybees have a special role to play in stabilizing crop productivity. The wheat-based cropping systems that are getting intensified might affect diversity and population of pollinators. It is preferable to have higher populations of pollinators. However, application of insecticides/fungicides may affect the honeybees, their diversity, and population. Reports by agricultural agencies in Pampas point out that herbicide-tolerant soybean has induced farmers to apply greater quantities of chemicals. This trend has affected diversity and population of pollinator bees (Le Feon et al., 2014). The wild species of bees too have been affected by the intensification of Pampas farming systems. The agricultural sky that harbors pollinators has changed significantly. Le Feon et al. (2014) state that despite aerial pollution, honeybees belonging to 33 morphospecies could be detected. Among them, wild species such as *Lasigblossum* (*Dialictus*) dominated the Pampas sky. In order to preserve and improve the diversity of honeybee and other pollinator insect species, there is an urgent need to reduce the use of harmful chemicals, during crop production. Clearly, Pampa's aerospace is home to pollinators that directly determine seed set/yield. So, let us not affect Pampa's sky and its biotic components deleteriously. Adoption of disease/insect pest tolerant cultivars reduces use of harmful chemicals. It reduces drift of pesticides/fungicides into the atmosphere. Use of helikites bird scarers too reduces the use of chemical baits.

7.3.7 AVES AND THEIR ROLE IN PAMPAS CROPLAND

Aves are an integral component of the Pampas sky. There is a general opinion that large-scale mechanization and expansion of agricultural crop production in Pampas has affected the biodiversity of birds, insects, and crop species (Solbrig, 1997, 2005). Large farms replaced hitherto small ones and affected the biodiversity. Bird (grain destroyers) and insect pests got accentuated too. Generalized observation of the Pampas sky aimed at knowing the bird diversity has been made, even in recent years. For example, Corbet (2012) lists several birds traced in different provinces within the Pampas. A few of them are related more to agricultural activities in the ecosystem. There are birds that are commensals and those that are pests on crops. There are predators that thrive on pest birds. There are bird species that serve as excellent pollinators of major crops cultivated in the Pampas (see Wikipedia, 2020b, c). Codesido et al. (2016) have observed the changes that occur in bird population as a consequence of agricultural land use. They say, a few rare birds have been retained in the crop belts despite changes in the vegetation. A few specialist species were confined to woods and grasslands, but many species did expand the foraging area. At the same time, caution about possible extinction of a few species of Pampa birds has also been expressed.

Medan et al. (2011) have observed that introduction of intensive agriculture in the erstwhile grasslands of Pampas has clearly affected birds in the sky. A few novel species of birds got introduced (attracted) into cropland. It seems there were increases of bird species that served as pollinators. Obviously, insect pollinators play a vital role in the pollination. Pollinator population will improve the seed set. *Therefore, not just the soils and crops, agricultural sky and its manifestations too got changed in the Pampas as intensive agriculture took roots.*

Now, let us consider birds that function as biological control agents in the Pampas sky. Raptors are the commonly traced "birds of prey" above the cropland in the Pampas biome. An examination of flooding Pampas has shown that there are few genera and at least 16 species of raptors that inhabit the natural vegetation regions, grasslands, and cropped fields. Type of land cover and pests (prey birds/insects) encountered may affect the species of raptors and owls, above the farms. Raptor species, such as Chimango Caracara (*Milvago chimango*) are frequently encountered in the sky. Chimango caracara is dominant species in the Flooding Pampas, particularly in the agricultural sky above the crops. Other species are Southern Caracara (*Caracara pancus*), American Kestrel (*Falco sparverious*), Hawk (*Rupornis magnirostris, Accipeter striatis*). Long-winged Harrier (*Circus buffoni*) were less abundant in this area. Owls traced commonly are the Burrowing owl (*Athene cunicularia*), Barn owl (*T. alba*), and short-eared owl (*Asio flammeus*). These predatory birds are known to keep the small birds (crop pests) that feed on crop's panicles and grains under threshold. They also reduce population of several insect species. So, agricultural sky above regions with predominance of predatory birds could be classified as "Predator bird rich sky" (Baladron and Bo, 2017; Lupoa, 2019).

One of the observations suggests that Pampas' vegetation has undergone significant changes due to intensification of agricultural crop production. Crop production areas have also extended into areas that hitherto supported natural vegetation. This has affected the predatory bird population. Row crop production in Pampas has affected few species of birds. It has affected avian abundance and diversity. No doubt, pest control functions of predatory birds need to be improved (Gavier-Pizarro et al., 2012).

Studies on abundance and diversity of bird species above the grassland of Pampas has revealed that land scape and its crop matrix plus the size of vegetational patches affect the bird population (Pratelli et al., 2018). Clearly, cropping systems, insect/bird pests, and predatory birds interact in a given farm. In the Pampas agrarian regions, land use, and cropping intensity seems to affect the bird species that inhabit the area. Phifer et al. (2019), for example, assessed the bird population and its diversity in an agrarian zone that supported pasture, annual crops, and major cereals. The area also had Eucalyptus plantation. It seems that bird species diversity was low in predominantly plantation zones. Native grassland, annual vegetables, and cereals supported greater population of birds. Further, bird species assemblages were distinct and different depending on the crops.

Bird pests in the Pampas sky can be deleterious to wheat/legume cultivating farms. Sunflower is among most susceptible crop species to bird pests. No doubt, Pampas aerospace/vegetation supports several species of birds that are pests on crops. They feed on grains and reduce the harvest size. Calamari et al. (2018) have examined the influence of landscape, environmental changes, and agricultural expansion on the pest species of birds. These are birds that affect crop yield by consuming/destroying the panicles of cereals. Firstly, changes in climate definitely affect bird population and diversity. Climate also affects the extent of damage that birds cause to the crops. Calamari et al. (2018) further state that population density of birds could be influenced by the individual atmospheric parameters, such as the changes in the temperature regime, etc. The pest birds are tightly linked to any changes in the climate or cropping systems adopted in the Pampas agroecosystems. Climate effects are actually severe on major pest bird species, such as the Eared dove and Monk Parakeet. These bird species affect grain productivity.

Managing pest birds in the Pampas aerospace could be important, at least in certain locations, although not all over the plains.

7.3.8 WIND AND SOLAR POWER GENERATION IN PAMPAS

Pampas agricultural region is actually endowed with different sources of power generation. Hydroelectric is the major renewable energy. It accounts for 68.6% of annual power generation in Argentina (not Pampas region). Wind power through wind parks amounts to 131 MW, that is, 21% of the total power generation, annually. Solar power generation is about 6.2 MW. It is equivalent to 1.0% of total power generation. Biomass/fossil fuel contributes 9.3% and geothermal 0.1% of the total power generation in Argentina.

Records indicate that Argentina initiated wind/solar generation in a significant way in 1994. Wind power generation facilities were located in seven provinces, such as La Pampa, Santa Cruz, Buenos Aires, La Roijas, Nuequen, and San Juan. Wind power farms with low level windmills occur in about 18 farms. They are connected to a grid (Invierta en Argentina, 2012; CADER, 2009). Rufin (2017) opines that larger investment in wind power is necessary. Argentina could be a major wind power generating nation in Latin America. Most of these mills that utilize the agricultural sky efficiently are private companies (Azzopardi, 2017). Incidentally, wind power generation has been in vogue for the past century. Farmers have generally adopted windmills of vintage design for a long time. However, in recent years, light-weight modern windmills have been installed. Farmers utilize such windmills for their own use. And, farmers lend the electricity generated to neighboring farms. Solar power generation facilities occur in two places in the La Pampas and San Juan. They generate small quantities of electric power that gets distributed via a grid, to different farms in the region (Invierta en Argentina, 2012).

7.3.9 USAGE OF AIRPLANES, DRONES, AND AEROSTATS

The piloted airplanes although costlier have been commissioned to study the geographical features, topography, and cropping pattern adopted in the Pampas. The airplane campaign is usually brisk. Aerial imagery is often insufficient regarding clarity and details. Aerial robotics have just entered the Argentinian Pampas. There are very few published data about the usage of agricultural drones in Pampas (see Krishna, 2018). Agricultural drones are highly suitable for consistent usage over the large private farming companies. Large-sized farms need digital data for adopting precision farming. This makes drone aircrafts a useful proposition. The cost of surveillance and of farms using the aerospace is relatively low. Also, aerial robots can offer data about the crop's progress and biomass/grain formation trends. The usage of tethered and untethered aerial vehicles, such as parafoils, tethered aerostats, and blimps are feeble. Perhaps, they are still in the military barracks monitoring the activities of military camps. These low-cost aerial vehicles may soon be among top priorities above the Pampas agrarian regions for the evaluation of crops. Blimps, autonomous parafoils and small drone aircrafts may all gain acceptance by farmers in the Pampas. The copter drone aircrafts fitted with aerial spray bar for the application of plant protection chemicals are perhaps best bets for the large private farms of the Pampas. They would utilize the agricultural sky both to collect data about infestation of insects/ diseases and help in spreading the pesticides/fungicides. Pampas is primarily a flat prairie vegetation. It supports several fields with crop species and grassy vegetation. So, introduction of drone aircrafts or tethered/untethered aerial robotics should be considered on priority.

Pampas and other agrarian regions of Argentina are among the major producers of food grains, especially wheat, soybean, sorghum, sunflower, and maize. They generate food grains through a series of large and small-scale private farms. Such farms utilize variety of farm vehicles and gadgets. Drones are among the recent entrants into Pampas. They could revolutionize the use of aerial space above the Pampas cropping zones. They could help farm companies in rapid and efficient management of farm operations (Meyerhoff, 2019). Recent initiatives in Pampas (Argentina) and other South American agrarian regions aim at utilizing the "Sprayer drones" in the agricultural sky. Sprayer drones could make spraying fertilizer formulations, pesticides, and fungicides over the vast wheat/soybean intercrops much easier. They would use the aerial space efficiently in distributing the chemicals swiftly, using the spectral data and precision agricultural principles. Such drones are usually fitted with variable-rate applicators (Grassi, 2019).

In Pampas, about 15–20% of agrarian region is managed by foreign-based large farm companies (Viglizzo, 2002). They adopt advanced methods, such as GPS-guided traction, fertilizer application, weeding, and sprinkler irrigation fitted with variable-rate nozzles. Harvesting is done using GPS-aided precision combine harvesters. Obviously, agricultural sky is utilized efficiently. During recent years, agricultural drones are being inducted to carry out spectral analysis of crops and utilize such data in the precision vehicles.

Drones are best suited in an agrarian region such as Pampas mainly because, Pampas farms are mostly large-sized farms. In future, sprayer drones may become popular with Pampas farming companies. They are among highly efficient aerial robotic vehicles that could make a mark in this agrarian region (Quest UAV, 2020; Krishna, 2018, 2020 a, b). Private farms in Argentina are said to be modernizing the agricultural methods. Satellite-mediated aerial observation of natural vegetation and crops is already in vogue. Drones with ability for close-up imagery of crops that supply spectral data about crops are preferred. Such spectral data usually pertains to crop's growth status (NDVI), crop's water stress index (CWSI) and disease/pest attack, if any (Meyerhoff, 2019; Precision Hawk, 2020; see Krishna, 2018, 2020a; Stickles, 2019). The aerial space above crops in the Pampas could also be efficiently used to get spectral data using the robotic parafoils tethered aerostats and helikites (see Krishna, 2020b; Allsopp Helikites Ltd., 2017). We should note that aerial drones reduce cost on crop scouting and farm surveillance.

At present, there is a trend to utilize helicopter and multicopter drones fitted with spray bar and variable-rate applicators (e.g., HSE, RMAX). It is conspicuous within the US Great Plains, in Northeast China, Japan and other regions. This is to spray the rice/wheat fields using sprayer drones. Soon, sprayer drones would become a common sight in the agricultural skies, wherever crop production is intense (Bolton, 2016; RMAX, 2015; DJI, 2016, 2017; Krishna, 2020a). Hence, the suggestion here is to use aerial drones (i.e., sprayer drones) to eradicate disease/insect pests in the Pampas crop production zones. Since vast crop production zones in the Pampas are actually controlled by large consortia and companies, adoption of copter drones to eradicate insect pests using drones is easier. Next, tethered aerostats and helikites are efficient in obtaining aerial imagery. They also provide digital data about crop's pest status. Therefore, it is possible that soon the Pampas' agricultural sky will also be dotted with helikites/aerostats. Overall, we will be using an *abiotic* factor (aerial robotic drone) to eradicate aerial insect pests (biotic factor) in the Pampas sky.

7.3.10 USAGE OF SATELLITE DATA AND GUIDANCE OF FARM VEHICLES

In Pampas, agricultural sky has been utilized effectively. Satellites monitor the salient features of terrain, topography, soil resources, water, natural, and agricultural vegetation. Satellites are also adopted to monitor water resources, such as the rivers, dams, canals lakes, irrigation lines, etc. At present, GPS-guided farm vehicles are in vogue in most of the large farms. The tractors used for deep ploughing, ridging, planting, and fertilizer inoculation are usually fitted with GPS receivers. Farming companies adopt autonomous ground vehicles wherever feasible. Satellite imagery is utilized for monitoring and forecasting weather-related events. Satellite imagery is apt to monitor floods, drought, and large-scale infestation by insect pest and disease-causing microbial pathogens. Satellite techniques are also utilized to forecast grain yield. Burning crop residues and forest plantations are often monitored using satellite imagery. Another possibility is to adopt satellite-guided aerial vehicles, such as the robotic parafoils and blimps to carry large farm cargo. This aspect has not been tested yet for adoption. Satellite imagery is in vogue to map the major changes in the natural vegetation, its biodiversity, and cropping systems. Satellite imagery lacks in high resolution if compared with the low-flying drone aircrafts. Yet, certain minor details could be monitored using satellite imagery. The sky should be clear without haze or clouds and resolution offered by imagery should be within the requirements.

7.4 THE AGRICULTURAL SKY ABOVE THE CROPPING EXPANSES OF EUROPEAN PLAINS

This section deals with the nature of agricultural sky and its impact on the farming zones of Europe. The vast stretch of European Plains extending from western France (Bay of Biscayne) till Russian steppes (Urals) is agriculturally a highly productive zone. Geographically, the European food grain producing zones extend between 5° W and 140° E longitude and 38° N and 55° N latitude. It is predominantly a prairie vegetation interspersed with timber and fruit trees, vineyards, and fiber crops. Food crops grown here are mostly the cereals, such as wheat, maize, barley, millets (Italian millets), legumes (peas, chickpea, and lentils) and pastures for domestic animals. It nourishes a large population of humans and domestic animals. For the past three millennia, the European Plain has been a stable agricultural belt, providing the nourishments to human population (Kusters, 2000; Plate 7.1A). European Plains are flat land occurring at about 153 m.a.s.l. (Shahgedanova, 2008). The European Plains are endowed with fertile Chernozem soils of black or brown color. They are rich in organic matter which is essential for better crop productivity. The French Plains are most intensively cultivated. The German, Polish, and Hungarian Plains support vast stretches of cereals predominantly wheat and legumes, such as pea, lentils, and chickpeas. In the east, "The Great Steppes" in Ukraine, Russia, Kazakhstan, Turkmenistan, and Uzbek too are agriculturally productive zones. They support prairie vegetation including food crops (Wichereck and Bossier, 1995 Dickenson, 1964; Moon, 2012, 2013; FAO, 2005; Wikipedia, 2013a, b; Krishna, 2015, Kosivart, 2013a, b).

7.4.1 THE AGROCLIMATE ABOVE THE EUROPEAN PLAINS

The European sky bestows the plains with generally a temperate ambient climate. The plains experience low temperatures of 3°C–10°C from November to next March. Yet, the cropping season that lasts from April till October is slightly warmer at 14°C–25°C. The precipitation period peaks around May–July/August. The European Plains are indeed a vast area that could be demarcated into subregions. The French Plains experience agroclimate that is congenial for cereal/legume production. Northern region of France experiences relatively colder weather pattern. The annual mean temperature is 15°C. In southern France, temperature during crop production reaches 23°C–25°C. Precipitation ranges from 850 to 1270 mm, annually. Droughts too are felt periodically. The North German Plains experience cold temperate climate. Southern German cropping belts experience slightly warmer temperate climate. The annual mean temperature in the German Plains is 9°C. Cold front usually affects the crops during December till March. The average ambient temperature in winter reaches 2°C. The annual precipitation received ranges from 600– to 800 mm. Berenyi et al. (2020) have reported that climate change effects have been observed in the precipitation pattern within the European agrarian regions. In the southern European lowlands, it includes the changes of precipitation. Precipitation changes are felt in terms of quantity and spatial patterns. Such alterations in precipitation may require appropriate risk management actions. The Central European Plains in Poland, Czech, and Hungary experience a slightly warmer temperature. The cropping season extends mainly into warmer months of April till October. Mean temperature during winter months of January–March could be. −5°C to −3°C. The warmest periods in August show up 18°C–22°C, on an average. Average annual precipitation in the Central Plains is about 525 mm. In the east, agroclimate of Ukrainian Plains and Russian steppes is congenial for cereal production. The temperature ranges from 0°C to 30°C. It experiences frosty and snowy period. Crops suffer during this period. Mean summer temperature ranges from 20°C to 24°C. Annual precipitation ranges from 500 mm–600 mm. The diurnal period varies from 8.5 h to 15 h. It depends on the location and season. The weather parameters stipulate the seeding time. Delays or mismatch often leads to reduction of biomass and grain productivity.

The European Plains have experienced climate change effects. During the past century from 1901 to 2005, natural vegetation, crops, fauna, and human inhabitations have experienced a rise in ambient temperature of 0.4°C–0.9°C, depending on the location (Olesen and Bindi, 2002, 2014; Jones and Moberg, 2003; Kjellström et al., 2007). It seems droughts have become frequent owing to higher ambient temperature. At several locations, the agricultural sky experiences 1°C–4°C higher temperature during cropping season (Olesen and Bindi, 2014). Forecasts suggest that during the next few decades agricultural sky may experiences higher temperature ranging 1°C–6°C. During past few years, European Plains have experienced extreme cold temperature. This comes immediately after an uncongenial summer and precipitation pattern. Severe cold has affected vegetable and fruit tree crops. In some locations, prolonged extreme low temperatures, frost, and snow have damaged crops entirely. Otherwise, it has reduced productivity markedly (Blaskovic, 2018; Plate 7.1A and B).

About half a decade ago, Olesen and Bindi (2014) studied the impacts of climate change on agriculture. They have assessed crop production methods adopted at that time. Then, forecasted the necessary changes in agronomic procedures required to obtain optimum productivity. Theyreported winters could be severe in Northern European Plains. Warming may induce expansion of cropping zones into Northern Plains (Plate 7.1A). We may see a marginal increase in cereal productivity in the Northern Plains if temperature rises slightly. Further, they have forecasted that the Northern and Western European agrarian regions would see intensification of crop production. Southern European cropping zones may experience warmer climate. Changes in CO_2 emission, temperature, and rainfall will affect the crop production levels. Heat stress, droughts, and low rainfall could be common in southern European Plains (Olesen and Bindi, 2002, 2014; Cuscar et al., 2018; Kjellström, 2004; Kjellström et al., 2007; European Commission, 2017). Overall, agricultural policies that suit each region too may be necessary.

Climate change scenarios that may result after considering a 2°C temperature rise in the plains have been simulated using several models. Agroclimate researchers have reported that agricultural activity in eleven different zones within the European Plains may alter. Generally, climate change pattern seems to differ in northern and southern European farming belts. Farmers have to be informed about the remedial measures accordingly. Climate change affects a series of human activities related to agriculture. European agrarian plains have experienced frequent droughts and heat waves, during past 3 years since 2018. High temperature, low precipitation, wildfire, and droughts have occurred. This phenomenon is supposedly a part of the wider atmospheric disturbances that occurred in the entire Northern Hemisphere. They say, jet streams (winds) have been weaker and feeble. Therefore, it allowed hot air to linger. Also, it has resulted in heat waves. Researchers at the Royal Netherlands Meteorological Institute and World Weather Attribution have surmised that global warming has doubled the frequency of heat waves in the European crop production zones (Wikimedia, 2020). Incidentally, in Northern European conditions, a heat wave is defined as the occurrence of summer temperatures of at least 25°C or more, for 5–7 consecutive days with a couple of days showing ambient temperature above 30°C. Temperature and moisture dearth-driven droughts have been perceived all through the European Plains (Ljungquist, 2019). Such extreme weather patterns have reduced productivity of spring wheat and maize in the Western Plains of Europe (van der Velde et al., 2012; Vogel et al., 2019).

PLATE 7.1 A Maize crop under optimum conditions of agrarian sky.

Note: A lush green maize crop adjoining an apple orchard in the northern European fertile chernozem plains. Agrarian sky is congenial to crop production. Now, compare vegetation and its survival during winter because of snow clad fields in the same region of Netherlands.

PLATE 7.1B Snow fall on Pastures and Forest vegetation.

Note: Snow fall has uncongenial effects on pasture grasses, crops, forest plantations, natural vegetation, and other biotic aspects in agrarian regions. Water sources may get clogged.

Source: Sharath Kowligi, Eindhoven, Nord Brabant, Netherlands.

Droughts due to scanty precipitation are not uncommon to European agrarian regions. For example, in German Plains, droughts are often accompanied with wildfires and dry conditions. They damage crops and natural vegetation (Hockenos, 2020). Farmers are asked to grow more diversified crops. This is to overcome climate change effects. In addition, forest fire and GHG emissions show marked increase in Europe. In some locations, agricultural sky can be more uncongenial. For example, in parts of German plains, winter rain, and snow too got feeble. This reduced the soil moisture available for crops in the next season. Therefore, a combination of drought in summer, lack of rain, and snow in winter can be devastating on yield levels that farmers can expect. In future, we should be able to characterize the European agricultural sky beforehand and inform farmers. Agrarian pockets prone to experience uncongenial sky could be mapped. It may help farmers to adjust the cropping pattern and investments. Clearly, we have to characterize and classify agricultural sky, just like we did for soils in a region.

Analysis of monthly and three monthly precipitation index in the Eastern European Plains for the past 50 years has revealed that severe droughts are less frequent (Cherenkova et al., 2015; Folland et al., 2014) . In the wheat growing regions of Ukraine and Russia, the drought events, if any, are influenced by quasi-biennial oscillation (Cherenkova and Kononova, 2009; Cherenkova et al., 2015). Spring wheat production on the black earth could get affected by mild droughts. Clearly, classifying the agricultural sky over the cropping zones of Eastern Europe including Russian steppes is helpful to farmers. They could adopt cropping systems to suit the nature of the sky.

There are several reports that incorporate forecasts about climate change in the European sky. Probable consequences on land use are highlighted. Clearly, atmospheric factors and their interactions with GHG have influenced agriculture. Agricultural productivity in European farms too is affected by GHG emissions. It also affects the spread of natural vegetation. No doubt, land-use pattern in European agrarian regions is under the influence of GHG (Downing et al., 2000; Marachi et al., 2005; Olesen and Bindi, 2002, 2014; Rounsevell et al., 2005). On a sector basis, about 62% of GHG emitted in 1990 was contributed by fuel combustion, 15% by transport vehicles and 9.0% by agriculture. At present, that is, in 2020, it is 54% by fuel combustion, 25% by transport and 10% agriculture. On country basis, Germany at 20.9% (906 kilotons of CO_2 equivalents), France at 10.8% (464 kt CO_2 eq), Spain at 8.5% (340 kt of CO_2 eq), United Kingdom at 11.3% (470 kt of CO2 eq), Italy at 9.0% (427 kt of CO_2 eq), and Netherlands 4.3% (193 kt of CO_2 eq) are the major GHG emitters (European Parliament, 2020). Overall, in the European Plains, GHG emissions have reduced from 1200 as base in 1990 to 78 in 2017. We have no clear idea about the consequences of GHG decrease on crop productivity in each small agrarian portion of European Plains.

European agrarian soils are prone to emit GHG at various levels (see Ge and Friedrich, 2020). The GHG that is generated accumulates in the atmosphere. This phenomenon modifies the gaseous composition of European sky. The GHG emission in Europe, like other regions is dependent on the land-use pattern. They say, soil factors such as moisture, temperature, nutrient dynamics, and primarily the soil textural traits affect the extent of GHG emissions. Cropping pattern and fertilizer supply are of course important factors that affect GHG emissions. Major gases emitted from agricultural regions in the European Plains are nitrous oxide, nitric oxide, carbon dioxide, and methane. Pastures emitted higher quantities of NO_2 and CO_2. Forest soils were prone to emit NO. Northern European soil emitted methane (Shaufler et al., 2010).

Global warming due to GHG has affected agricultural regions in the European Plains in different ways. Dunne (2018), for example, states that a warmer sky may induce higher population of insect pests. Therefore, atmosphere may induce harmful pests leading to greater damage. Pest population, its activity and feeding increases due to warmer climate. Data also show that spread of insect/pathogens may get hastened due to climate change effects. In wheat producing nations of Europe, such as France, Netherlands, Germany, Poland, and Hungary, crop loss due to insect infestation has increased. It is actually traced to uncongenial agricultural sky and its factors, such as warmer temperature. Yield loss of 10–25% has been attributable to insect damage. It is abetted by climate change factors.

The agrarian sky may also bring in hailstorms, incessant prolonged precipitation leading to floods, soil erosion, and reduction in crop productivity. This is yet another consequence of climate change. For example, hailstorms in the Grapevine region within France and Germany resulted in severe loss of fruit yield. Fruit orchards like apples, apricots, peaches, and cherries got devastated due to storms (Wellsher, 2019).

Snow, ice, and rain caused storms which induce shift of topsoil. Topsoil gets shifted from one set of fields to other due to such atmospheric manifestations. They do affect soil structure and nutrient composition. They also affect water status of agrarian fields. Sometimes, they cause floods and stagnation thus affecting crop growth productivity.

In European agrarian regions, winds are caused by variations in atmospheric pressure. The agricultural sky experiences such storms of different strengths during March–October each year. Regions most frequently affected because of North Atlantic oscillations are in Western and Northern Europe. Clearly, the sky above these agrarian regions in the Europe could be classified as those "prone to storm." Radler et al. (2019) have reported that convective weather in the European agrarian region may induce hails, lightning, severe wind gusts, and storms in greater frequency in future years. In the southern Europe, for example, in Italy and Greece, agricultural zones are known to suffer floods and crop damage. This is attributed to lashing rains and hails storms during summer months. However, recent forecasts suggest that the occurrence of thunderstorms and dust storms could decrease in some parts of Southern Europe. Yet, the overall severity of thunderstorms destroying agricultural crops may not alter (Radler et al., 2019).

European agrarian zone does receive dust and fine particles from storms created in other locations. The Sahara desert is a key source of dust storms. Sahara dust is frequently emitted into the Mediterranean atmosphere and transported by the winds sometimes as far north as central Europe and Great Britain.

7.4.2 MICROBES IN THE AGRICULTURAL SKY

This section is about microbes that are encountered in the aerial space above the European farming zones. Microbes traced could be generally commensals that float or reside over the dust particles and transit from a location to other and settle on terrain. The turbulence in the wind usually kicks up soil, dust, and organic particles that may harbor microbes capable of potentially diverse functions. Microbes could transform nutrients found on the organic debris, sand particles or even that found in moisture in the atmosphere. Microbes could be from the important groups like biological nitrogen fixers (BNF), both symbiotic and free-living asymbiotic. The European agricultural sky no doubt mediates transit of variety of BNF organisms. Microbes that cause disease on vast stretches of European crop production farms are indeed detrimental. The focus here is to discuss atmosphere-mediated disease-causing microbes.

7.4.2.1 AGRICULTURAL SKY MEDIATES DISSEMINATION OF PLANT PATHOGENIC MICROBES INTO THE PLAINS

There are several studies that deal with dust and sandstorms in North Africa (fringes of Sahara) and its consequences on movement of microbes. The microbes traced in the dust and sand particles traverse the long distances from Africa into Southern European cropping belts (Franzen, 1995; Hervas et al., 2009; Kubilay et al., 2000; LovePilot and Martin, 1996; Meola et al., 2015 Nicolaisen, 2017; Rodriguez et al., 2001; Prospero et al., 1970). Investigations have shown that pathogenic microbes, such as rust fungi migrate from Asia into European farms or those from African continent into Caribbean and North America via sand and dust storms. No doubt, the sky above the sandy deserts plays a vital role in the dispersal of microbes in general. Often, our focus has been to understand the spread of plant pathogens and causes for pandemics, such as rust fungi on wheat grown all over the world. The global sky plays a vital role in disseminating the microbes.

Airborne microbes are encountered in the canopy of crops from just above it and till different heights in the troposphere. The European sky too is a repository innumerable microbial species just like any other agrarian region. They could be commensals, beneficial organisms (e.g., nitrogen fixing bacteria) or pathogens (e.g., blotch causing bacteria, rust fungi, viruses, etc.). There are more investigations directed to understand airborne rust fungi. Let us consider a few examples. Airborne fungi spread into long distances in the European agrarian regions. The spores and other disease-causing propagules are disseminated through strong winds. Since control of spread is not easy, they can be devastating on the entire agrarian region, wherever host crop is cultivated. The economic losses could be devastating, particularly, on crops that are planted over large regions (Lucas et al., 2018).

Let us consider an example of fungus that attacks wheat grown in almost all regions of the globe. The aggressive strains of *P. graminis tritici* (*Pgt*) are easily picked in most wheat fields within, say, Europe, and in other continents (Meyer et al., 2017). Further, we may note that commercial genotypes of wheat cultivated in Europe are all susceptible to airborne rust fungi. Airborne rust may actually adopt different routes as they spread the disease. The microbes survive the aerial dissemination period and affect different wheat belts in different continents, including Europe. Targeted surveillance of *Pgt* spores at different field locations is necessary to quantify the disease-causing propagule. Such studies can help us in taking accurate remedial measures (Meyer et al., 2017). For example, long-term quantification of Ug99, a strain of *Pgt* has shown that rust fungi indeed travel as airborne spores from locations in India, Pakistan to other continents. No doubt, microbes utilize agricultural sky in different ways. Here, a pathogen resides in air and translocates swiftly to perpetuate itself as a pathogen on the crops. We can perhaps classify the agricultural sky above such cropping zones as "infested with pathogen's spores" highly contagious or "detrimental to cereal crops." Farmers may then plant appropriate crops and genotypes that tolerate or escape the devastation caused through the agricultural sky. Periodic enumeration of pathogen spores is useful. During recent years, aerial robots such as small aircraft drones with facility to collect air samples from above the crop's canopy and at different altitudes have been utilized.

The periodic sampling and assessment of fungal spores or other pathogens is useful in controlling the spread of the disease (see Krishna, 2020a).

Now, we should realize that epidemiology and rapid spread of wheat rust diseases in European locations is dependent on the two major factors that operate in the agricultural sky, namely, wind and rain. The fungal propagules transit short distances in air whenever intermittent gusts of winds blow over the wheat fields or through rain droplets. In France, for example, *Pgt* is said to spread rapidly in the wheat belt via wind (Sache, 2000, Madden, 1992, Aylor, 1990) and rain (Hirst and Stedman, 1963, Rowell and Romig, 1966, Sache, 2000). Again, a routine check of the atmosphere above the wheat region in southern France will be useful to judge the possible epidemics of wheat rust. Short distance dissemination via sky means survival of spores is not a major problem. This is unlike long distance spread of *Pgt* which requires proportionately longer duration of survival. In the European sky, black rust (*P. graminis*), brown rust (*Puccinia recondita*), and yellow rust (*P. striformis*) are all airborne fungi. They are traceable above the wheat canopy. They are almost endemic microbes in agrarian sky. Diurnal studies of spore release of rust fungi into atmosphere and short distance travel clearly indicates that this phenomenon is important epidemiologically. Airborne concentration of rust fungal spores seems to peak at around 11.30 A.M. in French Plains that support wheat crop (Hirst, 1953, 1961). We need to bestow greater attention in identifying the disease aiding characteristics of the agricultural sky. Perhaps, we can classify it for farmers to consult and adopt remedial measures. Agricultural sky as an entity like soil or crops need to be analyzed separately. Further, we may note that rust fungi that attack other crops, such as wheat, barley, maize, etc. too are airborne. The sky (wind, rain, temperature) plays a vital role in their dissemination. Powdery mildews too are wind dispersed disease (Hau and de Vallavielle-Pope, 2018).

In the past 150 years, the concentration of carbon dioxide in the atmosphere has risen from 280 parts per million (ppm) to 410 ppm. This is a mixed news for farmers. Any change in familiar weather patterns caused by the atmospheric warming is bound to be disruptive. But more carbon dioxide means more fuel for photosynthesis and therefore enhanced growth—sometimes by as much as 40%. In temperate zones, rising temperatures may bring milder weather and a longer growing season. What is not clear, though, and not much investigated, is how rising CO_2 levels will affect the relationship between crops and the diseases that affect them. Jarroudi et al. (2018) have discussed the use of computer-based modeling of wheat disease incidence and its spread due to atmospheric factors. The focus is on assessing disease establishment, epidemics and winter wheat grain yield loss. They say, factors related to agricultural sky, that is, atmosphere above the crop canopy is vital. Aerial factors, such as wind, precipitation and temperature could affect the drift of fungal spores from one location to other. Therefore, fungicide sprays should be carefully timed to thwart the spread of rust disease.

Researchers from Poland, Slovakia, and Czech have found that wind, mountain ranges and seasons have impact on the aeromycological aspects in the region. Air currents and terrain seem to affect the spread of fungal spores and disease initiation (Pusz et al., 2017). A recent study by Nicolaisenet al. (2017), however, has shown that pathogenic fungi above the wheat crops in Netherlands and other nations of European Plains could be uniformly distributed in the sky. Transmission of fungal pathogens (e.g., *Pgt*) in the sky is so rapid that location within 500 km did not have effect on spore distribution. The vastness of cropping belt could be overcome by the drifting spores. Further, they say, not only the rust fungi even other pathogens such as bacterial spores could get distributed rapidly in the surrounding regions.

Suffert et al. (2008) have dealt with the role of agricultural sky, along with its factors, such as wind, dust and rainstorms on drift of spores of plant pathogens. They say, it is becoming imperative that researchers could use the knowledge about wind and its currents and dissemination of propagules of fungal pathogens, usefully. For example, in the open skies of Europe, spores of rust fungi could easily be utilized to cause epidemics in a region. Wind aids the "agroterrorism" practices followed if other conditions too are congenial (Suffert et al., 2008). Agricultural sky is then detrimental to crop yield. Hence, in such cases, periodic monitoring of agricultural sky for spores of virulent pathogens is essential. For example, monitoring of rust fungi on wheat, barley, mildew or *Phytophthora* for disease may become mandatory. European region that produces wheat and other cereals in vast regions need to be extra alert regarding any wind-aided "agroterrorism."

We may note that wind in the agricultural sky is highly beneficial if it induces drift of pollen and enhances seed set. It is a detriment if it allows transit of pathogens. So, classifying agricultural sky as detriment or beneficial is useful to farmers. They have to time their agronomic procedures to suit the situation. McDonald and Stuckenbrock (2016) further caution us that airborne fungi and other plant pathogens mutate rapidly to overcome crop's genetic resistance built by researchers. In addition, monocropping stretches of cereals across European Plains make it easy for propagules of a pathogen to find host and spread. The wheat crop grown across the European Plains may show low levels of genetic diversity. This fact may make it susceptible to aerial spores of different rust fungi.

7.4.3 AGRICULTURALLY IMPORTANT AERIAL INSECTS

7.4.3.1 AERIAL INSECT PESTS IN THE EUROPEAN PLAINS

Innumerable insect species have been identified as pests on crops cultivated in the European Plains. Among them, a large number are aerial insects. They may dwell within and above crop's canopy. Soil dwelling insects have not been discussed here because, it is not in the purview of this volume. Insect caused damage could differ based on crops (cultivars) cultivated and insect pest species that inhabit the zone. Discussing all major crops cultivated in the European agrarian region and the insect species that attack them needs a few separate volumes. The published literature on these topics is vast. Therefore, only wheat that is grown widely in the European plains is considered as an example. Several pest species that infest the wheat crop cause only a mild or insignificant damage to the crop. There are others that cause epidemics and devastate the crop. There are insect species that attack the crop plus act as vectors for viral and/or bacterial pathogens. There are also predators and parasitoid insects that attack and destroy the pests (insects). Since 1950s, insecticide application to control aerial insects has increased enormously. Dosages of pesticides applied have been large during the period 1970 till 2000. During recent years, insecticide application on wheat has marginally reduced. This is attributed to adoption of IPM procedures (Miller and Pike, 2001). There are reports and manuals depicting pests and their control measures for several other crops cultivated in Europe. For example, there is an extended report by Meissle et al. (2010) dealing with pests affecting maize production in the European Plains. Similarly, an IPM manual by Cook (2019) for pests of *Brassica* (oilseed crop). There are also reports about simulation of pest attack and crop damage. Such simulation may help in understanding the dynamics of pest population and resulting crop loss (Donatelli et al., 2017). Overall, agricultural sky plays a vital role during adoption of various procedures related to IPM.

There are regions in the European Plains where strict adherence to commitments of GHG emissions and warming (i.e., 2°C stipulated in the

Paris Agreement) are in vogue. Despite it, insect pest incidence and crop loss could be significantly higher. The European wheat producing zones are among the most productive zones. Here, forecasts suggest that farmers may lose over 16 million tons of grains yearly to insect attack (Deutsch et al., 2018). The warmer climate in temperate agrarian belt may not induce better crop growth. It may also induce higher insect population and their spread. Metabolically active insects may cause greater damage to European wheat. Aerial insects that spread rapidly may have to be thwarted on a timely basis.

There are over 24 pests that attack wheat grown worldwide. Several of them are aerial insects. Similarly, there are crop species cultivated in the European Plains that are attacked by several insect species. Some of the insect pasts are endemic to European farming region. The three important aerial pests that attack wheat in different agrarian region of the world, including the various parts of European Plains are the Russian aphids (*Diuraphis noxia*), Hessian fly (*Mayetiola destructor*) and Sawfly (wheat sawfly-*Cephus cinctus*) (Crespora-Herrara et al., 2015). Agricultural researchers have tried to develop cultivars with innate genetic resistance to these pest species. The population of these insect species in the agricultural sky is important. If it exceeds a certain limit, then, crop genotypes may experience losses. Developing genetic resistance will allow farmers in avoiding aerial sprays of pesticides. Monitoring population of pest too could be useful. It helps us to judge resistance manifested by cultivars. In fact, there are several reports about the economic thresholds for insect pest caused wheat grain yield loss in Europe. However, we should note that monitoring the insect population in the sky above wheat canopy is equally important, if not more. The build up of pest population in the sky could be an indicator to crop loss. Remedial measures could be adopted accordingly (see Wetzel and Wellso, 1987).

There is also a trend in the European agrarian regions to use insect predators as biological control agents (parasitoids/predators). Releasing insect predators has often increased flowering and seed set with concomitant reduction of pest population. The biological control agent may keep the pest under threshold population levels for longer durations. In the wheat belt, forecasts show that insect predators/parasitoids may result in pest control. As a result, we may expect a 10–30% grain yield increase in the fields where biological control is adopted (Jeanneret et al., 2016).

7.4.3.2 INSECTS AS POLLINATORS IN EUROPEAN AGRARIAN SKY

Europe is home to an amazing variety of insects that pollinate agricultural crops and wild plants. This diversity is essential for a healthy nature and vegetation. However, pollinators are in serious decline. *Losing insect pollinators in the aerospace would be a major risk for nature and our own existence* (European Commission Environment, 2018). A compilation of reports from several European nations indicates that in the European agrarian regions, major insect pollinators are the bees, moths, a few species of beetles, and wasps. The domesticated Western Honeybee (*A. mellifera*) is supposedly the most prolific pollinator above the cropped fields. Forecasts indicate that insect pollinators may face extinction if remedial measures are not adopted. At present, there is a "EU pollinator initiative." It prescribes methods to improve the diversity and population of pollinators. Major reasons for reduction in bees and other pollinators are the use of pesticides and fungicides. Disappearance of native vegetation (flora) and monocropping trends during crop production too have affected diversity of insect pollinators. Certain insectivorous species of birds like Yellow legged hornets have been devastating on honeybee diversity in European farms. They say one out of ten bees or butterfly species are threatened with extinction (European Parliament, 2019). Kunast et al. (2020) consider that pollination is accomplished through biotic factors, such as bees by symbiosis. Further, we may note that insect-aided pollination in the open fields is a phenomenon of great consequences. *It takes place in the agricultural sky*. At the same time, we may note that atmospheric and other abiotic factors that operate in the European agricultural sky too affect the pollinators and their activity. For example, climate change and excessive GHG may affect pollinator activity. Reports by United Nations agencies dealing with ecosystems too caution us that a healthy pollinator population is essential to achieve higher grains, vegetable, and fruit yield (Agence France-Presse, 2016).

Basically, about 60% of the global crop production seems to be provided by crop species that *do not need* pollinator activity. About 35% of the food grain generation is dependent on pollinator activity. Further, pollinators (e.g., bees, moths, butterflies) are essential for 130 crop species cultivated. About 30 crop species are highly dependent on pollinator activity in the sky above their canopy, 27 crop species depend moderately on pollinators, 21 crop species depend only slightly on pollinators and for seven of the crops evaluated pollinator activity is not needed (Klein et al., 2007; Kunast et al., 2020).

An evaluation in 2011 has shown that population of insect pollinators in the European farms has stabilized. About 400 species of insects have been identified as pollinators in this cropping belt. During 1970s, excessive use of chemical pesticides did decrease the diversity of pollinator insects, such as bees, butterflies, and moths. Of course, at present, 150 bee species seem to dominate the role of pollination in cropped fields, within United Kingdom (Little, 2017). Clearly, deterioration of atmosphere through pesticides has been thwarted. Agricultural sky seems to have regained congenial traits. So, it allows pollinators, to survive better and operate efficiently.

Pollinator bees are vital biotic factors operating in the agricultural sky above the European crop production zones. Reports from FAO of the United Nations (Italy) reveal that as acreage of cross-pollinated crops increased in the European Plains, bee population and diversity too has kept pace. There has been a 30% increase in the population of bees in the European Plains (Majewska and Majewski, 2017). Bee population increased significantly in areas sown to brassicas and sunflower. The large area covered by these two crops are entirely pollinated by bees. Therefore, bee number and activity are important. In some regions of Europe, sunflower seed set gets affected if bee population during flowering, head maturation, and seed set is below threshold. Clearly, if we manage the bee population in the agricultural sky properly, then, yield of many cross-pollinated crops could be maintained at optimum levels or even enhanced.

Majewska and Majewski (2017) point out that beekeeping provides certain advantages to European farmers. Products, such as honey, propolis, royal jelly, beeswax, and bee venom are derived. Bees perform the function of pollination of crop plants. Plant pollination is the most important benefit provided by bees. It is estimated that these insects, by pollinating crop plants, improve grain produced by 10–100 times (Klein et al., 2007; Lautenbach et al., 2012). An important study about bees was conducted by researchers from different countries of the European Union. They aimed at understanding the role of bees in pollinations. Nieto et al. (2014) estimated the diversity of bee species involved in pollination in the European agrarian region. The species richness was higher in Southern European regions, particularly, Italian, Balkans, and Iberian peninsulas. The bee diversity seemed to decrease as we traversed toward the north of plains. Many bee species that were endemic to European farmland were threatened with extinction. About 9.2% of the bee species were actively involved in the pollination of crops. Still, many insect species that pollinated cereals, brassicas, legumes, fruit crops were threatened with extinction. On a comparative basis, the number of threatened bee species was much less than other animal species in the same regions. They have reported that over 1103 species of bees were traced in the European agrarian region.

The agricultural sky above the European crops is thronged by several bee species. However, the population of over 150 species of bees were declining. We need to maintain population levels and protect those bee species that are threatened (Nieto et al., 2014). Major problems are excessive use of pesticides by farmers and farming companies that adopt intensive crop production tactics. If there are other means by which we can encourage bee population and their diversity, we should then adopt them promptly. The ultimate result from encouraging bees over European Plains is after all beneficial to crop productivity plus honey production.

No doubt, bees are considered most prolific in carrying pollen grains and inducing pollination across most of the agricultural regions of the world. In addition, there are innumerable other insect species that serve to carry pollen and induce pollination. Hover flies, flies, beetles, bugs, moths, butterflies, and wasps are other insect pollinators traced in the European Plains (An-Taisce, 2018; Free, 1993; Kunast et al., 2020). Pollination is a prerequisite for seed set. Therefore, every insect species that contributes to pollination is important to farmers in Europe. Walton (2020) opines that moths that operate in the night may be vital to pollination of several crops grown in the European regions. Their evaluations in the Norfolk region of England and in the mainland Europe suggest that there are over 150 moth species that are active nocturnally. They effectively increase pollination and seed set of crops such as *Brassica* etc. There are now several initiatives supported by the European Commission. They aim at assessing pollinators and their role in farmland. They mostly aim at preserving and enhancing biodiversity of pollinators of crops (Fitzpatric, 2019).

7.4.4 AVES AND THEIR ROLE IN AGRICULTURAL CROPPING ZONES OF EUROPE:

Bird Diversity, Bird Population, Birds as Predators, Pests, and Pollinators

Crop loss due to bird pest can be severe in many of the subregions of European agrarian plains. We should note here that crop loss due to different kinds of pests are relatively high in the European agrarian regions. Crop loss could be ordinarily around 15–25% of the produce. If the infestation is severe, it ranges from 50% of the produce in the case of wheat to sometime even 80% (e.g., cotton), depending on the crop species (Oerke, 2005). It is possible that agricultural sky mediates innumerable pest species of birds. Crops, such as wheat, maize, soybean, cotton, pulses, and oilseeds generally encounter a range of different bird pests during a season. We have not provided much attention in understanding the role of agricultural sky in aiding birds that are pests of crops. They say, bird pests often induce greater loss on cash crops like fruits in orchards.

Generally, the European Starlings, pigeons, sparrows, and a few other species that thrive on crops are considered as pests. They may affect the grain yield. However, in nature, birds could be insectivorous thus reducing the pest population. There is no doubt that the European sky harbors hundreds of bird species that are pest on crops. They could be affecting the panicle and grain development of major cereal and legume crops. Crops such as sunflower are susceptible to bird that caused damage particularly during seed development.

Biological control using birds of prey is an environmentally safe bet. It does not involve the application of excessive pesticide nor the use of poisonous baits, to reduce bird pests. Birds of prey, that is, predatory birds and insectivorous birds are introduced or left to flourish in the farms. They may regulate the pests to low threshold levels and avoid loss of grains.

Raptors and owls are among the common birds of prey traced in the southern European farm and forest zones. When used for biological control to control pest birds, these birds of prey could be active during daytime (e.g., raptors, eagles, vultures) or during night (e.g., owls). The diversity of birds of prey is high in the forests of Southern Europe. For example, in Greece, we can detect about 286 species of eagles, vultures, and owls in the sky above farms/forests (Dadia Forest Reserve, 2020). Predatory birds (*Tyto alba, Elanus axillaris, Falco tinnunculus, Falco cenchroides, Bubo bengalensis,* and *Buteo rufinus*) are among the commonly listed biological control of rodents. Barn owls (*T. alba*) are the most cited species. However, we still need quantitative studies that accurately determine the role of birds of prey in controlling rodent pests. The efficiency of birds in controlling the rodents that destroy crops in the field needs attention (Labuschagne et al., 2016).

Farmers in France prefer to use IPM procedures that reduce the use of pesticides to control larvae that attack their apple orchards. To do this, they are opting for birds such as Tits (*Parus major*) to reduce leaf-eating larvae on apple trees. Commercial orchards too are preferring to release birds that reduce pest damage. This is followed instead of using high dosages of pesticides (Mols and Visser, 2007). Obviously, here we are trying to use a biotic factor in the sky above trees to reduce the pest damage.

Farmers could utilize the agricultural sky more efficiently by adopting insectivorous birds. Several countries in the Mediterranean Southern Europe are employing insectivorous birds, to reduce apple pest damage. Yet, it only forms part of integrated control of pests (Benayas et al., 2017). In Southern France, for example, great tits are the preferred insectivorous species that are released into orchards (Bouvier et al., 2005). In Italy, trials show that Tortricid moths affecting grape vineyards could be controlled using the birds of prey (Loriatti and Lucchi, 2016). Rubene et al. (2019) state that birds utilize both visual and olfactory systems to detect insect attacked crops.

Benayas et al. (2020) further state that regulation of agricultural pests by using their natural enemies represents an alternative to chemical pesticides. They have in fact evaluated insectivorous birds as pest regulators in woody crops located in Central Spain. Such an evaluation was conducted for four consecutive years. Predation experiments were conducted with greater wax moth (*Galleria mellonella*), sentinel caterpillars, and food consumption by birds was estimated. Population of insectivorous birds, such as great tit (*Parus major*) and sparrows (*Passer domesticus* and *Passer montanus*) increased over time in the vineyard and fruit orchards by the fourth year. In fact, there are clear suggestions to use predatory birds instead of poisons to reduce insect and bird pests.

There are reports about birds, their population, and diversity in parts of Europe (e.g., United Kingdom). They report that in the United Kingdom, it has either stabilized or marginally increased during 2009–2015 (Little, 2017). Birds of prey and those involved in pollination within crop fields need greater attention. We have to harness the advantages from agricultural sky better with regards to bird species in farms.

7.4.5 WIND POWER GENERATION USING EUROPEAN AGRICULTURAL SKY

Wind energy is an attractive alternative to fossil fuels. It is plentiful, renewable, widely distributed, and clean. It produces no GHG emissions. Europe remains a global leader in the generation of wind energy (European Commission, 2019). At present, out of the total power generated in Europe, about 50% is contributed by fuel burning, 27% by nuclear energy generating installations, 12% by hydroelectric power , 8% by wind energy sources, and

3% by others. European agencies could achieve higher quantities of wind energy. In the near future, about 30% of the total energy could be contributed through wind power (European Commission. 2019; Wikipedia, 2020d).

European sky provides wind energy to farmers through ground and low-altitude wind power generation methods (e.g., Windmills; see Plate 7.2). High-altitude wind power via the tethered aerostats is also a clear possibility (Canale et al., 2009; Frost and Sullivan Company, 2015). Wind power generated by the low-altitude ground level windmills has been in use in the European Plains since few centuries. Wind power is used to grind food grains to get flour, generate electricity, and operate lift irrigation machines. It is a phenomenon related to the agrarian sky. No doubt, characteristics of wind that blows over the cropped field needs greater attention.

Wind that blows above the crop's canopy is among the major renewable resources in the entire European agrarian region. Wind energy installations usually exploit wind resource at ground levels (e.g., traditional windmills) or low altitude (100 m to 200 m altitude). High-altitude winds that occur, at say, 750–1000 m above the ground surface too are utilized. The wind energy potential actually increases with higher altitude. At present, there are high-altitude wind generation equipment that are kite-based or tethered aerostat-based. Helium balloon hoisted at different altitudes are used to generate high-altitude wind energy. Incidentally, at high-altitudes wind is stable and blows at uniform speeds. It may depend on "jet streams" of wind that occur at high altitudes. Experimental evaluations and simulations have shown that windspeed increases by an average 30% between 110 m and 750 mm altitude. Therefore, high-altitude wind energy generation is feasible in the farm sector (O'Gairbhith, 2019; Wind Europe, 2020).

Now, let us consider trends in wind power generation and its utilization in the European Union. Total installed capacity for power in the European Union (not just plains region) in 2017 was 169.3 GW (giga watts) and is expected to be 239 GW, in the year 2020. Within 1 year in 2017, about 15,680 MW of wind power facility was installed. Wind power generation has only gained importance during recent years. Mainly because, it is clean, free of GHG emissions, less costly, and renewable. The European farming community requires electric power to run the farm and conduct various procedures. The demand for wind power is said to increase further, in the next few years. At present, wind power contributes about 11.6% electricity generated in the European Union. Wind power is perhaps destined to be important source of energy for European agrarian regions. More so, in the plains that can harness wind power better.

Let us consider an example from the Central Plains of Europe. Here, wind power is being encouraged as an alternative to nuclear power. In Romania, expansion of nuclear power was slowed down. Instead, wind power was considered as an option along with other power sources. Advantages of wind power are (a) small-scale forms; (b) decentral, supplemental power in windy areas, and (c) alternative for individual homeowner and small villages. The disadvantages are (a) highly variable source, (b) relatively low efficiency (30%), (c) more power than needed is produced when the wind blows, and (d) feeding into the grid or efficient energy storage is thus required. Romanian Energy Program has opted for planned installation of wind power stations with a total capacity of 550 MW until 2010. They would add a few more wind power stations to reach a cumulative of 3000 MW. It is said that such clean wind power systems can replace two larger nuclear power stations. Plus, wind power is cheaper with less initial capital requirements (Wenish and Pladerer, 2003). In a few nations of the European Plains, wind power generated using ground and low-altitude turbines is acting as lifeline. Particularly, in nations that are in transition from coal/nuclear sources to wind power or other cleaner sources.

A few leading wind power generating countries located in the European Plains are Germany (126,000 GWh), Spain (54,212 GWh), France (34100 GWh), Italy (20,200 GWh), Poland (15,000 GWH), Netherlands (11,458 GWh), Belgium (8118 GWh), and Romania (6750 GWh) (Wikipedia, 2020d; European Court of Auditors, 2018, 2019). At present, wind power covers only a small fraction of the total power consumed by the European farming community. On a country basis, wind power covers 26% of the electricity requirements in Germany, 21% in Spain, 12% in Netherlands, 10% in Belgium, 12% in Netherlands (Plate 7.2), 9% in Poland, 11% in Romania, 7% in Italy, 7% in France, 2% in Hungary and 1% in Czech. Major increases in the wind power generation since 2009 has been noticed in counties, such as Germany, Spain, United Kingdom, France, and Italy.

No doubt, wind is an important abiotic factor of agricultural sky. It is influenced by the climate change effects. Reports about climate change scenarios in the European Plains clearly state that historically, there has been an increase in ambient temperature. In future, global temperatures could increase by 1.5°C–2°C. Simulation models indicate a shift in the Atlantic jet (wind) toward north. It is supposed to lead to increased surface winds in the North European region. It also leads to a reduction in winds in the Southern Europe. Therefore, Scott Hosking et al. (2017) opine that potential to produce wind power is greater in the Northern European Plains. It may decrease gradually as we move to Southern Plains. Further, they report that decrease in wind power generation in the Southern Europe could be small or negligible if compared with present levels.

Ban et al. (2012) have suggested that it is useful to build maps that depict the potential of wind energy generation in various parts of the European Plains. Such maps will help in installing power generation equipment in the European agrarian regions based on wind power generation potential.

There are a few different methods using which we can generate wind power. Kite energy is among the recent methods tested and is yet to be adopted in large scale. Kite energy could become an important source of power to farmers in European plains (KITEnrg S.R.L. 2019; Plate 7.3). Kite energy may help individual farms or group of them situated in villages. Kite technology involves exploitation of high-altitude strong winds that occur in the troposphere. They use light-weight turbines or propellers fitted to kites. The energy generated is usually channeled to farms using electric cables attached to tethers.

Farmers in the European agrarian region could in due course explore and utilize wind energy generated, using high-altitude kites. The tethered kites that reach the "jet streams" of high altitude utilize strong and steady wind currents (Blackman, 2009). Such kite power could help farmers in a small village to run several gadgets.

7.4.6 SOLAR ENERGY GENERATION IN EUROPEAN AGRARIAN REGIONS

As stated earlier, solar energy is bestowed via the agricultural sky within the European farming zones. It is a clean energy source that does not emit GHG such as CO_2. Hence, a few agencies in the European farm belt are trying electrification of agrarian regions, using solar power (SolarPower Europe, 2019). Such solar power is to be utilized to energize irrigation pump sets, farm gadgets, farm buildings, and small agro-industries. It is said, such a trend helps in satisfying the long-term goals of European Commission to achieve decarbonization. Solar power offers a cleaner agricultural sky for crops to thrive. Economically, solar power is affordable. It helps in achieving "Greener Europe" by lessening environmental problems. Therefore, adopting solar energy in farm belt helps in at least two ways, firstly, it helps in obtaining electricity at lower cost, and secondly, it helps in avoiding GHG. Solar power helps in keeping the sky above crops a bit cleaner. Solar cells usage during farming depends on sunshine, reception of radiation energy during seasons of the ear, and cloudiness of the atmosphere.

Solar energy is generated using photovoltaic cells (solar PV cells). Such solar panels convert sun's energy that impinges on them is converted into electric energy. It is truly a phenomenon related to agricultural sky if it happens to be an agrarian farm. During recent years, interest in solar energy has gained. Solar panels have creeped into areas that erstwhile was farmland. There are also regions in Europe where solar farms are replacing cropland. Sometimes permanently or during off-season, the solar cells and energy derived depends on crop, season, cloud cover, and quantum of energy impinging on solar panels. A recent report about southern European region states that agricultural land is being converted to solar farms by placing large number of solar panels that cover hundreds of acres in area (Katalin et al., 2009; Ministerie van Landbaw Natuur en Voedselkwalitelt, 2020).

PLATE 7.2 A windmill in operation in agrarian regions near Eindhoven, North Brabant, Netherlands.
Source: Mrs Roopashree, B.N., Eindhoven, Nord Brabant, Netherlands.

PLATE 7.3 A Parafoil Kite meant for wind power generation.
Source: Kite Energy S.R.L., See KITEnrg S.R.L. 2019.

Maps depicting location of solar farms suggests that they are often located closer to electric substations. Locating PV solar cell farms closer to previous electric substations helps in selling the energy to power grids. Also, often these solar farms are located closer to agrarian farming pockets. The electricity generated is easily channeled to crop production farms. In any case, solar farms are situated in farming zones where at least a 1000 ha or more is easily available to install the solar panels. Sometimes, erstwhile farmland has been bought on 25–30 years lease, to use it for solar energy generation (Ministerie van Landbaw Natuur en Voedselkwalitelt, 2020). Reports suggest that solar energy generation is opted during recent years. It is actually a resurgence of solar energy that was actually abandoned since a few years. No doubt, production of solar energy helps us in using the energy bestowed via agricultural sky better than leaving it just for fallow for many years.

Germany is among major solar energy generating nations. German agrarian plains are bestowed with enormous potential to generate solar power because of bright sunshine. Cloudy days are reasonably small in number. German plains encountered a spurt in installation of solar power farms during early 2010. This trend continued till a few years. There was a dip in such expansion of solar energy during 3–4 years between 2014 and 2016. About 8% of total electric energy generated in Germany is contributed by solar energy. There are about 1.8 million solar array and the total capacity is 49 MW, annually. German solar power lobby envisages to increase solar power generation further, particularly, in the rural belt. German Plains added 4 MW solar power generation capacity in just 1 year in 2019. Solar power generation in the German farms seems to cost about 3.75 Eurocents per KWh, if the sun's radiant energy is intense and satisfactory (Wehrmann, 2020). There are forecasts that nations, such as France, Spain, Germany, and Poland are set to further improve the solar energy generation records, in 2020 (Whitlock, 2020). Farmers in the European Plains from France to Ukraine may depend on solar energy. Solar energy helps in providing farmers a continuous supply of power to farms at low cost. It is cleaner with no GHG emissions. Hence, in future, it could get preferred more than at present.

Let us consider another country from the European Plains. Romania is a central European nation that generates sizeable quantity of food grains. Romania receives satisfactory levels of sun's radiation in all seasons of the year. Solar radiation of 1300 –1500 kWh/m^2 is a good reason to develop solar power generation installations. Romania has made efforts to develop the solar energy equipment since 1979. Hot water systems as well as systems for drying and industrial application have been installed. Some of the major advantages and disadvantages listed about solar energy in Romania are that there is no pollution of atmosphere. Solar power is clean without GHG emissions. Low efficiency of 5–15% is a clear disadvantage. Initial capital requirements are high for installations. Adequate storage batteries are needed (Wenish and Pladerer, 2003). Overall, agrarian sky in Central Plains could be utilized to farmer's advantage via solar cells.

We may notice that in several farms, solar generation and agrarian crop production are conducted simultaneously. Solar panels are installed quickly in the off-season. This is to harness as much solar energy in the off-season. Some requirements of dual power generation options are accurate solar panel orientation. Solar panel orientation is determined using computer programs and controls. Solar panels are placed at a height above the crop's canopy. This allows movement of agricultural vehicles without obstruction while the panels above them. The obstruction to photosynthetic radiation if any due to solar panels has to be compensated.

7.4.7 UNMANNED AERIAL VEHICLES ABOVE THE EUROPEAN AGRARIAN REGIONS

Reports by European Commission clearly indicate that agricultural drones are making a debut. They are still to spread rapidly into European agrarian regions. The unmanned aerial vehicles (UAV) have not been adopted routinely to assess cereal fields, pastures, grapevine yards, as well as forests (European Commission, 2017). Farmers are yet to use the agricultural sky efficiently by adopting drones. At present, European farmers are testing drones to assess the growth of crops in their farms. They are yet to use them routinely for detecting droughts, diseases, and weeds. Some of these aspects are still in experimental farm stages. However, if we consider the production of drones with various capabilities relevant to farming, European Drone Industries have produced about 7 million drone units, in just 1 year in 2017. Obviously, drones are making it to European sky. One of the lacunae in adopting drones above European farms is said to be connected with farmer's use of computers and mobile phone-aided drone control packages. Farmer's ability to collect useful data depends on their ability to make versatile use of computers/mobile phones. A few interesting aspects of adopting drones in European farms relates to aerial seeding, spraying pesticides/fungicides, detecting weeds and spraying herbicides, crop monitoring, etc. Aerial seeding, for example, involves flying copter drones close to soil surface and then, fire seeds with force, so that they enter soil layer. Such use of agricultural sky is immensely useful. It reduces costs on farm labor. It improves accuracy of farm procedures enormously. It also reduces farm drudgery to almost nill during seeding (European Commission, 2017; Schimpf and Diamond, 2019).

There are indeed several established drone companies in Europe. Several start-up units that supply a range of models of drones are making rapid progress. Most of them are suitable for use above farms. They all help in using the agricultural sky, efficiently. A few examples of UAV models and their manufacturers are *e-Bee* (Sense Fly Inc, Switzerland), *mD*-series (Microdrone GMBH, Germany), Smart Planes (Skelleftea, Sweden), Threod Drones (Threod Inc, Vimsi, Estonia), Wingtra (Wyss Zurich Project, Switzerland), C-Astral Aerospace (Adjovscina, Slovenia), DV wing (Drone Volt Inc. Roissy, Paris, France), Novodrone (NOMAD AGR, Sevilla, Spain), MAVT Torino, Italy etc. (See Krishna, 2017, 2018, 2020a).

All along the European Plains, each nation supports several agricultural UAV manufacturing companies. Several models have found acceptance with farmers. The UAVs are utilized to conduct surveillance of farmers' installations, crops and farm activity on a daily basis. UAVs fitted with multispectral sensors, Lidar, and sometimes probes to detect atmospheric gases are common in the European Plains. They are utilized to monitor the crop growth rates using NDVI, leaf chlorophyll, LAI, crop's water stress index (CWSI) and visible aerial imagery (Table 7.2). Most, if not all, grain producing companies may soon be depending on UAV aircrafts, tethered aerostats or blimps to collect data about their crops periodically and also to develop appropriate and authentic forecasts about biomass and grain yield formation. The spectral data and digital surface models (DSM) could be utilized in precision farming vehicles. This situation is a good example how farmers can utilize the agricultural sky above the crops to their advantage. There are now several companies that manufacture "sprayer drones." These sprayer drones fitted with GPS facility and variable-rate applicators can spray the fluid fertilizer formulations, plant protection chemicals, and herbicides on to crops with great precision, also at an astonishingly rapid pace, using the airspace. Aerial robots help farmers in many ways. They also reduce the farm drudgery that otherwise is necessary during scouting the crops and conducting the agronomic procedures.

TABLE 7.2 Aerial Robotics Above European Agrarian Plains: A Few Examples of Efficient Use of Agricultural Sky to Monitor Crops in a Farm.

UAV model/ manufacturer	Purpose	References
Drone Aircrafts E-Bee/SenseFly Inc., Cheseaux-Louisanne, Switzerland	Monitoring farm installations periodically. Obtaining aerial visual imagery of crops, progress of agronomic, procedures crop's growth, water status (CWSI), using infra-red sensors. Assessing nutrient requirements mainly nitrogen, using leaf chlorophyll index. Mapping disease/pest affliction using spectral signatures. Mapping weeds in fields. Assessing panicle maturity etc.	SenseFly Inc. (2016)
Phantom-4 DJI Inc. Guongdong, China	To conduct aerial survey of long-term ecological sites. To obtain spectral data showing the diversity of botanical species and fauna. To study natural resources, mainly, soil type and water. The multispectral sensors are utilized to obtain NDVI data which is indicative of biomass accumulation	DJI (2016)
Scout B1/ Aeroscout Gmbh, Horw, Switzerland	It is utilized to scout over forest cover. Assess plantation for growth and biomass accumulation using spectral images and LIDAR data. It has been utilized to obtain phenomics data of field crops	Aeroscout (2018), Krishna (2020a)
Parafoils and Aerostats SUSI 62/ FU-Berlin, Linz am Rhein, Germany	Utilized to obtain high resolution visual and multispectral imagery of agricultural terrain, soil types, and vegetation. To obtain NDVI, CWSI and leaf chlorophyll index of elite lines of plant breeders, and to monitor experimental stations and farm installations	Thamm et al. (2013), Thamm and Judex (2006)

TABLE 7.2 *(Continued)*

UAV model/ manufacturer	Purpose	References
PIXY /INRA Auzeville and French Institute for Development, Montpellier France	Evaluation of wheat crop's response to fertilizer N inputs at INRA, Auzeville, Toulouse, France, using UAVs (autonomous Parafoil, e.g., Pixy ABS) fitted with high resolution visual and multispectral sensors	Lelong et al. (2016), Pudelko et al. (2012)
Helikite Allsopp Helikites Ltd., Hampshire, United Kingdom	Utilized to scare bird pests, conduct aerial surveillance and collect crop growth data	Allsopp Helikites (2017)
Aerostat Aerdrum Ltd, Belgrade, Serbia	Aerial photography of crop land and other natural features	Mijatovic (2014)

Note: The above list considers UAV manufacturers located in European continent. However, many of these models are exported to other nations where they are effectively utilized above the agrarian regions. For example, *E-Bee* is a compact, small fixed-winged drone model popular with farmers. They adopt it to collect visual and multispectral data about large fields with food grain crops. Conversely, we should note that European farmers in the plains do adopt several models manufactured in North America, China and Fareast. A good example is RMAX – a drone helicopter produced in Japan but utilized effectively in European agrarian regions also in North American Plains. Similarly, 's M-Agras, Phantom etc.

Diaz-Delgado et al. (2019) state that a quadcopter such as DJI's Phantom-4 is of great utility, particularly while acquiring useful data from experimental ecological sites. They could be used to collect close-up visual and multispectral data about vegetation, water resources, and effects of ambient weather on crops. Long-term ecological experiments were assessed, using quadcopter fitted with multispectral sensors.

The major advantages of using robotic parafoils, tethered aerostats, and helikites is that firstly, they provide vantage points in the agricultural sky. This allows sensors to pick accurate visual photography and digital data. This is not possible with ground vehicles. The parafoils can accomplish scouting the large fields at a pace much faster than the human skilled farm technicians. They are less costly than other means of scouting. They are relatively slower in covering large fields compared with rapid speed of fixed-winged drone aircraft or a copter drone. Yet, they are faster than human skilled technicians covering the fields slowly using ground vehicles and noting down the readings at each point. They offer data that are relatively more accurate than human skilled workers in the field.

Parachutes, both circular type and foils are useful in observing the crop fields from the sky. The RAM-AIR parachutes (parafoils) with necessary GPS connectivity and autonomous navigation have been adopted over the European Agricultural sky. For example, in Germany, Thamm (2011), Thamm et al. (2013), and Thamm and Judex (2006) have clearly demonstrated the utility of autonomous parafoils. The sensors placed on the parafoils (SUSI 62) provided digital data that was utilized to mark the "management blocks" during land preparation, planting, and while conducting agronomic procedures. Similarly, in Poland, Pixy ABS which is an autonomous RAM-AIR parachute was effectively adopted to conduct regular surveillance of farms. They are also utilized to monitor the crops in the agricultural experimental station. They were adopted to collect useful data about the progress and performance of several genotypes of wheat and brassica in the fields. Pudelko et al. (2012) and Lelong et al., 2016) state that "Pixy ABS" is low-cost aerial robot. It is efficient for use in the agricultural sky (Table 7.2). These autonomous RAM-AIR parachutes are versatile. They could be powered using a petrol engine or electric batteries. The atmosphere above the crop field must be calm and without turbulence if the parafoil has to perform flawlessly. These parachutes do not crash unlike the small drone aircrafts. In the absence of power or erroneous computer commands, they float and land slowly.

There are now several other aerial vehicles and their models. They are being briskly evaluated in the European sky. Tethered aerostats fitted with full complement of sensors, such as the visual (R, G. B) and multispectral sensors plus the Lidar are preferred. Sometimes, they even carry probes to detect chemical constituents. They are useful to detect the gaseous composition of atmosphere above the crops. Aerostats stay afloat in the sky above the crops for longer period because they are helium-filled balloons (aerostats). They are best utilized when continuous data about the crop's performance is required, also when experimental stations are to be monitored all through the season. In fact, such tethered aerostats with sensors, for example. "Helikites" have been utilized to surveillance crops from the sky. Helikites obtain useful digital data and imagery of crops. Helikite bird scarers are effective during sowing and during seed and maturation of grains (Allsopp Helikites Ltd., 2017).

7.5 THE AGRICULTURAL SKY ABOVE SAHELIAN WEST AFRICA

The West African Sahel is characterized by flat plains and undulating land with sparse natural vegetation contributed by perennial grasses. This natural expanse is interspersed with shrubs and trees placed at a distance. The agrarian belt in Sahel experiences semiarid climate (Koppen's classification). It extends from 15°W to 15°E latitude and 5°N to 25°N longitude. The Atlantic coast in Senegal and Morocco forms the western edge. In the east, the West African Sahelian region ends up near the Lake Chad. The Sahel is 5400 km in length from west to east and 1000 km in breadth (Fig. 7.1). The total area is 11, 78,850 km^2.

West African Sahel is a vast area endowed with highly sandy soils known as sandy Alfisols (Oxisols). They are marginal in fertility and low in moisture retention traits (FAO, 1997). However, in the present context, we are actually concerned with the "agricultural sky" just above vast semiarid agrarian belt. The Sahelian agricultural sky is harsh on crops. The temperature, precipitation, relative humidity, wind, and topsoil dust/sand particles are uncongenial to major crops, such as sorghum, pearl millet, cowpea, groundnuts, and agroforestry nurseries. Irrigation supplements from major rivers, such as the Niger, Volta, Benue, Bandama, Oueme are relatively feeble. Most of the agrarian region is rainfed. Hence, agricultural sky is an important factor that affects crops right from seed germination, seedling establishment, and until final harvest (Krishna, 2003, 2008, 2015). *The agricultural sky has partly guided evolution of apt crop species and cropping systems. It has no doubt influenced natural vegetation and its perpetuation immensely. Perhaps, "Sahelian sky" has imposed severest of natural selection pressures on crops. The present landraces have evolved experiencing the harsh sky for several millennia. Landraces and elite genotypes of crops that we encounter are adapted to sandy soils found in the Sahelian farming enterprises. It is perhaps the "Agricultural sky" that imposes several more constraints to optimum crop yield. Therefore, we need to analyze the Sahelian aerospace with this perspective in focus.* There is no doubt that we need to devise agronomic procedures to overcome agricultural sky-mediated constraints affecting the cropping systems. At present, cropping systems popular with farmers in the Sahel are crop–fallow (cereal–fallow), pearl millet-based cropping systems, groundnut-fallow systems in the "Groundnut Basin" of Senegal, maize–groundnut rotations in the Guinea region, pearl–millet, and groundnut rotations in Sudanian region of West Africa, and Sorghum-based regions of Southern Sahel (see Aregheore, 2009; Krishna, 2015, Manlay et al., 2002a, b).

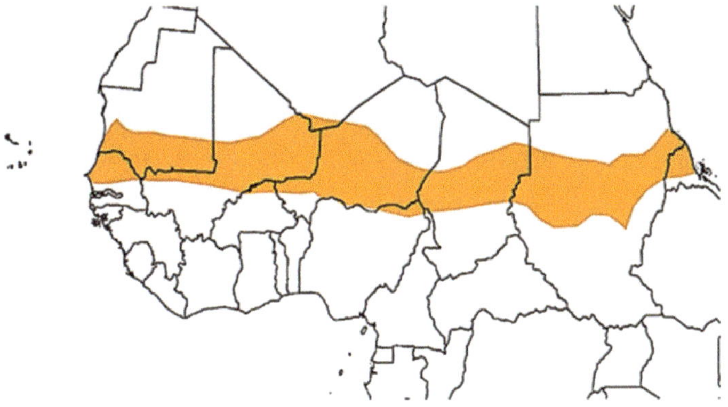

FIGURE 7.1 The Sahelian West Africa.

Note: We should recognize that the Sahelian terrain, soils, water resources, crops and sky all have their share of influence on the food grain generation. This is in addition to human toil. The suggestion here is that 'Agricultural sky' is important. The 'Sahelian Agricultural sky' bestows both positive effects and harsh detrimental effects through its various factors, on the agricultural farms. *Perhaps, sky is harsher on crops than soils!* The agricultural sky above the Sahelian cropping belt needs greater attention than provided now by the researchers, farmers and agricultural engineers. The agricultural sky provides useful resources to crops, but equally so, the crop suffers due to several factors related to sky (e.g., harsh temperature, storms, wind and dust storms, lack of precipitation, droughts, diseases, insects, pollution etc.). The intensity/duration of detriments suffered by crops due to agricultural sky may be much more than that, due to other aspects of farming such as soil, water or crop genotype. We have to study and understand the *'Sahelian sky'* better than we know today. Sahelian region experiences certain harsh weather parameters like low precipitation, erratic precipitation trends, droughts that occur periodically, dust storms that affect atmosphere, and improper crop establishment. They lead to low grain productivity. Basically, the Sahelian sky is different from that countered over other agrarian belts such as Great Plains of USA, Pampas, the Cerrados, European Plains, Indo-Gangetic Plains, Northeast Chinese plains etc. We could compare the sky above each of these cropping zones. Often, we have just studied and rarely compared the contrasting features and factors that operate in agricultural sky.

Source: FAO of the United Nations, Rome Italy, http://www.fao.org/3/x6543e/x6543e01.htm

7.5.1 AGRO-CLIMATE OF WEST AFRICAN SAHEL

The agroclimate is a major aspect of the agricultural sky in any agrarian region. In the Sahelian West Africa too, climate changes make it difficult for farmers to attain higher productivity. The semiarid weather, particularly, parameters, such as ambient temperature, precipitation pattern, relative humidity, wind and dust storms, frequent droughts reduce the productivity of crops. Often, even optimal levels of productivity are not attained because of vagaries of Sahelian weather. In the fringes of Sahara, where crops could still survive and perpetuate, the agricultural sky offers a harsh climate. In addition to difficult weather pattern, farmers experience high intensity dust storms such as Harmattan. Agricultural sky with its harsh atmosphere is an important portion that affects the cropping pattern. It dictates the extent of farm inputs and human labor that can go into farming enterprises. Subsistence farming too is affected by similar reasons related to agricultural sky. We should note that in addition to harsher agricultural weather, agricultural sky also harbors several other detrimental factors, such as the disease, pests, and bird damage that reduces grain yield. The already poor fertility of sandy alfisol is further depleted of topsoil and nutrients due to frequent winds of higher speed. High precipitation events too induce soil erosion and reduce fertility in the rooting region of crops. No doubt, agricultural sky above the Sahelian crop offers a difficult proposition to farmers. Despite it, we have not realized that we should analyze several of these agricultural sky-related factors comprehensively and then prescribe to farmers. We should first classify and subclassify the "agricultural sky" and then devise integrated approaches. This way we can answer several factors that operate above the crop's canopy. *We have been concentrating more on soil and crops! leaving the agricultural sky to play havoc on natural vegetation, crops, and human inhabitation in the Sahel.* Where feasible, we have to adapt the crops and farming systems to overcome the constraints imposed by the agricultural sky.

The agricultural weather-induced crop losses can be devastating to farmers in Sahel. First, we need to be alert to the vagaries perpetrated by the agricultural sky per se. Agricultural sky is as important as soils and genetic nature of crops, if not more. The Saharo-Sahelian region experiences high sunshine and ambient temperature. The average temperature is 26°C + 1.7–2.8°C. The ambient temperature during May/June can reach 53°C, during day time. In the Sahel, temperatures range from 15°C to 35°C during cropping season that lasts from July till November. Precipitation pattern and temperature are important agricultural sky-related parameters. They partially decide the length of cropping season. Usually, it lasts from May June till October/November. The diurnal period in the Sahel ranges from 11.5 to 12 h day^{-1}. As such, agricultural sky bestows optimum photosynthetic radiation for crops. It is interrupted feebly only during cloudy period or during the dust storms. The hanging dust and sand do reduce the light and radiation reaching the crop canopy during storms. Precipitation is a major agricultural sky-related factor in the Sahelian West Africa. The rainy period lasts for 90–100 day, extending from June/July till September end. Most parts of the Sahel receive

400–600 mm annual precipitation (Bertelesen and Traore, 2002; Buerkert et al., 2002; FAO, 2005; Coura Badiane, 2001; Tappan et al., 2004; Kamara et al., 2009). It is interesting to note that West African agrarian regions are classified as Saharo-Sahelian, Sahelo-Sudanian, Sudanian, Sudano-Guinean, etc. based on the precipitation (Rian et al., 2009; Coura Badianeet al., 2001). The agricultural sky above each region needs to be characterized too and data compiled. It does not suffice if just weather parameters are measured.

Precipitation is an important natural phenomenon originating in the Sahelian sky. Precipitation is the major source of water for the agrarian region. The total quantity of precipitation, its pattern during crop season and use efficiency are important. Precipitation affects the size of biomass and grain harvest. Precipitation is generally low in the Sahel. There are locations receiving about 200–400 mm precipitation during a crop season. It is low. It leaves a deficit in water requirement for higher yield. Some regions with 400–500 mm precipitation may support better crop yield. Use of surface water for irrigation is relatively feeble. Ground water has not been explored to any great extent. Although, ground water is potentially a good source of water to cropping belts of Sahel. Farmers adapt to precipitation pattern by selecting appropriate crops species, cropping systems, timing of sowing, several agronomic procedures, and harvesting (Aquastat, 2013; see Krishna, 2008, 2015). Precipitation regulates the biotic factors to a great extent in Sahel not just the crop.

7.5.2 WIND, DUST STORMS, AND HARMATTAN IN SAHEL

Wind is an important agricultural sky-related phenomenon in the Sahel. It has both useful and detrimental effects on the terrain, soils, their fertility, crops, and other biotic factors. In the Sahel, fields receive nutrient deposits due to a few different atmospheric phenomena. Atmospheric deposits, thunderstorms, and dust storms add nutrients derived from one location to other. Knowledge about the quantity of topsoil and sand particles shifted, then, placed at a new field is important. Nutrient carrying capacity of sand and dust/sand particle is a vital factor. Of course, nutrient carrying capacity of sand particles depends on the geographic location, cropping, and nutrient management trends practiced in the location (see Krishna, 2008; Romheld et al., 2000; Sterk et al., 1998; UNCCD, 2001). For example, Romheld et al. (2000) have reported that Harmattan received from Sahara on an average adds 3.0 kg N, 1.0 kg P and 0.7 kg K per ha yr^{-1} into fields located in the western Niger (Sahel). To avoid loss of sand/dust received into fields, usually, farmers are asked to develop good mulches made of crop residue. This procedure entraps nutrient-laden sand particles. It avoids further loss to a second location (Keitichi and Toshiyuki, 2002).

Wind is an atmospheric phenomenon that induces topsoil loss. The high intensity winds caused by Harmattan occur for short periods. Winds just prior to storms or intense precipitation events are detrimental to topsoil. Winds remove the sand in large quantities along with the essential nutrients. Wind affects ridges and furrows laid out prior to seeding. Wind-mediated sand shifts and dislodging of seeds/seed hills from its proper location are also experienced.

Wind-mediated soil erosion and other ill effects on soil fertility could be controlled using suitable agronomic procedures. The harshness of the agricultural sky could be curtailed to a certain extent. One of the methods to thwart topsoil drifts is to adopt crop residue application (Michels et al., 1995). Wind-mediated soil erosion can be severe. Winds at speeds of < 12–19 km may not be detrimental (Michels et al., 1995). However, *Harmattan* winds cause soil erosion, carry topsoil, and deposit sand particle embedded with nutrients. So, they deplete topsoil fertility. Harmattan winds sometimes coincides with storms that occur during May/June. However, the effects of Harmattan are severe when it occurs regularly during November–April (Michels et al., 1995, Sterk et al., 1998; Stahr and Herman, 1996). Again, the intensity and period for which the agrarian region suffers due to Harmattan phenomenon is important. Sterk et al. (1996, 1998) report that the enrichment ratio of particles for nutrients, such as C, N, P, and K are important. Overall, winds and Harmattan are major agricultural sky-related phenomena. They influence field soils, their fertility, seed germination, and crop stand. In some cases, dust and sand articles shifted and deposited into fallow field reached 24 t ha^{-1} yr^{-1} (Bielders et al., 2000a, b). As stated earlier, mulching is an important method. It reduces the severity of this aerial phenomenon. At the same time, mulching adds to soil fertility (see Bielders et al., 2000a, b).

7.5.3 GREENHOUSE GAS EMISSIONS IN SAHEL

Sahelian agrarian region has been experiencing negative nutrient balance since few decades. It is caused by several natural factors, such as topsoil loss due to wind and water forces. Crops have been extracting nutrients in high quantities than that replenished through fertilizers or farmyard manure.

Emissions into sky is rampant due to warm tropical weather. Loss of C and N as GHG emissions are intense during cropping season and during fallow. Fallows are mostly unprotected because cover crops are not practiced (Stoorvogel and Smalling, 1990; Bationo et al., 2005). Crop residue recycling procedures are insufficient in many locations. They do not overcome the rapid loss of Soil C into sky via soil respiration (Bationo et al., 2004).

Sahel is a large expanse of natural vegetation and crops. It is an important geographical region that affects the global carbon cycle. Its sandy oxidative soils induce massive CO_2 emissions. Soil microbial activity too contributes to soil respiration rather conspicuously. Large-scale changes from natural savannah vegetation to crops or barren fallows induces loss of CO_2 to atmosphere (Tieszen et al., 2004, 2011).

The GHG emissions have been reported from several agrarian locations within the Sahel. The West African Savanna vegetation as such interacts with other aspects, such as the soil, water bodies, and atmosphere. The gaseous exchange phenomena are among the conspicuous avenues of emissions of CO_2, CH_4, N_2O_2, NO_2, N_2, VOC, and water vapor. According to Grote et al. (2009), natural vegetation and crops are major factors that emit GHG into the sky. Reports suggests that in some locations, for example, in agrarian fields of Burkina Faso, the emissions were 20 μg NO_2 m^{-2} h^{-1} after a rainfall event. However, on a warm day, NO_2 emission reached 150 μg m^{-2} h^{-1}. Soil respiration leads to emission of CO_2 at a rate of 10–350 mg C m^{-2} h^{-1}. We should note that such GHG emissions are location-specific. They depend immensely on the soil, crop species cultivated, agronomic procedures, and seasonal changes in the ambient atmosphere (Grote et al., 2009).

Fire that affects natural vegetation and crops can result in loss of organic C. Loss of organic C occurs as CO_2, CH_4, VOC. Fires also contaminate the atmosphere with carbonaceous particulate matter. Historical evidence obtained through the occurrence of charred pearl millet grains suggest that fire has affected crops in the Sahel. Fires have occurred periodically in the thicker savannas and cropped fields since ages in the Sahel (Salzman, 2000, Goethe University, 2009; Ballouche, 2004; Diouf et al., 2012). Man-made or natural fires generally result in massive loss of CO_2 though emissions. However, the ash portion adds to mineral nutrient pool of the sandy oxisols found in the Sahel. In fact, recent farming trends in the Sahel stipulate short or long fallow to refurbish fertility. Then, addition of chemical fertilizer (N, P, and K), burning of scrub natural vegetation (e.g., *Chromolena odorata, Tridax procumbens, and Imperata* spp.) and addition of FYM to replenish organic matter and nutrients (Azeez et al., 2007).

Salzman (2000) and Salzman and Walker (1998) state that anthropogenic activity, such as forest clearing, cropping, and fossil fuel burning have affected the agricultural weather in Sahel since ages. Climate change effects are predominantly an agricultural sky-related factor. Its manifestation however has far-reaching effects on the several natural phenomena on terrain and water bodies (see Cooke, 2013; Descroix et al., 2009). Climate change affects faunal activity, the natural vegetation and crops. As such, Sahelian West Africa suffers from uncongenial atmospheric parameters. Any severe change can be still more severe to crop. Farmers need to be alert about the phenomenon related to agricultural sky. The decade 2000–2009 is approximately 1°C warmer in West Africa.

Reports state that during 1880s till 2010, the temperature in the Sahelian region fluctuated away from normal by −1°C to + 2°C (Appinsys, 2010). Jalloh et al. (2013a, b) state that weather models forecast 1°C–4°C increase in temperature in the next few decades. This causes fluctuations in the precipitation pattern, natural vegetation, crops, and even fauna. Well-directed efforts are needed to conserve the biotic diversity.

Shepard (2019) states that Sahel is among the most vulnerable cropping zones that are affected by vagaries of climate and its changes. The climate change affects the efficient use of already a shrinking natural resource base relevant to farming. In addition, the demand for cereals (millets), legumes, and oil seeds are increasing because of an estimated human population increase of 2.8% annually. Low and erratic precipitation patterns due to climate change have resulted in droughts and flash floods. Both of them are detrimental to optimum crop yield. Clearly, we should consider the agricultural sky as a composite of factors that determine natural vegetation and crop productivity. We should bestow greater attention to adapt or mend farms accordingly. A simple example would be, to make crop residue application on top of sandy soil is mandatory. This will reduce the loss of topsoil fertility, lessen the effects of high soil temperatures, improves fertilizer efficiency. It helps to retain nutrients and moisture better. Methods that reduce calamitous consequences due to strong storms created by intense precipitation events and wind/dust storms need greater attention. Methods that harness wind power better in the Sahel would be a boon to farming regions. Both low and high energy wind power generation need greater attention. Methods to improve seed germination and seedling establishment need to be explored. Sand shifts often submerge seed hills with high temperature sand particles that destroy the plant stand.

Droughts are frequently experienced in the Sahel. It is partly attributable to phenomena that occur in the agricultural sky above the cropped zones. Droughts are caused by paucity of rainfall. Uncongenial pattern and long gaps in rain fall events induces droughts. Droughts reduce biomass formation by natural vegetation and crops. Droughts and dust bowls may occur together. They may affect soil fertility, crops, and the general atmosphere. Drought induces sufferings for human population, domestic animals, natural vegetation and crops simultaneously. Dust fluxes in the Sahel are mostly caused by the large dust pools available in the Sahara. Dusts are also caused locally if the Sahel region suffers longer period of absence of rainfall (Multiza et al., 2010).

7.5.4 MICROBES AND AGRICULTURAL SKY ABOVE SAHELIAN AGRARIAN REGION

7.5.4.1 HARNESSING ATMOSPHERIC NITROGEN FIXATION IN SAHELIAN AGRARIAN REGION

Harnessing the Sahelian atmospheric N through biological nitrogen fixation is an important aspect. We need to improvise the performance of N-fixing microbes by inoculating them on crops or soil. Dissemination of N-fixing microbes via the agricultural sky (wind, dust, precipitation) is also important. Both, in natural expanses and in farms. There are a few new techniques to apply formulations containing N-fixing microbes. For example, we could utilize sprayer drones loaded with liquid/granular formulation of microbial inoculants (see Krishna, 2018, 2020a, b). Whatever be the methodology adopted, Sahelian legume fields should possess optimum population of N-fixing microbes for nodulation. This way, farmers can harness atmospheric N better.

Cowpea is an important legume crop. It is native to upper reaches of river Niger, in West Africa. Several species and biotypes of cowpea occur in this region (see Krishna, 2008, 2015). The harsh agricultural sky of the Sahelian West Africa must have played vital role during the evolution of the crop. Particularly, biotic factors, such as fungal, bacterial, and viral pathogens. Abiotic atmospheric factors too might have played important role during domestication of cowpea.

Cowpea has coevolved with symbiotic N-fixing *Bradyrhizobia*. Obviously, cowpea has evolved to harness benefits from the agricultural sky, better. Cowpea allows for conversion of atmospheric N (20–40 kg N ha) into the biomass. *Bradyrhizobium* is a soil bacteria capable of utilizing the atmospheric N. This process adds to soil N needed for the crop, including the succeeding crops, if any. *Bradyrhizobium* is disseminated through sand dust storms, precipitation, runoff, and farm vehicles. Perhaps, Sahelian dust storms do carry *Bradyrhizobium* cells to long distances and deposit them on soils. Legume nodulation gets supported by dust storms that carry inoculum in the sky.

In the Sahel, cowpea is mostly cultivated as intercrop with hardy cereals, such as pearl millet or sorghum. Both of these cereals too have been influenced immensely by the agricultural sky, and its abiotic, and biotic factors. There are several asymbiotic soil microbes that fix atmospheric N while living in a free state in the soil. Such microbes too get disseminated to different locations via the dust storms, precipitation, etc. In the present context, we are also concerned with the various disease-causing microbes that affect cowpeas.

The agricultural sky anywhere including the Sahel is generally endowed with higher levels of N_2 (inert gas). The soil air too is a good repository of N_2. These sources must be converted to NH_4 though microbial flora in the soil. The extent of N fixed into soil organic matter often varies with variety of factors, such as the soil type, its properties, crop species, and its genotype *Bradyrhizobium* species, and N fixation rates. A few other atmospheric traits, such as the humidity, temperature, pH, redox potential also affect the BNF. For example, Bado (2002) states that peanut grown in Burkina Faso fixes between 8 and 23 kg N ha depending on the location and cultural practices. N^{15} assays have shown that about 27–34% plant N was contributed by the conversion of atmospheric N by the *Bradyrhizobium* species. Naab (2001) has pointed out that our approach should be to select the efficient strain of *Bradyrhizobium*, appropriate peanut or cowpea genotype, and appropriate soil fertility practices. This will help in converting atmospheric N better than at present. Atmospheric N is also fixed by the asymbiotic free-living microbes, resident in sandy soils of West Africa. In all, the net potential of biological N fixation in the sandy soils of Sahel needs to be assessed periodically. This will help in ascertaining the soil N available to crops in the same season or the next one. For resource poor farmer in the Sahel, harnessing atmospheric N is indeed a good idea, particularly if the aim is to build soil N fertility of their fields at low economic costs. No doubt, agricultural sky has a positive effect on vegetation, crops, soil N stocks, and productivity.

There is need to expand legume-based cropping expanses to harness atmospheric N like they did in Cerrados (Brazil) and Pampas (Argentina). Perhaps, cereal-based rotations that include cowpea and groundnut and inter-crops with cereals need to be accentuated to high levels. Planting density and area under legume need a boost if atmospheric N has to be garnered better than that at the present levels (Sanginga, 2003, N'diaye, 1988, Dakora et al., 1987, Alvey et al., 2000, 2001; Bagayoko et al., 2000; Frank et al., 2004). Indeed, we already have such a situation in the peanut basin of Senegal. Groundnut being a legume possess the ability for fixation of atmospheric N through its symbiosis with *Bradyrhizobium*. It helps to add atmospheric N to soils at the rates ranging from 15 to 60 kg N ha^{-1} $season^{-1}$ (Dakora et al., 1987; Kaleem, 1989). A few other reports suggest that in a cropping season, groundnut basin converts and contributes upto 53 kg N of atmospheric N into soil N pool. It is easily usable by the succeeding crops (Sanginga, 2003; Shahendeh et al., 2004). A cowpea crop grown on the sandy soils of Sahel is said to convert 9–37 kg atmospheric N ha^{-1} $season^{-1}$.

The sandy soils of Sahel are supposedly low in phosphorus availability. Major crops, such as pearl millet, sorghum and cowpea are known to derive a degree of benefit in terms of higher P uptake, due to symbiosis with these zygomycetous mycorrhizal fungi. The arbuscular mycorrhizae do occur in the sandy Oxisols of Sahel (Krishna, 1986). There is perhaps little knowledge about the aerial spread of asexual spores and propagules (infective hyphae) of arbuscular mycorrhizae. Some reports state that dispersal of pathogenic fungi takes place via the Sahelian dust (dust storms). Mycorrhizal spores/propagules too may get dispersed via aerospace but is relatively less represented in the Saharan/Sahelian dusts (Elberti et al., 2007; Griffin et al., 2001). However, there are clear indications that soil dust lifts from cropped fields and transits a distance and then settles in other locations. So, they do carry spores/hyphae of arbuscular mycorrhizae. This phenomenon needs further data to substantiate.

7.5.5 AIRBORNE PLANT PATHOGENS IN THE SAHELIAN REGION

There are several reports stating that Sahelian/Saharan dust laden with microbes (fungal spores, bacteria, archae, virus particles, insect eggs, and organic matter particles) do traverse long distance on the atmospheric dust. They have been traced in the farmlands within Brazil, Caribbean, and even US Great Plains (Elberti et al., 2007; Griffin et al., 2001; Schutze et al.,

1981; Kellog et al., 2004). Short distance transmission of fungal spores has also been regularly recorded.

It is possible to trace spores of Ascomycetes and Basidiomycetes fungi in large numbers, at altitudes from few meters to 2000 m above ground level (Griffin, 2004). Whereas zygomycota spores could be traced at only feeble intensities (Elberti et al., 2007; Goudie and Middleton, 2001; Gregory,1971, 1973, 1978; Griffin et al., 2001). Fungal spores traced in the atmosphere are resistant to vagaries in weather parameters. They do withstand harsh climatic conditions, such as high winds, ambient temperature, desiccating conditions, wetness, and even cold tempeatures of winter. Several of these fungal spores may actually be pathogenic on crops. They could induce infections at new sites as they land on crop fields located at different distances from where they were first blown into atmosphere. Brown and Hovmeller (2002) have quoted as evidence the aerial dispersal of sugarcane rust (*Puccinia melanocephala*) from sugarcane fields in Cameroon (eastern Sahel) to Domnican Republic in Caribbean island. The uredospores of rust fungi (*P. melanocephala*) floats in the sky. The cyclonic winds are known as the major vectors for rust spores. Uredospores get lifted and blown across the Atlantic Ocean into North America. They initiate sugarcane rust disease in the new location.

Let us consider a different exmple of plant pathogenic fungus. Secondary spread of several crop diseases are mediated via wind and floatation in the agricultural sky (e.g., *Sclerospora graminicola*- downy mildew fungus) (Williams, 1984; Gilijamse et al., 1997). A few of the plant pathogenic fungi are pleomorphic and produce spores that tolerate harsh conditions prevalent in the atmosphere, particularly during airborne transit. Often the fungal spores are dislodged and blown based on the wind currents prevalent at the spot. A few members of ascomycetes produce asci that are released into air with pressure. They get dispersed to short distances away. It helps them in dispersal although only for short distance. Atmospheric conditions, such as relative humidity, dryness, and wind speeds are essential characteristic to be noted while studying the fungal spore dispersal and disease spread.

There are several diseases that occur on staple cereals, legumes, and vegetable crops grown in Sahel. Many of them have airborne stages that require the mediation of agricultural sky. The crop's canopy and atmosphere above the crop fields play a vital role in the dissemination of pathogenic organisms. Pearl millet and sorghum that are major cereals in the Sahel are affected by several disease. A few examples of diseases which need aerial space, wind and/or aerial insect vectors are as follows:

Fungal diseases: Downy mildew (*S. graminicola*), leaf spots (*Cercospora pennisiti*), ergot (*Claviceps purpurea*), rust (*Puccinia substriata*), and smut (*Moeszioyces pennisiti*).

Bacterial diseases: *Psuedomonas syringae, Xanthomonas campestris.*

Viral diseases: Maize dwarf mosaic virus, maize streak virus, wheat streak mosaic virus, etc.

There is clear need to understand the aerobiology of the causal agents. Regulating epidemiology and yield loss due to the disease-causing agent should be our major aim. Since many of these disease-causing agents are primarily disseminated via agricultural sky, we need to monitor farm sky, perhaps all through the year. In the Sahel, there is crop destruction caused by abiotic factors, such as dust storms, high temperature, drought, soil erosion, etc. In addition, there are detrimental biotic factors such as diseases. *Therefore, "agricultural sky" could be more detrimental to crops, than soils or water. It would be correct to say, 'Sahelian sky' is more dangerous and suppressive to crop yield than sandy soils with poor soil fertility or precipitation that is erratic."* Farmers should pay more attention to Sahelian sky and manage crops, accordingly.

Majority of the studies on aerial dispersal of plant pathogens (phytopathogens) have primarily focused on fungi. This is because, over a thousand species of fungi are responsible for 70% of all known plant diseases (Behzad et al., 2018). In the Sahel, downy mildew is among major fungal diseases that affect the cereals, such as pearl millet and sorghum. Downy mildew is an aerial disease. It affects the shoot system, foliage, and panicles. The photosynthetic efficiency of foliage gets reduced, crop gets stunted, foliage formation is markedly reduced and the seeds in the panicle get deformed (crazy tops) (Williams, 1984; Werder and Manzo, 1992; Mbaye, 1992, Gilijamse et al., 1997). Its infective stage during secondary spread is airborne. The agricultural sky can play a vital role in the dissemination and spread of this dreaded crop disease in the Sahelian region.

Pearl millet is a dominant cereal crop in the Sahel, along with sorghum (Mason et al., 2015). These crops suffer from downy mildew attack. The primary infection is via the oospores released in the soil. The airborne oospores are counted as a source of disease inoculum. Here, we may note that, agricultural sky and its atmospheric parameters play a vital role in the dissemination of asexual spores. Such sexual spores that are essential for secondary infection. Airborne inoculum in the form of sporangia/zoospores can remain viable in the air for a few hours. After that the sporangia may desiccate and become noninfective. We may note that over 35,000 sporangia are produced per cm^2 on the foliage. These could be transmitted through air. Airborne sporangia are released with a force, so that it gets lodged at about 2.5–5 m away from the point of liberation in the field. Sporangial liberation occurs continuously all through the day/night. Optimum temperature is 24°C and relative humidity is 85–90%. Darkness can aid dissemination of fungal spores. *Airborne sporangia infect the seedlings*. In fact, it is the secondary infection of crop that really deleterious to it and destroys the foliage of the crop and grain formation (Gilijamse, 1997; Gilijamse et al., 1997, Gilijamse and Jeger, 2002).

Sorghum is a hardy cereal grown in the Sahel. It is cultivated in regions that are slightly wetter on an average, particularly if we consider the zones that support pearl millet. Farmers in areas with relatively higher precipitation prefer sorghum cultivation. Sorghum too is affected by a series of fungal, bacterial, and viral pathogens that cause disease. They induce loss of biomass plus grains (Pron et al., 2020). Sorghum downy mildew is common in the Sahel. It becomes severe if the conditions are congenial during seedling stage. Humid climate builds disease. The wind-blown asexual spores spread the disease as secondary inoculum. There are reports that sorghum disease, such as anthracnose caused by *Colletotrichum sublineola* could be detected using aerial methods based on spectral analysis. Unmanned aerial vehicles seem to help in assessing diseases spread rapidly and in monitoring the fields (Pugh et al., 2018). Such methods could be adopted in the Sahel too. Maize cultivated in the Sahel is also susceptible to downy mildew disease that spreads aerially (Kutama et al., 2010). It is also attacked by several viral diseases that are transmitted by aerial insects (Ruf et al., 1995).

There are at least eight viral diseases that affect cowpeas cultivated in drier Sahel as well as in wet tropics of West Africa (Thottappilly and Rossel, 2008). They are transmitted aerially by insects, such as beetles, aphids, and flies (e.g., white fly Bemisia sp.). Insects residing in the crop's canopy and above it transmit the disease-causing virus. They are cowpea yellow mosaic comovirus, Southern bean mosaic virus, Cowpea aphid borne mosaic virus, Cowpea potty virus, Cucumber mosaic virus, Cowpea golden mosaic virus, Sun hemp mosaic virus, and Cowpea mild mottle virus. The general suggestion is to grow plant cultivars that are resistant to viruses and/or practice aerial sprays of pesticides that control the vector (Ambang et al., 2009).

Groundnut is an important cash crop in Senegal, Mali, and Burkina Faso. This legume crop is affected by several disease that are airborne (e.g., groundnut rust). There are also a few viral diseases that are transmitted by aerial insect vectors. For example, *Aphis craccivora* is a vector that transmits groundnut rosette virus. It is an aerial insect that proliferates rapidly. Aphids move across groundnut fields swiftly by flocking. Wind gusts too may spread the vector and the virus. Groundnut rosette virus is actually endemic to Sahelian West Africa. It reaches epidemic proportions when weather parameters are congenial for the proliferation and spread of the vector *A. craccivora*. There are two types of groundnut rosette disease. One that causes chlorotic leaves and other that produces green whorl.

Groundnut rosette virus has caused devastating epidemics in the Northern Nigeria during mid-1970s. Again, in mid-1980s, groundnut rosette epidemics occurred in many of the Sahelian nations. The insect vector that is active in the agricultural sky above groundnut canopy plays a vital role in causing groundnut virus epidemics.

7.5.6 INSECTS IN THE SAHELIAN AGRICULTURAL SKY

The agricultural sky above the Sahelian crops is an abode to several insect pests. They cause mild-to-severe devastations to crops. Most of the crops and their varieties grown by farmers do suffer from insect attacks. Foliage and grain yield losses are significant. This is despite strong international programs that have developed cultivars resistant to insect pests. Several insect species attack major crops of West African Sahel, such as pearl millet, sorghum, cowpea, cotton, groundnut, and vegetables. Pearl millet is a staple cereal grown all over the Sahelian West Africa from Senegal to Chad. In this region, pearl millet suffers from defoliators, panicle and grain-eating pests, and a few other species. Aerial pests are among chief destroyers of grain yield.

Fall armyworm has become a dreaded pest in the Sahel and other parts of Africa. Since 2016, it has spread into cropping zones of entire Sahelian West Africa. The average destruction due to this aerial pest is reported at 19% during 2020. It attacks all the major crops grown in the Sahel, such as pearl millet, maize, sorghum, small millets. (Ndumi Ngumbi, 2020). Stem borers such as *Chilo partellus* are also severe on sorghum and maize grown in the Sahel. The legume pod borer affects cowpeas severely. Cowpea is consumed by over 200 million Africans. Aerial pests such as pod borers reduce grain yield by 20–80% in Nigeria, Niger and Burkina Faso (Ndumi Ngumbi, 2020).

Most of the farmers in the Sahelain West Africa realize that pearl millet miner is a dreaded pest (Nwanze, 1989, 1991, Nwanze and Sivakumar, 1990,

Gahukar, 1984, 1988, Gahukar et al., 1986, Oumarou et al., 2019). It is among major biotic detriments caused via the agricultural sky. It affects the foliage, panicle, and grains. It migrates through the air. The adult flies from one location to other. During past decades, that is, for 28 years, this aerial pest has consistently caused 43–82% damage to grain yield in the Sahelain nations. Oumarou et al. (2019) has conducted a thorough survey about the farmer's awareness regarding deleterious effects from this pest. Farmers do realize that pest attack and its devastation is influenced by several of abiotic and biotic factors operating in the sky/canopy of the crop. Pest control methods need to be applied carefully and integrated with other procedures.

Climate change affects several aspects of a cropped field, impartially. Climate change effects have their specific effects on both the crops and insect pests (also pathogens). Certain climate change parameters and their combinations may affect the crops severely. A few of them may induce better growth and productivity. For example, a small increase in temperature induces better crops in colder regions. Climate change affects pests, their survival and destructive activity. A few combinations affect pests/disease severely so that crops may perceive the impact of pests/disease only at low levels. Other combinations of climate change factors may affect the pest less or accentuate them. In such cases, pest population increases thereby bringing about devastation. We know that temperature increase is among the major climate change factors affecting agrarian regions. Relative humidity and wind too may affect both the crops and pests. Coffrey and Farmer (2014) have published a list of several crops and their major pests/disease. They have clearly shown that certain combinations like high temperature–high humidity, high temperature–low humidity, high temperature–low precipitation, etc. can have detrimental effect on crops plus pests. Some combinations affect only the crop or others only the pest. At the bottom-line, agricultural sky has its unique effects on crops plus their pests/disease that occur in an agrarian region. We need to be aware of the details about such interactions in nature. A basic knowledge about crops and pests in the Sahel region is of course mandatory (Gahukar, 1984, 1987; Nwanze, 1989, 1991; Nwanze and Youma, 1995). Further, during recent years, farmers in the Sahel zone opt for IPM procedures. While applying IPM procedures, of course, due attention should be given to climate change effects bestowed via agricultural sky both on the crops and pests (Hausmann et al., 2020).

Cowpea is an important legume crop of Sahelian West Africa. Cowpea is affected by a few aerial insect pests. The Sahelian sky supports pests that attack legume crops. Sahelian ambient atmosphere (sky) is harsh on the pests' survival itself during offseason. Pod borers, pod sucking bugs, and defoliators do affect the crop yield. The aerial space above the cowpea canopy is congenial during crop season. However, it is harsh on pests that should survive and complete the life cycle. It is interesting to note that harsh agricultural sky is overcome by the pest by taking refuge in the alternative hosts. For example, in the case of cowpea, alternative hosts, such as *Cajanus cajan*, *Crotalaria retusa*, and *Rynchosia memnosa* are collateral hosts during the period when aerial space is harsh (Debire et al., 2005). In fact, Debire et al. (2005) provide a list of several alternative hosts that support the leaf sucking pests of cowpea. This is an example wherein Sahelian agricultural sky is harsh on crops.

At this juncture, we should pay attention to insect's migratory trends in the agricultural sky. Florio et al. (2020) state that migratory habits of insect pests have direct impact on food security in the Sahel region. In fact, they opine that windborne migration is a key strategy played out by crop pests encountered in the Sahel. Several pests were recorded in the Sahelian sky at altitudes ranging from 40–200 m above the crop's canopy. Night-time migration is also in vogue. Interestingly, Florio et al. (2020) report that pests migrate southwards from Sahel during dry season and return journey occurs as wet conditions and crops establish in the region. Some of these migrations are said to occur irrespective of the availability of crops.

Agricultural sky and its biotic factors might have played a vital role in the evolution of major pulse crop of the Sahel, namely, cowpeas. Agricultulral sky and its factors do affect both the crop and aerial pests. So, several aerial pests have evolved along with cowpea. Aerial pests do regulate crop's health and yield. Human efforts to select cowpea cultivars well adopted and resistant to aerial pest also needs consideration. Cowpeas are attacked by a series of aerial pests, such as the aphids (*Aphis fabae*), pod-borers (*Mauca vitrata*), pod sucking insects (*Riptotortus* sp., *Nezara viridula*) , blister beetles (*Mylabris* sp), and viral vectors like white fly (*Bemisia* sp). No doubt, abiotic (atmospheric) and biotic factors affect the severity of diseases. When conditions are congenial, epidemics that devastate the crop are also common in the Sahel region (Imrie, 2000; IITA, 2000; IPM CRSP, 2000).

Let us consider the devastation caused by locusts to Sahelian crops. *Locusts are aerial phenomenon.* They could become prominent detriment channeled on to crops via the agricultural sky. There are indeed several pests that attack the subsistence crops in the Sahel. Several of them are transboundary pests. The causal agents (insects) are often migratory. The Food and Agricultural Organization of the United Nations, Italy, suggests that such transboundary pests (e.g., locusts) could become severe. Sometimes they spread rapidly like the pandemics. They cause huge losses to Sahelian farmers. As such subsistence farmers suffer feeble crop stand and low grain productivity, even during normal years. They say, such transboundary pests are aerial. They get disseminated due to movement of humans, goods and other products of importance to different Sahelian nations. Most of the aerial pests disseminate rapidly through the agricultural sky. A few examples are locusts (short-horned grasshopper), army worms (e.g., fall armyworm), panicle and head miner *(Heliocheilus albipunctella),* cowpea pod sucking bugs, etc. Locust clouds are among excellent examples of pandemic insect pests. They migrate through the sky.

Let us consider locusts that are common to Sahelian agrarian regions (FAO, 2005, 2020). Locusts above crop fields is an excellent example that suggests how "agricultural sky" can be a nasty detriment on crops. Such devastation on crops is rapid and immense. Perhaps, there are several regions in the world, wherein, agricultural sky poses greater detriment on crops than soils phase of the farm world. In such locations, tackling sky (aerial factors) is more difficult than soil factors.

Locusts (*Locusta migratoria migratoroides*), also called short-horned grasshoppers are among the major biological detrimental factors in the West African sky. The African migratory locust has two phases. One is "solitary" phase when we encounter them individually. The other is "gregarious" when they group together as nymphs and fly from one region to other in large groups or clouds in the agricultural sky. The gregarious adults are dreaded pests that affect Sahelian crops, such as pearl millet, sorghum, cowpea, groundnut, vegetables. Locust nymphs are known to travel up to 24 m in search of vegetation. Swarms of adult locusts usually fly during the day and rest on the ground during night. These African migratory locusts are known to fly close to ground and destroy the crops. In comparison, the "Desert Locusts" fly at higher altitudes in the agricultural sky. When locusts land, they cause greater havoc by destroying crops and natural vegetation in a matter of hours. African migratory locusts erupt during August till October. This is the period of the year when precipitation is at peak and crops are green and leafy. Since 1890s, several different locusts, such as the Senegalese short-horned grasshopper, red locusts, brown locusts, and desert locusts have caused plagues in the Sahelian region. Locust plagues have occurred at different intervals. Locusts are a threat to Sahelian food security if they take to sky and migrate.

No doubt, agricultural sky is dangerous if it harbors migrating locusts. Regarding the present situation, Dongui (2020) warns that locusts could be affecting the entire food generating regions of East Africa and Sahel. Desert locusts are most severe detriments to crops cultivated in the Sahel. Their speed in the agricultural sky is high. Their population increases rapidly when rainfall is congenial during June–October in the Sahel. They are voracious feeders. Their appetite is enormous. They consume foliage and seeds double their body weight. A swarm with 200 billion short-horned grasshopper (locusts) feeds on food grains equivalent to requirement of 84 million people in a day. Locust control programs involving pesticide application, biological control using sprays of *Metarhizium* sp. fungus and birds are practiced. Over 500 million US$ are spent on locust control programs by FAO of the United Nations, Rome, Italy, each year.

Locusts are regularly monitored in the agricultural sky of sub-Saharan Africa. The desert locusts travel large distances spanning the continent from east to west of Sahara. They migrate from Kenya/Ethiopia to cropping zones of Sudan, South Sudan, Somalia, Pakistan, and India if they fly to east. They migrate to West African cropping zones in Chad from Libya and then on to Niger, Burkina Faso, Mali, Senegal, etc. The bulk of swarms usually get created in a short time of 2 weeks. Flocks develop during early seedling stages of the crop, in July/August (Anami, 2020). Food and Agricultural Organization, Italy regularly monitors and prepares reports about locusts for dissemination to farmers. Each year, locust swarms induce resource competition. Farmers in sub-Sahara have to take precautionary measures. Desert locust swarms are detected using satellite imagery, local tribal information channels, drones, and aircraft sorties. Reports suggest that worst attacks on crops have occurred in Kenya, Sudan, Chad, and West African Sahelian crop belts during June in 2020. Locust swarms have utilized agricultural sky rather efficiently and migrated from Kenya in east till the shores of Senegal in West Africa, within a month's period. *Here, agricultural sky mediates a disaster on crops.*

Desert locusts may occur in devasting proportion even until November. It depends on the rainfall pattern and availability of greenery/crops in the fields. For example, in 2012, desert locust migrations occurred from their breeding zones in Sudan, Libya, and Algeria to Niger, Mali, and Mauritania. It occurred in late October and November. During this period, they affected foliage, panicles, and grains. Sprouting vegetable crops too were devastated by the locust swarms (Hicks, 2012).

The desert locust (*Schistocerca gregaria*) swarms (i.e., large group or clouds) transit throughout the Sahelian agrarian regions. They also move into Middle East and Mediterranean farming regions (Sanchez-Zapata et al., 2007). Locust spikes, that is, plagues occur periodically in Sahel. They have been noticed in North Africa (Egypt) and Sahel since the times of Pharaohs (Nevo,1996). So, the agricultural sky in sub-Sahara has supported the phenomenon of locust swarms for over 5000 years. It affects food production strategies of farmers in the Sahel. They say desert locusts that migrate in the Sahel also encounter migratory predatory birds. Some of these predatory bird species travel from long distances in Europe each season. High populations of such predatory "Black Kite" are noticed in many regions of the Sahel. For example, in Mali and Mauritania, farms affected with locusts may also show up over 500–600 black kites within 2 km radius, particularly during November month. It coincides with panicle maturation and seed set. Regarding locusts, they report that there are two kinds of populations. One set affects trees, shrubs, and other types of natural vegetation. A second set affects pastures, grasses, and cereal food crops and vegetables. Generally, there is not much overlap among the two types of desert locusts (Sanchez-Zapata et al 2007). There is a tendency to spray the field crops to reduce locust attack. Farmers could also use biological control methods. Sanchez-Zapata et al. (2007) believe that if predatory birds (e.g., Black kite- *Milvus migrans*) from European regions are in good number affecting the build up of locusts, then, use of pesticide chemical could be reduced or even abandoned. Insect pest resistant cereals, legumes, vegetables, and cash crops could be planted. Our interest, based on the context of this book, would be to decipher if sowing large areas of resistant cultivars of, say, pearl millet, cow pea, or sorghum will affect the population of these insect pests in the agricultural sky. Perhaps, growing resistant cultivars will reduce pest population hovering into those fields. This would be a good example of a field crop affecting the biotic characteristics of agricultural sky and vice versa that is, a susceptible cultivar accentuating pest population of agricultural sky is also a clear possibility!

Breeding food grain crops and trees for resistance to insect pests is not a new aspect. There are several programs running concurrently in the Sahelian region. They aim at developing genetically resistant food crop species. We should note that in addition to inherent (genetic) resistance to insect pests, there are several agronomic procedures adopted by the Sahelian farmers that affect pest activity above the crop's canopy. The pest population in the agricultural sky is also dictated by the cropping intensity, its stage, cropping systems adopted, ratios of intercrops, for example, pearl millet and cowpea intercrops, etc. Gahuakar (1989) has clearly shown that activity and population of a major pest such as *H. albipunctella, or* thrips such as *Megalurothrips sjostedti* and *Frankliniella* sp. are affected by the intensity of intercrops (Pearl millet: Cowpea rows). The population of stem borers (*Acigona ignefusalis*) and spike worm (*H. albipunctella*) are also affected by millet: cowpea rows sown in the field. This is an example of crop species affecting the biotic composition of the agricultural sky. Not just the insect pests, it is said that even population of the pathogenic fungi, such as *Tolyposporium* sp. and *Puccinia* spp. in the agricultural sky is affected by intercrop species and their intensity.

Biological control of crop pests in Sahel is an idea worth attempting. It could be primarily an aerial phenomenon. It involves perhaps relatively lower cost compared with chemical pest control methods. It keeps pest at threshold. Pearl millet head miner (MHM) (*H. albipunctella)* is an important aerial pest that attacks the panicles possessing grains at maturation. MHM causes 30–40% loss of grains (Bhatnagar, 2011; Gahukar and Ba, 2019). Recently, biological control of MHM has been attempted in the cereal farms situated in Burkina Faso, Mali and Niger (Payne et al., 2011, Ba et al., 2013; Baoua et al., 2013) using parasitoides, namely, *Habrobracon hebetor*. The pest and parasitoid both are part of the aerial phenomenon in the agrarian Sahel. Parasitoids are released within and above the crop's canopy. Results have shown that *H. hebetor* population increases. Consequently, it decreases the pest-MHM. Parasitization levels may reach 97%, if the atmospheric conditions are congenial. Such biological control adopting parasitoides usually keeps the major pest under threshold levels (Baoua et al., 2013). Say, often at low levels of < 2–5% pest attack.

Biological control of pests that attack and flourish within vegetable farms in Sahel is again a useful proposition. Once established, the predators could be endemic to the agrarian zone. Predators (bugs) keep the parasitic insects under threshold levels. During recent years, tomatoes grown in Niger and Burkina Faso are being attacked by tomato pinworm (Brevault et al., 2014; Haougui et al. (2017). It is causing yield loss. Garba et al. (2020), for example, experimented with predatory mirid bugs (Heteroptera Miridae, *Nesidiocoris tenuis, N. volucer, and N. callani*) to achieve the control of tomato pinworm, namely, *Tuta absoluta*. Garba et al. (2020) suggest that as a matter of policy whenever crop pests are noticed in high intensity, it is preferable to search for natural enemies. In this case, natural enemies for tomato pinworm have been identified (Cherif et al., 2019). Incidentally, both insect pest and natural enemy used for biological control are part of sky above tomato fields.

Pollination of crops through aerial insect species is an important ecosystem service. It is counted among the important beneficial effects offered to crops through "agricultural sky." Most of the cultivated crops in the Sahel derive benefits from insect pollinators. Such pollinators hover above the crop's canopy. Insect pollinators ensure sexual reproduction of crops. They also stabilize grain productivity (seed set) in the Sahelian crops (Aizen et al., 2009; Stein et al., 2017; Intergovernmental Science-Policy Platform on Biodiversity (IPBES) (2012). Further, Stein et al. (2017) state that bee-aided pollination is an important aspect of production of cash crops, such as cotton and sesame. This is in addition to important role they play in pollination of food crops cultivated in Sahelian West Africa. Bee population, namely, wild, semi-wild, and honeybees (*A. mellifera*) multiplied in situ help in pollination, generating seeds, improving genetic diversity of crops, stabilizing yield by inducing better seed set, and honey collection in farmland. Sahelian farming belts and natural vegetation zones are endowed with several wild, semi-wild, and honeybee species. They need to be encouraged by developing natural vegetation with diverse plant species. Crop diversity has to be enhanced to provide better foraging fields for bee population. Preparing a list of plant species visited by bees and useful bee species is again good step (Wannarka, 2018). It helps farmers to locate plant (crop) -pollinator combinations properly. This way, we can improve the benefits from the agricultural sky in the Sahel.

Let us consider an interesting aspect of bees and their role in pollinating tree crops useful to Sahelian human population. Shea is an important agroforestry crop. Shea trees offer nuts and wood. Its fruits and nuts are processed and consumed as a source of fat. It is used by the local population resident in Sahelian region spanning from Senegal till Chad/Uganda covering 1 million km^2. These Shea trees are known to benefit immensely from bees that fly across the orchards. Regions with greater diversity of vegetation support better honeybee activity. Hence, Shea trees benefit more when plant diversity is held without dwindling (Stout et al., 2020).

Tamarind (*Tamarindus indicus*) is an important tree. It grows well-scattered in the Sudano-Sahelian region of West Africa. These trees are pollinated by several aerial insect species, such as wild bees, honeybees, wasps, and dipteran flies. *T. indicus* is also pollinated through wind, but it is relatively feeble. Studies in Burkina Faso by Diallo et al. (2014) suggest that long distance pollinators, such as *Xylcopa olivaceae* and *Megachille* sp. both belonging to Hymenoptera are important. Short distance pollinators, such as honeybees (*A. mellifera*), *Trigona* sp., of Hymenoptera and Dipterans, such as *Syrphida* sp. and *Bombylius* sp. are important pollinators of *T. indicus*. Certain wasps like *Polistes fastidious* also occupy the sky above *T. indicus* canopy. They aid in pollination. Generally, it is preferable to diversify natural vegetation and crops in the area where *T. indicus* is to be pollinated by the bees.

In Gambia, farmers realize that sky above crops have to be populated with pollinator insects like wild bees, and honeybees (*A. mellifera*). Reports suggest that a good population of honeybees above crop fields can result about 40% higher yield than that if they were not present. As such, there are no diseases that affect honeybees. Problems, if any, are caused by farmers applying high dosages of pesticides. Natural habitats of honeybees need to be restored. Genetic diversity of natural vegetation and crops need to be maintained. Further, Bee colonies need to be developed in farms on a mandatory basis (Africa BEECause, 2020).

We may note here that climate change factors do affect bee activity. Sometimes, erratic bee activity and low population in the air can reduce crop yield. For example, yield of cotton/sesame got reduced by 40% of the normal levels. Stein et al. (2017) believe that for resource poor farmers with < 2 ha land holdings and crop stand not so good, achieving higher rates of pollination and seed set via insect pollinators (bees) is important. Locating beehives in farms is indeed a good idea. Field experiments have revealed that bee-aided pollination improved seed set, its quantity (yield) and quality. In the case of cotton, increase in bolls were 47% compared with control plots. In the case of sesame, grain yield improved by 57% above control (Stein et al., 2017). Food and Agricultural Organization, Italy reports that insect pollination economic value (IPEV) in the West African Sahel is relatively high at 5.6 billion US$ (See Stein et al., 2017; Gallai et al., 2009; Fang, 2016). This is because, field crops, such as pearl millet, sorghum, cowpea, cotton, sesame, depend on pollination services of aerial insects. This is despite wind-aided pollination processes that occur naturally.

Sahelian bees also produce honey and wax. This is in addition to pollination-related benefits from bees in the agricultural sky. The honey produced in 2013 ranged from 190 tons by Mauritania, to 650 t by Cote de Ivoire, 840 t by Sierra-Leonne, 900 t by Guinea, 1050 t by Chad, 3150 t by Senegal and 5300 t by Morocco (Africa BEECause, 2020). Overall, agricultural sky harboring honeybees and other pollinators is a boon to Sahelian farmers. We should avoid any disturbance to their activity and ecosystem services.

7.5.7 AVES IN THE SAHELIAN AGRICULTURAL SKY

Birds that occupy Sahelian farms thrive under, within, and above the crop's canopy. Birds traced in natural vegetation and sparsely distributed shrubs/trees offer several ecosystematic services. A few bird species are severe pests on crops, others operate from the agricultural sky as predators. These predatory birds (e.g., falcons, raptors, owls) reduce the damage of crops by pest birds. There are bird species that serve to pollinate the Sahelian crops. There is general opinion that large birds and predator species are dwindling, particularly in terms of diversity and population in the Sahelian sky (Thiollay, 2015). We have to characterize the birds, their activity, and identify beneficial/detrimental effects. Classification of such portions of Sahelian sky with regard to birds is definitely useful to farmers, naturalists, and environmental specialists.

In the Sahel, *Quelea quelea* is an important aerial phenomenon. It is a bird pest operating in the sahelian sky. *Q. quelea* feed on the dry grass seeds and cereal food grains. It is a granivorous bird indeed. *Q. quelea* is relatively a small bird that occurs in flocks of large numbers. They destroy cereal food crops cultivated in the Sahelian West Africa. They are also common in the tropical savanna/guinea regions.

Now, let us consider wetland rice (*Oryza barthii*) and damage to it is inflicted by birds such as *Quelea*. Firstly, the agricultural sky above the rice fields does support a series of bird species capable of foraging on rice. Among them, *Q. quelea* seems to cause perceptible damage to rice crop. This species is almost endemic to aerial space and crop canopy since past several centuries (DeMey et al., 2011). The weaver birds may cause up to 13.2–15.2% grain yield loss. The population of pest and its intensity of feeding is important. Sometimes, the time tested *Quelea* control measures are not effective, particularly if the intensity of damage is higher. Economically, this aerial pest that occurs in Senegal's riverine regions may cause 1.0 million Euros loss per season to rice farmers (DeMey et al., 2011).

It is generally believed that *Q. quelea* primarily prefers grass seeds. However, in the absence of wild grasses in the vegetation, it feeds on larger cereal grains. Even field trials conducted for five decades have shown that other than grass seeds, *Quelea* sp. prefer grains of pearl millet, sorghum, rice, and maize (Ward, 1966, Yousufu et al., 2004; DeMey et al., 2011). This bird pest's preference also changes based on season and stage of crops. *Q. quelea* may face food scarcity during the period prior to rainy season. During this period fields are totally dry. Vegetation is sparce. Field trials in an area supporting diverse grass species and several cereal food grain crops in the Borno state of Nigeria indicate their preference for small-seeded wild grasses. They preferred 10 different wild grass species and five food grain crops in an experimental area in the Eastern Sahel. (Yousufu et al., 2004).

There are about 21 taxa of birds that are identified as pests of pearl millet, sorghum, wheat, rice, cowpeas, tomatoes, pepper, and guava (Ezealor and Giles, 2010). Birds, such as red-billed weaver (*Q. quelea*), golden sparrow (*Passer luteus*), and Ruppell's weaver (*Ploceus galbula*) are among commonly encountered pests in the West African Sahel and in other regions of the African continent (Ward, 1966, Williams, 1954; Magor and Ward, 1972; Manikowski, 1984). The US Agency for International Development (USAID) has published a list of several bird pests that attack the Sahelian food crops. Red-billed Quelea (*Q. quelea*) is prominent. Other bird species causing damage to cereal crops in West Africa include spur-winged goose (*Plectropterus gambensis*), knob-billed goose (*Sarkidiornis melanota*), village weaver (*Ploceus cucullatus*), black-headed weaver (*Ploceus melanocephalus*), red-headed *Quelea* (*Quelea erythrops*), and golden sparrow (*Passer luteus*) (Coffrey and Farmer, 2014; CABI, 2020). *Quelea* is among the most abundant bird pests encountered in Sahelian West Africa. Flocks are large with 60–70,000 birds per flock. About 30,000 nests could be found ha^{-1} (McWilliam and Cheke, 2004). In addition, reports suggest that climate change effects are affecting not just the crops, but it is altering the bird pest population and its activity. The damage caused by red-billed weavers are often attributable to large flocks of millions of birds feeding the grain crops swiftly. Ward (1964, 1971, 1972) has reported that several types of control measures have been adopted to regulate *Quelea*. Knowledge about their migratory routes is useful. It helps in avoiding food crops in those areas is a possibility (Ward, 1971, 1973). It seems that control methods should be effective in avoiding build up of *Quelea*. Pesticides and baits are other methods that has potential to reduce *Quelea* in large numbers. Each year, over 1000 million birds get sacrificed with a primary aim to reduce flock size. Avoiding crop's availability during dry season seems important (Ward, 1971, 1972, 1978; Ward et al., 1979). These control measures have been adopted for five to six decades. Bird scaring is usually labor intense and costly. However, in the recent years, helikites with bird scarer voices are becoming more common. Perhaps, they could be cheaper and efficient to regulate *Quelea* in the Sahel. It needs to be examined. Predatory birds too have been adopted to reduce bird pests. Maintaining optimum population of predatory birds is important to keep *Quelea* at threshold levels in the farms.

Usually, a large flock of bird pest *Quelea quelea* supports greater diversity of predatory birds too (McWilliam and Cheke, 2011; Zwarts et al., 2009).

Spraying pesticides such as Queltox (Fenthion) reduces *Quelea* population. This procedure of aerial sprays has been practiced for 40 years now (McWilliam and Cheke, 2004). Such chemical spraying affects nontarget birds too. Birds of prey, such as falcons, raptors, owls, and passerines are affected by the pesticide application (Zwarts et al., 2009). Mulie and Keith (1993) found that insecticides applied to control locust reduced grasshopper population by four times compared with unsprayed control. Further, it did affect several bird species found in the cropped fields. For example, application of fenitrothion reduced the three most common pest birds found in the Sahelian farms. Regarding raptors which are predatory birds, Keith and Bruggers (1998) have shown that application of common insecticides, such as malathion and chlorpyrifos affects raptor population, only mildly. Application of zinc phosphide too reduced raptors, only marginally. However, fenthion applied to control *Q. quelea* was detrimental to several nontarget species such as raptors. Secondary poisoning through dead *Quelea* birds has to be avoided. Aerial sprays may also affect organisms in the waterbodies of the region.

7.5.8 WIND AND SOLAR ENERGY GENERATION THROUGH AGRARIAN SKY

Wind and solar power sources are generally considered simultaneously as important renewable resources for power generation. They have greater relevance in areas not served well with natural resources for hydroelectric power or facility for nuclear energy. In the Sahelian agrarian regions too, we opt for solar/wind energy resources together as a method to improve power generation.

Solar power is a natural resource enjoyed profusely by the Sahelian region. It is derived via the agricultural sky. At present, there are several initiatives supported by various agencies that aim at harnessing the Sahelian sky better, by building solar energy parks in the Sahelian cropping zones. There are farms that support both crop production and solar panels that generate electricity for farm households. For example, a program named "Desert to Power." It aims at generating 10,000 MW of solar energy in the Sahelian West Africa. It is envisaged to provide electric power generated via solar panels to over 250 million people residing in the Sahelian nations.

Most of them are farmers who need in situ solar power generation facilities. It could be several small-scale generation units. But, the focus is to support agricultural farms and field activities (Africa50, 2017, 2020).

Policy makers believe that in the Sahelian Africa, like elsewhere in other continents, a combination of power generation resources is apt. For example, in the Sahel, a few nations are adopting a good mix of biomass-based energy sources, hydroelectric power, solar power, and wind power (Sterle and Brecha, 2018). There are hybrid solar-cum-biomass generation plants. Also, hybrid solar and wind power generation systems are installed (Zozen Boiler, 2020). Sterle and Brecha (2018) have opined that a good mix of renewable energy sources, such as wind, solar, and others could be opted. A mix of renewable energy source is feasible in most of the Sahelian nations.

Reports by International Energy Agency suggest that electrification in Sahelian zone is low at just 45% compared with other regions of Africa. There are innumerable farms that still suffer due to lack of electric power. Forecasts state that, in near future, Sahelian zone may encounter severe dearth of electric power. The reported supply of electric power to 45% in 2018 is still low compared with other parts of the world. Methods such as hybrid power generation using hydro, wind, and solar may still fall short of requirements (Sakyi, 2020). It is said that Sahelian nations, such as Senegal, Burkina Faso, and Niger are too dry to generate hydroelectric power. However, the agricultural sky is endowed with sumptuous supply of sunshine. Therefore, solar energy generation is a good proposition. Hybrid power generation utilizing solar and wind power seems a better idea. Forecasts suggest that about 60% of electricity requirements of Sahelian farms in countries, such as Senegal, Burkina Faso, and Niger could be covered. It could involve hydrowind and solar energy systems (Sakyi, 2020). Further, Leuven (2020) states that solar and wind power generation is increasing worldwide. These energy sources are becoming cheaper. This trend induces Sahelian nations to embark on spreading solar/wind power generation. This helps to keep climate targets in sight, but also poses certain challenges.

Wind power generation seems a good idea in the Sahel/Sahara region. Here, the vegetation is feeble. Crop production trends too are not intense. Precipitation levels are insufficient on many years. Simulations by wind power specialists has shown that it could help in increasing availability of power in the Sahel. Regarding solar power, Graham (2018) states that placement of solar power panels could increase power generation. We may note that high-altitude wind power generation plus solar power too could be useful in the Sahel/Sahara.

Senegal is a major producer of cereal food grains, peanuts, and cowpea. This is in addition to firewood in the western Sahel. It is supported by a mix of renewable resources, such as wind and solar energy. It utilizes both fuel wood and hydropower to a large extent. Recently, they have traced oil and gas. Yet, the nation has committed to producing wind and solar energy in the rural/agrarian regions. The parklands that support pearl millet, groundnut, and agroforestry also possess vast stretches of solar panels. At present, only half the population of urban zone and farmers are provided with electrification (Hedley, 2019). So, they have opted to utilize the agricultural sky efficiently. The aim is to generate solar and wind power. Solar photovoltaic installations are being developed in many locations in Senegal.

Bowlus (2019) states that wind has played havoc with agrarian regions. It induces dust storms. It spreads crop diseases, locusts, and other detriments. Climate change effect that creates erratic winds and rainfall pattern actually accentuates the problems. Yet, they say, wind in Sahel could be turned into an advantage by resorting to wind power generation. Senegal has also initiated wind power projects that utilize seashore wind. The offshore wind with higher speed aids wind power generation. The Atlantic wind, particularly, high-altitude ones could be a boon to farmers in Sahel.

Morocco in Western Sahel enjoys favorable conditions regarding wind and solar energy generation. Yet, in 2013, only 5.5% of total power was generated via wind power. Policy makers of Morocco believe that in near future, wind power could be a major contributor to energy budget in that area. The wind speed in 90% of the potential wind power generation belts is 5.3 ms^{-1}. In the north, wind is congenial at 8.3–11 ms^{-1}. Therefore, it is matter of time before the resources from agricultural sky and seashores are utilized more efficiently. This is to serve the farmers in the interior regions of Morocco. At present, Morocco has both private and government agencies involved in wind power generation. It seems they aim at increasing wind power contribution to 2000 MW by 2020 (Wikipedia, 2020a, b).

Morocco has agrarian regions that is endowed with highest insolation rates (3000 h. per year) both in agrarian regions and in the dry deserts. Therefore, the potential for solar (photovoltaic) power generation is high. They say, developing larger solar projects can improve solar power generation. About 37% of total power generation in Morocco could be from solar panels placed in rural areas (Wikipedia, 2020a, e).

Mauritania is endowed with wind and sunshine. It is sumptuous to generate power for farm sector plus rural households. At present, fossil fuel provides for large share of total installed capacity at 263 MW. While wind and solar energy plus other renewable energy sources add 117 MW annually. The intended expansion and intensification of farm production needs further addition to installed capacity of power generation. Therefore, it would be useful to enhance wind and solar energy generation. Solar/wind sources are clean and reduce GHG emissions (USAID, 2020). This way, agricultural sky over the farming regions could be utilized more efficiently than at present. There are possibilities for placing solar panels and cultivating crops simultaneously, within a farm, by shrewdly allocating land area.

Mali is a country with agrarian region that supports hardy crops, such as pearl millet, cowpea, and a few vegetables. Large patches of dry regions support agroforestry crops. The agrarian sky is good enough to be exploited for generating wind and solar energy. Recent initiatives suggest that Mali is developing solar parks able to generate 500 MW, for its farm belt. Similarly, Togo in West Africa has initiated solar energy parks in the northern portion of the nation (Afrik21, 2019). Clearly, in these semiarid Sahelian regions, potential to generate solar energy and supplement it with wind power is to be exploited better.

Burkina Faso with its primary emphasis on higher food grain generation is embarking on solar and wind energy generation. This region receives sunshine that needs to be utilized to support the production of pearl millet, groundnut, rice, cowpea, and other crops. Recently, solar panels that could generate 155 MW power were installed (Takoulen, 2019). At present, access to electricity is low. They wish to increase the power supply to most farming communities through wind and solar energy.

Let us now consider the potential of wind power generation in Niger. Niger is a Sahelian nation endowed with relatively meagre forest regions. Yet, we may note that 90% of energy utilized by farmers in this region is actually derived from fuel wood burning. It is not a situation feasible perpetually. Fuel wood burning results in GHG emissions. Biomass loss from the agrarian ecosystem is severe. Hence, several agencies are examining the potential of wind power generation. As stated earlier, wind power is cleaner. In Niger, so far, wind power generation trials have been conducted with wind turbines located at 80 m altitude from the ground surface. We have no idea about usefulness and efficiency of high-altitude wind power generation, using tethered aerostats and turbines placed at 2000 m altitude. High-altitude wind power generation is stable and wind speeds too are higher.

Garba and Ozer (2011) have evaluated the wind characteristics and their potential to generate electricity. They found that wind speed fluctuated based on height from say 10 m above ground till 80 m altitude. On yearly scales, wind speed ranged from 3.2 to 3.4 m s^{-1}. The wind speed changes were attributed to diurnal and seasonal fluctuation in ambient conditions. Electricity generated was not high. It needed improvement. Perhaps, search for high-altitude wind power is a better bet in agricultural sky above the cropping zones of Niger.

Although not West African Sahelian nations, there are other regions in the sub-Sahara that are actively considering harnessing the agricultural sky better via wind and solar energy. For example, Sudan in the east intends to develop better installed capacity of wind power. They wish to utilize it for farming (UNDP, 2020). Egypt is situated on the other end of Sub-Sahara and enjoys similar atmospheric conditions, if we consider wind speeds and sunshine hours. They say, the region has envisaged to develop world's largest solar energy park that extends to 37 km^2. It is composed of 7.0 million photovoltaic cells (panels). It is supposed to generate 1.5 GW power when complete. Such projects will in due course turn out to be most efficient utilizers of rural agrarian sky.

Low and high-altitude wind power generation in the Sahelian agricultural belt should be a good idea. High-altitude wind power generation using tethered aerostats needs to be examined. Agricultural communities and small townships supporting farming in the surrounding regions need to adopt it. Low-altitude wind power generation using the windy periods too is a possibility. Wind power and solar power could improve the usage of agricultural sky above Sahelian region. It is thought that a single high-altitude tethered aerostat attached with turbine can suffice to help a small village of 15–20 households. Multipurpose aerostats are attached with power generator, sensor to obtain clear imagery of terrain, its soil features, soil erosion, if any crops below. Tethered aerostats could also be fitted with bird scarers. Tethered aerostats are a great proposition to resource poor Sahelian farmers.

7.5.9 AERIAL VEHICLES IN SAHELIAN AGRARIAN REGIONS

Usage of agricultural drones by the farmers, particularly, to obtain digital data, farm/land imagery or to spray plant protection chemicals is rudimentary, if not nil. However, aerial surveillance of Sahelian vegetation and monitoring the changes has been attempted. Unmanned aerial vehicles (UAVs) have not been adopted in the Sahel to any extent. We can utilize the agricultural aerospace above the crop canopy efficiently by adopting UAVs. Several kinds of UAVs and their models are available. They need to be tested, evaluated, and adopted as quickly as possible to derive advantages. The UAVs such as autonomous small drone aircrafts (both fixed-winged and copters) are suitable for use above cropping expanses, all through the Sahel. Blimps can provide detailed images of natural vegetation, topography, crops, and water resources periodically as they float above the crops. Tethered aerostats with several attachments, such as set of cameras, bird scarers, and surveillance cameras too have not made entry into agrarian sky. In future, they could be adopted to monitor terrain, soil, crops and pests/disease on crops, water resources, droughts, floods, or soil erosion. Kites have only been examined as a possible alternative for resource poor farmers. Kite attached with multispectral sensors have provided data about cereal/groundnut growth, viral disease status, and foliage/biomass accumulation (Gerard, 2016). Aerostat networks spanning the Sahelian region from east to west should be helpful in monitoring locusts, spread of major disease, drought and its effects. Aerostat networks could forewarn the farmers using mobile phones. Such networks are in place to detect disease/pest on wheat grown in USA, Australia, etc. We have to replicate the same idea and procedures of using aerostats with ability for detailed collection of spectral data of crops.

Satellite has been used predominantly to collect digital data about the weather, its various parameters, and to develop appropriate forecasts. Otherwise, satellite techniques are utilized to monitor natural vegetation, water resources, and seasonal changes in cropping belts. The sparse natural vegetation, its botanical diversity, and biomass changes have been mapped using satellite imagery. Landsat and other series of satellites have been utilized to monitor the progress of agricultural weather. The crop planting trends, seedling establishment, soil crusts, sand drifts, loss of seedlings to high soil temperatures, soil erosion, and droughts have been monitored using satellite imagery. Locusts, other insect pests, and disease progress, if any, are monitored using satellites. Aerial imagery has been utilized by agricultural agencies to monitor crop growth, biomass accumulation trends, and grain formation and also to offer yield forecasts in different subregions of the Sahel. However, the use of farm vehicles that are GPS tagged seems still feeble.

7.6 AGRICULTURAL SKY AND ITS IMPACT ON INDO-GANGETIC PLAINS

The Indo-Gangetic Plains are regions that support intense farming of field crops. The prairie vegetation generates cereal food grains (wheat, rice, millets), pulses, and oilseeds. Cash crops, such as sugarcane and cotton are also predominant. This agrarian region generates food grains and other crops good enough for 500 million people living in the plains.

The Indo-Gangetic Plains are large expanse of alluvium. The moderately fertile agricultural expanse is formed due to deposition of silt by three major rivers, such as Indus in the western region, Ganges and Yamuna that flow to the east. Each river is fed by several tributaries emanating from snowy Himalayan Mountain ranges. The plains possess alluvial soils whose texture varies from sandy to clayey. Soil could be calcareous and alkaline. Soils are classified as light brown Mollisols or Vertic Inceptisols (see Krishna, 2003, 2015, IRRI, 2013b). The farming enterprises in the Indo-Gangetic plains are supplied with water derived through precipitation and melting of snow (Srivastava et al., 2014; Pal et al, 2009).

The Gangetic belt is grouped into four natural regions: *Bhabar, Terai, Bangar, and Khadar.*

Bhabar is a region that occurs adjacent to foothills of Himalaya (mountain range). It is a narrow strip of flat terrain of 6.0–15.5 km width. Bhabar area runs immediately below mountain slopes. The terrain is composed of boulder, pebbles, and gravel that are generated by the flat flowing rivers. Soils are highly porous. The region is endowed with large quantities of underground water and streams.

Terai region is situated south of Bhabar. The soils are created due to deposition of alluvium that is generated relatively recently. Silt depositions are recent, geologically. This region is moist due to water from snow melt and precipitation. It supports thick forest area and intense cropping. Wheat, rice, legumes, and oilseed *Brassica* are the major crops. The agricultural sky bestows a humid tropical climate that supports high biological diversity of flora and fauna.

Bangar is a region containing older alluvium. It forms a terrace in the flood plains of Ganga and Yamuna in the central and eastern region. It also has lateritic deposits.

Khadar is a lowland region traced below the Bangar in the entire length of central and eastern Gangetic Plains. The alluvium is fresh due to recent deposits of silt and silty clay by the major rivers and their tributaries.

The agrarian sky above the Indo-Gangetic Plains is as important as soil resources, if not more, in terms of crop productivity. The agrarian sky above the plains too may vary. The aerospace above crops needs to be characterized in greater detail. It must be demarcated and mapped, for the benefit of farmers. The terrain and cropping pattern of Indo-Gangetic Plains are perhaps influenced to a good extent by the manifestations of agricultural sky that needs to be deciphered. The Gangetic portion of the plains has been classified into eight agro-ecoregions and 14 sub-ecoregions (Abrol, 1998; see Biswas and Narayanaswamy, 1997; Krishna, 2003, 2015; Gajbhiye and Mandal, 2007: Sehgal, 1990).

The cropping systems in the Indo-Gangetic Plains have evolved based on several factors related to soils, atmosphere, biotic constraints (insect pests, diseases) human requirements, economics, etc. At present, rice–wheat plus a short duration pulse/fodder/vegetable is the dominant system. The productivity of rice–wheat system has improved perceptibly since 1980–1981 (Abrol, 1998; Krishna, 2003, 2015; IRRI, 2013b, Koshal, 2014). The Gangetic region supports diverse crop species, shrubs, and trees. The insect pests, aves, and microbial diversity detected in the agricultural sky is large (see Krishna, 2015; Pande et al., 2006). The Gangetic Plain is demarcated into Trans-Gangetic Plains, Upper Gangetic Plains, Middle-Gangetic Plains, and Lower Gangetic Plains. In all, the Gangetic Plains include 172 administrative districts of North India. The agricultural sky above each subregion needs greater attention by researchers and farmers. The drier regions may receive only 400–500 mm precipitation annually. Most of the regions receive precipitation between 800 mm and 1200 mm annually. High rainfall zones receiving 1600 mm occur in the east. Several traits, such as humidity, wind speed, and temperature too vary in the subregions.

The average productivity of wheat in the Gangetic Plains improved from 3000 kg ha^{-1} in 2000–2001 to 3600 kg ha^{-1} in 2012. It has further improved to 3900–4200 kg in 2020. Analysis of atmospheric constraints, soils, water, and cropping systems adopted reveals that farmers may have to negotiate climate change effects too. The biotic factors in the aerospace need greater attention while deciding the agronomic procedures. Farming systems need to adapt itself to changing demand of crop produce (Koshal, 2014).

The Indus agrarian region lies predominantly in Pakistan. No doubt, both the terrain and agricultural sky manifest variation. They need further careful study and evaluation. The Indus agrarian belt includes regions that are irrigated by the tributaries, such as Sutlej, Jhelum, and

Rabi. The tributaries of Indus are the major water resources for agrarian regions of Northwest India. The Indus region in Pakistan includes 10 agro-ecoregions (Chaudhry and Rasul, 2004; IRRI, 2013a, b; Malik et al., 2012). They are Indus Delta, Southern Irrigated Plain (Lower Sindh), Sandy desert (Thar, Nara and Dadu), Northern Irrigated Plains (Punjab including Jhelum and Sutlej), Barani Lands, Wet Mountains region, Northern dry mountains, Western Dry Mountains, Western Plateau, and Suleiman Piedmont. The Indus Plains support a large posse of field crops, vegetables, plantations, forest species, etc. (see NIAB, 2013). In Pakistan, major crops, namely, wheat, rice, sugarcane, millets chickpea, pigeon pea, and sunflower are cultivated in areas with assured rainfall/ irrigation. Brassicas and pulses too negotiate vagaries in terrain, soils, and agricultural sky. Guar occupies dry regions. The agricultural sky and its factors obviously have their share of influence on the crop species and cropping systems adopted in the Indus belt. No doubt, agricultural sky above each subregion of Indo-Gangetic Plains requires detailed analysis.

7.6.1 AGROCLIMATE OF INDO-GANGETIC PLAINS

Soils encountered in the Indus Plains are grouped into 38 great groups (112 subgroups). Mollisols, Inceptisols, and Alfisols are the most common soil types used, to cultivate the crops. Soils experience storms and floods that induce erosion. Wind caused erosion is also seen in the farming belts of Indus region (Ijaz, 2013). Precipitation pattern varies from as low as 400 mm annually in drier regions of Gangetic Plains to as high as 1600 mm in the eastern region of Gangetic Plains (West Bengal).

Heat waves are grouped into categories as defined: Moderate heat wave = five consecutive days with daily maximum temperature ≥ 35°C and < 40°C; mild heat wave = five consecutive days with daily maximum temperature ≥ 30°C and < 35°C (Rasul et al., 2014).

Climate change has been causing a drastic change in weather patterns both in summer and winter. It adversely affects the crop yields. Large variability has been observed in the precipitation and thermal regimes. Some recent examples in Pakistan and other parts of the plains are stated below.

Agricultural experts have diagnosed the causes of reduction in wheat yield in Ludhiana Province of India while the visible crop condition was the best. They pointed out that the occurrence of mild heat wave for 13 days having above normal (2°C–3°C) temperatures in early spring (at reproductive stage) had caused 28% reduction in the grain yield of wheat. In Sindh and southern Punjab of Pakistan, February in 2006 was 2°C–4°C warmer than normal. It resulted in significant yield reductions. Wheat was in the grain formation phase. High temperatures accelerated the development as the required heat units were met relatively quickly. The mean daily temperatures during the last week of February and first decade of March 2010 remained 3°C–6°C above normal. Early maturity did not allow the young wheat grains to grow to their normal weight, size, and starch contents. Due to higher night temperatures during 2003, the respiration overruled the photosynthesis causing reduction in net gain. Rice grain yield declined 10% for each 1°C increase in minimum temperature.

PLATE 7.4 Rice in Indo-Gangetic Plains.

Source: Conservation Agriculture Project, Food and Agricultural Organization of the United Nations, Rome, Italy, http://www.fao.org/conservation-agriculture/case-studies/indo-gangetic-plains/en/

Note: The farmers in the Indo-Gangetic Plains have adopted different farming tactics to improve rice productivity, during kharif season.

Farming regions of Indo-Gangetic Plains have been invaded with new crops, variety of implements, farm vehicles, and methods since Neolithic period. Farmers have adopted concepts that have offered impressive grain yield gains. However, some reports state that during past 30 years incessant farming has led to soil fatigue, loss of fertility, build up of pests, and diseases and climate change too have affected the crops (Food and Agricultural Organization of the United Nations, 2015). During past 30 years, since the adoption of rice–wheat copping system in the Northern Plains of India, grain yield has increased by 3.5% each year (Plate 7.4). However, grain yield stagnation too has been noticed. Adoption of conservation tillage, direct seeding, precision farming, and efficient N management has improved the condition in this region. *While soil, water, crop, and management methods have received due attention, we may notice that "agricultural sky" with high degree of variations and complexity of factors has not been studied to any great extent. Perhaps, increase of food production in the Indo-Gangetic Plains depends on managing agricultural sky in a better way. Matching cropping systems based on conditions prevalent in the agricultural sky also needs due consideration.*

There are clear evidences that agricultural sky and its manifestations, such as atmospheric temperature, solar radiation, diurnal periods, cloudy days, and precipitation pattern have immense influence on the biomass and grain productivity. Agricultural sky has impact on both the cereals, rice and wheat cultivated in the Indo-Gangetic Plains of India. They have shown that the late planting of wheat affects the crop through several factors (abiotic and biotic) that operate in the agricultural sky. A delay in planting of wheat after rice by 1 week reduces yield almost by a ton grains ha^{-1} (Agarwal and Mall, 2000; Agarwal et al., 1994; see Table 7.3).

According to Agarwal and Malla (2000), higher grain yield potential of rice–wheat system is noticed in the Western Indo-Gangetic Plains (Punjab, Haryana, Uttarakhand, and Uttar Pradesh) (17–18 t ha^{-1}) compared with Bihar (16. t ha^{-1}) and West Bengal (13.4 t ha^{-1}). This is attributable to higher solar radiation and temperature. Perhaps, a late planted wheat encounters detrimental factors operative in the ambient atmosphere. Also, crop's ability to negotiate ambient factors may be inadequate (see Table 7.3).

TABLE 7.3 Influence of Ambient Atmospheric Factors That Manifest Above Cereal Crops, on Grain Yield (t/ha).

Planting time	Rice–Wheat	Rice	Wheat
Optimum Planting time	16.70	9.88	6.82
Optimum Planting of Rice but Late planting of Wheat	15.85	9.88	5.97

Note: The reduction in grain yield of the rice–wheat system is attributable to yield reduction of wheat (not rice) due to late planting. Late planting means altered diurnal periods, precipitation, relative humidity, and even wind/biotic activity in the ambient air. Asynchrony between crops (its growth pattern) and natural factors occurs that reduces wheat grain yield. Several abiotic factors of agricultural sky, such as daily temperature, solar radiation, diurnal pattern, precipitation quantity and pattern, and wind get altered. Similarly, biotic factors, such as pathogens, insects, aves, and microbes could have altered effect on late sown wheat. Indo-Gangetic Plains of India generates 95% of the nation's bread wheat (*Triticum aestivum*). Also, about 4% of macaroni wheat is produced in this farming belt (Directorate of Wheat Development, 2015).

Source: Excerpted from Agarwal and Malla (2000); Also see Pande et al. (2006).

7.6.2 GREENHOUSE GAS EMISSIONS FROM AGRARIAN REGIONS OF INDO-GANGETIC PLAINS

Satellite-aided observations of the agricultural sky above the crops cultivated in the Indo-Gangetic Plains have consistently shown that GHG emissions are high. It is much higher than most other cropping belts of the world (Wang et al., 2020; Kuttippurath et al., 2020a, b). Indo-Gangetic Plains emit higher quantity of NH_3 gas during April–September/October. It is due to warm climate and rampant fertilizer N supplied during rice production. The NH_3 fluxes are high in the Indo-Gangetic Plains at 0.4 t km^{-2} $month^{-1}$. Ammonia emissions are greatest during rice phase of the rice–wheat cropping system. Ammonia gas can combine with sulfuric acid and/or nitric acid in the atmosphere to form NH_4^+ compounds. Sulfur content in the atmosphere tends to be low. Particulate fraction too is relatively low.

Farmers in the Indo-Gangetic Plains adopt intensive cropping practices. In a year, they cultivate two major cereals, or a cereal and legume or a cereal-legume and short duration vegetable/pasture grass, etc. They also add high quantities of fertilizer N to the cereals. This is in order to reap best grain yield possible in the given soil/environment. If fertilizer N inputs are higher, then they are prone to accumulation and loss into atmosphere via GHG. Fertilizer N is mostly lost as NH_3, NO_2, and sometimes as N_2 gas. GHG emissions from the agricultural zones in the Indo-Gangetic Plains are relatively higher. The GHG affects the atmospheric quality. It induces the formation of smog and dust storms, etc. Actually, application of fertilizer N at optimum rates is essential. Farmers should strike right balance between fertilizer N input and N uptake by the cereal crops. According to researchers at CIMMYT, Mexico, Doody (2019), and Sapkota (2018), the optimum fertilizer N supply for the two major cereals, that is, rice is 120–200 kg N ha^{-1}. For wheat that follows in rabi in the Indo-Gangetic Plains, it is 50–185 kg N ha^{-1}. These suggestions are based on yield goals. And, it simultaneously considers the possible amounts of N emissions as NH_3 from rice fields, and nitrous oxides from wheat. Obviously, nutrient inputs should match the crop's needs as exactly as possible. If not, we have to experience GHG emissions into agrarian sky at a greater intensity. Further, it is interesting to note that farmers have been adopting at least 10 different types of land preparation methods, ploughing intensities, and residue cycling. This is in order to reduce both emissions and fertilizer supply to cereals. Indeed, a wide choice is available for farmers. They have to adopt the best agronomic procedure to reduce GHG.

Further, Sapkota et al. (2015) have evaluated the usefulness of Conservation Agricultural (CA) practices in the Indo-Gangetic Plains. It supposedly enhances cereal production efficiency. At the same time, reduces detriments to agrarian sky, by reducing the GHG emissions. Their field evaluations indicate that cost of production reduces if we adopt conservation practices. The conservation agriculture (CA) methods improve irrigation productivity by 66–100% over conventional practices. The CA practices moderate temperature which caused ill effects on crops. It actually reduces canopy temperature by 1°C–4°C. It seems that heat and water stress effect get reduced if CA practices are adopted. Most importantly, CA reduces GHG emissions by 15% if we compare it with conventional farming procedures.

In the Eastern Gangetic Plains, rice phase of the rice–wheat cropping system emanates larger quantities of GHG into the agricultural sky. They report that soil factors and fertilizer inputs govern the extent of GHG emissions. Agronomic procedures, such as tillage, puddling, and residue retention have an important role in regulating the GHG emissions from rice fields. Low puddling and high retention of residue is best in terms of GHG emissions. This procedure induces lowest amounts of GHG. Low crop residue retention with moderate or high intensity puddling induces high quantities of GHG emission (Khairul Alam et al., 2016). In the Northwestern Indo-Gangetic Plains, emissions (CO_2, NO_2, and CH_4) are increasing. The GHG is increasing during the rice phase of rice–maize rotation. Kumar et al. (2018) state that as a result, rice/maize productivity has reduced by 40% between 1998 and 2011.

No doubt, in the Indo-Gangetic Plains, rice is among the major emitters of GHG. Hence, agronomic procedures that reduce GHG emissions are always welcomed by farmers. Tillage is supposedly an important procedure that has influence on the emissions from soil/crop. Let us consider an example. Pandey et al. (2012) have studied GHG emissions from rice that is grown under conventional and no-tillage systems. They found that no-tillage reduces emissions of CH_4, CO_2, and N_2O compared with conventional tillage. No-tillage does reduce rice yield but that gets paid-off considering the advantages bestowed in terms GHG emissions. And, no-tillage involves less expenditure to farmers. They say, no-tillage prior to rice sowing and reduced tillage prior to wheat clearly reduce GHG. Hence, agricultural sky above rice–wheat system is conserved to its pristine clean conditions. The GHG levels may not enhance to detriment the crop.

GHG emissions spoil the agricultural sky. They can induce global warming. Ambient temperature too increases. The GHG also pollutes the atmosphere above crops. As a consequence, productivity reduces. Now, there are certain agronomic procedures that reduce the emissions of GHG, such as N_2, CH_4, etc. There are urease and nitrification inhibitors that reduce emissions. Several natural and artificially synthesized chemicals have been evaluated in the Indo-Gangetic Plains. A few examples are neem cake, thiosulphate-coated calcium carbide, neem oil-coated urea, dicyandiamide (DCD), and urease inhibitors like hydroquinone. They have reduced the emissions. Reports suggest that 5–31% of GHG emissions could be reduced using urease/nitrification inhibitors during rice/wheat production (Malla et al., 2004). There is need to adopt application of nitrification inhibitors so that GHG emissions into agricultural sky is reduced.

There are several computer-based models and simulations that help researchers/farmers to decide exact quantities of inputs so that GHG emissions are low or nil. For example, InfoRCT (Information on the use of resource conserving technologies) is computer-based simulation program. It considers biophysical, agronomic, and socioeconomic aspects of farming in the Indo-Gangetic Plains. It prescribes inputs accurately avoiding excessive use of fertilizers and other inputs. It also provides a comparative analysis of various alternatives available to famers and extent of GHG that results due each set of procedures adopted. It highlights the global warming potential of various sets of procedures, so that farmers can select the best one. Such simulations could help us in reducing GHG in the Indo-Gangetic Plains (Pathak et al., 2011).

Since past decade, rice farmers in India have tried to adopt "System of Rice Intensification (SRI)." There is also "Modified System of Rice Intensification (MSRI)." These systems may involve higher inputs of fertilizer N. Proportionate GHG emissions are to be expected. For example, they report that under MSRI/SRI, N_2O emission increased by 22.5% over traditional systems. Whereas CH_4 and other GHG emissions were reduced by 61%. Under MSRI/SRI, the global warming potential reduced (Jain et al., 2014). It seems fertilizer N management is an important aspect of intensification of rice production in the plains.

7.6.3 DUST STORMS, HEAT WAVES AND DROUGHT IN THE INDO-GANGETIC PLAINS

Dust bowls are important abiotic phenomenon related to the agrarian sky. It can be a havoc above the large stretches of cropped land in the Indo-Gangetic Plains. Dust bowls, heat waves, and strong winds that blow during summer have reduced the natural vegetation in many locations within the Indo-Gangetic Plains. As such, the semiarid climate, hot summers, droughts, and scanty precipitation have supported only dry deciduous forests, hardy agro-forestry species, shrubs, and grasses. This type of vegetation does trap dust particles. Such vegetation reduced the movement of dust/haze all across the cropped plains. The dust storms that occur are different from the "*aandhi.*" It differs from regular heavy precipitation events that occur in the plains. During recent years, the dust bowl situation and haze has become a common event during summer season. The agricultural sky above the Indo-Gangetic Plains has definitely changed. Particularly, regarding its climatic parameters, dust and toxic haze (Jain, 2019; Sarkar et al., 2019).

Weather experts state that dust above cropped land is not considered toxic if it is natural and not contaminated with toxic carbonaceous particles. Crop burning, industrial activity, wind-blown soil particles, and transport vehicles generate the smog. Mineral elements traced in the dust/haze/smog above Indo-Gangetic Plains are magnesium, zinc, lead, iron, manganese, and copper. Organic fraction (carbon particles) could be a dominant fraction in the dust/haze depending on the season. Crop burning is an important activity after the harvest of cereals/sugarcane. This activity of farmers generates large quantities of carbonaceous particles that localize over the plains at low altitudes. The dust can obstruct the photosynthetic activity of the next crop too if it lasts for longer period.

The quality of air above the crop's canopy deteriorates as dust particles collect at higher density. The size of the particle also plays a role in clogging the air and reducing crop's stomatal activity. Even human respiratory system gets affected during recent years. Larger particles of > 10 μg have shown an increase in the agricultural sky over Western Indo-Gangetic Plain. Sudden spikes in quantity of dust particles are seen just prior to monsoon. Wind speeds too control the dust movement.

Weather stations in the Gangetic Plains have consistently recorded dusty summer, since past three decades. Hanging dust particles cause reduction in photosynthetic activity. They reduce crop productivity in the entire Indo-Gangetic Plains, that is, from Sind/Punjab in Pakistan till Eastern Bihar/West Bengal in India. They have recorded 500 μg/m dust in summer against a permissible 100 μg/mm during dusty period. It seems crustal dust accounts for 40% of particulate matter traced over the Indo-Gangetic Plains.

Overall, most of the recent reports indicate that agricultural sky above the Indo-Gangetic Plains has been highly detrimental during certain months/ seasons. The intensity of dust particles and size of the particles too have increased. Plus, its dust is getting contaminated with toxic emissions and GHG emissions due to human activity. Agrarian sky in this area needs attention by the policy makers, agricultural experts, weather experts, and farmers, particularly if we wish to reduce the crops loss due to environmental deterioration.

In addition to dust bowls, contamination of atmosphere occurs due to crop burning. Indo-Gangetic Plains are also experiencing enhanced ambient temperature. This could be attributed to GHG emissions. They say, during the past century, ambient temperature recorded above the crop canopy has increased uniformly by 1°C. *The cocktail mix of dust, smog/toxicity, and higher temperature could be devastating on crops and natural flora if droughts too occur frequently.* Eventually, the large patches of agricultural terrain in the Indo-Gangetic Plains may actually experience desertification and loss of vegetation. Generally, dust storms, crop burning events, and haze that cover large areas of the Indo-Gangetic Plains are monitored using satellite imagery. The dust/haze that occurs over Pakistan and India are easily monitored via satellites that carry spectroradiometers of moderate to high resolution. LANDSAT eight images are frequently consulted to forecast dust, haze, clouds, and precipitation patterns in the plains (Teotia et al., 1980; see Plate 7.5).

Dust storms that occur in the fertile Indo-Gangetic Plains during March–May/June induce deterioration of air quality. The dust from Arabian Peninsula reaches most parts of Western Plains covering areas in Pakistan, Northwest India and extending to Eastern Plains (see Plate 7.6). Hence, to monitor the air quality, farming agencies have installed air quality monitoring and early warning system. It helps to inform farmers of imminent dust storms that get initiated at small or long-range locations.

Bioaerosols are generated in large quantities within the Indo-Gangetic Plains. Bioaerosol is generated when crop residues are burnt. It is an integral part of carbon cycle in the plains. In addition, forest fires that occur periodically during summer months also induce formation of bioaerosols (Rajput et al., 2018). The carbonaceous particulate matter varies in size and distribution in the sky above crop belts. Intense burning induces thick emissions. It reduces interception of photosynthetic radiation, particularly if the crop is in vegetative phase.

Droughts due to paucity of rainfall are common in some parts of Indo-Gangetic Plains. There are at least four reasons for droughts in the Indo-Gangetic Plains. They are (a) delay in the onset of monsoons, (b) variability in monsoon rainfall, (c) long breaks in the precipitation pattern during mid-season, and (d) spatial variations in the precipitation pattern. Droughts often occur in combination with heat waves during summer months. The uncongenial ambient temperature causes greater damage to the kharif season crop. There are reports dealing with spatiotemporal changes in excessive heat and soil moisture within the Indo-Gangetic Plains. The Upper Middle Indo-Gangetic Plains are mostly prone to droughts. Frequency of droughts is relatively high in 40–50% of the area (Nath et al., 2017). Yet, we may realize that it is this area of Upper Middle Indo-Gangetic Plains that contributes 18–20% of cereal grains produced within in the plains. Clearly, selection of well-adapted cereal genotypes and excellent drought management methods are essential. The role of agricultural sky in causing droughts needs attention.

PLATE 7.5 Dust and haze over the agrarian regions of Indus Plains and Delta covering regions in Pakistan and India.

Note: Dust and haze are phenomena that occur in the agricultural sky. Such dusts could reduce interception of photosynthetic radiation by important cereal and vegetable crops. They also create smog and conditions if dust storms, or haze appears in conjunction with moisture and pollutants. Such situations of smog/fog are becoming common above the farming regions of Indo-Gangetic Plains. Some reports attribute dust, smog, and pollution to periodic crop burning tactics that farmers adopt. Here, agricultural sky could be a source of detriment to optimum crop production/grain yield. We have to reduce GHG emission and crop burning in the Indo-Gangetic Plains.

Source: Allen, J. Aqua MODIS Rapid response team, Earth Observatory, Goddard Space Observatory, National Aeronautics and Space Agency (NASA), USA. https://earthobservatory.nasa.gov/images/16826/dust-storm-in-the-indus-valley

Agricultural sky in the Indo-Gangetic Plains also induces soil deterioration if the wind speeds are high. Procedures, such as cover crops, timely ploughing, and contouring are essential. Wind-aided erosion seems to be gaining in prominence as the topsoil gets blown out of the cropped fields. The wind erosion could be rampant during summer fallows if the fields are kept unplanted with forage crops or pastures. Overall, we should note that agrarian sky can play a havoc with crops cultivated in the Indo-Gangetic Plains. We should not neglect the sky which happens to be an important portion of any agroecosystem.

PLATE 7.6 Hanging fog and haze over central and eastern regions of Indo-Gangetic plains.
Source: Schmaltz, J. National Aeronautics and Space Agency, Visible Earth, a Catalogue of Satellite Images. https://www.visibleearth.nasa.gov/images/87273/haze-over-the-indo-gangetic-plain/87273t

7.6.4 CROP BURNING, ATMOSPHERIC HAZE, AND POLLUTION IN INDO-GANGETIC PLAINS

There are indeed too many research reports, journal articles, and several treatises that help us in understanding the atmospheric phenomenon above the Indo-Gangetic Plains. Here, only a few examples relevant to the context have been quoted and discussed. Ravindran et al. (2019) have reported that pollutants increase coinciding with crop burning. In all, about 16 different pollutants were easily traceable in the agricultural sky above the crop land in the Indo-Gangetic Plains. Several authors have reviewed the trends in NH_3 emissions within the Indo-Gangetic Plains. They have also used detailed observations aimed at understanding the cropping systems and its relation to GHG emissions. They have concluded that intensive agricultural practices involving enhanced application of fertilizer N is a major cause of NH_3 emissions. It is said that Indo-Gangetic Plains with their large expanses of rice–wheat cropping system is an important hotspot for NH_3 emissions. However, it has been found that there is a decreasing trend in NH_3 emissions, during the past 3–4 years. Yet, there are spikes in NH_3 emissions during kharif. During this period, farmers in the plains tend to supply higher levels of fertilizer N. At this juncture, we may note that atmospheric NH_3 reacts with SO_2 and other NOx to form aerosols that are toxic to crops/humans.

As stated earlier, farmers in the Indo-Gangetic Plains burn crop residues after harvesting crops. The crop burning trends peak around October/November (particulate matter 207 ± 87 μg m^{-3}) immediately after the harvest of rice. There is a second peak of emissions of carbonaceous material. This is due to crop burning during April and May (particulate matter 111 μg m^{-3}), that is, after harvest of wheat. Such emissions that occur during summer are difficult to handle. This is because heat waves, dust storms, and carbon emissions all of them get accentuated. The crop burning procedures followed by farmers in the Indo-Gangetic Plains emit reactive nitrogen (NH_3, N_2O, NO_2) and VOC into atmosphere (Bray et al., 2019). Further, they have reported that rice burning emits 416 kg NH_3 day^{-1}, 231,621 kg NOx day, and 48,762 kg N_2O day^{-1}. Wheat burning emits 71,940 kg NH_3 day^{-1}, 94,750 kg NOx day^{-1}, and 40,950 kg N_2O day^{-1}. Such GHG emissions could have detrimental effects on crop productivity, during following seasons. Satellite-aided observations regarding crop residue burning in the Indo-Gangetic Plains too suggest that emissions could be rampant immediately after rice/wheat harvest (Badarinath et al., 2006).

Ramanathan and Ramana (2005) have examined the relevance of atmospheric brown (dust) clouds on the crops generated in the foothills of Himalayas and Indo-Gangetic Plains. They say, aerosols of extensive proportions can be traced in the plains. Dusty sky and aerosols are detrimental to crops. A dusty sky may also reduce the photosynthetic efficiency.

Analysis of climatic trends in the Indo-Gangetic Plains indicate that fogs occur periodically between December and February, each year. The winter temperature ranging from 5°C to 20°C induces fog formation. It results in emissions due to fertilizer N application and crop burning. During winter, such fog causes smog and pollution of the atmosphere. Atmospheric pollution has become a common factor in the Indo-Gangetic Plains since 1970 (Kutty et al., 2019; Saraf et al., 2010).

Fog that occurs on Indo-Gangetic Plains during winter months has been mapped using satellite imagery, year after year. For example, during 2000–2010, the area covered by fog extended on an average into 578,820 km^2. Atmospheric factors, such as relative humidity, wind speed, migration of fog above the plains have also been repeatedly mapped. Air temperature of 12°C–13°C seems to induce fog/smog, rampantly. The toxic smog firstly reduces the photosynthetic activity of major cereal – wheat. It also induces pollution effects such as toxic deposits on foliage.

Haze above the Indo-Gangetic Plains gets created due to a phenomenon known as "temperature inversion.". Sources for such haze are urban activity, such as transport, industrial emissions, large-scale crop burning, if any, and firewood. The air temperature at different layers of atmosphere becomes congenial for formation and dispersal of haze/pollutant. Such a phenomenon has its share of detrimental influence on crop productivity.

Understanding the long-term trends of O_3 contamination over the croplands of Indo-Gantetic Plains is also important (Lal et al., 2012). Ozone above the cropping belts of Indo-Gangetic Plains is considered an important form of atmospheric pollution. Ozone is directly toxic to crops. Ozone is a phenomenon emanating in the agricultural sky. It induces substantial loss of crop's productivity. It is supposedly a major concern in the Indo-Gangetic Plains during recent years. Ozone concentrations display considerable variation if we considered the sky above the large patches of rice–wheat cropping system of the Indo-Gangetic Plains. Agricultural researchers envisage regular monitoring of precursors, such as NOx, CO, and volatile organic compounds (VOC) that induce formation of ozone. Later, they develop methods that reduce the formation of NOx and O_3 (Singh and Agarwal, 2016;). Reports state that O_3 generated in the Indo-Gangetic Plains is actually drifting into agricultural sky above the crop belts of neighboring countries. "Bad Ozone" is said to affect crops and humans detrimentally (Ghosh, 2019).

There are indeed several research reports about the ozone formation above the crops cultivated in the Indo-Gangetic Plains. Further, they have studied the biochemical and physiological effects of ozone that lead to low yield. Several of the rice/wheat genotypes utilized by the farmers in the Indo-Gangetic Plains are susceptible. In fact, Indo-Gangetic Plains is considered as one of the "hot spots" where ozone pollution is accentuated. Researchers at the University of Eastern Finland opine that developing ozone tolerant rice/wheat varieties would be a good idea. In fact, among the rice varieties cultivated in the plains, only seven of them showed any semblance of ozone tolerance (University of Finland, 2015).

Researchers at the Indian Institute of Tropical Meteorology, Pune, believe that over 12% of the crop productivity is lost due to ozone effects on cereals in the Indo-Gangetic Plains. Otherwise, grain harvest in this region is good enough to feed the population in situ in the plains. Agronomic procedures that reduce GHG and precursors (NOx) of O_3 are suggested (Priyadarsini, 2014). Assessment of intensity of air pollution and components of pollutants encountered by the crops is essential. The duration for which the pollutants act on crop's physiology is also important. These aspects have to be studied periodically. The causes must be deciphered as accurately as possible. Agricultural sky here can be devastating if neglected even for a while, say a couple of years. In this regard, let us consider a report by Burney and Ramanathan (2010). They state that rice–wheat cropping system that extends into vast regions within the Indo-Gangetic Plains has undergone a lot of changes in climate and pollution aspects since 1960s till data. Agricultural sky has been detrimental to both the staple cereals. Short-lived climate pollutants (SLCP) in the air have caused greater detriment to air quality. Due to it, wheat crop yield has reduced by 36% and rice by 10% if we compared the possible yield levels without SLCPs. The SLCPs, tropospheric ozone, and black carbonaceous soot are the major pollutants. They reduced the grain productivity in the Indo-Gangetic Plains (Burney and Ramanathan, 2014).

One of the important concerns about air pollution in the Indo-Gangetic Plains is the rampant variation in the concentration and activity of various pollutants. Particularly, those generated and distributed in the aerospace above crops. According to Nazari (2019), pollutants, such as black carbon (BC), brown carbon (BrC), Ozone (O_3), nitrogen oxides (NO, NO_2 and N_2O), carbon monoxide, and ultraviolet particulate matter vary depending on the location within the plains. The influence of each pollutant and their specific combinations is actually influenced by the atmospheric parameters, such as relative humidity, temperature, wind speed, and its direction. Hence, monitoring pollutants and their influence on crops seems almost mandatory in the plains. In fact, periodic evaluation of atmospheric pollutants is being conducted in the Indo-Gangetic Plains by several agencies (see Gagoi et al., 2020).

7.6.5 MICROBES IN THE AGRICULTURAL SKY

Airborne particles that are biological in origin, such as bacteria, fungal spores, pollens, mites, dead tissues are called "bioaerosols" (Kakde, 2018). Bioaerosols can be a source of several plant diseases. Of course, such aerosol may also cause disease on farm animals and farmers. Bioaerosols encountered in the crop belts generally possess spores of fungi belonging to Zygomycetes, Deuteromycetes, several species of bacteria, viral particles. etc. Here, our focus is on bioaerosols and their role in crop production in the Indo-Gangetic Plains. Most commonly traced fungi that could be of detriment to crops are *Alternaria, Aspergillus, Cladosporium, Erysiphe, Puccinia, Penicillium, Mucor, Rhizopus, Helminthosporium*, etc. One of the earliest and conspicuous effort was made with wheat rust diseases in the plains of Ganges and Indus by Mehta (1952). Airborne rust fungi that affect cereals (e.g., wheat) were detected. Their dissemination pattern in the plains was deciphered (see Bhardwaj, 2017; Hodson, 2018).

The atmosphere above the crop's canopy harbors several genera of fungi. Many of them are regular pathogens on plant species traced in the natural vegetation and on food crops. In fact, initiation, spread, and development of fungal disease on food crops is linked with the surface air above crop's canopy. Intermixing of bioaerosols induces several different types of fungal infections. In the Indo-Gangetic Plains, aeromicrobiologists have traced several species of fungi that are potential plant pathogens. Genera, such as *Phytophthora, Puccinia, Pyricularia* are well known for their ability to cause disease on crop plants. They are traced most frequently in the bioaerosols above the Indo-Gangetic Plains. If *Pyricularia* affects rice, *Phytophthora* species are known to affect potatoes and other vegetables. *Puccinia* is a well-known rust fungus that is a deadly pandemic organism affecting wheat crop, all over the world, including Indo-Gangetic Plains. It is interesting to note that in one crop season, air samples from over 4000 locations within the Indo-Gangetic Plains showed occurrence of plant pathogenic fungi. About 69 different fungi were detected that could attack crops and cause disease. About 39 different crop species cultivated the Indo-Gangetic Plains were affected by the fungal spores traced in the bioaerosols in the agricultural sky above the crop's canopy. There could be several more fungi traced in the bioaerosols over the cropland in the Indo-Gangetic Plains. They may all be commensalistic in nature. They may ride over the particles from one location to another and thrive on organic particles, soil, or other organic debris on ground surface (see Yadav, 2020).

Reports suggest that in addition to the pathogenic fungi, innumerable bacterial and viral propagules are traced in the agricultural sky. In the Indo-Gangetic Plains, they may migrate in the air currents.

At the same time, we should note that beneficial microbes, such as biological nitrogen fixers (*Azotobacter* species, *Azospirillum*, *Bradyrhizobia*) and mycorrhizal fungi both ecto- and endomycorrhizal species could be traced in the bioaerosols. Mycorrhizal fungi may migrate long distance within the plains to cause new symbiotic associations. The dust storms, wind, and rain may aid transport and dispersal of symbiotic fungi traced on the surface of soil. Perhaps, this aspect of aeromicrobiology of Indo-Gangetic Plains requires greater attention than provided, at present.

7.6.5.1 AIRBORNE PATHOGENS OF CROPS CULTIVATED IN THE INDO-GANGETIC PLAINS:

The intention here is to explain how important the agrarian sky is with reference to crop diseases and their manifestation. Wheat rusts are being utilized in the discussion to substantiate the fact that it is the agrarian sky that needs greater attention. Particularly, if we want to manage many of the dreaded pandemics of crops.

Meyer et al. (2017) have studied the infectious diseases of staple cereal crops. They have clearly shown that airborne spores of fungi are among the most dreaded detriments to crops. They take full advantage of the light-weighted spores that levitate, float, and transit long distances in the agrarian sky, on sea surfaces, and general terrain. They cause pandemics and reduce grain yield. For example, Meyer et al. (2017) have proved that wheat rust fungi, mainly the brown and yellow rusts of wheat are spread through the continental wind currents. The weather parameters, such as the temperature, relative humidity, sunshine and atmospheric pollution, if any, may all be congenial for these obligate pathogenic fungi to spread faster. Meyer et al. (2017) have traced several routes thorough which the wheat rust fungi affect the crops cultivated. Knowledge about rust fungi of wheat and even other cereals, such as maize (*P. recondita*), barley, sorghum (Joshi and Palmer, 1970; Singh et al., 2000; Hodson, 2018), is essential. *It is the agrarian sky that plays detrimental role.* Agrarian sky aids the havocs that these aggressive pandemic fungi are known to cause. Since the spores of these obligates pathogenic fungi are known to mutate to more virulent versions, crop's resistance may break down causing damage. The ambient atmosphere and the UV radiation may induce the mutation to greater virulence (Le Roux and Rijkenberg, 1987). Again, it is the sky that needs to be monitored carefully. Remedial measure, if possible, should be adopted at the earliest. In the Indo-Gangetic Plains, repeated infections are often induced via atmosphere (wind) from southern Indian hills, Nepalese cropping zones, etc. The source of inoculum is collateral host (grasses, millets etc.).

Indo-Gangetic Plains support intensive farming of crops. This necessitates the careful management of disease/pests. Also, other constraints may thwart higher productivity. The agricultural sky above the crop canopy is among the most important aspects. It can induce diseases/pests. Farmers have been trained to assess the soils, manage water and other inputs efficiently. Factors that operate in the agrarian sky, however, seems neglected. Therefore, it needs greater attention in future. Several constraints to higher crop yield actually operate through the agricultural sky. Perhaps, sky above crops is detrimental to a greater extent than other components of farming, like soils and water. Abiotic and biotic factors of the agrarian sky, both could be detrimental to wheat yield. Here, we are concerned with diseases of crops cultivated in the Indo-Gangetic Plains that get disseminated through aerial factors.

Wheat has been cultivated in the Indo-Gangetic Plains for several centuries now, year after year. This has induced several airborne fungal/bacterial diseases. Assured presence of host and collateral hosts has helped the pathogenic microbes (fungi, bacteria/viruses). Pathogenic microbes have carved out a niche in the Indo-Gangetic Plains. The aerial parameters, such as the atmospheric conditions, wind, insect vectors, and collateral hosts have all aided the build up of disease. The agricultural sky anywhere in the plains is congenial to the development of a posse of fungal diseases. Airborne spores of pathogens are also known to disseminate to short/long distances from the plains. Fungal spores of pathogens generated in the plains (and in southern hills) are known to traverse long distance. Then, induce rust disease in different continents (Europe, West Asia,). The uredospores of *Puccinia graminis tritici Pgt* are known to traverse long distance through wind, storms, and dust (Brown and Hovmeller, 2002; see Nagarajan and Joshi, 1985; Nagarajan and Singh, 1990).

Major diseases of wheat encountered in the Indo-Gangetic Plains during recent years are stripe rust (*P. recondita*), leaf rust (*P. triticina*), stem rust (*P. graminis tritici*), karnal bunt (*Tilletia indica*), loose smut (*Ustilago tritici*), head scab or head blight (*Fusarium graminearum*), leaf blight (*Alternaria triticina*), and powdery mildew (*Blumeria graminis*) (Anikster et al., 1997; Keller et al., 2014; Joshi and Palmer, 1970; Saari and Prescott, 1985; Vikaspedia, 2020). The aerial space (i.e., troposphere) above the wheat canopy aids in the dissemination of several fungal diseases through wind, insects, aves, farm animals, and vehicles. The dynamics of fungal spores and routes they adopt to reach new wheat cropping belts have been delineated. These fungal pathogens are prone to mutate to virulence, periodically, and reach epidemic proportions. Therefore, we should monitor the disease foci and fungal spore density in the atmosphere above crops.

The tropical sky, which is humid, warm, and breezy can transmit several disease-causing organisms that affects rice crop cultivated in the Indo-Gangetic Plains. Agricultural sky and its component factors, such as wind, aerial insect vectors, storms. could induce large-scale spread of rice diseases in the plains. Major rice diseases are bacterial blight (*Xanthomonas oryzae*), rice blast (*Pyricularia oryzae*), sheath blight (*Rhizoctonia solani*), brown spot (*Helminthosporium oryzae*), false smut (*Ustilaginoides viresense*), and rice tungro disease (Rice tungro virus-RTV) (Sehgal, 1990; Sehgal et al., 1988, 2001). Aerial sprays to reduce pest and vectors are essential.

In addition to rice–wheat, Indo-Gangetic Plains also support rice–legume or wheat–legume rotations. Legumes are also intercropped with dryland cereals. Lentils are gaining in area in the Indo-Gangetic Plains (Tikoo et al., 2005; Singh and Singh, 2014). Here, we should note that legumes, such as pigeon pea, lentil, and cowpea are susceptible to a set of fungal pathogens that cause powdery mildews. Ambient atmosphere plays a vital role in the dissemination and establishment of powdery mildew. The powdery mildew can be a severe detriment to biomass and grain formation of lentil (or even other field legumes). A few of the diseases such as powdery mildew are worldwide in distribution. They may spread rapidly through the wind currents. It is not just the spread of the spores of *Erysiphe* sp. but even the establishment of infection on the leaf (entire canopy) is regulated, aided, and enhanced by appropriate conditions in the agricultural sky. The spores are produced and released when the temperature is congenial. Germination occurs on leaf tissue when the relative humidity is > 50% and temperature is relatively higher than normal in the ambient atmosphere (Singh et al., 2013). Recent reports from the Indo-Gangetic Plains suggest that lentils and other legumes grown are getting affected by powdery mildew. Lentils, for example, suffer total loss of foliage and grain yield becomes insignificant (Pande et al., 2008; Saxena and Khare, 1998). The control of powdery mildews that are airborne infections on legume in the Indo-Gangetic Plains are also dependent on aerial methods. Fungicidal spray at appropriate growth stages should be done repeatedly at intervals. It halts the establishment of disease foci on the leaves. Sulfur-containing fungicides such as benomyl are preferably sprayed through the atmosphere. Aerial methods that involve trapping of spores in the atmosphere, periodically, immediately after establishment of seedlings seems a good idea. Agricultural drones fitted with multispectral sensors are able to supply spectral data and will be useful. The spectral data have to be compared with known patterns of the powdery mildew affected and healthy (no disease) check plots. Quantification of powdery mildew attack in different areas of the Indo-Gangetic Plains and selection of lentil genotypes tolerant to powdery mildew disease should also be possible. We need to develop proper data banks of spectral imagery of disease and check plots.

Researchers at Cambridge, United Kingdom and CIMMYT, Mexico have mapped out the entire aerial routes taken by the different wheat rust fungi (Bhattacharya, 2017; Hodson, 2018; Meyer et al, 2017), particularly the uredospore dissemination across different continents. It is interesting to note that virulent strains of *Pgt* (Ug99) that is affecting wheat crop in Iran may not affect the crop that thrives in the Indo-Gangetic Plains (India and Pakistan regions). This is attributed to the routes that uredospore's take during long distance dispersal. There is at present chance of spores from wheat crop grown in Yemen to reach the Indo-Gangetic Plains, but feebly to Nepal. Uredospores from Nepal getting spread to Indian wheat belt too are feeble. It means, firstly, agricultural sky is the source of rust disease propagules. Further, if we dwell into details, then it is the wind currents and general pattern of worldwide uredospore dissemination routes adopted in the agricultural sky that is important. Here again, we can classify the agricultural sky as "prone to wheat rust" of a definite kind emanating from a specific location in the world. For example, agricultural sky above wheat belt in the Indo-Gangetic region is free from strains that affect the crop in Kenya or Iran. But the same region is prone to wheat rust strains from Southern Hills of India.

There are several diseases that affect sugarcane, oilseeds, and vegetables grown in rotations/intercrops in the Indo-Gangetic Plains. They too are susceptible to many fungal and viral diseases. Their dispersal is mediated through the agricultural sky. There are rusts, powdery mildews, downy mildews, blotches, smuts whose spores get transmitted through the wind currents. Viral diseases that need aerial vectors also cause damage to these crops produced in this region. Aphids and flies are among prominent aerial vectors that transmit viral disease of these crops.

7.6.6 INSECTS IN THE AGRICULTURAL SKY ABOVE INDO-GANGETIC PLAINS

Geographically, Gangetic Plains can be demarcated into Trans-Gangetic Plains, Upper Gangetic Plains, Middle-Gangetic Plains. Aerial insect pests do vary based on the subregion and cropping systems adopted in the plains. The insect population, diversity, and devastation caused may be affected by the agricultural sky and its manifestations. We need to understand this aspect in greater detail. The Indo-Gangetic Plains have been utilized to raise several different food crops. The cropping system followed has been dynamic. It is based on natural resources, food habits, and farmer's economic aspirations. At present, major cropping systems are rice–wheat, maize–wheat, rice–mustard–rice, cotton–wheat, pearl millet–wheat, etc. Pests traced in the plains include aphids, stem fly, shoot fly, pod borer, army worms, boll worms, thrips, etc. (Koshal, 2019). Farmers try to avoid the crops that may get affected by insect pests. Aerial sprays of pesticides too are applied based on the pests traced above the canopy.

Wheat is the major cereal crop cultivated in the Indo-Gangetic Plains since ages. During past five decades, beginning in the 1970s, the rice–wheat cropping system has gained acceptance. The cropping intensity too increased resulting in certain problems, such as insect pests. Aerial pests, such as aphids, pink borer, and fall army worms are common on wheat. Aphids that suck the sap are really dreaded aerial pests of wheat. There are a group of 11 different aphid species that attack the wheat crop throughout the season. However, only a cluster of three aphid species, namely, *Rhopalosiphum maidis, R. padi, Sitobion avenae*, and *S. miscanthi* seem to cause most of the damage to wheat crop. These aphid species are common in the Western Gangetic Plains. They may cause 3–20% grain yield reduction if unchecked. Aphids that attack wheat also transmit viral and fungal disease (Jasrotia and Kumar, 2019).

Wheat cultivated in the Indus Valley (Pakistan) is also exposed to onslaughts by insect pest via the agricultural sky. For example, Commonwealth Agricultural Bureau reports and describes the biology and nature of damage caused by several genera of aerial pests of wheat. A few of them are aphids, cut worms, shoot fly, etc. There are several lists of pests published. Among them, *Sesamia inferens, Sitobion avenae, Myzus persicae, Atherigona naqvii, Mythimna separata, Agrostis ipsilon* are common. They could reach devastating proportions. Aphids affect wheat in several locations of Indus Plains. For example, *A. craccivora* that affected cowpea seems to switch hosts and affect wheat grown in the plains (Islam et al., 2015; Akhtar and Parveen, 2002). Special attention to thwart the pests that operate through the agricultural sky seems a good idea. Reports from International Maize and Wheat Center in Mexico state that in the Indus Plains of Pakistan wheat is a major crop along with maize and is affected by Fall army worm. It becomes severe and devastating in many locations within the plains of Pakistan (CIMMYT, 2018).

The enhanced acceptance of rice–wheat cropping system in the Indo-Gangetic Plains has meant larger area under rice during kharif season. The tropical conditions and moisture induce greater activity of insect pests. Again, humid tropical agricultural sky plays host to large number of insect pests. Several of them are devastating at times. They reduce rice crop yield, perceptibly. Major rice pest that transmits aerially above the rice canopy are leaf folder (*Cnaphalocrosis medinalis*), leaf hoppers, such as the brown leaf hopper (*Nilaparvatha lugens*), white backed leaf hopper (*Sogatella furcifera*), green leaf hopper (*Nephotetix virescens*), and rice hispa (*Dicladispa armigera).* Aerial sprays are excellent prophylactic measures to reduce the pest population and its deleterious effects (Sehgal et al., 1988, 2001). We should also adopt IPM methods that specifically thwart the aerial insects affecting rice crop. The role of agrarian sky, exclusively with regard to spread of pest needs attention.

Mustard is an important oilseed crop of Indo-Gangetic Plains. It is grown often in rotation with major cereal crops such as wheat/rice. Insects play a vital role as pollinators during flowering and seed set stage. However, at this juncture, we are concerned with another set of insects traced often in the agricultural sky. They are called aphids (*Lipaphis erysimi).* They are among the most dreaded pests of Brassicas grown in the Indo-Gangetic Plains. These aerial insects attack *Brassica campestris* (mustard), *B. rapa* (Sarson) and *B. juncea* (raya) grown in North India. Generally, aphids travel only short distances from the focus of infection. At best, they move 5–10 m within the cropped field. However, researchers have traced the movement of aphid from *Brassica* fields in Bengal, to central region of the plains into Allahabad region. Also, aphids are known to migrate away from Western foothills of Himalayas into the plains and proliferate (Sharma, 2019; Maheshwary, 2019). Aphid's migration has been noticed from 100 m to 1000 m above the ground surface. Wind currents seem to play an important role during the transit of these sucking insects. Migrant aphids (winged) begin appearing right in October. Aphids transit mainly between December till next March. They cause the damage to the crop. Low temperature seems to induce insect to migrate away from the foothills of Himalayas. Yield losses attributable to aphids range from 39 to 91%. The devastation caused by this aerial pest depends on the weather pattern and cropping systems followed. The intensity of pest population is important. In a year, about 30–35 overlapping generations are possible. Aphid population around 25 per 10 m^2 on the canopy causes yield reduction of 1.25 t ha^{-1}. Clearly, eradication prior to migratory trends of aphids is a good idea. Pesticide spray schedules should match the migratory trends and avoid the spread.

Locusts have attacked crops and natural vegetation traced in the Indo-Gangetic plains, since several decades. During past five decades, severe attacks by locust swarms have been reported in India and Pakistan in 1962, 1978, and 1993. The most recent upsurge of locusts that occurred in April 2020 utilized the agrarian sky rather efficiently to migrate and feed on crops across different geographical locations. This desert locust (*Schistocerca gregaria*) attack was confined to Southwest Pakistan and Northwest Indian cropping zones. Reports indicated that a single swarm of 1 km size contained over 80 million grasshopper individuals. There were swarms of 3–5 km detected during the same period in 2020, particularly in the border locations between India and Pakistan in Punjab. Desert locusts are known to be efficient in flight. So, they are dreaded aerial pests if they do get accentuated. The most recent upsurge in Pakistan and Northwestern India devastated over 25,000 ha of major cereals, such as wheat, maize, millets, such as pearl millet, sorghum, and legumes (India Today, 2020). The general migratory pattern of desert locusts includes Southwest Pakistan to India in eastern journey. In western transit, locusts usually cover countries, such as Kenya, Ethiopia, Somalia, then on to Sudan and Chad, and later to Sahelian West Africa. The most recent upsurge noticed in Indus belt was caused by cyclones that occurred in Kenya and Ethiopia. Cyclonic winds pushed the insect into Indus belt of Pakistan and India (India Today, 2020). There are also reports that cyclones, such as Mekunu and Luban that occurred in Yemen, Oman, and adjoining areas literally, induced the locust movement into western Indo-Gangetic Plains (Biswas, 2020). In Western Indo-Gangetic Plains locusts have devastated cash crops, such as sugarcane, cotton. These above examples make it clear that the agricultural sky can become a host to dreaded migratory pests. Agricultural sky induces devastation on greenery, including valuable agricultural crops. Therefore, here, we have to bestow greater alertness and research time to study and develop control methods. Locust control measures should either reduce or totally eliminate locust problems. It is the agrarian sky above the plains that is dangerous and detrimental to crops.

7.6.6.1 INSECTS AS POLLINATORS OF CROPS IN THE INDO-GANGETIC PLAINS

In the Indian agricultural belts, forests and mountainous regions we can trace a few honeybee species. They could be effective as pollinators of crops. A few bee species prefer restricted type of vegetation for their normal pollinating activity. A few migrate periodically. Reports suggest that *A. mellifera,* the commonly known European honeybee is most commonly traced in agrarian regions of the plains. *Apis dorsata,* also called rock-bees too are common in the farms of Indo-Gangetic Plains. These bees usually build colonies on rocks, buildings, tall forest trees, mountains, etc. *Apis cerapa indica* is smaller than *A. dorsata* and *A. mellifera*. They can be domesticated. *A. flores* are still smaller sized bees. In addition to *Apis*, we can also find *Trigona* species of honeybees that are useful.

Pollination is an essential aspect of crop production. Pollination through bees and other insect species, birds, bats, is necessary, particularly in the case of open-pollinated plants (see Roubik, 1995). The agricultural sky above the natural vegetation and adjoining cropped fields supports the activity of several insect species, including honeybees that serve as excellent pollinators. Several crop species that are open pollinated depend on honeybees and other insects for pollination. Their productivity may depend on pollinator activity to a large extent. There are indeed several reports about insect pollinators that operate in the Indo-Gangetic Plains. For example, it was found that crop species that honeybees forage include cereals, such as wheat, brassicas, coriandrum, and litchi. Forest species such as *Myrica, Rumex*, and *Erigeron* are also visited frequently by the pollinators. Similarly, Chauhan et al. (2017) have listed species of crop plants and those from natural vegetation in the North Indian plains. They say, about 55 plant species including several crops and weeds were foraged by the honeybees. A list of plants mentioned includes forest species, such as *Eucalyptus, Prosopis, Syzygium cumini.* Pollen from a few other plant species traced in the honey were brassicas, *Pimpinella tomentosa, Xanthium strumarium,* and *Ziziphus* sp. Reports from central regions of Indo-Gangetic Plains suggest that European honeybee *A. mellifera* and *A. dorsata* are frequently encountered in the crop fields. *A. dorsata* seems to be dominant in some locations. *Apis dorsata* seem to visit flowers more frequently and provide good pollination efficiency (Udikeri and Chandra, 2019).

Pollinator population and activity in the agricultural sky is no doubt important for proper seed set and productivity. In most of the major agrarian regions, intensification and preference to only very few species of food crops and fruits have induced loss of biodiversity, in terms of both flora and fauna. In the Indo-Gangetic Plains, during the past six decades (1947–2010), monocropping trends have gained ascendency. Even among food grain crops, small grain cereals, millets, and sorghum, it has dwindled. In its place, rice/wheat have found acceptance. This trend seems to have affected several insect species including honeybees. Honeybees are facing decline. There is loss of diversity attributable to pesticide application (Smith et al., 2019; Hitaj et al., 2018; Kremen et al., 2002; Pannure, 2016).

There are, in fact, several reports about the need for pollinator diversity, not just crop diversity. They say, introduction of cross-pollinated crops induces better pollinator diversity. Let us consider an example. Pigeon pea (*C. cajan*) is a legume common to Indo-Gangetic Plains. Several insect pollinators are supposed to operate in pigeon pea fields. For example, Singh et al. (2017) have reported that insects belonging to hymenoptera were the most common pollinators. Insect belonging to Lepidoptera and Diptera too function in the agrarian sky. They are pollinators of pigeon pea crop but at low frequency. Honeybees dominated the pollinator activity in legume fields of Indo-Gangetic Plains. About 66% of pollinator insects belonged to hymenopteran honeybees. In India, 52–73% of pollination in the agrarian regions is accomplished by honeybees. Flies account for 19%, bats 6.5%, wasps 5%, beetles 5%, birds 4%, and butterflies 4% of pollinators (Abrol, 2009; Pannure, 2016).

7.6.7 AVES IN THE RICE–WHEAT BELT OF INDO-GANGETIC PLAINS

There are over 1065 species of birds recognized as feeders on crops grown in the Gangetic Plains within India. Several of them may gain the status of severe pests when the season and the crop stage coincide with their migratory and feeding habits (Avibase, 2020). There are several others that are useful as insectivorous biocontrol agents in the wheat–rice belt of Gangetic Plains. A few of them are also bee-eaters. Therefore, they reduce bee population. We should note that during past five decades, use of pesticides and herbicides has increased perceptibly in the Gangetic regions. This agronomic procedure involving agrarian sky does affect the bird population. Bird pests may get reduced. However, biocontrol birds too suffer due to rampant use of chemical pesticides. Hence, we may have to use the agricultural sky above the crop canopy shrewdly and reap the best benefits out of birds, pesticides, and bee pollinators. Usually, farmers in the Gangetic Plains are offered an assortment of suggestions that help them in reducing the insect pests. They could adopt both biocontrol birds and pesticides to maximize returns. There are also treatises and websites that list the bird species traced in the Indus Valley and other locations within Pakistan (Roberts, 1992).

Rice belt in the Indo-Gangetic Plains is large. They report that 1300 species of birds could be traced in the plains. At least about 351 species utilized lowland rice fields regularly for their foraging (Sunder and Subramanya, 2010). The spread of rice cultivation zones seems to have provided good habitat to some of the species of birds. At least 64 species of birds that thrive on rice crop has been experiencing increase in population. Bird pests that reduce rice yield has been studied intensely, during past decade. However, there are several bird species that have useful role in the lowland rice fields. It is clear that proper maintenance of agricultural sky above the natural wetlands and lowland rice fields in the Indo-Gangetic Plains needs special attention. We have to reduce the use of toxic chemicals and also to improve population of biocontrol birds and preserve species that are getting extinct due to loss of habitat.

Indo-Gangetic grasslands are also home to several bird species that are decreasing in population. A few are threatened with extinction. This has been attributed to loss of habitat. Excessive use of land for farming is a possible reason. Also, farm procedures that involve the use of plant protection chemicals in higher quantities affect birds. Recent reports from Pakistan, Western and Eastern Gangetic Plains and Plains of Terai in Nepal suggest that babblers, weavers, grey Prinia are becoming scarce in the wetter areas. This induces imbalance in ecosystem functions. There are indeed several reports depicting the useful role of wetland birds in regulating the insect pest population in the rice fields of Indo-Gangetic Plains. Conservation of biocontrol birds is essential.

The agrarian sky in the Indo-Gangetic Plains is no doubt an abode to several species of birds. These bird species display different roles in the agroecosystem. A series of about 14 reports by the Ministry of Agriculture, India states that about 63 bird species have been recognized as regular pests on crops. About 52 species out of the 63 are severe pests on cereal crops. About 14 species attack millets and pulse grain crops. Fruit crops were attacked by 23 species. In addition, there are birds that feed excessively on small grain cereals, such as sorghum, pearl millet, panicum, and setaria. Most commonly traced bird pests are Grey Partridges, Blue rock Pigeons, House Sparrow, Parakeet, Weavers, Munnia, and Doves. Mostly, these birds form their nests close to farmland. Hence, they become pest quickly when the crop matures (DTE Staff, 2019; Varghese et al., 2008). Further, the above report states that 46 bird species were beneficial. Since these bird species destroyed insect pests and rodents within the cropped fields. These species could be selectively encouraged as biological control agents.

Bird pest control is also an important activity related to agricultural sky above the crops. We are yet to utilize the sky above crops shrewdly to avoid birds from the cropped fields. Bird scaring is a better proposal to control bird species that are prone to devour grains of wheat, pulses, etc. There are several methods, each has its advantages and lacunae. The efficiency of sound guns decreases once the bird species get accustomed to it. Baiting is not to be practiced if bird species needs to be conserved in the agroecosystem. However, helikites with taped sounds of predatory birds, such as eagles, falcons, and vultures are effective. Similarly, many farmers in Pakistan (Indus Valley) seems to utilize reflector plastic tapes. They deter bird pests from alighting on the mature crops. To quote an example, in the Indus Valley of Punjab in Pakistan, farmers have utilized low-cost technology such as plastic reflector tapes. Hafeez et al. (2008) have reported that hanging reflector ribbons of 65–100 cm above the crop's canopy was effective. It deterred birds from approaching the cropped fields. The four main bird pest species such as crow (*Corvus splendens*), Parakeet (*Psittacula krameria*), Bank Myna (*Acridothres ginginianus),* Common myna (*A. tristis*) attack wheat and maize grown in Punjab, Pakistan. They could be effectively diverted from crops, using reflectors (Hafeez et al., 2008). There are also several suggestions regarding use of repellents, to reduce loss of grains in the Indian subcontinent (see Kale et al., 2012).

Biocontrol of insect pest using insectivorous birds is a good idea. It avoids the use of pesticides. In the Indus Valley of Pakistan, where cotton/wheat cropping system is popular, bird species that are insectivorous are highly preferred. These bird species are natural biocontrol agents of pests (insects) that attack wheat and cotton. Reports from Punjab (Pakistan) state that several bird species are insectivorous. They are useful to cotton/wheat farmers as biocontrol agents. Hussain and Afzal (2005) have reported that there are at least 32 bird species that are useful as pest control agents. Most of them, that is, about 55% are Passerines. Indian warblers, jungle babblers, yellow-throated sparrow, jungle sparrow, doves, etc. are commonly traced biocontrol agents (Roberts, 1992; Khan et al., 1980). The population density of birds that act as biocontrol agents is moderately high. This is despite rampant use of pesticides. Some of them are honeybee eaters. So, they may not be of great use if the farmers are depending on insect-aided pollinators. Further, Hussain and Afzal (2005), have pointed out that, since 1954, pesticide use on wheat/cotton has increased enormously. This procedure has reduced the population of biocontrol bird species (Hussain, 1999).

7.6.8 WIND AND SOLAR POWER IN INDO-GANGETIC PLAINS AND AGRICULTURAL CROP PRODUCTION

Utility of sky above the shorelines, mountains, and agricultural plains for wind energy is important. Wind power generation was initiated in India in 1952. Reports by Tata Energy Research Institute, New Delhi suggest that Indo-Gangetic Plains have good potential to increase the wind energy generation. The potential for wind energy in India (not Indo-Gangetic belt) is estimated at 2000 GW per year. Installed capacity of wind energy in India was 7850 MW in 2006 but now in 2020, it is 37,669 MW (Wikipedia, 2020f). The emphasis here is only on wind farms above agricultural farms. At present, wind farms in the agricultural plains are still feeble. The windy plains in North India are of course a good candidate to improve wind energy production. Wind energy generation through low- and high-altitude turbines could be enhanced. In fact, during the recent years, wind turbines are being installed at a rapid rate in the plains. Aerostats with turbines placed at high altitude could be harnessed in individual large farms or groups of farms. Several wind energy generation companies operate in the Indian subcontinent. Some are localized in the Indo-Gangetic plains. A few examples of wind power generating companies in India are Vesta India ltd., Suzlon Energy Ltd., Enercon India Ltd., Wind world India ltd., Inox Wind Ltd., Indo-Wind Energy Ltd. etc. They serve the vast population in farming regions of North India (Santosh, 2020). No doubt, wind energy generating companies benefit the farmers in solving their energy requirements.

Solar energy generation in India is receiving great impetus. The total generation of solar energy reached 20,000 MW in 2017 (MEDA, 2020; Prateek, 2018; Phukan, 2014). Solar energy contributes only small portion of the total energy generated in India. It is only 1.67% of the total. However, there has been rapid increase in the installed capacity and power generation in the subcontinent. Solar power improves the way we utilize the agrarian sky, particularly when solar panels are placed in greater numbers in the farm belt. However, there is need to maximize the land-use efficiency by carefully arranging the panels in off-season. In many of the locations of the plains, the present trend is to adopt hybrid power generation systems. It involves both solar and wind power generation system. They connect them into a grid for use by farms.

7.6.9 AERIAL VEHICLES AND AGRICULTURE IN INDO-GANGETIC PLAINS

7.6.9.1 USE OF DRONES IN THE AGRARIAN SKY

Aerial vehicles, especially the totally autonomous ones seem to promise the efficient use of aerospace above the crop fields. The small drone aircrafts have already invaded the sky above crops in the Indo-Gangetic Plains. However, they are still in rudimentary stages. They have just started helping farmers with spectral data. Some farms do adopt drones even for spraying pesticides. There are also several other types of aerial vehicles, such as the parafoils, helikites, blimps, and aerostats, even kites that can enhance our ability to utilize agricultural sky in the Indo-Gangetic Plains. Arial vehicles also help us in knowing disease progress and insect attack, if any. Like soil health, aerial health too should be monitored above the rice/wheat agroeco-system in the Indo-Gangetic Plains. We can then reduce the loss of grain yield. Drones may also aid in fore-warning farmers with dire difficulties that they could face, due to climate change processes, torrential rains, drought, soil erosion, etc.

Sylvester (2018) reports that in the rice growing regions of South Asia, including the Indo-Gangetic Plains, drones have been tested and adopted to deliver the required data about the crop. They have been used to develop sharp maps with variety of characters of the crop being depicted in them. Yield goals and forecasting possible final yield is an essential exercise during farming. Yield forecast and final harvest data has been reported to many farm agencies, using drones. In case of crops like maize, number of plants and cobs have been counted, using aerial imagery from drone's cameras (see Murali, 2018). During the season, drones' aerial imagery and spectral data have been effectively used to inform farmers and agencies about the crop's nutritional status, disease/pest attack if any, and even crop's water stress index (CWSI) (See Krishna, 2015, 2018). Farmers in the plains may eventually receive images of their crops (e.g., rice, wheat, sorghum, maize or legumes) into their computers prior to deciding on the agronomic procedures required (Sylvester, 2018). Small drone air crafts have also been utilized to assess the status of sugarcane and other plantations in the Indo-Gangetic Plains. Typically, fully autonomous drones are used to provide aerial imagery of plantations, along with GPS tags. Such data could be utilized on the GPS connected ground vehicles in the farms. Reports suggest that in the Indo-Gangetic Plains, imagery provided via drones have been accurate to the extent of 80–90% when ground reality maps of disease/pest or water status or NDVI are verified (Sylvester, 2018; see Krishna, 2018, 2020a, b; Equinox Drones, 2020). Since farm holdings are small, drones that are inexpensive may be needed. Small areas of 1–2 km^2 could be imaged to ascertain the crop's status. Larger cooperatives may adopt sophisticated drones, so that aerial space above crops is utilized more efficiently than now. Agricultural cooperatives may use sophisticated drones for aerial imagery, surveillance, and even for spraying plant protection chemicals. *However, they say, in the Indo-Gangetic Plains drone-aided spraying of harmful plant protection chemicals is still illegal. Regulations for use of drones to spray using aerospace above crops, is yet to be decided.* Development of software to suit the crops, such as wheat, rice, maize cajanus, cicer, cowpea, and other crops, such as cotton, sugar cane is essential. This is because, agronomic prescriptions depend immensely on the accuracy of softwares. Sylvester (2018) states that the quality of software is an important determinant during evaluation of aerial images. As a result, agricultural drones could be effective in assessing the crop's NDVI at various stages. Such data could be useful in grain/fruit yield forecasting the plains.

A few other aspects that need attention prior to popularization of drone technology are accessibility of drones (i.e., cost) by farmers, flight range, and area to be assessed. Legal aspects for usage of airspace above farms, weather and its vagaries if any, farm operations that can be accomplished by drones (e.g., spraying), etc., also need consideration. Farm connectivity with drone companies or agricultural service agencies is necessary.

Wheat crop cultivated in the Indo-Gangetic Plains is exposed to diseases, such as brown rust, streak rust, streak mosaic virus, and stripe rust. The fungal/viral disease occur at varying intensities. They can reach devastating proportions if they are not detected early in the season. Prophylactic sprays are to be conducted. Agricultural sky needs to be utilized efficiently to first monitor the spore movement into wheat belt. Drones offer excellent aerial imagery about the disease. Drones could also be utilized to assess spore density in the atmosphere. It seems, during the past year, that is, in 2015, wheat farmers in Punjab and Haryana could adopt drones to thwart wheat stripe rust crisis by quickly identifying fields and spraying fungicides (Gupta, 2016).

Agricultural drones are expected to solve problems that currently affect farming enterprises in the Indo-Gangetic Plains and other parts of India, particularly they can aid in monitoring crop health rapidly. Crop spraying can be accomplished with greater efficiency if drones are used in the sky above crops. Hence, long-term plan is to manufacture drones which can carry 20 liters of pesticides for bigger farms (Singh, 2019).

A recent report states that drones with spray bars and pesticide (Organophosphorus) tanks are being utilized to overcome locust swarms, particularly in the airspace above crops in Punjab, Haryana, Rajasthan, and Uttar Pradesh (Equinox Drones, 2020; Krishna, 2020a, b; Kulkarni and Phadnis, 2020; Sinha, 2020). Locusts attack crops (wheat, rice, legumes) in the Indo-Gangetic Plains covering areas both in Pakistan and India. They transit swiftly covering about 150 km a day in the agricultural sky. Recent reports suggest that Desert locusts that breed in the Sind region of Pakistan can destroy crops rapidly. Locust swarms could reach Indian wheat belt in few days. They fly rapidly in the agricultural sky (Haq, 2020). So, to monitor and deter locusts, these nations are buying large number of sprayer drones (Bhardwaj and Jadhav, 2020).

Management of water resources (i.e., irrigation scheduling) is an important aspect for farmers in the Indo-Gangetic Plains. There are areas that are prone to drought. Some locations suffer severe flooding. Agricultural sky could be utilized efficiently to overcome these difficulties. In each of these situations, drone's aerial imagery is helpful. Small totally autonomous drones are handy in judging the crops' water stress index (CWSI) from a close range. They can transit or hover at low altitude (5–10 ft.) above the crop's canopy. Then, pick spectral data useful in assessing the water status of the canopy (leaf tissue). Farmers in the plains could arrange irrigation schedules after analyzing the spectral data and even practice precision irrigation if they have facility. Satellite imagery has been utilized to assess crop's water status, drought effects if any or flooded zones. However, resolution of aerial imagery from satellites is only in few meters say 5 or 10 m. Drones offer resolution in the range of 1–2 cm (see Niu et al., 2019).

Forest fires, burning crop residue and smog are among phenomena common to most locations of Indo-Gangetic Plains. These are related to agricultural sky. These aspects could be efficiently monitored using drone aircrafts (IANS, 2017; Synergia Foundation, 2017).

In order to serve the farmers in the Indo-Gangetic Plains, drone companies need to produce them in large number and make them cost-effective. Low-cost farm drones are necessary if they have to become popular with famers in the Indo-Gangetic Plains (See Krishna, 2018, 2020a).

7.6.9.2 USE OF SATELLITES

There are satellite-based rice monitoring (SRM) initiatives, integrated remote sensing, crop modeling and ICT tools. They generate and provide near-real time and accurate information on rice growth, yield, as well as damage caused by abiotic and biotic stresses. The "RIICE technology" is capable of providing accurate and timely village level information about rice. Satellite imagery provides details about planted areas, including information on the start of the season and its variability. Satellite images depict geographical features of rice belt. They also help in forecasting yield and the impact of any disaster on specific rice growing areas (Sylvester, 2018).

The satellite-based rice monitoring system deploys three software modules, namely, MAPscape-Rice (Nelson et al., 2014), Rice Yield Estimation System (Rice-YES) (Setiyono et al., 2017), and ORYZA Crop Growth Model (Li et al., 2017).

In the case of natural disasters such as those resulting from flood and drought, a map of the impacted area, particularly the rice area is generated by the system. The SRM provides more detailed information on the flood impacted rice area in contrast to the conventional information on the flood-affected area, more generally.

KEY WORDS

- **agricultural sky**
- **agroclimate**
- **dust storms**
- **european plains**
- **great plains**
- **greenhouse gas**
- **indo-gangetic plains**
- **sahel**
- **solar power**
- **wind**

REFERENCES

Abrol, D. P. Plant-Pollinator Interactions in the Context of Climate Change—An Endangered Mutualism. *J. Palynol.* **2009,** *45*, 1–25.

Abrol, I. J. Sustaining Rice-Wheat Cropping System Productivity in the Indo-Gangetic Plains. *The 4th JIRCAS International Symposium on Sustainable Agriculture Development Compatible with Environmental Conservation in Asia*; ICRISAT: Hyderabad, India, 1998; pp 155–165.

Aeroscout. High Resolution Laser Ranging of Earth's Forests and Topography. *Aeroscout: Unmanned Aerial Technology, Newsletter* March 2018, pp 1–4. http://www.aeroscout.com (accessed July 21st, 2021).

Africa BEECause. Bee Keeping for Sustainable Income Generation. Why Are Honeybees Important? 2020, pp 1–4. https://africabeecause.org/?page_id=56/ (accessed July 24th, 2021).

Africa50. AfDB, GCF and Africa50 Support Solar Energy Programme for the Sahelian Region, 2020, pp 1–3. http://africanreview.com/energy-a-power/renewables/afdb-gcf-and-africa-50-support-solar-energy-programme-for-the-sahel-regionsola/ (accessed July 9th, 2021).

Africa50. Africa50 Joins Investors to Power the World's Largest Solar Park. Smart Energy International, 2017, pp 1–6. https://www.smart-energy.com/renewable-energy/africa50-joins-investors-to-power-the-worlds-largest-solar-park/ (accessed July 30th, 2021).

Afrik21. Mali: Solar Development Project, 2019, pp 1–3. https://africa-energy-portal.org/news/mali-ntpc-appointed-consultant-500-mwp-solar-project-development/ (accessed July 24th, 2021).

Agarwal, P. K.; Kalra, N.; Singh, A. K.; Sinha, S. K. Climatically Potential Yields, and Optimal Management Strategies. Rice-Wheat Consortium of Indo-Gangetic Plains. *Field Crops Res* **1994,** *38*, 73–91.

Agarwal, P. K.; Mall, R. K. Potential Yields of Rice-Wheat System in the Indo-Gangetic Plains of India. Rice-Wheat consortium Series of papers No. 10, 2000, pp 1–16. https://www.researchgate.net/publication/235925367/ (accessed July 22nd, 2021).

Agence France-Presse. Decline of Bees and Other Pollinators Threaten Crop Output—UN Body, 2016. https://www.rappler.com/science-nature/environment/123852-decline-bees-pollinators-threaten-crop-output/ (accessed June 28th, 2021).

Agritech Tomorrow. Nine Ways Your Agricultural Drones Revolutionize Your Farm or Ranch, 2019, pp 1–10. https://www.agritechtomorrow.com/article/2017/10/9-ways-agricultural-drones-revolutionize-your-farm-or-ranch/10311/ (accessed July 20th, 2021).

Aizen, M. A.; Garibaldi, L. A.; Cunningham, S. A.; Klein, A. M. How Much Does Agriculture Depend on Pollinators? Lessons from Long-Term Trends in Crop Production. *Ann. Bot-London* **2009,** *103*, 1579–1588.

Akhtar, M. S.; Parveen, S. Studies on Population of Wheat Aphids on Wheat Crop in New Crops, Lahore. Punjab University. *J. Zool* **2002,** *17*, 14–22.

Allsopp Helikites Ltd. GIS, Geomatics, Surveying and Inspection Helikites Balloons, 2017, pp 1–15. http://www.allsopp.co.uk/index.php?mod=page&id_pag=63 (accessed Nov 21st, 2017).

Alvarez, R. Soil Organic Carbon Stock in the Pampean Soils. Change Associated with Rotation and Tillage, 2012, pp 1–3. http://www.istro2012.congressos-rohr.info/programa/_4_1_Roberto_Alvarez.ht (accessed Aug 1st, 2021).

Alvarez, R. Taboada, M. A.; Gutlerrez Boem, A.; Fernandez, P. L.; Prystupa, P. Topsoil Properties as Affected By Tillage Systems in the Rolling Pampas Region of Argentina. *Soil Sci. Soc. Am. J.* **2009,** *73*, 1242–1250.

Alvey, S.; Bagayoko, M.; Neumann, G.; Buerkert, A. Cereal/Legume Rotations Affect Chemical Properties and Biological Activities in Two West African soils. *Plant Soil* **2001,** *231*, 31–44.

Alvey, S.; Bagayoko, M.; Neumann, G.; Buerkert, A. Cereal-Legume Rotation Effects in Two West African Soils Under Control Conditions. *Plant Soil* **2000,** 231–234.

Ambang, Z.; Ndongo, B.; Amayana, D.; Djile, B.; Ngoh, J. P. and Chewachong, G. M. Combined Effect of Host Plant Resistance and Insecticide Application on the Development of Cowpea Viral Diseases. *Aust. J. Crop Sci.* **2009,** *3*, 162–172.

Anami, L. Alert Over Renewed Desert Locust Swarms Heading into West Africa, 2020, pp 1–5. https://www.theeastafrican.co.ke/tea/news/east-africa/alert-over-renewed-desert-locust-swarms-heading-into-west-africa-1443952 (accessed July 24th, 2021).

Anikster, Y.; Bushnell, W. R.; Eilam, T.; Manisterski, J.; Roelfs, A. P. *Puccinia recondita* Causing Leaf Rust on Cultivated Wheats, Wild Wheats, and Rye. *Can. J. Bot.* **1997,** *75*, 2082–2096.

An-Taisce. Natures Way: Pollinators in Ireland. The National Trust of Ireland, Dublin, 2018, pp 1–12. https://www.duhallowlife.com/ird-duhallow-life-raptor-life/natures-way-pollinators-ireland (accessed July 20th, 2021).

Appinsys. Sahel, Africa. Global Warming Science, 2010, pp 1–3. http://www.appinsys.com/globalwarming/RS_Sahel.htm (accessed July 21st, 2021).

Aquastat. Computation of Long-Term Annual Renewal of Water Resources by Country: Senegal. Food and Agricultural Organization of the United Nations, Rome, Italy, 2013, pp 1–3 http://www.fao.org/nr/aquastat/ (accessed Dec 23rd, 2013).

Archer, C. L.; Daran, L. D.; Rife, L. Airborne Wind Energy: Optimal Locations and Variability, 2013, pp 1–8. https://doi.org/10.1016/j.renene.2013.10.044/ (accessed July 19th, 2021).

Aregheore. *Nigeria: Country Pasture/Forage Profiles*; Food and Agricultural Organization: Italy, Rome, 2009; pp 1–42.

Arnolds, R. Will Kite Power Fly High as the Next Renewable Energy Solution. Earth.com, 2019, pp 1–3. https://www.earth.com/news/kite-power-renewable-energy/ (accessed July 27th, 2021).

Avery, M. L. Bird in Pest Management University of Nebraska-Lincoln, USDA National Wildlife Centre, Nebraska, USA, 2002, pp 1–5 https://digitalcommons.unl.edu/cgi/viewcontent.cgi?article=1452&context=icwdm_usdanwrc (accessed July 22nd, 2021).

Avibase. Avibase-Bird Checklists of the World-Gangetic Plains. Birdlife International, 2020, pp 1–14. https://avibase.bsc-eoc.org/checklist.jsp?region=INgg&list=howardmoore/ (accessed July 20th, 2021).

Aylor D. E. The Role of Intermittent Wind in the Dispersal of Fungal Pathogens. *Annu. Rev. Phytopathol.* **1990,** *28*, 73–92.

Aylor, D. E.; Schmale, E. J.; Shields, E. J.; Newcombe, M.; Nappo, C. J. Tracking the Potato Late Blight Pathogen in the Atmosphere, Using Unmanned Aerial Vehicles and Lagrangian Modelling. *Agric. Forestry Meteorol.* **2008,** *151*, 251–260.

Azeez, J. O.; Adetunji, M. T.; and Adebusuyi, M. Effect of Residue Burning and Fertilizer Application on Soil Nutrient Dynamics and Dry Grain Yield of Maize (*Zea mays*) in an Alfisol. *Nigerian J. Soil Sci.* **2007,** *17*, 71–80.

Azzopardi, T. Argentina Awards Contracts to Four More Wind Projects, 2017, pp 1–6. https://www.windpowermonthly.com/article/1453400/argentina-awards-contracts-four-wind-projects (accessed July 27th, 2021).

Az-Zorita, M.; Buschiazzo, D. E.; Peinemann, N. Soil Organic Matter and Wheat Productivity in the Semi-Arid Argentine Pampas. *Agron. J.* **1999,** *91*, 276–279.

Ba, M. N.; Baous, I. B.; N'Diaye, M.; Dabire-Binso, C.; Sanon, A.; Tamo, M. Biological Control of the Millet Head Miner *Heliocheilus albipunctella* in the Sahelian Region by Augmentative Releases of the Parasitoid Wasp *Habrobracon hebetor*: Effectiveness and Farmer's Perceptions. *Phytoparasitica* **2013,** *41*, 569–576.

Babar, S. A. Spectral Reflectance to Estimate Genetic Variation for In-Season Biomass, Leaf Chlorophyll, and Canopy Temperature in Wheat. *Crop Sci.* **2006,** *46* (3), 1046–1049.

Badarinath, K. V.S.; Chand, R. K.; Prasad, V. K. Agricultural Crop Residue Burning in the Indo Gangetic Plains: A Study Using IRS P6 AWiFS Satellite Data. *Curr. Sci.* **2006,** *91*, 1085–1089.

Badner, J. *Low-Level Wind Shear: A Critical Review*; National Weather Service: Washington, DC, 1979; pp 1–8. https://archive.org/stream/lowlevelwindshea00badn/lowlevelwindshea00badn_djvu.txt/ (accessed July 26th, 2020).

Bado, B. V. Roles des legumueses sur la fertilie des sols ferrugineaux des zones guineene et sudanienne du Burkina Faso, 2002, p 197. https://www.collectionscanada.gc.ca/obj/s4/f2/dsk3/QQLA/TC-QQLA-20487.pdf (accessed July 21st, 2021).

Baethgen, W. E.; Magrin, G. O. Assessing the Impacts of Climate Change on Winter Crop Production in Uruguay and Argentina Using Crop Simulation Models. In *Climate Change and Agriculture. Analysis of Potential International Impacts*, 1995; pp 207–230.

Bagayoko, M.; Buerkert, A.; Lung, G.; Bationo, A.; Romheld, V. Cereal/Legume Rotation Effects on Cereal Growth in Sudan-Sahelian West Africa: Soil Nitrogen, Mycorrhiza and Nematodes. *Plant Soil* **2000,** *218*, 103–116.

Baladron, A. V.; Bo, M. S. Relative Abundance, Habitat Use and Seasonal Variability of Raptor Assemblages in the Flooding Pampas of Argentina. *J. Raptor Res.* **2017,** *51*, 38–49. https://doi.org/10.3356/JRR-15-56.1 (accessed July 1st, 2021).

Balaskovitz, A. Tribal Leaders Hope Large South Dakota Project Becomes Models for Others, 2020, pp 1–18. https://nativenewsonline.net/business/tribal-leaders-hope-large-south-dakota-solar-project-becomes-model-for-others/ (accessed June 1st, 2020).

Baldi, G.; Paruelo, J. M. Land Use and Land Cover Dynamics in South American Temperate Grasslands. *Ecol. Soc.* **2008,** *13*, 1–10.

Ballouche, A. Fire and Burning in West African Holocene Savannah Paleo-Environment. Anthropogenic and Natural Processes in Environmental Changes. In *Proceedings of conference on environmental catastrophe in Mauritania, the desert and coast*, 2004; p 15 http://www.at.yorku.ca/c/a/m/u/05 (accessed Aug 4th, 2004).

Ban, M.; Perkovic, L.; Duic, N.; Penedo, R. Estimating the Spatial Distribution of High-Altitude Wind Energy Potential in Southeast Europe. *Elsevier Energy* **2012,** *57*, 24–29. https://doi.org/10.1016/j.energy.2012.12.045/ (accessed July 29th, 2021).

Baoua, I.; Oumarou, N.; Amadou, L.; Payne, W. A. Estimating Effect of Augmentative Biological Control on Grain Yields from Individual Pearl Millet Heads. *J. Appl. Entomol.* **2013,** *138* (4), 1–7. DOI: 10.1111/jen.12077/; https://www.researchgate.net/publication/256744121 (accessed July 14th, 2021).

Bationo, A.; Kihara, J.; Waswa, B.; Ouatara, B.; Vanlauwe, B. Technologies for Sustainable Management of Sandy Sahelian Soils. Food and Agricultural Organization of the United Nations, 2005, pp 1–19. www.fao.org/docrep/010/ag125e/AG125E42.htm (accessed July 12th, 2021).

Bationo, A.; Traore, Z.; Kimetu, J.; Bagayoko, M.; Kihara, J.; Bado, V.; Lompo, M.; Tabo, R.; Kaola, S. *Crop Productivity's Affected by Cropping System*; CTA Workshop Wageningen, Netherlands, 2004; p 209.

Baumhardt, R. L. Wind Water and Growing Season: Cropping System Selection Pressures in the Southern Great Plains. In *Proceedings of Dynamic Cropping System Symposium*; Colorado State University: Fort Collins, CO, USA, 2003; pp 114–123.

Baumhardt, R. L.; Salinas-Garcia, J. Mexico and the Unites States Southern Great Plains. In *Dry Land Agriculture*; Peterson, Ed.; *Agron. Monographs* **2005,** *23*, 1–85.

Bayer, C.; Martin-Neto, L.; Mielniczuk, C. N.; Sandoi, L. Changes in Soil Organic Matter Fractions Under Subtropical No-Tillage System. *Soil Sci. Soc. Am. J.* **2001,** *65*, 1473–1478.

Behzad, H.; Minata, K.; Gojobori, T. Global Ramifications of Dust and Sandstorms in Microbiota. *Genom. Biol. Evol.* **2018,** *10*, 1970–1987. DOI: 10.1093/gbe/evy134/ (accessed July 21st, 2021).

Bello, D. *The Sky Is the Limit for Wind Power*; Scientific American, 2012; pp 1–3. https://www.scientificamerican.com/article/no-limit-for-wind-power/ (accessed July 25th, 2021).

Benayas, R. J. M.; Meltzer, J.; Heras-Bravo, D. L.; Cavuela, L. Potential of Pest Regulation by Insectivorous Birds in the Mediterranean Woody Crops, 2017, pp 1–12. DOI: 10.1371/journal.pone.0180702.ecollection/ (accessed July 19th, 2021).

Berenyi, A.; Barholy, J.; Pongracz, R. Analysis of Precipitation Patterns and Extremes in European Lowlands. European Commission GU, Hungary, 2020, pp 1–2. https://doi.org/10.5194/egusphere-egu2020–963/ (accessed July 25th, 2021).

Berg, L. K.; Riihimaki, L. D.; Qian, Y.; Yan, H.; Huang, M. The Low-Level Jet Over the Southern Great Plains Determined from Observations and Reanalysis and Its Impact on

Moisture Transport. *J. Clim.* **2015,** *28* (17), 6682–6706. DOI: 10.1175/JCLI-D-14-00719.1 (accessed July 22nd, 2021).

Bert, F. E.; Satorre, E. H.; Toranzo, F. R.; Podsta, G. P. Climatic Information and Decision-Making in Maize Crop Production Systems of the Argentinean Pampas. *Agric. Syst.* **2006,** *88*, 180–204.

Bertelesen, M.; Traore, G. The NRM inerCRSP for West Africa: Context, Challenges and Approach, 2002, pp 13–15. http://www.oired.vt.ed/projects/current/projectynthesis.pdf (accessed July 22nd, 2021).

Bhardwaj, M.; Jadhav, R. India Buys Drones, Specialist Equipment to Avert New Locust Attack, 2020, pp 1–7. https://in.reuters.com/article/india-locusts-idINKBN20D1XB/ (accessed July 30th, 2021).

Bhardwaj, S. C. Growing with Wheat and Barley Rusts for Three Decades. *Indian Phytopatol.* **2017,** *70*, 22–31.

Bhatnagar, V. S. Conservation and Encouragement of Natural Enemies of Insect Pests in Dryland Subsistence Farming: Problems, Progress and Prospects in the Sahelian Zone. *Recent Adv. Res. Trop. Entomol.* **2011,** *8*, 791–795.

Bhattacharya, S. Wheat Rust Back in Europe. *Nature* **2017,** *542*, 145–146. https://www.nature.com/news/polopoly_fs/1.21424!/menu/main/topColumns/topLeftColumn/pdf/nature.2017.21424.pdf/ (accessed July 22nd, 2021).

Bielders, C. Michels, K.; Rajor, J. On Farm Evaluation of Wind Erosion Control Technologies, 2002a, p 5. ICRISAT.org. CCER.ht (accessed August 4th, 2021).

Bielders, C.; Rajor, J. Amadou, M.; Skidmore E. On-Farm Evaluation of Wind Erosion Under Traditional Management Practices, 2002b, p 6. ICRISAT.Org. CCER.htm (accessed Aug 4th, 2021).

Biswas, P. Locust Attack: How They Arrived, Seriousness of the Problem, and Ways to Solve It. *Indian Express* May 29th, 2020, pp 1–3. https://indianexpress.com/article/explained/explained-how-locusts-came-what-next-6432162/ (accessed July 22nd, 2021).

Biswas, T. D.; Narayanaswamy, G. *Sustainable Productivity of Rice-Wheat System*; Indian Society of Soil Science: New Delhi Book No 18, 1997; pp 1–82.

Blackman, C. Kites Flying in High Altitude Winds Could Provide Clean Electricity. *Phys.org*, 2009, pp 1–7. https://phys.org/news/2009–06-kites-high-altitude-electricity.html (accessed July 29th, 2021).

Blaskovic, T. Extreme Cold Results in Severe Agricultural Damage Across Europe, Food Prices Rising, 2018, pp 1–4. https://watchers.news/2018/03/05/extreme-cold-results-in-severe-agricultural-damage-across-europe-food-prices-rising/ (accessed July 24th, 2021).

Blumsack, S.; Richardson, K. Cost and Emissions Implications of Coupling Wind and Solar Power. *Smart Grid Renew. Energy* **2012,** *3*, 308–315.

Bolton, A. Japan Develops a Drone to Patrol Farmland and Destroy Insect Pests, 2016, pp 1–3. https://www.cnet.com/news/japan-develops-a-drone-to-patrol-farmland-and-destroy-insect-pests/ (accessed June 10th, 2020).

Borade, G. Facts About Wind ErosionBuzzle.com, 2012, pp 1–2.

Bouvier, J. C.; Toubon, J. F.; Boivain, T.; Sauphanor, B. Effects of Apple Orchard Management Strategies on the Great Tit (Parus Major) in Southern France. *Environ. Toxicol. Chem.* **2005,** 1–18. DOI: 10.1897/04–588r1.PMID:16398122/ (accessed June 19th, 2020).

Bowlus, J. V. Senegal Surfs Offshore Winds to Power West Africa. Global Political Monitor, 2019, pp 1–6. https://www.geopoliticalmonitor.com/senegal-surfs-offshore-winds-to-power-west-africa/ (accessed July 11th, 2020).

Bray, C. D.; Betty, W. H.; Aneja, V. P. The Role of Biomass Burning Agricultural Emissions in the Indo-Gangetic Plains on Air Quality in New Delhi, India. *Atmos. Environ.* **2019,** *218*, 1–16.

Brévault, T.; Sylla, S.; Diatte, M.; Bernadas, G.; Diarra, K. *Tuta absoluta* Meyrick (Lepidoptera: Gelechiidae): A New Threat to Tomato Production in Sub-Saharan Africa. *Afr. Entomol.* **2014,** *22*, 441–444.

Brewer, M. J.; Elliot, N. L. Biological Control of Cereal Aphids in North America and Mediating Effects of Host Plant and Habitat Manipulations. *Annu. Rev. Entomol.* **2004,** *49*, 219–242.

Brewster, K. A.; Degelia, S. K. Availability and Variability of Potential PV Solar and Wind Power Production in Oklahoma. In *Proceedings of 7th Conference on Weather, Climate and New Energy Economy*; American Meteorological Society Paper No. 33: New Orleans, Louisiana, USA, 2016; pp 1–9.

Brown, J. K. M.; Hovmeller, M. S. Aerial Dispersal of Pathogens on the Global and Continental Scles and Its Impact on Plant Disease. *Science* **2002,** *297*, 537–541.

Buerkert, A.; Piepho, H. P.; Bationo, A. Multi-Site Time Trend Analysis of Soil Fertility Management Effects on Crop Production in Sub-Saharn West Africa. *Exp. Agric.* **2002,** *38*, 163–183.

Burkart, M. R.; Jones, D. E. Agricultural Nitrogen Contributions Through Arachis Hypogea in the Gulf of Mexico. *J. Environ. Qual.* **1999,** *28*, 850–859.

Burney, J.; Ramanathan, V. Recent Climate and Air Pollution Impacts on Indian Agriculture. *Proc. Natl. Acad. Sci.* **2014,** *111*, 16319–16324.

Buschiazzo, D. E.; Zobeck, T. M.; Abascal, A. Wind Erosion Quantity and Quality of an Entic Haplustoll of the Semi-Arid Pampas of Argentina. *J. Arid Environ.* **2006,** *69*, 29–39.

Buschiazzo, D. E.; Zobeck, T. M.; Silvia A. Wind Erosion in Loess Soils of the Semi-Arid Argentine Pampas. *Soil Sci.* **1999,** *164*, 133–138.

Buschiazzo, D.; Teddy, Z.; Silvia, A. Wind Erosion in Loess Soil of the Semi-Arid Argentinian Pampas, 1998, pp 1–2. http://are.usda.gov/research/publications/publications.htm?seq_no_115=86320 (accessed April 12th, 2013).

Bynum, E.; Knutson, A.; Swart, J.; Jungman, M. *Managing Insects and Mite Pests of Texas Small Grains*; Texas A and M Agrilife Extension, TX, USA, 2019; pp 1–21.

CABI. *Quelea quelea* (Weaver Bird). Commonwealth Agricultural Bureau, United Kingdom 2020, pp 1–28. https://www.cabi.org/isc/datasheet/66441/ (accessed July 16th, 2021).

CADER. Camara Argentina de Energias Removables- Estado de la Industrial Eolica e Argentina, 2009, pp 1–3. https://www.cader.org.ar/category/informes-y-estudios/ (accessed July 7th, 2021).

Calamari, N. C.; Canavelli, S.; Ceezo, A.; Dardanelli, S.; Bernardos, J. N.; Zaccagnini, M. E. Variations in Pest Bird Density in Argentinian Agroecosystems in Relation to Land Use and/or Cover, Vegetation Productivity and Climate. *Wildlife Res.* **2018,** *45*, 668–678. DOI: https//doi.org/10.1071/WR17167/ (accessed July 21st, 2021).

Canale, M.; Fagiana, L.; Milanese, M. High Altitude Wind Energy Generation Using Controlled Power Kites. *IEEE Trans. Control* **2009,** *18*, 279–293.

Charlet, L. D.; Gavloski, J. Insects of Sunflower in the Northern Great Plains of Northern: Arthropods of Canadian Grasslands-2. In *Inhabitants of a Changing Landscape*; Floats, K. D., Ed.; Biological Survey of Canada, 2011; pp 159–178.

Chaudhry, Q.; Rasul, G. Agroclimatic Classification of Pakistan. *Sci. Vision* **2004,** *9*, 59–66.

Chauhan, M. S.; Farooqui, A.; Trivedi, A. Plants Foraged By Bees for Honey Production in Northern India. The Diverse Flora of India Is Implication for Apiculture. *Acta Palaeobotanica* **2017,** *57*, 119–132.

Cherenkova, E. A.; Kononova, N. K. Dangerous Meteorological Drought Over European Russia in the 20-th Century and Atmospheric Circulation Processes Relationship. *Izvestiya Akademii Nauk Seriya Geograficheskaya* **2009,** *1*, 73–82 (in Russian).

Cherenkova, E.; Semen ova, I.; Bardin, M.; Zolotokrylin, A. N. Drought and Grain Crop Yields over the East European Plain under Influence of Quasibiennial Oscillation of Global Atmospheric Processes. *Int. J. Atmos. Sci.* **2015,** *2015*, Article ID 932474, 11 pp. http://dx.doi.org/10.1155/2015/932474/ (accessed June 25th, 2021).

Cherif, A.; Mansour, R.; Barhoumi-Attia, S.; Zappalà, L.; Grissa-Lebdi, K. Effectiveness of Different Release Rates of Trichogramma cacoeciae (Hymenoptera: Trichogrammatidae) against Tuta absoluta (Lepidoptera: Gelechiidae) in Protected and Open Field Tomato Crops in Tunisia. *Biocontrol Sci. Technol.* **2019,** *29*, 149–161.

Cherubuni, A.; Papini, A.; Vertechy, R.; Fontana, M. Airborne Wind Energy Systems-A Review of the Technologies. *Renew. Sustain. Energy Rev.* **2015,** *51*, 1461–1476 https://doi.org/10.1016/j.rser.2015.07.053/ (accessed July 25th, 2021).

CIMMYT. Pakistan Warned Against Deadly Wheat Pest, 2018, pp 1–2. https://www.pakissan.com/2019/02/25/pakistan-warned-against-deadly-wheat-pest/ (accessed July 23rd, 2021).

Civeira, G. Estimation of Carbon Inputs to the Soil from Wheat in the Pampas Region of Argentina. *Czech J. Genet. Plant Breed.* **2011,** *47*, S39–S42.

Codesido M.; Gozalez-Fisher, C.; Bilenca, D. Agricultural Land Use, Avian Nesting and Rarity in the Pampas of Central Argentina. *Emu J. Ornithol.* **2016,** *112*, 46–54.

Coffrey, P.; Farmer, A. *Agricultural Adaptation to Climate Change in the Sahel: Expected Impacts on Pests and Diseases Afflicting Selected Crops*; United States Agency for International Development: Washington, DC, 2014; p 88.

Cook, S. *The Insect Pests of Oil Rape: Biology and Potential for Control by IPM*; Rothamsted Research, BBSRC: Great Britain, 2019; pp 1–71.

Cooke, K. Warming Climate Could Affect West Africa River Levels. Climate News Network, 2013, pp 1–3. http://www.climatenewsnetwork.net2013/07/warmer-climate-will-hit-volta-river-levels/ (accessed July 10th, 2021).

Corbet. Birding in the Buenos Aires Area, 2012, pp 1–8. https://fatbirder.com/world-birding/south-america/argentine-republic/buenos-aires/ (accessed June 2nd, 2020).

Coura Badiane. Senegal's Trade in Groundnut Economic, Social and Environmental Implications. TED Case Studies No. 646, 2001, pp 1–32 http://www.Senegalcasestudy.htm (accessed July 22nd, 2021).

Crawford, H. S.; Jennings, D. T. Predation by Birds on Spruce Budworm *Choristoneura fumiferana*: Functional, Numerical and Total Responses. *Ecology* **1989,** 152–160.

Crespora-Herrara, I.; Singh, R. P.; Sabroui, A.; El-Bhoussani. Resistance to Insect Pests in Wheat-Rye and *Aegilops speltoides* Translocation and Substitution Lines. *Euphytica* **2015,** *2019* (215), 123.

Cuscar, J. C.; Feyen, Z.; Iberrata, D.; Seria, A. Climate Impacts in Europe. European Commission: JRC Science for Policy Report, Seville, Spain, 2018; p 95.

Dadia Forest Reserve. Predatory Birds: Beauty Can Fly. Dadia Leukimi-Soufli Reserve Area, Greece, 2020, pp 1–3. https://en.wikipedia.org/wiki/Dadia_Forest (accessed July 19th, 2021).

Dakora, F. D.; Aboynga, R. A.; Mahama, Y.; Apaseku, J. Assessment of N Fixation in Groundnut (*Arachis hypogeae*) and Cowpea and Their Relative N Contribution to a Succeeding Maize Crop in Northern Ghana. *MIRCEN J.* **1987,** *3*, 389–399.

Debire, C.; Antoine, S.; Heve, B.; Kouahou, F. B. Alternative Host Plants of *Clavigralla tomentosicollis* Stall (Hemiptera, Coreidae) the Sucking Bug of Cowpea in Sahelian Zone of Burkina Faso. *J. Entomol.* **2005,** *2*, 9–16.

Deckers, J. A.; Nactergaele, F. O.; Spaargeren, O. C. *World Reference Base for Soils Resource-Introduction*; Acco, Leauven/Ammersfort, 1998; p 165.

Degelia, S. K.; Mueller, D. M.; Brewster, K. A. Variability and Economics of a Hybrid Wind and Solar Power System in Oklahoma. University of Oklahoma School of Meteorology Capstone Project Final Report, 2014. ftp://ftp.caps.ou.edu/users/kbrews/solar/Hybrid Wind&Solar_20140602.pdf (accessed July 22nd, 2021).

DeMey, Y.; Demont, M.; Diagne, M. Estimating Bird Damage to Rice in Africa: Evidence from the Senegal River Valley. *J. Agric. Econ* **2011,** *63* (1), 175–200.

Descroix, L.; Mahe, G.; Lebel, T.; Favreauu, G.; Gautier, E.; Olivry, J. C.; Albergel, J.; Amogu, O.; Cappelareae, B. Dessaoussi, R.; Diedhiou, A.; Le Beton, E.; Mamadou, I.; Sigmnou, D. Spatio-Temporal Variability of Hydrology Regimes Around the Boundaries Between Sahelian and Sudanian Areas of West Africa: A Synthesis. *J. Hydrol.* **2009,** *375*, 90–102.

Deutsch, C. A.; Tewksbury, J. J.; Tigchelaar, M.; Battisti, D. S.; Merrill, S. C.; Huey, R. B.; Naylor, R. L. Increase in Crop Losses to Insect Pests in a Warming Climate. *Science* **2018,** *361* (6405), 916–919. DOI: 10.1126/science.aat3466 (accessed July 22nd, 2021).

Diallo, B. O.; Ouedraogo, M.; Chevalier, M.; Joly, H. J.; Hassaert-McKey, M.; McKey, D. Potential Pollinators of *Tamarindus indica* (Caesalpiniaceae) in Sudanian Region of Burkina Faso. *Afr. J. Plant Sci.* **2014,** *8*, 528–536.

Diaz-Delgado, R.; Onodi, G.; Kroel-Dulay, G.; Kertesz, M. Enhancement of Ecological Field Experimental Research by Means of UAV Multi-Spectral Sensing. *Drones* **2019,** *3*, 1–7. https://doi.org/10.3390/drones3010007 (accessed July 20th, 2021).

Dickenson, R. E. *Germany: Regional and Economic Geography*; London, 1964; pp 1–84.

Diouf, A.; Barbier, N.; Lykke, A. M.; Couteron, P.; Deblauwe, V.; Mahamane, A.; Saadou, M.; Bogaert, J. Relationships Between Fire History, Edaphic Factors and Woody Vegetation Structure and Composition in a Semi-Arid Savanna Landscape (Niger, West Africa). *Appl. Veg. Sci.* **2012,** *15*, 488–500.

Directorate of Wheat Development. Status Paper on Wheat. Ministry of Agriculture, Government of India, 2015, p 180. https://www.nfsm.gov.in/StatusPaper/Wheat2016.pdf (accessed July 22nd, 2021).

DJI. Above the World. *UAS Magazine*, 2016, p 238. http://www.uasmagazine.com/articles/1591/dji-explains-new-book-of-UAV-captured-images (accessed July 20th, 2021).

DJI. DJ1 MG-1S Agricultural Wonder Drone, 2017. https://www.youtube.com/watch?v=P2YPG8PO9JU (accessed August 18th, 2021).

DMZ Aerial Inc. Unmanned Aerial Vehicles and Scouting, 2013, p 10. http://www.dmzaerial.com/uavscouting.html (accessed Aug 25th, 2021).

Dolper, R. A.; Linz, G. M. Black Birds. World Life Damage Management. United States Department of Agriculture. Technical Series paper, 2016, *12*, 1–16.

Donatelli, M.; Magarey, R. D.; Bregeglio, S.; Wilocquet, L.; Whish, J. P. M.; Savary, S. Modelling the Impacts of Pests and Diseases in Agricultural Systems. *Agric. Syst.* **2017,** *155*, 213–234.

Dongui, Q. In West Africa at Risk of Desert Locust Invasion-FAO. Udegbunam, O. (Ed.), 2020, pp 1–8. https://www.premiumtimesng.com/agriculture/agric-news/394804-west-africa-at-risk-of-desert-locusts-invasion-fao.html/ (accessed July 12th, 2021).

Doody, A. *Optimum Nitrogen Fertilizer Rates for Rice and Wheat in the Indo-Gangetic Plains*. International Centre for maize and Wheat: Mexico, 2019; pp 1–4. https://wheat.org/tag/greenhouse-gas-emissions/ (accessed July 24th, 2021).

Downing, T. E.; Harrison, P. A. Butterfield, R. E.; Lonsdale, K. G. Climate Change, Climate Variability and Agriculture in Europe. An Integrated Assessment, Research Report No 21, Brussels, Belgium: Commission of the European Union. Contract ENV4-CT95–0154, 2000, p 455.

DTE Staff. Birds Impacting Agricultural Crops a Major Concern. DowntoEarth, 2019, pp 1–5. https://www.downtoearth.org.in/news/agriculture/birds-impacting-agricultural-crops-a-major-concern-64588/ (accessed August 21st, 2021).

Duncan, E. Airborne Wind Power. Popular Science Series, 2016, pp. 1–5. https://issuu.com/emilyeduncan/docs/draft_1_pages (accessed July 22nd, 2021).

Dunne, D. Rise in Insect Pests Under Climate Change to Hit Crop Yields Study Says. Carbonbrief.org, 2018, pp 1–8. https://www.carbonbrief.org/rise-in-insect-pests-under-climate-change-to-hit-crop-yields-study-says/ (accessed July 17th, 2021).

Easterling W. E. Adapting North American Agriculture to Climate Change in Review. *Agric. Forest Meteorol.* 1996, *80*, 1–53.

Editors. The Pampas. Encyclopaedia Britannica, 2020, pp 1–3. https://www.britannica.com/place/the-Pampas/ (accessed June 4th, 2021).

Elberti, W.; Taylore, P. E.; Andrae, M. O.; Poschl. Contribution of Fungi to Primary Biogenic Aerosols in the Atmosphere: Wet and Dry Discharges Spores, Carbohydrates and Inorganic Ions. *Atmos. Chem. Phys.* **2007,** *7*, 4569–4588.

Engel, R.; Dusenbury, M.; Miller, P.; Lemke, R. First Check of Nitrous Oxide Emissions Under Cropping Systems Adopted for the Northern Great Plains. *Proc. Western Nutr. Manage. Conf., Lake City, Utah, USA*, **2005,** *6*, 25–31.

Equinox Drones. Importance of Drone technology in Indian Agriculture, Farming. Equinox Drones Pvt ltd, Bangalore, India, 2020, pp 1–8 https://www.equinoxsdrones.com/blog/importance-of-drone-technology-in-indian-agriculture-farming/ (accessed July 30th, 2021).

European Commission. Wind Energy Generation. SETIS Section on Wind Energy, 2019, pp 1–4. https://setis.c.europa.eu/technologies/wind-energy/ (accessed July 30th, 2021).

European Commission. Sustainable Adaptation of Special EU Farming Systems on Climate Change. A: Baseline Reports for the 4 Main Climate Risk Regions. *Life* 15, CCA/DE/000072, AgriAdopt, 2017, p 105 https://agriadapt.eu/wp-content/uploads/2017/04/Baseline-report.pdf/ (accessed July 24th, 2021).

European Commission–Environment. EU Pollinators: Small, Precious and in Need of Protection, 2018, pp 1–7. https://ec.europa.eu/environment/nature/conservation/species/pollinators/index_en.htm (accessed July 21st, 2021).

European Court of Auditors. Electricity Production from Wind and Solar Photovoltaic Power Within European Union: Background Paper, 2018, pp 1–17. https://www.eca.europa.eu/en/Pages/DocItem.aspx?did=45188/ (accessed July 22nd, 2021).

European Court of Auditors. Special Report No 8/2019 Wind and Solar Power for Electricity Generation Significant Action Needed If EU Targets To Be Met, 2019, pp 1–2. https://www.eca.europa.eu/en/Pages/DocItem.aspx?did=50079/ (accessed July 31st, 2021).

European Parliament. Greenhouse Gas Emitted by Countries and Sectors, 2020, pp 1–22. https://www.europarl.europa.eu/news/en/headlines/society/20180301STO98928/greenhouse-gas-emissions-by-country-and-sector-infographic/ (accessed July 27th, 2021).

European Parliament. What's Behind the Decline in Bees and Other Pollinators? 2019, pp 1–4 https://www.europarl.europa.eu/news/en/headlines/society/20191129STO67758/what-s-behind-the-decline-in-bees-and-other-pollinators-infographic/ (accessed July 20th, 2021).

Eversemeyer, M. G.; Burleigh, J. B. A Method of Predicting Epidemics Development of Wheat Leaf Rust. Phytopathology **1970,** *60*, 805–811.

Eversemeyer, M. G.; Kramer, B. C. Epidemiology of Wheat Leaf and Stem Rust in the Central Plains of the USA. *Annu. Rev. Phytopathol.* **2000,** *38*, 491–513.

Ezealor, A.; Giles, R. H. Vertebrate Pests of Sahelian Wetland Agroecosystem: Perceptions of Indigenes and Potential Strategies. *Int. J. Pest Manage.* **2010,** *43*, 97–104. DOI: 10.1080/096708797228762 (accessed July 24th, 2021).

Fabrizzi, P. K. Soil Carbon and Nitrogen Organic Fractions in Degraded Versus Non-Degraded Mollisols in Argentina. *Soil Sci. Soc. Am. J.* **2003,** *67*, 1831–1841.

Fang, O. Early Exclusion of Overwintering Cotton Bollworm Moths from Warming Temperatures Accentuates Yield Loss in Wheat. *Agric. Ecosyst. Environ.* **2016,** *217*, 89–98. http://www.fao.org/ag/agll/carbonsequestration/docs/csequestration/docs/cseqtropicaldrylands.hm

FAO. Abiotic Disturbance: Storms. Food an Agricultural Organization of the United Nations, 2005, pp 1–17. http://www.fao.org/forestry/abiotics/71634/en/ (accessed July 17th, 2021).

FAO. Plant Pest and Diseases—FAO in Emergencies. Food and Agricultural Organization of the United Nations, Rome, Italy, 2020, pp 1–6. http://www.fao.org/emergencies/emergency-types/plant-pests-and-diseases/en/ (accessed July 15th, 2021).

FAO. Soil Map of the World. Technical Paper ISRIC, Wageningen Netherlands Revised Reprints, 1997, p 140.

FAO. Soil Map of the World. Technical Paper ISRIC, Wageningen Netherlands Revised Reprints, 1997, p 140.

Farage, P.; Pretty, J.; Ball, A. Carbon Sequestration in Tropical Dry Land Agroecosystem. *Soil Tillage Res.* **2005,** *94* (2), 457–472.

Fitzpatric, U. Protecting Farmland Pollinators. EIP-Agri, European Commission, Brussels, Belgium, 201, pp 1–4. https://ec.europa.eu/eip/agriculture/en/find-connect/projects/protecting-farmland-pollinators/ (accessed June 21st, 2021).

Florio, , J.; Veru, L.; Dao, A. B.; Diallo, M. B.; Sanogo, Z. L.; Samaka, D. B.; Huestis, D. L.; Yossi, O. B.; Talamas, E. C. D.; Chomoro, L. C.; Frank, J. H.E.; Biondi, M. F.; Morkel, C. G.; Bartlett, C. H.; Linton, Y M ; Krajacich, B. J.; Smith, C. S.; Lehman, T. Massive Windborne Migration of Sahelain Insects: Diversity, Seasonality, Altitude and Direction, 2020, pp 1–32. DOI: https://doi.org/10.1101/2020.02.28.960195/ (accessed July 22nd, 2021).

Folland, C. K. Hannaford, J.; Bloomfield, J. P. Multi-Annual Droughts in the English Lowlands: A Review of Their Characteristics and Climate Drivers in the Winter Half Year. *Hydrol. Earth Syst. Sci. Discussions* **2014,** *11*, 12933–12985.

Food and Agricultural Organization of the United Nations. Conservation Agriculture the Key to Food Security. *Fact Sheet* **2015,** *6*, 1–3.

Frank, A. C.; Schulz, S.; Oyewole, D.; Bako, S. Incorporating Short Season Legumes and Green Manure Crops Into Maize-Based Systems in the Moist Guinea Savannah of West Africa. *Exp. Agric.* **2004,** *40*, 463–479.

Franzen, L. G. The Saharan Dust Episode of South and Central Europe, and northern Scandinavia, March 1991. *Weather* **1995,** *50* (9), 313–318.

Free, J. B. *Insect Pollination of Crops*; Academic Press: London, 1993; p 684.

Frost and Sullivan Company. High Altitude Wind Energy-Visionary Outlook, 2015, p 73. https://www.reportlinker.com/p03327515/High-Altitude-Wind-Energy-Visionary-Outlook.htm/ (accessed July 22nd, 2021).

Fryerear, D. W.; Sutherland, P. L.; Davis, G.; Hardee, G.; Dollar, M. Wind Erosion Estimation with RWEQ and WEQ. In *Sustaining the Global Farm*; Stott, D. E., Mohtar, R. H., Steinhardt, G. C., Eds.; United States Department of Agriculture-ARS National Soil Erosion Research Laboratory: Purdue, Indiana, 2001; pp 760–765.

Fuente, F. B.; Lenardis, A. E.; Suarz, S. A.; Gil, A.; Chersa, C. M. Insect Communities Related Wheat and Coriander Cropping Histories and Essential Oils in the Rolling Pampas, Argentina. *Eur. J. Agron* **2006,** *24*, 385–395.

Gagoi, M. M.; Jayachandran, V. N.; Vaishya, A.; Babu, S. N.S.; Satheesh, S. K.; Krishna Moorthy, K. Airborne In Situ Measurements of Aerosol Size Distribution and Black Carbon Across Indo-Gangetic Plains During SWAMI and RAWEX. *Atmos.Chem. Phys.* **2020,** *20*, 8593–8610.

Gahukar R. T. Problems and Perspectives of Pest Management in the Sahel: A Case Study of Pearl Millet, TCrop. *Pest Manage.* **1988,** *34*, 3538.

Gahukar R. T.; Guèvremont T. H.; Bhatnagar, V. S.; Doumbia Y. O.; Ndoye M.; Pierrard G. A Review of the Pest Status of the Millet Spike Worm, Rhaguva albipunctella (De Joannis) (Noctuidae: Lepidoptera) and Its Management in the Sahel. *Insect Sci. App.* **1986,** *7*, 457–463.

Gahukar, R. T. Insect Pests of Pearl Millet in West Africa: A Review. *Trop. Pest Manage.* **1984,** *30*, 142–147.

Gahukar, R. T. Pest and Disease Incidence in Pearl Millet Under Different Planting Density and Intercropping. *Agric. Ecosyst. Environ.* **1989,** *20*, 69–74. https://doi.org/10.1016/0167–8809(89)90040–6/ (accessed July 15th, 2021).

Gahukar, R. T. Population Dynamics of Sorghum Shoot Fly, Atherigona soccata (Diptera: Muscidae), in Senegal. *Environ. Entomol.* **1987,** *16* (4), 910–916.

Gahukar, R. T.; Ba, M. N. An Updated Review of Research on *Heliocheilus albipunctella* (Lepidoptera: Noctuidae) in Sahelian West Africa. *J. Integr. Pest Manage.* **2019,** *10*, 1–19. DOI: 10.1093/jipm/pmz003 (accessed July 14th, 2021).

Gajbhiye, K. S.; Mandal, C. *Agroecological Zones, Their Soil Resources and Cropping Systems*; National Bureau of Soil Survey and Land Use Planning. Nagpur India, 2007; pp 1–32.

Gallai, N.; Salles, J.-M.; Settele, J.; Vaissière, B. E. Economic Valuation of the Vulnerability of World Agriculture Confronted with Pollinator Decline. *Ecol. Econ.* **2009,** *68*, 810–821.

Garba, M. A.; Ozer, P. Assessment of Wind Power Energy Potential in Niamey, Niger. First International Conference on Energy, Environment and Climate Changes, 2011. https://orbi.uliege.be/bitstream/2268/97730/1/ICEC2011_Full_GarbaAbdou%26Ozer.pdf (accessed July 21st, 2021).

Garba, M.; Streito, J. C.; Gauthier, N. First Report of Three Predatory Bugs (Heteroptera: Miridae) in Tomato Fields Infested by the Invasive American Tomato pinworm *Tuta absoluta* in Niger: An Opportunity for Biological Control? *Phytoparasitica* **2020,** *48*, 215–229.

Garcia-Prechac, R.; Ernst, O.; Siri-Prietog and Terra, J. A. Integrating No-Tillage Into Crop-Pasture Rotations in Uruguay. *Soil Tillage Res.* **2004,** *77*, 1–13.

Gavier-Pizarro, Calamari, N.; Thompson, J.; Sonia, B.; Laura, B. C.; Solari, M.; Andrea, J. D.; Bernardos, J. N.; Zaccagnini, M. E. Expansion and Intensification of Row Crop Agriculture in the Pampas and Espinal of Argentina Can Reduce Ecosystem Service Provision by Changing Avian Density. *Agric. Ecosyst. Environ.* **2012,** *154*, 44–55. https://doi.org/10.1016/j.agee.2011.08.013/ (accessed July 25th, 2021).

Ge, M.; Friedrich, J. Four Charts Explain Greenhouse Gas Emissions by Countries and Sectors, 2020. https://www.wri.org/blog/2020/02/greenhouse-gas-emissions-by-country-sector/ (accessed June 26th, 2020).

Gerard, B. It's a Bird, It's a Plane No, It's a Super Scientist. SAT Trends, ICRISAT Newsletter, 2016, p 14. http://www.icrisat.org/what-we-do/satrends/01dec/1.htm (accessed August 3rd, 2021).

Ghosh, S. 'Bd Ozone' from Indo-Gangetic Plains May Be Affecting Entire Indian Sub-Continent. *The Wire*, 2019. https://thewire.in/environment/indo-gangetic-plain-bad-ozone/ (accessed July 6th, 2021).

Giles, K.; Hein, G. L.; Pearis, F. *Areawide Pest Management of Cereal Aphid in Dryland Wheat Systems of the Great Plains of USA*; University of Nebraska: Lincoln, USA, 2008; pp 1–12.

Gilijamse, E. *Downy Mildew in South Niger*; Joint Report of WAU Department of Phytopathology, DFPV, Niamey, Niger, 1997; pp 1–23.

Gilijamse, E.; Frinking, H. D.; Hager, M. J. Occurrence and Epidemiology of Pearl Millet Downy Mildew, Sclerospora Graminicola in Southwest Niger. *Int. J. Pest Manage.* **1997,** *43*, 279–287.

Gilijamse, E.; Jeger, M. J. Epidemiology and Control of Pearl Millet Downy Mildew *Sclerospora graminicola* in Southwestern Niger. In *Advances in downy Mildew Research Spencer-Phillips*; Kluwer Academic Publishers: Dordrecht, Netherlands, 2002; pp 189–193.

Glass, B. In Soaring 'SuperTowers' Aim to Bring Mobile Broadband to Rural Areas (By Steadler, T.). ITU News 2018, pp 1–7. https://news.itu.int/soaring-supertowers-aim-to-bring-mobile-broadband-to-rural-areas/ (accessed July 15th, 2021).

Glatzle, A. Compendio para el manejo de pastures en el Chaco. Proyeceto Estacion Experimental Chaco Central (MAG-GTZ) GTZ. *Ellector* 1999, 188–190.

Goethe University. Holocene Vegetational History of Northeast Nigeria-Sahel: Manga Grass Lands, 2009, pp 1–3. http://www.aaf.stadiumdigitale.uni-frankfurt.de/index.php/en/research/results/88 (accessed August 4th, 2021).

Goudie, A. S.; Middleton, N. J. Saharan Dust Storms: Nature and Consequences. *Earth-Sci. Rev.* **2001,** *56*, 179–204.

Graham, K. We Can Make the Sahara Desert Green Again Through Clean Energy. *Science*, 2018, pp 1–4. http://www.digitaljournal.com/tech-and-science/science/we-can-make-the-sahara-desert-green-again-through-clean-energy/article/531421/ (accessed July 9th, 2021).

Grassi, M. J. Three Agricultural Spray Drone Models That Promise To Be a Breakthrough (Part One). *Meister Media Worldwide*, 2019, pp 1–2. https://www.precisionag.com/in-field-technologies/drones-uavs/three-agricultural-spray-drone-models-that-promise-to-be-breakthroughs-part-one/ (accessed June 6th, 2021).

Gregory, P. H. Distribution of Airborne Pollen and Spores and Their Long-Distance Transport. *Pure Appl. Geophys.* **1978,** *116*, 309–315.

Gregory, P. H. Airborne Microbes: Their Significance and Distribution. *Proc. R. Soc. London* **1971,** *177*, 469–483.

Gregory, P. H. *The Microbiology of the Atmosphere*; Leonard Hill Books: Aylesbury, 1973; p 287.

Griffin, D. W. Terrestrial Microorganisms at an Altitude of 20,000 m in Earth's Atmosphere. *Aerobiologia* **2004,** *20*, 135–140.

Griffin, D. W.; Garrison, V. H.; Herman, J. R.; Shinn, E. A. African Desert Dust in the Caribbean Atmosphere: Microbiology and Public Health. *Aerobiologia* **2001,** *17*, 203–213.

Gross, S. *Renewables, Land Use, and Local Opposition in the United States*; Foreign Policy at the Brookings Institute: USA, 2020; pp. 1–24.

Grote, R.; Lehman, E. Brummer, C.; Nicoloas, B.; Szarzynski, J.; Kuntsmann, H. Modelling and Observation of Biosphere-Atmosphere in Natural Savannah in Burkina Faso, West Africa. *Physiol. Chem. Earth* **2009,** *34*, 251–260.

Gupta, A. *Drones Help Indian Farmers Fighting Stripe Rust Crisis*; Kansas State University: USA, 2016, pp 1–5. https://www.thequint.com/news/environment/drones-farmers-rising-stripe-rust-crisis-wheat-rice-kansas-punjab-bihar-climate-change-kansas-state-university/ (accessed July 30th, 2021).

Hafeez S.; Khan T. H.; Khan T. N.; Shahbaz M.; Ahmed M. Use of Reflector Ribbon as a Pest Birds Repellent in Wheat and Maize Crop. *J. Agric. Soc. Sci.* **2008,** 4, 92–94.

Hall, A. J.; Rebella, C. M.; Ghersa, C. M.; Cullot, J. H. Field Crop Systems of the Pampas. In *Ecosystems of the World: Field Crops Ecosystem*; Pearson, E. J., Ed.; Amsterdam: The Netherlands, 1992; pp 413–450.

Halvorson, A. D.; Del Grasso, S. J. Nitrogen Source and Placement Affect Nitrous Oxide Emissions from Irrigated Corn in Colorado. *Better Crops* **2012,** *96*, 7–9.

Halvorson, A. D.; Del Grasso, S. J.; Alluvione, F. Nitrogen Rate and Source Effects on Nitrous Oxide Emissions from Irrigated Cropping Systems in Colorado. *Better Crops* **2009,** *93*, 16–18.

Halvorson, A. D.; Del Grasso, S. J.; Jantalia, C. P. Nitrous Oxide Emissions from Several Nitrogen Sources Applied to a Strip-Tilled Corn Field. In *Proceedings of Fluid-Form, Fluid Fertilizer Foundation*; Scottsdale, AZ, 2011, pp. 1–6.

Haougui, A.; Garba, M.; Dan Mairo, M.; Adamou, B.; Oumarou, S.; Gougari, B.; Kimba, A.; Abou, M.; Delmas, P. Geographical Distribution of the Tomato Borer, *Tuta absoluta* Meyrick (Lepidoptera, Gelechiidae) in Niger. *Schol. Acad. J. Biosci.* **2017,** *5* (2), 108–113.

Haq, Z. Inter-Ministerial Group Set Up to Tackle Locust Invasion Across States, 2020, pp 1–4 https://www.hindustantimes.com/india-news/interministerial-group-set-up-to-tackle-locust-invasion-across-states/story-xbWsSmF2FkGiqKiaTjNlSN.html/ (accessed July 23rd, 2021).

Hau, B.; de Vallavielle-Pope, S. Wind Dispersed Diseases. In *The Epidemiology of Plant Diseases*; Jones, G., Ed.; Springer: Heidelberg, Germany, 2018; pp 323–347.

Hausmann, B.; Some, B.; Schuch, D. Advancing Together: Overview of the Crop Proliferation in West Africa as of 2020. Collaborative Crop Research Program-CCRP. McKnight Foundation, 2020, pp 1–31. https://www.ccrp.org/resources/advancing-together-%C2%B7-overview-of-the-ccrp-portfolio-in-west-africa-as-of-2020/ (accessed July 14th, 2021).

Heathcoat, R. L. *Perception of Desertification in the Southern Great Plains—A Preliminary Enquiry*; United Nations University: Tokyo, Japan, 1980; p 134.

Hedley, T. Senegal Commits to Renewable Energy, 2019, pp 1–5. https://www.africaoilandpower.com/2019/09/03/senegal-comments-to-renewable-energy/ (accessed July 11th, 2021).

Helbaum, N. The Bases for a Classification of World Agriculture. *Prof. Geogr.* **2010,** *9*, 2–17.

Hervas, A.; Camarero, L.; Reche, I.; Casamayor, E. O. Viability and Potential for Immigration of Airborne Bacteria from Africa That Reach High Mountain Lakes in Europe. *Environ. Microbiol.* **2009,** *11* (6), 1612–1623.

Hicks, C. Sahel Locust Invasion Threatens Niger and Mali. *The Guardian*, 2012, pp 1–3. https://www.theguardian.com/global-development/2012/jun/11/sahel-locust-invastion-niger-mali/ (accessed July 11th, 2021).

Hirst, J. M. Changes in Atmospheric Spore Content: Diurnal Periodicity and the Effects of Weather. *Trans. Br. Mycol. Soc.* **1953,** *36*, 375–393.

Hirst, J. M. The Aerobiology of Puccinia graminis Uredospores. *Trans. Br.Mycol. Soc.* **1961,** *44*, 138–139.

Hirst, J. M.; Stedman, O. J. Dry Liberation of Fungal Spores by Raindrops. *J. Gen. Microbiol.* **1963,** *33*, 335–344.

Hitaj, C.; Hunt, K.; Smith, D. J. Honeybees on the Move. Pollination and Honey Production. In *Proceedings of the Conference 2017 Agriculture and Applied Economics Association Annual Meeting*, Washington, DC, USA, 2018, pp. 1–8. https://www.researchgate.net/publication/330686236_Honeybees_on_the_move_Pollination_services_and_honey_production (accessed July 23rd, 2021).

Hockenos, P. Red alarm: Europe's Next Drought Is Already Upon Us, 2020, pp 1–3. https://energytransition.org/2020/04/red-alarm-europes-next-drought-is-already-upon-us/ (accessed June 24th, 2021).

Hodges, S. C. *Environmental Guidelines for Plant Nutrient Used*; The University of Georgia Extension Service: Athens Georgia, USA, 1991; pp. 1–5.

Hodson, D. A Global Wheat Rust Monitoring System. Study Identifies Likely Senerios for Global Spread of Devastating Crop Disease. RustTracker.org., 2018, pp 1–17.

Hudson, J. C. Agriculture. In *Encyclopaedia of the Great Plains*; Wishart, D. J., Ed.; 2012; pp. 1–8.

Hussain, I.; Afzal, W. Insectivorous Birds and Their Significance in Cotton-Wheat Based Agroecosystem of Punjab, Pakistan. *Pak. J. Zool.* **2005,** *37*, 133–143.

Hussain, T. Pesticide Use and Its Impact of Crop Ecologies: Issues and Options. Working Paper Series No. 42. Sustainable Development Policy Institute (accessed SDPI), Islamabad, Pakistan, 1999, p 114.

IANS. In a First, Uttarakhand Monitors Forest Fires with Drones, 2017, pp 1–8. http://www.saharasamay.com/regional-news/uttarakhand-news/676611474/in-a-first-uttarakhand-monitors-forest-fires-with-drones.html/ (accessed September 30th, 2021).

IITA. Cowpea International Institute for the Tropical Agriculture, Ibadan, Nigeria, 2000, p 16. www.iita.org/info/impact/Cowpea.pdf/ (accessed July 22nd, 2021).

Ijaz, S. S. Organic Matter Status of Pakistan Soils and Its Management. Pakkissan, 2013, pp. 1–3. http://www.pakkissan.com/english/advisory/organic.farming.matterr.status.of-pakistan.shtml (accessed July 23rd, 2021).

Imrie, B. Cowpea. International Institute for Tropical Agriculture, 2000, pp 1–71. www.rirdc.go.au/pub/handbook/cowpea. Pdf/ (accessed July 23rd, 2021).

India Today. Worst Attack in 27 Years: Swarms of Locusts Destroy Crops in Several States, 2020, pp 1–3. https://www.indiatoday.in/india/story/

worst-locust-attack-in-27-years-swarms-destroy-crops-across-north-india-1681782–2020–05 25#: ~:text=Locust%20swarms%20have%20travelled%20from, insect%20attacks%20on%20their%20crops (accessed August 5th, 2021).

Intergovernmental Science-Policy Platform on Biodiversity IPBES. The Regional Report for Africa on Pollinators and Pollination and Food Production. IPBES-Executive summary, Kaulalumpur, Malaysia, 2012, pp 1–43. https://www.cbd.int/gbo/gbo4/African-Pollinators-en.pdf/ (accessed July 23rd, 2021).

Invierta en Argentina. Renewable Energy Sector in Argentina, 2012, pp 1–8. https://energypedia.info/wiki/Wind_Energy_Country_Analysis_Argentina (accessed July 7th, 2021).

IPM CRSP. Significant Achievement of the IPM CRSP in Uganda, 2000. www.aaec.t.edu/pmcrsuganda/annual20%report/highlight.htm) (accessed July 23rd, 2021).

IRRI. Rice in India International Rice Research Institute, Manila, India, 2013b, pp 1–4. http://www.irri.org/inex.php?option-com-k27view=itm&id=8744:rice-in-india&lang=en (accessed October 15th, 2021).

IRRI. Rice in Pakistan. International Rice Research Institute, Manila, Philippines, 2013a, pp 1–3. http://www.irri.org/inex.php?option-com-k27view=itm&id==12077:rice-in-pakistan&lang=en/ (accessed Aug 2nd, 2021).

Islam, S. U.; Inayathullah, M.; Jan, S.; Ibrahim, M.; Shah, S. J. A. Aphis Craccivora Koch on Wheat in Pakistan. *Int. J. Farm. Allied Eng.* **2015,** *4*, 86–88.

Jain, N. Poor Air Quality Over Northern India from Desert Storms in May 2018. Mongabay Series, Environment and Health, 2019, pp 1–8. https://india.mongabay.com/2019/05/poor-air-quality-over-northern-india-from-dust-storms-in-may-2018/ (accessed September 6th, 2021).

Jain, N.; Dubey, R.; Dubey, D. S.; Singh, J. Khanna, M.; Pathak, H.; Bhatia, A. Mitigation of Greenhouse Gas Emission with System of Rice Intensification in the Indo-Gangetic Plains. Paddy and Water Environment. Springer, 2014; pp 1–16. DOI: 10.1007/s10333–013–0390–2 (accessed September 13th, 2021).

Jalloh, A.; Faye, M. D.; Roy-McCaulay, H.; Sereme, P.; Zougmore, P.; Thomas, T. S.; Nelson, G. C. Overview. In *West African Agriculture and Climate Change*; International Food Policy Research Institute: Washington, DC, 2013a; pp 1–30.

Jalloh, A.; Faye, M. D.; Roy-McCaulay, H.; Sereme, P.; Zougmore, P.; Thomas, T. S.; Nelson, G. C. Summary and Conclusions. In *West African Agriculture and Climate Change*; International Food Policy Research Institute: Washington, DC, 2013b, pp 1–30.

Jarroudi, M. E.; Kouadi, L.; Tycheoon, B.; Jarroudi, M. E.; Jung, J.; Bock, C.; Delfosse. Modeling the Fungal Diseases of Winter Wheat: Constraints and Possible Condition, 2018, pp. 1–35 (June). https://www.intechopen.com/books/advances-in-plant-pathology/modeling-the-main-fungal-diseases-of-winter-wheat-constraints-and-possible-solutions/ (accessed July 23rd, 2021).

Jasrotia, P.; Kumar, R. Management of Major Insect Pest of Wheat Under Field and Storage Conditions. Indian Institute for Wheat Research, Karnal, 2019, pp 1–3. https://www.iiwbr.org/project-management-of-major-insect-pests-of-wheat-under-field-and-storage-conditions/ (accessed September 4th, 2021).

Jeanneret, P.; Begg, J.; Gosme, M.; Alomar, O.; Reubens, B. Landscape Features to Improve Pest Control in Agriculture, 2016, p 1–6. https://www.thesolutionsjournal.com/article/landscape-features-improve-pest-control-agriculture/ (accessed July 27th, 2021).

Jiang, X.; Lau, N.; Heldt, I. M.; Ploshay, J. J. Mechanism of the Great Plains Low-Land Jet Is Simulated in and AGCM. *J. Atmos. Sci.* **2007,** *64*, 532–541.

Jones, P. D.; Moberg, A. Hemispheric and Largescale Surface Air Temperature Variations: An Extensive Revision and an Update Till 2001. *J. Climatol.* **2003,** *16*, 206–223.

Joshi, L. M.; Palmer, L. T. Epidemiology of Stem Leaf and Stripe Rusts of Wheat in Northern India. *Plant Dis. Rep.* **2019,** *57*, 8–12.

Jossi, F. Industry Report: Midwest and Great Plains Lead Wind Energy Expansion, 2017, pp 1–8. https://energynews.us/2017/04/19/midwest/industry-report-midwest-and-great-plains-lead-wind-energy-expansion/ (accessed July 24th, 2021).

Kakde, U, B. Fungal Bioaerosols: Global Diversity, Distribution, and Its Imapct on Human Beings and Agriculture Crops. *Bionano Front.* 2018, *5*, 323–329.

Kale, M.; Balfors, B.; Mortberg, U.; Bhattacharya, P.; Chakane, S. Damage to Agricultural Yield Due to Farmland Birds, Present Repelling Techniques and Its Impact: Am Insight from the Indian Perspective. *J. Agric. Technol.* **2012,** *8*, 49–62.

Kaleem, F. Z. Assessment of Nitrogen Fixation by Legumes and Their Relative N Contribution to a Succeeding Maize Crop. Annual Report 1988/89 Nynakpal Agricultural Experimental Station Tamale, Ghana, 1989; pp 23–28.

Kamara, A. Y.; Ekeleme, F.; Chikoye, D.; Omogui, L. O. 2009 Planting Date and Cultivar Effects on Grain Yield in Dry Land Corn Production. *Agron. J.* **2009,** *101*, 91–98.

Kansas State University. Project Using Drones to Detect Emerging Insect Pests, Disease in Crops. Department of Entomology, Kansas State University at Salinas and Manhattan, KS, USA, 2015. https://www.ksnt.com/news/kansas-state-researches-using-drones-to-protect-crops/ (accessed April 5th, 2017).

Katalin, B.; Ioannis, K.; Nigel, T.; and Jaeger-Weldan, A. Solar Photovoltaic Electricity Generation: A Lifeline for the European Coal Regions in Transition. European Commission: Belgium, 2009, pp 1–8. DOI:10.3390/su11133703 (accessed July 2nd, 2021).

Keith, J. O.; Bruggers, R. L. Review of Hazards to Raptors from Pest Control in Sahelian Africa. *J. Raptors Res.* **1998,** *32*, 151–158.

Keitichi, H.; Toshiyuki, W. Sustainable Soil Fertility Management by Indigenous and Scientific Knowledge in the Sahel Zone Niger. In *Proceedings of 17th World Congress of Soil Science Symposium No. 15 Paper No. 1251*; 2002; pp 611–612.

Keller, M.; Bergstrom, G.; Shields, F. Aerobiology of Fusarium graminearum. *Aerobiologia* **2014,** *30*, 1–8. DOI: 10.1007/s10453–013–9321–3 (accessed July 16th, 2021).

Kellog, C. A.; Griffin, D. W.; garrison, V.; Kealy Pak, K.; Royall, N.; Raymond, R. S.; Shinn, E. A. Characteristics of Aerosolized Bacteria and Fungi for Desert Dust Event in Mali, West Africa. *Aerobiologia* **2004,** *20*, 99–110.

Khairul Alam, M.; Biswas, K.; Bell, R. W. Greenhouse Gas Implications of Novel and Conventional Rice Production Technologies in the Eastern Gangetic Plains. *J. Clean Prod.* **2016,** *112*, 3977–3987.

Khan, A. R.; Qauyoom, M. A.; Khan, M. H. Insectivorous Behaviour of Some Faisalabad Birds. *Pak. J. Entomol.* **1980,** *2*, 21–26.

KITEnrg S. R.L. Automatically Controlled Airfoils to Harvest High Altitude Wind Energy. Kite Energy S.R.L. Demerghitta, Italy, 2019, pp 1–4 http://www.kitenergy.net/technology-2/key-points/ (accessed July 30th, 2021).

Kjellström, E. Recent and Future Signatures of Climate Change in Europe. *Ambio* **2004,** *33*, 193–198.

Kjellstrom, E.; Barring, L.; Jacob, D.; Jones, R.; Landerink, G.; Schar, C. Variability in Daily Maximum and Minimum Temperatures: Recent and Future Changes Over Europe. *Climatic Change* **2007,** *81*, 249–265.

Klein, A.; Vaissiere, B. E.; Cane, J. H.; Steffan-Dewenter, I.; Cunningham, S. A.; Kreman, R.; Tschrantke, T. Importance of Pollinators in Changing Landscapes for World Crops. *Proc. R. Soc. B* **2007,** *274*, 303–313.

Knodel, J. J.; Charlet, L. D.; Gavloski, J. *Integrated Pest Management of Sunflower Insect Pests in the North Great Plains*; North Dakota State University Extension Service: Fargo, North Dakota, USA, 2015, pp. 1–20.

Knodel. J. J.; Shrestha, G. Pulse Crops. Pest Management of Wireworms and Cutworms in the Northern Great Plains of USA and Canada; *Ann. Entomol. Soc. Am.* **2018,** *111*, 195–204.

Koshal, A. K. Analysis of Pest Detection in Cropping Systems of Indo-Gangetic Plains Through Remote Sensing and GIS, 2019, pp 1–14. https://www.geospatialworld.net/article/pest-detection-in-cropping-systems-of-indo-gangetic-plains-through-remote-sensing/ (accessed July 23rd, 2021).

Koshal, A. K. Changing Scenario of Rice-Wheat System of Indo-Gangetic Plains of India. *Int. J. Sci. Res. Pub.* **2014,** *4*, 1–13.

Kosivart. Ukraine. Climate, 2013a, pp 1–5. http://www.kosivart.com/eng/index.cfm/do/ukraine.climate/ (accessed July 23rd, 2021).

Kosivart. Ukraine. Land, 2013b, pp 1–3. http://www.kosivart.com/eng/index.cfm/do/ukraine.land/ (accessed July 19th, 2021).

Kremen, C.; Williams, N. M.; Thorp, R. W. Crop Pollination from Native Bees at Risk from Agricultural Intensification. *Proc. Natl. Acad. Sci. USA* **2002,** *99*, 16812–16816. https://doi.org/10.1073/pnas.262413599 PMID: 12486221/ (accessed July 24th, 2021).

Krishna K. R. *Aerial Robotics in Agriculture: Parafoils, Blimps, Aerostats and Kites*; Apple Academic Press Inc.: Palm Bay, FL, 2020b, p 416.

Krishna, K. R. *Agricultural Drones: A Peaceful Pursuit*; Apple Academic Press Inc.: Waretown, NJ, 2018; p 425.

Krishna, K. R. *Agricultural Prairies: Natural resources and Crop Productivity*; Apple Academic Press Inc.: Waretown, NJ, 2015; p 499.

Krishna, K. R. *Agrosphere: Nutrient Dynamics, Ecology and Productivity*; Science Publishers Inc. Enfield: New Hampshire, 2003; p 346.

Krishna, K. R. *Peanut Agroecosystem: Nutrient Dynamics and Productivity*; Alpha Science Publishing: Oxford, 2008; p 298.

Krishna, K. R. *Unmanned Aerial Vehicle Systems in Crop Production*; Apple Academic Press Inc.: Palm Bay, FL, 2020a, p 689.

Krishna, K. R. Vesicular Arbuscular Mycorrhizae and Phosphorus Nutrition of Pearl Millet in West Africa. Pearl Millet Improvement Program. International Crops Research Institute for the Semi-arid tropics, Patancheru 502324, Andhra Pradesh, India, Internal Report, 1986, pp 1–33.

Kubilay, N.; Nickovic, S.; Moulin, C.; Dulac, F. An Illustration of the Transport and Deposition of Mineral Dust Onto the Eastern Mediterranean. *Atmos Environ.* 2000, *34* (8), 1293–1303.

Kukal, M.S.; Irmak, S. *Climate-Driven Crop Yield and Yield Variability Impacts on the US Great Plains Agricultural Production*; Scientific Reports, 2018; pp 1–12. DOI:10.1038/s41598–018–21848–2/ (accessed Aug 3rd, 2020).

Kulkarni, V.; Phadnis, A. *Centre Allows Use of Drones to Control Locusts*; Agribusiness: New Delhi, 2020. https://www.thehindubusinessline.com/economy/agri-business/

centre-allows-use-of-drones-to-control-locusts/article31643470.ece/ (accessed July 23rd, 2021).

Kumar, V.; Jana, S.; Bhardwaj, A.; Deepa, R.; Sahu, S. K.; Pradhan, P. K.; Sirdas, S. A. Greenhouse Gas Emission, Rainfall and Crop Production Over North India. *Open Ecol. J* **2018,** *11*, 47–61. DOI: 10.2174/1874213001811010047 (accessed July 12, 2021).

Kunast, C.; Riffel, M.; deGraeff, R.; Whitmore, G. *Pollinators and Agriculture: Agricultural Productivity and Pollinator Protection*; European Crop Protection Agency, 2020; p 96.

Kunkel, K. E.; Stevens, L. E.; Sun, L. *Regional Climate Trends and Scenarios of the U. S. National Climate Assessment. Part 4: Climate of the US Great Plains*; National Oceanic and Atmospheric Administration: Washington, DC, 2013; p 91.

Kusters, H. 2000. *Northern Europe: Germany and Surrounding Regions*; Kipple, K. E., Ornelas, K., Eds., Vol. 2; The Cambridge World History of Food, 2000; pp 1226–1231.

Kutama, A. S.; Aliyu, B. S.; Emechebe, A. M. State of Sorghum Downy Mildew in Maize in the Sudan and Sahel Savanna Agro-Ecological Zones of Nigeria. *Bayero J. Pure Appl. Sci.* **2010,** *3* (1), 233–237.

Kuttippurath, J.; Singh K. A.; Dash, S. P.; Malik, N.; Clerbaux, M.; Van Denme, M.; Clarrise, P.; Raj, K.; Varikoden A. H. Record High Levels of Atmospheric Ammonia Over India: Spatial and Temporal Analysis, 2020a, pp 1–12. https://www.sciencedirect.com/science/article/pii/S0048969720335063/ (accessed July 7th, 2021).

Kuttippurath, J.; Singh, K. A.; Dash, S. P.; Mallick, N.; Damme, M. V.; Clarisse, P.; Cobeur, E.; Rai, K. Abishek, K.; Varikoden, H. Record High Levels of Atmospheric Ammonia Over India: Spatial and Temporal Analysis, 2020b, pp 1–22. https://doi.org/10.1016/j.scitotenv.2020.139986/ (accessed July 12th, 2021).

Kutty, S. G.; Dimri, A. P.; Gultene, I. Climatic Trends in Fog Occurrence Over the Indo-Gangetic Plains. *Int. J. Climatol.* **2019,** 1–8. https://doi.org/10.1002/joc.6317/ (accessed July 7th, 2021).

Labuschagne, L.; Swanepod, L.; Taylor, P. I.; Bilmani, I. R. 2016 Are Avian Predators' Effective Biological Agents for Pest Management in Agriculture Systems. *Biol. Control* **2016,** *101*, 1–14. DOI: 10.1016/j.biocontrol.2016.07.003/ (accessed July 19th, 2021).

Lal.; Ghude, S. D.; Kulkarni, S. H.; Jena, C.; Tiwari, S.; Srivasava, M. K. Tropospheric Ozone annd Aerosol Long Term Trends Over the Indo-Gangetic Plain (IGP), India. *Atms. Res.* **2012,** *116*, 82–92.

Lautenbach, S.; Seppelt, R.; Liebscher, J.; Dormann, C. F. Spatial and Temporal Trends of Global Pollination Benefit. *PLoS ONE* **2012,** *7* (4), e35954. DOI: 10.1371/journal.pone.0035954 (accessed July 23rd, 2021).

Le Feon, V.; Pogio, S. L.; Torretta, J. P.; Bertrand, C.; Gonzalo, A.; Molina, R.; Burrel, F. Diversity and Life-History Traits of Wild Bees (Hymenoptera) in Intensive Agricultural Landscapes in the Rolling Pampa Argentina. *J. Nat. History* **2014,** *50*, 1175–1196 https://doi.org/10.1080/00222933.2015.1113315/ (accessed August 8th, 2021).

Le Roux, J.; Rijkenberg, F. H. J. Pathotypes of Puccinia Graminis Striiformis with Increase Virulence for Sr 24. *Plant Dis.* **1987,** *71*, 1118–1119.

Lelong, C. C.; Burger, P.; Jubein, G.; Roux, B.; Labbe, S.; Baret, F. 2016 Assessment of Unmanned Aerial Vehicles Imagery for Quantitative Monitoring of Wheat Crop in Small Plots. *Sensors* **2016,** *8*, 3557–3585.

Leuven, K. U. Hydropower Plants to Support Solar and Wind Energy in West Africa, 2020, pp 1–5. https://nieuws.kuleuven.be/en/content/2020/hydropower-plants-to-support-solar-and-wind-energy (accessed July 9th, 2021).

Levitan D. High Altitude Wind Energy Huge Potential-and Hurdles. YALE Environment 360 2012, pp 1–8. https://e360.yale.edu/features/high_altitude_wind_energy_huge_potential_and_hurdles/ (accessed July 25th, 2021).

Ley, E. L.; Buchmann, S.; Stricht, L. *Great Plains Steppe and Shrub Province, Including Oklahoma and Texas*; North American Pollinator Campaign: Washington, DC, 2018; pp 1–24.

Li, T.; Angeles, O.; Marcaida III, M.; Manalo, E.; Manalili, M. P.; Radanielson, A.; & Mohanty, S. From ORYZA2000 to ORYZA (v3): An Improved Simulation Model for Rice in Drought and Nitrogen Deficient Environments. *Agric. Meteorol.* **2017,** *237*, 246–256.

Linz, G. M.; Homan, H. J. Use of Glyphosate for Managing Invasive Cattail (*Typha spp.*) to Disperse Blackbird (Icteridae) Roosts. *Crop Protect.* **2011,** *30*, 98–104.

Linz, G. M.; Homan, H. J.; Bleir, W. J. Blackbird Densities and Sunflower Damage in North Dakota and South Dakota 1996–1998. *Proc. Sunflower Res. Workshop-1999.* **2001,** *21*, 136–129.

Linz, G. M.; Homan, H. J.; Werner, S. J.; Hagy, H. M.; Bleier, W. J. Assessment of Bird-Management Strategies to Protect Sunflowers. *BioScience* **2011,** *61*, 960–997.

Little, J. The Decline of Pollinators in the UK Not According to the Latest Report, 2017, pp 1–2. https://cropscience.bayer.co.uk/blog/articles/2017/08/the-decline-of-pollinators-in-the-uk-not-according-to-the-latest-report/ (accessed July 20th, 2021).

Ljungquist, F. C. European Warm Season Temperature and Hydroclimate Since 850 CE. *Environ. Res. Lett.* **2019,** *14*, 084015 (accessed July 21st, 2021).

Lloyd, M. Crosswind Kite Power. *J. Energy* **1980,** *3*, 10–113.

Loriatti, C.; Lucchi, A. J. Semi-Chemical Strategies for Tortricid Moth Control in Apple Orchards and Vineyards in Italy. *Chem. Ecol.* 2016, pp 1–14. DOI: 10.1007/s10886–016–0722-y/ (accessed June 19th, 2021).

Loring, K. Soil Conservation and Allotment Act-1935 Major Acts of Congress, 2004, pp 1–3.

Lothar Beyer. Soil Geography and Sustainability of Cultivation. In *Soil Fertility and Crop Production* Krishna, K. R., Ed.; Science Publishers Enfield: New Hampshire, 2002; pp 33–64.

LovePilot, M. D.; Martin, J. M. Saharan Dust Input to the Western Mediterranean: Eleven Years Record in Corsica. *Environ. Sci. Technol. Lib.* **1996,** *11*, 191–199.

Lucas, G. B.; Lee Cambell, C.; Lucas, L. T. *Diseases Caused By Airborne Fungi*; Springer: Heidelberg, Germany, 2018; pp 192–245.

Lupoa, L. J. Birds That Are Considered Pest Birds, 2019, pp 1–2. https://www.thespruce.com/natures-beauty-spoiled-by-pest-birds-2656529/ (accessed July 19th, 2020).

Lyons, C. The Year of Locust in Great Plains, 2018, pp 1–8. https://www.historynet.com/1874-the-year-of-the-locust.htm/ (accessed July 17th, 2021).

Lyons, D. J.; Smith, J. A. Wind Erosion and Its Control. Institute for Agriculture and Natural Resources. University of Nebraska, Neb Guide G1537, 2010, pp 1–8 https://extensionpublications.unl.edu/assets/pdf/g1537.pdf/ (accessed July 2021).

Madden, L. V. Rainfall and the Dispersal of Fungal Spores. *Adv. Plant Pathol.* **1992,** *8*, 39–79.

Magor, J. I.; Ward, P. Illustrated Descriptions, Distribution Maps and Bibliography of the Species of Quelea (Weaverbirds: Ploceidae). Tropical Pest Bulletin 1, Centre for Overseas Pest Research, London, 1972; p 23.

Maheshwary, R. Tracing the Footsteps of Aphids to Save India's Mustard Crops. *Res. Matters*. 2019, pp 1–5 https://researchmatters.in/news/tracing-footsteps-aphids-save-india%E2%80%99s-mustard-crops/ (accessed August 5th, 2021).

Majewska, B. M.; Majewski, J. Importance of Bees Pollination of Crops in the European Union Countries. In *Proceedings of the 2016 International Conference 'Economic Science for Rural development;* LLU ESAF: Poland, 2017; pp 114–119.

Malik, S. M.; Awan, H.; Khan, N. Mapping Vulnerability to Climate Change and Its Repercussions on Human Health in Pakistan. *Global Health* **2012,** *8*, 31–35.

Malla, G.; Bhatia, A.; Pathak, M.; Prasad, S.; Jain, N.; Singh, J. Mitigating Nitrous Oxide and Methane Emissions from Soil in Rice-Wheat System of the Indo-Gangetic Plains with Nitrification and Urease Inhibitors. *Chemosphere* **2004,** *58*, 141–147.

Manikowski, S. Birds Injurious to Crops in West Africa. *Trop. Pest Manage.* **1984,** *30* (4), 379–387.

Manlay, R. J. Masse, D.; Chotte, J. C.; Fellr, C.; Kaire, M.; Fardoux, J.; Pontanier, R. 2002b Carbon, Nitrogen and Phosphorus Allocation in Agroecosystems of a West African Savannah 2. The Soil Component Under Semi-Permanent Cultivation. *Agric., Ecosyst. Environ.* **2002b,** *88*: 215–232.

Manlay, R. J.; Chotte, J.; Masse, D.; Laurent, J.; Feller, C. Carbon, Nitrogen and Phosphorus Allocation in Agroecosystems. Of a West African Savanha 3 Plant and Soil Components Under Continuous Cultivation. *Agroecosyst. Environ.* **2002a,** *88*, 249–264.

Marachi, G.; Sirotenko, O.; Bindi, M. Impacts of Present and Future Climate Variability on Agriculture and Forestry in the Temperate Regions. *Eur. Clim. Change* **2005,** *70*, 117–135.

Marchigiani, R.; Gordy, S.; Cipolla, J. Adams, R. A.; Evans, D. C.; Stey, S.; Galwankar, S.; Russell, S.; Marco, A. P.; Kmans, N.; Bhoi, S.; Papdimos, D. J. Wind Disasters A Comprehensive Review of Current Management Strategies. *Int. J. Crit. Illness Sci.* **2013,** *3*, 72–79.

Martin, M. J. How to Choose a Wind Turbine, 2019, pp 1–3. https://homeguides.sfgate.com/choose-wind-turbine-78770.html/ (accessed July 28th, 2021).

Mason, S. E.; Maman, N.; Pale, S. Pearl Millet Production in Semi-Arid West Africa: A Review. *Exp. Agric.* **2015,** *51*, 501–521.

Mbaye, D. F. Les Maladies du mil au Sahel. CiLLS/INSAH, Bamako, Mali, 1992, pp 42–53.

McDonald, B. A.; Stuckenbrock, E. H. Rapid Emergence of Pathogens in Agroecosystem: Global Threats to Agricultural Sustainability and Food Security. *Phil. Trans. R. Soc. B* **2016,** *371*, 20160026. DOI: http://dx.doi.org/10.1098/rstd.20160026.

McGlashen, A. Protecting Crops with Predators Instead of Poisons, 2018, pp 1–5. https://ensia.com/features/predatory-birds/ (accessed July 19th, 2021).

McRae, T. C. Sunflower looper-*Rachiplusia nu*, 2011, pp 1–5. https://beetlesinthebush.com/2011/11/13/sunflower-looper-rachiplusia-nu/ (accessed August 2nd, 2021).

McWilliam, A. N.; Cheke, R. A. A Review of the Impacts of Control Operations Against Red-Billed Quelea (*Quelea quelea*) on Non-Target Organisms. *Environ. Conserv.* **2004,** *31*, 130–137.

MEDA. Small Wind Energy and Hybrid Systems Program, 2020. https://www.mahaurja.com/meda/en/off_grid_power/small_wind_solar_hybrid/ (accessed July 26th, 2021).

Medan, D.; Hodara, K.; Toretta, J. P.; Fuenta, E. 2011 Effect of Agriculture on the Vertebrate and Invertebrate Diversity in the Pampas of Argentina. *Biodiversity Conserv* **2011,** *20*, 3077–3100.

Mehta, K. C. *Further Studies on Cereal Rusts of India, Part II. Scientific Monographs 18*; Indian Council Agri. Res.: New Delhi,1952; p 244.

Meissle, M.; Mouran, P.; Musa, J.; Bigler, F. 2010 Pests, Pesticides Use and Alternative Options in European Maize Production. Current Status and Future Prospects. *J. Appl.*

Entomol. **2010,** *134*, 357–375. DOI: 10.1111/j.1439–0418.2009.01491.x/ (accessed July 28th, 2021).

Meola, M.; Lazzaro, A.; Zeyer, J. Bacterial Composition and Survival on Sahara Dust Particles Transported to the European Alps. *Front. Microbiol.* **2015,** *6*, 14–54.

Mersereau, D. Why the Weather on the Great Plains Is So Extreme, 2017, pp 1–3. https://www.mentalfloss.com/article/502515/why-weather-great-plains-so-extreme/ (accessed July 26th, 2021).

Meyer, M.; Cox, J. C.; Hitchings, M. D.T, Burgin, L.; Hort, M. C.; Hodson, D. P.; Gilligan. S. Quantifying Airborne Dispersal Routes of Pathogens Over Continents to Safeguard Global Wheat Supply. *Nature Plants* **2017,** *3*, 780–786.

Meyerhoff, H. Argentine Gets Smarter in the Sustainable Agriculture, 2019, pp 1–3 https://www.forbes.com/sites/sap/2019/10/22/argentina-grows-a-more-sustainable-and-efficient-agricultural-sector-with-machine-learning-and-geospatial-technology/#24645edf63bc/ (accessed July 23rd, 2021).

Michelina, R. O.; Irrutia, C. B. Susceptibility of Soil to Wind Erosion in La Pampa Province, Argentina. *Arid Soil Res. Rehab.* **1995,** *9*, 227–234.

Michels, K.; Shivakumar, M. V.K.; Allison, B. E. Wind Erosion Control Using Crop Residue. 1. Soil Flux and Particles. *Field Crops Res.* **1995,** *40*, 101–110.

Mijatovic, A. *Compact Aerial system: Specifications*; Aero Drum Ltd. Technical Specifications, 2014; pp 1–2. http://www.rc-zeppelin.com/compact-aerial-photography-systems.html (accessed July 23rd, 2021).

Miller, R. H.; Pike, K. S. *Insects in Wheat-Based Systems*; Food and Agricultural Organization of the United Nations: Rome, Italy, 2001; pp 1–6 (accessed June 29th, 2021).

Ministerie van Landbaw: Natuur en Voedselkwalitelt. Spain: Agricultural Land Turning Into Solar Farms, 2020, pp 1–6. https://www.agroberichtenbuitenland.nl/actueel/nieuws/2020/03/03/spain-agricultural-lands-turning-into-solar-farms/ (accessed July 1st, 2021).

Mols, C. M.M.; Visser, M. E. Great Tits (*Parus major*) Reduce Caterpillar Damage in Apple Orchards, 2007, pp 1–9. DOI: 10.1371/journal.pone.0000202 (accessed June 19th, 2021).

Moon, D. *The Plough That Broke the Steppes. Agriculture on Russian Grasslands. Agriculture and Environment on Russian Grassland. 1700–1994*; Oxford University Press: Oxford, 2013; p 313.

Moon, D. The Russian Steppes. An Environmental History 1700–1994 CRCEES Working Papers WP2008/071–15, 2012. http://www.gla.ac.uk/media/media_209913_en.pdf/ (accessed July 21st, 2021).

More, C. Great Plains States Dominate Renewables Generation in 2018, 2019, pp 1–5. https://www.utilitydive.com/news/great-plains-states-dominate-renewables-generation-in-2018/549846/ (accessed May30th, 2021).

Mortenson, E. Oregon Firm Developing Airborne Wind Energy System, 2016, pp 1–5. https://www.capitalpress.com/state/oregon/oregon-firm-developing-airborne-wind-energy-system/article_4e1227e1–4a47–5261–907a-76008e16d9fd.html (accessed July 31st, 2021).

Mostacelli, G.; Pazos, M. S. Soils of Argentina-Nature and Use, 2000, p 17 http://www.elsitioagricola.com/articulos/mostacelli/soilsofargentina-natureanduse.asp (accessed August 19th, 2021).

Motyka, M. Renewable Energy Industry Primed for Continued Growth: 2020 Renewable Energy Industry Outlook, 2020, pp 1–7. https://www2.deloitte.com/us/en/pages/energy-and-resources/articles/renewable-energy-outlook.html/ (accessed June 1st, 2021).

Mulie, W. C.; Keith, J. O. The Effect of Aerially Applied Fenitrothion and Chlorpyrifos on Birds in the Savanna of Northern Senegal. *J. Appl. Ecol.* **1993,** *3*, 536–550. DOI: 10.2307/2404193 (accessed July 16th, 2021).

Multiza, S.; Heslop, D.; Pittauerova, D. Fischer, H.; Meyer, I. Stuut, J. B.; Zahle, M.; Mollenhauer, G.; Collins, J. A.; Kuhnert, H.; Schultz, M. Increase in West African Dust Flux at the Onset of Commercial Agriculture in the Sahel Region. *Nature* **2010,** *466*, 226–288.

Murali, A. How Drones Are Paving a New Path for Precision Agriculture in India, 2018, pp 1–12. https://factordaily.com/drones-for-precision-agriculture-in-india// (accessed July 24th, 2021).

MyFields. Russian Wheat Aphid, 2020, pp 1–8. https://www.myfields.info/book/cereal-aphids/ (accessed July 19th, 2021).

Naab, J. B. The Role of Legume-Maize Rotations in Sustainable Intensified Maize Seed Farming Systems in West and Central Africa, 2001, pp 269–280 http://iita.org/info/wecmansection%2001.pdf (accessed July 220th, 2021).

Nagarajan, S.; Joshi, L. M. Epidemiology in the Indian Subcontinent. In *The Cereal Rusts, Vol. 2, Diseases, Distribution, Epidemiology, and control*; Roelfs, A. P.; Bushnell, W. R., Eds.; Academic Press: Orlando, FL, 1985; pp 371–402.

Nagarajan, S.; Singh, D. V. Long Distance Dispersion of Rust Pathogens. *Annu. Rev. Phytopathol.* **1990,** *28*, 139–153.

Nazari, H. *Trend Analysis of Air Pollutants Concentration and Meteorological Parameters Variations at Rural Site in the Indo-Gangetic Plains*; University of Gothenburg: Germany, MS Thesis, 2019, p 115.

NASA. Drought in Southern South America. NASA-Godard Space Centre, 2009, pp 1–2 http://earthobservatorynasa.gov/naturalaards/View.php?id=37239 (accessed July 24th, 2021).

Nath, R.; Nath, D.; Li, Q.; Chen, N.; Cui, X. Impact of Droughts on Agriculture in the Indo-Gangetic Plains, India. *Adv. Atmos. Sci.* **2017,** *34*, 335–346.

N'diaye, A. Groundnut in Senegal: Production Related Operations and Research. First ICRISAT Regional Groundnut Meeting for West Africa. ICRISAT Sahelian Centre, Niamey, Niger, 1988; pp 65–68.

Ndumi Ngumbi, E. Africa's Most Notorious Insects—The Bugs That Hit Agriculture the Hardest, 2020, pp 1–8. https://theconversation.com/africas-most-notorious-insects-the-bugs-that-hit-agriculture-the-hardest-83107/ (accessed July 15th, 2021).

Nelson, A.; Setiyono, T.; Rala, A. B.; Quicho, E. D.; Raviz, J. V.; Abonete, P. J.; Maunahan, A. A.; Garcia, C. A.; Bhatti, H. Z. M.; Villano, L. S. Toward an Operational SAR-Based Rice Monitoring System in Asia: Examples from 13 Demonstration Sites Across Asia in the RIICE Project. *Remote Sens.* **2014,** *6*, 10773–10812.

Nevo, D. The Desert Locust, *Schistocerca gregaria*, and Its Control in the Land of Israel and the Near East in Antiquity, with Some Reflections on Its Appearance in Israel in Modern Times. *Phytoparasitica* **1996,** *24*, 7–32.

NIAB. Research Progress of Soil Science Division, Faisalabad, Pakistan, Nuclear Institute for Agriculture and Biology, 2013, pp 1–10. http://www.niab.org.pk/soil.htm/ (accessed August 30th, 2021).

Nicolaisen, M. Fungal Communities Including Plant Pathogens in Near Surface Air Are Similar Across Northwestern Europe. *Front. Microbiol.* **2017,** 1–12. https://doi.org/10.3389/fmicb.2017.01729/ (accessed July 24th, 2021).

Nieto, A.; Roberts, S. P.M.; Kemp, J.; Rasmont, P.; Kuhlmann, M.; García Criado, M.; Biesmeijer, J. C.; Bogusch, P.; Dathe, H. H.; De la Rúa, P.; De Meulemeester, T.; Dehon, M.; Dewulf, A.; Ortiz-Sánchez, F. J.; Lhomme, P.; Pauly, A.; Potts, S. G.; Praz, C.; Quaranta, M.; Radchenko, V. G.; Scheuchl, E.; Smit, J.; Straka, J.; Terzo, M.; Tomozii, B.; Window, J.; Michez, D. *European Red List of Bees*; European Commission: Luxemberg, Europe, 2014; p 96.

Niu, H.; Zhao, T.; Wang, D.; Chen, Y. Evapotranspiration Estimation with UAVs in Agriculture: A Review. *Preprints* 2019070124, 2019, pp 1–7. DOI: 10.20944/preprints201907.0124.v1 (accessed July 24th, 2021).

North Dakota Energy Centre. Solar. North Dakota Energy Resource, Bismark, ND, USA, 2018, pp 1–3. https://www.energynd.com/resources/solar/ (accessed May 30th, 2020).

NREL. *North Dakota: Wind*; National Renewable Energy Resources: Washington, DC, 2018, pp 1–3. https://www.nrel.gov/docs/fy00osti/28054.pdf/ (accessed July 7th, 2021).

Nwanze, K. F. 1989 Assessment of yield loss of sorghum and pearl millet due to Stem Borer Damage. International Journal of Pest Management, 35(accessed 2), 137–142.

Nwanze, K. F. Components for the Management of Two Insect Pests of Pearl Millet in Sahelian West Africa. *Insect Sci. Its App* **1991,** *12* (5–6), 673–678.

Nwanze, K. F.; Sivakumar. M. V.K. Insect Pests of Pearl Millet in Sahelian West Africa-II. Raghuva albipunctella De Joannis (Noctuidae, Lepidoptera): Distribution, Population Dynamics and Assessment of Crop Damage. *Int. J. Pest Manage.* **1990,** *36*, 59–65.

Nwanze, K. F.; Youma, O. Panicle Insect Pests of Sorghum and Pearl Millet. In *Proceedings of an International Consultative Workshop*, 4–7 Oct 1993, ICRISAT Sahelian Center, Niamey, Niger. Nwanze, K. F., Youm, O., Eds. *International Crops Research Institute for the Semi-Arid* Tropics, 1995; p 187. http://oar.icrisat.org/549/ (accessed July 15th, 2021).

O'Gairbhith, C. Assessing Viability of High-Altitude Wind Resources in Ireland, 2019, pp 1–9. http://carbontracking.com/reports/High_Altitude_Wind_Resource_in_Ireland.pdf/ (accessed July 30th, 2021).

Oerke, E. C. Crop Losses Due to Pests. *J. Agri. Sci.* **2005,** *144*, 31–43. https://doi.org/10.1017/S0021859605005708/ (accessed July 24th, 2021).

Olesen, J. E.; Bindi, M. Consequences of Climate Change for European Agricultural Productivity Land Use and Policy. *Eur. J. Agron.* **2002,** *16*, 230–262.

Olesen, J. E.; Bindi, M. Agricultural Impacts and Adaptations to Climate in Europe, 2014, pp 1–15. https://www.pik-potsdam.de/avec/peyresq2005/talks/0926/bindi/literature/olesen_bindi_2004.pdf/ (accessed July 24th, 2021).

Oumarou, N.; Baoua, I.; Saidou, A. A.; Amadou, L.; Stern, D. Farmer's Perception of Pearl Millet Hed Miner *Heliocheilus albipunctella*, a Major Millet Pest in the Sahel. *Tropicultura* **2019,** *37*, 1–8. DOI: 10.25518/2295–8010.255 (accessed July 24th, 2021).

Otegui, M. E.; Nicolini, M. G.; Ruiz, R. A.; Dodds, P. A. Sowing Date Effects on Grain Components for Different Maize Genotypes. *Agron.* **1995,** *87*, 29–33.

Otto, C. R.V, Roth, C. L.; Carlson, B. L.; Smart, M. D. Land-Use Change Reduces Habitat Suitability for Supporting Managed Honeybee Colonies in the Northern Great Plains. *Proc. Natl. Acad. Sci.* **2016,** *113*, 10430–10435. https://doi.org/10.1073/pnas.1603481113/ (accessed July 20th, 2021).

Pal, D. K.; Bhattacharya, T.; Srivastava, P.; Chandran, P.; Raj, S. K. Soils of the Indo-Gangetic Plains: A Historical Perspective. *Curr. Sci.* **2009,** *96*, 1193–1999.

Palma, R. M.; Saubidet, M. I.; Rimolo, M.; Utsumi, J. Nitrogen Losses by Volatilization in a Corn Crop with Two Tillage Systems in Argentine Pampa. *Commun. Soil Sci. Plant Analys.* **1998,** *29*, 2285–2879.

Pande, S. Gupta, R. K.; Dahiya, S. S.; Singh, S.; Singh, U. P.; Chauhan, Y. S.; Jat, M. L.; Singh, S. S.; Sharma, H. C.; Rao, J. P.; Chauhan, P. Re-Introduction of Extra Short Duration Pigeon Pea in Rice-Wheat Cropping System of the Indo-Gangetic Plains. In *Rice-Whet Consortium and International Crops Research Institute for the Semi-Arid tropics*, Patancheru, India, Technical Bulletin No. 9, 2006; pp 1–48.

Pande, S.; Sharma, M.; Rao, J. N. Etiology, Biology and Management of Diseases of Food Legumes. In *Food Legumes for Nutritional Security and Sustainable Agriculture*; Khakrawal, M. C., Ed.; Indian Society of genetics and Plant Breeding: New Delhi, India, 2008; pp 1–15.

Pandey, D.; Agarwal, M.; Bohra, J. S. Greenhouse Gas Emissions from Rice Crop with Different Tillage Permutations in Rice–Wheat System. *Agric. Ecosyst. Environ.* **2012,** *159*, 133–144.

Pannure, A. Bee Pollinator Decline: Perspective from India. *Int. J. Nat. Resour. Appl. Sci* **2016,** *3*, 1–10.

Parton, W. J.; Gutmann, M. P.; Merchant, E. R.; Hartman, M. D. Adler, P. R.; McNeal, F.; Lutz, S. M. Measuring and Mitigating Agricultural Greenhouse Gas Production in the US Great Plains-1870–2000, 2011, pp 1–33. https://www.pnas.org/content/112/34/E4681 (accessed August 7th, 2021).

Paruelo, J. M.; Sala, O. E. Effect of Global Change on Maize Production in the Argentinean Pampas. *Clim. Res.* 1993, *3*, 151–167.

Pathak, H.; Saharawat, Y. S.; Gathada, M. W.; Ladha, J. Impact of Resource Conserving Technologies on Greenhouse Gas Emissions Within Indo-Gangetic Plains, 2011, pp 1–12. https://doi.org/10.1002/ghg.27 (accessed July 12th, 2021).

Payne W.; Tapsoba, H.; Baoua, I. B.; Malick, B. N.; N'Diaye, M.; and Dabiré-Binso, C. On-Farm Biological Control of the Pearl Millet Head Miner: Realization of 35 Years of Unsteady Progress in Mali, Burkina Faso and Niger. *Int. J. Agric. Sustain.* **2011,** *9* (1), 186–193.

Pearson, C. J. *Field Crop Ecosystem*; Bhattarai, E. S., Ed.; Ecosystems of the World; Elsevier Science Publications Inc.: Netherlands, 1992; p 576.

Peiretti, R. A. The Development and Future of Direct Seed Cropping Systems in Argentina. In *Sustaining the Global Farm. Selected Papers from the 10th International Soil Conservation Organization Meeting Held at Purdue University*; Stott, D. E., Mohtar, R. H., Steinhardt, G. C., Eds.; Purdue University: West Lafayette, IN, 1998; p 234.

Perkins, C. Wind Power Is Looking Up to Clouds. Society for the Science and Public, 2014, pp 1–3. https://www.sciencenewsforstudents.org/article/wind-power-looking-to-the-cloud (accessed May 27th, 2020).

Peterson, G. A.; Westfall, D. G. Managing Precipitation Use in Sustainable Dryland Agroecosystem, 2005, pp 1–24. https://doi.org/10.1111/j.17348.2004.tb00326.x/ (accessed March 31st, 2020).

Phifer, C. C.; Knowlton, J. L.; Webster, C. R.; Flaspohler, D. J.; Licata, J. A. Bird Community Responses to Afforested Eucalyptus Plantations in the Argentine Pampas. *Biodiversity Conserv.* 2019, pp 1–30. DOI 10.1007/s10531–016–1126–6/ (accessed June 1st, 2020).

Phukan, R. S. Solar Energy in India: Pros, Cons and Future, 2014. https://www.mapsofindia.com/my-india/india/scope-of-solar-energy-in-india-pros-cons-and-the-future

Piesley, R. K.; Saunders, M.; Luck, G. W. A Systematic Review of the Benefits and Costs of Bird and Insect Activity in Agroecosystems, 2015, pp. 1–18. DOI: 10.1007/s40362–015–0035–5 (accessed July 21, 2021).

Plumer, B. These Maps Show the Best Places to Put Solar and Wind Power, 2013, pp 1–6. https://www.washingtonpost.com/news/wonk/wp/2013/07/15/these-maps-show-the-best-places-to-put-solar-and-wind-power-its-not-where-you-think/ (accessed July 30th, 2021).

PNNL. When the Wind Comes Sweeping Down the Plains. Atmospheric Science and Global Change Research highlights. Pacific Northwest National Laboratory, 2016, pp 1–8. https://www.pnnl.gov/science/highlights/highlight.asp?id=4108/ (accessed July 28th, 2021).

Popper, D. E.; Popper, F. J. The Great Plains: From Dust to Dust, 2004, pp 1–7. http://www.planning.org/25anniversary/planning/1987dec.htm (accessed July 23rd, 2021).

Podesta, G.; Bert, F.; Rajagopalan, B.; Apipattanavis, L.; Weber, E.; Easterling, W.; Katz, R, Letson, D.; Menendez, A. Decadal Climate Variability in the Argentine Pampas. Reginal Impacts of Possible Climate Scenarios on Agricultural Systems. *Clim. Res.* **2009,** *40*, 199–210.

Portela, S. A.; Ruillo, A. E.; Sasal, M. C.; Mary, B.; Jobaggy, B. G. Fertilizer Versus Organic Matter Contribution in Nitrogen Leaching of the Pampa. 15N Application in Lysimeters. *Plant Soil* **2009,** *289*, 265–277.

Powell, L. *Great Plains Birds*; Bison Books: New York, 2019; p 224.

Prasad, M. M. Genetic Analysis of Indirect Selection for Winter Wheat Grain Yield Using Spectral Reflectance Indices. *Crop Sci.* **2007,** *47* (4), 1416–1419.

Pratelli, M. G.; Isacch, J. P.; Cardoni, D. A. Species-Area Relationship of Specialist Versus Opportunistic Pampas Grassland Birds Depend on Landscape Matrix. Ardeola **2018,** *65* (1), 3–23.

Prateek, S. Solar Power Generation in India Jumped 86% in 2017, 2018, pp 1–7. https://mercomindia.com/solar-power-generation-india/ (accessed July 26th, 2021).

Precision Hawk. Drone-Based Aerial Intelligence in Precision Agriculture, 2020, pp 1–22. https://www.precisionhawk.com/hubfs/PrecisionHawk%20PrecisionAnalytics%20Agriculture%20Solution%20Brief%202019.pdf/ (accessed July 6th, 2021).

Priyadarsini, S. Ozone Killing Enough Crops to Feed Millions of Poor. Panos South Asia Climate Change, 2014, pp 1–3. http://www.natureasia.com/en/nindia/article/10.1038/nindia.2014.148/

Pron, L. K.; Adamou, H.; Issa, A.; Abdoulkadri, A.; Issa, K.; Bibata, O. A.; Magill, C. Survey of the Prevalence and Incidence of Foliar and Panicle Diseases of Sorghum Across Production Fields of Niger. *Plant Pathol. J.* **2020,** *19*, 106–113.

Prospero, J. M.; Bonatti, E.; Schubert, C.; Carlson, T. N. Dust in the Caribbean Atmosphere Traced to an African Dust Storm. *Earth Planet. Sci. Lett.* **1970,** *9*, 287–293.

Pudelko, R.; Stuckzynski, T.; Borzecka-Walker, M. The Suitability of an Unmanned Aerial Vehicle (UAV) for the Evaluation of Experimental Fields and Crops. *Zemdirbyste Agric.* **2012,** *99*, 431–436.

Pugh, N.; Han, X.; Collins, D.; Thomasson, A.; Corpe, D.; Chang, A.; Jung, J.; Isakiet, T.; Prom, I. K.; Carvalho, G.; Gates, I.; Vree, A.; Bagnall, G.; Rooney, W. Estimation of Plant Health in a Sorghum Field Infected with Anthracnose Using a Fixed-Wing Unmanned Aerial System. *J. Crop Improv.* **2018,** *32*, 861–877. DOI: https//doi.org/10.1080/15427528.2018.1535462/ (accessed July 24th, 2021).

Pukett, H. L.; Brandle, J. R.; Johnson, R.J.; Blackenship, B. E. Avian Foraging Patterns in Crop in Crop Field Edges Adjacent to Woody Habitat. *Agric. Ecosyst. Environ.* **2009,** *131*, 9–15.

Pusz, W.; Weber, R.; Dancewicz, A.; Kita, W. 2017 Analysis of Selected Fungi Variation and Its Dependence on Season and Mountain Range in Southern Poland—Key Factors in Drawing Up Trial Guidelines for Aeromycological Monitoring. *Environ. Monitor. Assess.* **2017,** *189*, 526–541.

Putnam, E. S.; Oplinger, D. R.; Hicks, B. R.; Durgan, D. M.; Noetzel, R. A.; Meronuck, J. D.; Doll; Schulte, E. E. Sunflower. Alternative field Crops Manual, 2012, pp 1–15. https://hort.purdue.edu/newcrop/afcm/sunflower.html (accessed June 17th, 2021).

Quest UAV. Drones in South America. South American Real Estate News, 2020, pp 1–2. https://www.gatewaytosouthamerica-newsblog.com/drones-in-south-america-agriculture/ (accessed July 6th, 2021).

Radler, A. T.; Pieter, H. G.; Faust, E.; Sausen, R.; Pucik, T. Frequency of Severe Thunderstorms Across Europe Expected to Increase in the 21st Century Due to Rising Instability. *Clim. Atmos. Sci.* **2019,** *2*, 30–36. https://doi.org/10.1038/s41612–019–0083–7/ (accessed July 25th, 2021).

Rajput, P.; Chauhan, A. S.; Gupta, G. Bioaerosols Over the Indo-Gangetic Plain: Influence of Biomass Burning Emission and Ambient Meteorology. In *Environmental Contaminants*; Gupta, T., Agarwal, A., Agarwal, R., Labhsetwar, N., Eds. Springer: Singapore, 2018; pp 93–121. https://doi.org/10.1007/978–981–10–7332–8_5/ (accessed July 28th, 2021).

Ramanathan, V.; Ramana, M. V. Persistent, Widespread and Strongly Absorbing Haze Over the Himalayan Foothills and the Indo-Gangetic Plains. *Pure Appl. Geophys* **2005,** *162*, 1609–1626.

Ramsperger, B.; Peinemann, N.; Stahr, K. Input and Characteristics of Aeolian Dust in the Argentinean Pampas. *Journal of Arid Environment* **1998,** *39*, 467–476.

Rasul, G.; Chaudhry, Q. Z.; Mahmood, A.; Hyder, K. W. Effect of Temperature Rise on Crop Growth & Productivity. *Pak. J. Meteorol.* **2014,** *8*, 53–62.

Ravindran, K.; Mor, S.; Singh, V.; Gehlawat, T. K.; Dhankar, R.; Mor, S. S.; Beig, G. Realtime Monitoring of Air Pollutants in Seven Cities of North India During Crop Residue Burning and Their Relationship with Meteorology and Transboundary Movement. *Sci. Total Environ.* **2019,** *690*, 717–729. https://doi.org/10.1016/j.scitotenv.2019.06.216/ (accessed July 24th, 2021).

Reed, D. K.; Pike, K. S. *Biological Control Exploration in Brazil, Argentina, and Chile for Natural Enemies of Russian Wheat Aphid,* Diuraphis noxia. USDA-ARS Trip Report, 27 Oct.–19 Nov. 1990. Stillwater, OK, USA, USDA-ARS, Plant Science Research Laboratory, 1990; p 32.

Reinhardt, C.; Ginzel, B. *Farming in the 1930s*; Ginzel Group: Nebraska, 2003, pp 1–3. https://livinghistoryfarm.org/farminginthe30s/life_02.html/ (accessed July 19th, 2021).

Reynolds, M. Phenotyping Approached for Physiological Breeding and Gene Discovery in Wheat. *Ann. Appl. Biol.* **2012,** *155*(3), 309–320.

Rian, S.; Xue, Y.; MacDonald, G. M. Toure, M. B.; Yu, Y.; De Sales, F. Levine, P. A. Doumbia, S.; Taylor, C. E. 2009 Analysis of Climate and Vegetation Characteristics Along the Savannah-Desert Ecotone in Mali Using MODIS Data. *GIScі. Remote Sens.* **2009,** *46*, 424–450.

RMAX. RMAX Specifications. Yamaha Motor Company, Japan, 2015, pp 1–4. http://www.max.yamaha-motor.Drone .au/specifications (accessed July 28th, 2021).

Robbins, C. S.; Bruun, B.; Zim, H. S. *Birds of North America: A Guide to Field Identification*; Golden Press: New York, 1983, 360 pp.

Roberts, T. J. *The Birds of Pakistan. Passeriformes*; Oxford University Press, **1992,** p 617.

Rodrıguez, S.; Querol, X.; Alastue, A, Kallos, G.; Kakaliagou, O. Saharan Dust Contributions to PM10 and TSP Levels Is Southern and Eastern Spain. *Atmos Environ.* **2001,** *35* (14), 2433–2447.

Romheld, D.; Marschner, H.; Becker, K.; Schecht, E.; Hulsebusch, S.; Buerkert, A. Effect of Site Factors on the Efficiency of Soil Amendments and on Nutrient Fluxes in Subsistence-Oriented Cropping of the West African Sahel, 2000, pp 1–24. http://www.troz.uni-hohenheim.de/research/sfb308/Ebcont/b13 (accessed July 7th, 2021).

Rosenweig, C.; Iglesius, A.; Yang, X. B.; Epstein, R. P.; Chevian, A. Climate Change and Extreme Weather Events Implications for Food Production, Plant Disease and Insects. University of Nebraska Licln, NASA Paper 24, 2011, pp 1–16 https://digitalcommons.unl.edu/nasapub/24/ (accessed July 24th, 2021).

Roubik, D. W. Pollination of Cultivated Plants in the Tropics. *FAO Agric. Serv. Bull.* **1995,** *118*, 1–194.

Rounsevell, M. D.A.; Ewert, F.; Reginster, I.; Leomans, R.; Carter, T. R. Future Scenarios of European Agricultural Land Use II Projecting Changes in Crop Land and Grassland. *Agric. Ecosyst. Environ.* **2005,** *107*, 177–135.

Rowell, J. B.; Romig, R. W. Detection of Urediospores of Wheat Rusts in Spring Rains. *Phytopathology* **1966,** *56*, 807–811.

Rubene, D. Leidefors, M.; Ninkovic, V.; Eggers, S. Low, M. Disentangling Olfactory and Visual Information Used By the Field Foraging Birds. *Ecol. Evol.* **2019,** *11*, 545–552. DOI: 10.1002/ece3,4773.eCollection2019 jan (accessed July 19th, 2021).

Ruf, J. C.; Marchand, J. L.; Peterschmitt, H. M. Maize Streak, Maize Stripe and Maize Mosaic Virus Diseases in the Tropics of Africa and Islands in the Indian Ocean—A Dossier. *Agric. Dev.* **1995,** Special Issue 1–15.

Rufin, C. Wind Energy in the Latin America: Realizing the Potential. ReVista Harvard review of Latin America, 2017, pp 1–6. https://archive.revista.drclas.harvard.edu/book/wind-energy-latin-america (accessed July 7th, 2021).

Sa, J. C.M.; Cerri, C. C.; Dick, W. A.; Lal, R.; Venske Filho, S. P. Piccolo, M. C.; Feigi, B. E. Organic Matter Dynamics and Carbon Sequestration Rates for a Tillage Chrono Sequence in a Brazilian Oxisol. *Soil Sci. Soc. Am. J.* **2001,** *65*, 1486–1489.

Saari, E. E.; Prescott, J. M. World Distribution in Relation to Economic Losses. In *The Cereal Rusts*, Vol. 2, *Diseases, Distribution, Epidemiology, and Control*; Roelfs, A. P., Bushnell, W. R., Eds.; Academic Press: Orlando, FL, 1985; pp. 259–298.

Sache, I. Short Distance Dispersal of Wheat Rust Spores by Wind and Rain. *Agronomie* **2000,** *20*: 757–787.

Sakyi, P. K. A. Harmonizing West Africa's Increased Electricity Generation with Climate Objectives, 2020, pp 1–5. https://www.ghanaweb.com/GhanaHomePage/features/Harmonizing-West-Africa-s-increased-electricity-generation-with-climate-objectives-994504/ (accessed July 9th, 2021).

Salzman, U.; Waller, M. The Holocene Vegetational History of Nigerian Sahel Based on Multiple Pollen Profiles. *Rev. Palaeobot. Palynol.* **1998,** *100*, 33–72.

Salzmann U. Are Savannas Degraded Forests. A Holocene Pollen Record from the Sudanian Zone of NE Nigeria. *Veg. History Palynol.* **2000,** *100*, 39–72.

Sanchez, N. E.; DeWysiecki, M. L. Grasshoppers of the Argentine Pampas. Encyclopaedia of Entomology. Capinera, J. L., Ed.; 2008, pp 1–3. DOI: https://doi.org/10.1007/978-1-4020-6359-6_1169.

Sanchez-Zapata, J. A.; Donazer, J. A.; Delgado, A.; Forera, M. G.; Ceballos, O.; Hiraldo, F. Desert Locust Outbreaks in the Sahel. *J. Appl. Ecol.* **2007,** *44*, 323–329.

Sanginga, N. Role of Biological Nitrogen Fixation in Legume-Based Cropping Systems: A Case Study of West Africa Farming Systems. *Plant Soil* **2003,** *25*, 25–39.

Santosh, K. Top 10 Best Energy companies in India, 2020, pp 1–8. https://www.worldblaze.in/top-10-best-wind-energy-companies-in-india/ (accessed July 26th, 2021).

Sapkota, T. B. Cost-Effective Opportunity for Climate Change Mitigation in Indian Agriculture. *Sci. Total Environ.* **2018**. Doi: 10.1016/j.scitotenv.2018.11.225/ (accessed July 22nd, 2021).

Saraf, A. K.; Berry, A. K.; Das, J.; Rawat, W.; Sharma, K.; Jain, S. K. Winter Fog Over the Indo-Gangetic Plains: Mapping and Modelling Using Remote Sensing and GIS. *Nat. Hazards* **2010,** *58*, 199–2020.

Sarkar, S.; Chauhan, A.; Kumar, R.; Singh, R. P. Impact of Deadly Dust Storms (May 2018) on Air Quality, Meteorological, and Atmospheric Parameters Over the Northern Parts of India. *GeoHealth* **2019,** 3. https://doi.org/10.1029/2018GH000170 (accessed September 6th, 2020).

Saskatchewan Natural Resources Department. Insects on Crops. Saskatchewan Agriculture. Canada, 2020, pp 1–12. https://www.saskatchewan.ca/business/agriculture-natural-resources-and-industry/agribusiness-farmers-and-ranchers/crops-and-irrigation/insects/ (accessed July 19th, 2021).

Saxena, M.; Khare, M. N. Diseases of Lentil and Their Management. In *Diseases of Field Crops and Their Management*; Thind, T. S., Ed.; National Agricultural Technology Information Centre: Ludhiana, India, 1998; pp 271–284.

Schimpf, M.; Diamond, E. Digital Farming. Friends of the Earth Europe. European Commission, Belgium, 2019, pp 1–8. http://www.foeeurope.org/sites/default/files/gmos/2020/foee-digital-farming-paper-feb-2020.pdf (accessed July 23rd, 2021).

Schutze, L.; Jainicke, R.; Pietrick, H. Saharan Dust Transport Over with Atlantic Ocean. In *Desert Dust*; Pewe, T. L., Ed. *Geol. Soc. Am.* **1981,** *187*, 87–100.

Scott Hosking, J.; Macleod, D.; Phillips, I.; Watson, P.; Shuckburgh, E. F.; Mitchell, D. Changes in European Wind Energy Generation Potential Within 1.5°C Warmer World. *Environ. Res. Lett.* **2017,** *13* (5), 1–18, 054032. DOI:10.1088/1748–9326/aabf78 (accessed June 29th, 2020).

Sehgal, J. L. *Agroecological Regions of India Technical Bulletin National Bureau of Soil Survey*; India Publication No. 24: Nagpur, 1990, pp 1–73.

Sehgal, J. L.; Sharma, P. K.; Karale, R. L. Soil Resource Inventory of Punjab Using Remote Sensing Technique. *J. Indian Soc. Remote Sens.* **1988,** *16*, 39–47.

Sehgal, M.; Jeswani, M. D.; Kalra, A. Management of Insects, Diseases and Nematodes Pests of Rice-Wheat in the Indo-Gangetic Plains. *J. Crop Product.* **2001,** *4*, 167–226.

SEIA. *Solar Industry Research Data*; Solar Energy Industries Association, 2019; pp 1–6. https://www.seia.org/solar-industry-research-data (accessed July 30th, 2021).

SenseFly Inc. *Drones for Agriculture*; SenseFly Inc.- Parrot Company, Cheaseux-Louisanne, Switzerland, 2016; pp 1–9. http://www.sensfly.com/applications/agriculture.html (accessed July 24th, 2021).

Setiyono, T. D.; Quicho, E. D.; Romuga, G. C. Rice Yield Estimation System (Rice-YES) User Manual. *Int. Rice Res. Inst. Los Baños, Philippines Remote Sens.* **2017,** *10* (2), 293–305.

Shahendeh, H.; Bilanton-Knewtson, S. J.; Doumbia, M.; Hons, F. M.; HosnerL. R. Nitrogen Dynamics in the Tropical Soils of Mali, West Africa. *Biol. Fertility Soils* **2004,** *39*, 258–268.

Shahgedanova, M. L. *The Physical Geography of Northern Eurasia*; Oxford University Press Inc.: New York, 2008; p 535.

Sharma, K. How Tiny Pests Traverse Long Distances to Attack Mustard Crops. DownToEarth, 2019, pp 1–3. https://www.downtoearth.org.in/news/agriculture/how-tiny-pests-traverse-long-distance-to-attack-mustard-crops-64559/ (accessed July 24th, 2021).

Shaufler, G.; Kitzler, B.; Schindlbacher, A.; Skiba, U.; Sutton, M. A.; Zechmeister-Boltenstern Greenhouse Gas Emissions from European Soils Under Different Land Use: Effects of Soil Moisture and Temperature. *Eur. J. Soil Sci.* **2010,** 1–17. https://doi.org/10.1111/j.1365–2389.2010.01277/ (accessed July 24th, 2021).

Shepard, D. Global Warming: Severe Consequences for Africa. Africa Renewal, 2019, pp 1–9. https://www.un.org/africarenewal/magazine/december-2018-march-2019/global-warming-severe-consequences-africa (accessed July 27th, 2021).

Shi, Y.; Thomasson.; Murray, S. C.; Puch, N. A.; Rooney, W. L.; Shafian, S.; Rajan, N.; Rouze, G.; Morgan, C. L.S.; Neely, H. L.; Rana, A.; Bagvthiannan, M. V.; Herrickson, J.; Bowden, E.; Vilsack, J.; Olsenholler, J.; Bishop, M. P.; Sheridan, R.; Putman, E. B.; Popescu, S.; Burks, T.; Cope, D.; Ibrahim, A.; McCutchen, B. F.; Baltensperger, D. D.; Vent, R.; Vidrine, M.; Yang, C. Unmanned Aerial Vehicles for High Throughput Phenotyping and Agronomic Research. *PLOS One* **2016,** 1–15. http://dx.doi.org/10.1371/journal.pone.0159781 (accessed July 29th, 2021).

Singh, A. A.; Agarwal, S. B. Tropospheric Ozone Pollution in India: Effects on Crop Yield on Product Quality. *Environ. Sci. Pollut. Res.* **2016,** *24*, 1367–4382.

Singh, A. Agricultural Drone: The Shape of Indian Agriculture to Come, 2019, pp 1–6. https://smefutures.com/agricultural-drones-shape-indian-agriculture-come/ (accessed July 30th, 2021).

Singh, A. K.; Bhatt, B. P.; Singh, K. M.; Kumar, A.; Bharathi, R. C. Dynamics of Powdery Mildew (*Erysiphe trifoli*) Disease of Lentil Influenced By Sulphur and Zinc Nutrition. *Plant Pathol. J.* **2013,** *12*, 71–77.

Singh, I.; Shankar, U.; Abrol, D. P.; Mondal, A. 2017 Diversity of Insect Pollinators Associated with Pigeon Pea (Cajanus cajan) and Their Impact on Crop Production. *Int. J. Curr. Microbiol. Appl. Sci.* **2000,** *6*, 528–538.

Singh, K. M.; Singh, A. K. Lentil in India: An Overview, 2014, pp 1–8. https://www.researchgate.net/publication/266968749/ (accessed July 24th, 2021).

Singh, R. P.; Huerta-Espina, J.; Roelfs, A. P. *The Wheat Rusts. Food and Agricultural*; Organization of the United Nations: Rome, Italy, 2000; pp 1–38 http://www.fao.org/3/y4011e0g.htm/ (accessed July 24th, 2021).

Sinha, S. Using Drones Allowed for Combating Desert Locust Menace to Aerial Imagery, 2020; pp 1–6. https://timesofindia.indiatimes.com/india/using-drones-allowed-for-combating-desert-locust-menace-to-aerially-spray-pesticides/articleshow/75926781.cms (accessed July 30th, 2021).

Smart, M. D.; Pettis, J. S.; Euliss, N.; Spiva, M. S. Land-use change in the Northern Great Plains Region of the US Influences the Survival and Productivity of Honeybees. *Agric. Ecosyst. Environ.* **2016,** *230*, 139–149.

Smith, J. C.; Ghosh, A.; Hijmans, R. S. Agricultural Intensification Was Associated with Crop Diversification in India (1947–2014). *PLOS One* **2019,** *14* (12), 1–17, e0225555 https:/doi.org/10.1371/journal.pone.o225555/ (accessed July 24th, 2021).

Soil Survey Staff. *Keys to Soil Taxonomy*, 5th ed.; USDA-NRCS: Washington, DC, 1992; p 541.

Soil Survey Staff. *Keys to Soil Taxonomy*, 7th ed.; USDA-NRCS: Washington, DC, 1996b; pp 644–693.

Soil Survey Staff. *Keys to Soil Taxonomy*, 8th ed.; USDA-NRCSL: Washington, DC, 1998, p 326.

Soil Survey Staff. *Soil Survey Laboratory Information Manual. Soil Survey Investigation Report 45*. USDA-NRCDL: Lincoln, Nebraska, 1995; p 305.

Soil Survey Staff. *Soil Survey Laboratory Information Manual. Soil Survey Investigation Report 42*. USDA-NRCD: Lincoln, Nebraska, 1996a; p 693.

SolarPower Europe. How Solar-Based Electrification Can Accelerate European Decarbonization, 2019, pp 1–6. https://www.solarpowereurope.org/how-solar-based-electrification-can-accelerate-european-decarbonisation/ (accessed July 1st, 2020).

Solbrig O. T.; Viglizzo, E. T. Sustainable Farming in Argentine Pampas: History, Society, Economy and Ecology, 2011, pp 1–12 https://www.researchgate.net/publication/268297020_Sustainable_Farming_in_the_Argentine_Pampas_History_Society_Economy_and_Ecology/ (accessed July 24th, 2021).

Solbrig, O. T. The Dilemma of Biodiversity Conservation: Agricultural Expansion in the Pampas. Revista, 2005, pp 1–5. https://archive.revista.drclas.harvard.edu/book/dilemma-biodiversity-conservation (accessed July 27th, 2021).

Solbrig, O. T. Towards a Sustainable Pampa Agriculture: Past Performance and Prospective Analysis, 1997, p 52. https://www.wzw.tum.de/public-html/lattanzi/Lit/Solbrig%201997.pdf (accessed July 27th, 2021).

Sorghum and Millet Innovation Laboratory. Pearl Millet Production Problems in West Africa. Feed the Future Innovation Lab for Collaborative Research on Sorghum and Millet. Kansas State University-Hays Experimental Station, Manhattan, KS, USA, A Report, 2002, pp 1–17.

SPP Market Monitoring Unit, 2015: State of the Market Report: Fall 2015, September–November 2015. Presented at the 37th Conference on Radar Meteorology, Norman, OK, September 14–18, 2015. American Meteorological Society, Paper 14A.3, p 65. https://www.spp.org/documents/34155/qsom_2015fall.pdf (accessed July 24th, 2021).

Srivastava, P.; Pal, D. K.; Aruche, M. Soils of the Indo-Gangetic Plains. *Elsevier Earth Sci. Rev.* **2014,** *140*, 54–71.

Stackman, E. C.; Christensen, C. M. Aerobiology in Relation to Plant Pathogens. *Bot. Rev.* **1946,** *12*, 205–253.

Stahr, K.; Herman, L. Origin, Deposition and Composition of Dust and Consequences for Soil and the Site Properties with Special Reference to Semi-Arid Regions of West Africa. In *Wind Erosion in Niger: Implications and Control Measures in a Millet-Based Farming System*; Buerkert, B., Allison, B. E., von Oppen, M., Eds.; Kluwer Academic Publishers: Dordrecht, The Netherlands, 1996, pp 56–65.

Stary, P.; Gerding, M.; Norambuena, H.; Remaudière, G. Environmental Research on Aphid Parasitoid Biocontrol Agents in Chile (Hym.; Aphididae; Hom.; Aphidoidea). *J. Appl. Entomol.* **1993,** *115*, 292–306.

Stein, K.; Coulibaly, D.; Stenchly, K.; Goetze, D.; Porembski, S.; Linder, A. Konate, S.; Linsenmair, E. K. Bee Pollination Increases Yield Quantity and Quality of Cash Crops in Burkina Faso, West Africa. *Sci. Rep.* **2017,** *7*, 1–12, 17691. DOI: 10.1038/s41598–017–17970–2/ (accessed July24th, 2021).

Sterk, G.; Herman, L.; Bationo, A. Wind-Blown Nutrient Transport and Soil Productivity Changes in Southwest Niger. *Land Degrad. Dev.* **1996,** *7*, 325–335. https://doi.org/10.1002/(SICI)1099–145X(199612)7:4<325::AID-LDR237>3.0.CO;2-Q/ (accessed July 24th, 2021).

Sterk, G.; Stroosnijder, I.; Raats, P. A.C. Wind Erosion and Control Techniques in Sahelian Zone Niger, 1998, pp 1–14 https://www.researchgate.net/publication/40138628_Wind_Erosion_Processes_and_Control_Techniques_in_the_Sahelian_Zone_of_Niger (accessed July 24th, 2021).

Sterle, S.; Brecha, R. How West Africa Can Expand Power Supply and Meet Climate Goals. *Conversation,* 2018, 1–4. https://www.agri4africa.com/how-west-africa-can-expand-power-supply-and-meet-climate-goals/ (accessed July 29th, 2021).

Stickles, N. Farmers Are Using AI to Spot Pests and Catch Diseases- and Many Believe It Is the Future of Agriculture. *AI Revol.* **2019,** 1–3. https://www.businessinsider.in/science/news/farmers-are-using-ai-to-spot-pests-and-catch-diseases-and-many-believe-its-the-future-of-agriculture/articleshow/71975166.cms (accessed July 26th, 2021).

Stoorvogel, J. J.; Smalling, E. M.A. Assessment of Soil Nutrient Depletion in Sub-Saharan Africa. 1983–2000 Volume 1 Main Report. The Winand Staring Centre, Wageningen, The Netherlands, 1990, p 342.

Stout, J.; Delaney, A.; Marshall, T.; Vickerey, J. Birds, Bees and Butter- Biodiversity Crucial for Shea Production in West Africa. Trinity College, University of Dublin, Ireland, 2020, pp 1–7. https://www.enn.com/articles/63724-birds-bees-and-butter-biodiversity-crucial-for-shea-production-in-west-africa/ (accessed July 28th, 2021).

Straub, R. W. Red-Winged Blackbird Damage to Sweet Corn in Relation to Infestations of European Corn Borer. *J. Econ. Entomol.* **1989,** *82*, 1406–1416.

Suffert, F.; Latxague, E.; Sache, I. *Plant Pathogens as Agroterrorist Weapons: Assessment of the Threat of European Agriculture and Forestry*; Springer, 2008; pp 1–13. DOI: 10.1007/s12571–009–0014–2/ (accessed July 23rd, 2021).

Sullivan, D. G.; Fulton, J. P. Shaw, J.; Bland, G. Evaluating the Sensitivity of an Unmanned Thermal Infrared Aerial System to Detect Water Stress in a Cotton Canopy. *Trans. ASABE* **2007,** *50*, 1955–1962.

Sunder, K. S.G.; Subramanya, S. Bird Use of Rice Fields in the Indian Subcontinent. *Int. J. Waterbird Biol.* **2010,** *33*, 44–70.

Sylvester, G. E-Agriculture in Action: Drones for Agriculture. Food and Agricultural Organization of the United Nations, Rome Italy and International Telecommunication Union, Bangkok, Thailand, 2018; pp 1–128.

Synergia Foundation. Drones in Uttarakhand, 2017, pp 1–4. https://www.synergiafoundation.org/insights/analyses-assessments/drones-uttarakhand/ (accessed July 30th, 2021).

Taboada, M. A. Soil Structural Behaviour in Flooded and Agricultural Soils of the Argentine Pampas Doctoral Thesis Submitted to Devant L'institut National Polytechnique De Toulouse, France, 2006, p 135.

Takoulen, J. M. Burkina Faso: Six Solar Power Plants to Increase to Energy Supply by 155 MW. Afrik21, 2019, pp 1–3. https://www.afrik21.africa/en/burkina-faso-six-solar-power-plants-to-increase-energy-supply-by-155-mw/ (accessed July 10th, 2021).

Tappan, G. G.; Sall, M.; Wood, E. C.; Cushings, M. Ecoregions and Land Cover in Senegal. United States Geographical Service EROS Data Centre. Centre de Salvi Ecologique Dakar Senegal, 2004, pp 1–22.

Techy L.; Schmale D. G.; Woolsey C. A. Coordinated Aerobiological Sampling of a Plant Pathogen in the Lower Atmosphere Using Two Autonomous Unmanned Aerial Vehicles. *J. Field Robot.* **2010,** *27*, 335–343.

Teotia, H.S, D'Hoore, J.; Goombeer, R. Soil and Land Use Distribution Over a Part of North Plains (Indo Gangetic Plains) of India Based on the Optical Interpretation of Landsat-2 Multi-Spectral Satellite Imagery. *Pedologie* **1980,** *30*, 19–42.

Thamm, H. P. SUSI 62 A Robust and Safe Parachute UAV with Long Flight Time and Good Payload. *Int. Arch. Photogrammetry Remote Sens. Spatial Inform. Sci.* **2011,** *38*, 19–24.

Thamm, H.P.; Judex, M. The Low-Cost Drone. An Interesting Tool for Process Monitoring in a High Spatial and Temporal Resolution. In *International Archives of Photogrammetry, Remote sensing, Spatial information Science. ISPs commission 7th Mid-term symposium. Remote Sensing: From Pixels to Process*. Enchede, The Netherlands **2006,** *36*, 140–144.

Thamm, H. P.; Menz, G.; Becker, M.; Kuria, D. N.; Misana, S.; Kohn, D. The Use of UAS for Assessing Agricultural Systems in AN Wetland in Tanzania, in the Wet-Season, for Sustainable Agriculture and Providing Ground Truth for Terra Sar X Data. ISPRS International Archives of the Photogrammetry, Remote Sensing and Spatial Information Sciences. XL-1/w2, 2013; pp 401–406. DOI: 10.5194/isprsarchives-XL-1-W2–401–2013 (accessed July 20th, 2021).

Thiollay, J. M. Severe Decline of Large Birds in Northern Sahel of West Africa: A Long Term Assessment. *Bird Conserv. Int.* **2015,** *16*, 353–365.

Thomson, L. J. Predicting the Effects of Climate Change on Natural Enemies of Agricultural Pests. *Biol. Control* **2010,** *52*, 296–306.

Thottappilly, G.; Rossel, H. W. Virus Diseases of Cowpea in Tropical Africa. *PEST Manage.* **2008,** *38*, 337–342.

Tieszen, L. L.; Tappan, G. G.; Tan, Z.; Tachie-Obeng, E. 2011 Land Cover Change, Biogeochemical Modelling, of Carbon Socks and Climate Change in West Africa. In *The Proceedings of Conference on "Africa and Carbon cycle-The CarboAfrica Project Accra*, Ghana, 2011; pp 75–83.

Tieszen, L. L.; Tappan, G. G.; Toure, A. Sequestration of Carbon in Soil Organic Matter in Senegal—An Overview. *J. Arid Environ.* **2004,** *59*, 409–425.

Tikoo, J. L.; Sharma, B.; Mishra, S. K.; Dikshit, D. K. Lentil (*Lens culinaris*) in India: Present Status and Future Prospectives. *Indian J. Agric. Sci.* **2005,** *75*, 539–562.

Timmer, J. Wind Power Prices Now Lower Than the Cost of Natural Gas. ARS Technica, 2019, pp 1–3. https://arstechnica.com/science/2019/08/wind-power-prices-now-lower-than-the-cost-of-natural-gas/ (accessed July 28th, 2021).

Tomas. Climate Change Effects on Physiology and Population Processes of Hosts and Vectors That Influence the Spread of Hemipteran-Born Plant Viruses. *Global Change Biol.* **2008,** *15*, 1884–1894.

Trigo, E.; Cao, S.; Malach, V., Villarcal, F. Innovating in the Pampas: Zero-Tillage-Soybean Cultivation in Argentina. In *The Case of Zero-Tillage Technology in Argentina*. International Food Policy Research Institute: Washington, DC, 2009; p 237.

Trostle, C. In-Field Correlation of Bradyrhizobium Nodulation with Soil Parameters and Peanut Yield in West Texas Precision Agriculture Initiatives for Texas High Plains Texas Agricultural Experimental Station at Lubbock, TX, USA Annual Report, 2002, pp. 1–4.

UCSUSA. Renewable Energy and Agriculture: A Natural Fit, 2008, pp 1–4. https://www.ucsusa.org/resources/renewable-energy-and-agriculture/ (accessed July 30th, 2021).

Udikeri, A.; Chandra, U. Pollination Efficiency of Different Insects on Phalsa, Grewia Subaequalis D. C. *J. Entomol. Zool. Stud.* **2019,** *7*, 1061–1065.

UNCCD. Global Alarm: Dust and Sandstorm from the World's Dryland. United Nations Convention to Combat Desertification, United Nations Buildings, Bangkok, Thailand,

2001, p 343. https://www.droughtmanagement.info/literature/UNCCD_global_alarm_dust_and_sandstorms_2002.pdf (accessed July 13th, 2021).

UNDP. Solar for Agriculture: Empowering Sudan's Farmers. ESI Africa, Africa's Power Journal, 2020, pp 1–7. https://www.esi-africa.com/industry-sectors/generation/solar/solar-for-agriculture-empowering-farmers-in-sudan/ (accessed July 10th, 2021).

University of Finland. Use of Ozone Tolerant Cultivars to Enhance Indi's Food Security. *Science Daily*, 2015, pp 1–8. www.sciencedaily.com/releases/2015/08/150831085613.htm/ (accessed July 28th, 2021).

United States Department of Geology. *The Great Plains and Prairies*; United States Department of Geology: Washington, DC, 2013, pp 1–8. http://country studies.us/united-states/geography-17 (accessed July 20th, 2021).

USAID. *Mauritania. Power Africa Fact Sheet*; US Agency for International Development: Washington, DC, 2020, pp 1–3. https://www.usaid.gov/powerafrica/mauritaniap (accessed July 10th, 2020).

USDA. Focus on Crop Lands to Northern Plains. United States Department of Agriculture Climate Hub, 2019, pp 1–3. https://www.climatehubs.usda.gov/hubs/northern-plains/topic/focus-croplands-northern-plains/ (accessed July 21st, 2021).

USDA-NRCS. Insects and Pollinators: Pollinators by the Numbers. Natural Resources Conservation Service, US Department of Agriculture, 2016, pp 1–32. www.nrcs.usda.gov/wps/portal/nrcs/main/national/plantsanimals/pollinate/ (accessed July 23rd, 2021).

USGCRP. *Great Plains: Global Climate Impacts in the United States of America*; United States Global Change Program: Washington, DC, 2009a; pp. 1–4.

USGCRP. Global Climate Change Impact in United States of America. In *United State Global Change Research Program*; Karl, T. R., Melillo, J. M., Peterson, T. C., Eds.; Cambridge University Press: New York, 2009b; p 596.

USGCRP. *Overview: Great Plains*; United States Global Change Program: Washington, DC, 2012; pp 1–4.

USEPA. *Climate Change: Great Plains Impacts and Adaptation*; United States Environment Protection Agency: Washington, DC, 2012, pp 1–3. http://www.epa.gov/climatechange/impacts-adaptation/greatplains.html (accessed January 11th, 2013).

van den Bosch, R. Informe sobre el control biológico de los áfidos de los cereales en Chile. *Agric. Técn.* (accessed Chile) **1976,** *36*, 141–145.

van der Velde, M.; Tubieloo, F. N.; Vrielling, A.; Bouraoui, F. Impacts of Extreme Weather on Wheat and Maize in France. *Clim. Change* **2012,** *113*, 751–765. DOI 10.1007/s10584–011–0368–2/ (accessed July 26th, 2021).

Varenhost, A. *Soybean Aphids in South Dakota*; South Dakota State University Extension Service, 2019; pp 1–5.

Varghese, S.; Narayanan, S. P.; Raj, V. P.; Vajja, H. P. Analysis of Wetland Habitat Changes and Its Implications on Avifauna in Select Districts of the Indo-Gangetic Plains of Uttar Pradesh, India between 1972 and 2004. Conference paper, 2008, pp 1–15. https://www.researchgate.net/publication/315912598/ (accessed July 21st, 2021).

Viglizzo, E. F. *La sustantabilidad Ambiental del agro Pampeano Edisiones*; Institut Nacional Technologie Agriculture: Buenos Aeries, Argentina, 2002; pp. 135.

Viglizzo, E. F. Argentina Pampas- Millennium Ecosystem Assessment: The Provision of Ecosystem Services and Human Well-Being in the Pampas of Argentina, 2005, pp 1–4. http://www.milleniumassessment.org/en/SGAArgentinePampas.aspx (accessed August 4th, 2021).

Viglizzo, E. F. Ecological and Environmental Footprint of Agricultural Expansion in Argentina. *Global Change Biol.* **2011,** *17*, 959–973.

Vikaspedia. Diseases of Wheat, 2020, pp 1–18. https://vikaspedia.in/agriculture/crop-production/integrated-pest-management/ipm-for-cerels/ipm-strategies-for-wheat/wheat-diseases-and-symptoms/(accessed July 5th, 2021).

Vogel, E.; Donat, M. G.; Alexander, A. V.; Meinhausen, M.; Ray, D. E.; Karoly, D. E.; Meinhausen, N.; Frieler, K. The Effects of Climate Extremes on Global Agricultural Yields. *Environ. Res. Lett.* **2019,** *14*, 05410.

Walton R. *Moths Have a Secret But Vital Role as pollinators in the Night*; University of California, 2020, pp 1–5. ps://phys.org/news/2020–05-moths-secret-vital-role-pollinators.html (accessed June 20th, 2020).

Wang, T.; Song, W.; Xu, Z.; Lin, M.; Xu, T.; Liao, W.; Yin, L.; Cai, W.; Kang, L.; Zhang, H.; Zhu, T. Why Is the Indo-Gangetic Plains Region the Largest NH3 Column in the Globe During Pre-Monsoon and Monsoon Seasons. *Atmos. Chem. Phys* **2020,** *20*, 8727–8736.

Wannarka, M. Bee.otany a Non-Profit Organization Helping Pollinators One Flower at a Time, 2018, pp 1–2. https://beeotany.org/megan-wannarka/ (accessed July 21st, 2021).

Ward, P. The War Against the Quelea Bird. *New Sci.* **1964,** *22*, 736–738.

Ward, P. Distribution, Systematics and Polymorphism of African Weaver Bird *Quelea quelea. Ibis* **1966,** *108*, 37–39.

Ward, P. The Migration Patterns of *Quelea quelea* in Africa. *Ibis* **1971,** *113*, 275–297.

Ward, P. New Views on Controlling Quelea. *Span* **1972,** *15*, 136–137.

Ward, P. A New Strategy for the Control of Damage by Quelea. *Pest. Abstracts* **1973,** 97–106.

Ward, P. The role of the crop among red-billed queleas *Quelea quelea. Ibis* **1978,** *120*, 333–337.

Ward, P.; Pant, N. C.; Roy, E.; Dorow, E.; Betts, E.; Whellan. Regional Strategies for the Control and Other Migrant Bird Pests in Africa. *Phil. Trans. R. Soc. Lond. B Biol. Sci.* **1979,** *287*, 289–300.

Warrick, R. A. The Possible Impacts on Wheat Production of a Recurrence of the 1930s Drought in the U. S. Great Plains. *Clim. Change* **1984,** *6*, 5–26.

Wehrmann, B. Solar Power in German- Output, Business and Perspectives, 2020, pp 1–4. https://www.cleanenergywire.org/factsheets/solar-power-germany-output-business-perspectives/ (accessed July 21st, 2021).

Wellsher, K. France to Declare Natural Disaster After Storms Through, 2019, pp 1–3. https://www.theguardian.com/world/2019/jun/16/france-to-declare-natural-disaster-after-storms-rip-through-crops/ (accessed June 25th, 2020).

Wenish, A.; Pladerer, C. Energy Situation and Alternatives in Romania. Compagne per la Reforme delle anca Mondiale. Vienna, Austria, 2003, pp 1–9.

Werder, J.; Manzo, S. K. Pearl Millet Diseases in West Africa. In *Sorghum and Millet Diseases—A Second World Review*; Milliano, W. A. J., Frederickson, R. A., Bengston, G. D., Eds.; International Crops Research Institute for the Semi-Arid Topics: Patancheru, AP, 1992; pp 109–114.

West, J. S. Aerobiology and Air Sampling in Plant Pathology. *Alergologia Immunologia* **2012,** *9*, 80–81.

West, J. S.; Atkins, S. D.; McCartney, A.; Fitt, B. D. L. Detecting Airborne Inoculum to Forecast Arable Crop Diseases. *Appl. Aspects Aerobiol*; The Association of Applied Biologists, Wellesbourne, UK, 2008, pp. 1–6.

Wetzel, T.; Wellso, S. G. Economic Thresholds for Central European and North American Wheat Insect. *Great Plains Entomol* **1987,** *20*, 51–60.

Whetter, J. Canola Growth Stalled Under a Shroud of Smoke, 2018, pp 1–4. https://www.country- guide.ca/crops/canola-growth-stalled-under-a-shroud-of-smoke/ (accessed July 25th, 2021).

Whitlock, R. Further European Solar Records to Follow in 2020 Says Solargis, 2020, pp 1–5. https://www.renewableenergymagazine.com/pv_solar/further-european-solar-records-to-follow-in-20200513/ (accessed July 21st, 2021).

Whittelesey, D. Major Agricultural Regions of the Earth. *Ann. Assoc. of Geogr.* **2009,** *26*, 234–239.

Wichereck, S.; Bossier, M. O. Impact of Exploitive Farming Practices on Soil Water and Crop Quality. A Need for Remedial Measures, 1995, pp 1–8 (accessed July 25th, 2021).

Wikimedia. 2018 European Heat Wave, 2020, pp 1–9. https://en.wikipedia.org/wiki/2018_Euroepan_heat_wave/ (accessed June 26th, 2021).

Wikipedia. Environment of Argentina, 2013a, pp 1–3. http://www.en.wikipedia.org/wiki/Environment_of_Argentina (accessed July25th, 2021).

Wikipedia. Climate of Argentina, 2013b, pp 1–4 http://www.en.wikipedia.org/wiki/Environment_of_Argentina/ (accessed July 25th, 2021).

Wikipedia. Wind power in the United States. 2019, pp 1–8. https://en.wikipedia.org/wiki/Wind_power_in_the_United_States/ (accessed July28th, 2021).

Wikipedia. Airborne Wind Turbine, 2020a, pp 1–7. https://en.wikipedia.org/wiki/Airborne_wind_turbine/ (accessed July 25th, 2021).

Wikipedia. List of Birds of Argentina, 2020b, pp 1–8. https://en.wikipedia.org/wiki/list_birds_of_Argentina/ (accessed July 30th, 2021).

Wikipedia. Pampas: Bird in Agrarian Regions, 2020c, p 15. en.wikipedia.org/wiki/Pampas/ (accessed July 30th, 2021).

Wikipedia. Wind Power in the European Union, 2020d, pp 1–7. https://en.wikipedia.org/wiki/Wind_power_in_the_Euoepan_Union/ (accessed July 29th, 2021).

Wikipedia. Renewable Energy in Morocco, 2020e, pp 1–8. https://en.wikipedia.org/wiki/Renewable_energy_in_Morocco (accessed July 10th, 2021).

Wikitpedia. IGP Wind Power in India, 2020f, pp 1–23. https://en.wikipedia.org/wiki/Wind_power_in_India/ (accessed July 26th, 2021).

Williams, J. G. The Quelea Threat to Africa's Grain Crops. *East Afr. Agric. J.* **1954,** *19*, 133–136.

Williams, R. J. Downy Mildew of Tropical Cereals. *Adv. Plant Pathol.* **1984,** *3*, 1–103.

Wilson, J.; Henshaw, H. W. Fifty Common Birds of Farm and Orchard. United states Department of Agriculture: Washington, DC. Farmer's Bulletin No 513, 1913, pp 1–32.

Wind Europe. Wind Energy in Europe in 2019: Trends and Statistics, 2020, pp 1–24. Windeurope.org/ (accessed July 29th, 2021).

Wishart, D. The Great Plains Region. In *Encyclopaedia of the Great Plains*; University of Nebraska Press: Lincoln, Nebraska, 2004; pp 13–18.

Wooten, M. UGA Scientists Use Robots and Drones to Accelerate Plant Genetic Research, Improve Crop Yield. UGA Today, 2017, pp 1–3. http://news.uga.edu/releases/articles/robots-and -drones-improve-crop-yield/ (accessed July 26th, 2021).

Yadav, S. Bioaerosol Impact on Crop Health Over India Due to Emerging Fungal Diseases: An Important Missing Link. *Environ. Sci. Pollut. Res.* **2020,** pp 1–8. DOI: 10.1007/s11356–020–08059-X (accessed July 25th, 2021).

Yates, V.; Nakatsu, C. H.; Miller, R. V.; Pillai, S. V. Aerobiology of Agricultural Pathogens. In *Manual of Environmental Microbiology*; American Society of Microbiology Press, 2015; pp 1–7. DOI: 10.1128/9781555818821.ch3.2.8/ (accessed July 25th, 2021).

Yousufu, S. D.; Yakubu, Y.; Maize, B. N. 2004 The Food of Quelea During Dry Season in Borno State of Nigeria. *Pak. J. Biol. Sci.* **2004,** *7*, 620–622.

Zhang, J.; Huang, Y.; Yuan, L.; Yang, G.; Chen, I.; Zhao, C. Using Satellite Multispectral Imagery for Damage Mapping of Armyworm (*Spodoptera frugidera*) in Maize at a Regional Scale. *Pest Manage. Sci.* **2015,** 1–3. DOI: 10.1002/ps.4003 (accessed August 18th, 2021).

Zotarelli, L.; Eleno, T.; Bodey, R. M.; Sugeno, U.; Bruno, A. Role of legumes in the Nitrogen Economy of Cereal Production in Crop Rotations Under Conventional and No Tillage. *Proceedings of the 17th World soil Science Conference*; Bangkok, Thailand, 2002, *866*, 1–6.

Zozen Boiler. Biomass Generation Plant for West Africa, 2020, pp 1–5. https://www.dreikugelapotheke-bad-hindelang.de/1625540893/western-sahara-type-boiler---zozen-boiler.html/ (accessed July 25th, 2021).

Zuniga, E. Biological Control of Cereal Aphids in the Southern Cone of South America. In *World Perspectives on Barley Yellow Dwarf*; Burnett, P. A. Ed.; International Centre for Maize and Wheat: Mexico, DF, 1990; pp 362–367.

Zwarts, R.; Bijlsma, J.; Van der Kemp; Wymenga, E. *Living on the Edge: Wetlands and Birds in a Changing Sahel*; KNNV Publishing: Zeist, Netherlands, 2009; p 564.

Index

B

G

H

I

L

M

N